T0222059

PROCEEDINGS OF THE 2ND INTERNATIONAL HOSPITALITY & TOURISM CONFERENCE 2014, PENANG, MALAYSIA, 2–4 SEPTEMBER 2014

Theory and Practice in Hospitality and Tourism Research

Editors

Salleh Mohd Radzi, Norzuwana Sumarjan, Chemah Tamby Chik, Mohd Salehuddin Mohd Zahari, Zurinawati Mohi, Mohd Faeez Saiful Bakhtiar & Faiz Izwan Anuar
Faculty of Hotel and Tourism Management, Universiti Teknologi MARA, Malaysia

CRC Press
Taylor & Francis Group
Boca Raton London New York Leiden

CRC Press is an imprint of the
Taylor & Francis Group, an **informa** business

A BALKEMA BOOK

Organized by
Faculty of Hotel & Tourism Management
Universiti Teknologi MARA (Shah Alam), Malaysia

Co-organized by
Faculty of Hotel & Tourism Management
Universiti Teknologi MARA (Penang), Malaysia

Cover photo: Dewan Agong Tuanku Canselor, UiTM Shah Alam, Copyright: Rizal Osman Architect, Kuala Lumpur, Malaysia

CRC Press/Balkema is an imprint of the Taylor & Francis Group, an informa business

© 2015 Taylor & Francis Group, London, UK

Typeset by V Publishing Solutions Pvt Ltd., Chennai, India
Printed and bound in Great Britain by CPI Group (UK) Ltd, Croydon, CR0 4YY

Published by: CRC Press/Balkema
        P.O. Box 11320, 2301 EH Leiden, The Netherlands
        e-mail: Pub.NL@taylorandfrancis.com
        www.crcpress.com – www.taylorandfrancis.com

ISBN: 978-1-138-02706-0 (Hbk)
ISBN: 978-1-315-73735-5 (eBook PDF)

Theory and Practice in Hospitality and Tourism Research – Radzi et al. (Eds)
© *2015 Taylor & Francis Group, London, ISBN 978-1-138-02706-0*

Table of contents

Hospitality and tourism marketing

Tourism management

Technology and innovation in hospitality and tourism

Foodservice and food safety

Relevant areas in hospitality and tourism

Preface

Dear Distinguished Delegates and Guests,

Welcome to the second *International Hospitality and Tourism Conference 2014*, organized by the Faculty of Hotel and Tourism Management, Universiti Teknologi MARA on September 2–4, 2014.

The relevance of the 2nd International Hospitality and Tourism Conference (IHTC) 2014 theme, "Theory & Practice in Hospitality and Tourism Research", is reflected in the diverse range of papers that have been submitted for publication. Altogether there are 110 contributed papers included in the proceedings and this demonstrates the popularity of the IHTC Conference in sharing ideas and findings with a truly international community. Thank you to all who have contributed to producing such a comprehensive 2nd IHTC conference and proceedings and thus contributed to the body of knowledge within the Hospitality and Tourism industry.

This conference presents an opportunity to:

1. explore contemporary and future research issues in hospitality and tourism industry and
2. exchange ideas and information on the "state-of-the-art" research in hospitality and tourism. The conference will also offer a unique networking opportunity to students, academia and practitioners.

It has been an honour for me to have the chance to edit the proceedings. I have enjoyed considerably working in cooperation with the committee to call for papers, review papers and finalize papers to be included in the proceedings. Refereeing papers for an international conference such as 2nd IHTC is a complex process that relies on the goodwill of those researchers involved in the field. The refereeing process for 2nd IHTC papers is optional but the number of papers submitted to the refereeing process has increased by nearly 15% compared to those submitted for 1st IHTC. I would like to thank all these reviewers for their time and effort. Without this commitment it would not be possible to have the important 'referee' status assigned to papers in the proceedings. The quality of these papers is a tribute to the authors and also to the reviewers who have guided any necessary improvement.

We are very grateful and acknowledge the contribution of various parties who have been involved directly and indirectly—the organizing committees, students, reviewers, faculty members, speakers, sponsors, and the participants—for realizing the success of this conference.

Finally, we wish you success in your presentation; enjoy fruitful discussions and a pleasant stay in Penang, Malaysia.

Theory and Practice in Hospitality and Tourism Research – Radzi et al. (Eds)
© 2015 Taylor & Francis Group, London, ISBN 978-1-138-02706-0

Acknowledgements

A. Parasuraman (University of Miami, USA)
Ahmad Azmi Mohd Ariffin (Universiti Kebangsaan Malaysia, Malaysia)
Amrul Asraf Mohd Any (Universiti Malaya, Malaysia)
Artinah Zainal (Universiti Teknologi MARA (Shah Alam), Malaysia)
Aslinda Mohd Shahril (Universiti Teknologi MARA (Shah Alam), Malaysia)
Azdel Abdul Aziz (Universiti Teknologi MARA (Shah Alam), Malaysia)
Azila Azmi (Universiti Teknologi MARA (Penang), Malaysia)
Azni Zarina Taha (Universiti Malaya, Malaysia)
Bany Ariffin Amin Noordin (Universiti Putra Malaysia, Malaysia)
Chemah Tamby Chik (Universiti Teknologi MARA (Shah Alam), Malaysia)
Fadzilah Mohd Shariff (Universiti Teknologi MARA (Shah Alam), Malaysia)
Faiz Izwan Anuar (Universiti Teknologi MARA (Shah Alam), Malaysia)
Hashim Fadzil Ariffin (Universiti Teknologi MARA (Shah Alam), Malaysia)
Inoormaziah Azman (Universiti Teknologi MARA (Shah Alam), Malaysia)
Johanudin Lahap @ Wahab (Universiti Teknologi MARA (Penang), Malaysia)
Kamril Juraidi Abdul Karim (Universiti Teknologi MARA (Shah Alam), Malaysia)
Kasyif Hussain (Taylor's University, Malaysia)
Khairil Wahidin Awang (Universiti Putra Malaysia, Malaysia)
Lim Lay Kian (Universiti Teknologi MARA (Shah Alam), Malaysia)
Mohamad Abdullah Hemdi (Universiti Teknologi MARA (Shah Alam), Malaysia)
Mohd Faeez Saiful Bakhtiar (Universiti Teknologi MARA (Shah Alam), Malaysia)
Mohd Hafiz Mohd Hanafiah (Universiti Teknologi MARA (Shah Alam), Malaysia)
Mohd Raziff Jamaluddin (Universiti Teknologi MARA (Shah Alam), Malaysia)
Mohd Salehuddin Mohd Zahari (Universiti Teknologi MARA (Shah Alam), Malaysia)
Mohd Syaquif Yasin Kamaruddin (Universiti Teknologi MARA (Shah Alam), Malaysia)
Mohhidin Othman (Universiti Putra Malaysia, Malaysia)
Muhammad Naim Kamari (UNITAR International University, Malaysia)
Noor Azmi Ahmad (Universiti Teknologi MARA (Shah Alam), Malaysia)
Nor'Ain Othman (Universiti Teknologi MARA (Shah Alam), Malaysia)
Nor Asmalina Mohd Anuar (Universiti Teknologi MARA (Shah Alam), Malaysia)
Nor Azah Mustapha (Universiti Teknologi MARA (Shah Alam), Malaysia)
Norfezah Md Nor (Universiti Teknologi MARA (Penang), Malaysia)
Norliza Aminudin (Universiti Teknologi MARA (Shah Alam), Malaysia)
Norzuwana Sumarjan (Universiti Teknologi MARA (Shah Alam), Malaysia)
Nuraisyah Chua Abdullah (Universiti Teknologi MARA (Shah Alam), Malaysia)
Nur'Hidayah Che Ahmat (Universiti Teknologi MARA (Penang), Malaysia)
Rafidah Aida Ramli (Universiti Teknologi MARA Penang, Malaysia)
Rahmat Hashim (Universiti Teknologi MARA (Shah Alam), Malaysia)
Raja Puteri Saadiah Raja Abdullah (Universiti Teknologi MARA (Shah Alam), Malaysia)
Rosita Jamaluddin (Universiti Putra Malaysia, Malaysia)
Roslina Ahmad (Universiti Teknologi MARA (Shah Alam), Malaysia)
Rozila Ahmad (Universiti Utara Malaysia, Malaysia)
Saidatul Afzan Abdul Aziz (Universiti Teknologi MARA (Shah Alam), Malaysia)
Salamiah A. Jamal (Universiti Teknologi MARA (Shah Alam), Malaysia)

Salleh Mohd Radzi (Universiti Teknologi MARA (Shah Alam), Malaysia)
Salim Abdul Talib (Universiti Teknologi MARA (Shah Alam), Malaysia)
Vikneswaran Nair (Taylor's University, Malaysia)
Wan Edura Wan Rashid (Universiti Teknologi MARA (Puncak Alam), Malaysia)
Yuhanis Abdul Aziz (Universiti Putra Malaysia, Malaysia)
Zulhan Othman (Universiti Teknologi MARA (Shah Alam), Malaysia)
Zurinawati Mohi (Universiti Teknologi MARA (Shah Alam), Malaysia)

Theory and Practice in Hospitality and Tourism Research – Radzi et al. (Eds)
© 2015 Taylor & Francis Group, London, ISBN 978-1-138-02706-0

Organizing committee

PATRON

Tan Sri Dato' Prof. Ir. Dr. Sahol Hamid Abu Bakar
FASc, DPMS, SSAP, DJMK, DSPN, DJN, DSM, BCN

ADVISOR

Assoc. Prof. Mohamad Abdullah Hemdi, PhD

CONFERENCE CHAIR

Assoc. Prof. Mohd Salehuddin Mohd Zahari, PhD

COMMITTEES

Assoc. Prof. Salleh Mohd Radzi, PhD
Assoc. Prof. Zafrul Hj Isa
Raziff Jamaluddin
Johanudin Lahap @ Wahab, PhD
Rafidah Aida Ramli, PhD
Zurinawati Mohi, PhD
Hashim Fadzil, PhD
Mohd Hafiz Mohd Hanafiah
Mohd Raziff Jamaluddin
Ahmad Fitri Amir
Azdel Abdul Aziz
Zulhan Othman
Inoormaziah Azman
Mohd Faeez Saiful Bakhtiar
Mohd Syaquif Yasin Kamaruddin
Mohd Noor Ismawi Ismail
Hamizad Abdul Hadi
Roslina Ahmad
Noradzhar Baba
Zulhan Othman
Asmalina Anuar
Flora Shumin Chang Abdullah
Mohd Zulhilmi Suhaimi
Noor Azmi Ahmad
Hazwani Ab Hamid
Siti Ruhanis Mohamad Nazer

Hospitality management

Theory and Practice in Hospitality and Tourism Research – Radzi et al. (Eds)
© *2015 Taylor & Francis Group, London, ISBN 978-1-138-02706-0*

The contribution of internship in developing industry relevant management competencies among hotel and tourism management students

A.H. Ahmad Ridzuan & N.C. Ahmat
Faculty of Hotel and Tourism Management, Universiti Teknologi MARA, Penang, Malaysia

A.A. Azdel
Faculty of Hotel and Tourism Management, Universiti Teknologi MARA, Shah Alam, Malaysia

ABSTRACT: Among the goals of higher education is to prepare students for their future career. Traditionally, universities and colleges education emphasises on teaching theoretical knowledge, but industry demand practical skills and practices. This disparity somehow create imbalance between what industry wants and what academician are offering. Hospitality and tourism employers reported that they faced scarcity of suitably qualified employees in both management and skilled workers. With that, internship has become a major factor in the curricula and has been actively pursued as a viable procedure for bridging the theoretical world of academia and the practical world of industry. It is said that education of these potential future employees need to be more aligned with the industry in order to compete in the rapidly changing business practical skills and management competencies besides strengthen the networking with the opportunity to screen potential employees.

Keywords: Hospitality, internship, management competencies

1 INTRODUCTION

Hospitality career education program is offered to train individuals to work in the vast sector of hospitality industry which demand a significant amount of young trained professional. Due to fast paced changes in the hospitality industry with great emphasis on employability skills, there is a need to determine the level of competencies that graduates in hospitality management program should possess before moving on to workplace. With the increasing number of hospitality graduates, questions have emerged regarding industry expectations pertaining to the generic managerial competencies. Do hospitality education programs adequately prepare their students with the employability skills? If not, can the program be improved or changed to meet the demands of the employers? What is at stake is significant or else, colleges and universities may be failing to prepare graduates to meet the demands and expectations of the industry (Robinson, 2006). The combination of theory with practical is viewed as an essential component in hospitality and tourism management education. Internship, which originated as apprenticeship and then evolved into academic based experience have long been regarded as important experiences. This

further raised a critical question whether there is any development in hospitality student's management competencies before and after the internship program in order to measure the effectiveness of the internship program. However, limited attempts have been made to provide empirical support to claim that a practical internship experience develops management competencies in hotel and tourism students especially in Malaysia hospitality education context. This paper will investigate the contribution of internship in developing industry relevant management competencies among hospitality and tourism students.

2 REVIEW OF LITERATURE

2.1 *Management competencies*

Previous scholars defined manager competencies as traits, skills, and values of effective managers (Boyatzis, 2008; Mihail, 2006) or in other words of saying, competencies are a combination of observable and applied knowledge, skills, and behaviours that create a competitive advantage for an organization. However, Quinn, Faerman Thompson, and McGrath (2003) argued that competency suggest both the possession of knowledge and the behavioural capacity to act appropriately and to develop competencies, one person must both be introduced to knowledge and have the opportunity to practice their skills.

Be that as it may, the main objective of management education is to provide the industry with high calibre graduates equipped with the most relevant management competencies (Christou, 2002; Hansson, 2001). With that, it is important for hotel and tourism educators to incorporate a wide range of learning experiences that can contribute to the development of essential competencies (Breen, Walo, & Dimmock, 2004). Concurrently, recent years has witnessed a significant emphasis being placed in understanding the most critical management competencies for those who working in the hospitality and tourism industry, with particular attention devoted to examining the skills base of graduate hotel managers (Raybould & Wilkins, 2005). This further supported by Jauhari (2006) who commented on the need for curricula to have a heavy focus on leadership and competency development in order to meet the industry expectations and needs.

2.2 Competing Values Framework (CVF)

Originally developed by Quinn and Rohrbaugh (1983), the first value dimension related to organizational focus from internal (micro) to external (macro), while the second value dimension related with organizational structure for an emphasis on stability to an emphasis on flexibility. Both dimension of CVF classified four models and each one contains a different set of effectiveness criteria. The four models are Human Relation model, Open System model, Rational Goal model, and Internal Process model.

Since hospitality managers need transferable generic competencies, a framework by Quinn, Faerman, Thompson, and McGrath (1990) entitled "Competing Values Framework" (CVF) in Figure 1 was deemed suitable to be discussed in this paper to identify and explain significant management competencies.

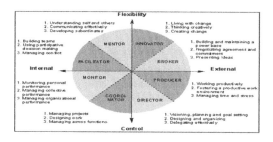

Figure 1. Competing Values Framework. Source: Quinn et al. (1990).

The model has a sound theoretical and research base and offers an opportunity to examine key managerial competencies based on organizational outcomes such as maximization of output, adaptation and change, human relations and consolidation and continuity. Quinn et al. (1990) scrutinizing 8 managerial roles and 24 management competencies linked to each role by using CVF. The CVF conceptualizes that each of the organizational outcomes are crucial parts that make up the larger construct, namely organizational and managerial competencies (Quinn, et al., 1990). The CVF provides managers with the opportunity to increase effectiveness by developing and utilizing a range of often competing competencies, depending on the situation. It is worth mentioned that the CVF framework was popularly adopted by previous scholars in the hospitality researches (Breen, et al., 2004; Walo, 2001).

2.3 Competencies required by hospitality and tourism graduates

Velo and Mittaz (2006) discovered the competencies needed by hospitality management graduates to cope with challenges and the skills found to be the utmost critical to break into emerging international markets are flexibility, openness, and cultural awareness. Earlier study by Raybould and Wilkins (2005) pointed out managers rated skills associated with interpersonal, problem solving, and self-management skill domain as the most important. In the same tone, few scholars noted that competencies required by the hospitality and tourism graduates are more on transferrable generic skills such as leadership, communication, and human resource management (Kay & Russette, 2000; Walo, 2001) and those skills are essential to ensure their success across departments, organizations, and industry.

2.4 Developing competencies through internship

Beggs, Ross, and Goodwin (2008) elaborated that internship courses provide learning opportunities for undergraduates to experience professional practice and activities associated with knowledge application thus give opportunity to close the gap between learned theory and practical reality. The internship is an integral part of a student's career development and may have numerous positive impacts on students (Lam & Ching, 2007). It helps the students to develop the critical core skills demanded by global markets including communication, time management, self-confidence, and self-motivation which further supported Walo (2001) who positively affirmed that students have gained managerial competencies after the internship programme. She claimed that for pre-internship, skills may have been developed either through coursework in the program or the students have already worked at an operational level previously hence able to apply the skills. For post-internship, the level of competence were increased because the students became more confident in interpersonal communication, presenting information by writing effectively, understanding self and others, and personal productivity and motivation. However, her findings revealed that students still least confident on presenting ideas, motivating others, controlling and participative decision making. As well, her study identified that students are most competent in Mentor and Monitor roles while least competent in Broker and Innovator roles (Quinn, et al., 1990; Walo, 2001).

Additionally, tourism and hospitality employers are encouraged to structure the orientation and training of new graduates to take advantage of their strong mentor competencies and to further develop coordinator competencies through job work experience. Into the bargain, there is a need for on-going evaluation by hospitality and tourism educators regarding the students' competency development to ensure it fits with the industry needs and requirement (Breen, et al., 2004; Hansson, 2001).

3 METHODOLOGY

A causal cross-sectional study was conducted to determine whether the students' management competencies are developed after they had undergone the internship. Besides, researchers compared the mean scores of the students' responses to the 24 management competencies before and after internship. The sampling frame was limited to semester six students comprises of four departments in the Faculty of Hotel and Tourism Management, Universiti Teknologi MARA (UiTM) Pulau Pinang namely Diploma in Hotel Management, Diploma in Foodservice Management, Diploma in Tourism Management, and Diploma in Culinary Arts. The majority of these students were expected to undertake the internship subject for their final year in Diploma program. Statistic from the Industrial Training Department showed the number of senior hospitality students going for internship in the four different programmes were 270 students. The first round of the survey was distributed to all students

who enrolled in a third year core unit in semester five, before they left the campus to start internship in semester six.

Purposive sampling method was used as researchers believed the respondents who were belongs to the targeted groups are able to answer the survey precisely. A self-administered questionnaire comprised of two parts was developed. Part one adopted items from Quinn et al. (1990) 113 competency statements whereby each student was asked to self-assess himself/herself in relation to each competency statement. Likert scale was used to rate each of the competency statements ranging from 1 "Strongly Disagree" to 6 "Strongly Agree". The 113 competency statements were presented in the same order as set out by Quinn et al. (1990). Part two collected basic demographic data of the respondents including gender and intention to work in the industry after graduation. The students were informed of the purpose of the study before the surveys were distributed and they were also been informed that a same questionnaire will be given to them at the completion of their internship.

4 ANALYSIS AND FINDINGS

Of the 270 surveys distributed to the students, 200 surveys were returned given a response rate of 74 per cent. The internal consistency of the item was assessed by computing the Cronbach-Alpha reliability scale with the result of 0.987 for all items which are excellent (Hair, Money, Samouel, & Page, 2007). Based from the frequency statistics, 75 per cent from the total respondents are female (n = 150) while another 25 per cent are male (n = 50). Findings worth noted that almost majority of the students have the intention to work in hotel industry after graduation (n = 178; 89%) and only 11% (n = 22) refuse to work in the field. Somehow, this result is contradicted with a study done by Mohd Zahari, Hanafiah, Othman, Jamaluddin and Zulkifly (2010) who found that students became less interested and less committed to accept any job in the hotel industry even after few years they had studying the program and had undergone their practical training.

Result in Table 2 revealed the output from paired sample t-tests which was conducted to compare the students' pre and post-internship mean scores for the 24 management competencies and eight managerial roles. It was found that students' pre-internship mean scores ranged from 4.5250 to 4.0720. These relatively high scores for pre-internship possibly due to majority of this group of students had the opportunity to apply or practice those skills. Alternatively, those skills may have been developed through coursework in their diploma program.

Table 1. Distribution of the students.

No.	Departments	Number of Practical Students
1.	Hotel Management (HM110)	93
2.	Tourism Management (HM111)	41
3.	Foodservice Management (HM112)	46
4.	Culinary Arts (HM115)	90

Table 2. Comparison of students' pre and post-test managerial competencies.

Management competency/ Role	Student intern status n=200	Mean scores (M)	Differences in mean scores
Director Role	Pre-intern	4.4093	+0.2734
	Post-intern	4.6827	
Taking initiative	Pre-intern	4.3600	+0.298
	Post-intern	4.6580	
Goal setting	Pre-intern	4.4340	+0.236
	Post-intern	4.6700	
Delegating effectively	Pre-intern	4.4338	+0.2862
	Post-intern	4.7200	
Producer Role	Pre-intern	4.3727	+0.3326
	Post-intern	4.7053	
Personal productivity & motivation	Pre-intern	4.3690	+0.369
	Post-intern	4.7380	
Motivating others	Pre-intern	4.2970	+0.339
	Post-intern	4.6360	
Time & Stress management	Pre-intern	4.4520	+0.29
	Post-intern	4.7420	
Coordinator Role	Pre-intern	4.2692	+0.3251
	Post-intern	4.5943	
Planning	Pre-intern	4.3963	+0.2612
	Post-intern	4.6575	
Organizing	Pre-intern	4.1890	+0.339
	Post-intern	4.5280	
Controlling	Pre-intern	4.2225	+0.375
	Post-intern	4.5975	
Monitor Role	Pre-intern	4.2562	+0.3348
	Post-intern	4.5910	
Reducing information overload	Pre-intern	4.3130	+0.307
	Post-intern	4.6200	
Analyzing information & critical thinking	Pre-intern	4.2130	+0.395
	Post-intern	4.6080	
Presenting information & writing effectively	Pre-intern	4.2425	+0.3025
	Post-intern	4.5450	
Mentor Role	Pre-intern	4.3613	+0.2724
	Post-intern	4.6337	
Understanding yourself & others	Pre-intern	4.5250	+0.187
	Post-intern	4.7120	
Interpersonal communication	Pre-intern	4.3390	+0.295
	Post-intern	4.6340	
Developing subordinates	Pre-intern	4.2200	+0.335
	Post-intern	4.5550	
Facilitators Role	Pre-intern	4.2632	+0.3965
	Post-intern	4.6597	
Team building	Pre-intern	4.1613	+0.4162
	Post-intern	4.5775	
Participative decision making	Pre-intern	4.4260	+0.348
	Post-intern	4.7740	
Conflict management	Pre-intern	4.2025	+0.425
	Post-intern	4.6275	
Innovator Role	Pre-intern	4.2011	+0.3768
	Post-intern	4.5779	
Living with change	Pre-intern	4.2333	+0.3984
	Post-intern	4.6317	
Creative thinking	Pre-intern	4.2230	+0.321
	Post-intern	4.5440	
Managing change	Pre-intern	4.1470	+0.411
	Post-intern	4.5580	
Broker Role	Pre-intern	4.2248	+0.3757
	Post-intern	4.6005	
Building & maintaining a power base	Pre-intern	4.3050	+0.361
	Post-intern	4.6660	
Negotiating agreement & commitment	Pre-intern	4.2975	+0.35
	Post-intern	4.6475	
Presenting ideas	Pre-intern	4.0720	+0.416
	Post-intern	4.4880	

*significance at the p = .000 level

Furthermore, for pre-internship, the competencies in which students perceived they were most confident are Understanding Self and Others, Time and Stress Management, and Goal Setting. The competencies in which they were least confident are Presenting Ideas,

Managing Change, and Team Building. On Likert scale of 1 to 6, the students' lowest pre-internship mean score was 4.0720 for Presenting Ideas and the highest mean score was 4.5250 for Understanding Self and Others. Ironically, the results are perfectly aligned with Walo (2001). Moreover, students' pre-internship perceived they were most competent in the Director, Producer, and Mentor Roles whilst least competent in the Innovator, Broker, and Monitor Roles. The students' highest pre-internship mean score was 4.4093 for Director Role and the lowest mean score was 4.2011 for Innovator Role.

Conversely, students' post-internship mean scores ranged from 4.7740 to 4.4880 on the Likert scale of 1 to 6. It shows that students perceived a relatively high level of competence for the 24-management competencies after internship. Additionally, for post-internship, the competencies in which students perceived they were most confident are Participative Decision Making, Time and Stress Management, and Personal Productivity and Motivation whereas they were least confident on Presenting Ideas, Organizing, and Creative Thinking. On Likert scale 1 to 6, the students' lowest post-internship mean score was 4.488 for Presenting Ideas and their highest mean score was 4.7740 for Participative Decision Making. Students' post-internship perceived they were most competent in the Producer, Director, and Facilitator Roles. They perceived to be least competent in the Innovator, Monitor, and Coordinator Roles. The students' highest post-internship mean score was 4.7053 for the Producer Role and the lowest mean score was 4.5779 for Innovator Role.

When comparing both pre- and post-data, students' post-internship mean scores were higher than their pre-internship mean score for all 24 competencies. A comparison of students' pre- and post-internship mean scores found those competencies indicating the greatest numerical difference were Conflict Management (+0.4250), Team Building (+0.4162), and Presenting Ideas (+0.4160). The management competencies indicating the least numerical difference in mean scores were Understanding Self and Others (+0.187), Goal Setting (+0.236), and Planning (+0.2612). In addition, the greatest numerical difference of (+0.3965) occurred in the Facilitator Role with the least numerical difference of (+0.2724) for Mentor Role. The low difference in the Mentor Role is due to a low after internship in students' mean scores for the competency in Understanding Self and Others.

5 CONCLUSION AND RECOMMENDATIONS

Based from thorough analysis and findings discussed, it is confirmed that the hospitality students' management competencies has developed after they return from their internship. Practical training clearly helps the students to improve their various skills as needed by the industry players. When comparing among all 24 management competencies, it shows consistency of the students' high competency level on time and stress management before and after the internship. These portrayed that the hospitality students are able to manage their time and stress level working in the field with the ability to cope with the nature of the business and perhaps this reason contribute to the large percentage of them decide to remain in the industry after they graduated. Though, something needs to be done to improve the least competent presenting ideas skill among students as it also shows constant result for pre- and post-intern and not to mention, soft skill is among the important attributes that a young trained professional should possess in order to work in the industry.

The development in training and education should accommodate the expected industry growth and future competitive demands. Most importantly, the curricula must reflect the changing needs of the industry to ensure the courses are not only relevant but at the same time producing graduates with skills and knowledge needed as future managers. The university has taken the step by implementing outcome-based education and student-centred learning based curriculum to produce a more quality and marketability graduates to cater for the industry. It is believe that both academicians and industry players have their own target to achieve. Hence, there is a need for educational strategies to foster the development of management skills and competencies within Malaysia. Future researches could expand the study to examine the attributes of management competencies from the industry players' perspectives and also could enlarge the sample by targeting at different faculties to investigate the result.

REFERENCES

Beggs, B., Ross, C.M., & Goodwin, B. (2008). A comparison of student and practitioner perspectives of the travel and tourism internship. *Journal of Hospitality, Leisure, Sport and Tourism Education, 7*(1), 31–39.

Boyatzis, R.E. (2008). Competencies in the 21st century. *Journal of Management Development, 27*(1), 5–12.

Breen, H., Walo, M., & Dimmock, K. (2004). Assessment of tourism and hospitality management competencies: A student perspective. *Southern Cross University, ePublications, SCU School of Tourism and Hospitality Management.*

Christou, E. (2002). Revisiting competencies for hospitality management: Contemporary views of the stakeholders. *Journal of Hospitality & Tourism Education, 14*(1), 25–32.

Hair, J.F., Money, A.H., Samouel, P., & Page, M. (2007). *Research methods for business.* New York: John Wiley & Sons, Ltd.

Hansson, B. (2001). Competency models: Are self-perceptions accurate enough? *Journal of European Industrial Training, 25*(9), 428–441.

Jauhari, V. (2006). Competencies for a career in the hospitality industry: An Indian perspective. *International Journal of Contemporary Hospitality Management, 18*(2), 123–134.

Kay, C., & Russette, J. (2000). Hospitality-management competencies identifying managers' essential skills. *Cornell Hotel and Restaurant Administration Quarterly, 41*(2), 52–63.

Lam, T., & Ching, L. (2007). An exploratory study of an internship program: The case of Hong Kong students. *International Journal of Hospitality Management, 26*(2), 336–351.

Mihail, D.M. (2006). Internships at Greek universities: An exploratory study. *Journal of Workplace Learning, 18*(1), 28–41.

Quinn, R., Faerman, S., Thompson, M., & McGrath, M. (1990). *Becoming a master manager: A competency frame-work.* New York.

Quinn, R., Faerman, S., Thompson, M., & McGrath, M. (2003). *Becoming a master manager: A competency frame-work* (2nd ed.). USA: John Wiley & Sons.

Quinn, R.E., & Rohrbaugh, J. (1983). A spatial model of effectiveness criteria: towards a competing values approach to organizational analysis. *Management science, 29*(3), 363–377.

Raybould, M., & Wilkins, H. (2005). Over qualified and under experienced: Turning graduates into hospitality managers. *International Journal of Contemporary Hospitality Management, 17*(3), 203–216.

Robinson, J.S. (2006). *Graduates' and employers' perceptions of entry-level employability skills needed by agriculture, food and natural resources graduates.* University of Missouri, Columbia.

Velo, V., & Mittaz, C. (2006). Breaking into emerging international hotel markets: Skills needed to face this challenge and ways to develop them in hospitality management students. *International Journal of Contemporary Hospitality Management, 18*(6), 496–508.

Walo, M. (2001). Assessing the contribution of internship in developing Australian tourism and hospitality students' management competencies. *Asia Pacific Journal of Cooperative Education, 2*(2), 12–28.

Zahari, M., Hanafiah, M., Othman, Z., Jamaluddin, M., & Zulkifly, M. (2010). Declining interest of hospitality students toward careers in hotel industry: Who's to be blamed. *Interdisciplinary Journal of Contemporary Research in Business, 2*(7), 269–287.

Theory and Practice in Hospitality and Tourism Research – Radzi et al. (Eds)
© *2015 Taylor & Francis Group, London, ISBN 978-1-138-02706-0*

An examination of current compensation and performance appraisal practice among hotel employers in Malaysia: A preliminary study

J. Lahap, S.M. Isa, N.M. Said, K. Rose & J.M. Saber
Universiti Teknologi MARA, Penang, Malaysia

ABSTRACT: This paper seeks to examine the current compensation and performance appraisal practice among hotel employers in Malaysia. It was known that effective Human Resource Management practice is crucial to any organization ergo act as a catalyst to other sub-department in hotel establishment. The main activity of the department is to recruit, select, remunerate and appraise employees. Moreover, HRM department also performs tasks such as planning, controlling and organizing human resources. Therefore, to have effective and efficient HRM practice is inescapable. The finding of the study reveals that the practice of compensation and performance appraisal among hotel employers in Malaysia meets the industry requirement and standard, however, needs some improvement.

Keywords: Human Resource Management practices, compensation, performance appraisal, hospitality and tourism

1   INTRODUCTION

The top three leading contributors to the inflow of foreign exchange in the Malaysian economy is the Tourism sector, it was estimated that the business were believed to worth MYR 60.6 billion in 2013 (approximately USD19.5 billion) (Tourism Malaysia, 2012). The hospitality industry is also considered part of the Tourism sector and it can be categorized into three segments; transportation services, accommodation and tourist destination (Hayes & Ninemeier, 2009). Kuala Lumpur was ranked as the seventh most visited destination in the world (Euromonitor Report, 2010). The development and growth of the Hospitality and Tourism industry in Malaysia promotes Kuala Lumpur to be the top tenth Best Shopping Cities in the world by CNN's Travel Survey (Prime Minister Department, 2013). This study is developed as a means to examine the practice of Human Resources Management practice that concentrate on factors such as compensation and performance appraisal. Lodging business provide accommodation for those who are away from homes and in many circumstances the industry uses human to deliver hotel products and services. Therefore, to employ the best service deliverer is seen to be vital. A number of reviews of literature suggest that there are many factors that can enhance employee job performance. Some of the strategies are working environment (culture, belief), physical environment (facilities, equipment), rules and regulations

(policies, procedures) and Human Resources Management practices (Lahap, O'Mahony & Sillitoe, 2010). Proper Human Resources Management practice can alter employee's intention to leave their job, influence the level of job satisfaction and employees' commitment towards their employer (Arnett, Laverie & McLane, 2002). An excellent Human resource management is primarily able to reduce problem relating to employees' performance, employees' job dissatisfaction, employees' turnover rate and employees' absenteeism (Lahap, O'Mahony & Dalrymple, 2013). For many years, the Malaysian Tourism Industry had shown a dramatic improvement, thus enables it to compete with the neighbor countries. However, a recent study reports that the monetary compensation paid by hotel employers to hotel employees in the sector is known to be unsatisfactorily (Lahap, et al., 2010). In addition, hotel employees work for long hours and received only a day off in a week. These factors have caused many hotel employees refuses to perform their job effectively, to the extent quit their job. Furthermore, working in the hotel sector is enervated, therefore, absenteeism and employee turnover were known to be the main issue (Gounaris, 2008; Phonsanam, 2010).

2   LITERATURE REVIEW

In the 21st century, the need to match compensation and organizational strategy has become

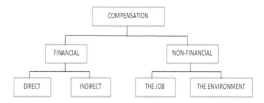

Figure 1. Component of total compensation program (Wayne, 2008, p. 242).

essential. It was known that there are two types of compensation program namely; direct and indirect financial compensation. Fundamentally, compensation practice can be divided into two categories that are financial and non-financial. As the name implies, the financial compensation involved all benefits in terms of monetary given to employees. Financial rewards can be further divided into two categories that are direct reward and indirect reward. Wages, commission, bonuses are all known as direct reward. Meanwhile, non-financial compensation refers to the satisfaction received by the employees from the job environment itself. Non-financial can be divided into two categories; the job and the job environment (Wayne, 2008). Examples of indirect rewards are; medical leave, life insurance, retirement plan, social security and paid vacation. For the non-financial reward, it refers to the skill and knowledge gained through the job surrounding. The component of compensation system is depicted in Figure 1.

The cost of living in Malaysia nowadays encouraged many job seekers to be more demanding in choosing employment. Hence, the need to find a appropriate job that will help them to survive is significant. Therefore, we could summarize that if an effective and efficient compensation practice is implemented by top management on a continuous basis; employees will tend to perform well in their work (Frye, 2004; Gounaris, 2008; Nebel, 1991).

2.1 Direct financial

Direct financial compensation can be defined as pays that a person received in a form of salary, wage, bonus and commission (Wayne, 2008). Direct financial often been classified as something that has to do with monetary incentives (Phonsanam, 2010). A compelling compensation system can encourage employees not only to perform well, but also boost their productivity and strengthen their job performance (Lin, 2000). Gould-Williams (2003) proposed that compensation is not just a return for labor service but it is also a crucial tool that managers could use to encourage and shape employee behavior.

2.2 Indirect financial

Mondy (2008) define indirect financial as the satisfaction that a person received from the job itself or from the psychological or physical environment in which the person works (Phonsanam, 2010). A good leader [managers], should not assume that the monetary compensation act as the only factor that boost employee morale, but, they must also acknowledge the true meaning of non-monetary compensation.

2.3 Compensation in the service sector (hospitality industry)

In Crick and Spencer (2011) works postulated that hotels could have a higher employees' retention rates if they are provided with good financial incentives. In addition, they also asserted that employees tend to quit their jobs, if hotel employer concentrate solely on the pay itself. Furthermore, hotels in Malaysia should offer different types of incentives to attract the best talent available. To support that contention it was also suggested that hotels employers should clearly communicate organization's compensation policies and disseminate their goals to employees on what to achieve (Phonsanam, 2010).

2.4 Performance appraisal system

This literature review will start with differentiating the term 'Performance Management' and 'Performance Appraisal'. Generally, Performance Management (PM) and Performance Appraisal (PA) are intertwined, however, different. Performance management is defined as a continuous process that includes several stages such as; planning, monitoring, developing, appraising and rewarding employees (Hayes & Ninemeier, 2009). For that reason, in order for organizations to maintain productivity, they must take a good care of their performance management program (Ahmad & Ali, 2004). Some of the common methods for performance appraisals are; Behaviorally Anchored Rating Scales (BARS), Behavioral Observation Scale (BOS), Management by Objective (MBO), Peer Evaluation, Upward Assessments and 360-Degree Appraisal (Hayes & Ninemeier, 2009). Besides that there are also methods used such as Rating Scales, Critical Incident, Essay Method, Work Standard, Ranking Method, Forced Distribution and Result-Based System (Wayne, 2008).

2.5 Performance appraisal in the service sector (hospitality industry)

In service industry, conflict exists in the process of performance appraisal and it can't be avoided. Citing from Bretz, Milkovich and Read (1989) review

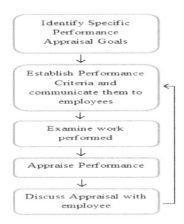

Figure 2. The performance appraisal process (Wayne, 2008, p. 213).

on Fortune 100 best companies in the US have found that the most highlighted issue pertains to performance apprasial practice raised by American employees are; 1) the accepted performance appraisal system by those who are being rated, 2) fairness during the appraisal process, 3) employees' perception towards the results (fairness), 4) types of response given by the employer, and 5) the effectiveness of the performance appraisal program.

3 METHODOLOGY

In this study, the respondents selected in this study were those who work in the hotel industry. The questionnaire and email were randomly distributed to the selected population. The researcher distributed 200 printed questionnaires and 100 emails, which made up of 300 questionnaires. The majority of the questionnaires were distributed within the 4 and 5 stars hotels in Klang Valley. This strategy was employed due to the density of well-structured hotel exists in Klang Valley. As a result, 229 responses were successfully gathered which representing 76.3 percent response rate. Using the 5 point Likert's Scale to measure the responses (Strongly Disagree = 1 and Strongly Agree = 5). The questionnaire consists of two sections adapted from Khatri (2000) and Tsaur and Lin (2004).

4 FINDINGS

This section will be divided into two segments which are; (i) descriptive analysis (demographic) and (ii) principal component analysis (compensation and performance appraisal). From the analysis, it was found that out of the 229 of total respondents, 124

Table 1. Respondents' gender.

	Gender		
	Female	Male	Total
Count	124	105	229
Percentage (%)	54.1%	45.9%	100.0%

Table 2. The distribution of respondents' age range by gender.

Age Range (years old)		Gender		Total
		Female	Male	
18-20	Count	4	3	7
	Percentage %	1.8%	1.3%	3.1%
21-25	Count	35	33	68
	Percentage %	15.3%	14.4%	29.7%
26-30	Count	35	30	65
	Percentage %	15.3%	13.1%	28.4%
31-35	Count	22	12	34
	Percentage %	9.6%	5.2%	14.8%
36-40	Count	11	13	24
	Percentage %	4.8%	5.7%	10.5%
41-45	Count	13	8	21
	Percentage %	5.7%	3.5%	9.2%
46-50	Count	2	4	6
	Percentage %	.9%	1.7%	2.6%
51-55	Count	1	2	3
	Percentage %	.4%	.9%	1.3%
> 56	Count	1	0	1
	Percentage %	.4%	.0%	.4%
	Total	124	105	229
		54.1%	45.9%	100.0%

of them are female, which represent 54.1 percent and 105 respondents are male that represent 45.9 percent of the whole sample population. It shows that female pose the highest percentage and dominated the sample for this study. The distribution of respondents by gender is presented in Table 1.

4.1 Age range

In this analysis, it was found that the highest hotel employees' age range is within 21–25 with the total of 29.7 percent of the sample. The second highest hotel employees' age range is within 26–30 with the total of 28.4 percent. The smallest percentage of the hotel employees' age range is 0.4 percent of the age range above 56. Table 2 depicts the age range of the hotel employees.

4.2 Length of employment with current employer

Based on the data collected, it shows that the majority of the respondents have worked with their current employers for more than 2 years, which represents 21.4 percent of the population. Moreover,

Table 3. The distribution of respondents' length of employment with their current employer by gender.

Length of Employment		Gender		Total
		Female	Male	
0-5 months	Count	10	12	22
	%	4.4%	5.2%	9.6%
6-12 months	Count	18	10	28
	%	7.8%	4.4%	12.2%
> 1 years	Count	23	18	41
	%	10%	7.9%	17.9%
> 2 years	Count	25	24	49
	%	11%	10.4%	21.4%
> 3 years	Count	23	15	38
	%	10%	6.6%	16.6%
> 4 years	Count	7	7	14
	%	3%	3.1%	6.1%
> 5 years	Count	10	3	13
	%	4.4%	1.3%	5.7%
6 years above	Count	8	16	24
	%	3.5%	7%	10.5%
	Total	124	105	229
		54.1%	45.9%	100.0%

Table 4. The result of the unrotated factor extraction from the 4 questions representing compensation.

Component	Eigenvalues	Percentage of Variance	Cumulative Percentage
1	1.0	62.4%	62.4%
2	1.0	16.1%	78.5%
3	1.0	12.6%	91.1%
4	1.0	8.9%	100.0%

Kaiser-Meyer-Olkin Measure of Sampling Adequacy = 0.76 Bartlett's Test of Sphericity = 280.325, Significance = .00. Source: Data analysis 2013.

Table 5. The result of varimax rotated factor matrix for compensation.

Item	Compensation	LD
1	The financial reward practices of my company consist with my expecting.	0.83
2	Promotion is based primarily on seniority (reverse-coded)	0.72
3	My company has elaborate/comprehensive flexible benefits scheme.	0.85
4	My company constantly reviews and updates the range of benefits to meet the needs of employees.	0.75
	Reliability Cronbach's Alpha = 0.80	

Source: Data analysis (2013); LD: Factor Loading.

it was also found that the lowest distribution of respondents' length of employment with their current employer is more than 5 years, which represent 5.7 percent of the sample. Table 3 shows the overall distribution of the hotel employees' length of employment with their current employer.

4.3 Principal Component Analysis (PCA)

The analysis focused on 7 questions that representing two dimensions, which are; i) compensation (4 questions) and ii) performance appraisal (3 questions). Raw data had undergo three types of analyses and these are; i) principle component analysis ii) reliability and validity testing and iii) the Cronbach's Alpha value to assess the consistency of each questions. Based on Hair, Anderson, Tatham, and Black (2004), there are two possible multivariate methods which can be used to identify the importance of each dimension. It was deemed important for multivariate analysis to be conducted in this study as a means to explore the significant among the questions.

4.4 Compensation

Based on the analysis, the Kaiser-Meyer-Olkin value of 0.76 exceeded the recommended value of 0.60 and Bartlett's Test of Sphericity and, therefore, it is statistically significant (Hair, et al., 2004). This shows that there is a high degree of interrelationship between the questions within the scope of compensation. The unrotated factor solution extracted shows Eigenvalues is equal to 1. According to Table 4, these four components account for 100 percent of the explained variance with the first factor explaining 62.4 percent of the variance.

The pattern loadings, factor structure and factor interpretation are shown in Table 5. The components were defined by the variables with significant factor loadings of 0.60 and above. Reliability tests on each of the factors indicated a Cronbach's Alpha coefficient above 0.80. Thus, all the 4 questions are all accepted. Based on the analysis, it was found that the financial rewards in the employees' hotel consist with their expectation (0.83) and their promotion is based on seniority (0.72). Respondents reported that their employer has elaborate/comprehensive flexible benefits scheme (0.85). It was also found that, their employer constantly reviews and updates the range of benefits to meet their needs (0.75). Therefore, the highest question loading (LD) is the item 3 that has LD = 0.85 and it can be proposed that the employer in the hotel industry provides a comprehensive benefits scheme for their employees.

4.5 Performance appraisal

Based on the data analysis, the Kaiser-Meyer-Olkin value of 0.70 exceeded the recommended value of 0.60 and Bartlett's Test of Sphericity and hence, it

Table 6. Result of the unrotated factor extraction from the 3 questions representing performance appraisal.

Component	Eigenvalues	Percentage of Variance	Cumulative Percentage
1	1.0	71.0%	71.0%
2	1.0	16.2%	87.3%
3	1.0	12.7%	100.0%

Kaiser-Meyer-Olkin Measure of Sampling Adequacy = 0.70
Bartlett's Test of Sphericity = 208.959, Significance = .00
Source: Data analysis 2013

Table 7. The result of varimax rotated factor matrix for performance appraisal.

Item	Performance Appraisal	LD
1	Employee could know the performance appraisal result by the formal feedback system.	0.82
2	Performance appraisal includes the supervisor setting objectives and goals of employees for the period ahead in consultation with them.	0.86
3	My company emphasizes job relevant criteria in the appraisal systems.	0.85
	Reliability Cronbach's Alpha = 0.79	

Source: Data analysis (2013); LD: Factor Loading

is statistically significant. (Hair, et al., 2004) There is a high degree of interrelationship between the questions within the scope of the Performance Appraisal. The unrotated factor solution extracted shows the Eigenvalues are equal to one and according to Table 6, these three factors account for 100 percent of the explained cumulative percentage, with the first factor explaining 71.0 percent of the variance.

The pattern loadings, factor structure and factor interpretation are shown in Table 7. The dimensions were defined by the variables with significant factor loadings of 0.60 and above. Reliability tests on each of the factors indicated that the Cronbach's Alpha coefficient was above 0.79. This means that the three questions were all accepted. Based on the analysis, the respondents' reports that their employer inform the performance appraisal results to them using a formal system (0.82). Other than that, it was also found that, their performance appraisal includes the setting of objectives and goals by the supervisor for a period ahead when consultation with them (0.86). In addition, the employers also do emphasize job relevant criteria in the performance appraisal (0.85). Therefore, the respondents' are all agreed that their employer have developed an effective performance appraisal system based on their feedback in this study.

5 DISCUSSIONS AND CONCLUSION

Based on the analysis, it was found that hotel employees' financial reward consists of expectation, promotion based on seniority and employee

has a comprehensive flexible benefits scheme statistically significant. This is in concurrent with Gounaris (2008) contention that compensation has a direct affect to employee job performance and satisfaction. As for the performance appraisal practices, it was found that the employer inform the performance appraisal result to hotel employees by using formal feedback system. Furthermore, it was also found that hotel employers include setting up objectives and goals before the performance appraisal is conducted. The most important analysis in performance appraisal was that it was found that hotel employer' emphasis job relevant criteria in appraising employees in which Crick and Spencer (2011) asserted as the main ingredient to excellent performance appraisal program. In this study, from the academic perspective, revealed that the compensation and performance appraisal practice in the selected sampling (Klang Valley) met the standard of best practices. Practically, this study helps hotel employer to acknowledge the importance of compensation and performance appraisal system as part of the Human Resource Management practice. To sum up, it is worth noting that this study enhances our understanding of the current Human Resource Management practices predominantly towards the compensation and performance appraisal practice.

REFERENCES

Ahmad, R. & Ali, N.A. (2004). Performance appraisal decision in Malaysian public service. *International Journal of Public Sector Management, 17*(1), 48–64.

Arnett, D.B., Laverie, D.A. & McLane, C. (2002). Using job satisfaction and pride as internal-marketing tools. *The Cornell Hotel and Restaurant Administration Quarterly, 43*(2), 87–96.

Crick, A.P. & Spencer, A. (2011). Hospitality quality: New directions and new challenges. *International Journal of Contemporary Hospitality Management, 23*(4), 463–478.

Euromonitor Report. (2010). Euromonitor international's top 100 city destination ranking *Passport Travel Tourism. Euromonitor International.*

Frye, M.B. (2004). Equity-based compensation for employees: firm performance and determinants. *Journal of Financial Research, 27*(1), 31–54.

Gould-Williams, J. (2003). The importance of HR practices and workplace trust in achieving superior performance: a study of public-sector organizations. *International Journal of Human Resource Management, 14*(1), 28–54.

Gounaris, S. (2008). Antecedents of internal marketing practice: Some preliminary empirical evidence. *International Journal of Service Industry Management, 19*(3), 400–434.

Hair, J., Anderson, R., Tatham, R. & Black, W. (2004). *Multivariate data analysis* (5th ed.). Madrid: Prentice Hall.

Hayes, D.K. & Ninemeier, J.D. (2009). *Human resources management in the hospitality industry*: John Wiley & Sons.

Jr., B., Milkovich, R.D. & Read, G.T.W. (1989). Comparing the performance appraisal practices in large firms with the directions in research literature: Learning more and more about less and less. *Cahrs Working Paper Series*, 414.

Khatri, N. (2000). Managing human resource for competitive advantage: A study of companies in Singapore. *International Journal of Human Resource Management, 11*(2), 336–365.

Lahap, J., O'Mahony, G.B. & Sillitoe, J. (2010). *Developing a service delivery improvement model for the Malaysian hotel sector*. Paper presented at the Asia—Euro International Conference 2010, Taylors University, Malaysia.

Lahap, J.B., O'Mahony, G.B. & Dalrymple, J. (2013). *Improving service quality in Malaysian hotels: Exploring the people dimension*. Paper presented at the Global Business and Technology Association International Conference, Helsinki, Finland.

Lin, W. (2000). *Study on influence of compensation system and organizational climate on job satisfaction and job performance*. Master's thesis, Graduate Institute of Industrial Engineering, Da Yeh University.

Mondy, R. (2008). *Human Resource Management* (10th ed.). New Jersey: Pearson Prentice.

Nebel, E.C. (1991). *Managing hotels effectively: Lessons from outstanding general managers*. Canada: John Wiley & Sons, Inc.

Phonsanam, S.T. (2010). *Total compensation practices and their relationship to hospitality employee retention*. Master of Hospitality Administration, University of Nevada, Las Vegas. Retrieved from http://digitalscholarship.unlv.edu/thesesdissertations/561/

Prime Minister Department. (2013). Economic transformation programme annual report 2012.

Tourism Malaysia. (2012). *Tourism Report: Malaysian Tourism Promotion Board*. Retrieved from www.corporate.tourism.gov.com.my.

Tsaur, S.-H. & Lin, Y.-C. (2004). Promoting service quality in tourist hotels: The role of HRM practices and service behavior. *Tourism management, 25*(4), 471–481.

Wayne, M. (2008). *Human Resource Management* (10th ed.). New Jersey: Pearson Prentice Hall.

Theory and Practice in Hospitality and Tourism Research – Radzi et al. (Eds)
© 2015 Taylor & Francis Group, London, ISBN 978-1-138-02706-0

Six Sigma as a source of service delivery improvement methodology for the Malaysian hotel sector

J. Lahap
Universiti Teknologi MARA, Penang, Malaysia

B. O'Mahony & J. Dalrymple
Swinburne University of Technology (SUT), Melbourne, Australia

ABSTRACT: This paper is presented as to examine the significance of Six Sigma methodology approach in the Malaysian hotel industry. It was known that there are many strategies that had been developed over the years to improve service delivery in the hospitality industry. The industry by norm is known to use human to deliver services. Therefore, seeking the best method to increase employee job performance is vital. From the review of the literature it was found that there are several hospitality establishments used Six Sigma to enhance their service delivery efficiency. In Six Sigma, 6σ symbolizes a specific number, which is 3.4 Defects Per Million Opportunities (DPMO), where opportunity is understood as any possible source of error in product, process, or service. There are three principles that builds Six Sigma methodology and they are: a) teamwork, b) Statistical Control Process (SPC) and c) shared vision. Despite of that, however, this paper will only focus on a single element of the approach that is shared vision.

Keywords: Six Sigma, teamwork, statistical process control shared vision, hospitality, job performance

1   INTRODUCTION

Tourism is an important industry that contributes at least 9 percent of the world's GDP, employs over 255 million people worldwide and represents 8 percent of the global workforce (World Travel & Tourism, 2012). The industry makes a significant contribution to the economic development of many countries, with some countries reliant on tourism as a catalyst for growth and development. This is the case in Malaysia where the tourism industry has been hosting large numbers of tourists arriving for business or vacation purposes as well as transient passengers' en-route to other destinations. In 2013, the hospitality and tourism sector contributed MYR 65.44 billion (US$ 20 billion) towards Malaysia's Gross Domestic Product (GDP) and it was ranked second, after the manufacturing sector, in terms of overall economic contribution (Tourism Malaysia, 2012). Intense competition has been a feature of the tourism industry for some time and this has led to the development of sophisticated business operation strategies. However, the hotel sector has traditionally been slow to adopt these innovations (Fyall & Spyriadis, 2003; Renaghan, 1993). O'Mahony (2006) notes, for example, that instead of improving service and developing distinctive products that satisfy the needs of their customers, hotels tend to concentrate on improving revenue through increased sales, generally by lowering prices. The review of the literature found that the Six Sigma methodology has been widely used in manufacturing businesses (Gutiérrez, Lloréns-Montes & Sánchez, 2009). However, there are number of service organizations that have embraced the product improvement methodology (Pearlman & Chacko, 2012). Kivela and Kagi (2009) stated that Six Sigma is a product improvement approach that is also well accepted by service oriented businesses. Although the methodology is designed specifically to improve productivity, the concept gained serious interest from service organizations as a method to improve the quality of services. This was due to the effectiveness of the methodology in improving products and services as well as work systems (Johns, Tyas, Ingold & Hopkinson, 1996; Kivelä & Chu, 2001; Oh & Jeong, 1996; Pearlman & Chacko, 2012; Qu, 1997).

2   LITERATURE REVIEW

Six Sigma was developed and named by Dr. Mikel Harry a Senior Engineer from the Motorola Corporation, in 1975. He designed the approach after his search for ways to reduce defects and improve production (Pearlman & Chacko, 2012). The purpose of the methodology is to improve employee performance, work processes, productivity and the quality of the product and, at the same time, reduce the cost of production (Bhote & Bhote, 1991). The main reason for the development of the Six Sigma framework was the continuous improvement in the manufacturing of complex devices involving large numbers of parts with a high probability of defects (Gutiérrez, et al., 2009). Linderman Schroeder, Zaheer and Choo (2003, p. 195) offered the following definition of Six Sigma:

> *"Six Sigma is an organized and systematic method for strategic process improvement and new product and service development that relies on statistical methods and the scientific method to make dramatic reductions in customer defined defects rates"*

In Six Sigma, 6σ symbolizes a specific number, which is 3.4 defects per million opportunities (DPMO), where opportunity is understood as any possible source of error in product, process, or service (Breyfogle III, 2003). This philosophy proposed continuous improvement in the firm as a means of increasing the efficiency of job processes (free from defects) (Pande, Cavanagh & Neuman, 2002). Therefore, by implementing Six Sigma, firms are able to decrease their defect rates in work processes, improve the quality of products, satisfy customers, reduce costs, improve efficiency, and increase productivity (Breyfogle III, 2003; Pande, et al., 2002).

2.1 The dimensions of Six Sigma (σ)

Teamwork is one of the major pillars in the Six Sigma methodology (Breyfogle III, 2003; Lloréns-Montes & Molina, 2006; Lowenthal, 2002; Pande, et al., 2002). The continuous improvement proposed in this philosophy was developed through different tasks assigned to teams of workers. The success of an improvement program depends on cross-functional teams (Dedhia, 2005; Pande, et al., 2002). Teamwork was a key factor in Six Sigma's success because team members are the main carriers of the philosophy. In Six Sigma, roles such as 'Champions', 'Master Black Belts', 'Black Belts' and 'Green and Belts' are explicitly established. According to Pande et al. (2002) and supported by Gitlow (2005), people that work in the organization can be categorized into; a) Champions (Executive Committee)—obtain resources and eliminate barriers, b) Master Black Belts (Top management)—have important abilities and deep knowledge of Six Sigma methodology, c) Black Belt (full time agent)—improvement projects, and d) Green Belt (Employee)—belong to the improvement program

but this group of employees only has part time contribution to a task. These groups of employees in the organization have different task and roles. Thus, teamwork is seen to be pivotal to Six Sigma implementation. On the other hand, Six Sigma team members are trained to improve employees' abilities, teamwork, statistical methods and tools (Gitlow, 2005; Lee & Choi, 2006; Ravichandran, 2006). Six Sigma offers a solid statistical methodology for experimentation and research (De Mast, 2006). In fact, the definition given by Linderman et al. (2003) indicates that the initiatives for improvement were grounded in powerful statistical methods. Hence, another distinctive aspect of the Six Sigma approach is its strong statistical foundation (Eckes, 2001; McAdam & Lafferty, 2004). Pearce and Ensley (2004, p. 260) defined Shared Vision as '... a common mental model of the future state of the team.' This definition represents a capacity for sharing the future image of what is desired by firm members, developing common commitment to this future image and establishing some principles for pursuing it. Shared vision was among the most important ideas concerning leadership in the twentieth century (Harrington, 1999). Employees' ability to share a future image toward which they can direct their efforts enables the achievement of a series of significant advantages for the organization. Through Shared Vision, the relationship between professional improvement, lifelong learning and long term commitment can be stimulated (Senge, 1992). These factors have significant implications for organizational performance.

2.2 The application of Six Sigma to service industry and hospitality sector

In general, the majority of studies on quality management and improvement programs originate in, and are designed to improve, product quality in the manufacturing sector (Sureshchandar, Rajendran & Anantharaman, 2001). However, scholars and specialists have posited that some principles could also be implemented successfully in the service sector (Kivela & Kagi, 2009; Pearlman & Chacko, 2012).

3 METHODOLOGY

In Part A, the demographic profiles of respondents will be presented (gender/designation). In Part B, six questions were adapted to assess Shared Vision and, again, a 7 point Likert type scale was used and the questions relating to Shared Vision were taken from the work of Gutiérrez et al. (2009). The Klang Valley was identified as the most appropriate geographical sampling area because the region includes the main Central Business District (CBD) of Kuala

Lumpur, the capital city of Malaysia, as well as the surrounding cities Putra Jaya and Selangor. The region is also recognized as the most popular choice for business, transitory and vacation tourists with more than 10.7 million tourist visits or 41.2 percent of overall international tourists in Malaysia (Tourism Malaysia, 2012). There are approximately 40-50 four and five star hotels in this region. Hence, the Klang Valley was seen to be the most appropriate geographic area for this research. Four and Five star international hotels were chosen as the target population because these establishments are recognized as having well defined management systems, organizational structures and large, well-staffed departments. The raw data underwent three different types of analyses and these were: i) principal component analysis (PCA), ii) reliability and validity testing and iii) the Cronbach Alpha value to assess consistency. Those methods had the capacity to identify the most important questions (or variables) of each dimension and to explain the variation between variables. Four hundreds questionnaires were distributed to four and five star hotels with 312 usable responses collected, which represents a 78 percent response rate.

4 FINDINGS AND DISCUSSIONS

By conducting cross tabulations between department and gender it was found that the front office department represents 50.3 percent of respondents and the food and beverage department represents 39.1 percent of the total respondents in the sample. In addition, there were respondents that represented departments directly related to the front office or the food and beverage department. This group of respondents constituted 10.6 percent of the total respondents. Overall, it was found that female respondents represented the highest percentage of respondents (60.9%) of the overall sample.

However, as the hotel industry workforce, in this study in particular, is dominated by female employees. The distribution of respondents by department is presented in Table 1.

Table 1. Respondent distribution by departments.

Departments			Gender		
---	---	---	Male	Female	
Front Office	Count		77	80	157
	%		24.7%	25.6%	50.3%
Food & Beverage	Count		37	85	122
	%		11.9%	27.2%	39.1%
Others	Count		8	25	33
	%		2.6%	8.0%	10.6%
	Count		122	190	312
	Total		39.1%	60.9%	100.0%

5 DESIGNATION

The analysis of data found that employees from the front office department constituted 38.6 percent of the total respondents, with front desk assistants or receptionists constituting the highest percentage of these respondents.

Respondents in the food and beverage department represented a total of 40.1% of the sample and waiters and waitresses from the coffee house represented (9.9%), bartenders represented (4.5%), and banqueting staff represented (9.3%) of these respondents.

Table 2. Respondents designation.

Front Office Department				
Front Office Supervisor	Count	3	1	4
	%	1.0%	.3%	1.3%
Assistant Front Office Supervisor	Count	8	16	24
	%	2.6%	5.1%	7.7%
Receptionist/Front Desk Assistant	Count	13	29	42
	%	4.2%	9.3%	13.5%
Telephone Operator	Count	5	8	13
	%	1.6%	2.6%	4.2%
Bell Captain	Count	10	0	10
	%	3.2%	.0%	3.2%
Bellman	Count	18	0	18
	%	5.8%	.0%	5.8%
Porter	Count	4	0	4
	%	1.3%	.0%	1.3%
Doorman	Count	5	0	5
	%	1.6%	.0%	1.6%
Total	21.3 %	17.3%	38.6%	
Food & Beverage Department				
Food & Beverage Supervisor	Count	13	12	25
	%	4.2%	3.8%	8.0%
Assistant Food & Beverage Supervisor	Count	5	14	19
	%	1.6%	4.5%	6.1%
Captain	Count	3	4	7
	%	1.0%	1.3%	2.2%
Waiter/ess	Count	11	20	31
	%	3.5%	6.4%	9.9%
Bartender	Count	14	0	14
	%	4.5%	.0%	4.5%
Banquet Waiter/ess	Count	0	29	29
	%	.0%	9.3%	9.3%
Total	14.8 %	25.3	40.1%	
Others				
	Count	10	57	67
	%	3.2%	18.3%	21.5%
	Total	3.2%	18.3%	21.5%
Grand Total	122	190	312	
	39.1 %	60.9%	100.0%	

6 SHARED VISION

The Kaiser-Meyer-Olkin value of 0.79 exceeded the recommended value of 0.60, and Bartlett's Test of Sphericity is statistically significant (Hair, Anderson, Tatham & Black, 2004). This showed that there is a high degree of interrelationship among the questions within the dimension of shared vision. The unrotated factor solution extracted one factor with an Eigenvalue greater than one.

According to Table 3, this factor accounted for 63.9 percent of percentage variance. In this test only one factor was generated from the analysis. Hence six questions were reduced into one factor. The pattern loadings, factor structure and factor interpretation are shown in Table 4. The dimensions were defined by the variables with significant factor loadings of 0.60 and above. Reliability tests on each of the factors indicate a Cronbach Alpha coefficient above 0.88. This means that the 6 questions can be accepted.

In shared vision respondents reported that there is a clear vision guiding the strategic goals and mission in their organization (0.69). They also asserted that the leadership of the company shares a common vision of the organization's future with hotel employees (0.86). Furthermore, different departments shared the same ambition and vision as other departments (0.83). It was also revealed that these hotel employees were enthusiastic about pursuing the goals and mission of the whole organization (0.85), and they reported that the shared vision in their organization was appropriate (0.80). Finally, they agreed with management on what was important for their organization (0.73). The evaluation of results found that the factor loading of "shared vision" was 0.6, which is above the recommended value (Hair, et al., 2004).

Thus, it can be concluded that shared vision plays a significant role in improving hotel employees' performance. Shared Vision was found to be significant to employee performance as a means of achieving personal and organizational goals and therefore, it can be suggested that Shared Vision is a significant dimension contributing to high employee performance. This is in line with several researchers' contention that an organization should clearly explain to employees the reason for the organizations existence (Gutiérrez, et al., 2009; Kivela & Kagi, 2009) (Lahap, O'Mahony & Dalrymple, 2013; Pearce & Ensley, 2004; Pearlman & Chacko, 2012). They added that employees should also be informed of the reason why they do the job and the implications of doing the job. Therefore, it can be suggested that shared vision in this study is well-understood by hotel employee in this sample. Moreover, top management knows the importance of shared vision towards their employee. This study enriches the body of knowledge towards the field of service improvement methodology.

7 CONCLUSION

Despite of the success of this study, few limitation were also faced. Time, money and sample selection constraint were the main concern in this research. The duration of 5 months of data collection limits the researcher to get more respondents. The sample selection was also found to be a barrier to this research. Therefore, Klang Valley was chosen in this study. This study was done and acts as a pilot study to assist towards bigger sample such as, the Northern and Southern part of Peninsula Malaysia as well as Sabah and Sarawak. To sum up, bigger sampling and ample time will be allocated towards future study. Overall, this study had successfully achieved its main objective that was to examine the significance of Six Sigma methodology for the Malaysian hotel Industry.

Table 3. Result of the unrotated factor extraction from the 6 questions (variables) representing Shared Vision (SV).

Factor	Eigenvalues	Percentage of Variance	Cumulative Percentage
1	3.837	63.9	63.9

Kaiser-Meyer-Olkin Measure of Sampling Adequacy = 0.79: Bartlett's Test of Sphericity = 1136.664, Significance = .00

Table 4. Result of Varimax rotated factor matrix for Shared Vision (SV).

No	Factor One	LD
1.	In the organization, there is clear vision guiding the strategic goals and missions	0.69
2.	The leadership of the company shares a common vision of the organization's future with me	0.86
3.	Our department shares the same ambitions and vision as other departments	0.83
4.	People in our department are enthusiastic about pursuing goals and missions of the whole organization	0.85
5.	I think the shared vision in the organization is appropriate	0.80
6.	I agree with management on what is important for our organization	0.73
	Reliability Cronbach's Alpha = 0.88	

LD: Factor Loading

REFERENCES

Bhote, K.R. & Bhote, A.K. (1991). *World class quality: Using design of experiments to make it happen* (2nd ed.). New York: American Management Association.

Breyfogle III, F.W. (2003). *Implementing six sigma: Smarter solutions using statistical methods*. New York: John Wiley & Sons.

De Mast, J. (2006). Six Sigma and competitive advantage. *Total Quality Management and Business Excellence, 17*(04), 455–464.

Dedhia, N.S. (2005). Six sigma basics. *Total Quality Management & Business Excellence, 16*(5), 567–574.

Eckes, G. (2001). *The Six Sigma Revolution*. New York: John Wiley and Sons.

Fyall, A. & Spyriadis, A. (2003). Collaborating for Growth: The international hotel industry. *Journal of Hospitality and Tourism Management, 10*(2), 108–123.

Gitlow, H.S. (2005). *Design for six sigma for green belts and champions: Applications for service operations— foundations, tools, DMADV, cases, and certification.* Englewood Cliffs, New Jersey: Prentice Hall.

Gutiérrez, L.J.G., Lloréns-Montes, F. & Sánchez, Ó.F.B. (2009). Six sigma: From a goal-theoretic perspective to shared-vision development. *International Journal of Operations & Production Management, 29*(2), 151–169.

Hair, J., Anderson, R., Tatham, R. & Black, W. (2004). *Multivariate data analysis* (5th ed.). Madrid: Prentice Hall.

Harrington, A. (1999). The best management ideas. *Fortune, 104*, 152–154.

Johns, N., Tyas, P., Ingold, T. & Hopkinson, S. (1996). Investigation of the perceived components of the meal experience, using perceptual gap methodology. *Progress in Tourism and Hospitality Research, 2*(1), 15–26.

Kivela, J. & Kagi, J. (2009). Applying Six Sigma in foodservice organizations. *Turizam: znanstveno-stručni časopis, 56*(4), 319–337.

Kivelä, J.J. & Chu, C.Y.H. (2001). Delivering quality service: Diagnosing favorable and unfavorable service encounters in restaurants. *Journal of Hospitality & Tourism Research, 25*(3), 251–271.

Lahap, J.B., O'Mahony, G.B. & Dalrymple, J. (2013). *Improving service quality in Malaysian hotels: Exploring the people dimension.* Paper presented at the Global Business and Technology Association International Conference, Helsinki, Finland.

Lee, K.-C. & Choi, B. (2006). Six Sigma management activities and their influence on corporate competitiveness. *Total Quality Management & Business Excellence, 17*(7), 893–911.

Linderman, K., Schroeder, R.G., Zaheer, S. & Choo, A.S. (2003). Six Sigma: A goal-theoretic perspective. *Journal of Operations management, 21*(2), 193–203.

Lloréns-Montes, F.J. & Molina, L.M. (2006). Six Sigma and management theory: Processes, content and effectiveness. *Total Quality Management and Business Excellence, 17*(04), 485–506.

Lowenthal, J.N. (2002). *Guía para la aplicación de un proyecto Seis Sigma*. Madrid: FC Editorial.

McAdam, R. & Lafferty, B. (2004). A multilevel case study critique of six sigma: Statistical control or strategic change? *International Journal of Operations & Production Management, 24*(5), 530–549.

O'Mahony, G.B. (2006). *Understanding the impact of wine tourism on post-tour purchasing behaviour.* Victoria University. Footscray Park, Melbourne Australia.

Oh, H. & Jeong, M. (1996). Improving marketers' predictive power of customer satisfaction on expectation-based target market levels. *Hospitality Research Journal, 19*(4), 65–85.

Pande, P.S., Cavanagh, R.R. & Neuman, R.P. (2002). *Las claves de Seis Sigma: La implantación con éxito de una cultura que revoluciona el mundo empresarial*: McGraw-Hill Interamericana de España.

Pearce, C.L. & Ensley, M.D. (2004). A reciprocal and longitudinal investigation of the innovation process: The central role of shared vision in product and process innovation teams (PPITs). *Journal of Organizational Behavior, 25*(2), 259–278.

Pearlman, D.M. & Chacko, H. (2012). The quest for quality improvement: Using six sigma at Starwood hotels and resorts. *International Journal of Hospitality & Tourism Administration, 13*(1), 48–66.

Qu, H. (1997). Determinant factors and choice intention for Chinese restaurant dining: a multivariate approach. *Journal of Restaurant & Foodservice Marketing, 2*(2), 35–49.

Ravichandran, J. (2006). Six-Sigma milestone: An overall Sigma level of an organization. *Total quality management, 17*(8), 973–980.

Renaghan, L. (1993). The international hospitality industry: Organizational and operational issues. In P. Jones & A. Pizam (Eds.), *International Hospitality Marketing* (pp. 165–172). London: Pitman Publishing Limited.

Senge, P.M. (1992). La Quinta disciplina. *Granica*, 56.

Sureshchandar, G., Rajendran, C. & Anantharaman, R. (2001). A conceptual model for total quality management in service organizations. *Total quality management, 12*(3), 343–363.

Tourism Malaysia. (2012). *Tourism Report: Malaysian Tourism Promotion Board.* Retrieved from www.corporate.tourism.gov.com.my.

World Travel & Tourism. (2012). World travel and tourism report:Tourism economic research.

Theory and Practice in Hospitality and Tourism Research – Radzi et al. (Eds)
© 2015 Taylor & Francis Group, London, ISBN 978-1-138-02706-0

Assessing employees' performance in Integrated Service Recovery Strategies (ISRS): The role of self-efficacy

F. Farook, C. Mason & T. Nankervis
Swinburne University of Technology, Victoria, Australia

ABSTRACT: This conceptual paper aims to assess front-line employees' performance in Integrated Service Recovery Strategies (ISRS) by incorporating the self-efficacy theory. Service recovery is a vital tool for service business following service failure. Integrating customer and employee in the recovery strategies alleviates the effectiveness of recovery management. Self-efficacy is incorporated in the framework to capture employees' service performance using SERVPERF. This study will employ the mixed method technique for data collection; combining online questionnaire and interview. Front-line employees and middle managers of hotels are the samples for the study. Front liners are the first and primary contact that customers deal with throughout their service encounters. Hence they are the most relevant personnel approach in order to obtain the data and answer the research question. The results of the study will contribute to the existing body of knowledge of services marketing; particularly in service recovery which lacks in studies focusing on employees.

Keywords: service recovery, self-efficacy, hotel, service performance

1 INTRODUCTION

Hotel development is rapidly increasing in Malaysia, especially the development of international chain hotels. In an effort to provide service of the international standard, hotels should be aware of service failure given the nature of the service, which is intangible and inseparable. Service failure must be recovered for several reasons such as avoiding negative word of mouth, enhancing customer satisfaction, building customer relationship, loyalty and impact on profits (Bitner, Booms & Tetreault, 1990). Service recovery is the attempt by the service provider to correct mistakes and bring the customer to the state of satisfaction after the failure. Traditional service recovery literature mainly emphasizes recovery satisfaction from the customers' perspective while studies on employee satisfaction are relatively scant (Lin, 2010).

An integrated approach to service recovery gives the organization a new opportunity to maintain dissatisfied clients, recover employee and learn to prevent or reduce the likelihood of future error (Johnston & Michel, 2008; La & Kandampully, 2004; Michel, Bowen & Johnston, 2009). Employees, especially the front-liners, experience job dissatisfaction, stress and demotivation when service failure occurs (Yoo, Shin & Yang, 2006). They have a high level of interaction with the customers on a daily basis and when failure happens, they have

to rectify the failure and are bound to experience similar effects as customers would (Lin, 2010). The integration of employees and customers in the recovery process ensures a more effective and efficient recovery management by service providers. The concept allows recovery to be managed proactively by achieving the benefits of successful service recovery to the customers and employees. Utilizing Albert Bandura (1977) self-efficacy theory, front-line employees' performance in ISRS will be assessed based on the SERVPERF.

2 LITERATURE REVIEW

2.1 *Service recovery strategies*

Zero defects in service are an unattainable goal (Cranage, 2004; Hart, Heskett & Sasser Jr, 1990). A failure in the service process at any point usually destroys the customer's perception of the whole experience (Cranage, 2004). Service failures are inevitable and occur in both the process and the outcome of service delivery (Lewis & McCann, 2004). Lewis and McCann (2004) added that the inseparability of production and consumption means that failures occur at the point of consumption. Service failure can put companies out of business if attention is not paid to this problem.

Thus, businesses have a second chance to correct the mistake with service recovery and bring the customer to the state of satisfaction after the failure. Successful service recovery has significant benefits: it enhances customers' perception of the service and organization, leads to positive word-of-mouth publicity, enhances customer satisfaction, builds customer relationship, loyalty and impacts on profits (Bitner et al., 1990). While failure is inevitable, it is critical for an organization to establish service recovery strategies as part of customer retention strategies. According to the literature, there are various strategies for recovery, for example apology and correcting the mistake (Lewis & McCann, 2004). According to Cranage (2004), the type of recovery strategy depends on the severity of failure. Bell and Zemke (1987) suggested five strategies for service recovery: apology, urgent reinstatement, empathy, symbolic atonement and follow-up, while Bitner et al. (1990) proposed acknowledgement, explanation, apology and compensation. An apology is sufficient for a less serious failure, while a managerial intervention is necessary for a more serious failure (Cranage, 2004). A good recovery can turn angry, frustrated customers to loyal ones. It can create more goodwill than if things had gone smoothly (Hart et al., 1990).

2.2 Integrated Service Recovery Strategies (ISRS)

Service recovery literature has always focused on a single dimension, either investigating the customer or the employee. In fact, there are more studies on customer satisfaction with service recovery compared to employee recovery satisfaction (Lin, 2010). The need to investigate internal recovery is evidenced by that fact and with the new concept called Integrated Service Recovery Strategies (ISRS), the recovery can be managed in a more efficient manner. ISRS is an integrated concept of managing failure by proactively preventing failures, efficiently recovering and learning from mistakes and successfully maintaining long-term relationship (Santos-Vijande, Díaz-Martín, Suárez-Álvarez & Río-Lanza, 2013).

Tensions among customer recovery, process recovery and employee recovery caused service recovery to fail, reinforcing that effective recovery management requires an integrated approach (Michel et al., 2009). ISRS would assist businesses in developing practices that allow firms to perform the recovery procedures in a proactive, relational and strategic manner (Santos-Vijande et al., 2013). Santos-Vijande et al. (2013) also added that ISRS is a holistic view of service recovery which reinforces advisability of considering service recovery as a comprehensive management system. This study seeks to extend the existing work (Santos-Vijande et al., 2013) by implementing ISRS in the hotel industry and focusing on employees as compared to from B2B and in the manufacturing industry.

2.3 Theoretical development of ISRS

According to Santos-Vijande et al. (2013), there are three dimensions in ISRS:

a) Failure detection—internal and external orientation (customer and/or employees)
b) Failure analysis—service shortcomings are recognized collectively and the willingness to work to avoid them becoming tangible
c) Response to failure—rapid response, fair outcome, employee empowerment, learning innovation.

Failure detection falls under the category of recovery information. Smith, Karwan, and Markland (2009) coined the term 'accessibility' in his seven dimensions of integrated recovery system but limit this dimension to capturing the voice of the customers only. Hence, ISRS includes customer and employees information pertaining failure detection to enable the gathering of complete data on why failure happens. Next is failure analysis where Santos-Vijande et al. (2013) believes that ISRS should include the analysis and collective assessment of information available on the failure in order to establish proactively the necessary measure for improvement and to respond as efficiently as possible when failure occurs. Johnston and Michel (2008) include the analysis and interpretation of information on failure within the 'process recovery' dimension and uses 'system intensity' and 'comprehensiveness' as a measure of the magnitude of the resources devoted to monitoring and controlling failures and gathering information about all potential recovery activities. However, it is essential to share and analyze information on service failure given the potential complexity of the organizational response necessary to achieve service recovery and value creation so as to foster joint interpretation of its implications, reach a consensus on service improvement priorities and achieve long-term improvements in performance (Tax & Brown, 1998).

The following ISRS element is response rate which fall under the category of recovery action and presented in two fold orientation—external and internal. The external orientation is targeted to customer, primarily to front office (front-line) corrective actions that seek to maintain or even increase client satisfaction and avoid damaging the intention to repurchase (Kau & Loh, 2006) while the internal dimension corresponds with back office operations that seek to transform the knowledge generated within the firm into service improvements, innovations and allow employee recovery

(La & Kandampully, 2004). Santos-Vijande et al. (2013) proposed external oriented response into three factors called rapid response, fair outcome and employee support while internal orientation refers to learning innovation. Rapid response is similar to Procedural Justice in Perceived Justice Theory. Customers' value speedy recovery rates response time as crucial is service recovery (del Río-Lanza, Vázquez-Casielles & Díaz-Martín, 2009; La & Kandampully, 2004). Fair Outcome is similar to Distributive Justice and La and Kandampully (2004) propose that compensation is the second crucial recovery action. Although speedy recovery is important, customers expect compensation for the damages. Subsequently, employee support is comparable to Interactive Justice and DeWitt and Brady (2003) mentioned that it is important to empower employees to prevent failures and to fix them rapidly if they re-occur. Nevertheless, responsibility or empowerment has to be backed up by training; to encourage them (employees) in their work and gives them confidence to use the discretion they have received. Santos-Vijande et al. (2013) also added that employees need to have minimum level of authority to detect, resolve any quality-related problem especially front-line employees. This dimension should contribute to employees' satisfaction after the failure since it is the employee who can make decisions to rectify the process and compensate client immediately (Santos-Vijande et al., 2013). Learning and Innovation is the external orientation of response rate in ISRS which refers to using the failure and recovery as an opportunity to constantly innovate service delivery system to a higher standard. While correcting the mistake, providing the compensation and empowering the employees are all important, a firm should not neglect the importance of change in organization to prevent future failures, to minimize the risk of problems and reduce cost of any lack of quality (Santos-Vijande et al., 2013). Furthermore, Nonaka (1991) affirms that successful firms are those that consistently create new knowledge, disseminate it throughout the organization and embody it rapidly in new technologies and product while Nevis, DiBella, and Gould (1995) stresses that service failure should be seen as opportunities for learning. When a firm is able to learn from mistakes and offer improved or new service, it can have long-term results (La & Kandampully, 2004; Slater, 2008; Vos, Huitema & Lange-Ros, 2008). Service firms need to keep up with ever changing trends to sustain the business and this dimension distinguishes the integrated recovery system introduced by Smith et al. (2009) which is more internally focused but does not explicitly take into account the learning innovation arising from the management of failures. To reiterate, the concept of integrated service recovery strategies is viable for businesses to create a proactive and efficient service recovery management especially in the service business where failure occurs on a daily basis.

2.4 Self-efficacy

The concept of self-efficacy evolves from Albert Bandura's social cognitive theory (Bandura, 1977). Self-efficacy is a theory based on the strong beliefs that psychological procedures serve as means of creating and strengthening expectations of personal efficacy (Bandura, 1977). According to (Bandura, 1977) a person's abilities, attitudes and cognitive skills comprise of what is known as 'self-system'. The system decides how we perceive, behave and response in given situations and self-efficacy is an essence of this self-system. An efficacy belief influences peoples thinking, feeling, motivation and act (Bandura, 1995, p.2). Developing self-efficacy requires the combination of cognitive and physical effort over a period of time. Realistic and achievable goals must be established for one to be fully efficacious. Perceived self-efficacy refers to beliefs in one's capabilities to organize and execute the courses of action required to produce given attainments (Bandura, 1995, p.3). The strength of people's conviction in their own effectiveness is likely to affect their willingness to even try to cope with given situations (Bandura, 1977). If one perceives that they can achieve or succeed in an activity, then the action will lead towards achieving it. Perceived self-efficacy serves as directive influence on choice of activities and settings as well as affect coping efforts, through expectations of success once the choice is initiated (Bandura, 1977). This is followed by efficacy expectation. Efficacy expectation is defined as the conviction that one can successfully execute the behaviour required produce the outcome (Bandura, 1977) while outcome expectancy is defined as a person's estimate that a given behaviour will lead to certain outcomes. Efficacy expectation is a major factor of people's choice of activities, how much effort they will expend and how long they will sustain effort in dealing with stressful situation (Bandura, 1977); the expectations vary on several dimensions that have important performance implications such as magnitude, generality and strength. Therefore, it is clear that self-efficacy explains how motivation comes about as well as the entire process of action and performance in a rather systematic manner.

2.5 Self-efficacy and ISRS

Since the self-efficacy theory was introduced, it has received a lot of attention from scholars and the theory has been applied to various research involving the assessment of people with certain situations such as travelling by automobile, using elevators and escalators, climbing stairs to high levels, dining in restaurants, shopping in supermarkets and venturing alone in public spaces (Bandura, 1982). Besides that, there were also studies in other field or disciplines. One example is the study by Karatepe, Arasli, and Khan (2007) examining the effect of self-efficacy on job performance, job satisfaction and affective organizational commitment among hotel employee. The result showed that highly self-efficacious employees perform more effectively in the workplace.

In the context of this study, self-efficacy is incorporated in the implementation of ISRS in hotel industry to encapsulate the efficacy factors within the employees and middle managers. Self-efficacy influences peoples thinking, feeling, motivation and act. Hence, understanding the employees and middle managers efficacy beliefs toward ISRS enables the hotel to implement ISRS effectively. Self-efficacy development involves physical and cognitive effort, over time before the person becomes fully efficacious and it is worthwhile assessing the efficacy factors in relation to ISRS, an emerging concept.

Middle managers often need to intervene in the recovery process upon customers demand. Incorporating self-efficacy in the process of implementation of the ISRS in hotels would enable the researcher to examine the acceptance and perception of ISRS among middle managers and employees. Apart from that, empirical evidence shows that firms with ISRS can achieve a more valuable client base since customer satisfaction, loyalty and perceived value are enhanced.

2.6 Service performance

The SERVPERF model emerged from SEVQUAL introduced by Parasuraman, Zeithaml, and Berry (1985) to measure service quality. The SERVQUAL model received criticism by scholars particularly the way it measures expectation among customers. Therefore Cronin Jr and Taylor (1992) argued that service quality measurement should be based on performance. Since then, SERVPERF has gained a lot of attention by service marketing scholars and many studies has utilised SERVPERF in order to measure service quality. In reference to this study, modified SERVPERF will be used to measure employees' performance with regards to the implementation of ISRS in the hotel industry. The existing scale measurement is replicated from the study by Yilmaz (2009). His study revolves around measuring hotel service quality performance from the customers' perspective in Turkey. For the purpose

of this study, the element of tangible is removed for the reason that the measurement is based on employees' perception of service performance in ISRS.

3 RESEARCH CONTRIBUTION

This study will contribute to the existing knowledge of service marketing literature generally and the service recovery area specifically. Previous studies on service recovery mainly revolve around customer satisfaction, loyalty and profitability and lacks in focusing on employees, especially frontline employees. Service failure and service recovery affect both the customers and employees. Hence, focusing on employees would create the balance in the service recovery literature. Apart from that, this study would improve the understanding of the effectiveness of integrated service recovery strategies, specifically in the hotel industry.

By assessing the employees and middle managers performance with ISRS, the acceptability and suitability of the new concept can be captured. Apart from that, hotel management will have a clear understanding of the approach to implement ISRS specifically by focusing on the self-efficacy factors identified from the result. When ISRS is incorporated in the hotel's recovery strategy, the outcome would ensure reduced level of stress, dissatisfaction and frustration among employee and hence directly contribute to proactive and efficient recovery management. Following this, customers would experience improved service, reducing the risk of service failure as well as guaranteed effective service recovery. Finally, the outcome of effective recovery management is important in ensuring the sustainability of the tourism industry which is crucial to the Malaysian economy.

4 CONCLUSION

An integrated approach to service recovery gives the organisation a new opportunity to maintain dissatisfied clients, recover employee and learn to prevent or reduce the likelihood of future error (Johnston & Michel, 2008; La & Kandampully, 2004; Michel et al., 2009). The benefit and importance of an integrated service recovery strategy is evidently clear and this study will shed light towards the effectiveness of ISRS in hotel industry based on front-line employees' service performance by incorporating self-efficacy theory in the framework of the study. The result will provide theoretical as well as managerial implications towards the betterment of the hospitality and tourism industry in Malaysia.

REFERENCES

Bandura, A. (1977). Self-efficacy: Toward a unifying theory of behavior change. *Psychological Review, 84*(2), 191–215.

Bandura, A. (1982). Self-efficacy mechanism in human agency. *American Psychologist, 37*(2), 122–147.

Bandura, A. (1995). *Exercise of personal and collective efficacy in changing societies* (A. Bandura Ed.). USA: Cambridge University Press.

Bell, C.R. & Zemke, R.E. (1987). Service breakdown, the road to recovery. *Management Review, 76*(10), 32–36.

Bitner, M.J., Booms, B., H. & Tetreault, M.S. (1990). The service encounter:Diagnosing favorable and unfavorable incidents. *Journal of Marketing, 54*(1), 71–84.

Cranage, D. (2004). Plan to do it right: Plan for recovery. *International Journal of Contemporary Hospitality Management, 16*(4), 210–219.

Cronin Jr, J.J. & Taylor, S.A. (1992). Measuring Service Quality: A Reexamination and Extension. *Journal of Marketing, 56*(3), 55–68.

del Río-Lanza, A.B., Vázquez-Casielles, R. & Díaz-Martín, A.M. (2009). Satisfaction with service recovery: Perceived justice and emotional responses. *Journal of Business Research, 62*(8), 775–781.

DeWitt, T. & Brady, M.K. (2003). Rethinking Service Recovery Strategies: The Effect of Rapport on Consumer Responses to Service Failure. *Journal of Service Research, 6*(2), 193–207.

Hart, C.W.L., Heskett, J.L. & Sasser Jr, W.E. (1990). The profitable art of service recovery. *Harvard Business Review, 68*(4), 148–156.

Johnston, R. & Michel, S. (2008). Three outcomes of service recovery: Customer recovery, process recovery and employee recovery. *International Journal of Operations & Production Management, 28*(1), 79–99.

Karatepe, O.M., Arasli, H. & Khan, A. (2007). The Impact of Self-Efficacy on Job Outcomes of Hotel Employees: Evidence from Northern Cyprus. *International Journal of Hospitality & Tourism Administration, 8*(4), 23–46.

Kau, A.-K. & Loh, E.W.-Y. (2006). The effects of service recovery on consumer satisfaction: a comparison between complainants and non-complainants. *Journal of Services Marketing, 20*(2), 101–111.

La, K.V. & Kandampully, J. (2004). Market oriented learning and customer value enhancement through service recovery management. *Managing Service Quality, 14*(5), 390–401.

Lewis, B., R. & McCann, P. (2004). Service failure and recovery: Evidence from the hotel industry. *International Journal of Contemporary Hospitality Management, 16*(1), 6–17.

Lin, W.-B. (2010). Relevant factors that affect service recovery performance. *The service industries journal, 30*(6), 891–910.

Michel, S., Bowen, D. & Johnston, R. (2009). Why service recovery fails: Tensions among customer, employee, and process perspectives. *Journal of Service Management, 20*(3), 253–273.

Nevis, E.C., DiBella, A.J. & Gould, J.M. (1995). Understanding organizations as learning systems. *Sloan Management Review, 36*(2), 73–85.

Nonaka, I. (1991). The knowledge-creating company. *Harvard Business Review, 69*(6), 96–104.

Parasuraman, A., Zeithaml, V.A. & Berry, L.L. (1985). A Conceptual Model of Service Quality and Its Implications for Future Research. *Journal of Marketing, 49*, 41–50.

Santos-Vijande, M.L., Díaz-Martín, A.M., Suárez-Álvarez, L. & Río-Lanza, A.B.D. (2013). An Integrated Service Recovery System (ISRS): Influence on Knowledge-Intensive Business Services Performance. *European Journal of Marketing, 47*(5), 934–963.

Slater, S.F. (2008). Learning how to be innovative. *Business Strategy Review, 19*(4), 46–51.

Smith, J.S., Karwan, K.R. & Markland, R.E. (2009). An Empirical Examination of the Structural Dimensions of the Service Recovery System. *Decision Sciences, 40*(1), 165–186.

Tax, S.S. & Brown, S.W. (1998). Recovering and learning from service failure. *Sloan Management Review, 40*(1), 75–88.

Vos, J.F.J., Huitema, G.B. & Lange-Ros, E. d. (2008). How organisations can learn from complaints. *The TQM Journal, 20*(1), 8–17.

Yilmaz, I. (2009). Measurement of Service Quality in the Hotel Industry. *Anatolia: An International Journal of Tourism and Hospitality Research, 20*(2), 375–386.

Yoo, J.J.-E., Shin, S.-Y. & Yang, I.-S. (2006). Key attributes of internal service recovery strategies as perceived by frontline food service employees. *International Journal of Hospitality Management, 25*(3), 496–509.

Theory and Practice in Hospitality and Tourism Research – Radzi et al. (Eds)
© 2015 Taylor & Francis Group, London, ISBN 978-1-138-02706-0

Revisit the legal duty of hoteliers towards contractual entrants and invitees in hotels

N.C. Abdullah
Faculty of Law, Universiti Teknologi MARA, Shah Alam, Malaysia

ABSTRACT: In all jurisdictions, the standard of duty of care is the issue that arises as a result of injury of any entrant in any hotels. In the complicated common law concept of duty of care adopted in many Western jurisdictions, a guest in a hotel who pays to use the accommodation is classified as contractual entrant; but he is an invitee when he uses common restaurant in the hotel. Hence, the common law concept causes confusion and is often being criticized for its ambiguous nature. Ironically, many decided cases in Malaysia, which is well-known for its strict adherence of common law; illustrates that the determining factor in deciding the liability of hotelier is whether the injury is a result of unusual danger which is frequently associated with the concept of invitee attached to a paying guest, in the absence of discussion on the distinction between invitees and contractual entrants. This paper discusses the complicated task of the judiciary to determine the duty of care owed by hotelier as a result of the different classification of entrants and the different standards of duty of care owed to these categories in the current context of hospitality industry whereby hotels no longer only provide the traditional services of only food, beverages and accommodation. This paper argues that though statutory provisions of duty of care is not available in Malaysia, sometimes, the judiciary tends to apply the standard duty of care established in the case of *Donoughue v. Stevenson*, though at times, technical issues of classification of entrants and different standards of duty of care are mentioned in the judgments. The paper concludes that the exiting framework can be better strengthened with the application of a standard duty of care which is applicable to all lawful entrants to the hotel.

Keywords: hoteliers, duty of care, contractual entrant, invitees, unusual danger

1 INTRODUCTION

Under the common law, the standard of duty of care differs from a contractual entrant, invitee and licensee although these different groups of entrant, contractual entrant, invitee and licensee are legal entrants to the inn. The contractual entrants are divided into two categories, i.e. main purpose entrant and ancillary purpose entrant. As for the main purpose entrant, an occupier has a duty to ensure that the premise is safe for the purposes of the contract as reasonable care and skill on the part of anyone can make them. On the other hand, for the ancillary purpose entrant, an occupier has a duty to ensure that the premise is in all respect reasonably safe for the purposes for which the other party was invited to use them. Invitees are also divided into two categories, i.e. the legally authorized entrants and business visitors. However, the occupier owes the same duty of care to both these two categories of invitees i.e. a duty to take reasonable care to prevent damage from 'unusual danger', of which the occupier knows or ought to know and

of which the entrant does not know. The feature of a contractual entrant is that an entry is made lawful or permitted by the contract itself—and not, for example, by virtue of conferring a business benefit on the occupier (as in the case of an invitee) or by virtue of the occupier's express or implied permission (as in the case of a licensee).

2 COMPLICATED CONTRACTUAL DUTY AND DEBATABLE CONCEPT OF 'UNUSUAL DANGER'

The nature of the duty owed to contractual entrants turns on the express of implied terms of the contract. In the absence of an express term, an implied term may vary depending on the type of contract. Where there is no express term or well-settled standard for the type of contract in question, the term that is usually implied is that the premises are as safe as reasonable care on the part of anyone can make them. It is now clear, despite some earlier suggestions to the contrary, that an occupier is not normally considered to have given an absolute guarantee of the safety of the premises. Nevertheless, the duty is more than an ordinary duty of reasonable care, and has been described rather as a duty to see that care is taken. Thus, the occupier is liable not only for the negligence of himself and his servants but also of that of independent contractors and previous occupiers. In the case of *Maclenan v. Segar* [1917] 2 K.B. 325, that plaintiff is a guest in a hotel who was injured in a fire which was caused by the negligence of contractors employed by a previous owner to the remodel the kitchen area of the hotel. The nature of the duty owed to her was expressed in the following terms: "Where the occupier of premises agrees for reward that a person shall have the right to enter and use them for a mutually contemplated purpose, the contract between the parties (unless it provides to the contrary) contains an implied warranty that the premise are as safe for that purpose as reasonable care and skill on the part of anyone who can make them. The rule is subject to the rule that the defendant is not to be held responsible for defects which could not have been discovered by reasonable care or skill on the part of any person concerned with the construction, alteration, repair or maintenance of the premises it matters not whether the lack of care or skill be that of the defendant or his servants, or that of an independent contractor or his servants, or whether the negligence takes place before or after the occupation by the defendant of the premises."

In *Culvert v. Stollznow* (19821 1 N.S. W.L. R. 175) the New South Wales Court of Appeal held that a restaurant patron was to be treated as an invitee so far as the structural condition of the premises was concerned, albeit that his contract for the provision of food and drink would contain a term, implied in fact, that he had a right to remain on the premises during the meal. Drinkers in hotels are also to be treated as invitees instead of contractual entrants *(Whiteman v. Boyd* [1962] N.S.W.R. 328). However, in *Branninen v. Harrinnton* (1921) 37 Times L.R. 349, a restaurant customer is considered a contractual entrant. The Court appears to have thought that the relevant distinction was not between contracts whose main purpose was the provision of accommodation on the part of the premises where the plaintiff was injured, and those where the use of the premises was merely incidental to some other contract. Rather a distinction should be drawn between contracts by virtue of which the plaintiff could be said to have 'bought' the right to enter the premises and those which merely entitled him to remain on or use the premises. Even a theatregoer or sporting spectator would not be entitled to the higher duty as the main purpose of their contracts is not accommodation in the premises but the viewing of a spectacle. However, it may be queried whether the distinction between a contractual right to enter and a contractual right to remain draws the line at the right point. It would seem that there is a further requirement, namely that use of the premises was central to or the main purpose of the contract, rather than merely ancillary or incidental to it. (Winfield & Jolowicz on Tort, 1984; J.G. Fleming, The Law of Torts, 1987) Furthermore, bearing in mind that the rationale for the imposition of a higher duty on the occupier towards contractual entrants is that the occupier is being paid for granting permission to enter, it is no doubt necessary that the entry should be for a purpose beneficial to the contemplated purpose or at any rate that it should be for a mutually contemplated purpose.

It seems that even those who do qualify as contractual entrants may be disentitled to the higher duty if the injury occurs on a part of the premises to which the public is entitled to resort, since here the permission to enter does not derive from the contract. Thus, it has been said in *Calvert v. Stollznow* [19821 1 N.S.W.L.R. 181: "a paying guest in a hotel may be a contractual entrant upon those portions of the premises reserved for paying guests, because he has paid to enter them. But he may be an invitee in the hotel's public restaurant. Similarly, a visitor to the theatre may be an invitee in the foyer, although a contractual entrant to the auditorium to which his ticket admits him. He may be an invitee in the lavatories if they are open to the public; but otherwise if they are reserved for holders of a ticket to the performance."

3 SURVIVAL OF THE 'UNUSUAL DANGER' CONCEPT

It appears that the Malaysian Courts often classify the legal entrant who enters into premise which benefits the occupier as invitee and hence, applied the concept of unusual danger. In the Malaysian case of *Lee Lau & Sons Realty Sdn Bhd v Tan Yah & Ors* [1983] 2 MLJ 51, the appellants operated a brick-making factory. They also owned a forklift machine which was under the management of one Tan Kam Sing. While repairs were being affected to the forklift, Tan Kam Sing (the deceased) was injured and he later died from his injuries. In this case, the court started to discuss a bit on the extent of duty of care by occupier. Here, the court decided that the invitor's duty is confined to protection against unusual dangers, that is, those not usually found in carrying out the task which the invitee has in hand and the onus of pleading and proving 'unusual' danger which the appellants knew or ought to have known rests on the respondents. Thus, this is in line with the case of Lau Tin Sye pertaining to the standard of care used.

In the Malaysian case of *Takong Tabari v. Government of Sarawak & Ors* [1996] 5 MLJ 435, an explosion occurred in Public Bank which caused deaths and injuries to the customers inside the premise. The court held that as the occupier, Public Bank owed a duty to the deceased, an invitee, to prevent damage or injury to the invitee due to unusual dangers on the premises which it knew or ought to know and which the invitee did not know. Since Public Bank, through its staff, did not take any reasonable or reasonably sufficient measures to meet its duty as the occupier towards the deceased, the plaintiff therefore had established, on a balance of probabilities, the liability of Public Bank as the occupier of the premises. One of the issues here is whether there was a breach of duty by the Public Bank as an occupier. The court elaborates little on the invitee. Before making any finding of liability as an occupier, it must be determined whether the deceased was an invitee or lawfully in the premises at the material time. He was in fact a customer of Public Bank. He was in the premises for a business purpose of material benefit to the occupier, Public Bank. There was in fact a common interest between the deceased and Public Bank. Accordingly, the deceased could not be described as anything else but an 'invitee' under common law. This shows that an invitee is a person who has a common interest with the occupier. Like any other cases, the court did not dispute the category of entrant of the plaintiff.

In this case, the court here explains about the unusual danger or unusual risk. In respect of the meaning of 'unusual risk', the judge quoted the case of *London Graving Dock Co Ltd v Horton* [1951] AC 737, which said that an 'unusual' risk is one which is not usually found in carrying out the task which the invitee has in hand. In the present case, the premise was used for banking business. Surely, no customer would expect to find gas, an admittedly dangerous thing, in such a place. Thus, the danger was unusual to the deceased and unknown to him at that material time. Accordingly, due to lack of any action taken by the staff of Public Bank even with the smell prevailing for about a month, the Public Bank through its staff did not take any reasonable or any reasonably sufficient measures to meet its duty as the occupier towards the deceased, an invitee. This case is more comprehensive in the sense that it illustrates the standard of care required towards an invitee which is to take reasonable care to prevent injury arising from unusual danger.

4 ABOLISHMENT OF COMMON LAW CATEGORIES

In US, the courts moved away from the rigid common law standard of care based on categories of entrants. In the case of *O'Leary v. Coenen* (251 N.W.2d 746, 749–50 (N.D. 1977), the plaintiff, an insurance salesperson, was bitten by the defendant's dog while approaching the defendant's farm to carry out business. The plaintiff's presence was uninvited and unexpected. The North Dakota Supreme Court held that "the status of an entrant, a licensee or an invitee is no longer solely determinative of the duty of care owed such entrant." The court unanimously held it would generally apply a single standard of reasonable care under all the circumstances in premise liability cases. In accordance with ordinary principles of negligence, this standard includes foreseeability of plaintiff's presence, likelihood and seriousness of the injury and the burden of avoiding the risk of injury. In the case of *Clayton W. Williams, Jr., Inc. v. Olivo* (912 S.W.2d 319; Tex. App.-San Antonio 1995), an employee of an independent contractor sued Clayton Williams, Jr., Inc., the operator of an oil and gas lease for personal injuries to his back when he fell from a pipe rack and landed on a piece of equipment left lying on the ground. The court held that the evidence was sufficient to support the verdict under a simple negligence theory because Clayton Williams and its supervisor retained some right to control Olivo's work as an employee of a sub-contractor.

This case demonstrates how the court is ready to apply a duty of ordinary care and submission of jury issues under a simple negligence jury submission. It is applicable in all premises liability suits without consideration of whether the claim was one of negligent activity or premise condition. The courts in US indicate that they are taking a major step toward abolishing the common law categories. Common law categories simply do not reflect the issues presented in present premise liability suits, some of which would have been unimaginable when such categories were judicially created. Abolishing the common law categories and adopting a duty of reasonable care as applied under the circumstances will bring simplicity, more predictability and hopefully greater fairness to the citizens of US.

5 GENERALISED STANDARD OF REASONABLE CARE

In Canada, it is doubtful that this highly structured responsibility ever worked very well. Any attempt to place entrants into a limited number of fixed categories could produce difficulties, anomalies and uncertainties (McDonald, D.C. & Leigh, L.H., 1965).

However, during the twentieth century, the classical doctrine of occupiers' liability fell away, leaving an archaic set of rules which was not compatible with modern notions of civil responsibility. By virtue of the case of *Donoghue v. Stevenson* ([1932] A.C. 562, 101 L.J.P.C. 119, [1932] All E.R. Rep. I.), the tort of negligence extended its reach to virtually all activities and conduct, and imposed a generalized standard of reasonable care for the safety of others. After 1932, it became difficult to defend the different standards of care owed to the contractual entrant as compared to the licensee. The degree of proximity, which was the touchstone of the obligation of reasonable care under *Donoghue v. Stevenson*, did not seem to differ greatly and this clearly created an added pressure on the established doctrine. The substantive principles of the tort began to be re-constructed by the courts and re-adjusted to provide much greater protection for injured persons. There was also an increasing reluctance to exonerate defendants on the grounds that the plaintiff had some knowledge of the danger which caused the injury. On a more practical level, societal changes during the twentieth century increased the difficulty in applying the classical doctrine. The increasing variety of land use created a plethora of problems, both in the definition of 'occupier', and in the classification of visitors into the four categories.

This appeared to be also the approach which is sometimes taken by the Malaysian judiciary. In Chang Fah Lin v United Engineers (M) Sdn Bhd [1978] 2 MLJ 259, the court did not use concealed danger as the standard of care but instead applied the concept of injury which could reasonably be foreseen. In this case, the plaintiff was in the process of fixing zinc sheets on the roof at a height of twenty five feet from the ground when he slipped and fell down sustaining severe injuries. No safety belts were used because none was provided. The court decided that as between the first defendant (United Engineers Bhd) and the plaintiff, there was an invitor/invitee relationship and therefore the first defendant were bound to take such reasonable care as would avoid the risk of injury to such person as they could reasonably foresee might be injured by failure to exercise such care. This is different from the earlier decided cases and subsequent cases because the standard of care for invitee is usually the duty to prevent the unusual danger from affecting the invitees. The court did not explain on what it means by injury which could be reasonably be foreseen. After stating the duty of the defendant, the court proceeded to explain that the defendant is the occupier and has overall occupation and control over the premise. This case is different in the sense that the occupier owes a duty to prevent injury which could reasonably be foreseen only. Hence, the test is objective one which is whether an ordinary man would foresee the injury if he is the occupier.

6 STATUTORY DECLARATION OF STANDARD OF CARE

The UK Occupiers' Liability Act in section 3(1), replaces the common law classification system and its standard of care with a generalized standard of common care without distinguishing specifically between permitted entrants and others. The occupier owes to persons on the premises and to the owners of property on the premises, such care as in all the circumstances of the case is reasonable to see that the person or property is reasonably safe. This duty of reasonable care is common to all occupiers' liability legislation in other jurisdictions. The generalized obligation of care under section 3(1) of the Act is essentially the same standard of care dictated by the negligence doctrine. The effect of imposing this standard is to raise the standard of care owed to entrants who would be treated at common law as licensees and trespassers and to lower the standard of care owed at common law to contractual entrants.

Where a person willingly assumes the risks of entering premises, section 4(1) of the Occupiers' Liability Act in Ontario substitutes a lesser duty on the occupier to not create a danger with deliberate intent to do harm and to not act with reckless disregard. Section 4(3) of the Act is a deeming provision that provides that a person who enters certain types of premises that are outlined in section 4(4) is deemed to have willingly assumed the associated risks. Those premises outlined in section 4(4) include such property as rural premises and recreational trails. The trial judge held that the lesser duty of care did not apply because the premises did not come within one of the categories listed in section 4(4). He noted that although the premises contained recreational trails, the concrete wall was not on one of the trails. The Court of Appeal allowed the appeal and dismissed the action. The Court of Appeal sensibly stated that it would make little sense to impose a lesser standard when users remained on the trail, but to impose a higher standard when they veered off of it. The trail was being used by Ms. Schneider for recreation and it met the definition of recreational trail, thus bringing it within the provisions of section 4.

7 THE WAY FORWARD AND CONCLUSION

As discussed earlier, although statutory provisions of duty of care is not available in Malaysia, sometimes, the judiciary tends to apply the standard duty of care established in the case of *Donoughue v. Stevenson*, though at times, technical issues of classification of entrants and different standards of duty of care are mentioned in the judgments. In the complicated context of a hotel where many other non-traditional services are provided, it is submitted that the exiting framework can be better strengthened with the application of a standard duty of care which is applicable to all lawful entrants to the hotel. In applying the statutory standard, it is suggested that a wide range of factors and circumstances can be adopted from experiences of other jurisdictions (Osborne, P.H., 1985). Hence, an essential condition of liability is a foreseeable risk of injury which is significant to prompt action from a reasonably prudent person. (*Preston v. Canadian Legion of British Empire Service League.* Kingsway Branch 1981 ABCA 105). In determining whether there is such a risk, the courts have imposed on the occupier a duty to make reasonable inspection of the premises. The sufficiency of the care, in light of such a risk, is determined by a consideration of the kind of premises, (*Flint v. Edmonton Country Club* (1980), 26. A.R. 391 (Q.B.)). the class of visitors, the gravity of the danger *(Meier v. Qualico Developments Ltd.* 1984 ABCA 289), the likelihood of injury or damage, the burden of removing or mitigating the danger, the practice of occupiers in such circumstances (*Epp v. Ridgetop Builders* (1979) 7 C.C.L.T. 291 (Alta. Q.B.), the unexpected nature of the danger, the weather conditions where relevant *(Lyster v. Fortress Mountain Resorts Ltd.* 1978 688 (AB QB), the purpose of entry, the degree of warning and in certain cases, special assurances of safety given to the visitor.

ACKNOWLEDGMENT

The author wishes to thank the Research Management Institute (RMI) UiTM Shah Alam for funding this project entitled "Re-visiting the Concept of Contractual Entrants and Invitees in Hotels," under the Research Intensive Fund (RIF).

REFERENCES

Atkinson, M.C. (1968). Occupiers' Liability Some Developments and Criticisms. University of Tasmania Law Review, 3(1): 82–114.

Bohlen, F. Studies in the Law of Torts. (Indianapolis, Bobbs-Merrill Company, 1926) p. 163, 190.

Buckley, R.A. (2006). Occupiers' Liability in England and Canada. Common Law World Review, 35(3), 197–215.

Charlesworth & Percy on Negligence (7th ed., London, Sweet & Maxwell, 1983) p. 425.

Clerk & Lindsell on Torts (15th ed., London, Sweet & Maxwell, 1983) p. 592.

Faridah Hussain & Nuraisyah Chua Abdullah. (2013). Uplifting the Standard of Duty of Care towards Guests: An overview of recent developments in selected countries. Elsevier, 101, 488–494.

J.G. Fleming, The Law of Torts (7th ed., Sydney, Law Book Co., 1987) p. 423.

Keener, T.H. (1997). Can The Submission of A Premises Liability Case Be Simplified. Texas Tech Law Review, 28(4): 1161–1174.

Marsh, N.S. (1953). The History and Comparative Law of Invitees, Licensees and Trespassers. Law Quarterly Review, 69(2): 182–199.

McDonald, D.C. & Leigh, L.H. (1965). Law of Occupiers' Liability and Need for Reform in Canada. University of Toronto Law Journal, 16(1): 56–88.

Nurchaya Talib. (2010). Law of Torts in Malaysia (3rd ed.). Petaling Jaya: Sweet & Maxwell Asia.

Osborne, P.H. (1985). The Occupiers Liability Act of Manitoba. Manitoba Law Journal, 15(2): 177–197.

Recent Developments, Torts-Abrogation of Common Law Entrant Classes of Trespasser, Licensee and Invitee. (1972). Vanderbilt Law Review, 25(3): 623–640.

Strenkowski, E.A. (1979). Tort Liability of Owners and Possessors of Land—A Single Standard of Reasonable Care Under the Circumstances Towards Invitees and Licensees, Arkansas Law Review, 33(1): 194–210.

The Law Reform Commission. (1988). Occupiers Liability. Canberra: Australian Government Publishing Service.

Winfield & Jolowicz on Tort (12th ed., London, Sweet & Maxwell, 1984) p. 202–203.

Theory and Practice in Hospitality and Tourism Research – Radzi et al. (Eds)
© 2015 Taylor & Francis Group, London, ISBN 978-1-138-02706-0

A snapshot of work-family conflict among hotel managers in Malaysia

M.F.S. Bakhtiar, N. Sumarjan, S. Tarmudi & S.M. Radzi
Faculty of Hotel and Tourism Management, Universiti Teknologi MARA, Shah Alam, Malaysia

I.M. Ghazali
Faculty of Hotel and Tourism Management, Universiti Teknologi MARA, Terengganu, Malaysia

ABSTRACT: Authors of papers to proceedings have to type these in a form suitable for direct photographic reproduction. The nature of work within hotel industry is known to have caused managers sacrificing their time with families and personal lives to make way for relentless work demands. For such reason, it is imperative to assess the influence of work variables and non-work variables towards work-family conflict among managers within the hotel industry. This paper unveils such understanding through the perspective of hotel executives in Malaysia. Through self-administered questionnaire, 51 valid responds were gathered among executives of five stars hotels in Kuala Lumpur to unravel the relationship between all variables. Several formulated hypotheses were tested following data collection. Findings revealed that work variable and non-work variable had a negative influence and shows insignificant relationship towards work-family conflict.

Keywords: Work variable, non-work variable, work-family conflict, hotel industry

1 INTRODUCTION

Contribution of competent and loyal employees towards organizational profits has been affirmed by various researches. It is clear that the hotel unable to achieve excellent operations without excellent employees (Enz & Siguaw, 2000). These understandings are suggestive of the fact that hospitality employees are the most significant resource within the organization. Being a part from the hospitality spectrum, hotel managers are responsible to ensure that the attitudes and behaviors of their employees are consistent with expectations of their customers and the organization to be successful.

Since the hotel industry is a people-oriented business, hotel managers are expected to be able to cope with the effects of role stress and work-family conflict while not affecting their job outcomes. Because employees can play different roles at work and at home, many organizations are increasingly concerned with the management of individuals' career and family roles (Mulvaney, O'Neill, Cleveland & Crouter, 2007). Work-family conflict occurs once individuals are required to cope with conflicting role pressures between job and family simultaneously (Ballout, 2008). To balance between both pressures is challenging; hence for most employees, careers involve a trade-off between individual success and family success.

The nature of hotel operations is 24 hours a day and 7 days a week requiring staffing and services continuously. Because of the demands of being a "24/7" industry, hotel companies often have norms that encourage their employees, especially executives, to work as many hours as possible, including weekends and holidays (Ismail, 2002). Turnover among hotel managers is primarily caused from having to work long hours while not having minimal time to spend with families (Stalcup & Pearson, 2001). Other work attributes including non-standard and irregular hours shift have also been well-established to correlate with lower marital quality and divorce, children with more problem behaviors, as well as increased life-conflict (Presser, 2004).

2 LITERATURE REVIEW

A comprehensive review is made on the studied variables including work variable (WV), non-work variable (NWV) and work-family conflict (WFC); this is further supported relevant theoretical framework.

2.1 *Work Variable (WV)*

Work is considered as central to human existence, providing the necessities for life, sources of identity, opportunities for achievement, and determining standing within the larger community. Demand of work placed on individuals employed in the organization has gradually increased (Farrell & Geist-Martin, 2005). Through their studies in related hospitality employees; about half from total respondents believed that they spent most their life at work while 78 percent of them reported that their primary cause of stress were contributed by work. From the above contemplation, there is noteworthy to study about job related stress since many researchers also believed that this negative emotional state may extend and contributing towards work-family conflict (Fu & Shaffer, 2001).

An extensive literature revealed that a group of five elements being most relevant to affect one's wellbeing at workplace including number of work hours; schedule inflexibility; work stressor; job involvement; and social support at work. Based on Anderson and Ungemah (1999) findings, the work hours' programs variables are becoming more popular with both large and small companies; employers realize they are good for employee's morale and for business. Second elements involved is schedule inflexibility which refers to the inability to alter one's work schedule to meet work and non-work pursuits, including the family (Golden, 2009) and will cause "time bind", a term means an imbalance between the demand of work and demand of family or personal life (Hochschild, 1997). Next, work stressor may refer to any characteristic of the workplace that poses a threat to an individual. This may be related to either work demand a person cannot meet or the lack of sufficient resources to meet work needs (Interactive, 2011). Forth element; job involvement was defined as the degree to which a person identifies psychologically with the job, and the importance of the job to the person's self-image and self-concept (Uygur & Kilic, 2009). Ballout (2008) noted that job involvement is one potential work-related variable that may cause work-family conflict. Another work-related variable that relates to work-family conflict is job social support have pointed out that social support from the non-work domain, such as the partners or family members is of great importance in reducing work-family conflict (Sloan, Newhouse & Thompson, 2013).

2.2 Non-Work Variable (NWV)

One's wellbeing may be affected by various personal and social factors in addition to those encountered at work. Non-work stress can be defined as a real or imagined imbalance between the demands on the home and the family's ability to meet those demands (Jones & Fletcher, 1996). This form of pressure may exert an individual's emotional state; meanwhile excessive and uncontrollable stress may contribute towards work-family conflict in the long run (Bazana & Dodd, 2013). Through literature review, various antecedents were reviewed and a group of three elements being most relevant to affect one's wellbeing outside of work namely gender, parental demands as well as working spouse (Kim & Ling, 2001).

Gender is perhaps the socio-demographic characteristic most often included in studies of work-family conflict and career success (Ramadoss & Rajadhyaksha, 2012). Traditional gender roles assumed that men are primarily responsible for family financial support by working, while women are primarily responsible for childcare and household duties (Loscocco, 2000). Next, a study by Ballout (2008) indicates that parental demand can affect work-family conflict and subsequently success in careers. Parents tend to experience more work family conflict than non-parents as they need to spend greater amount of time at home taking care of their children (Kim & Ling, 2001). Working spouses is another element that might be expected to experience more work-family conflict than those who are not working, because they often encounter dual demands from work and family activities, and social expectations may pressurized them to do good job in both work and family domains (Ballout, 2008). Therefore, working spouse who think that their work is very important will highly committed to their work will spent more time to work, thus, forcing their husband to share home care responsibilities (Martire, Stephens & Townsend, 2000). On the other hand, Judge, Cable, Boudreau and Bretz (1995) said having a non-working spouse may be associated with a high rate of career progress; while (1992) found in their cross-cultural study examining the antecedents and outcomes of work-family conflict in a sample of working women in Singapore showed that spouse support did reduces work-family conflict.

2.3 Work-Family Conflict (WFC)

Work-family conflict is a form of inter-role conflict in which pressures from the work role are contrary with pressures from the family-role, and this conflict effect on the quality of both work and family domains (Thomas & Ganster, 1995). Bragger Rodriguez-Srednicki, Kutcher, Indovino and Rosner (2005) defined work-family conflict as interference of work demands with one's abilities to perform his or her family responsibilities. Work-family conflict is divided into three parts such as job-spouse conflict, job-parent conflict and job-homemaker conflict as married couples usually play more than one role in the family.

Employment can have an effect on working women's marital relationship, as her job may limit her from spending time with her spouse. Stoner, Hartman and Arora (1990) reported that female entrepreneur who experiences lower levels of marital happiness tend to have higher level of work family conflict. A negative relationship between work family conflict and marital satisfaction for Singapore women professionals was reported by Aryee (1992), and found to be consistent with Holahan and Gilbert's (1979) finding on dual career couples.

Study by Bragger et al. (2005) indicated that parents experience greater work family conflict. While Pleck, Staines and Lang (1980) parents with young children will be much affected with work-family conflict. According to Kim and Ling (2001), hours worked correlates most strongly with job-parent conflict for Singapore women entrepreneurs, similar results obtained from Aryee (1992).

Work stressor also has an impact on job-homemaker conflict. According to Barling and MacEwen (1992), work stressors will results to strains such as worry and preoccupation with the work can reduce the level of concentration, making household chores more difficult. Parent with older children will be less affected with job-homemaker conflict since the children can help them out with household chores (Kim & Ling, 2001). Greater number of older children parents have, more help is available for doing household chores.

2.4 Relationship between work variable, non-work variable and work-family conflict

Time pressure which can be measured by the number of hours worked and schedule flexibility is one of the work characteristics that lead to work-family conflict. Jones and Fletcher (1996) confirmed in their research that work stressors affect family life in the form of unpleasant moods that have spilled over from work to family. According to Kahn, Wolfe, Quinn, Snoek and Rosenthal (1964) they suggested that work-family conflict occurs when demands from work and family are mutually incompatible to some degree. Job stressors and job involvement would predict work-family conflict which would in turn predict family distress and depression (Frone, Russell & Cooper, 1992). Emotional demands from work and home deal with the intentional or overt displays of certain emotions expected of employees and family members (Edwards & Rothbard, 2000). Moreover, according to Wilson (2009), work family conflict occur when individuals experience various pressures regarding how or what emotions to display at work and at home.

3 RESEARCH METHODOLOGY

This research utilized self-administered questionnaires. The samples of the study were from 133 executives of five stars hotels in Kuala Lumpur. The data were later analyzed to test two hypotheses formulated for the study.

4 FINDINGS

4.1 Response rate

A total of 19 from 24 five star rated hotels agreed and participated in the study; a total of 51 questionnaires were returned and coded from 133 issued, represents 38 percent response rate. According to Baldauf Reisinger and Moncrief (1999) and Tomaskovic-Devey, Leiter and Thompson (1994), surveys of organizations typically receive substantially lower return rates than surveys of individuals, which is with 15 percent return rates sometimes reaching a level of acceptability for organizational surveys. Besides, Visser, Krosnick, Marquette, and Curtin (1996) mentioned that some studies with low response rates, even as low as 20 percent, are able to yield more accurate results than studies with response rates of 60 to 70 percent.

4.2 Cronbach's alpha reliability analysis of study variables

In order to ensure that the gathered data were reliable, the measurement of each variable and its sub scales were assessed for reliability. The Cronbach's alpha coefficient value for all variables in the study revealed a range of coefficient value from 0.63 to 0.71 accordingly. The dependent variable of work-family conflict had a reliability coefficient of 0.71 constant with the number of independent variables. The independent variable of work variable showed that its dimensions of meaningfulness had a coefficient value of 0.66, and non-work variable had a reliability coefficient value of 0.63. The results also indicated that the reliabilities of the scales used in this study were commonly accepted and considered to be the criterion for demonstrating the internal consistency of a reliable scale (Hinton, Brownlow, McMurray & Cozens, 2004).

4.3 Bootstrap/INDIRECT Analysis

The objectives of this study are to examine the relationship among work variable, non-work variable and work-family conflict. The result from bootstrap analysis is used to examine all the relationship outlined through developed hypotheses.

The Bootstrap/Indirect analysis was applied to test the relationship between independent (WV) and (NWV) with dependent variable (WFC). It was found that the relationship for both work and non-work variables being insignificant with work-family conflict.

H1: Work variable significantly influence hotel managers' work-family conflict.

Results gained in the study revealed that work variable had a negative influence on hotel managers' work-family conflict. It was earlier proposed that there was a positive influence of work variable on managers' work-family conflict and based on the findings, it shows that the scores of work variable are $B = .002, p = 0.9898$. Hence, hypothesis (H1) is not supported. However, non-work variable shows insignificant relationship towards quality of work life ($B = 0.400, p = 0.35$). The analyses were different from the following hypothesis:

H2: Non-work variable significantly influence hotel managers' work-family conflict.

Hypothesis two identifies whether non-work variable have an influence on hotel managers' work-family conflict. Based on findings, the bootstrap result of non-work variable was also shows insignificant relationship with work-family conflict. It was found that there was no statistically significant difference at the .05 level in non-work variable on executives' work-family conflict. The non-work variable scores ($B = -.028, p = 0.9154$). Hence, hypothesis (H2) is not supported.

As mentioned in Weer (2006), the supportiveness of an organization's environment for non-work role participation did mitigate the negative impact of non-work emotional energy demands on work engagement. In addition, Thomas and Ganster (1995) found that employees who worked for organizations that were supportive of their involvement in family life experienced a greater sense of control over their work and family responsibilities, which in turn reduced their perceptions of conflict and strain and increased their job satisfaction.

5 CONCLUSION

Findings of this study suggest some important implications for hotels' operator in Malaysia, particularly in Kuala Lumpur. The findings indicate that both work variable and non-work variable insignificant relationship towards work-family conflict. Result gained is supported by Nadeem and Abbas (2009) whereby job satisfaction is negatively interrelated with work to family interference, but it is not significant from the study on the impact of work life conflict on job satisfaction of employees

in Pakistan. Study by Lapierre et al. (2008) suggest that workforces tend to experience less work-family conflict when view their employing organization as family-supportive. In line with Allard, Hass and Hwang (2011), whereby in a study of Swedish fathers found that if they perceived the organization as family-friendly, they were better able to combine work and family and experience less conflict.

Alternatively, non-work role (parental demands and working spouse) did not influence the executives' work-family conflict as they can separate their personal life from work. Since Malaysians give high importance to the family, they tend to block the work-family conflict from influencing their quality of non-work (family) (Md-Sidin, Sambasivan & Ismail, 2010). Besides, Koekemoer and Steyl (2011) study found that the work interferes most negatively on family life. From the industry perspective, employees that aware of their responsibilities is believed to be able to deliver quality service hence directly enhance the customers' satisfaction. Thus, managers' performance towards the organization may be unaffected. Furthermore, the industry could retain talented managers in the workforce longer.

The current study has some limitations; it was conducted in Kuala Lumpur focusing on 5 star hotels only. Hence, the findings could not be generalized to the whole industry in Malaysia. Besides, this study focused on work variable, non-work variable and its implication towards work-family conflict. It is suggested that future studies looking at inter-role conflicts and how it influence QWL.

ACKNOWLEDGEMENTS

This research is funded by Universiti Teknologi MARA through Research Acculturation Grant 600-RMI/RAGS 5/3 (140/2013).

REFERENCES

Allard, K., Haas, L. & Hwang, C. (2011). Family-supportive organizational culture and fathers' experiences of work–family conflict in Sweden. *Gender, Work & Organization, 18*(2), 141–157.

Anderson, S. & Ungemah, D. (1999). Variable work hours: An implementation guide for employers. *Oregon Department of Environmental Quality*.

Aryee, S. (1992). Antecedents and outcomes of work-family conflict among married professional women: Evidence from Singapore. *Human relations, 45*(8), 813–837.

Baldauf, A., Reisinger, H. & Moncrief, W.C. (1999). Examining motivations to refuse in industrial mail surveys. *Journal of the Market Research Society, 41*(3), 345–353.

Ballout, H.I. (2008). Work-family conflict and career success: the effects of domain-specific determinants. *Journal of Management Development, 27*(5), 437–466.

Barling, J. & Macewen, K.E. (1992). Linking work experiences to facets of marital functioning. *Journal of Organizational Behavior, 13*(6), 573–583.

Bazana, S. & Dodd, N. (2013). Conscientiousness, work family conflict and stress amongst Police Officers in Alice, South Africa. *Journal of Psychology, 4*(4), 1–8.

Bragger, J.D., Rodriguez-Srednicki, O., Kutcher, E.J., Indovino, L. & Rosner, E. (2005). Work-family conflict, work-family culture, and organizational citizenship behavior among teachers. *Journal of Business and psychology, 20*(2), 303–324.

Edwards, J.R. & Rothbard, N.P. (2000). Mechanisms linking work and family: Clarifying the relationship between work and family constructs. *Academy of Management Review, 25*(1), 178–199.

Enz, C.A. & Siguaw, J.A. (2000). Best practices in human resources. *Cornell Hotel and Restaurant Administration Quarterly, 41*(1), 48–61.

Farrell, A. & Geist-Martin, P. (2005). Communicating social health perceptions of wellness at work. *Management Communication Quarterly, 18*(4), 543–592.

Frone, M.R., Russell, M. & Cooper, M.L. (1992). Antecedents and outcomes of work-family conflict: Testing a model of the work-family interface. *Journal of Applied Psychology, 77*(1), 65–78.

Fu, C.K. & Shaffer, M.A. (2001). The tug of work and family: Direct and indirect domain-specific determinants of work-family conflict. *Personnel Review, 30*(5), 502–522.

Golden, L. (2009). Work hours and inflexibility: The costs to work-life.

Hinton, P.R., Brownlow, C., McMurray, I. & Cozens, B. (2004). *SPSS Explained.* New York: Routledge Inc.

Hochschild, A.R. (1997). *The time bind: When work becomes home and home becomes work.* New York: Macmillan.

Holahan, C.K. & Gilbert, L.A. (1979). Interrole conflict for working women: Careers versus jobs. *Journal of Applied Psychology, 64*(1), 86–90.

Interactive, H. (2011). Stress in Workplace. *American Psychological Association.* Retrieved from http://204.14.132.173/news/press/releases/phwa-survey-summary.pdf

Ismail, A. (2002). *Front office operations and management.* Delmar: Cengage Learning.

Jones, F. & Fletcher, B.C. (1996). Taking work home: A study of daily fluctuations in work stressors, effects on moods and impacts on marital partners. *Journal of Occupational and Organizational Psychology, 69*(1), 89–106.

Judge, T.A., Cable, D.M., Boudreau, J.W. & Bretz, R.D. (1995). An empirical investigation of the predictors of executive career success. *Personnel psychology, 48*(3), 485–519.

Kahn, R.L., Wolfe, D., Quinn, R., Snoek, J. & Rosenthal, R. (1964). *Organizational stress.* New York: Wiley.

Kim, J.L.S. & Ling, C.S. (2001). Work-family conflict of women entrepreneurs in Singapore. *Women in Management review, 16*(5), 204–221.

Koekemoer, F.E. & Steyl, J.M.E. (2011). Conflict between work and nonwork roles of employees in the mining industry: Prevalence and differences between demographic groups. *SA Journal of Human Resource Management, 9*(9), 277–291.

Lapierre, L.M., Spector, P.E., Allen, T.D., Poelmans, S., Cooper, C.L., O'Driscoll, M.P., Sanchez, J.I., Brough, P. & Kinnunen, U. (2008). Family-supportive organization perceptions, multiple dimensions of work-family conflict, and employee satisfaction: A test of model across five samples. *Journal of Vocational Behavior, 73*(1), 92–106.

Loscocco, K. (2000). Age integration as a solution to work–family conflict. *The Gerontologist, 40*(3), 292–301.

Martire, L.M., Stephens, M.A.P. & Townsend, A.L. (2000). Centrality of women's multiple roles: Beneficial and detrimental consequences for psychological well-being. *Psychology and Aging, 15*(1), 148–156.

Md-Sidin, S., Sambasivan, M. & Ismail, I. (2010). Relationship between work-family conflict and quality of life: An investigation into the role of social support. *Journal of Managerial Psychology, 25*(1), 58–81.

Mulvaney, R.H., O'Neill, J.W., Cleveland, J.N. & Crouter, A.C. (2007). A model of work-family dynamics of hotel managers. *Annals of Tourism Research, 34*(1), 66–87.

Nadeem, M.S. & Abbas, Q. (2009). The impact of work life conflict on job satisfactions of employees in Pakistan. *International Journal of Business and Management, 4*(5), 63–83.

Pleck, J.H., Staines, G.L. & Lang, L. (1980). Conflicts between work and family life. *Monthly Labor Review, 103*, 29–32.

Presser, H. (2004). Employment in a 24/7 Economy: Challenges for the Family. In A. Crouter & A. Booth (Eds.), *Work-Family challenges for low-income parents and their children* (pp. 83–106). Mahwah: Lawrence Erlbaum.

Ramadoss, K. & Rajadhyaksha, U. (2012). Gender differences in commitment to roles, work-family conflict and social support. *Journal of Social Science, 33*(2), 227–233.

Sloan, M.M., Newhouse, R.J.E. & Thompson, A.B. (2013). Counting on coworkers race, social support, and emotional experiences on the job. *Social Psychology Quarterly, 76*(4), 343–372.

Stalcup, L.D. & Pearson, T.A. (2001). A model of the causes of management turnover in hotels. *Journal of Hospitality & Tourism Research, 25*(1), 17–30.

Stoner, C.R., Hartman, R.I. & Arora, R. (1990). Work-home role conflict in female owners of small businesses: An exploratory study. *Journal of small business management, 28*(1), 30–38.

Thomas, L.T. & Ganster, D.C. (1995). Impact of family-supportive work variables on work-family conflict and strain: A control perspective. *Journal of Applied Psychology, 80*(1), 6–15.

Tomaskovic-Devey, D., Leiter, J. & Thompson, S. (1994). Organizational survey nonresponse. *Administrative Science Quarterly, 39*, 439–457.

Uygur, A. & Kilic, G. (2009). A study into organizational commitment and job involvement: An application towards the personnel in the central organization for Ministry of Health in Turkey. *Ozean journal of applied sciences, 2*(1), 113–125.

Visser, P.S., Krosnick, J.A., Marquette, J. & Curtin, M. (1996). Mail surveys for election forecasting? An evaluation of the Columbus Dispatch poll. *Public Opinion Quarterly, 60*(2), 181–227.

Weer, C.H. (2006). *The impact of non-work role commitment on employees' career growth prospects.* Doctor of Philosophy, Drexel University. Retrieved from http://scholar.google.com.my/scholar?hl=en&as_sdt=0,5&q=The+Impact+of+Non-Work+Role+Commitment+on+Employee%E2%80%99+Career+Growth+Prospects

Theory and Practice in Hospitality and Tourism Research – Radzi et al. (Eds)
© 2015 Taylor & Francis Group, London, ISBN 978-1-138-02706-0

A preliminary study on boutique hotels in the city of Kuala Lumpur

A.S. Arifin, S.A. Jamal, A.A. Aziz & S.S. Ismail
Faculty of Hotel and Tourism Manageemnt, Universiti Teknologi MARA, Shah Alam, Selangor, Malaysia

ABSTRACT: This paper discusses about boutique hotels in the centre of the city of Kuala Lumpur and provides a definition, overview and the characteristics of boutique hotels and the responses of their guests. This study also investigates factors influencing guest experience in these boutique hotels. To investigate these experiences, the methodology includes site observation, a rigorous literature review, questionnaire surveys and an analysis of online guest reviews on the hotel website. The findings suggest that the density of boutique hotels is skewed to the Kuala Lumpur city centre. The results also indicate that the existence of boutique hotels is welcomed by local guests, with a majority of them giving positive responses. The results reveal the following three dimensions of guest experience: (i) physical environment; (ii) interaction with employees; and (iii) interaction with other guests. The findings offer important implications for industry marketing.

Keywords: Boutique hotels, guest experience, Malaysia, preliminary study

1 INTRODUCTION

In the past twenty years, market change, shifting production and varying guest trends have encouraged product differentiation in the hotel industry (Timothy & Teye, 2009). Thus, product differentiation puts pressure on hotels to improve their performance, anticipate change, and develop new structures. Rogerson (2010) emphasised that guests currently search for new and unique experiences that are different from traditional hotels. These guests deliberately search for accommodations that are distinctly different in appearance and experience from brand hotels (Albazzaz et al., 2003). Because of this search for unique experiences, boutique hotels were invented in the early 1980s, and today, the boutique hotel sector is growing worldwide.

Boutique hotels are an example of customer experience differentiation because guests are treated with personalised service. The definition of the boutique hotel as a small-scale, design-conscious operation with individuality (McNeill, 2009) has been controversial lately. Nevertheless, the design and uniqueness of boutique hotels are the focus of a differentiation strategy used to compete with mainstream hotel chains. The unique architecture, buildings and furniture are the driving forces of a boutique hotel's development (Aggett, 2007).

In today's service economy, guests seek unique events that create unforgettable and long-lasting experiences. Services can cater to guests' unique personal tastes and increase the value of a hotel's products (Oh, Fiore & Jeoung, 2007). Erdly and Kesterson-Townes (2002) acknowledged that the hospitality industry is focusing on delivering customised experiences to its guests. Customised experiences may be shifting in certain sectors of the hospitality industry, such as the accommodation sector, from a differentiation of services to guest experiences and a memorable stay.

Previous researchers have examined boutique hotels all over the world, covering the United States (Gao, 2013; Walls, 2013), the United Kingdom (Aggett, 2007; Lim & Endean, 2009), South Africa (Rogerson, 2010), Turkey (Erkutlu & Chafra, 2006), Australia (Lwin & Phau, 2013), Thailand (Rompho & Boon-itt, 2012), and Singapore (Henderson, 2011). Though a significant number of boutique hotels exist throughout Malaysia, research in this area is still very limited.

This study investigates the factors that influence guest experience in the boutique hotels in Kuala Lumpur. This study identifies the factors that inspire guests to stay in boutique hotels, which should explain what guests want in a boutique hotel.

The paper includes four sections. The first section examines the definition of boutique hotels, which includes the current definition. The second section discusses the development of boutique hotels in Malaysia, and the third section examines the characteristics of boutique hotels there. Finally, the last section discusses the factors that influence guests to stay at a boutique hotel.

2 LITERATURE REVIEW

2.1 *Defining boutique hotels*

The universal definition of boutique hotel is still being debated among scholars (Gao, 2012). The evolution of the definition can be observed from 2005 to 2009. The definition involves uniqueness (Aggett, 2007; Mcintosh & Siggs, 2005), personalised service (Aggett, 2007; Erkutlu & Chafra, 2006; Mcintosh & Siggs, 2005), home-comfort, involved emotions, and an individual design (Lim & Endean, 2009). Boutique hotels are also furnished in a stylistic and sometimes a thematic manner (Erkutlu & Chafra, 2006). Moreover, scholars agreed that location (Aggett, 2007; Lim & Endean, 2009) is important to the hotel's success.

A boutique hotel may only have a maximum number of 100 rooms (Aggett, 2007; Henderson, 2011). With smaller rooms than mainstream hotels, it is easy for a boutique hotel to deliver personalised service; it is also easy for guests and employees to have the integrity in their exchange (Henderson, 2011). Chan (2012) argued that it is not possible for properties with larger accommodations to deliver the same level of personalised service as a boutique hotel. Because larger properties have more employees, relations between individual guests and employees are reduced. These employees cannot deliver the high degree of personalised service that is used as a selling point by boutique hotels. Furthermore, mainstream hotels lack the flexibility and degree of employee empowerment required to deliver the highest level of personalised service that distinguishes a boutique hotel (Henderson, 2011).

2.2 *An overview of boutique hotel*

The development of boutique hotels in the 1980s is widely attributed to Blakes in London (Brights, 2007; McDonnell, 2005). The concept spread across the Atlantic and stateside. In 1984, Ian Schrager and Steve Rubell opened Morgans Hotel in New York; they used the term "boutique" to describe their first hotel. By using a remarkable design, they portrayed their hotel as "theatrical magic and glamorous mystique". Since the 1980s in the boutique hotel industry, travellers seek information to experience their destination (Freund de Klumbis & Munsters, 2005). Hotels may understand their guests from the experiences for which their guests search; hotels may gain from the hotel's experiential view. For example, hotels could provide intangible qualities for guests, facilitating feelings, emotions, imagination, knowledge and beneficial experiences. Thus, a boutique hotel can attract and satisfy customers (Tidtichumrernporn, 2012), achieve more sustainable returns and charge premium prices (Pine & Gilmore, 1999).

Table 1. Distribution of boutique hotels according to States in Malaysia for 2013.

No	State	Operators
1.	Melaka	7
2.	Johor Bharu	6
3.	Kuala Lumpur	13
4.	Sarawak	3
5.	Perak	4
6.	Sabah	3
7.	Putrajaya	1
8.	Penang	9
9.	Selangor	5
10.	Negeri Sembilan	1
11.	Pahang	6
12.	Kelantan	1
13.	Kedah	4
	Total	63

Source: Tripadvisor.com

2.3 *Boutique hotels in Malaysia*

Malaysia has many traditional hotels classified as economy, mid-scale, upscale, upper-upscale, and luxury. Hotels that can be classified as luxury and upper-upscale typically provide fine services and amenities, large rooms, and the best location for a hotel or resort. Most of the upscale hotels emphasise on quality and comfort and are used by business travellers. By contrast, mid-scale and economy hotels cater to the basic needs of short-term business and budget-minded travellers; these properties provide a low price and minimum services and amenities (Albazzaz, et al., 2003).

The boutique hotels in Malaysia are individual in their design features (Lim & Endean, 2009). Table 1 shows the distribution of boutique hotels according to states in Malaysia in 2013.

A boutique hotel has an individual quality in culture, art and history and has likely incorporated contemporary individual luxuries on its premises. Moreover, boutique hotels in Malaysia are constructed from private mansions such as Chymes in Penang, former shop houses such as Courtyard @ Heeren in Melaka and historical buildings such as Anggun Boutique Hotel in Kuala Lumpur. An individual style exists for each property. In addition, personalised service between guests and employees can create a good experience that affects guest satisfaction (Liljander & Strandvik, 1997), conveys guest loyalty (Mascarenhas, Kesavan & Bernacchi, 2006) and instils confidence (Flanagan, Johnston & Talbot, 2005). According to Tripadvisor.com, Malaysia have 63 boutique hotels.

3 METHODOLOGY

The methodology for this paper includes site observation, literature reviews, a structured questionnaire survey and an analysis of guest comments through online review sites such as Tripadvisor, Orbitz, Travelocity, Expedia and booking.com. Thirty questionnaires were pilot to tested Postgraduates from the Faculty of Hotel and Tourism Management, Universiti Teknologi MARA. These Postgraduates had stayed in boutique hotels. This sample includes representatives from all departments in the faculty, namely, Hospitality Management, Tourism Management, Gastronomy and Food Service.

The questionnaire was designed to measure the respondents' level of agreement or disagreement with various statements or items developed. Each scale has a minimum of "1" and a maximum of "7". A score close to "7" means a very strong attitude in favour of the statement, while a score close to "1" means a very strong attitude against the statement. All data were analysed through the Statistical Package for the Social Sciences software (SPSS) version 21.0.

4 RESULTS AND DISCUSSION

From the literature review and site observation, 63 boutique hotels exist in Malaysia and 21 percent are located in the city of Kuala Lumpur. This high density of boutique hotels in the country is located within this vicinity. In Kuala Lumpur, the average number of guestrooms was 29, and hotels with a total number of rooms of 10 to 50 will be the focus of this study.

The pilot survey was conducted in March 2014 and involved postgraduates from the Faculty of Hotel and Tourism Management, Universiti Teknologi MARA. The respondents are from all the departments as follows: Hospitality Management (28 percent), Tourism Management (25 percent), Gastronomy (17 percent) and Food Service (30 percent). The majority of respondents was female (80%), local travellers (96.7%), below 30 years old (76.7%) and holds a degree in hospitality and tourism management (98.7%). The main purpose of their previous travel was leisure and vacation (70%), and most of them have stayed one to two days in the hotel during their previous holidays (60%).

The preliminary findings suggest that a majority of the respondents (89%) are aware of the existence of boutique hotels, and more than half (56%) look forward to staying in boutique hotels in the future. Table 2 summarises the degree of expectation that potential local guests have for each variable.

The online interview results recorded various positive comments from respondents whereby many respondents emphasised that one of the

Table 2. Degree of experience expectancy of variables.

Variable	Percentage of expectancy
Unique character	57
Personalised	57
Homely	73
Quality	40
Value added	77

important features of a boutique hotel is the unique design offering a distinctive stay.

"We recently stayed in the Eiffel room and had a fantastic time. The room was new and had a fun twist to it." – Guest A

"The decor is made from the personal collection of the owner. It is a very tasteful collection of old Chinese antiques. So you'll certainly take away something to remember from this stay." – Guest B

In the literature, personal service was also stressed as part of a boutique hotel's character (Aggett, 2007; Henderson, 2011). This observation corresponds to some statements from the respondents.

"Friendly staff are helpful and dedicated to our travel needs" – Guest C

"The staffs are excellent! The staff showed us maps with great recommendations for wherever we wanted to go" – Guest D

The findings also suggested that the location of the boutique hotel either in the city centre or near historical places determine different types of travellers. Some comments regarding the location of the boutique hotels are as follows:

"We spent a long weekend to visit the main sights in the heritage zone. The hotel building is a renovated old heritage building in the heart of the core heritage zone. It is a small, boutique hotel and is within walking distance of all the main sights." – Guest B

"The location is great, within walking distance of all the entertainment spots in Bukit Bintang and the local food streets. It also has convenient access to the monorail stations, which are only a five to ten minute walk." – Guest E

5 CONCLUSION

It can be concluded that different characteristics offered by boutique hotels have created different experiences to different guests. Examining what potential guests expect and how actual guests react to and benefit from their experiences can provide important information for the product development and marketing of boutique hotels. The findings of this preliminary study suggest that any effort to improve the guest experience would help boutique hotels enhance their market reputation and influence their marketing strategies.

The findings suggest that favourable employee interaction, such as appropriate, respectful, and helpful conduct, influences a guest's experience. Additionally, it can be concluded that online reviews play an important role as a modern tool of marketing and promotion. According to Reuters (2012), 93 percent of consumers read online product reviews and other user-generated content before purchasing. This figure is consistent with this study whereby many guests used online review sites to give their opinions, share experiences and provide suggestions and recommendations.

ACKNOWLEDGMENT

This study was made possible by the continuous support from Universiti Teknologi MARA, UiTM-Grant No: 600-RMI/DANA 5/3/RIF (517/2012).

REFERENCES

Aggett, M. (2007). What has influenced growth in the UK's boutique hotel sector? *International Journal of Contemporary Hospitality Management, 19*(2), 169–177.

Albazzaz, A., Birnbaum, B., Brachfeld, D., Danilov, D., Kets de Vries, O. & Moed, J. (2003). Lifestyles of the rich and almost famous: The boutique hotel phenomenon in the United States *High Tech Entrepreneurship and Strategy Group Project. Fontainebleau: Insead Business School.*

Brights, D. (2007). The definition of boutique hotels. *Article Dashboard.*

Chan, C. (2012). Lodging subsector report: Boutique hotels.

Erdly, M. & Kesterson-Townes, L. (2002). Experience rules. *IBM Business Consulting Services' Vision for the Hospitality and Leisure Industry, circa 2010.*

Erkutlu, H.V. & Chafra, J. (2006). Relationship between leadership power bases and job stress of subordinates: example from boutique hotels. *Management Research News, 29*(5), 285–297.

Flanagan, P., Johnston, R. & Talbot, D. (2005). Customer confidence: the development of a "pre-experience" concept. *International Journal of Service Industry Management, 16*(4), 373–384.

Freund de Klumbis, D. & Munsters, W. (2005). Developments in the hotel industry: design meets historic properties. *International Cultural Tourism*, 162–184.

Gao, L. (2012). An Exploratory study of the boutique hotel experience: Research on experience economy and designed customer experience. *International Journal of Hospitality and Tourism*, 186–204.

Gao, L. (2013). *An exploratory study of the boutique hotel experience: Research on experience economy and designed customer experience.* Purdue University.

Henderson, J.C. (2011). Hip heritage: The boutique hotel business in Singapore. *Tourism and Hospitality Research, 11*(3), 217–223.

Liljander, V. & Strandvik, T. (1997). Emotions in service satisfaction. *International Journal of Service Industry Management, 8*(2), 148–169.

Lim, W.M. & Endean, M. (2009). Elucidating the aesthetic and operational characteristics of UK boutique hotels. *International Journal of Contemporary Hospitality Management, 21*(1), 38–51.

Lwin, M. & Phau, I. (2013). Effective advertising appeals for websites of small boutique hotels. *Journal of Research in Interactive Marketing, 7*(1), 18–32.

Mascarenhas, O.A., Kesavan, R. & Bernacchi, M. (2006). Lasting customer loyalty: a total customer experience approach. *Journal of Consumer Marketing, 23*(7), 397–405.

McDonnell, C.J. (2005). Boutique hotels are getting new interest. Business First of Buffalo, from http://bizjournals.com/buffalo/stories/2005/01/31/focus4.html

Mcintosh, A.J. & Siggs, A. (2005). An exploration of the experiential nature of boutique accommodation. *Journal of travel research, 44*(1), 74–81.

McNeill, W.H. (2009). *The rise of the West: A history of the human community*: University of Chicago Press.

Oh, H., Fiore, A.M. & Jeoung, M. (2007). Measuring experience economy concepts: tourism applications. *Journal of travel research, 46*(2), 119–132.

Pine, B.J. & Gilmore, J.H. (1999). *The experience economy*. Boston, Ma: Harvard Business Press.

Reuters, T. (2012). The global consumer report 2012; A Clearwater consumer team report Retrieved February 24, 2013, from http://www.clearwatercf.com/documents/sectors/Final_Version_PDF.pdf

Rogerson, J.M. (2010). *The boutique hotel industry in South Africa: definition, scope, and organization.* Paper presented at the Urban Forum.

Rompho, N. & Boon-itt, S. (2012). Measuring the success of a performance measurement system in Thai firms. *International Journal of Productivity and Performance Management, 61*(5), 548–562.

Tidtichumrernporn, T. (2012). *Lifestyle segmentation for soutique accommodation in relation to the service quality and customer.* Paper presented at the The 17th Asia-Pacific Decision Sciences Institute Conference on July 2012.

Timothy, D.J. & Teye, V.B. (2009). *Tourism and the lodging sector.* Oxford: Butterworth-Heinemann.

Walls, A.R. (2013). A cross-sectional examination of hotel consumer experience and relative effects on consumer values. *International Journal of Hospitality Management, 32*, 179–192.

Theory and Practice in Hospitality and Tourism Research – Radzi et al. (Eds)
© 2015 Taylor & Francis Group, London, ISBN 978-1-138-02706-0

Blue Ocean strategies in hotel industry

N.C. Ahmat, R. Abas & A.H. Ahmad-Ridzuan
Faculty of Hotel and Tourism Management, Universiti Teknologi MARA, Penang, Malaysia

S.M. Radzi & M.S.M. Zahari
Faculty of Hotel and Tourism Management, Universiti Teknologi MARA, Shah Alam, Malaysia

ABSTRACT: In the face of intense competition, the merit of Blue Ocean strategy towards the creation of value innovation and the ability to exploit the untapped market for firm's survival is indubitable. Some outstanding companies have proven that Blue Ocean strategy contributed significantly towards the organizational growth and success. However, despite its long existence and its worth, the scantiness of the strategy adoption is eminent. In light of this, it is worth diagnosing the characteristics of Blue Ocean strategies in the hotel industry to enable the hoteliers to synthesize the relevant factors associated with the strategy for a long-term sustainability of the industry. It is also imperative to know to what extent the execution of such strategy will have an impact on hotel financial performances.

Keywords: Blue Ocean strategy, hotel industry

1 INTRODUCTION

In today's saturated industries, competing head-to-head has resulted to a bloody 'red ocean' of rivals fighting over a shrinking profit pools (Kim & Mauborgne, 2005a; Koo, Koo & Luk, 2008). Therefore, employing the right strategy is vital for a firm's long-term survival particularly for the hotel industry (Awang, Ishak, Mohd Radzi & Taha, 2008). According to Yang (2012a), hotel operators should continuously provide new offerings in an uncontested marketplace through the evolution of a great number of modifications of room-associated amenities and the development of hotel products, which consecutively result in greater customers' satisfaction, acceptance, and loyalty. This strategy known as the blue ocean strategy emphasizes on "value innovation" with the aim of reducing company cost, yet increasing values to customer hence making competition irrelevant. With supply exceeding demand in the hotel industry, competing for a share of contracting market will not be sufficient enough to sustain high performance. For that reason, companies need to go beyond competing through grasping on new profit and improving the growth opportunities for their business.

2 THE CONCEPT OF BLUE OCEAN STRATEGIES

Blue Ocean reflects the unknown or the unexploited market space. Numerous studies conducted on Red Ocean strategies indicated that as companies try to outperform their rivals to grab a greater share of existing demand, stiff competitions in a bloody red ocean is resulted. This contradicted with the Blue Ocean Strategy (BOS) where companies need to think beyond competing to seize new profit and increase growth opportunities. It is undeniable that regardless of the approach used, each strategy involves both opportunities and risks. It is important to clarify the aim of the Blue ocean strategy is not to outperform the rivals in an existing industry but rather to create a new market space of a Blue Ocean by doing things that others disregard. Apart from that, tapping the niche areas, and reengineering the product and service provisions are the cornerstone in making the competition irrelevant.

Prior to discussing BOS, it is worth to distinguish the traits of ROS and BOS as illustrated by Kim and Mauborgne (2005a) in Table 1. It is apparent that BOS is rather a silent mover and a trend setter than becoming a follower. Kim and Mauborgne (2005b) further established various tools and framework to formulate and execute the Blue Ocean Strategy (BOS) interalia include the Strategy canvas, the Four-Actions Framework and the Eliminate-Reduce-Raise-Create grid. The strategy canvas is an alignment of key competing factors and the offering values to the customers. The vertical axis comprises of the values or the benefits rendered to the customers within the variables or the key competing factors as technology and product offerings and the competition currently invested

Table 1. Different traits of Red Ocean and Blue Ocean strategy.

Red Ocean	Blue Ocean
Compete in existing market space	Create uncontested market space
Beat the competition	Make the competition irrelevant
Exploit existing demand	Create and capture new demand
Make the value-cost trade-off	Break the value-cost trade-off
Align the whole system of a firm's activities with its strategic choice of differentiation or low cost	Align the whole system of a firm's activities with its strategic choice of differentiation and low cost

Source: Kim and Mauborgne (2005a).

in which is positioned at the horizontal axis. The diagnosis enable firm to position it and measures its relative performance against the rivals across the industry. What distinguishes BOS with other management strategies then? As BOS emphasizes on value innovation, the strategy enables firm to ploy and galvanize the right actions to maneuver by shifting the focus from competitors to alternatives and from customers to non-customers.

The Four-action framework is an analytical tool where the management team systematically establishes a new value curve through series of insightful questions by reducing factors below the standard; eliminating factors that the company disregard, raising factors well below the standard and creating the untapped factors. Reconstructing the buyer value elements according to Yang (2012b), enables hotel operators to create customer value, innovative product and service offerings for customers, thus resulting to an uncontested marketplace.

The Eliminate-Reduce-Raise-Create grid on the other hand, is an extension of the four action framework, which can be presumed as another version of SWOT analysis with the Eliminate factors replacing the Weaknesses, Reduce substitutes the Threats, Raise to replace the Strengths and Create to replace the Opportunities. Apart from manipulating those factors as organization normally does when performing the SWOT analysis, BOS allows organization to pursue product and service differentiation at a low cost through the eliminate-reduce factors to break the value-cost tradeoff.

3 SUCCESSFUL COMPANIES UNDER BLUE OCEAN STRATEGIES

It is said that Air Asia managed to avoid Red Ocean by competing with Malaysia Airlines (MAS) and other regional airlines by examining the factors that industry take for granted and factors that are important for customers (Ahmad, 2010; Ahmad & Neal, 2006). With the four-action framework proposed in the Blue Ocean Strategy,

Air Asia implemented various strategic moves to ensure the competition with Malaysia Airlines (MAS) and regional companies are irrelevant. This includes elimination of over the counter booking system, free food/beverage on the plane, seating class booking system. The company also managed to reduce 'luxury' facilities provided by Airport Lounge, inflight attendance service, and seat quality. In addition, raising or increasing the flight frequency and shifting the focus on several key destinations, creating online booking system, and creating point-to-point travel system (Source: http://EzineArticles.com/1031603) proved that Air Asia demonstrates a strong value creation to benefits customers whilst reducing the company cost. Implicitly, the introduction of the low cost carrier with its strategic moves, the alluring tagline of *"now everyone can fly"* and the cheaper price connote that people from all walks of life are given the chance to experience what was once considered as "luxury" at a value for money price. With the successful implementation of Blue Ocean Strategy, Air Asia has diversified their business into Tune Hotel and Tune Money maintaining the same concept, the Blue Ocean marketplace (INSEAD, 2007; airasia.com website).

Other examples of successful company that implemented Blue Ocean Strategy include Apple and Google. Apple is currently the world's most outstanding telecommunication brand as competitors find it hard to imitate the products. Despite rumours over imitation of Apple iPhone application by Samsung Galaxy SIII, to date, there is no other company could outperform Apple in terms of security, antivirus, application and whatnot. Even the rumour is true, but with its current performance and credibility, it is not impossible for Apple to recreate another new Blue Ocean. In addition, the death of Steve Job somehow helped the company to boost its profit as customers are willing to queue up just to get the brand new hand phone. Past studies on the Blue Ocean Strategy proved the effectiveness of the Blue Ocean strategy implementation in various multinational companies including Apple, Canon, IBM, Mc Donalds and Air Asia (to name a few). In short, the Blue Ocean will remain the engine of growth and prospects in most established market spaces and it is consistent across time regardless of the industry while Red Ocean are shrinking readily (Kim & Mauborgne, 2005b). Nonetheless, despite all the success stories proven by the aforementioned companies, lack of studies provides the information pertaining to the impact of the Blue Ocean Strategy on their financial performances. With that, it is important to look at the impact of such strategy towards the financial aspect of the company.

4 THE ATTRIBUTES OF BLUE OCEAN STRATEGIES IN HOTEL INDUSTRY

In assessing the growth strategies of three Chinese Domestic Hotel Companies, Qin, Adler and Cai (2012) asserted that the essential elements of their strategies were innovative positioning, keeping cost low, rapid expansion, continuous innovation, focus on quality consistency, extensive training and several indigenous Chinese cultural operational practices. Except for the later, these attributes are the key elements in the Blue Ocean Strategy. According to authors also aggregating and integrating with the indigenous operation practices could create a huge competitive advantages for the three hotels which allow them to outperform other domestic hotels and compete successfully with their foreign counterparts.

Yang (2012b) further identified the attributes of Blue Ocean Strategies in selected hotels in Taiwan as laid out in Table 2. Although the findings cannot be generalized to all contextual setting, the input somehow is influential precedent in determining the intrinsic values to be considered when adapting BOS in hotel setting.

Parvinen, Aspara, Heitanen, and Kajalo (2011), pointed out that different approaches to BOS work in different contexts and context combinations. Like other industry, the attributes of BOS in hotel industry may vary from one hotel to another due to differences in business scale, scope, spatial and cultural differences. As hotel is a labor-intensive industry, other internal and external factors such as staff turnover rate can be considered. Other attributes for example hotel theme, product and services offering, marketing intelligence and type of delivery system employed may profoundly impact the company's financial and non-financial performance. The most important thing is for organization to consider the factors that they have control over and those they could not or beyond control.

These attributes however, should dwell within the three main features of BOS, namely focus, divergence, and offering gripping tagline to create value innovation and enable firm to occupy a strong branding image to the customers. As competition is unavoided, firms need to find a way to reconstruct the market boundaries by focusing on various issues as the alternative industries, the business dyads, the network chain, complementary products and service offerings, functional or emotional appeal to buyers and time. At the level of sales management, the strategic seeking of uncontested space by creating totally new network roles, value creation logics and benefits for customers does indeed facilitate profitable growth (Parvinen et al., 2011).

Table 2. Characteristics of BOS in Taiwan hotel.

Elimination	Removal of travel barriers for prospective visitors to Taiwan
	Remove constraints on foreign capital investment
Reduction	Bargaining power (over room capacity; change of target market in a region)
	In-room equipment and amenities
	Stabilize use of guestroom products and services
Creation	Establish a regional brand for the Asia Pacific Rim
	New market segmentation to attract mainland tourists
	Market integration among regional hotels
	The development of hotel packages to accompany cultural tourism
Raise	Brand development (Branding expansion, brand differentiation)
	Market segmentation (expand the marketing territories, re-structure distribution channels)
	The provision of products and services (innovation in in-room amenities, understand guest preferences, differentiate offerings, upgrade in-room electronics equipment)
	Cross-industry strategic alliances
	Developing customized travel packaged

Source: Yang (2012a).

5 FORMULATING AND EXECUTING BLUE OCEAN STRATEGIES IN HOTEL INDUSTRY

Understanding of the basic strategy canvas will help the company to start looking at how to formulate the Blue Ocean strategy. Company must first break the competition and unlock the uncontested market space. It is important to justify the reasons why companies must change from Red Ocean to Blue Ocean. It is mainly because of the increase in price wars among hotels that boost up the intensity level of competition. With that, it will leads to shrinking of profit margin and market shares. Anecdotal evidence shows that in most of the industries, the supply is now overtaking the demand. The concept introduced by the companies becomes more similar while differentiation becomes harder. Take the example of hotel industry, there are more and more hotels being built from time to time. Albeit hotel providers claimed that they had launched a new concept, somehow the basic things are still the same, providing product and services like rooms for the customers to stay. The only way that differentiates one hotel over the other is the element of value innovation inside their strategic move. For example, improving on the technology used like room internet to the customers could be one of the value innovation elements or targeting students (youngsters) to stay in hotel rather than middle or high income customers.

Furthermore, Qin et al. (2012) postulated that the more brand expansion is performed, the more hotels are able to create new customer value. Creating or adding value to hotel offerings would beneficially reinforce revenue management and strengthen market positions in the marketplace. The continuous of new product offering in an uncontested marketplace will leads to the evolution of a great number of modifications of room-associated amenities and the development of hotel products which in turn result in greater satisfaction, acceptance and loyalty on the part of the customers. The value characteristics of the Blue Ocean Strategy of the selected hotels are creating added-valued offerings for existing customers and aligning all hotel products, services and activities with the strategic choice of differentiation. The innovation component in this hospitality setting would include restructuring market segmentations, rebranding, and reformulating pricing strategies to capture new demand (Yang, 2012a).

On top of that, hotel operators should be able to visualize various strategies from different angles the visual awakening is where firm is able to see where the company stands; visual exploration in where organization visualizes how others perceive it; visual strategy fair where firm projects where it could be in the short and long run; and visual communication where firm is able to draw a map of every competitors by plotting who is the pioneer, migrator and the settler. Identifying the equivalent of the swimmer, snorkeler and diver enables firm to predict which company is the true rival and how it could differentiates it from the snorkeler and diver. Identifying the attributes of BOS also enable firm to reach beyond its existing demand. This enable firm to scrutinize the three tiers of non-customers at which the "soon to be noncustomers is the first tier, followed by the second tier or those who consciously refuse your market and the third tier or those "unexplored" noncustomers in distant market". The attributes also helps firm to analyze them with any or all of the analytical tools and frameworks so as to design strategies to save the first two tiers and to attract the third one.

In addition to that, the sequence of BOS, which comprises of the buyer utility, the price, the cost and the adoption hurdles in realizing the business idea must be strictly adhered to ensure a commercially viable BOS idea. Likewise, Parvinen et al. (2011) pointed out that linking to performance, at the end of the day to cash flow, is governed by the fit between the context and selected approach to BOS. Therefore, the ability of firm to understand the combination of right approach is vital for the firm's performance and survival.

It is good to realize that the hotel industry will continuously grow where market will expand, operations will improve and hotel players will come and go. Nevertheless, with the overabundance of industries existed today namely 3D televisions, Ipad, Ipod, Nanotechnology and lots more, we could projected that in twenty years from now there will be many unknown industries will likely exist then. The same goes to hotel industry milieu, nowadays we could see various types of concepts being introduced such as green hotel, Syariah Compliant hotel, boutique hotel and library hotel. Just few decades ago none of these concepts have ever existed. Perhaps in the future, let say twenty decades from now there will be many unknown concepts being created.

6 CONCLUSION

It is a challenge to explore the ability to create new market space that is uncontested. This requires firm to steer away from complacency of the current product and service offerings and management practices and to move ahead to strategize by understanding the business environment and configuration. With that, an empirical study is crucial to examine the Blue Ocean Strategy tools and framework from hotel industry perspectives. It is also imperative to scrutinize the impact of implementing Blue Ocean Strategy in hotel industry towards hotel performances particularly the objective or financial performance.

REFERENCES

Ahmad, R. (2010). AirAsia Indeed the Sky's the Limit! *Asian Journal of Management Cases, 7*(1), 7–31.

Ahmad, R. & Neal, M. (2006). AirAsia The Sky's the Limit. *Asian Journal of Management Cases, 3*(1), 25–50.

Awang, K.W., Ishak, N.K., Mohd Radzi, S. & Taha, A.Z. (2008). Environmental Variables and Performance: Evidence from the Hotel Industry in Malaysia. *International Journal of Economics and Management, 2*(1), 59–79.

Kim, W.C. & Mauborgne, R. (2005a). *Blue ocean strategy: How to create uncontested market space and make competition irrelevant.* Boston, Massachusetts: Harvard Business Press.

Kim, W.C. & Mauborgne, R. (2005b). Value innovation: a leap into the blue ocean. *Journal of business strategy, 26*(4), 22–28.

Koo, L., Koo, H. & Luk, L. (2008). A pragmatic and holistic approach to strategic formulation through adopting balanced scorecard, SWOT analysis and blue ocean strategy–a case study of a consumer product manufacturer in China. *International Journal of Managerial and Financial Accounting, 1*(2), 127–146.

Parvinen, P., Aspara, J., Hietanen, J. & Kajalo, S. (2011). Awareness, action and context-specificity of blue ocean practices in sales management. *Management Decision, 49*(8), 1218–1234.

Qin, Y., Adler, H. & Cai, L.A. (2012). Successful growth strategies of three Chinese domestic hotel companies. *Journal of Management & Strategy, 3*(1), 40–54.

Yang, J.-t. (2012a). Identifying the attributes of blue ocean strategies in hospitality. *International Journal of Contemporary Hospitality Management, 24*(5), 701–720.

Yang, J.-T. (2012b). Thinking outside the hotel box: Blue Ocean strategies for hotels in Taiwan. *International Journal of Contemporary Hospitality Management, 28*(10), 9–11.

Theory and Practice in Hospitality and Tourism Research – Radzi et al. (Eds)
© 2015 Taylor & Francis Group, London, ISBN 978-1-138-02706-0

Work variables, non-work variables and quality of work life: The Malaysia hotel executives' insights

S.M. Radzi, N. Sumarjan & M.F.S. Bakhtiar
Faculty of Hotel and Tourism Management, Universiti Teknologi MARA, Shah Alam, Malaysia

I.M. Ghazali
Faculty of Hotel and Tourism Management, Universiti Teknologi MARA, Terengganu, Malaysia

ABSTRACT: Issues regarding the quality of work life have long been investigated and it is gaining greater attention from various industries. This paper fills the gap in understanding the Work Variable (WV), Non-Work Variable (NWV) towards Quality of Work Life (QWL) among hotel executives in Malaysia. The objectives of this study are to examine whether work variable and non-work variables have an influence on hotel executives' quality of work life. This research utilized self-administered questionnaires and result revealed that the work variable had a positive influence on hotel executives' quality of work life while non-work variable shows insignificant relationship towards quality of work life.

Keywords: Malaysia hotel industry, non-work variable, quality of work life, work variable

1   INTRODUCTION

Looking at the nature of the hotel industry, which is very challenging and demanding, it is understood why the T industry needs to provide a good quality of work life (QWL) in order to attract and retain capable employees. Sirgy et al. (2001) explained QWL does not only affect job satisfaction but also satisfaction in other life domains such as family life, leisure life, social life, financial life, and etc. Therefore, the focus of QWL is actually beyond the job satisfaction itself. It does involve the effect of the workplace on satisfaction with the job, satisfaction in non-work life domains, and satisfaction with overall life, personal happiness, and subjective well being. There is some evidence showing that a happy employee is a productive, dedicated and loyal employee (Greenhaus et al., 1987). Study also affirmed the contribution of competent and loyal employees toward organizational profits (Hinkin and Tracey, 2000). In fact, no hotel can have excellent operations without excellent employees (Enz and Siguaw, 2000). These inferences are suggestive of the fact that hospitality employees are the most significant resource within the organization. Hotel managers are supposed to ensure that the attitudes and behaviors of employees are consistent with the expectations of their customers and the organization as a key to be successful in this global competitive environment. Work and family are both important to many people and this has been long realized by previous study (Whitely and England, 1977).

Nevertheless, the issue of work-family balance and conflict has continued to attract popular media and serious research attention worldwide. Kandasamy and Ancheri (2009) observed that despite acknowledging employees' start job with expectations that has been found to influence their QWL, a definition founded and formulated on the basis of employee expectation is conspicuously lacking, in the hospitality context. On the other hand, this industry found that it is difficult to attract and retain their human resources. In the light of the foregoing concerns in the hotel industry, Yu (1999) stated it becomes a challenge for the management to motivate employees to stay on the job, by ensuring measures at enhancing the working condition to counter turnover and to design measures that can enhance the working condition in reducing turnover. Research finding has acknowledged the impact of improved quality of work life (QWL) on reduced turnover (Ference, 1982). It is essential to harmonize the hospitality employees' quality of work life induct a quality then into the work lives of hospitality employees. Thus, this study aims to investigate whether work variable and non-work variable have an influence on quality of work life of five star hotels' executives.

2   LITERATURE REVIEWS

An extensive review will focus on the variables under study mainly work variable, non-work variable and quality of work life to further define and support a theoretical framework

2.1 Work Variable (WV)

Anderson and Ungemah (1999) mentioned that variable work hour's programs are becoming more and more popular with companies both large and small, as employers realize they are good for employee morale and good for business. They explained that due to some reason such as the increase in two-career families, the rise in single-parent families and the huge number of working women including mothers of school-age children have forced companies to reconsider their work schedules. Employers have been pleasantly surprised to find the variable work hours have many benefits. Companies are realizing that everyone wins by providing the flexibility to accommodate family needs, leisure activities and other obligations. According to Kim and Ling (2001), work variables are number of hours worked, work schedule inflexibility and work stressor. They explained that time pressure can be measured by the number of hours worked and schedule flexibility. They also discovered that past research studies have shown that long hours worked and work schedule inflexibility are related to high work-family conflict. On the other hand, job characteristics and work role pressure have a positive relationship with work-family conflict (Kim and Ling (2001). Work stressors can produce strain symptoms such as anxiety, frustration, tension and irritability (Greenhaus and Beutell, 1985). Ballout (2008) noted that job involvement is one potential work-related variable that may cause work-family conflict. Moreover, individuals who are highly involved in their jobs or careers may devote more time and effort to the work role than to family role. According to Adams et al. (1996), job involvement has got to do with the degree of importance people assign to work involvement. Another work-related variable that relates to work-family conflict is job social support (Ballout, 2008). Adams et al. (1996) and Bernas and Major (2000) pointed out that social support from the non-work domain, such as the partners or family members is of great importance in reducing work-family conflict.

2.2 Non-Work Variable (NWV)

The non-work variables are gender, parental demands, and working spouse. Gender is perhaps the socio-demographic characteristic most often included in studies of work-family conflict and career success (Ballout, 2008). The evidence suggests that there are gender differences in work-family conflict due to social role differences between working men and working women. Traditional gender roles assumed that men are primarily responsible for family financial support by working, while women are primarily responsible for childcare and household duties (Loscocco, 2000). In his research, Ballout (2008) indicated that parental demand can affect work-family conflict and subsequently success in careers. Due to the increased demands of spending a great amount of time at home taking care of their children, investing the required time and energy at work to support their career advancement and success, employed parents tend to experience more family interfere work conflict and work interfere family conflict than employed non-parents. On the other hand, non-parents often have more flexibility in managing their time and personal life, and fewer familial responsibilities, which less likely to experience family-to-work conflict.

2.3 Quality of Work Life (QWL)

Sirgy et al. (2001) mentioned that a new measure of QWL was developed based on need satisfaction and spillover theories. The measure was designed to meet the needs of an employee to capture the extent to which the work environment, job requirements, supervisory behavior, and ancillary programs in an organization. They further explained that QWL differs from job satisfaction whereby job satisfaction is construed as one of many outcomes of QWL. Abo-Znadh and Carty (1999) noted that QWL efforts not only on how people can do work better, but also on how work may cause people to be better. Moreover, QWL in an organization also concerns about participation of workers regarding problem solving and decision making. Higher quality of work life would then correlate with lower work-to-family interference (Cheung and Tang, 2009 & Aziz et al. 2011).

2.4 Relationship between work variable and quality of work life

Time pressure which can be measured by the number of hours worked and schedule flexibility is one of the work characteristics that lead to work-family conflict. As time is a limited resource, the more time a women entrepreneur spends on her business, the less time she will have with her family (Kim and Ling, 2001). Jones and Fletcher (1996) confirmed in their research that work stressors affect family life in the form of unpleasant moods that have spilled over from work to family. Job involvement may lead to work interfering with family, which in turn leads to less time and energy devoted to family roles, thereby making it more difficult to comply with pressures associated with family roles (Ballout, 2008). Ballout (2008) also mentioned that supervisor and perceived organizational support could cause employees to report less spillover from work to family and to show greater job satisfaction and better performance. He further added that both work-specific variables and work-family conflict affect individuals' perceptions of career success.

2.5 *Relationship between non-work variable and quality of work life*

Ballout (2008) proposed that non-work variables to affect individuals' perceived career aspiration and success. In particular, he expected that such demographic variables do affect the way individuals make allocation investments to family responsibilities and career aspirations. According to Ashforth et al. (2000), social identity theory focuses on the choice made by individuals in role transitions involving home, work, and other places and postulates that roles can be arrayed on a continuum, ranging from high segmentation to high integration. While Ruderman et al. (2002) explored the benefits of multiple roles for managerial women and found that women managers who integrated careers with family and other non-work roles were successful and satisfied in engaging in both managerial and personal roles.

2.6 *Relationship between work variable, non-work variable and quality of work life*

As suggested by Ballout (2008), organizations that have supportive cultures can reduce work-family conflict of their employees by offering work-family support programs built around flexibility, teamwork, and cooperation. Further, when employees do experience greater conflict in the workplace that intrudes them into the home, their disposition and attitudes relating to career roles will be affected. Then employees may become preoccupied by managing the imbalance between work and home, if this is the case. Lack of balance will cause a detrimental effect on how committed employees feel toward their career competences.

3 METHODOLOGY

This research utilized self-administered questionnaires. The samples of the study consist of 133 executives that have been recruited from five stars hotels in Kuala Lumpur. The data were later analysed to test several hypotheses formulated for the study.

4 FINDINGS AND DISCUSSION

4.1 *Response rate*

Out of 24 five star rated hotels, only 19 hotels have agreed to participate. A total of 51 questionnaires

were returned out of 133 questionnaires and finally coded. This represents 38 percent response rate. According to Baldauf et al. (1999) and Tomaskovic-Devey et al. (1994), surveys of organizations typically receive substantially lower return rates than surveys of individuals, which is with 15 percent return rates sometimes reaching a level of acceptability for organizational surveys. Besides, Visser et al. (1996) mentioned that some studies with low response rates, even as low as 20 percent, are able to yield more accurate results than studies with response rates of 60 percent to 70 percent.

4.2 *Cronbach's alpha reliability analysis of study variables*

In order to ensure that the gathered data were reliable, the measurement of each variable and its sub scales were assessed for reliability. The Cronbach's alpha coefficient value for all variables in the study revealed a range of coefficient value from 0.63 to 0.66 accordingly. The dependent variable of quality of work life had a reliability coefficient of 0.65 constant with the number of independent variables. The independent variable of work variable showed that its dimensions of meaningfulness had a coefficient value of 0.66, and non-work variable had a reliability coefficient value of 0.63. The results also indicated that the reliabilities of the scales used in this study were commonly accepted and considered to be the criterion for demonstrating the internal consistency of a reliable scale (Hinton, Brownlow, McMurray & Cozens, 2004).

4.3 *Bootstrap / INDIRECT analysis*

The objectives of this study are to examine the relationships among work variable, non work variable, and quality of work life. The result from bootstrap analysis is used to examine all the relationships outlined in research objectives and to answer the rest of the hypotheses in the study.

Bootstrap/Indirect analysis was conducted to test the relationship between the independent and dependent variable. It was found that by simply looking to the bootstrap result on path (a) in Table 4.1, the result indicates that only work variable has a positive and significant relationship with quality of work life *(B = 1.194, p < .01)*. The analyses support the following hypothesis:

H1: Work variable significantly influence hotel executives' quality of work life

Results gained in this study revealed that work variable had a positive influence on hotel executives' quality of work life. It was earlier proposed that there was a positive influence of work variable on executives' quality of work life and based on

Table 1. Summary of Bootstrap results (5000 bootstrap samples, $N = 51$).

Independent variable (IV)	Dependent variable (DV)	Direct effect (c')	Indirect effect (a*b)	95% CI for a*b	Total effect (c)
Work variable	Quality of work life	1.194*	.0003	−.1236, .0728	1.194*
Non-work variable	Quality of work life	.437	−.0034	−.1749, .0721	.400

Independent variable *(IV)*; Dependent variable *(DV)*. In both analyses, the effect of the other independent variable were controlled for **p < .001. *p < .05.

the findings, it shows that there was indeed a significant positive and significant relationship with scores *(B = 1.194, p < .01)*. However, non-work variable shows insignificant relationship towards quality of work life ($B = 0.400$, $p = 0.35$). The analyses were different from the following hypothesis:

H2: Non-work variable significantly influence hotel executives' quality of work life

It was found that there was no statistically significant difference at the .05 level in non-work variable on executives' quality of work life. According to Table 4.1, the scores of non-work variable towards quality of work life were $B = 0.400$, $p = .35$.

As expected by Weer (2006), the supportiveness of an organization's environment for non-work role participation did mitigate the negative impact of non-work emotional energy demands on work engagement. In addition, Thomas and Ganster (1995) found that employees who worked for organizations that were supportive of their involvement in family life experienced a greater sense of control over their work and family responsibilities, which in turn reduced their perceptions of conflict and strain and increased their job satisfaction.

5 CONCLUSION

Findings of this study suggest some important implications for hotels' operator in Malaysia particularly in Kuala Lumpur. From the study, the result indicates that only work variable has positive and significant relationship with quality of work life whereby non work variable shows insignificant relationship towards quality of work life. This indicates that executives did depend on their work life such as number of hours work, work schedule inflexibility, work stressor, job involvement and job social support to enhance their quality of work life. They show that they are passionately interested in dealing with things that are related to their jobs. On the other hand, non-work role (parental demands and working spouse) did not influence the executives' quality of work life as they can separate their personal life from work. For the industry, when the

employees are aware of their responsibilities and could deliver good service, indirectly it will enhance the customers' satisfaction. Thus, this will not affect performance of executives towards the organization they served. Furthermore, the industry could retain their talented executives in the workforce longer.

This current study has some limitations that deserve to be mentioned. This study was conducted in Kuala Lumpur focusing on 5 star hotels only. Hence, the findings could not be generalized to the entire industry in Malaysia. The current study focused on work variable, non-work variable and its implication towards quality of work life. It is suggested that future studies to look into inter-role conflicts and how it influence QWL.

ACKNOWLEDGEMENTS

This research is funded by Universiti Teknologi MARA, under the Research Acculturation Grant Scheme (RAGS) (600-RMI/RAGS 5/3 (140/2013).

REFERENCES

Abo-Znadh, S. & Carty, R. (1999). An exploration of selected staff and job characteristics, and their relationship to quality of work life, among staff nurses in medical/surgical units in two tertiary care hospitals in Saudi Arabia. Doctoral Dissertation, George Mason University.

Adams, G., King, L. & King, D.W. (1996).Relationships of job and family involvement, family social support, and work-family conflict with job and life satisfaction. *Journal of Applied Psychology, 81* (4), 411–20.

Anderson, S. & Ungemah, D. (1999).Variable work hours: an implementation guide for employers. Oregon Department of Environmental Quality.

Aryee, S. (1992). Antecedents and outcomes of work-family conflict among married professional women: evidence from Singapore. *Human Relations, 45* (2), 813–37.

Ashforth, B., Kreiner, G. & Fugate, M. (2000). All in a day's work: boundaries and micro role transitions. *Academy of Management Review, 25* (3), 472–91.

Aziz, R.A., Nadzar, F.M., Hussaini, H., Maarof, A., Radzi, S.M. & Ismail, I. (2011). Quality of work life of librarians in government academic libraries in the Klang Valley, Malaysia. The *43* (3), 149–158.

Baldauf, A., Reisinger, H. & Moncrief, W.C. (1999). Examining motivations to refuse in industrial mail surveys. *Journal of the Market Research Society, 41*, 345–353.

Ballout, H.I. (2008). Work-family conflict and career success: the effects of domain-specific determinants. *Journal of Management Development, 27* (5), 437–466.

Barnett, R.C. & Baruch, G.K. (1987). Determinants of fathers' participations in family work. *Journal of Marriage and The Family, 49* (1), 29–40.

Bernas, K.H. & Major, D.A. (2000). Contributors to stress resistance: testing a model of women's work-family conflict. *Psychology of Women Quarterly, 24* (2), 170–8.

Cheung, F.Y. & Tang, C.S. (2009). Quality of work life as a mediator between emotional labor and work family interference. *Journal of Business Psychology, 24*, 245–255.

Enz, C.A. & Siguaw, J.A., (2000). Best practices in human resources. *Cornell Hotel and Restaurant Administration Quarterly, 41* (1), 48–61.

Ference, E.A., (1982). Human resources development: toward a definition of training. *Cornell Hospitality Quarterly, 23* (3), 25–31.

Greenhaus, J.H. & Beutell, N.J. (1985), Sources of conflict between work and family roles. *Academy of Management Review, 10* (1), 76–88.

Greenhaus, J.H., A.G. Bedian & K.W. Mossholder: (1987). Work experiences, job performances, and feelings of personal and family well being. *Journal of Vocational Behavior, 31*, 200–215.

Hinkin, T.R. & Tracey, B.J., (2000). The cost of turnover: putting a price on the learning curve. *Cornell Hotel and Restaurant Administration Quarterly, 41* (3), 14–21.

Hinton, P.R., Brownlow, C., McMurray, I. & Cozens, B. (2004). *SPSS Explained*. New York: Routledge Inc.

Jones, F. & Fletcher, B.C. (1996). Taking work home: a study of daily fluctuations in work stressors, effects on moods, and impact on marital partners. *Journal of Occupational and Organizational Psychology, 69*, 89–106.

Kandasamy. I & Ancheri. S. (2009). Hotel employees' expectations of QWL: A qualitative study. *International Journal of Hospitality Management, 28*, 328–337.

Kim, J.L.S. & Ling, C.S. (2001). Work-family conflict of women entrepreneurs in Singapore. *Women in Management Review, 16* (5), 204-221.

Loscocco, K. (2000). Age integration as a solution to work-family conflict. *The Gerontologist, 40* (3), 292-300.

Martel, J.P. & Dupuis, G. (2006). Quality of work life: Theoretical and methodological problems, and presentation of a new model and measuring instrument. *Social Indicators Research, 77*, 333–368.

Martire, L.M., Stephens, M.P. & Townsend, A.L. (2000). Centrality of women's multiple roles: beneficial and detrimental consequences for psychological well-being. *Psychology and Aging, 15* (1), 148–56.

Md-Sidin, S., Sambasivan, M. & Ismail, I. (2010). Relationship between work-family conflict and quality of life. *Journal of Managerial Psychology, 25* (1), 58–81.

Morton, S.M.B., Bandara, D.K, Robinson, E.M. & Carr, P.E.A. (2003). In the 21st Century, what is an acceptable response rate? *Australian and New Zealand Journal of Public Health, 36* (2), 106–108.

Ruderman, M., Ohlott, P., Panter, K. & King, S. (2002). Benefits of multiple roles for managerial women. *Academy of Management Journal, 45* (2), 369–86.

Sirgy, M.J., Efraty, D., Siegel, P. & Lee, D. (2001). A new measure of quality of work life (QWL) based on satisfaction and spillover theories. *Social Indicators Research, 55*, 241–302.

Thomas, L. & Ganster, D. (1995). Impact of family-supportive work variables on work-family conflict and strain: A control perspective. *Journal of Applied Psychology, 80*, 6-15.

Tomaskovic-Devey, D., Leiter, J. & Thompson, S. (1994). Organizational survey non-response. *Administrative Science Quarterly, 39*, 439-457.

Weer, C.H. (2006). The Impact of Non-Work Role Commitment on Employees' Career Growth Prospects. Unpublished Doctor of Philosophy Dissertation. Drexel University.

Whitely, W. & England, G.W. (1977). Managerial values as a reflection of culture and the process of industrialization. *Academy of Management Journal, 20*, 439-53.

Yu, L., (1999). *The International Hospitality Business*. The Haworth Hospitality Press, New York.

Theory and Practice in Hospitality and Tourism Research – Radzi et al. (Eds)
© 2015 Taylor & Francis Group, London, ISBN 978-1-138-02706-0

Knowledge strategic choices in implementing knowledge strategy: Case of Malaysian hotel industry

H. Ismail, S.M. Radzi, N. Ahmad & S.K.A. Nordin
Faculty of Hotel and Tourism Management, Universiti Teknologi MARA, Shah Alam, Malaysia

ABSTRACT: Malaysian Government has encourages majority industry players who contributing toward national's Gross National Income (GNI) to have the cognizance of managing knowledge asset. Since the business environment in Malaysia are highly competitive, all the industry players including hotel industry must consider knowledge as a vital asset in implementing business strategy. It is expected that Knowledge Management (KM) via the Knowledge Strategy (KS) implementation with the appropriate knowledge strategic choices enable hotels to achieve sustainable competitive advantages. Therefore, this proposed that the execution of KS will guide Malaysian hotel industry to manage the valuable asset of knowledge. 120 hotels located in Malaysia were agreed to participate in this study and the data were eligible to for analysis process. Exploratory Factor Analysis (EFA) was conducted and result has confirmed that hotels Malaysia have selected six appropriate knowledge strategic choices as to implement successful KS.

Keywords: Knowledge management, knowledge strategy, knowledge strategic choices

1 INTRODUCTION

Complex business environment requires business firm to put variety of efforts in order to survive and perform in a particular industry. However, due to the factors such as global competition, technological changes and sophistication of customer has make business firm struggle to remain competitive in the industry (Nordin, Radzi, Ismail & Ahmad, 2013). Hence, business firm's has begun to explore on how to attain sustainable competitive advantages (SCA) thus able to achieve superior business performance.

In doing so, business firm must look into the existence of resources in the organization. According to Hansen (2008), instead of looking to position the product or service in the market, business firm must look also into the buddle of assets which can be the vital sources of competitive advantages. This is where resource based view (RBV) play its role in determining competitive advantages and superior business performance.

Nevertheless, as strongly argued by Huang (2012), the existence of resources in most of the business firm are tangible which easy to be imitated by the competitor. Besides that, not all tangible resources are in line with the contention made by Barney (2001) which he argued that tangible resource must be *valuable, rare, inimi*table, and *no substitu*table in order to achieve SCA. This has led business firm to search a new alternatives as to remain competitive in the industry.

2 RESEARCH ISSUE

Today, Malaysia is the country that thrives to become successful towards the year 2020. As a result, Malaysian economic nowadays is seemed to produce a good impact to the whole country. Reported in Key Economic Indicator (2012) report, the Gross National Income (GNI) has shown a positive outcome which in 2012, the GNI was RM 237.1 with the increment of 5.7 percent compared to the previous year. One of the important industries that contribute to national's GNI is service industry. Mentioned by Kim (2011), since Malaysian government has focuses on industrialization, service industry growth as important engine for Malaysian's economic.

Since then, hotel industry has received the positive effect resulting from the growth of service industry (Khairil Wahidin, Nor Khomar, Radzi & Azni Zarina, 2008). The average occupancy rate from 2005 until 2012 were at a stable rate of 60 percent (Tourism Malaysia, 2005, 2012a). Additionally, Malaysian government has taken an initiative to register more hotels to cope with demand from tourist, which gradually increased yearly. Reported in Tourism Malaysia (2012b) tourist arrival statistic, there are about six million increment of number of tourist from 2007 until 2012. Based on this increment, the number of hotel registered recently is 2277 in 2012 from 1567 in 2007 (Ministry of Tourism Malaysia, 2007, 2012a).

Particularly in these two issues, there is a stiff competition in Malaysian hotel industry. This assumption made has supported the previous observation made Auzair (2011) where she contended that hotel industry in Malaysia is believed to be highly competitive. More specifically, in 2010 the number of hotel room supply is 105,849 (Economic Cencus, 2011) (Ministry of Tourism Malaysia, 2012b). Business firm ought to develop best strategies in sustaining competitive advantages enabling attainment of superior business performance. One of the primary competitive tools is the management of knowledge within the firm.

Knowledge Management (KM) is the best answer to understand knowledge strategy (KS) (Hallin, 2009; Zack, 1999). KS is still new concept in the business environment, thus KM implementation is still very conservative and passive. (Cheng, 2010; Musulin, Gamulin & Crnojevac, 2011). Therefore, this study proposed that the execution of KS will guide Malaysian hotel industry to manage the valuable asset of knowledge.

3 LITERATURE REVIEW AND HYPOTHESIS DEVELOPMENT

3.1 Knowledge based view

Researchers have called to refine the RBV to include intangible resource as factors that affect competitive advantage. (Sousa & Hendriks, 2006).

Knowledge can be defined as *"fluid mix of framed experience, values, contextual information, and expert insight that provides a framework for evaluating and incorporating new experiences and information. It originates and is applied in the minds of knower. In organizations, it often becomes embedded not only in documents or repositories but also in organizational routines, processes, practices, and norms"* (Davenport & Prusak, 2000, p. 4).

Knowledge is something difficult to be understood and there are no universally accepted of knowledge definition (Russ, Fineman & Jones, 2010).

Given that tangible assets such as land, capital and labor are easily to be copied, business firms have changed their mindset to recognize knowledge as the main asset within the organization (Sadaghiyani & Tavallaee, 2011). However, there is still unsolved problem regarding on how this knowledge can be created and sustained in the organization (Kiessling, Richey, Meng & Dabic, 2009). This has led to a rise KM initiatives within organization.

3.2 Knowledge management

Knowledge management literature revolves around the discussion on how to develop and sustain knowledge within organizations. (Smith, McKeen & Singh, 2010).

The main function of KM rests on the premise to locate, organize, transfer, and leverage the knowledge in the organization (e.g. Anderson, 2009; Civi, 2000; Lee, 2008). These function act as the key to avoid reinventing the wheel and to leverage increasing organizational knowledge for more well-defined and informed decision making (Ling, Yih, Eze, Guan Gan & Ling, 2008). The appearance of KM give an obvious answer to respond to the what, when, where and how this assets must exist in the organization. Aboud Zeid (2009) supported that KM must be the tool for managing the knowledge in the organization. Though beneficial, KM can be time consuming and temporary, where knowledge goes through incremental change over time (Sabherwal & Sabherwal, 2007). The implementation and execution of KS is affected by issues relating to the conceptualization of KM and the identification benefits associated with KM (Abdollahi, Rezaeian & Mohseni, 2008; Asoh, Belardo & Duchessi, 2008).

3.3 Knowledge strategy

The practice of KS in understanding the concept of KM is widely ignored and received little attention in research (Mitch & Samson, 2007). KS is the critical area of strategic choices for the firm, where it guide business firm to implement KM (Kasten, 2007). KS is a possible new method to manage knowledge since it can bring benefit to the firm when it comes to survive in highly competition business environment (Denford & Chan, 2011). KS is defined as "a set of strategic actions or choices made at high strategic level to identify the strategic knowledge assets, resources and capabilities and then orientating them toward achieving the organizational goals and improving the organizational performance" (Al-Ammary, 2008, p. 66). Particularly in this definition, knowledge strategic choices are essential to be determined first before implementing KS. KS choices are the knowledge activities that needs to be required by business firm prior to implement KS (e.g. Al-Ammary, 2008; Asoh, 2004; Chew, 2008; Zack, 1999). Table 1 summarized knowledge strategic choices that commonly found in business environment.

Implementation of strategic choices can be conducted once the knowledge strategic choices have been determined. Consequently, there are many typology of KS has been developed by several researchers. To determine KS in this study, knowledge strategic choice must be determined first. However, the balance of trade off must be specified first before selecting the suitable knowledge strategic choice. The reason is that business firm

Table 1. Knowledge Strategic Choice.

Knowledge strategic choice	Description
Internal knowledge source	Knowledge basically existed in the organization such as standard operation procedure, book, and employees skills
External knowledge source	Knowledge basically existed outside of the organization such as competitor information, vendors information and research studies
Knowledge exploitation	Utilization and transformation of knowledge existed in the organization
Knowledge exploration	Finding and create new knowledge either inside or outside of the organization
Explicit knowledge	Tangible knowledge in the organization
Tacit knowledge	Intangible knowledge in the organization
Knowledge codification	Knowledge that store, share, and use by using technology such as explicit knowledge
Knowledge personalization	Knowledge that store, share and use by using human or employees such as tacit knowledge
Deep knowledge base	Focus on the specialized knowledge and capabilities in the organization
Broad knowledge base	Focus on the multiple and extensive knowledge
Speed of learning	Refer how slow or fast organization learn their knowledge
Level of radical of learning	Refer to incremental exploiting knowledge and radical exploring knowledge

incapable to manage all the resources in one particular time (Porter, 1996) and they usually possess a limited intangible resource especially knowledge asset (Al-Ammary, 2008).

It is apparent to choose internal knowledge, external knowledge, knowledge exploitation, knowledge exploration, knowledge codification, knowledge personalization, deep knowledge, and broad knowledge as the main knowledge strategic choices in this study. Based on this argument, the following hypothesis has been proposed:

H1: There are knowledge strategic choices to implement KS in Malaysia hotel industry

4 METHODOLOGY

4.1 *Target population, sampling and data collection method*

Two hundred and six of four-star and five-star hotels in Malaysia were the target population in this study. This target population was obtained from Malaysia Accommodation Directory (2011) and used as sampling from this study. The main reason to select four and five stars hotels is because they have greater operational sophistication such as wider range of product or service, provide high personalized service, large investment and wider span of control among department managers (Patiar & Mia, 2008). This can be assumed that

they have similar common of strategic management characteristic (Şentürk, 2012).

Simple random sampling via the computer was the main sampling techniques in this study. Self-administered questionnaire was employed in this study by using mail postal method. The respondents were guaranteed their anonymity. Finally, 120 hotels were agreed to participate which equivalent to 58 percent response rate.

5 FINDING, DISCUSSION, AND CONCLUSION

In order to test the hypothesis the 25 knowledge activities that related to knowledge strategic choice were subjected to factor analysis to analyze the existence of knowledge strategic choices in hotels.

Using principal component analysis (PCA) with varimax rotation and Kaiser Normalization via the SPSS, seven factor components emerged that explained by 65.70 percent of variance. The Kaiser Meyer-Olkin measure, the KMO's value were 0.81 which indicate as great in term of its sampling adequacy (Field, 2013). While in term of Bartlett's test analysis, the items were highly significant at below -001 alpha levels. This result indicates that the factor analysis was appropriate to be conducted.

Furthermore, PCA result has confirmed seven factor solutions after rotation. However, the seventh factor was removed due to the one item only loaded into the factor.

As a result, six items were excluded due to the items loading load below than 0.55. Out of six factor solutions, factor component number one, two, three and six corresponded with Asoh (2004) contention regarding to the common knowledge strategic choices. However, factor component number four and five are inconsistent with Asoh (2004) components but there was no new interpretation made on those two components. This is because items that load onto components numbers 4 and 5 were actually represents similar aspect.

Ultimately, components one appear to represent knowledge exploration, factor two represent broad knowledge, factor three loaded as knowledge exploitation, factor four classified as knowledge personalization, factor five known as external knowledge and factor six was knowledge codification. Table 2 illustrates knowledge strategic choices as to implement KS particularly in Malaysian hotel industry. As the conclusion, hypothesis one was partially supported.

Zack (1999) revealed that business firm mostly focused on internal, external, exploitation, exploration and innovation as main knowledge strategic choices and classified that those business firms who are keen to internal knowledge and knowledge

Table 2. Exploratory Factor Analysis for Knowledge Strategic Choice.

Items	1	2	3	4	5	6
QKSb7	.77					
QKSb9	.72					
QKSb8	.71					
QKSb10	.66					
QKSb13		.77				
QKSb11		.77				
QKSb12		.66				
QKSa9			.76			
QKSa8			.61			
QKSa7			.60			
QKSb1				.74		
QKSa4				.58		
QKSb2				.57		
QKSb5					.73	
QKSb6					.70	
QKSb3					.55	
QKSa2						.85
QKSa1						.76
QKSa3						.63
Eigenvalues	8.03	1.93	1.67	1.52	1.17	1.08
% of Variance	32.13	7.72	6.74	6.07	4.68	4.33

exploitation are able to pursue conservative knowledge strategy (CKS) and those who are focusing external, exploration and innovation knowledge able to pursue aggressive knowledge strategy (AKS). While in study of Al-Ammary (2008), the author stressed that internal, external, codification, personalization, exploitation and exploration were appropriated knowledge strategic choices in implementing CKS and AKS. Whereas in study of Chew (2008), he also identified the same knowledge strategic choices and classified those business firms who focused on this knowledge activity were able to pursue CKS and AKS.

However, the retained four factor solutions were consistent and two factor solutions were not consistent with previous studies discussed above. Internal and deep knowledge were not successfully load at expected component. Therefore, these two particular factors have revealed that hotels were not likely to have these two knowledge strategic choice in implementing the KS. The findings of this study supports Porter (1996) and Al-Ammary (2008) propositions. They argued that in order to manage the knowledge successfully, business firm must require a balance trade off in choosing knowledge strategic choice.

As for the conclusion, hotel industries in Malaysia were likely to choose knowledge exploration, broad knowledge, knowledge exploitation, knowledge personalization, external knowledge and knowledge codification. Based on the chosen knowledge strategic choices, they might able to implement successful KS.

6 LIMITATION AND RECOMMENDATION

Despite of the contribution made from this study, there were two limitations posed. Firstly, the data were only collected in the period of the year 2012; it might not enough to tackle more on knowledge strategic choices as general. This is because knowledge strategic choices are likely to evolve due to the forces such as technology changes, government regulation, and uncertainty of world economy. Hence, future studies must require a longitudinal study.

Last but not least, a new measurement of knowledge strategic choices must be reconstructed as the current knowledge strategic choices measurement was constructed in 2004. Although the present study adapted the knowledge strategic choices measurement, it still can be used because strategy of managing knowledge in hotel industry yet to be recognized and therefore, if hotel industry in future becoming knowledge oriented firm, future researchers need to develop a new measurement of knowledge strategic choices so that it parallel with future business environment.

REFERENCES

Abdollahi, A., Rezaeian, A. & Mohseni, M. (2008). Knowledge Strategy: Linking Knowledge Resources to Competitive Strategy. *World Applied Sciences Journal, 4*(2), 08–11.

Aboud Zeid, E.-S. (2009). Alignment of Businses and Knowledge Management Strategy Encyclopedia of Information Science and Technology, 2nd Ed (pp. 124–129): IGI Global.

Al-Ammary, J.H. (2008). *Knowledge Management Strategic Alignment in the Banking Sector at the Gulf Cooperation Council (GCC) Countries.* Murdoch University Western Australia.

Anderson, K.K. (2009). *Organizational Capabilities as Predictors of Effective Knowledge Management: An Empirical Examination.* Nova Southeastern University, Ann Arbor.

Asoh, D. (2004). *Business and Knowledge Strategies: Alignment and Performance Impact Analysis.* State University of New York at Albany, Ann Arbor.

Asoh, D., Belardo, S. & Duchessi, P. (2008). Knowledge Strategic Alignment: Research Framework, Models, and Concepts Knowledge Management and Business Strategies: Theoretical Frameworks and Empirical Research (pp. 188–208): IGI Global.

Auzair, S.M. (2011). The Effect of Business Strategy and External Environment on Management Control System: A Study of Malaysian Hotels. *International Journal of Business and Social Science, 2*(13), 236–244.

Barney, J.B. (2001). Resource-Based Theories of Competitive Advantage: A Ten-Year Retrospective on the Resource-Based View. *Journal of Management, 27*(6), 643–650.

Cheng, X. (2010). *A Systematic Review of Knowledge Management Research in the Hospitality and Tourism Industry.* University of Nevada, Las Vegas.

Chew, K.C. (2008). *An Exploratory Case Study. Knowledge Management: Managing Organizational Knowledge Assets by Aligning Business Strategy, Knowledge Strategy, and Knowledge Management Strategy.* Golden Gate University.

Civi, E. (2000). Knowledge Management as a Competitive Asset: A Review. *Marketing Intelligence & Planning, 18*(4), 166–174.

Davenport, T.H. & Prusak, L. (2000). Working Knowledge: How Organizations Manage What They Know. *Ubiquity, 2000* (August), 2.

Denford, J.S. & Chan, Y.E. (2011). Knowledge Strategy Typologies: Defining Dimensions and Relationships. *Knowledge Management Research & Practice, 9,* 102–119.

Economic Cencus. (2011). Department of Statistics Malaysia Retrieved 5 May, 2012, from http://www.statistics.gov.my/portal/images/stories/files/LatestReleases/BE/BI/BE2011_PenginapanBI.pdf

Field, A. (2013). *Discovering Statistics Using IBM SPSS Statistics.* London: Sage Publication Ltd.

Hallin, C.A. (2009). *Exploring the Strategic Impact of Service Employees' Tacit Knowledge: The Development of an Indicator for Forecasting Economic Performance of Hotel Companies.* University of Stavanger, Norway.

Hansen, F.R. (2008). *The Utility of Strategic Management Knowledge for Strategic Management Practice: The Actors Perspective.* Capella University, United States, Minnesota.

Huang, H.I. (2012). *An Empirical Analysis of the Strategic Management of Competitive Advantage: A Case Study of Higher Technical and Vocational Education in Taiwan.*

—. (2007). *Knowledge Strategy and Its Influence on Knowledge Organization.* Paper presented at the North American symposium on knowledge organization, Arizona.

Key Economic Indicator. (2012). National Accounts Gross Domestic Product Retrieved 14 November, 2012, from http://www.statistics.gov.my/portal/download_Akaun/files/quartely_national/2012/SUKU_KEEMPAT/KDNK_Q412.pdf

Khairil Wahidin, A., Nor Khomar, I., Radzi, S.M. & Azni Zarina, T. (2008). Environmental Variables and Performance: Evidence from the Hotel Industry in Malaysia. *International Journal of Economics and Management, 2*(1), 59–79.

Kiessling, T.S., Richey, R.G., Meng, J. & Dabic, M. (2009). Exploring Knowledge Management to Organizational Performance Outcomes in a Transitional Economy. *Journal of World Business, 44*(4), 421–433.

Kim, S. (2011). Factor Determinants of Total Factor Productivity Growth in the Malaysian Hotel Industry: A Stochastic Frontier Approach. *Cornell Hospitality Quarterly, 52*(1), 35–47.

Lee, M.L. (2008). *A Qualitative Case Study Apporoach to Define and Identify Perceived Challenges of Knowledge Management for Casino Hotel Industry.* University of Nevada, Las Vegas.

Ling, T.N., Yih, G.C., Eze, U.C., Guan Gan, G.G. & Ling, L.P. (2008). *Knowledge Management Drivers for Organizational Competitive Advantages.* Paper presented at the International Applied Business Conference.

Malaysia Accommodation Directory. (2011). Malaysia Accommodation Directory *2010/2011.* Putrajaya Tourism Malaysia: Ministry of Tourism Malaysia.

Ministry of Tourism Malaysia. (2007). Malaysia Hotel Registered Retrieved 5 October, 2011, from http://www.hotels.org.my/appstorage/gallery/statistics_registered_hotels_31_dec_2007.pdf

Ministry of Tourism Malaysia. (2012a). Malaysia Hotel Registered Retrieved 5 May, 2012, from http://www.motour.gov.my/en/download/viewcategory/27-statistik-pelesenan.html

Ministry of Tourism Malaysia. (2012b). Number Of Room Retrieved 5 May, 2012, from http://www.motour.gov.my/en/download/viewcategory/27-statistik-pelesenan.html

Mitch, C.R. & Samson, D. (2007). Aligning Knowledge Strategy and Knowledge Capabilities. *Technology Analysis & Strategic Management, 19*(1), 69–81.

Musulin, J., Gamulin, J. & Crnojevac, I.H. (2011, 23–27 May 2011). *Knowledge Management in Tourism: The Importance of Tacit Knowledge and the Problem of Its Elicitation and Sharing.* Paper presented at the MIPRO, 2011, 34th International Convention.

Nordin, S.K.A., Radzi, S.M., Ismail, H. & Ahmad, N. (2013). Knowledge and Business Strategy Model. Paper presented at *the* Hospitality and Tourism: Synergizing Creativity and Innovation in Research (pp. 52–56). Shah Alam, Malaysia.

Patiar, A. & Mia, L. (2008). The Interactive Effect of Market Competition and Use of MAS Information on Performance: Evidence From the Upscale Hotels. *Journal of Hospitality & Tourism Research, 32*(2), 209–234.

Porter, M.E. (1996). What is Strategy. *Harvard Business Review.*

Russ, M., Fineman, R. & Jones, J.K. (2010). Conceptual Theory: What Do You Know?Knowledge Management Strategies for Business Development (pp. 1–22).

Sabherwal, R. & Sabherwal, S. (2007). How Do Knowledge Management Announcements Affect Firm Value? A Study of Firms Pursuing Different Business Strategies. *Engineering Management, 4*(3), 409–422.

Sadaghiyani, J.S. & Tavallaee, R. (2011). *Review Different Knowledge Strategy Models and Designing a New Model.* Paper presented at the International Conference on Innovation, Management and Service, Singapore.

Şentürk, F.K. (2012). A Study to Determine the Usage of Strategic Management Tools in the Hotel Industry. *Procedia-Social and Behavioral Sciences, 58*(0), 11–18.

Smith, H.A., McKeen, J.D. & Singh, S. (2010). Developing and Aligning a KM Strategy. *Journal of Information Science & Technology, 7*(1), 40–60.

Sousa, C.A.A. & Hendriks, P.H.J. (2006). The Diving Bell and the Butterfly: The Need for Grounded Theory in Developing a Knowledge-Based View of Organizations. *Organizational Research Methods, 9*(3), 315–338.

Tourism Malaysia. (2005). Average Occupancy Rate. Retrieved 5 May, 2012, from http://corporate.tourism.gov.my/images/research/pdf/2005/AOR/AOR_2004_2005.pdf

Tourism Malaysia. (2012a). Average Occupancy Rate. Retrieved 5 May, 2013, from http://corporate.tourism.gov.my/images/research/pdf/2012/AOR/AOR%20_2012.pdf

Tourism Malaysia. (2012b). Tourist Arrivals & Receipts to Malaysia. Retrieved 15 November, 2012, from http://corporate.tourism.gov.my/research.asp?page=facts_figures

Zack, M.H. (1999). Developing a Knowledge Strategy. *California Management Review, 41*(3), 125–145.

Theory and Practice in Hospitality and Tourism Research – Radzi et al. (Eds)
© 2015 Taylor & Francis Group, London, ISBN 978-1-138-02706-0

Quality practices and quality implementation: A proposed case study in Grand Bluewave Hotel, Shah Alam

N. Sumarjan, B. Syaripuddin, S.A. Jamal, C.T. Chik & Z. Mohi
Faculty of Hotel and Tourism Management, Universiti Teknologi MARA, Shah Alam, Selangor, Malaysia

ABSTRACT: The purpose of this study is to identify the relationship between TQM practices and quality implementation in Grand Bluewave, Shah Alam. This study will employ the seven MBNQA criteria (leadership, strategic planning, customer focus, measurement, analysis, and knowledge management, workforce focus, process management, and business results) as TQM practices. A quantitative method approach will be used. Questionnaires will be distributed to all operational employees in the hotel. Descriptive statistics, analysis and multiple regressions will be used to analysis the data. Practically, the outcomes of this study will provide the management team with some advantages in planning and organizing its resources to better enhance the quality process and system in the hotel. Theoretically, the finding from this study also hopefully should provide academicians the platform with valuable information on the current and the real situation of the hospitality industry regarding quality implementation, thus eliminating gaps between literatures and academic findings.

Keywords: Malaysia hotels, MBNQA, Total Quality Management (TQM), quality implementation, and quality practices

1 INTRODUCTION

The Economic Transformation Process (ETP) showed tourism sector has been classified as a catalyst for Malaysia economic and a major source of government revenue (Performance Management & Delivery Unit, 2012). These will benefit the local and the international hotel operators as they are welcomes to open new hotel in Malaysia to accommodate the increasing number of tourists' arrival (Ng, 2013).

However hotel operators cannot look this opportunity as a reason to be complacent. They have to remain competitive as Malaysia is surrounded by Thailand, Indonesia and Singapore that are offering a similar tourism product and services (Sumarjan, Arendt & Shelley, 2013). Quality implementation has become a top priority for companies around the world and plays an important role to gain competitive advantage and achieve organizational goals (Fotopoulos & Psomas, 2010; Prajogo & McDermott, 2005; Sadikoglu & Zehir, 2010; Sumarjan, et al., 2013; Talib, Rahman & Qureshi, 2013). Talib et al. stated that the most recommended and recognized approached is Total Quality Management (TQM) concept.

Total quality management (TQM) can be described as a management philosophy that helps organization to manage and improve overall performance and effectiveness towards achieving world-class status (Yusof & Aspinwall, 2000; Zhang, Waszink & Wijngaard, 2000). A lot of success stories with regard to TQM practices have been released and successfully published (Arumugam, Ooi & Fong, 2008; Gustafsson, Nilsson & Johnson, 2003; Karia & Asaari, 2006; Lagrosen, 2003; Miller, Sumner & Deane, 2009; Prajogo & McDermott, 2005; Pun, 2003; Sila, 2007; Yoo, Rao & Hong, 2006). Nevertheless, researchers found mixed results related to TQM (Nair, 2006; Kaynak, 2003). Nair (2006) stated TQM practices only contribute to certain company's performance while Kaynak (2003) revealed TQM practices (e.g. supplier quality management positively affecting process management, product and service design, and inventory management performance) had an indirect positive effect on market, financial as well as quality performance. Theoretically, TQM practices play an important part on the improvements of business performance but in reality, a lot of organizations fail in implementing their owned quality programmes (Rad, 2006). The mixed result and findings concerning the implementation and success of TQM practices poses the question of what factors or quality practices that lead to TQM success? Currently, investigation related to the relationship between TQM practices and quality implementation in the hotel industry, particularly in the Malaysia setting is still limited. Additionally, despite numerous studies that have been conducted, concentration on operational employee perspectives regarding quality practices and quality implementation are less known especially within hotel industry context (Sumarjan, et al., 2013). Gaining employee commitment is critical to ensure the success of quality implementation in organizations (Demirbag & Sahadev, 2008; Jackson, 2004). Therefore, this study proposes to investigate the relationship between quality practices and quality implementation from employee perspective. Operational employees have been chosen as they have direct contact with customers. Furthermore, employee involvement in quality management practices will reduce stress levels and resistance to change (Jarrar & Zairi, 2002) and affect employee commitment (Howard & Foster, 1999).

2 LITERATURE REVIEW

2.1 *Quality practice*

Stories about the successful implementation of TQM practices have been frequently highlighted and published (Arumugam, et al., 2008; Gustafsson, et al., 2003; Karia & Asaari, 2006; Lagrosen, 2003; Miller, et al., 2009; Prajogo & McDermott, 2005; Pun, 2003; Sila, 2007; Yoo, et al., 2006). Arumugam et al. (2008) who conducted a study in 122 Malaysia organizations certified with ISO 9001:2000 found TQM practices such as continual improvement, customer focus, process management, information analysis, people involvement, quality system improvement, supplier relationships and leadership, partially influenced quality performance. Gustafsson et al. (2003) found that even though TQM practices such as process orientation, customer satisfaction and employee management can improve firm performance, the relationship between quality practices and firm performance is dependent on firm size.

Karia and Asaari (2006) conducted a study to examine the impact of TQM practices on employees' work-related attitudes and found that quality practices such as customer focus, empowerment and teamwork, training and education, as well as continuous improvement and problem prevention improve the level of career and job satisfaction, enhance job involvement and promote greater organizational commitment. Miller et al. (2009) carried out an assessment of quality management practices in healthcare industry found eight quality practices: role of management leadership; customer focus; role of the physician; training; quality data reporting; employee relations; process management/training; and the role of quality department

were having a significant effect on TQM practices. Thus, to investigate the relationship between quality practices, this study will use the seven MBNQA criteria: strategic planning; leadership; business results; process management; workforce focus; customer focus; and measurement, analysis and knowledge management (National Institute of Standards and Technology (NIST), 2014). This MBNQA criterion is comprehensive and has been frequently cited in the literature to measure quality endeavor (Khoo & Tan, 2003; Lau, Zhao & Xiao, 2004; Prajogo & McDermott, 2005; Sumarjan, et al., 2013).

2.2 *Employee involvement in quality implementation*

According to Pun and Gill (2002), the implementation of quality management practices that involve employee comes in many different approaches. One of the approach is Malcom Baldrige National Quality Award (National Institute of Standards and Technology (NIST) (2014)), which come out with seven criteria: leadership; strategic planning; process management; workforce focus; customer focus; measurement, analysis and knowledge; and business results, that can investigate the integration of employees participation and total quality management (TQM) (Sumarjan, et al., 2013).

Sumarjan et al. (2013) who conducted a study to compare perceptions of Malaysian employee and Hotel Quality Managers (HQMs) in three- and four-star hotels by using two Malcolm Baldrige National Quality Award (MBNQA) criteria (leadership and workforce focus), found that failure to develop explicit quality objectives and policies by the management team and inefficient communication system as the main contributor for perceptions incongruence between managers and employees.

Yeh (2003) conducted a study to investigate the implementation of sustainable total quality management (TQM) implementation from the employee perspectives in a United States. This study used critical factors such as job characteristic, individual training and project involvement, social support, employees' self-efficacy, and organizational structure, found that organizational interpersonal support, employees' self-efficacy, and standardized organizational structure were factors that critically predicted employee's involvement in total quality management (TQM). On the other hand, factors like training and individuals' project involvement did not have direct effect employees' practices of total quality management. The author also revealed individuals' project involvement and training had indirect effect to employees' practices of quality management as they believe workloads increased as quality management is implemented.

Vanichchinchai (2012) conducted a study about the relationship between employee involvement, partnership management and supply performance in Thailand automotive industry found that employee involvement has significantly positive impact on firm's supply performance and external partnership management. At the same time, external partnership management intervene the positive relationship between employee involvement and firm's supply performance. Thus, from supply chain and quality management perspectives, barriers pertinent to employee issues like fear of change, internal resistance, lack of participation, inadequate skills and knowledge, and lack of a common perspective should be changed.

3 METHODOLOGY

A quantitative approach will be used to collect and analyze data. Grand Bluewave Hotel, Shah Alam will be chosen as this organization has received several quality achievement awards and already ISO 90001 certified (Crescent Rating, 2013). Operational employees from various department and functions that have been working for at least three months in this hotel and not holding top management position will be used as the sample. Dual language (English and Malay) questionnaire with a cover letter will be distributed to all respondents. This study will use the descriptive statistic and multiple regressions to answer the research question. Statistical Program for Social Analysis (SPSS), version 20 will be used for data coding, processing and analyzing.

4 CONCLUSION

Practically, the outcomes of this study will give the hotel management team some advantages to plan and organize for future quality practices in hotel. The findings from this study would also provide significant knowledge and information on quality practices and implementation within the operational workforce in creating better work environment in the workplace. Theoretically, the finding from this study also should provide academicians with valuable information on the current and the real situation of the hospitality industry regarding quality implementation, particularly from the employees' perspective, thus eliminating gaps between literatures and academic findings.

ACKNOWLEDGEMENTS

This research is funded by Universiti Teknologi MARA through Research Faculty Grant 600-RMI DANA 5/3/RIF (449/2012).

REFERENCES

Arumugam, V., Ooi, K.B. & Fong, T.C. (2008). TQM practices and quality management performance: An investigation of their relationship using data from ISO 9001:2000 firms in Malaysia. *The TQM Journal, 20*(6), 636–650.

Baird, K., Hu, K.J. & Reeve, R. (2011). The relationships between organizational culture, total quality management practices and operational performance. *International Journal of Operations & Production Management, 31*(7), 789–814.

Crescent Rating (2013). *Grand Bluewave Hotel Shah Alam*. Retrieved from http://www.crescentrating.com/malaysia-halal-friendly-accomodation/2383-grand-bluewave-hotel-shah-alam.html

Demirbag, M. & Sahadev, S. (2008). Exploring the antecedents of quality commitment among employees: An empirical study. *International Journal of Quality & Reliability Management, 25*(5), 494–507.

Fotopoulos, C.V. & Psomas, E.L. (2010). The structural relationships between TQM factors and organizational performance. *The TQM Journal, 22*(5), 539–552.

Gustafsson, A., Nilsson, L. & Johnson, M.D. (2003). The role of quality practices in service organizations. *International Journal of Service Industry Management, 14*(2), 232–244.

Howard, L.W. & Foster, S.T. (1999). The influence of human resource practices on empowerment and employee perceptions of management commitment to quality. *Journal of Quality Management, 4*(1), 5–22.

Jackson, P.R. (2004). Employee commitment to quality: Its conceptualisation and measurement. *International Journal of Quality & Reliability Management, 21*(7), 714–730.

Jarrar, Y.F. & Mohamed Zairi, M. (2002). Employee empowerment—A UK survey of trends and best practices. *Managerial Auditing Journal, 17*(5), 266–271.

Karia, N. & Asaari, M.H.A.H. (2006). The effects of total quality management practices on employees' work-related attitudes. *The TQM Magazine, 18*(1), 30–43.

Kaynak, H. (2003). The relationship between total quality management practices and their effects on firm performance. *Journal of Operations Management, 21*(4), 405–435.

Khoo, H.H. & Tan, K.C. (2003). Managing for quality in the USA and Japan: differences between the MBNQA, DP and JQA. *The TQM Magazine, 15*(1), 14–24.

Lagrosen, S. (2003). Exploring the impact of culture on quality management. *International Journal of Quality & Reliability Management, 20*(4), 473–487.

Lau, R.S.M., Zhao, X. & Xiao, M. (2004). Assessing quality management in China with MBNQA criteria. *International Journal of Quality and Reliability Management, 21*(7), 699–713.

Miller, W.J., Sumner, A.T. & Deane, R.H. (2009). Assessment of quality Management practices within the healthcare industry. *American Journal of Economic and Business Administration, 1*(2), 105–113.

Nair, A. (2006). Meta-analysis of the relationship between quality management practices and firm performance-Implications for quality management theory development. *Journal of Operations Management, 24*(6), 948–975.

National Institute of Standards and Technology (NIST) (2014). *Baldrige performance excellence program.* Retrieved from http://www.nist.gov/baldrige/

Ng, A. (2013). *More upscale hotels to open.* Retrieved from The Star Online: http://www.thestar.com.my/Business/Business-News/2013/10/21/More-upscale-hotels-to-open-Hospitality-sector-continues-to-attract-international-hotel-operators.aspx/

Performance Management & Delivery Unit (2012). *Annual Report 2012.* Retrieved from Economic Transformation Programme: http://etp.pemandu.gov.my/annualreport/upload/Eng_ETP2012_Full.pdf

Prajogo, D.I. & McDermott, C.M. (2005). The relationship between total quality management practices and organizational culture. *International Journal of Operations & Production Management, 25*(11), 1101–1122.

Pun, K.F. (2003). Exploring employee involvement and quality management practices: A review of the literature. *Asian Journal on Quality, 4*(2), 123–144.

Pun, K.F. & Gill, R. (2010). Integrating EI/TQM efforts for performance improvement: A model. *Integrated Manufacturing System, 13*(7), 447–458.

Rad, A.M.M. (2006). The impact of organizational culture on the successful implementation of total quality management, *The TQM Magazine, 18*(6), 606–625.

Sadikoglu, E. & Zehir, C. (2010). Investigating the effects of innovation and employee performance on the relationship between total quality management practices and firm performance: An empirical study of Turkish firms. *International Journal of Production Economics, 127*(1), 13–26.

Sila, I. (2007). Examining the effects of contextual factors on TQM and performance through the lens of organizational theories: An empirical study. *Journal of Operations Management, 25*(1), 83–109.

Sumarjan, N., Arendt, S.W. & Shelly, M. (2013). Incongruent quality management perceptions between Malaysian hotel managers and employees. *The TQM Journal, 25*(2), 124–140.

Talib, F., Rahman, Z. & Qureshi, M.N. (2013). An empirical investigation of relationship between total quality management practices and quality performance in Indian service companies. *International Journal of Quality & Reliability Management, 30*(3), 280–318.

Vanichchinchai, A. (2012). The Relationship between Employee Involvement, Partnership Management and Supply Performance: Findings from a developing country. *International Journal of Productivity and Performance Management, 61*(2), 157–172.

Yeh, Y.F. (2003). Implementing a sustainable TQM system: Employee focus. *The TQM Magazine, 15*(4), 257–265.

Yoo, D.K., Rao, S.S. & Hong, P. (2006). A comparative study on cultural differences and quality practices—Korea, USA, Mexico, and Taiwan. *International Journal of Quality & Reliability Management, 23*(6), 607–624.

Yusof, S.M. & Aspinwall, E. (2000). Total quality management implementation frameworks: Comparison and review. *Total Quality Management, 11(3),* 281–294.

Zhang, Z., Waszink, A. & Wijngaard, J. (2000). An instrument for measuring TQM implementation for Chinese manufacturing companies. *International Journal of Quality & Reliability Management, 17*(7), 730–755.

Theory and Practice in Hospitality and Tourism Research – Radzi et al. (Eds)
© 2015 Taylor & Francis Group, London, ISBN 978-1-138-02706-0

Organizational DNA and human resource practices: Its implication towards hotel performance

Z.N. Aishah & M.N. Syuhirdy
Faculty of Hotel and Tourism Management, Universiti Teknologi MARA, Penang, Malaysia

M.S.M. Zahari & S.M. Radzi
Faculty of Hotel and Tourism Management, Universiti Teknologi MARA, Shah Alam, Malaysia

ABSTRACT: It is undeniable that success of the organization's performance is well depended on the effectiveness of its structural management and operating protocols that exist within the organization itself. Organizational DNA represents a deliberate way of thinking about organizations and their patterns of leadership, culture, management and human resource practices within their related contexts. Organizational DNA is proven helps companies to recognize and expose hidden strengths and entrenched weaknesses thus aids managers to focus efforts on reinforcing what works in their organization and modifying what does not. This paper reviews the Organizational DNA, the types and its strengths and subsequently proposed to empirically investigate whether the patterns of human resource practices in hotel organization is aligned with which types of Organizational DNA and how far the effect of it towards hotel performance.

Keywords: Organizational DNA, human resource, practices, hotel, performance

1 INTRODUCTION

Many business scholars argue that the success of the organizational performance is well depend on the effectiveness of its structural management and operating protocols that exist within the organization itself (Kaipa & Milus, 2002; Spiegel, Gary, Flaherty Thomas & Srini, 2005). Honold and Silverman (2002) posit that the key to a company's operating destiny in modern business is strongly associated with four core elements of organization attributes namely leadership, culture, management and human resource practices. These elements are in fact considered the organizational building blocks or organizational genetic code which combine and recombine to express distinct identities or personalities and in today's circumstances it is called as Organizational DNA. Organizational DNA is proven that helps companies to recognize and expose hidden strengths and entrenched weaknesses, thus managers can focus efforts on reinforcing what works in their organization and modifying what does not (Neilson, Pasternack, Mendes & Tan, 2004).

The importance of Organizational DNA for the company is supported by a few case studies (Dehoff, Jaruzelski & Kronenberg, 2008; Honold & Silverman, 2002; Kaipa & Milus, 2002; Spiegel, et al., 2005). Kaipa and Milus (2002) asserted that the dynamic nature of the relationships between DNA elements could impact the outcomes in organizational transformation efforts. Organizational transformation will not be successful or less able to realize its full potential, if personal transformation leading to business transformation is not clearly connected with the change effort. Organizational DNA can be reconfigured by identifying and realigning the four building blocks to fixing the execution problems in their company (Spiegel, et al., 2005). This illustrates that any organization, across any industry, has its own DNA that drives its actions and its performances.

Organizational DNA represents a deliberate way of thinking about organizations and their patterns of leadership and management practices within their related contexts (Honold & Silverman, 2002). They proposed four distinct types of Organizational DNA; Factual DNA, Conceptual DNA, Contextual DNA and Individual DNA. Each type of DNA has been determined by four different manufacturing industries in Honold and Silverman (2002) studies in the late nineties. This help organization to discern their DNA type by looking at the organizational practices carefully and considering the alignment of DNA types in order to develop the organization in the most effective manner. Modifying or misalign one element somewhat obstructs the organization's performance.

According to the Honold and Silverman (2002) alignment model, the types of DNA is identified through organizational practices; leadership, management, human resource and everyday cultural practices. The alignment for each of the practices and the tools/instruments has been developed in determining the company's dominant organizational DNA. The understanding of the dominant type of organizational DNA is critical to undertake resolution of the organizational problems, relieve organizational stress as well as in selecting the appropriate consultants and other resources for organizational improvement (Honold & Silverman, 2002).

2 LITERATURE REVIEW

2.1 Organizational DNA

The DNA is an acronym for deoxyribonucleic acid that contains the genetic instructions used in the development and functioning of all known living organisms with the exception of some viruses. The DNA is vital for inheritance, coding for proteins and the genetic blueprint of life. Given the enormity of DNA's functions in the human body and its responsibility for the growth and maintenance of life, it is not surprising that the discovery of DNA has led to such a great number of developments in treating disease (Explore DNA, September 2010).

As of medical sciences, organizational DNA refers to the building blocks that combine and recombine of its components (leadership, culture, management and human resource practices) thus portray the identities or personalities of the organization (Neilson, et al., 2004). The Organizational DNA helps the organization identify and expose hidden strengths and entrenched weaknesses, reinforcing what works and modifying what does not. Prakash (2001) posits the organizational DNA is the vision and values upon which the organization is built and organization must have a strong foundation and certain core values that constitute the genetic make-up of the company. Spiegel et al. (2005) affirm that reconfiguring a company's organizational DNA through a structured market can increase institutional agility, accelerate readiness for external change and enable sustained their financial leadership. Results of online poll conducted by Booz Allen Hamilton Company show the utility industry is in a stage of ongoing volatility, which requires companies to adapt continuously, not only to changes in the business cycle, but also to the turbulence of deregulation therefore an organization need to reconfigure its DNA.

2.2 Organizational DNA alignment model

Prasad Kaipa, in 1989, developed the 3-dimensional modeling process called a DNA model. This DNA model consist a system or the dynamic nature of the relationships between the leadership, strategy, structure and culture and how the relationships could impact the outcomes in organizational transformation efforts. The basic premise is that organization transformation will not be successful, or fail to realize its full potential, if personal transformation leading to business transformation is not clearly connected with the change effort. Honold and Silverman (2002) on the other hands noted that organizational DNA represents a deliberate way of thinking about organizations and their patterns of leadership, culture, management and human resource practices within their related contexts. They claimed the organizational DNA replaces many alternatives, the initial approaches, organizational model, forms or practices such as teamwork, decision making and employee development. They subsequently proposed four types of Organizational DNA, namely Factual DNA, Conceptual DNA, Contextual DNA and Individual DNA.

Factual DNA speaks to the reality of our factual world and to organizations that are committed to knowing themselves and their environments through the collection of data. This information provides a constant barometer and continual assessment of performance. It is also crucial in the development of all policies and procedures of the organization. Conceptual DNA focuses on large motivating ideas that may take the form of major theories, visions and other conceptual devices. The use of these ideas is to orient the company to belief systems, ideologies and frameworks that often provide justification and guidance in the life and work. Contextual DNA deals with the environments in which the company functions. It directs the attention to the problems and issues the company face and the strategies it employs as its shape the organizations and the contexts in which they are situated in relation to each other. Individual DNA is about people who live in cooperation with each other, but also singularly, with own voices, wills, goals and interests. In human terms, they are inventive as they develop positive, appreciative, relationships within the organizations that will sustain in the fulfillment of deeper needs.

Each type of DNA developed by Honold and Silverman (2002) has been determined by four diverse midsized manufacturing using empowering practices for more than fifteen years and named them as an organizational DNA alignment model. This alignment model helps the organization to discern their DNA type by looking at the organizational practices carefully and considering the alignment of DNA types in order to develop the organization in the most effective manner. In this alignment model, the type of DNA is identified through core organizational practices like leadership, management, and everyday cultural and human resource practices. The alignment for each of the practices and the tools/instruments has been developed in determining the company's organizational DNA. The understanding on the organizational DNA types is critical to undertake resolution of the organizational problems, as well as in selecting the appropriate consultants and other resources for organizational improvement.

2.3 Human Resource Practice (HRP)

Due to the importance in helping organizations to achieve its goal faster, to compete with competitor effectively, and to survive in the long run in an unpredictable today's business environment, scholars have emphasized the impact of human resource practice of organizational performance (Chand, 2010; Tavitiyaman, 2009). A number of researchers have attempted to measure the effect of HR practices (HRP) on an organization's performance from its employees' satisfaction with its market value (Bamberger & Meshoulam, 2000; Becker, Huselid, Becker & Huselid, 1998).

Some have suggested several domains of human resource managements (HRM) practices in terms of influencing an organization's performance, including the hotel industry (Ulrich & Lake, 1990). In the hotel industry, Ulrich and Lake (1990) named six domains for effective HRM which are staffed, training and development, employee performance appraisal, employee performance rewards, organization design and communication. HR practices or functions include recruitment and selection, training and development, performance management and compensation management. When these functions are integrated into a system, it is identified as an HR practices system.

2.4 Organizational performance

Organizational performance is the outcome of the strategy that an organization implements. After managers implement strategic management to firms, managers must measure the organizational effectiveness by measuring performance data (Crook, Ketchen & Snow, 2003). Performance may vary depending on whether it is the customers' or stakeholders' viewpoints or during different time periods (Tse, 1991). Important trade-offs between

performance measures may occur depending on the strategy used, structure selected, and the relative competitive strength from which the firm implements its strategy (Tse, 1991).

In the hotel industry, Haktanir and Harris (2005) explored performance measurement practices in the context of independent hotels. Six measurements are: (1) business dynamics—concerned with decision-making and information flow in the departments of the hotel; (2) overall performance measures—identified the performance measures utilized by different departments in order to summarize the performance of the whole establishment; (3) employee performance measures—revealed the important role of human resources in providing rooms, food and beverage, and leisure services; (4) customer satisfaction measures—reflected the significance of understanding customer requirements and developing systems accordingly; (5) financial performance measures—identified the financial performance that is measured and utilized at different levels of the business and the rationale for utilizing such measures, and (6) innovative activity measures—identified the new activities, products and different ways of delivering service to customers and the measurement of their outcome. Bai (2001), Cho (2004) and Chand (2010) strongly emphasized the impact of HRP on hotel performance as it helping organization to achieve its goal fasters, to compete with competitor successfully, and to survive in the long run in an unpredictable business environment. There are two outcomes implemented for organizational performance. The first relates to an organization's behavioral performance and the other relates to an organization's financial performance.

Behavioral performance refers to the performance in job-related tasks and its measurement is appropriate for situations in which performance results are hard to measure and in which there is a clear cause-effect connection between activities and results (Botten & McManus, 1999). Donavan, Brown, and Mowen (2004) asserted that it is important for the motivational well-being of the service workers (e.g. satisfaction and commitment) because their willingness to commit to the organization, satisfy with their job will improve the business performance.

Bridoux (1997) views financial performance as "profit in excess of the cost of capital, depends upon the attractiveness of the industry in which the firm operates (industry-effect on performance) and the firm's competitive advantage." Financial performance indicators such as return on investment, total sales, profit before tax, net profit, and total asset have been widely used as a method for evaluating business performance and for comparing a firm with others in an industry.

3 ORGANIZATIONAL DNA, HUMAN RESOURCE PRACTICES AND HOTEL PERFORMANCE

In the hotel perspective, studies have focused on general aspects of leadership, cultural, management and human resource, On the leadership, Weerakit (2007) identifying the future leader's competencies and the impact of applying different styles of leadership on employees while Weber (2006) and Brown (2007) looked on the impact of culture to the leadership, organization management and employee performance rather than organization performance. In management practice, Cerviño and Bonache (2005) examined the best management practices that can be implemented in hotel and multiple corporate strategies have a strong impact on a hotel's financial performance. Similarly, human resource managements have also given the impact on the organizational performance (Chand, 2010; Cho, 2004; Tavitiyaman, 2009). Those preceding studies in actual fact just focused on a small part of organizational DNA elements in different size, location and type of the hotels. However, the dominant Organizational DNA used by hotel as proposed by Honold and Silverman (2002) has not been yet explored.

In relation to the above notion, among the four organizational DNA elements, human resource practices (HRP) is considered the most critical area in the hotel industry (Cho, 2004; Haynes & Fryer, 2000). This is because hotel organizations are normally well depending on their employees that fit to the organization to consistently deliver excellent service experiences to customers through human resource practices which directly determine the organizational performance (Davidson, 2003; Hoque, 2000). Chand (2010) strongly emphasized the impact of HRP on hotel performance through hiring practices, compensation basis and employees training and development as these core human resources practices attributes helping organizations to achieve its goal faster, to compete with competitors successfully and to survive in the long run in an unpredictable business environment. In other words, fail to maintain good hiring practices, compensation and employees training and development as part of HR practices will lead to ineffective organization's performance and employees' satisfaction with its market value (Bamberger & Meshoulam, 2000; Becker, et al., 1998). Thus HRP has been seen as one of the vital element in identifying the organizational DNA.

There have been many studies on HRP conducted across a range of industries industry however none looking at the type of organizational DNA particularly in hotel industry. In other word, to what extent the alignment of HR practices (hiring practices, compensation and employees training and development) with the type of Organizational DNA for the hotel industry to date is not available. In fact, how far the effect of Organizational DNA towards hotel's behavior and performance is also not known? With this gap, there is really a need to explore on these issues in the hotel industry perspective particularly in five star hotels owing to the complication of it HRP practices.

4 CONCLUSION

The understanding on the organizational DNA types is critical to undertake resolution of the organizational problems, as well as in selecting the appropriate consultants and other resources for organizational improvement.

REFERENCES

Bai, X. (2001). *Measuring hotel financial performance: The role of training.* Retrieved from Retrieved from Proquest Digital Dissertations (AAT 3075636).

Bamberger, P.A. & Meshoulam, I. (2000). *Human resource strategy: Formulation, implementation, and impact.* Oaks, CA: Sage Publications, Inc.

Becker, B.E., Huselid, M.A., Becker, B. & Huselid, M.A. (1998). *High performance work systems and firm performance: A synthesis of research and managerial implications.* Paper presented at the Research in personnel and human resource management.

Botten, N. & McManus, J. (1999). *Competitive strategies for service organisations.* West Lafayette, Indiana: Purdue University Press.

Bridoux, F. (1997). A resource-based approach to performance and competition: An overview of the connections between resources and competition. *Strategic Management Journal, 18*(1), 1–21.

Brown, E.M. (2007). *An examination of the link between organizational culture and performance: A study of three county public health departments.*

Cerviño, J. & Bonache, J. (2005). Hotel management in Cuba and the transfer of best practices. *International Journal of Contemporary Hospitality Management, 17*(6), 455–468.

Chand, M. (2010). The impact of HRM practices on service quality, customer satisfaction and performance in the Indian hotel industry. *The International Journal of Human Resource Management, 21*(4), 551–566.

Cho, Y.S. (2004). *Examining the impact of human resources management: A performance based analytic model.* University of Nevada, Las Vegas. Retrieved from Proquest Digital Dissertations (AAT 3144524).

Crook, T.R., Ketchen, J.D.J. & Snow, C.C. (2003). Competitive edge: A strategic management model. *The Cornell Hotel and Restaurant Administration Quarterly, 44*(3), 44–53.

Davidson, M.C. (2003). Does organizational climate add to service quality in hotels? *International Journal of Contemporary Hospitality Management, 15*(4), 206–213.

Dehoff, K., Jaruzelski, B. & Kronenberg, E. (2008). Innovation's Org DNA. *Booz & Company, Florham Park*, from http://www.orgdna.com/about.cfm

Donavan, D.T., Brown, T.J. & Mowen, J.C. (2004). Internal benefits of service-worker customer orientation: job satisfaction, commitment, and organizational citizenship behaviors. *Journal of marketing, 68*(1), 128–146.

Explore DNA. (September 2010). An overview of DNA functions, from http://www.exploredna.co.uk/

Haktanir, M. & Harris, P. (2005). Performance measurement practice in an independent hotel context: A case study approach. *International Journal of Contemporary Hospitality Management, 17*(1), 39–50.

Haynes, P. & Fryer, G. (2000). Human resources, service quality and performance: A case study. *International Journal of Contemporary Hospitality Management, 12*(4), 240–248.

Honold, L. & Silverman, R.J. (2002). *Organizational DNA: diagnosing your organization for increased effectiveness*. Palo Alto, CA: Davies-Black Publication.

Hoque, K. (2000). *Human resource management in the hotel industry: Strategy, innovation and performance*. London: Routledge.

Kaipa, P. & Milus, T. (2002). Mapping the organizational DNA: A living system approach to organization transformation. Campbell, CA: Kaipa Group.

Neilson, G., Pasternack, B., Mendes, D. & Tan, E.-M. (2004). *Profiles in Organizational DNA Research and Remedies*. USA: Booz Allen Hamilton.

Prakash, N. (2001). Mapping the organizational DNA. The Indian Express Limited.

Spiegel, E.A., Gary, N., Flaherty Thomas, J. & Srini, R. (2005). Test your company's DNA. *Electric Perspectives, 30*(2), 32.

Tavitiyaman, P. (2009). *The impact of industry forces on resource competitive strategies and hotel performance*. Oklahoma State University. Retrieved from Proquest Digital Dissertations (AAT 3372198).

Tse, E.C.-Y. (1991). An empirical analysis of organizational structure and financial performance in the restaurant industry. *International Journal of Hospitality Management, 10*(1), 59–72.

Ulrich, D. & Lake, D. (1990). *Organizational capability: Competing from the inside out*. New York: John Wiley & Sons, Inc.

Weber, T.J. (2006). *Performance oriented cross-cultural management research: Examining the impact of national culture on the practice-performance relationship*. University of North Carolina. Retrieved from Proquest Digital Dissertations (AAT 3207366).

Weerakit, N. (2007). *Leadership competencies required for future hotel general managers' success in Thailand*. Retrieved from ProQuest (AAT 3275114).

Theory and Practice in Hospitality and Tourism Research – Radzi et al. (Eds)
© *2015 Taylor & Francis Group, London, ISBN 978-1-138-02706-0*

Organizational citizenship behaviors of hotel employees: The role of discretionary human resource practices and psychological contract

M.A. Hemdi & M. Hafiz
Faculty of Hotel and Tourism Management, Universiti Teknologi MARA, Shah Alam, Malaysia

F. Mahat
Faculty of Hotel and Tourism Management, Universiti Teknologi MARA, Penang, Malaysia

N.Z. Othman
KFCH International College, Selangor, Malaysia

ABSTRACT: The aims of this study were to investigate the relationship between discretionary HR practices and OCBs (OCBO and OCBI), and to clarify the mediating effects of psychological contract fulfillment. 380 operational employees from 22 large hotels in Malaysia participated in this study. The results show that discretionary HR practices, particularly training and development, performance management, and participation and involvement have significant positive relationship with either OCBO or OCBI. Furthermore, psychological contract fulfillment was a significant mediating mechanism of the relationship between discretionary HR practices and OCBs. These findings suggest that practicing hotel managers should continue to place greater emphasis on HR practices in order to enhance feeling of psychological contract and to promote greater OCBs among hotel's operational employees.

Keywords: Discretionary HR practices, psychological contract, organizational citizenship behavior

1 INTRODUCTION

Tourism industry has been performing extremely well in the Malaysian economy. Between 2006–2009, revenue from the tourism industry increased 67.1% to RM53.4 billion and tourist arrivals increased 43.6% to 23.6 million. In 2012, a total of 25.03 million tourists arrived in Malaysia and spent RM60.6 billion. During the 10th Malaysian Plan (2011–2015) period, the target is to improve Malaysia's position to be within the top 10 in terms of global tourism receipts, contributing RM115 billion in receipts and providing 2 million jobs in the industry in 2015 (*Economic Planning Unit – 10th Malaysian Plan*, 2011). As a result of a steady increase in tourism over the years, Malaysia now has more hotel rooms than ever before. For example, as of May 2013 the total room supply has increased to 184,998 rooms as compared to 168,497 rooms in 2010(*Economic Planning Unit - 10th Malaysian Plan*, 2011).

For Malaysia to be competitive in the global tourism industry, service-oriented organizations such as hotels, an important sector in the tourism industry, need to be proactive in their human resources implementations in order to deliver high quality services to their customers (Chang, Gong & Shum, 2011). However, the understanding of how it is that HR practices impact employee attitudes and behavioral outcomes is still lacking, especially within the Malaysian hotel industry (Hemdi & Othman, 2014; Tang & Tang, 2012). According to Li, Frenkel and Sanders (2011), HR practices are the means through which employee perceptions, attitudes, and behaviors are shaped. Therefore, building on somewhat limited research conducted to date on discretionary HR practices and psychological contract, this study proposes to examine the employees' assessment of the quality of discretionary HR practices on their perceptions of psychological contract fulfillment and its impact on individual outcomes, specifically organizational citizenship behaviors (OCB).

2 LITERATURE REVIEW

2.1 *Discretionary HR practices*

Discretionary HR practices are those which the organization chooses to invest in, are typically more strategically focused, and are different from those which are transactional (Hayton, 2004).

Hayton (2004) conceptualized discretionary HR practices are those which imply optional investment in the human capital of the organization in programs and practices such as those included in the selection, training and development, participation and involvement, pay for performance, and performance management initiatives.

In contrast to discretionary HR practices, non-discretionary HR practices are administrative, are considered a cost of doing business, and its activities are typically associated with mandated compliance (Hayton, 2004). These more traditional HR practices are transactional in nature and are requirement of doing business (Hayton, 2004). Examples of transactional practices include timely administration and processing of payroll, enrollment and administration of benefits, processing worker compensation claims, etc. Both discretionary and non-discretionary HR practices are posited to influence employees' behavioral outcomes such as organizational citizenship behavior (OCBO and OCBI) and turnover intentions (Benjamin, 2012; Hemdi, Hanafiah & Tamalee, 2013) and organizational outcomes such as performance, customer commitment, and organizational commitment (Nadiri & Tanova, 2010).The core discretionary HR practices used in this study include pay for performance, staffing and selection, performance management, training and development, and employee involvement.

2.2 *Psychological contract*

According to Rousseau (1995), the psychological contract is individual beliefs, shaped by the organization, regarding terms of an exchange agreement between an individual and the organization. Recent research (Lee & Kim, 2010) considers psychological contract to be composed of beliefs regarding exchange relationships that are shaped by the employee's experience in "current" organization. MacNeil (1985) categorized psychological contract into relational and transactional contracts. Transactional contracts are short-term, purely of economic focus, characterized by limited involvement of both parties, and mostly job-task, and contract-oriented. Relational contracts are long-term and broad, as these entail exchanges of socio-emotional resources and largely focused on relationship and development. Researchers (Hoffman, Blair, Meriac & Woehr, 2007) suggest that the two types of contracts are on rather independent dimensions. The reported correlations in previous studies (Raja, Johns & Ntalianis, 2004) do not strongly support that the two types of contracts are bipolar. Psychological contract breach occurs when an individual perceives an organization to have failed to fulfill promised obligations (Conway

& Briner, 2005). Most research uses the term fulfillment as the opposite concept of breach (Lambert, Edwards & Cable, 2003). Studies within the hotel industry found that found that psychological contact have significant influence on turnover intentions (Hemdi & Rahim, 2011) and affective commitment (Hemdi & Rahman, 2010).

2.3 *Organizational citizenship behavior*

Organizational citizenship behavior (OCB) was defined as "individual behavior that is discretionary, not directly or explicitly recognized by the formal reward system, and that in the aggregate promotes the effective functioning of the organization" (Organ, 1988: pg.4). Organ (1988) proposed five-dimension framework OCB comprising altruism, courtesy, conscientiousness, civic virtue, and sportsmanship. Based on Organ's (1988) framework, Williams and Anderson (1991) proposed a two-dimensional framework of OCBO and OCBI. Coleman and Borman (2000) proposed an integrated model of OCB, in which the first category includes the interpersonal citizenship performance such as altruism and courtesy of Organ's (1998) framework and is similar to OCBI of Williams & Anderson (1991). The second category, organizational citizenship performance, includes Organ's (1998) sportsmanship, civic virtue, and conscientiousness pertaining to OCBO of Williams and Anderson (1991). A review by Podsakoff, Mackenzie, Paine, and Bachrach (2000) found that job attitudes, task variables, and leadership behaviors are strongly related to OCBs. Job satisfaction, organizational commitment, and perceptions of fairness appeared to be important determinants of OCBs. In addition, task feedback, task routinization, and intrinsically satisfying tasks were consistently related to OCBs. However, prior research did not adequately deal with HR practices as potential antecedents of OCBs. Nevertheless, based on the data obtained from hotels, Sun, Aryee, and Law (2007) found high-performance HR practices are positively related to service-oriented OCBs. Hemdi and Nasurdin (2008) found that hotel employees' perceptions of organizational justice have significant influence on OCB intentions. Based on the above discussion, the following hypotheses were proposed:

H1: Employee's perception of quality of discretionary HR practices will be positively related to OCBO and OCBI.

H2: Employee's perception of quality of discretionary HR practices will be positively related to psychological contract fulfillment.

H3: Psychological contract fulfillment will be positively related to OCBO and OCBI.

H4: Psychological contract fulfillment will mediate the relationship between employee's perception of quality of discretionary HR practices and OCBO and OCBI.

3 METHODOLOGY

3.1 Sample

The purpose of this study is to investigate whether employees' perception of discretionary HR practices and psychological contact fulfillment affects their behavioral, in the form of OCBs in the hotel industry. This study was correlational in nature. In this study, the unit of analysis was individual operational employee working in large hotels located in the state of Selangor, Kuala Lumpur, and Pulau Pinang, Malaysia. A total of 630 questionnaires were distributed to operational employees attached to 22 large hotels. A total of 411 questionnaires were returned. Of these, 31 responses were found to be non usable. Therefore, only 380 questionnaires (60.3%) were coded and analyzed.

3.2 Data analysis

Hypotheses were tested by using hierarchical multiple regression analysis with OCBs as the dependent variables. Baron and Kenny's (1986) mediation rules were followed for testing mediation effects of psychological contract fulfillment. To reduce the possibility of spurious statistical influence, this study also measured five demographic control variables: age, gender, marital, education, and organizational tenure. All the reliability coefficients for the measures were acceptable since they exceeded the minimum recommended level of 0.60 (Sekaran, 2000).

4 RESULTS

4.1 Hypotheses testing

Table 1 shows the summary of the regression analyses, direct and indirect relationships on OCBO.

From Table 1, none of the controlled variables significantly affect psychological contract or OCBO (M1). When discretionary HR practices were entered into the equation to be regressed onto OCBO (M2), discretionary HR practices variables were able to explain 39.0% (R^2 change = .39, F-change = 10.45, $p < .01$) of the observed variations on OCBO. Of the five HR practice dimensions, training and development ($\beta = .42, p < .01$) and participation and involvement ($\beta = .31, p < .01$) were positively and significantly contributes to the prediction of OCBO. Thus, hypothesis H1 was partially supported.

On the effect towards psychological contract fulfillment (M3), training and development ($\beta = .37$, $p < .01$), participation and involvement ($\beta = .33$, $p < .01$) and performance management ($\beta = .19$, $p < .01$) had positive and significant effects. Hence, H2

Table 1. Summary of the Regression Results on the Influence of Discretionary HR Practices and Psychological Contract on OCBO.

Predictors	M1	M 2 (OCBO)	M3 (PC)	M4 (OCBO)
Control Variables:				
Gender[a]	−.01	−.06	−.03	−.02
Age	−.15	.06	−.14	−.15
Marital Status[b]	.02	−.05	−.02	.02
Education	.05	.02	.06	−.00
Organizational Tenure	.02	−.01	−.01	−.00
Discretionary HR Practices:				
Staffing & Selection		.07	.09	.05
Training & Development		.42**	.37**	.08
Performance Management		.10	.19**	.06
Pay for Performance		.08	.10	.06
Participation & Involvement		.31**	.33**	.11
Psychological Contract				.34**
R^2	.02	.41	.46	.31
Adj. R^2	−.01	.39	.44	.29
R^2 Change	.02	.39	.44	.19
F-Change	.80	10.45**	19.06**	95.39**

Note: $N = 380$; *p < .05, **p < .01.

73

Table 2. Summary of the Regression Results on the Influence of Discretionary HR Practices and Psychological Contract on OCBI.

Predictors	M1	M2 (OCBI)	M3 (PC)	M4 (OCBI)
Control Variables:				
Gender[a]	−.03	−.03	−.03	−.05
Age	−.11	.06	−.14	−.14
Marital Status[b]	.05	−.06	−.02	.02
Education	.05	.02	.06	−.01
Organizational Tenure	.04	−.04	−.01	−.02
Discretionary HR Practices:				
Staffing & Selection		.12	.09	.13
Training & Development		.32**	.37**	.11
Performance Management		.31**	.19**	.09
Pay for Performance		.09	.10	.08
Participation & Involvement		.09	.33**	.11
Psychological Contract				.38**
R^2	.02	.39	.46	.36
Adj. R^2	.01	.37	.44	.28
R^2 Change	.02	.37	.44	.18
F-Change	.96	46.87**	19.06**	87.13**

Note: $N = 380$; *p < .05, **p < .01.

is partially supported. Further, psychological contract fulfillment was significantly affect OCBO ($\beta = .34$, $p < .01$), thus H3 was supported. On mediation effect (M4), the effect of both training and development and participation were found to be insignificant ($\beta = .08$, $p < .05$; $\beta = .11$, $p < .05$). This implies that psychological contract fulfillment fully mediated the relationship between training and development and participation and involvement and OCBO, in which partially supported hypotheses *H4*.

Table 2 shows the summary of the regression analyses on OCBI.

Again, none of the controlled variables affect psychological contract or OCBI in all the regression models (M1). Discretionary HR practices variables were able to explain 37.0% (R^2 change = .39, F-change = 46.87, $p < .01$) of the observed variations on OCBI (M2). Training and development ($\beta = .32$, $p < .01$) and performance management ($\beta = .31$, $p < .01$) were positively and significantly contributes to the prediction of OCBI. Thus, hypothesis H1 was partially supported. Psychological contract fulfillment was also significantly affect OCBI ($\beta = .38$, $p < .01$), thus H3 was supported. On the mediation effects of psychological contract, the effect of both training and development and performance management were found to be insignificant ($\beta = .11$, $p > .05$; $\beta = .09$, $p > .05$) implying that psychological contract fulfillment fully mediated the relationship between training and development and performance management towards OCBI. Thus, H4 was partially supported.

5 DISCUSSION AND FUTURE RESEARCH

The purpose of this paper was to examine hotel employees' assessment of the quality of discretionary HR practices on their perceptions of psychological contract and its impact on OCBs. Results from this study showed that discretionary HR practices in the form of training and development, performance management, and employee participation and involvement significantly influenced hotel employees' feeling of psychological contract fulfillment, and subsequently their extra-role behaviors. Hence, when hotel employees highly perceived that their organizations had provided them with extensive training program to enable them to do their jobs better and when they satisfied with training opportunities given to them, they will reciprocate with positive work performance behaviors in the form of both OCBO and OCBI. In addition, feeling of psychological fulfillment is achieved when employees were given enough opportunities in their career development through challenging job assignments, which subsequently translated in positive extra-role behaviors that can benefits the organization. The results also impliedly showed that proper management of employee performance was important in ensuring positive feeling of employees' psychological contract fulfillment and extra-role behaviors. Thus, performance appraisal processes and targets must be transparent to employees. They must be allowed to participate in goal settings and been given timely feedbacks.

In summary, the impact of investing in discretionary HR practices is an important topic both from an academic and practitioner perspective as it addresses the "value-added" not only with respect to the HR function, but in the ability to more clearly recognize the relationship between these discretionary HR practices and employee attitudinal (psychological contract fulfillment) and behavior outcomes. It is evident from the results of this study that training and development, performance management, and employee involvement practices directly influence psychological contract and OCBs.

Some limitations which necessitate caution in accepting the results of this study. First, we followed the traditional view that breach and fulfillment are on a single continuum. Future studies may need to investigate the effects of both breach and fulfillment on outcomes. Second, we collected data from the same respondents. Longitudinal and dyad data may be needed to minimize the effect of common method bias.

ACKNOWLEDGEMENT

The authors wish to thank Universiti Teknologi MARA (UiTM) for the financial support via the Research Incentive Faculty (RIF) grant [600-RMI/DANA 5/3/RIF (721/2012)] to carry out and publish this research.

REFERENCES

Baron, R.M. & Kenny, D.A. (1986). The moderator–mediator variable distinction on social psychological research: Conceptual, strategic and statistical considerations. *Journal of Personality and Social Psychology*, 51, 1173–1182.

Benjamin, A. (2012). Human resource development climate as a predictor of citizenship behavior and voluntary turnover intentions in banking sector, *International Business Research*, 5 (1), 110–119.

Chang, S., Gong, Y., Shum, C. (2011). Promoting Innovation in hospitality companies through human resource management practices. *International Journal of Hospitality Management*, 30 (4), 812–818.

Conway, N. and Briner, R. (2005), *Understanding Psychological Contracts at Work: A Critical Evaluation of Theory and Research*, Oxford University Press, New York, NY.

Economic Planning Unit – 10th Malaysian Plan 2011–2015. www.epu.gov.my/rmk10

Hayton, J. (2004). Strategic human capital management in SMEs: An empirical study of entrepreneurial performance, *Human resource Management*, 42 (4), 375–391.

Hemdi, M.A. & Othman, N.Z. (2014). Discretionary Human Resources Practices and Psychological Contract Fulfillment on Hotel Employees' Organizational Citizenship Behaviors. *Proceeding of the 5th. International Conference on Business and Economic Research (ICBER2014), Kuching, Malaysia, 24–25 March, 2014.* ISBN: 978-967-5705-13-7.

Hemdi, M.A., Hanafiah, M.H. & Tamalee, K. (2013). The Mediation Effect of Psychological Contract Fulfillment on Discretionary Human Resource Practices and Organizational Citizenship Behaviors of Hotel Employees. *Proceeding of The 4th International Conference on Business, Economics and Tourism Management (CBETM2013), 1–5, Jeju Island, Korea, 19–20 Oct 2013.* ISBN: 978-981-07-8017-3

Hemdi, M.A. & Rahim, A, R. (2011). The Effect of Psychological Contract and Affective Commitment on Turnover Intentions of Hotel Managers. *International Journal of Business and Social Science*, 2 (23), 76–88, *Special Issues December 2011*. ISSN 2219-1933 (Print), 2219-6021 (Online).

Hemdi, M.A. & Rahman, N. A,. (2010). Turnover of hotel managers: Addressing the effect of psychological contract and affective commitment. *World Applied Science Journal, 10 (Special Issues on Tourism & Hospitality)*, 1–13. ISSN: 1818-4952.

Hemdi, M.A. & Nasurdin, A.M. (2008). Investigating the influence of organizational justice on hotel employees' organizational behavior intentions and turnover intentions. *Journal of Human Resources in Hospitality and Tourism*, 7(1), 1–23.

Hoffman, B.J., Blair, C.A., Meriac, J.P. & Woehr, D.J. (2007). Expanding the criterion domain? A quantitative review of the OCB, *Journal of Applied Psychology*, 92 (2), 555–566.

Lambert, L.S., Edwards, J.R. & Cable, D.M. (2003). Breach of fulfillment of the psychological contract: A comparison of traditional and expanded views, *Personnel Psychology*, 56 (4), 895–934.

Lee, K.Y. & Kim, S. (2010). The effect of commitment-based human resource management on organizational citizenship behaviors: The mediating role of the psychological contract, *World Journal of Management*, 2 (1), 130–147.

Li, x., Frenkel, S.J. & sanders, K. (2011). Strategic HRM as process: How HR system and organizational climate strength influence Chinese employee attitudes, *The International Journal of Human Resource Management*, 22 (9), 1825–1842.

MacNeil, I.R. 1985. Relational contract: What we do and do not know, *Wisconsin Law Sociology Review,* 28: 55–69.

Morrison, E.W. 1996. Organizational citizenship behavior as a critical link between HRM practices and service quality. *Human Resource Management*, 35: 493–512.

Organ, D.W. (1988). *Organizational Citizenship Behavior: The Good Soldier Syndrome*. Lexington: Lexington Books.

Podsakoff, P.M., MacKenzie, S.B., Paine, J.B. & Bachrach, D.G. 2000. Organizational citizenship behaviors: A critical review of the theoretical and empirical literature and suggestions for future research. *Journal of Management*, 26: 513–563.

Raja, U., Johns, G., Ntalianis, F. (2004). The impact of Personality on psychological contract, Academy of Management Journal, 47 (3), 350–367.

Rousseau, D.M. (1995). *Psychological contracts in organizations: Understanding written and unwritten agreements*. Thousand Oaks, CA: Sage.

Sekaran, U. (2000). *Research Methods for Business: A Skill-Building Approach.* New York: John Wiley & Sons.

Tang, T.W. & Tang, Y.Y. (2012). Promoting service-oriented organizational citizenship behavior in hotels: The role of high-performance human resource practices and organizational social climate. *International Journal of Hospitality Management*, 31(3), 885–895.

William, L.J., Anderson, S.E., (1991). Job satisfaction and organizational commitment as predictors of organizational citizenship behaviors, *Journal of Management*, 17 (3), 601–617.

Theory and Practice in Hospitality and Tourism Research – Radzi et al. (Eds)
© *2015 Taylor & Francis Group, London, ISBN 978-1-138-02706-0*

A stakeholder approach to working conditions in the tourism and hospitality sector

A. Walmsley
Plymouth University, Plymouth, UK

S.N. Partington
Manchester Metropolitan University, Manchester, UK

ABSTRACT: Increasing concerns about the business and society relationship are reflected in concerns about the impacts of tourism development. However, an area that has seen relatively little attention in sustainable tourism is that of working conditions. From a theoretical standpoint few advances have been made. This paper explores the relationship between the owners and employees of tourism businesses interpreting stakeholder theory through the lens of traditional economic theory. It thereby aims to provide a basis for improving working conditions that extends beyond purely strategic and moral rationales.

Keywords: CSR, stakeholder theory, employment, working conditions

1 INTRODUCTION

'Employees are our most valuable asset'. This is the often-proclaimed mantra of executives in the tourism and hospitality sector. Despite such proclamations the reality of working conditions in the sector are almost invariably described as poor by scholars the world over. An early study from Japan (Tomoda, 1983), for example, described working conditions in the sector as 'unsatisfactory or even deplorable'. Other authors such as Wood (1997) reach similar conclusions a little over a decade later. Fast forward another decade and the conclusion of Baum's (2007) review of the status of human resources in tourism are succinctly summarized in the paper's title 'still waiting for change'.

This paper addresses the discrepancy between employer discourse and employment applying Blair's (1998) interpretation of stakeholder theory to the sector. It thereby provides a novel perspective with which to justify the legitimacy of concerns about poor employment practices in tourism. It strengthens the conceptual foundation of labour relations in the sector, which is required if scholarship and practice are to make progress in an area that has witnessed little conceptual development.

The paper is structured as follows. Initially, to set the scene for the ensuing theoretical discussion, the paper exposes current thinking on working conditions in the sector. Subsequently, the paper demonstrates how working conditions relate to the concepts of sustainable tourism and corporate social responsibility (CSR). Blair's (1998) justification of the relevance of stakeholder theory using economic theory is discussed before being applied to tourism. The paper concludes by summarizing the key points and suggesting avenues for further research.

2 WORKING CONDITIONS IN THE TOURISM SECTOR

The tourism sector is acknowledged by the International Labor Organization (ILO, April, 16, 2014), as one of the fastest growing sectors of the global economy that is labor intensive and has significant multiplier effect on employment in other related sectors. It offers varied opportunities for people in diverse sub-sectors and at different levels (Baum, 2007). However United Nation World Tourism Organization (UNWTO, April, 4, 2014) in its latest report emphasized that the world of work in tourism is generally not well known because reliable data are missing on tourism employment, hence their latest joint project with ILO titled 'employment and decent work in tourism' that attempts to address this gap. Non-Governmental Organizations (NGOs) in the UK such as Tourism Concern who advocates ethical tourism also identified the tourism sector '…as a fertile ground for exploitation of workers at the bottom of the tourism supply chain in countries all over the world', which echoes Wood's (1997) bleak assessment of the industry, in both developed and less developed countries.

Unfortunately the negative image of industry in relation to its working condition have continued to be highlighted by tourism scholars even in recent years (Baum, 2007; Janta, Ladkin, Brown & Lugosi, 2011; Lucas, 2004; Nickson, 2013).

The tourism sector in the main is often considered to be secondary labor market that faces low wages, insecure employment and poor opportunities. The Low Pay Commission when it started its research on working conditions before the implementation of the National Minimum Wage (NMW) in the UK identified the tourism sector (more specifically hospitality) as one of the industry's with the highest concentration of low paid workers in the UK. Authors such as Wilkinson and Sachdev (1998) at that time also highlighted that socially downgraded low paid jobs are generally occupied by disadvantaged people in the labor market and they argued that '...the low paid are not low paid because they lack skill: they are low paid because their skills are not recognized or valued'. This has been a continuous challenge faced by employees in this sector and has hindered progress in relation to improving pay practices. However, a number of recent studies have started advocating against these perceptions of 'unskilled' labour, highlighting the importance of soft skills in the tourism sector that should not be undervalued (Burns, 1997; Warhurst & Nickson, 2007).

Over a decade has passed by since the introduction of the NMW in the UK aimed to combat low pay, but Human Resource Management (HRM) research in relation to working conditions in the tourism sector is still topical. Some of the poor employment practices continuously highlighted by tourism scholars to name a few included long and unsociable working hours, low pay, minimal compliance with law, lack of training, lack of job security and career progression. In short, as powerfully presented by Wood (1997).

'Hospitality work is largely exploitative, degrading, poorly paid, unpleasant, insecure and taken as last resort or because it can be tolerated in the light of wider social and economic commitments and constraints'.

This view can still be seen to be relevant within current studies on experiences of migrant workers in the tourism sector that highlighted migrant workers plan to eventually move out of this sector because of the poor working conditions (Janta et al., 2011) supporting the notion proposed by Szivas and Riley's (1999) that '...tourism employment might play the role of "any port in a storm"; a refuge sector'. Even though a numbers of studies (Devine, Baum, Hearns & Devine, 2007; Janta et al., 2011) have shown the benefits migrant workers bring to the industry, we should not deny that there is an 'elephant in the room'. A large proportion of migrant workers are fulfilling vacancies in the industry mainly because the local employees are not willing to tolerate the poor working conditions. Therefore it is concerning to think that as long as the migrant workers are able to fill these vacancies, employers do not see the necessity to change their practices to improve working conditions in the sector. Indirectly, migrants may help to continue entrench certain employment practices that are favorable to employers (Janta et al., 2011) and as Baum (2007) asserts '...the widespread recruitment and use of migrant labor in the tourism industry of developed countries has acted to the detriment of real change within the sector's workplace'. In order not to end this section on this pessimistic note, we would like to propose that there are more debates that are needed under the sustainable tourism and CSR agenda to recognize and address this worldwide problem. It is to these that we now turn.

3 SUSTAINABLE TOURISM, CSR AND WORKING CONDITIONS

The notion of sustainable tourism is frequently adopted by tourism scholars as well as practitioners. Sustainable tourism, although variously defined, has its roots in the sustainable development movement which emerged in the late 1960s and which raised awareness of potential negative side-effects of a growth at all costs mentality. It brought to light competing demands of economic growth and environmental protection (Dresner, 2008) and sought a more equitable relationship between the two. In relation to tourism development Jafari, Smith and Brent (2001) emphasized the move from an advocacy to a cautionary platform where tourism was initially only acknowledged as a tool to promote economic development, with potential negative side-effects not being raised until much destruction of natural habitats had already occurred (Krippendorf, 1999).

Although sustainable tourism recognizes the legitimacy of social concerns, there is a strong emphasis on environmental protection in the sense of intergenerational equity which is evident in the ubiquitous definition of sustainable development provided in the so-called Brundtland Report (United Nations, 1987) from which sustainable tourism draws much inspiration. Thus, we would argue, that although not absent from the sustainable tourism agenda, the notion of social impacts of tourism, in particular the relationship between tourism development and working conditions have not been of primary concern.

By contrast, scholarly concern with the 'business and society relationship' has a longer history than that of sustainable development. Carroll (1999) in a review of the evolution of the CSR construct is one of many authors who have described the development of the relationship between business and society (Lee (2008). He takes early formal endeavours to explain this relationship back to the first half of the 20th century. The modern era of the social responsibilities of business is argued to have occurred with the publication of Bowen's (1953) *Social Responsibilities of the Businessman* (Carroll, 1999). At this time other definitions of CSR do not mention the environment at all but focus on the broader concept of social ends. This is understandable from a historical perspective where the environmental movement was really only beginning to emerge with works such as (Carson, 1962) *Silent Spring* and a decade later Meadows, Meadows, Randers, and Behrens (1972), *Limits to Growth* (Hussey & Thompson, 2000). It was only later, towards the end of the 20th century that distinctions between social and environmental dimensions of responsibility were formally voiced in, for example, Elkington's (1998) now famous phrase of the triple bottom line of: profit, people and planet (sometimes also referred to as economic, social and environmental impacts). Whereas the general discussions around the business and society relationship have therefore had a longer history of dealing with the impacts of business on society, in tourism, due to the emergence and pre-dominance of the concept of sustainable tourism which itself arose out of the sustainable development movement, the environmental dimension appears to have attracted more attention.

The above review serves to explain, at least in part, why working conditions have not featured as strongly as environmental concerns in tourism. This is not to argue that working conditions have been completely ignored as this is manifestly not the case, demonstrated by the appeals from academics discussed above. And yet, if Baum (2007) conclusions are valid then little has in fact changed in the last 20–30 years when it comes to working conditions in the sector. This lack of progress relates however not just to employment practices in tourism but also to the development of theory amongst tourism scholars. Thus, whereas the responsibility of business to its employees has received some attention drawing on the related concepts of CSR, social responsiveness, stakeholder theory and so forth, we argue that these discussions have remained marginal compared to debates around sustainable tourism.

4 EMPLOYEES AS STAKEHOLDERS OF THE TOURISM FIRM

Today, the nature of the business and society relationship is in the spotlight due to the economic turmoil brought about by the 'get rich quick at any costs' mentality that was seen to pervade the financial services sector, which has even be described as psychopathic (Boddy, 2011). This left little space for consideration of stakeholder needs beyond those of shareholders and senior management. More than twenty years after Freeman (2010) wrote his seminal work on the stakeholder approach to business, much of the business community seems to have only paid lip-service to it. The suggestion that businesses have responsibilities that extend beyond satisfying the needs of owners and shareholders can be found at the core of CSR (Carroll, 1999) and is where stakeholder theory and CSR overlap. Indeed, the often-quoted rebuttal of CSR as a concept that came from Friedman (1973) targets precisely the extent of a company's responsibility, arguing that first and foremost the firm is responsible to its shareholders. The kernel of Friedman's (1973) position is that managers (agents) are acting irresponsibly if they do not act in the best interest of the owners (principals). Any activity that detracts from shareholder value, such as corporate philanthropy, or paying employees more than necessary to retain them, is irresponsible (leaving aside momentarily the business case considerations often attributed to CSR). Traditional economic theory suggests that once the firm has met all its financial obligations any residual income, or profits, go to the owners of the firm as recompense for their assumption of financial risk.

Blair (1998) on the other hand contests the idea that economic value equates solely to profits, i.e. what is left to shareholders once all other financial obligations have been met. She would include payments to employees as stakeholders of the firm part of the firm's wealth creation. Wealth created by the firm is the sum of profits and payments to employees, not as traditionally viewed profits where payments to employees are regarded as a cost. This latter view will result in attempts to reduce the 'wage bill' as the residual profits will increase in line with a reduction in 'costs'.

To further substantiate the claim that wealth creation includes payments to employees Blair (1998) argues that risk inherent in business is also carried by employees. If a business fails the employees are at risk of losing their jobs. They may not have invested capital in the business, but they have invested their labor, and developed firm-specific human capital. This firm-specific human capital allows them to command a wage premium, but if they lose their jobs, they are likely to face a reduction in wages because, by definition, they cannot take their firm-specific human capital with them.

Blair (1998) was however writing in a very general, non-sector specific sense. One of the often-cited characteristics of tourism employment is its high level of labour turnover (Johnson, 1981; Riley & Ladkin, 2002; Walmsley, 2004). This provides employees with limited opportunities to develop firm specific human capital. In addition, tourism employment, certainly much if not all of it, does not rely on employees becoming experts in the use of firm-specific technology. While firm-specific skills are required, these may be relatively straightforward to acquire compared to the examples provided by Blair (1998) of a machine operator in an engineering or manufacturing firm. This condition of tourism employment represents a double-edged sword for the tourism employee. On the one hand it places downward pressure on tourism wages; on the other the tourism employee can find similarly, admittedly low-paid employment in other tourism firms.

If the risk of losing one's wage premium due to the acquisition of firm specific human capital is less applicable to tourism it may be argued that tourism employees do not bear as much risk inherent in losing one's job as employees in other sectors. All else being equal, the tourism employee has more chances of obtaining employment at similar levels of remuneration than employees in other sectors. But, and this is a considerable 'but' there is still the risk of losing one's job, and whether the employee is able to find alternative work will depend to a large degree on factors outside his/her control, specifically labor market conditions.

The tourism employee contributes to the wealth creation of the firm and carries some risk if the firm folds. As such, drawing on Blair's (1998) analysis, the employee is a legitimate stakeholder of the firm and should not simply be equated with a cost to be minimized. The reason this analysis is important is because it provides a foundation for employee relations in tourism, and by implication the improvement of working conditions, that does not draw on the two prevailing grounds for this to happen: a business case or a moral case. The scope of this paper does not permit a full discussion of the two so a brief overview will suffice to set the argument within its context.

Wry (2009) claims that 'business and society' scholars have taken two approaches to further their normative aims of improving firms' social performance: one strategic, the other moral. The first attempts to use the business case for socially responsible behavior in a strategic manner. Effectively, this view argues that what is good for society is good for business. The second approach takes a moral standpoint and suggests firms should behave responsibly for moral reasons, even if the behavior limits profit growth. Both approaches accept the economic imperative underlying firm behavior. The verdict on the business case for CSR is still out (Margolis & Walsh, 2003; Vogel, 2006).

Whether appeals to owner/managers' moral sensitivities is to have any notable impact is even less clear. This paper has thus attempted to address what Lee (2008) calls for when he suggests that future research needs to develop conceptual tools and theoretical mechanisms that go beyond existing approaches.

5 CONCLUSION

Concerns about tourism's impacts are not new. Neither are characterizations of poor working conditions in the sector. Rather than using the business case or a moral argument to achieve improvement in working conditions, Blair (1998) focuses on the economic imperative of business. Wealth creation itself is not questioned. Instead, what wealth creation means and who is entitled to it lies at the heart of Blair's (1998) analysis. The economic arguments provided by Blair (1998) have been applied to tourism. As demonstrated, because of the nature of tourism employment, the economic risks borne by tourism employees are weaker than those for other sectors. However, they still carry economic risks and the employer-employee relationship is therefore based on more than the neo-classic economic view of a sole labor-wage exchange. In other words, the traditional economic view of the firm that sets employers' and employees' interests against one another does therefore not hold. Maximization of shareholder value at the expense of working conditions is not legitimate. While it may be naïve to believe that this view on its own will improve working conditions, providing alternative vantage points of the business and society relationship is a step in addressing poor employment practices in the sector. As the literature indicates, there is still a long way to go. We recommend further conceptual work in the area of tourism employment, particularly with reference to the business and society relationship and related concepts such as stakeholder theory, sustainable tourism and CSR.

REFERENCES

Baum, T. (2007). Human resources in tourism: Still waiting for change. *Tourism Management, 28*(6), 1383–1399.
Blair, M.M. (1998). For whom should corporations be run?: An economic rationale for stakeholder management. *Long range planning, 31*(2), 195–200.
Boddy, C.R. (2011). The corporate psychopaths theory of the global financial crisis. *Journal of Business Ethics, 102*(2), 255–259.
Bowen, H.R. (1953). *Social responsibilities of the businessman.* New York: Harper & Brothers.

Burns, P.M. (1997). Hard-skills, soft-skills: undervaluing hospitality's 'service with a smile'. *Progress in Tourism and Hospitality Research, 3*(3), 239–248.

Carroll, A.B. (1999). Corporate social responsibility evolution of a definitional construct. *Business & society, 38*(3), 268–295.

Carson, R. (1962). Silent Spring. Greenwich, Connecticut: Fawcett Publications.

Devine, F., Baum, T., Hearns, N. & Devine, A. (2007). Cultural diversity in hospitality work: the Northern Ireland experience. *The International Journal of Human Resource Management, 18*(2), 333–349.

Dresner, S. (2008). *The principles of sustainability*. London: Earthscan Publications.

Elkington, J. (1998). Partnerships from cannibals with forks: The triple bottom line of 21st century business. *Environmental Quality Management, 8*(1), 37–51.

Freeman, R.E. (2010). *Strategic management: A stakeholder approach*. Boston: Cambridge University Press.

Friedman, M. (1973). The social responsibility of business is to increase its profits. *New York*.

Hussey, S. & Thompson, P. (2000). *Roots of Environmental Consciousness*: Routledge.

ILO. (April, 16, 2014). from http://www.ilo.org/global/industries-and-sectors/hotels-catering-tourism/lang-en/index.htm

Jafari, J., Smith, V. & Brent, M. (2001). The scientification of tourism. *Hosts and guests revisited: Tourism issues of the 21st century*, 28–41.

Janta, H., Ladkin, A., Brown, L. & Lugosi, P. (2011). Employment experiences of Polish migrant workers in the UK hospitality sector. *Tourism Management, 32*(5), 1006–1019.

Johnson, K. (1981). Towards an understanding of labour turnover. *Service Industries Journal, 1*(1), 4–17.

Krippendorf, J. (1999). *The holiday makers: Understanding the impact of leisure and travel*. London: Routledge.

Lee, M.D.P. (2008). A review of the theories of corporate social responsibility: Its evolutionary path and the road ahead. *International journal of management reviews, 10*(1), 53–73.

Lucas, R. (2004). *Employment relations in the hospitality and tourism industries*. London: Routledge.

Margolis, J.D. & Walsh, J.P. (2003). Misery loves companies: Rethinking social initiatives by business. *Administrative science quarterly, 48*(2), 268–305.

Meadows, D.H., Meadows, D., Randers, J. & Behrens III, W.W. (1972). The Limits to Growth: A Report for the Club of Rome's Project on the Predicament of Mankind (New York: Universe).

Nickson, D. (2013). *Human Resource Management for Hospitality, Tourism and Events*. Oxford: Routledge.

Riley, M. & Ladkin, A. (2002). *Tourism employment: Analysis and planning* (Vol. 6). Clevedon: Channel View Publications.

Szivas, E. & Riley, M. (1999). Tourism employment during economic transition. *Annals of Tourism Research, 26*(4), 747–771.

Tomoda, S. (1983). Working Conditions in the Hotel, Restaurant and Catering Sector: A Case Study of Japan. *Int'l Lab. Rev., 122 (2)*, 239–252.

United Nations. (1987). Report of the World Commission on Environment and Development: Our Common Future United Nations.

UNWTO. (April, 4, 2014). from http://statistics.unwto.org/en/project/employment-and-decent-work-tourism-ilo-unwto-joint-project

Vogel, D. (2006). *The market for virtue: The potential and limits of corporate social responsibility*: Brookings Institution Press.

Walmsley, A. (2004). Assessing staff turnover: a view from the English Riviera. *International Journal of Tourism Research, 6*(4), 275–287.

Warhurst, C. & Nickson, D. (2007). Employee experience of aesthetic labour in retail and hospitality. *Work, Employment & Society, 21*(1), 103–120.

Wilkinson, F. & Sachdev, S. (1998). *Low pay, the working of the labour market and the role of a minimum wage*. London: Institute of Employment Rights.

Wood, R.C. (1997). *Working in hotels and catering*. London: International Thomson Business Press.

Wry, T.E. (2009). Does business and society scholarship matter to society? Pursuing a normative agenda with critical realism and neoinstitutional theory. Journal of Business Ethics, 89(2), 151–171.

Theory and Practice in Hospitality and Tourism Research – Radzi et al. (Eds)
© *2015 Taylor & Francis Group, London, ISBN 978-1-138-02706-0*

Strategic information system and environmental scanning practices in Malaysian hotel organizations

H.M. Said & R.A. Latif
UNITAR International University, Selangor, Malaysia

N.K. Ishak
City University College, Selangor, Malaysia

ABSTRACT: Environmental Scanning is the acquisition and use of information about events or trends in an organizations external environment, the knowledge of which would assist management in planning for the organization next course of action. Environmental Scanning or known as ES is necessary for every organization yet this task is very difficult to conduct. Environment uncertainty is a key issue for strategic decision makers to manage continuously in a competitive market. It is seen as a threat more than an opportunity to some industries. Previous studies show that there are significant relationship between scanning the environment, strategic response and organizational performance. Changes in the environment had in the past effected on how organization conduct its businesses, thus have impact on their performance. Reports from the literature show that organizations do conduct some practice of ES on their business environment; some are informal while others are formal. For that reasons, organization needs to implement a systematic ES as part of its corporate planning activities. How well the organization use information from ES is also another key area to competitive advantage. How organizations organizes the scanning effort so that information to specific situations can be more readily available. Having objective information is crucial for organization's survival in the industry. The purpose of this paper is to discuss conceptually on how organization conduct ES. It will discuss the significant strategic information models in the literature to address the importance of obtaining information for decision making.

Keywords: Environmental scanning practices, external environment, strategic information scanning system, organizational competitiveness, Malaysian hotels

1 INTRODUCTION

1.1 *Background of the study*

Customer spending on tourism and hotel sector is closely related to the economic cycle. Spending on leisure activities such as travel and lodging tends to be one of the first things that consumers hold back in times of economic hardship. The travel and hotel industry is further affected by reduced demand from business sector due to critical incidents, economic crisis, technology advancement, political regulations and other changing external environment. Luxury hotels are among the most affected sector whereby hotel will experience reduced occupancy, reduced room rates, access rooms, and high operating cost.

The impact is very much related to the environment around the organization. Environment is viewed as a source of information, continually creating signals and message that organizations should attend to (Dill, 1962; Weick, 1979). The external business environment can be defined as customers, competition, technological, regulatory, economy and socio cultural (Daft, et al., 1988).

Scanning is a process of identifying, monitoring and evaluating major environmental changes in the industry (Fahey & King, 1977). The purpose of scanning is to develop broad strategies and long term policies based on information that has been gathered and interpreted from the environment. The content of scanning can be considered in terms of the scope (general or specific environment), range (geographical), source (magazines, trade exhibitions) and futurity (5 years, 10 years) (Thomas, 1980).

ES is also seen as a systematic process to collect information, interpret trends and events and provide feasible courses of action (Zhao & Merna, 1992). In short, ES can be defined as an approach to gathering relevant information from the external environment and converting the information into a meaningful knowledge for the executives to use in managing their organizations.

As reported in the previous study on corporate planning, a systematic ES could lead to organizational performances. Researches (such as Dev & Olsen, 1989; West, 1990) have found a positive relationship between environmental scanning, strategy and performances. In a dynamic situation confronting hospitality industry ES can be seen as becoming more important as a tool for corporate planning.

1.2 *Hotel industry in Malaysia*

Hotel sector are vital to the Malaysian economy as tourism industry is the 2nd contributor to the country's Gross Domestic Products (GDP). Tourist arrival into Malaysia for 2010 is about 24 million people. Hospitality industry at large strives in unpredictable business environments characterized by technological advancement, economic structure, intense competition, changing social preference and uncertainty of climate change. There are about 2,503 hotels in Malaysia (MOT, 2013) with average occupancy rate of 60 percent, the competition is very intense and organizations that can anticipate changes and address them before-hand can assist hotels to remain relevant to the market. In Kuala Lumpur city centre alone, there are about 180 five star hotel establishments.

In the Malaysian Economic Transformational Program (ETP), tourism has been earmarked as one of the National Key Economic Areas (NKEA). Malaysia is to focus on 12 NKEAs to boost the economy and achieve a high income status by 2020. These 12 NKEAs are the core of the Malaysian ETP that will receive prioritized government support including funding, top talent and national attention. In addition, policy reforms such as the removal of barriers to competition and market liberalization will be targeted at these NKEAs. This program will be the driver to stimulate economic activity that will contribute towards attaining high income, sustainability and completeness to the nation. The development of new luxury Hotel sectors under ETP is one of the driving forces for NKEA (12_National_Key_Economic_Areas, 2102).

1.3 *Research issue*

Environmental scanning systems are an important instrument for supporting managerial decision making, especially in turbulent times. The recent economic crisis provided a sustainable impulse for focusing earlier on emerging threats and opportunities. Although a rich body of knowledge exists, concepts remain unused in practice. Most often they lack applicability due to the lack of integrity on the information/data collected on which strategic decisions are made (Bischoff, et al., 2012). Past researches show that scanning firms demonstrate significant performance than non-scanning firms (Newgren, et al., 1984; Jain, 1984, Karim & Hussein, 2008).

Preservation of organizational memory becomes increasingly important to organizations as it is recognized that experiential knowledge and organizational learning is the key to competitiveness. Nevertheless, the information/data collected is not properly organized (Stein, et al., 1995). Mayer (2012) proposed five ways to improve today's Environmental Scanning system (1) design a more comprehensive information model (2) set up a collective learning process for interpreting information (3) use IS to enable management technique familiar to executives (4) design processes for more internationalization and (5) accelerate prototyping.

The business environment is increasingly becoming uncertain and complex. Environmental Scanning is a systematic way for organizations to detect changes and trends. Therefore, information literacy skills are required to conduct effective and efficient environmental scanning activities as it is an information intensive process (Zhang, et al., 2010). Good quality information can improve decision making, enhance efficiency and allow organization to gain competitive advantage (Karim & Hussein, 2008).

2 LITERATURE REVIEW

2.1 *Environmental scanning*

Environmental Scanning is the acquisition and use of information about events, trends and relationships in an organization's external environment, the knowledge of which would assist management in planning the organization's future course of action (Aguilar, 1967, Choo & Auster, 1993).

Environmental Scanning is a process that systematically surveys and interprets relevant data to identify external opportunities and threats. With the changes in economy, Environmental Scanning becomes the main driver for competitiveness (Raja Mejri, et al., 2013).

An organization gathers information about the external world, its competitors and itself. The company should then respond to the information gathered by changing its strategies and plans when the need arises (Choo, 1998).

It is a system using formalized procedures to provide management at all levels in all functions with appropriate information, based on data from both internal and external sources, to enable them to make timely and effective decisions for planning, directing and controlling the activities for which they are responsible (Lucey, 1991). The information processing approach seeks to understand and predict how organizations perceive stimuli, interpret them, store, retrieve and transmit information, generate judgments and solve problems (Larkey & Sproull, 1984).

2.2 Strategic information scanning system

It is a system using formalized procedures to provide management at all levels in all functions with appropriate information, based on data from both internal and external sources, to enable them to make timely and effective decisions for planning, directing and controlling the activities for which they are responsible (Lucey, 1991).

The information processing approach seeks to understand and predict how organizations perceive stimuli, interpret them, store, retrieve and transmit information, generate judgments and solve problems (Larkey & Sproull, 1984).

2.3 Underpinning theory

Aaker (1983) states planning requires an external analysis of the environment that relies upon information that has been gathered in an ad hoc, unsystematic way by those involved in the planning process. From his perspective a considerable amount of information is exposed to managers but is lost, dissipated or unused. This may be due to the scanning effort tends to be undirected, not partitioned among the participants and no vehicle to store and subsequently to retrieve and disseminate the information. To resolve this situation Aaker (1983) developed the 'Strategic Information Scanning System' (SISS) to enhance the effectiveness of the scanning effort and reserve much of the information now lost in the organizations. This system consists of six steps. Steps 1 and 2 specify information needs and sources, steps 3 and 4 identify the participants of the system and assign them to scanning tasks, finally steps 5 and 6 deals with the storage, processing, and dissemination of the information.

2.4 Review on strategic environmental information process models

2.4.1 Refined model on environment information process by Daft (1988)

This model is developed based on a formal six-step environmental scanning process lead to fulfil top management's requirements for strategic decision making. In this refined model, equal importance has been attached to the scanning steps starting from scanning needs identification to information evaluation and use: 1) scanning needs identification, 2) information acquisition, 3) information processing and synthesizing, 4) information organization and storage, 5) information distribution, and 6) information evaluation and use. Daft (1988) believed that senior manager's perceived strategic uncertainty is still proposed to have effect on frequency and rate of importance for environmental scanning, but is not limited to the steps of information collection.

2.4.2 Systematic approach to being information literate by Doyle (1994)

Doyle (1994) identified 10 steps required to perform an information task, which convey a total and systematic approach to being information literate. The steps are: 1) recognize the need for information, 2) recognize the need for accurate and complete information, 3) formulate questions based on needs, 4) identify potential sources of information, 5) develop successful search strategies, 6) access sources including computer-based and other technology, 7) evaluate information, 8) organize information for practical application, 9) integrate new information into an existing body of knowledge, and 10) use information in critical thinking and problem solving.

2.4.3 Information management model by Choo (2002)

Six distinct information management processes may be distinguished as 1) identifying information needs, 2) acquiring information, 3) organizing and storing information, 4) developing information products and services, and 5) distributing information and using information (Davenport 1993, McGee & Prusak 1993).

2.4.4 Environmental scanning process by Zhang, Majid and Foo (2010)

A six-step environmental scanning process is proposed based on Choo's (2010) information management model. The steps comprises of: 1) the formal environment scanning process starts with clearly defined scanning needs, 2) organizations actively obtained information through various sources, 3) the collected information is either stored for future use or processed and synthesized with the existing organizational knowledge, 4) after filtering (removing the irrelevant part of the information), repackaging (selecting information from different sources and merging it) or interpreting, 5) the processed environmental intelligence may be organized and stored in an organization knowledge repository for future utilization, and 6) disseminated directly to target users.

2.5 Research gap

Several models of Strategic Scanning Information System have been suggested by researchers, for example, Daft (1988) Refined Model on Environment information Process; Doyle (1994) Systematic Approach to being Information Literate, Choo (2002) Information Management Model, and Zhang, Majid and Foo (2010), the Environmental Scanning Process.

All these models have as a common agreement on the adopted methodologies for information management. Nevertheless, most of the publications did not bring an actionable knowledge about the mechanisms of implementing environmental scanning device (Caron-Fasan & Lesca, 2003). In spite the awareness of the importance of Environmental Scanning practice and its contribution to organizational performance; it remains unpunctual, unstructured and not holistic in approach (Raja Mejriet, et al., 2013).

3 RESEARCH METHODOLOGY

3.1 Research design

Qualitative research design will be deployed for this study. Qualitative method includes interview, non-participant observations and case study method. This technique is aimed at understanding the complex nature of the phenomena. In the context of this research, the case study is considered appropriate for providing a holistic approach to the study of strategic scanning information system in the hotel organizations.In creating the case study, several types of data and information will be collected, including general information about the hotel and specific information on the respondents. Case studies enable researchers to learn from practice and understand complex processes in its natural settings (Benbasat, et al., 1987).

3.2 Data gathering method

The qualitative research method will be used for collecting the core information for this research will be the semi-structured interview, a tool flexible enough to adapt to each of the variables that will be investigated, and also to pursuing unexpected responses and cues suggested by the respondents and via the on-site observation techniques as deemed necessary by the researcher throughout the research process.The methodology used is of major importance in obtaining in-depth information that are based mainly on the personal experience and practices of general managers, documentary evidence and on direct observation by the researcher (Yin, 2009).

3.3 Types of data collection

3.3.1 Primary data
Primary data will be collected via unstructured interview and observation with selected hotels in Malaysia.

3.3.2 Secondary data
Published information i.e. company information, directory of hotels, newspaper articles and press release are available from the internet sources and establishments such as Malaysian Hotel Owners Association directory, Tourism Information Centre Library, financial statements and prospectus from respective parent company of the selected hotels and other government authorities offices.

4 DATA ANALYSIS

The data obtained will be analyzed using the qualitative analysis method. Software called ATLAS. ti for computer-aided qualitative data analysis (CAQDA) will be employed. The analysis will show patterns of similarities and differences of ES practices among hotels in Malaysia. This study will also outline Strategic Information Scanning System for each hotel under study, and therefore, will extend further the literature on strategic information and environmental scanning practices in organizations.

5 CONCLUSIONS

Expected results on new knowledge will expand the new theoretical and empirical insights on environmental scanning and a model of strategic information scanning system practices in Malaysian hotel industry are expected to be derived from the proposed research. The knowledge in this field of research is crucial for hotels competitive advantage and survival of local hotels.It is hoped that the research will derive at the following prepositions:

P1: The more comprehensive the environment scanning system; the more structured will be the information on the external environment.

P2: The better the environment scanning structure; the more structured will be the information on the external environment.

P3: The more regular the environment scanning process; the more structured will be the information on the external environment.

P4: The more structured the information on external environment; the better the quality of information input for strategic decision making.

P5: The better the quality of information input for strategic decision making; the more competitive the organization.

REFERENCES

12 National Key Economic Areas. (2012). Retrieved from http://etp.pemandu.gov.my/annualreport2011/12_National_Key_Economic_Areas-@-12_National_Key_Economic_Areas.aspx

Aguilar, F.J. (1967). *Scanning the business environment.* New York, NY: MacMillan.

Choo, C.W. &Auster, E., (1993) Environmental Scanning by CEOs in two Canadian Industries. Journal of the American Society for Information Science, 44(4), pp 194–203.

Choo, C. (1998).The Knowing Organization: How Organizations Use Information for Construct Meaning, Create Knowledge and Make Decisions. New York: Oxford Press.

Davenport, Thomas H. & Lawrence Prusak. 1993. Blow Up The Corporate Library. *International Journal of Information Management* 13, no. 6 (Dec 1993): 405–412.

Jain, S.C. (1984). Environmental scanning in U.S. corporations. *Long Range Planning, 17*(2), 117-128.

Lee S. Sproul, Patrick D. & Larkey (1984). Advances in Information Processing in Organizations. Jai Press.

Lee Sproull. 1984. The nature of managerial attention. *Advances in Information Processing in Organizations,* Vol. 1: 9-27. Greenwich, CT: JAI Press.

Lucey, T. (1991; 1993), Management Information Systems. DP. Publications Ltd.

Masrek, M.N., Karim, N.S.A. & Hussein, R. (2008). The utilization and effectiveness of intranet: a case study at selected Malaysian organizations. *Communications of the IBIMA*, 4 (27), 200–206.

Mayer, J., Steinecke, N., Quick, R. & Weitzel, T. (2013). More applicable environmental scanning systems leveraging "modern" information systems. Information Systems and e-Business Management December 2013, Volume 11, Issue 4, pp 507–540

Raja Mejri & Mahmoud Zouaoui (2013), "Environmental Scanning and Its Perception by the Managers of the Tunisian Companies," IBIMA Business Review, 2013.

Xue Zhang, Shaheen Majid & Schubert Foo (2010). Environmental scanning: An application of information literacy skills at the workplace. Journal of Information Science, 36 (6) 2010, pp. 719–732

Theory and Practice in Hospitality and Tourism Research – Radzi et al. (Eds)
© 2015 Taylor & Francis Group, London, ISBN 978-1-138-02706-0

Structural relationships between career development learning, work integrated learning and employability: A structural equation modelling approach

H.F. Ariffin, R.P.S. Raja-Abdullah & N. Baba
Faculty of Hotel and Tourism Management, Universiti Teknologi MARA, Shah Alam, Malaysia

S. Hashim
Faculty of Technical and Vocational Education, Universiti Pendidikan Sultan Idris, Malaysia

ABSTRACT: This study investigates the structural relationships of career development learning and work integrated learning towards employability among hospitality graduates. Questionnaires that consist of career development learning, work integrated learning and graduate's employability was administered to 425 hospitality graduates from University Teknologi MARA (UiTM). Findings showed that the significant relationships between career development learning, workplace experience and graduates' employability upon graduation is confirmed in Structural Equation Modelling (SEM). It can be inferred from this that it is important for students to attain the career development learning and work integrated learning in order to be highly employable. Implications and suggestions for future research are also provided.

Keywords: employability, career development learning, works integrated learning, structural equation modelling

1 INTRODUCTION

For the past three decades, Malaysia's economy has grown at an annual rate of 6.6 percent, but the unemployment rates has increased from 2.6 percent in 1996 to 3.6 percent in 2003 as reported by Department of Statistics Malaysia, 1996–2003 (Ismail, 2011). In February of 2014, the unemployment rate in Malaysia was reported at 3.2 percent (Department of Statistics Malaysia (2014).

The analysis of unemployment rate was reported in Tenth Malaysia Plan with 30,000 graduates could not find job six months after graduation (Economic Planning Unit (EPU), 2010). Each year, 200,000 of the total 923,000 students were estimated to complete their study in various areas of field from institution of higher learning. Consistently, recent survey through Graduate Online Tracer Study (Ministry of Higher Education, 2010) (MOHE) reported that 42,955 of graduates were still unemployed after six months of their graduation. Statistics on recent graduate's employment status by MOHE Graduates Online Survey in 2008 indicated that there were 35.6 percent of first degree graduates in Tourism, Hospitality and Food and Beverage were unemployed (Abu Bakar, Jani & Zubairi, 2009). Most of all, labor market now is less predictable, changing more rapidly and more competitive

(Organization for Economic Cooperation and Develoment (OECD), 2004; Perrone & Vickers, 2003). Consequently, some fresh graduates are facing obscurity in getting a job. Thus, it is pertinent to highlight some effects of graduates' unemployment on the economy growth and towards graduates themselves. It appears that both unemployment and underemployment has contributed some negative effect which are increasing in burglary and armed robbery; psychological and financial stress; aggression, fear, anxiety and frustration; homelessness, wandering, vagrancy, insecurity; prostitution, kidnapping and drug addiction (Olowe, 2009).

Pool and Sewell (2007) suggests that it is essential for students to receive some education in career development learning to get better chance of securing occupations in which they can be satisfied and successful. The currents and future working environment also requires graduates to be "work-ready", equipped with work experiences (Hodges & Burchell, 2003) in order to be more competitive.

Smith, Brooks, Lichtenberg, Mcllveen, Torjul and Tyler (2009) propose that there is a need to provide evidence that career development learning can significantly give impact to the students' outcome. Therefore this study will cover the perceived relationship between career development learning and employability from the perspective of

Malaysian graduates specifically among hospitality students. As work integrated learning through internship training is one important part in hospitality program, the effect of internship on employability among graduates was also addressed.

2 LITERATURE REVIEW AND HYPOTHESES

2.1 Graduates' employability

Graduate employability (GE) recently has become a main issue for institutions because of the changing nature of the graduate labor market, mass participation in institutions of higher learning, pressures on student's finance, competition to recruit students and expectations of students, employers, parents and government (McNair, 2003). Lees (2002) mention that from the perspective of institutions of higher learning, employability is about to produce graduates who are able and capable which gives an impacts upon all areas of university life, in terms of the delivery of academic programs and extra curricula activities.

Consequently, Pool and Sewell (2007) refer employability as "having a set of skills, knowledge, understanding and personal attributes which makes a person more likely to choose and secure occupations in which they can be satisfied and successful". Based on their definition, they have developed a model that can be used to explain the concept of employability and known as "CareerEDGE". According to that model, career development learning and experience are construct as a "key" to choose and secure occupations in which the graduates has the opportunity to achieve satisfaction and success.

2.2 Career development learning

Career development learning (CDL) is associated with lifelong learning and related to: Learning about the content and process of career development or life/career management. The content of career development learning in essence represents learning about self and learning about the world of work. Watts (2008) indicates in his study, career development learning can help students to enlighten their future and expose the path they will choose, which they can build their employability and venture competences. He also asserts that career development learning may be organized variously to raise students' awareness of employability and how to get self-management in their studies and extra-curricular activities to optimize the employability. Pool and Sewell (2007) point out that graduate needs to grab the best chance of securing occupations in which they can be satis-

fied and successful by receiving some education in career development learning.

Also known as career education, Watts (2008) defines it as consisting of planned experiences designed to assist the development of: self-awareness—in terms of interest, abilities, value and others; opportunity awareness—knowing what work opportunities and what their requirement are; decision learning—decision-making skill; and transition learning—including job search and self-presentation skill. This formulation widely describes as DOTS models; decision-making learning (D), opportunity awareness (O), transition learning (T) and (self-awareness (S). Activities that help students to become more self-aware, to enable them to give real deliberation to the things that they take pleasure in doing, engrossed in, motivate them and suit their personalities should be included as proposed by DOTS model. Job market search needs to be learned by them in order to see what opportunities are available, how to present themselves effectively to the potential employers, and how to make decisions about their careers (Watts, 2008).

H1: There is a significant relationship between career development learning and graduates' employability.

2.3 Work integrated learning

Work integrated learning (WIL) is learning resulting from participation in a workplace community setting (Association of Graduate Careers Advisory Services (AGCAS), 2005). Most of the universities offer their students with work-integrated learning such as internship, practicum, practical placement, and industry based project, mentoring or vacation work in many of their academic programs. This program require students to implement their learning knowledge and reflecting upon the experience, knowing themselves and the world of work better in order to authorize them to enter and succeed in the world-of-work (AGCAS, 2005). Excellent work experience can boost learning and employability, so that work experience opportunities can be well-managed to be educationally valuable (Knight & Yorke, 2002).

H2: There is a significant relationship between workplace experience and graduates' employability.

3 METHODOLOGY

3.1 Participants and procedures

The respondents of this study consisted of Hotel Management, Tourism Management, Foodservice Management and Culinary Art Management fresh graduates from Faculty of Hotel and Tourism Management, UiTM. All respondents graduated in May 2011 both in degrees and diplomas and they were selected through systematic sampling. The rationale for choosing this group of students is due to their experiences as first time job seeker after graduate from university. In addition they were also selected because they have already working experiences during their practical or internship training in tertiary program. Data was collected via mail survey and self-administered questionnaires to 760 respondents however only 450 questionnaires were returned. Hence, only 425 questionnaires that were found useful and have been retained for further analysis.

3.2 Measures

A 22-items instrument used to measure career development learning was adopted from the outcome of career development learning, namely the DOTS model as listed by Watts (2008). The 24-items used to measure work integrated learning through internship training adapted from Muhamad, Yahya, Shahimi and Mahzan (2009), Dickerson& Kline (2008) and Singh & Dutta (2010). Employability was measured using 39-items adapted from Knight & York (2002) pertaining on aspect of employability that graduates should possess to enhance the employability. All of the items were measured by using the 5-point Likert Scale.

3.3 Data analyses

The demographic information was used to provide an overview of respondents' profile. Principal factor analysis was performed to reduce the number of factors or items from each variable. The final results from this factor analysis were then used for further investigation using Structural Equation Modelling (SEM) with AMOS 20.0 program. It aims to find the most optimal model or combination of the variables that fits well with the data on which it is built and serves as a purposeful representation of the reality from which the data has been extracted, and provides a parsimonious explanation of the data (Kline, 2011). In this study, the SEM technique was used to identify the influence of career development learning and work integrated learning to graduates' employability upon graduation.

4 RESULTS

4.1 Profile of sample

Majority of the respondents 70.8 percent ($N = 301$) were female while the remaining 29.2 percent ($N = 104$) were male. The ages of respondents involved were at a range from 22 and 26 years old. Most of the respondents are 24 years old with 36.2 percent ($N = 154$). Among of them, 69.4 percent ($N = 295$) are degree holder graduates and only 30.6

percent ($N = 130$) held a diploma certificate while 28.7 percent ($N = 122$) were graduates from Tourism Management program. It was found that all of the respondents had internship training experience or practical work experience in hospitality industry with duration of training for 3–6 months. The respondents' present status show that majority of graduates were employed with 76.2 percent ($N = 324$), 9.9 percent ($N = 42$) were unemployed and 13.9 percent ($N = 59$) furthering their study to the higher level.

4.2 Structural equation of hypothesized final model

Based on the modification index of CFA, the measurement model of exogenous and endogenous and the final model as the examination of the hypothesized model confirmed the constructs of GE, CDL and WIL of the hypothesized paths. In SEM, factor analysis and hypotheses are tested in the same analysis. SEM techniques also provide fuller information about the extent to which the research model is supported by the data. Goodness of fit indices for the 12 observed variables of CDL, WIL and GEB shows that the reading is good if it ranges from 0.188 to 0.941 for the significance standardized regressions weight. Standard error (SE) for each observation shows the goodness of fit and low level reading from .014 to .029, and estimate (Square Multiple Correlation) of observation shows the contribution level to the latent variable (.099 to 0.885). The standardized regression weight between CDL and GE is 0.525, and between WIL and GE is 0.296. The final model shows the model explained in a substantial portion of the variance in all the endogenous variables (square multiple correlations) that indicates the two exogenous variables (CDL and WIL) jointly explained 42.3 percent variance in GE. Finally, from the Structural Model the reading for GFI is at 0.90 (acceptable fit criteria) and RMSEA is less than .08. The measurement model has a good fit with the data based on assessment criteria such as GFI and RMSEA (Bagozzi & Yi, 1988). Table 1 summarizes the goodness of

Table 1. Summary of the goodness fit of CDL, WIL, GE and final model.

Model Fit Indicator	CDL	WIL	GE	Final Model
(x^2)		10.322	7.203	128.489	182.060
DF	3	2	3.4	49
CMIN/DF	3.441	3.601	3.779	3.716
P	.016	0.027	.000	.000
GFI	0.990	0.992	0.945	0.937
RMSEA	.076	.078	.080	.080

model fit of CDL, WIL, GE and the structural model. The structural models testing of endogenous variables (CDL and WIL) fulfill the GFI (GFI > 0.90) and RMSEA criteria (less than .08).

5 DISCUSSION

Based on the findings presented, it was observed that career development learning significantly influences graduates' employability with a fair correlation and standardized regression weight of 0.525 (p < .01), indicating that career development learning significantly predicted graduates' employability, thus supporting H1. This is consistent with Watts (2008) where career development learning able to assist students to make clear chosen career path which they can develop and build their employability and competence. Career development learning can help to optimize the employability by raising student's awareness and how to manage their studies and extra-curricular activities (Watts, 2008). Based on the result between career development learning dimension which is self/opportunities awareness and decision-making and transition learning, it was found that self/opportunity awareness and decision-making has more influence on graduates' employability with standardized regression weight of 0.780 compare to transition learning with standardized regression weight of 0.626. This result is also consistent with study done by McIlveen et al. (2011) which results of the survey indicates convergence of the career development-learning domains of self-awareness and opportunity awareness, but relatively less integration of decision-making and transition learning.

Consequently, it was observed that work integrated learning significantly influences graduate employability with a fair correlation and standardized regression weight of 0.296 (p < .01), indicating that work integrated learning significantly predicted graduates' employability, thus supporting H2. The result highlights that the more exposure on internship training helps students to be more aware on self-interest, abilities and value or knowing what opportunities available and what requirement need to be comply. They also feel that they are able to make decision about their career after internship training. This is in relation with (Brooks, Cornelius, Greenfield & Joseph, 1995; Taylor, 1988) who suggests that practical experience and exposure gained during internship program to be helpful in improving career decision making. Similarly, McIlveen et al. (2011) found that self-awareness and opportunity awareness are rated as most often present in work-integrated programmes.

6 IMPLICATIONS

It is imperative that the importance of career development in enhancing graduate employability after graduation such as self-awareness, opportunity awareness, decision-making and transition learning has been realized among students. By that, it will help them to enhance their employability chances upon graduation. For that reason, it is important for the students to take part in their career planning by making their own career goals and action plans to attain better career employability and employment chances in the future.

This study also can be used as a guideline for institutions to develop a better career development learning programs. Institutions play a big part to ensure the level of their students through the learning outcomes especially which related to career development. With the result of this study, university departments especially academics staff and career development practitioner may reflect upon which the extent to give to their students in regards to career learning process during their tertiary program. Teaching and learning in the curriculum also should be more effective to prepare graduates to link their potential skills and knowledge to meet demand required by employers.

Through this study, the industry also has information regarding students' perception of internship program so that they will know how to assist students in meeting their internship goals better. Both school and industry should work together closely to develop a well-organized and quality internship program in order to enhance the level of graduates' employability. Employers also can incorporate career development during any work experience by provision of career mentor; provide in-house speaker, rotate students through a range of role, skill development and training. The inclusion of career development opportunities within workplace experiences also can lead better matching of students and opportunities and help students to have a clear sense of whether they fit or not in any industry.

7 LIMITATIONS AND RECOMMENDATIONS

Results gained might not be able to generalize to the other sample due to the limited sampling frame used. For future research, a bigger and more diversified sample could be used from various types of institutions. Private and public universities have different implementation and career development planning for their students. Since this study focuses on one of the public university, it is worth to conduct and compare the result from both public and private institution students in the future research.

It is recommended that future research be conducted in long term period to cover both perception and expectation from students regarding the workplace experiences through internship training before and after the programs. It is also recommended that future research to study on level of understanding on career development learning in teaching and learning process among university employers and also in workplace supervision of internship training among employers.

8 CONCLUSION

Based on the findings, it can be concluded that career development learning is an important element that could assist graduates to secure their future work after graduation. Thus institutions of higher learning should be aware on the need of effective career development learning among students during their tertiary programs. Internship training program should be well planned and design systematically to make this program a good experiences for students. Lack of commitment from industry for better internship program created a negative perception from students about the effect of internship in influencing their learning for employability. Hence, both institution and industry should take appropriate actions in increasing and enhancing the learning condition and work integrated learning to help students to be more employable upon graduation.

REFERENCES

Abu Bakar, M., Jani, R. & Zubairi, Y.Z. (2009). *Overview of graduate employability of recent graduates: Some facts and figures.* Paper presented at the Seminar on Employability, the Ministry of Higher Education of Malaysia, Putrajaya.
Association of Graduate Careers Advisory Services (AGCAS). (2005). Careers education benchmark statement. Sheffield: Careers Education Task Group.
Bagozzi, R.P. & Yi, Y. (1988). On the evaluation of structural equation models. *Journal of the academy of marketing science, 16*(1), 74–94.
Brooks, L., Cornelius, A., Greenfield, E. & Joseph, R. (1995). The relation of career-related work or internship experiences to the career development of college seniors. *Journal of Vocational Behavior, 46*(3), 332–349.
Department of Statistics Malaysia. (2014). Labour Force Statistics: February 2014. Putrajaya: Department of Statistics Malaysia.
Dickerson, J.P. & Kline, S.F. (2008). The early career impact of the co-op commitment in hospitality curricula. *Journal of Teaching in Travel & Tourism, 8*(1), 3–22.

Economic Planning Unit (EPU). (2010). Tenth Malaysia Plan 2011–2015, from http://www.epu.jpm.my/RM10/html/english.htm
Hodges, D. & Burchell, N. (2003). Business graduate competencies: Employers' views on importance and performance. *Asia-Pacific Journal of Cooperative Education, 4*(2), 16–22.
Kline, R.B. (2011). *Principles and practice of structural equation modeling*: Guilford press.
Knight, P. & Yorke, M. (2002). Skills Plus: Turning the Undergraduate Curriculum., from http://www.heacademy.ac.uk/945.htm
Lees, D. (2002). *Graduate employability-literature review*: LTSN Generic Centre.
McIlveen, P., Brooks, S., Lichtenberg, A., Smith, M., Torjul, P. & Tyler, J. (2011). Career development learning frameworks for work-integrated learning *Developing Learning Professionals* (pp. 149–165): Springer.
McNair, S. (2003). Employability in Higher Education: Developing Institutional Strategy: Routledge-Falmer.
Ministry of Higher Education, G.T.S. (2010). Statistic of Malaysia Higher Education, from http://www.mohe.gov.my/web_statistik/satistik
Muhamad, R., Yahya, Y., Shahimi, S. & Mahzan, N. (2009). Undergraduate Internship Attachment in Accounting: The Interns Perspective. *International Education Studies, 2*(4), P49.
Olowe, O. (2009). Graduate Unemployment and its Resultant Effects on Developing Economies. *Available at SSRN 1457041*.
Organization for Economic Cooperation and Development (OECD). (2004). Career giudance: A handbook for policy maker, from http://www.oecd.org/dataoecd/53/53/34060761.pdf
Perrone, L. & Vickers, M.H. (2003). Life after graduation as a "very uncomfortable world": An Australian case study. *Education+ Training, 45*(2), 69–78.
Pool, L.D. & Sewell, P. (2007). The key to employability: developing a practical model of graduate employability. *Education+ Training, 49*(4), 277–289.
Singh, A. & Dutta, K. (2010). Hospitality internship placements: analysis for United Kingdom and India. *Journal of Services Research, 10*(1).
Smith, M., Brooks, S., Lichtenberg, A., McIlveen, P., Torjul, P. & Tyler, J. (2009). *Career development learning: maximising the contribution of work-integrated learning to the student experience: Australian Learning & Teaching Council Final project report*: University of Wollongong, Careers Central, Academic Services Division.
Taylor, M.S. (1988). Effects of college internships on individual participants. *Journal of Applied Psychology, 73*(3), 393.
Watts, A. (2008). *Career Development Learning and work-integrated learning: A conceptual perspective from the UK*. Paper presented at the National Symposium on Career Development Learning: Maximising the contribution of work-integrated learning (WIL) to the student experience, Melbourne.

Hospitality and tourism marketing

Theory and Practice in Hospitality and Tourism Research – Radzi et al. (Eds)
© 2015 Taylor & Francis Group, London, ISBN 978-1-138-02706-0

An investigation of customers' satisfaction as a mediating effect between hotel customer loyalty antecedents and behavioral loyalty

A.A. Azdel, M.F.S. Bakhtiar, M.S.Y. Kamaruddin & N.A. Ahmad
Faculty of Hotel and Tourism Management, Universiti Teknologi MARA, Shah Alam, Malaysia

N.C. Ahmat
Faculty of Hotel and Tourism Management, Universiti Teknologi MARA, Penang, Malaysia

ABSTRACT: Customer's satisfaction come in nature and individualized as each customer will face dissimilar experience. The attributes of customer loyalty will come after the experience that the customer gained and it will trigger them to be a loyal customer. This paper studies the antecedents from the perspective of the hotel's customer in Malaysia and sees whether it will lead to customer satisfaction and behavioral loyalty. The customers that had stayed and experienced the services at five star hotels in Kuala Lumpur were the target population. Bootstrap Analysis had been used to test customer satisfaction as the mediating variable between customer loyalty antecedents and behavioral loyalty. Most of the findings were statistically significant. This study had shown the consequence of understanding the customer loyalty antecedents as it can give an impact towards retaining the loyal customer.

Keywords: Customer loyalty antecedents, satisfaction, behavioral loyalty

1 INTRODUCTION

Customer's satisfaction come in nature and individualized as each customer will face dissimilar experience. The attributes of customer loyalty will come after the experienced that the customer gained and it will trigger them to be as a loyal customer. The customer will face different experience even though they are staying at the same hotel and received the same service. Different person conveys a different background, values, attitudes, and beliefs to the situation as each experience is derived through his or her individual lens (Knutson, Beck, Kim & Cha, 2009).

When the customer is gaining more experience over time, the cumulative satisfaction become more important and it will contribute to the loyalty. Furthermore, the experienced customer and the customer that have cumulative satisfaction will not defect just because one unsatisfying experience (Brunner, Stöcklin & Opwis, 2008).

The study that had been done by Lee and Back (2009) had mentioned that the loyal customer will give lots of advantages for the firm and perhaps for the hotel too as it can help the hotel to reduce their marketing and operation cost, increase their revenue, positive word of mouth and also it will reduce the possibility for the customer to switch for the other competitors.

Conversely, there is a distinction regarding this relationship as Mohsan, Nawaz, Khan, Shaukat and Aslam (2011) noted that customer loyalty is vulnerable because even if consumers are satisfied with the services they will be disloyal to the provider if they think they can get better value, convenience or quality elsewhere. Furthermore, the same authors also stated customer satisfaction and loyalty were the different things as satisfaction is essential but not a sufficient condition of loyalty.

Therefore, it raised an interest for the researcher to investigate the antecedents from the perspective of the hotel's customer in Malaysia and see whether it will lead to customer satisfaction and behavioural loyalty. Furthermore, this study will help the hotel industry to figure out the customer loyalty antecedents and make use of it to tackle this target customer. Hence, this study will be using different settings from the previous research.

2 LITERATURE REVIEW

2.1 *Hotel customer loyalty antecedents*

Customer loyalty can be defined from various definitions. Bowen and Shoemaker (2003) stated that customer loyalty is the measurement of the probability that the customer will revisit or come again to a property. The same authors also noted that customer loyalty is essential in hotel industry as this industry had matured and competition is strong. A study done by Shoemaker and Lewis (1999) stated that customer loyalty is not the same as customer satisfaction. Customer satisfaction will measure how well a customer's expectation is met by a given transaction. Meanwhile, customer loyalty measures how likely a customer to repurchase and engage in partnership activities.

Hence, this study had identified customer loyalty antecedents that will help the hoteliers or academia to view that those antecedents will give an input to them in executing their strategies, which similar dimension of customer loyalty antecedents in Azdel, Azizan, Bakhtiar and Ahmad (2014) study. The antecedents that had been review and discussed in this study are perceived value, familiarity and attachment.

Velazquez, Saura and Molina (2011) noted that customer perceived value will engage the evaluation of the customer between what is attain which is the result and benefits and what has been invested that means money, time and effort that the customer need to pay. Perceived value is important as it will give an impact as the customer will consider it when seek information about certain hotel, comparing with other hotels and also their willingness to buy. A study done by Cronin, Brady and Hult (2000) has some distinction as the authors mentioned that perceived value that the customer gained will be the better measurement of repurchase of a product or service than the customer satisfaction or the quality of the product or services.

Previous study by Williams and Vask (2003) mentioned that place attachment refers to the emotional and symbolic relationship that individual form with a certain place. Besides that, place attachment also can be referred to the experience of a long term affective bond to a particular geographic area as had been defined by Hay (1998) that a person will has a feeling of affection or a sense of belonging of that place if they lives in a particular place over an extended period of time.

According to Hummon (1992), people form emotional bonds to a place and also to other people and the attachment to a place reveal the embeddedness of individuals within their socio-physical environments. It had been agreed by Hidalgo and Hernandez (2001) as the authors stated that the place attachment will comprises of cognitive and emotional linkage of a customer to the place.

Familiarity is being defined by Moorthy, Ratchford and Talukdar (1997) as the customer perception on how much he/she knows about the place that means the hotel. Familiarity also undertake in the marketing as the customer will go or purchase product or service that they familiar (Moore & Lehmann, 1980).

Gursoy (2001) notified that besides using the product which means the internal source, customer familiarity can also be acquired from external sources that are word of mouth and advertising. Meanwhile, Gefen (2000) mentioned that familiarity will act as a mechanism in reducing uncertainty and it had been agreed by Bhattacherjee (2002) that familiarity had works as a risk reduction and consequently it had effect the customer's trust for the product and service.

2.2 Customer satisfaction and behavioural loyalty

Recent study done by Prayag and Ryan (2012) agreed that as satisfaction levels increase, the propensity of the customer to return and recommend will increase as well. The authors also stated word-of-mouth recommendations are critical for loyalty. Moreover, Choi, Cho, Lee, Lee and Kim (2004) in their study noted that the customer who feels satisfied with the service that they get will make a repeat purchase and recommend about the service provider to other people.

However, study done by Walsh, Evanschitzky and Wunderlich (2008) found that satisfaction with behavioural loyalty was only moderately significant as the authors indicated customers with weakly held satisfaction have a greater risk of defection than the customers that have more strongly held satisfaction. The statement showed if the customer felt only a little bit of satisfaction, the probability of them to return or recommend to others is slightly low.

3 METHODOLOGY

A quantitative approach was used to gather the data by distribution of questionnaire to the sample that has been determined. It involves gathering numerical data using structured questionnaires to collect primary from individuals. The research took place at the 5 stars hotels around Kuala Lumpur. The customers that had stayed and experienced the services at five star hotels in Kuala Lumpur were the target population. From the recent record, there are 25 hotels in Kuala Lumpur are 5-stars rating. The researchers choose the 25 hotels that involved in this study. A total of 169 valid samples were gathered and analyzed.

4 FINDINGS

Among the 169 respondents who answered the questionnaires, it appears that majority (57.4%) of them were female while almost half (42.6%) were men. Meanwhile, most of the of them (38.5%)

were in the age of 20 to 30 then followed by the respondents that were in age of 31–40 that contributed 30.8 percent.

The respondents in the age of 41 to 50 and 51 to 60 contributed to the same percent ages which were 13 percent respectively. The rest of the respondents were above the age of 60 or below 20 (4.1% and 0.6% accordingly). Almost half (47.9%) of them stayed at the five stars hotel for holiday purposes. Other than that, the respondents who stayed for the purpose of business contributed 45.0 percent.

Bootstrap analysis was used to assess the mediational model of Satisfaction as a mediator of the relationship between Customer Loyalty Attributes (Perceived value, Attachment, Familiarity) and Behavioural Loyalty. To prove the mediating effect from the variables, the 95 percent bootstrap confidence interval for the indirect effect must not include zero which means if there is zero value, "0" between the interval which are lower and upper, it indicates the mediator does not mediate between independent and dependent variable and was not significant.

In this analysis, Satisfaction does mediate the effect of Customer Loyalty Attributes on Behavioural Loyalty. The analysis for each dimensions showed Satisfaction does mediates between Perceived Value and Attachment on Behavioural Loyalty. However, it does not mediate between Familiarity on Behavioural Loyalty. As can be seen in Table 1, when customer loyalty attributes was controlled, both indirect and direct effects were significant. The mean of indirect effect from the analysis was positive and significant (a x b = 0.128), with a 95% bias corrected and accelerated confidence interval excluding zero (.078 to 0.167) and the direct effect c' (0.10) was significant (p <. 001). The respondents who had high level of customer loyalty were more likely to have satisfaction relationship (a = 0.268) while holding constant for customer loyalty attributes, the high level of satisfaction will indicate positivity of behavioural loyalty (b = 0.478).

When perceived value was controlled, the mean of indirect effect from the bootstrap analysis was positive and significant (a x b = 0.229), with a 95% bias corrected and accelerated confidence interval excluding zero (0.148 to 0.321). In the direct path, the respondents who indicated high level of perceived value were more likely to have satisfaction relationship (a = 0.471) while holding constant for perceived value, the respondents who had high level of satisfaction will more likely lead to the behavioural loyalty (b = 0.489). In the direct effect path (c' = .059), the satisfaction relationship was not significant and the respondents with high level of perceived value were less likely to have behavioural loyalty.

Whereby, when attachment was controlled, both indirect and direct effects were significant. The mean of indirect effect from the analysis was positive and significant (a x b = 0.140), with a 95% bias corrected and accelerated confidence interval excluding zero (.071 to 0.221) and the direct effect c (0.140) was significant (p < .05). The respondents who had high level of attachment were more likely to have satisfaction relationship (a = 0.288) while holding constant for attachment, the high level of satisfaction will indicate positivity of behavioural loyalty (b = 0.489).

On the other hand, when familiarity was controlled, the mean of indirect effect from the bootstrap analysis was significant (a x b = .005). However, with a 95 percent bias corrected the accelerated confidence interval was insignificant as the interval was including zero (-.046 to .053). This model showed that familiarity cannot stand alone to lead to the behavioural loyalty as familiar-

Table 1.   Summary of mediation result (5,000 bootstrap samples, $N = 169$).

Independent Variable (IV)	Mediating Variable (M)	Dependent Variable (DV)	Effect Of IV on M (a)	Effect of M on DV (b)	Direct Effects (c')	Indirect Effect (a*b)	95% CI for a*b	Total Effects (c)
Customer Loyalty Attributes	Satisfaction	Behavioural Loyalty	.27**	.49**	.10**	.128*	.078, .167	.228**
Perceived Value	Satisfaction	Behavioural Loyalty	.47**	.49**	.059	.229*	.148, .321	.289**
Attachment	Satisfaction	Behavioural Loyalty	.29**	.49**	.13**	.140*	.071, .221	.266**
Familiarity	Satisfaction	Behavioural Loyalty	.005	.49**	.102*	.005*	−.046, .053	.104*

IV, independent variable; DV, dependent variable; M, mediator. In both analyses, the effects of the other independent variables were controlled for **p.001, *p.05.

Note: 95% CI for a*b is significant if zero is not included in the interval.

ity need to be in a group of other attributes such as perceived value and attachment before it had an effect towards satisfaction and behavioural loyalty. Therefore, the respondents who had familiarity were less likely to have satisfaction relationship (a = .005) while holding constant for familiarity, the high level of satisfaction indicate positivity of behavioural loyalty (b = 0.489). Apparently, the direct effect c' (0.102) was significant.

5 CONCLUSION

The finding proves that satisfaction does mediate between customer loyalty attributes and behavioral loyalty. The result was supported by early research done by Spiteri and Dion (2004) as the authors conducted the study confined to veterinarians. The finding had recognized from the attributes of customer loyalty, the customer will ensure that they were satisfied with the service offered by the hotel before turn them as a loyal customer and have a behavior loyalty towards the particular hotel.

In a similar vein, the result provides acceptable evidence that satisfaction also does mediate between perceived value and attachment towards behavioral loyalty. It was similar with Ariff, Fen and Ismail (2012) as finding validated the configuration of customers' perceived value, satisfaction and loyalty. Satisfaction also had a significance effect towards attachment and behavioral loyalty and the result supports Lee, Kyle and Scott (2012) stated their findings as satisfied visitors at a festival will develop an affecting attachment towards the festival host destination and eventually they will become loyal to that destination.

Nonetheless, the result from the bootstrap analysis showed that satisfaction does not mediate the effect of customer's familiarity on behavioural loyalty. The result was opposed with the study done Casaló, Flavian and Guinalíu (2008) as the authors found out as customer familiarity increases it will influence their loyalty and the loyalty also acquire from the effect on satisfaction.

Hoteliers must keep in their minds that they cannot treat customer the same and anticipate receiving the same evaluation of satisfaction from the as different customer have their own expectation and level of satisfaction. Thus, the hoteliers need to understand the disparity between their customers and produce the best ways to create satisfaction among the customer.

Retaining customer is essential for the future in this industry and also within this globally competitive market as hotel industry have to hunt for way to communicate their exclusivity for their customer and gain loyalty from them. The management also needs to focus for their aim for prosperity in this industry and ensure that their mission is an ongoing basis. A warm welcome, serving with caring and responsive approach of the employee can turn the ordinary service into an outstanding and exceptional experience.

Therefore, it can be concluded that this study really assist in determining the attributes that trigger satisfied customer to be a loyal customer. As a result, this study should be able to help the hoteliers, as well as other service providers, who sensed that it is necessary to find out the best ways to retain the customer as this kind of customer will help them to bring the luxurious path to their business.

ACKNOWLEDGEMENT

This research is funded by Universiti Teknologi MARA, under the Research Acculturation Grant Scheme (RAGS) (600-RMI/RAGS 5/3 (146/2012).

REFERENCES

Ariff, M. S. B. M., Fen, H. S., Zakuan, N., Ishak, N. & Ismail, K. (2012). Relationship between customers' perceived values, satisfaction and loyalty of mobile phone users. *Review of Integrative Business and Economics Research, 1*(1), 126–135.

Azdel, A. A., Azizan, N. N., Saiful Bakhtiar, M. F. & Ahmad, N. A. (2014). Hotel customer loyalty antecedents and behavioral loyalty. *Hospitality and Tourism: Synergizing Creativity and Innovation in Research*, 265.

Bhattacherjee, A. (2002). Individual trust in online firms: Scale development and initial test. *Journal of management information systems, 19*(1), 211–242.

Bowen, J. T. & Shoemaker, S. (2003). Loyalty: A strategic commitment. *Cornell Hotel and Restaurant Administration Quarterly, 44*(5–6), 31–46.

Brunner, T. A., Stöcklin, M. & Opwis, K. (2008). Satisfaction, image and loyalty: New versus experienced customers. *European Journal of Marketing, 42*(9/10), 1095–1105.

Casaló, L., Flavián, C. & Guinalíu, M. (2008). The role of perceived usability, reputation, satisfaction and consumer familiarity on the website loyalty formation process. *Computers in Human Behavior, 24*(2), 325–345.

Choi, K.-S., Cho, W.-H., Lee, S., Lee, H. & Kim, C. (2004). The relationships among quality, value, satisfaction and behavioral intention in health care provider choice: A South Korean study. *Journal of Business Research, 57*(8), 913–921.

Cronin Jr, J. J., Brady, M. K. & Hult, G. T. M. (2000). Assessing the effects of quality, value, and customer satisfaction on consumer behavioral intentions in service environments. *Journal of retailing, 76*(2), 193–218.

Gefen, D. (2000). E-commerce: The role of familiarity and trust. *Omega, 28*(6), 725–737.

Gursoy, D. (2001). *Development of a travelers' information search behavior model.* Unpublished doctoral dissertation, University Libraries, Virginia Polytechnic Institute and State University.

Hay, R. (1998). Sense of place in developmental context. *Journal of environmental psychology, 18*(1), 5–29.

Hidalgo, M. C. & Hernandez, B. (2001). Place attachment: Conceptual and empirical questions. *Journal of environmental psychology, 21*(3), 273–281.

Hummon, D. M. (1992). *Community attachment: Local sentiment and sense of place.* New York: Plenum Press.

Knutson, B. J., Beck, J. A., Kim, S. & Cha, J. (2009). Identifying the dimensions of the guest's hotel experience. *Cornell Hospitality Quarterly, 50*(1), 44–55.

Lee, J.-S. & Back, K.-J. (2009). An examination of attendee brand loyalty: Understanding the moderator of behavioral brand loyalty. *Journal of Hospitality & Tourism Research, 33*(1), 30–50.

Lee, J. J., Kyle, G. & Scott, D. (2012). The mediating effect of place attachment on the relationship between festival satisfaction and loyalty to the festival hosting destination. *Journal of travel research, 51*(6), 754–767.

Mohsan, F., Nawaz, M. M., Khan, M. S., Shaukat, Z. & Aslam, N. (2011). Impact of customer satisfaction on customer loyalty and intentions to switch: evidence from banking sector of Pakistan. *International Journal of Business and Social Science, 2*(16), 230–245.

Moore, W. L. & Lehmann, D. R. (1980). Individual differences in search behavior for a nondurable. *Journal of consumer research, 7*(3), 296–307.

Moorthy, S., Ratchford, B. T. & Talukdar, D. (1997). Consumer information search revisited: Theory and empirical analysis. *Journal of consumer research,* 263–277.

Prayag, G. & Ryan, C. (2012). Antecedents of tourists' loyalty to Mauritius The role and influence of destination image, place attachment, personal involvement, and satisfaction. *Journal of travel research, 51*(3), 342–356.

Shoemaker, S. & Lewis, R. C. (1999). Customer loyalty: The future of hospitality marketing. *International Journal of Hospitality Management, 18*(4), 345–370.

Spiteri, J. M. & Dion, P. A. (2004). Customer value, overall satisfaction, end-user loyalty, and market performance in detail intensive industries. *Industrial Marketing Management, 33*(8), 675–687.

Velázquez, B. M., Saura, I. G. & Molina, M. E. R. (2011). Conceptualizing and measuring loyalty: Towards a conceptual model of tourist loyalty antecedents. *Journal of Vacation Marketing, 17*(1), 65–81.

Walsh, G., Evanschitzky, H. & Wunderlich, M. (2008). Identification and analysis of moderator variables: Investigating the customer satisfaction-loyalty link. *European Journal of Marketing, 42*(9/10), 977–1004.

Williams, D. R. & Vaske, J. J. (2003). The measurement of place attachment: Validity and generalizability of a psychometric approach. *Forest science, 49*(6), 830–840.

Theory and Practice in Hospitality and Tourism Research – Radzi et al. (Eds)
© 2015 Taylor & Francis Group, London, ISBN 978-1-138-02706-0

Impact of actual self-congruity and ideal self-congruity on experiential value and behavioral intention

J.M. Yusof, H.A. Manan, N.A.M. Kassim & N.A. Karim
Faculty of Business Management, Universiti Teknologi MARA, Shah Alam, Malaysia

ABSTRACT: This study is to test the effect of actual self-congruity and ideal self-congruity on experiential value and behavioral intention and to compare which of the type of self-congruity has a stronger effect on these two behavioral outcomes. Despite of much attention given to the concept, few, if any, have reported its effect on experiential value and in the context of destination image. Data were collected from tourists who visited island destination. The findings of the study provide evidence that, actual self-congruity and ideal self-congruity has a significant relationship with experiential value, but has no significant relationship with behavioral intention. Experiential value has an effect on behavioral intention and actual self-congruity appears to have a stronger effect on experiential value. This study provides an empirical support for the new framework on the importance of experiential value in the self-congruity framework in destination image.

Keywords: Actual self-congruity, ideal self-congruity, experiential value, behavioral intention, destination image

1 INTRODUCTION

1.1 *Tourism industry*

Tourism industry in Malaysia is one of the important industries, which has contributed significantly to the nation's economic growth and employment. Total tourists arrival in Malaysia in 2011 has reached RM24.7 million, an increase of 10 percent from previous year (Musa, Muhammad Kassim & Putit, 2012). The industry has been earmarked under the Economic Transformation Program (ETP) as the National Key Economic Area (NKEA) with the target set as "2020:36:168". This means that the industry has a target of 36 million tourists visiting the country and with the contribution of RM168 billion generated from the industry by year 2020. The attempts to draw in more tourists to the country appear to be an important agenda to the government. To address this issue, the government continues to upgrade the tourists' offerings and services in order to enhance the country's image and at the same time to improve tourists visiting experience.

In the marketing context, research supports the importance of the product, brand or store image. In tourism industry specifically, it has been acknowledged by researchers that image is an important factor of a consumer's destination process (Stepchenkova & Morrison, 2006). Much of the image studies in destination process have focused on factors that influence on consumer's decision process such as awareness, choice intention, and the like (Phau, Shanka & Dhayan, 2010). In particular, there has been little empirical evidence that has been provided in the theory of self-congruity that examines experiential value of the tourists. It is important to recognize that tourists think about the destination and how it matches their own image. Therefore, research efforts on understanding the self-congruity of tourists as well as their experiential value are imperative.

Much of the self-congruity studies have been in the product, brand, and store domain (e.g., Jamal & Goode, 2001; Sirgy & Samli, 1985; Sirgy & Su, 2000). Results from these studies indicated that image congruence affects consumer behavior both directly and indirectly through functional aspects of product, brand, or retail stores (Sirgy, Johar, Samli & Claiborne, 1991; Sirgy & Samli, 1985). Despite numerous self-congruity studies, there has been a dearth of literature examining self-congruity in tourist destination that particularly investigated actual self-congruity and ideal self-congruity that incorporate the concept of tourist' experience.

Hence, the present study proposed a model that incorporates actual and ideal self-congruity, experiential value and behavioral intention of tourists. By proposing and subsequently testing the structural relationships among the four constructs, this study is intended to achieve the following research objectives: (1) to investigate the effect of actual self-congruity and ideal self-congruity on experiential value and behavioral intention, (2) to examine the influence of experiential value on behavioral intention, (3) to investigate the role of experiential value in the self-congruity and behavioral intention relationship.

2 REVIEW OF LITERATURE

2.1 *Actual self-congruity and ideal self-congruity*

In either product, brand, or store context, actual congruity refers to the degree of match between a consumer or shopper's actual self-image and the image of the users or patrons of product, brand or store. It is the fit between how the consumers or shoppers see themselves in relations to the image of the users or patrons of the retail stores. Ideal self-congruity refers to the degree of match between a consumer ideal-self-image and the image of users or retail store patrons, or the fit between how consumers or shoppers would like to see themselves in relation to the image of users or patrons.

In the service industry, the actual self-congruity and ideal self-congruity have been found in studies of service industry such as destination, hotel, and also restaurant. Litvin and Kar (2004) examined the congruity between destination image and self-image and found that ideal self-congruity influences satisfaction and loyalty. Usakli (2009) also found support for self-congruity theory in destination image in that both the actual and ideal self-congruity has positive impact on behavioral intention.

Specifically, Graeff (1996) examined the congruity between brand image and self-image in the context of consumer user situation. It was found that the evaluations of the brands were affected more by the congruity of brand image and ideal self-image than actual self-image. Ericksen (1997) the existence of relationship between actual and ideal self-congruity on purchase intention. Similarly, Chang (2001) found support for the link between actual and ideal self-congruity with product evaluation. Jamal and Goode (2001) as well as Govers and Schoormans (2005) examined ideal and actual self-congruity as the types of self-congruity that influence product or brand preference. Meanwhile, Helgeson and Supphellen (2004) found that actual and ideal self-congruity has positive effect on brand attitude.

2.2 *Experiential value*

Experiential value has been explored mainly in the education context as an important factor that stimulates learning. The concept has nevertheless begun to receive a wide attention among marketing and retailing scholars as well as the practitioners. In fact, in recent times, the marketing and retailing literature has highlighted the critical role of experience of the consumers. To provide this experiential value to the consumers, retailers specifically are proposed to create a theatrical retailing environment, stress fun, excitement and promotion, as well as encourage greater customer participation in the retail shopping experience (Baron, Harris & Harris, 2001; Mathwick, Malhotra & Rigdon, 2001). Experiential value is defined as

> *"..... a perceived, relativistic preference for product attributes or services performance arising from interaction within a consumption setting that facilitates or blocks achievement of customer goals or purposes" (Mathwick, et al., 2001, p. 53).*

In the present research context, experiential value is viewed as the perceived benefits gained when engaging in the destination product offerings and services. For example, the aesthetic value could be obtained from the natural beauty of the destination; social bonding experienced from the interaction with family as well as other tourists, financial value is derived from the quality of services compared with the price paid; knowledge enrichment is derived when visited the destination and have the feeling of enjoyment of diverging to a new self.

2.3 *Behavioral intention*

Behavioral intention plays a vital role in a consumers' adoption and continuance use of service and product. As suggested in the theory of reasoned action and the theory of planned behavior, behavioral intention represents the degree of conscious effort that a person exert in order to perform a behavior en (Ajzen, 1991; Fishbein & Ajzen, 1975). It symbolizes the latter's expectations about a particular behavior. Zeithaml, Berry and Parasuraman (1996) in their examination of perceived service quality, suggested behavioral intention as willingness of customers to a) say positive things about the service provider, b) recommend the service provider to other consumers, c) remain loyal to it, e) spend more money, and e) pay a higher price. In a different context, Kim (2004) examined the role of perceived value and satisfaction with the products and services of online shopping in influencing behavioral intention. The aspects of intentions to purchase, revisit the website, search form information, say positive things, and recommend the website to others. Burton, Sheather, and Roberts (2003) acknowledged that customer experience is related to behavioral intention. They suggested that the more positive the customer's experience, the more likely he or she would reuse the service. In this study, behavioral intention is conceptualized to capture the intensity and likelihood to engage on activities in advocating as well as sharing of experience pertaining to the destination with others through words of mouth as well as social media. The two aspects are known as destination attachment and interpersonal engagement. On the basis of preceding findings, the following is hypothesized:

H1: Actual-congruity has a significant positive effect on experiential value

H2: Ideal self-congruity has a significant positive effect on experiential value

H3: Actual self-congruity has a significant positive effect on behavioral intention

H4: Ideal congruity has a significant positive effect on behavioral intention

H5: Experiential value has a significant positive effect on behavioral intention

3 METHODOLOGY AND DATA ANALYSIS

3.1 *Research design*

To test the hypothesis, questionnaires were distributed to visitors of island destinations to answer the questionnaire with regard to their perception of the current visit of the island destination. Thus, island destination was chosen as the research context for the current study.

3.2 *Sampling method and procedure*

Four hundred and twenty questionnaires were distributed using a survey methodology based on the convenience sampling. The study has used a survey approach with self-administered questionnaire distributed in island destinations at Tioman Island and Redang Island. The respondents were asked for information about their perception of island destination in terms of congruity between the destination image and their self-image and also the value they derived in terms of aesthetic, escapism, social value and economic value as a result of their visit at the island destination. Subsequently, they were also asked about specific behavior that may have occurred as a result of the perception. Finally, they were requested to provide demographic information about themselves. For a period of over four months, 364 questionnaires were collected and

finally there were 298 usable questionnaires to be included in the actual data analysis.

3.3 *Data analysis*

A two-phase of data analysis was used for the data analysis. As a general procedure, the data was first analyzed using exploratory data analysis (EFA) to identify the underlying structure of the constructs examined. Using varimax rotation the latent root criterion of 1.0 was used for factor inclusion, and a factor loading of 0.50 was used as the benchmark to include items in a factor. This analysis procedure was done to help to reduce or decrease multicollinearity or error variance correlations among indicators in the confirmatory factor analysis of the measurement model (Yoon & Uysal, 2005). Such errors should be avoided as much as possible in structural equation modeling procedures (Bollen, 1998). Eight factors with eigenvalue above 1.0 were generated which explained about 74 percent of the total variance. Each factor has yielded a reliability coefficient ranging from 0.80 to 0.90. Factor with Cronbach alpha value greater than 0.60, is considered as having good internal consistency (Hair, Babin, Money & Samouel, 2003; Malhotra, 2004).

Subsequently, the confirmatory factor analysis (CFA) was conducted to examine the psychometric properties of the measures. The measurement model test includes the estimation of convergent, unidimensionality and discriminant validity of the items in the measurement. The maximum likelihood was used as the estimation method for the analysis of this study. Indices such as Chi-square (χ^2), ratio of Chi-square to degrees of freedom, root mean square error of approximation (RMSEA), goodness of fit index (GFI), normed fit index (NFI), and comparative fit index (CFI) were adopted for model fit criteria. The structure equation modeling (SEM) was conducted after CFA to examine the relationships among the self-congruity, experiential value, and behavioral intention.

4 RESULTS

4.1 *Descriptive statistics*

Results of demographic profile of tourists found that the composition of gender were almost equal. Male were shown as 52% and female were 48%. They were mostly at the age from 21 years to 30 years, which constitutes about 66% of the respondents. In terms of marital status, the result shows that the respondents were mainly single. It is reflected by 70% of the respondents. Most of the respondents were from the European countries such as German and also France, which constitute about 35% of the total respondents.

4.2 Measurement model analysis

The first stage in the two-phase of data analysis is the estimation of the measurement model. One measurement model analysis was conducted to assess the measurement properties of the latent constructs. The measurement model consists of two-factor structure of self-congruity, four-factor structure of experiential value and two-factor structure of behavioral intention. The measurement model shows acceptable fit:

CFA: Chi-square = 1093, df = 426, (χ^2/df = 2.5), p < .001, root mean square error of approximation (RMSEA = .074), goodness of fit index (GFI = 0.8), normed fit index (NFI = 0.81), and comparative fit index (CFI = 0.87).

4.3 Validity

The assessment of the Cronbach alpha and the composite reliability has revealed good internal consistency and reliability. All the constructs have values exceeded the minimum benchmark of 0.70. (Hair, et al., 2003). The average variance extracted (AVE), which assesses the amount of variance captured by the constructs' measures relative to measurement error and the correlations among the latent constructs in the model has values range from 0.67-0.86. The values exceeded the minimum recommended level of 0.50, which indicates the convergent validity is achieved. Table 1 summarizes the result of the measurement model. The reliability of the variables was also achieved composite reliability value greater than the threshold value of 0.60 (Bagozzi & Yi, 1988) and Cronbach alpha greater than the cut-off value of 0.70 (Anderson & Gerbing, 1988). To test for discriminant validity, the correlation

matrix of latent variables was examined. The correlations between each latent variables were lower than 0.80, a cut-off point for discriminant validity, which in this measurement model it indicates discriminant validity was established (Yanamandram & White, 2006). In conclusion, it is reasonable to claim that all the measures used in this study possess adequate psychometric properties. The measures of the proposed constructs have achieved satisfactory reliability, convergent, and discriminant validity.

4.4 Structural model test

The structural model was tested to assess the hypothesized structural relationships of the four constructs (Refer to Figure 1). The results revealed that the structural model has a significant χ^2 value (χ^2 = 558, df = 181, χ^2/df = 3.08, p<.001) indicating inadequate fit of the data with the hypothesized model. Based on the suggestion by Hair et al. (1998), reliance on the chi-square test as the sole measure of fit in not recommended due to its sensitivity to sample size. Hence, alternative fit indices were used as the test for model fit. Based on the result of other fit indices (RMSEA = .06, GFI = 0.91, NFI = 0.94, and CFI = 0.96), it was shown that the model fits the data satisfactorily. Hence, the study's attempt to establish a plausible model that has statistical and explanatory power, which could permit confident interpretation of results, was thus fulfilled. Therefore, the model was accepted as reasonable and the hypothesis tests were interpreted. Table 1 illustrates in detail the results of the hypothesized model.

5 DISCUSSION AND CONCLUSION

The findings suggest that three of the hypotheses in the study were supported by the data. The hypothesis test results indicated that actual self-congruity and ideal self-congruity has a significant relationship with experiential value. The two constructs were found to have no effect on behavioral intention. In addition, experiential value also has a significant relationship with behavioral intention. In addition, actual self-congruity has a stronger effect on experiential value. Therefore, in the current study, experiential value not only has a direct effect on behavioral intention, but also mediates the actual self-congruity as well as ideal self-congruity and behavioral intention relationship. Hence, this result supports previous studies, which found that there is no direct relationship between self-congruity and consumer behavior. For example, several studied have found that such relationship was mediated by satisfaction (Jamal & Goode, 2001;

Table 1. Results of tested hypotheses.

Hypotheses and hypothesized paths		Standardized coefficient	t-value	Results
H1	Actual self-congruity →Experiential value	0.4	5.0	Supported
H2	Ideal self-congruity → Experiential value	0.36	4.5	Supported
H3	Actual self-congruity →Behavioral intention	(0.08)	(1.00)	Not supported
H4	Ideal self-congruity → Behavioral intention	(0.03)	(0.36)	Not supported
H5	Experiential value →Behavioral intention	0.84	6.8	Supported

Magin, Algesheimer, Huber & Herrmann, 2003) and functional congruity (Sirgy & Samli, 1985). In destination image, Yusof, Musa and Putit (2013) found that experiential value mediates the relationship with self-congruity and behavioral intention.

The findings of this study provide several managerial implications for actual self-congruity, ideal self-congruity, and experiential value strategies in the tourism industry. Firstly, apart from the emphasis by many marketers on the actual self-congruity and ideal self-congruity, experiential value initiative cannot be undervalued. This suggests that actual and ideal self-congruity and experiential value is important in the tourists' experiences that were translated into their behavioral intention. For this reason, tourism operators have to provide the destination environment that matches the actual as well ideal image of the consumers. As shown in the present study, more interesting features may be an effective way to offer experiential value to the tourists, which in turn enhances their behavioral intention. Since the experiential value is obtained from the exposure of the destination, the interesting features of the destination with the intention to uplift the tourists' experiential value has to continually be upgraded. Thus, the government should consider investing in more interesting features of the destination in order to attract tourists to come to the country.

REFERENCES

Ajzen, I. (1991). The theory of planned behavior. *Organizational Behavior and Human Decision Processes, 50*(2), 179-211.

Anderson, J. C. & Gerbing, D. W. (1988). Structural equation modeling in practice: A review and recommended two-step approach. *Psychological bulletin, 103*(3), 411-423.

Bagozzi, R. P. & Yi, Y. (1988). On the evaluation of structural equation models. *Journal of the Academy of Marketing Science, 16*(1), 74-94.

Baron, S., Harris, K. & Harris, R. (2001). Retail Theater The "Intended Effect" of the Performance. *Journal of Service Research, 4*(2), 102-117.

Bollen, K. A. (1998). *Structural equation models*. New York: Wiley.

Burton, S., Sheather, S. & Roberts, J. (2003). Reality or perception? The effect of actual and perceived performance on satisfaction and behavioral intention. *Journal of Service Research, 5*(4), 292-302.

Chang, C. (2001). TThe impact of personality differences on product evaluations. *Advances in consumer research, 28*(1), 26-33.

Ericksen, M. K. (1997). Using self-congruity and ideal congruity to predict purchase intention: a European perspective. *Journal of Euromarketing, 6*(1), 41-56.

Fishbcin, M. & Ajzen, I. (1975). *Belief, attitude, intention and behavior: An introduction to theory and research*. Addison-Wesley: Reading, MA.

Govers, P. C. & Schoormans, J. P. (2005). Product personality and its influence on consumer preference. *Journal of Consumer Marketing, 22*(4), 189-197.

Graeff, T. R. (1996). Image congruence effects on product evaluations: The role of self-monitoring and public/private consumption. *Psychology & Marketing, 13*(5), 481-499.

Hair, J. F., Babin, B., Money, A. H. & Samouel, P. (2003). *Essential of business research methods*: John Wiley & Sons.

Helgeson, J. G. & Supphellen, M. (2004). A conceptual and measurement comparison of self-congruity and brand personality-The impact of socially desirable responding. *International Journal of Market Research, 46*(2), 205-233.

Jamal, A. & Goode, M. M. (2001). Consumers and brands: A study of the impact of self-image congruence on brand preference and satisfaction. *Marketing Intelligence & Planning, 19*(7), 482-492.

Kim, M. (2004). *The Role of Self-and Functional Congruity on Online Retail Patronage Behavior*. (Unpublish Doctoral dissertation), University of Tennessee, Knoxville.

Litvin, S. W. & Kar, G. H. (2004). Individualism/collectivism as a moderating factor to the self-image congruity concept. *Journal of Vacation Marketing, 10*(1), 23-32.

Magin, S., Algesheimer, R., Huber, F. & Herrmann, A. (2003). The impact of brand personality and customer satisfaction on customer's loyalty: theoretical approach and findings of a causal analytical study in the sector of internet service providers. *Electronic Markets, 13*(4), 294-308.

Malhotra, N. K. (2004). *Marketing research: An applied orientation* (4th ed.). New Jersey: Prentice Hall.

Mathwick, C., Malhotra, N. & Rigdon, E. (2001). Experiential value: conceptualization, measurement and application in the catalog and Internet shopping environment*. *Journal of retailing, 77*(1), 39-56.

Musa, R., Muhammad Kassim, R.-n. & Putit, L. (2012). Disentangle the factorial structure of Ecotourism Experiential Value (EEV) and its effects on Total Ecotourist Experience Quality (TEEQ) and Eco-tourist Behavioural Intentions (EBI): An analysis of Taman Negara National Park. *Universiti Teknologi MARA*.

Phau, I., Shanka, T. & Dhayan, N. (2010). Destination image and choice intention of university student travellers to Mauritius. *International Journal of Contemporary Hospitality Management, 22*(5), 758-764.

Sirgy, M. J., Johar, J., Samli, A. C. & Claiborne, C. (1991). Self-congruity versus functional congruity: predictors of consumer behavior. *Journal of the Academy of Marketing Science, 19*(4), 363-375.

Sirgy, M. J. & Samli, A. C. (1985). A path analytic model of store loyalty involving self-concept, store image, geographic loyalty, and socioeconomic status. *Journal of the Academy of Marketing Science, 13*(3), 265-291.

Sirgy, M. J. & Su, C. (2000). Destination image, self-congruity, and travel behavior: Toward an integrative model. *Journal of travel research, 38*(4), 340-352.

Stepchenkova, S. & Morrison, A. M. (2006). The destination image of Russia: From the online induced perspective. *Tourism management, 27*(5), 943-956.

Usakli, A. (2009). *The Relationship between destination personality, self-congruity, and behavioral intentions*. (Unpublished Doctoral dissertation), University of Nevada, Las Vegas.

Yanamandram, V. K. & White, L. (2006). *Exploratory and confirmatory factor analysis of the perceived switching costs model in the business services sector.* Paper presented at the Australia and New Zealand Marketing Academy Conference (ANZMAC 2006), Brisbane, Queensland, 4-6 December 2006.

Yoon, Y. & Uysal, M. (2005). An examination of the effects of motivation and satisfaction on destination loyalty: a structural model. *Tourism management, 26*(1), 45-56.

Yusof, J. M., Musa, R. & Putit, L. (2013). Mediating role of ex-periential value in self-congruity and behavioral intention relationship. *International Journal of Business and Management Studies, 2*(3), 109-121.

Zeithaml, V. A., Berry, L. L. & Parasuraman, A. (1996). The behavioral consequences of service quality. *Journal of marketing, 60*(2), 31-46.

Theory and Practice in Hospitality and Tourism Research – Radzi et al. (Eds)
© 2015 Taylor & Francis Group, London, ISBN 978-1-138-02706-0

Corporate sponsorship and intention to sponsor sporting events in Malaysia

S. Tarmudi, I.S. Bahar, M.F.S. Bakhtiar, S.A. Jamal, N. Othman & I.R. Razak
Faculty of Hotel and Tourism Management, Universiti Teknologi MARA, Shah Alam, Malaysia

ABSTRACT: The involvement of corporate bodies in sponsoring sports events in Malaysia plays an important part in the success of sports events in attracting a large number of audiences and achieving their event objectives. This paper aims to determine the objectives of corporate sponsorship, the most important classification of corporate sponsorship as well as clarification about the relationship between corporate sponsorship objectives and intention to sponsor. Using self-administered questionnaire adapted from previous studies, several sports events in Malaysia sponsored by several corporate bodies in Kuala Lumpur will be selected as samples to accomplish the objectives of this study. This study will contribute to respective parties since there are very limited studies on corporate sponsorship and its intention to sponsor sports events in Malaysia particularly. Apart from that, corporate bodies sponsorship contribution in any level of sports events in Malaysia will help to increase the country's economic growth through tourism industry as well as to promote Malaysia as a tourist destination and to boost international audience.

Keywords: Corporate sponsorships, intention to sponsor, sports events, Malaysia

1 INTRODUCTION

Malaysia's tourism industry is the second largest foreign exchange earner after manufactured goods. It is also the seventh largest contributor to the Malaysian economy. The total of tourist arrival to Malaysia for the first half of 2013 was 12, 552,731; an increase of 7.9 percent from year 2012 with a total RM60.6 billion receipt (Tourism Malaysia, 2013) Major Events has contributed RM0.4 billion incremental Gross National Income (GNI) providing 8,000 jobs in sports, arts, cultural and lifestyle events, homegrown and home-hosted events.

In Malaysia, the importance of sports development and sports event tourism were emphasized during Budget 2013. As such, an allocation RM738 million was provided for youth and sports development. RM239 million were allocated for sports development in Budget 2014 while RM150 million to the Sports Trust Fund for the development of elite sports, medical treatment and dedicated government agency with the purpose of managing various events including sports events to generate Malaysia's tourism revenue especially in international events under Malaysia Convention and Exhibition Bureau (MyCEB). MyCEB is an agency under Malaysia's Ministry of Tourism and Culture. Major sports events in Malaysia include Formula Grand Prix, Le Tour de Langkawi and MotoGP to name a few.

In conjunction with Visit Malaysia 2014, the Ministry of Tourism and Culture has geared up high—impact programs to increase Malaysia's economic growth through tourism industry. Hosting international and major events is one of the strategies to promote Malaysia as a tourist destination in order to boost international audience such as the Malaysian Grand Prix series, various existing homegrown events were repackaged and clustered. Tourist expenditure of 916 million was received from 19 events in 2012. To ensure the successful of Visit Malaysia Year 2014, each and every Malaysian plays an important role which includes the role played by corporate bodies or private sectors.

Sport sponsorship is an exchange relationship involving sport organizers, corporations and others that play in between which is a basic form of sponsorship. These interdependent relationships are predicated on the understanding that all parties involved will benefit from the exercise with minimum risk According Thwaites (1995) it includes sponsoring major sporting events which have an advantage of being attended and watched by large amount of people that makes it easier to communicate with the target group in the market as well as use the opportunity of significant media coverage.

A number of corporations in Malaysia have been involved in sports sponsorship for quite a long time. They are Government Linked Companies (GLCs) such as Khazanah Nasional, Government

Linked Investments Companies (GLICs) such as PNB and corporate bodies or private sectors such as CIMB Bank, Nestles' Milo and 100plus.

The importance of sporting events was acknowledged where the government intends to expand sports tourism offerings in Malaysia beyond hosting events such as golf tourism included in Malaysian Economic Transformation Programme (ETP) in order to increase tourist arrival and Malaysia's GNI. The government of Malaysia has also recognized the impact of corporate bodies in sports development in the country. The world of sports without the support of corporations would collapse (Irwin, 1993). Thus, the purpose of this study is to examine the relationship between corporate sponsorship objectives; public awareness, corporate image and community involvement towards intention to sponsor sports events and to explore the relative importance and most influential corporate sponsorship objectives to the corporate sectors.

2 LITERATURE REVIEWS

2.1 *Corporate sponsorship objectives*

Corporate sponsorship objective is defined as the successfulness of an event being sponsored by a company and how useful it is and how well the end result can be evaluated (Abiodun, 2011). These objectives set the frame through which a sponsorship should be selected in order to reach those objectives (Ivarsson & Johansson, 2004).

Thwaites (1995) proposed that one of the features of sponsorship is its ability to contribute to a wide range of objectives at both corporate and brand level. Among the many related objectives would include increasing product or brand awareness, reinforce or alter market perception of both product and brand and identifying it with a particular market segment (Ivarsson & Johansson, 2004; Meenaghan, 1983). Through achievement of multiple objectives stated earlier is what sponsorship companies would expect (IEG, 2008). Meenaghan (1983) also stated in his studies about how a specific sport sponsorship may be selected is referred to the ability to fulfill objectives through the sponsorship. Companies must have apparent objectives and reasons for their actions at all times before committing themselves to sponsor an activity (Shank, 1999). It is also worth stating that Javalgi Traylor, Gross and Lampman (1994) suggest that the objectives tend to be unclear.

Corporate objectives are the basis for the list development of sponsorship benefits (Buhler, Chadwick & Nufer, 2009) that vary from organization to organization, from industry to industry, from large organization to small and from one author to another in terms of company's involvement in sport sponsorship (Buhler et al., 2009; Dolphin, 2003) (Abiodun, 2011). Corporate organizations view sponsorship relationship as an activity that will benefit both the corporate organizations, as well as the sport federations (Benadie, 2005). Javalgi et al. (1994) opined that sponsorship is the underwriting of a special event with the object of supporting organizational objectives by enhancing corporate image, increasing awareness of brands and leverage corporate reputation.

2.2 *Public awareness*

An organization engage in sponsorship is due to a number of factors. Sponsorship increases public awareness of the organizations product or brand, altered public perception, enhancing organizations products and brand image as well as enhancing the sales (Verity, 2002). According to Dolphin (2003), brand awareness will unsurprisingly boost from sponsorships. Profile of the corporate brand will then increase the value of the brand through brand awareness and sponsorships might also arouse the products and services. Corporate reputation is an intangible cost of corporation's main focus in business. Verity (2002) proposed that generating brand awareness accrues naturally from sponsorship. However Hoek, Gendall, Jeffcoat, and Orsman (1997) suggested that the benefits of sponsorship do not accrue automatically.

Mullin, Hardy, and Sutton (2007) included public awareness of a company under the corporate-related objectives while increase product/brand awareness is listed under product-related objectives. In IEG list of why companies sponsor, they listed 'to create awareness' as one of the reasons for companies to sponsor.

2.3 *Corporate image*

Hall (1993) reports that company reputation is the intangible resource most focused on by CEOs. Image and reputation can be seen as resources that enable a company to secure a competitive advantage (Amis, Slack & Berrett, 1999). Sport sponsorship is a useful tool to enhance or even to change the company's brand or the company's image status which has been proven by authors (Abiodun, 2011). Sponsorship might have an impact on the image of the corporate bodies (Dolphin, 2003). A specific brand or a company perception by a particular market segment can be influenced by sponsorships that might enhance the image of the corporate bodies. Meenaghan (1983) and K. Gwinner (1997) listed corporate image as one of the two most common objectives to sponsorship engagement where it can help to establish, strengthen or change brand image.

Corporate or brand positioning-related objectives are predominant goals in sponsorship arrangements because they directly relate to the alignment with the image of the sponsorship property (Fahy, Farrelly & Quester, 2004). Sponsorship is particularly useful in attaining brand awareness and brand attitude/image objectives as suggested by (Meenaghan, 1983). Hoek et al. (1997). K. P. Gwinner and Eaton (1999) also support this statement; they stated that sponsorship can enhance or change companies' reputation and brand image. Adjustment and strengthening image is what that makes company wants to be involved in sponsorship (Abiodun, 2011). For example, Coca-cola sponsored the Helsinki Cup to reinforce its relationship to youth.

Amis et al. (1999) pointed out that a sponsorship should produce an outcome that matches well with the image that the sponsor is trying to convey. Any sponsorship should therefore produce an image which is so superior that it clearly differentiates the firm from its competitors.

2.4 *Community involvement*

Community involvement has been stated as an important corporate objective. Sponsors involve themselves in improving the life of the community, either at local or national level in order to show their manner of good citizenship (Meenaghan, 1983). Organizations that participate in sponsorship deals with the aims of revolving around community support, and will give back to the public that supports them (Walker, 2007) and it is an excellent mechanism that will boost the community involvement (Mack, 1999). David (2001) stated that another way of introducing products and services directly to the market is through sponsorship and continuously talking about the improvement of community involvement.

Mullin et al. (2007) have also listed community involvement amongst the corporate objectives. David (2001) stated that companies get involved in sponsorship because of social responsibility. Companies' reputation in its community may be increased through sponsorship. Eventually, customers will perceived the image of the company as a caring and socially responsible. This is also supported by IEG (2008) who listed that showcase community responsibility is one of the reasons why companies sponsor. In fact, customers are willing to spent on the company's products and services based on corporate citizenship.

3 METHODOLOGY

A descriptive design using a quantitative research approach through cross-sectional study will be used. A self-administered questionnaire will be designed and distributed personally to the CEOs, managers or personnel in charge of sport sponsorship of the corporate bodies that has already sponsored sport events or currently involved in any level of sports event sponsorship in Kuala Lumpur will be chosen for data collection using convenience sampling technique. A total of valid respondents will be gathered and analyzed using Statistical Package for Social Science.

4 FINDINGS AND DISCUSSION

4.1 *Corporate objectives*

The aim of this paper is to gain a thorough understanding on companies' sponsorship selection criteria of sports events. Numerous research (Abiodun, 2011); Amis et al. (1999); (Mullin et al., 2007; Verity, 2002) have identified most important corporate objectives when choosing a sponsorship.

Corporate objectives are the main priority why companies have an intention to sponsor sporting events and eventually involve in sponsorship relationship (Dolphin, 2003). Initial literature analysis suggested that corporate involvement in sport sponsorship is growing. Public awareness, corporate image and community involvement were mostly listed in the top five importance ratings of sponsorship selection criteria and objectives of previous finding and results studies (Hartland, Skinner & Griffiths, 2005). Most sponsors are more interested to sponsor events that can increase their companies as well as products and services awareness to the public, enhancing its company image in addition to increasing involvement of the community in that particular event. The sponsors expected that sponsorship activities will manage to deliver its corporate objectives.

In contrast, previous studies listed marketing objectives and personal objectives among other sponsorship objectives (Ivarsson & Johansson, 2004). However, both objectives are listed as less important than corporate objectives. The data analysis will reveal the most important and the most influential corporate objective among the three dimensions that will create an intention to sponsor sports events. The study will also reveal Malaysia's corporate profile in sports events sponsorship such as sports mostly sponsored and its sports events sponsorship histories.

5 CONCLUSION

Overall, this study is assumed to reveal that all companies have their own corporate objective in selecting a sponsorship. However, it is revealed that not all companies share the same objectives or have constant sponsorship objectives. Therefore, the sponsorship selection criteria will be based on the most important objective that gives the best interest to the company. Less similarity between the organizers and potentials sponsors will somehow lessen the opportunity to receive sponsorship deals. Based on previous studies, most findings listed the three objectives discussed earlier as important objectives for a company in doing sponsorship. They were likely to sponsor sports events if the events fit its company's corporate objectives; hence event or sports organizers should take the opportunity to gain better understanding on its potential sponsors objectives before going into any sponsorship activities which eventually identify the relative importance of various issues and objectives in determining sponsorship participation of corporate organizations. Thus, this study should be valuable to sports event organizers, corporate sponsors and as well as towards the development of sports tourism in Malaysia respectively.

ACKNOWLEDGEMENTS

This research is funded by Universiti Teknologi MARA through Research Faculty Grant 600-RMI/RAGS 5/3 (138/2013).

REFERENCES

Abiodun, O. (2011). The Significance of Sponsorship as a Marketing Tool in Sport Events. *International Business. Finland.*

Amis, J., Slack, T. & Berrett, T. (1999). Sport sponsorship as distinctive competence. *European journal of Marketing, 33*(3/4), 250-272.

Benadie, S. (2005). Relationships in sport sponsorship: A marketing perspective.

Buhler, A., Chadwick, S. & Nufer, G. (2009). *Relationship marketing in sports*: Routledge.

David, J. (2001). Principles and Practice of marketing. *McGraw Publishing Company* (5).

Dolphin, R. R. (2003). Sponsorship: perspectives on its strategic role corporate communications. *An International Journal, 8*(3), 173-186.

Fahy, J., Farrelly, F. & Quester, P. (2004). Competitive advantage through sponsorship:A conceptual model and research propositions. *European journal of Marketing, 38*(8), 1013-1030.

Gwinner, K. (1997). A model of image creation and image transfer in event sponsorship. *International marketing review, 14*(3), 145-158.

Gwinner, K. P. & Eaton, J. (1999). Building brand image through event sponsorship: the role of image transfer. *Journal of advertising, 28*(4), 47-57.

Hall, R. (1993). A framework linking intangible resources and capabiliites to sustainable competitive advantage. *Strategic management journal, 14*(8), 607-618.

Hartland, T., Skinner, H. & Griffiths, A. (2005). Tries and conversions: are sports sponsors pursuing the right objectives? *International Journal of Sports Marketing & Sponsorship, 6*(3).

Hoek, J., Gendall, P., Jeffcoat, M. & Orsman, D. (1997). Sponsorship and advertising: a comparison of their effects. *Journal of Marketing Communications, 3*(1), 21-32.

IEG. (2008). Guide to Why Companies Sponsor Retrieved November 12, 2013 from http://stjude.org/SJFile/alsac_ieg_ guide_why_companies_sponsor.pdf

Irwin, D. (1993). In search of sponsors. *Athletic Management, 5*(2), 11-16.

Ivarsson, C. & Johansson, M. (2004). Sport sponsorship: as a promotional tool. *International Business and Economics Programme.*

Javalgi, R. G., Traylor, M. B., Gross, A. C. & Lampman, E. (1994). Awareness of sponsorship and corporate image: An empirical investigation. *Journal of advertising, 23*(4), 47-58.

Mack, R. W. (1999). Event sponsorship: An exploratory study of small business objectives, practices, and perceptions. *Journal of Small Business Management, 37*(3), 25-30.

Meenaghan, J. A. (1983). Commercial sponsorship. *European journal of Marketing, 17*(7), 5-73.

Mullin, B. J., Hardy, S. & Sutton, W. A. (2007). *Sport marketing* (Vol. 13): Human Kinetics.

Shank, M. D. (1999). *Sports marketing: A strategic perspective* (Vol. 341): Prentice Hall Upper Saddle River, NJ.

Thwaites, D. (1995). Professional football sponsorship-profitable or profligate? *International Journal of Advertising, 14*(2), 149-164.

Tourism Malaysia. (2013). Tourists Arrivals & Receipts to Malaysia Retrieved October 5, 2013, from http://corporate.tourism.gov.my/research.asp?page=facts_figures

Verity, J. (2002). Maximising the marketing potential of sponsorship for global brands. *European Business Journal, 14*(4).

Walker, J. R. (2007). *The restaurant: from concept to operation*: John Wiley and Sons.

Theory and Practice in Hospitality and Tourism Research – Radzi et al. (Eds)
© *2015 Taylor & Francis Group, London, ISBN 978-1-138-02706-0*

The time-satisfaction relationship of interpretation topics in Yehliu Geopark

C.H. Chu, Y.J. Guo, C.C. Chen, C.H. Hsu, Y.H. Wang & Y.C. Wang
Aletheia University, Taiwan, R.O.C.

ABSTRACT: One of the major missions of Yehliu Geopark is to convey educational and inspiring information of geology and environment to visitors. Unfortunately, in many cases, although the interpreters own plenty knowledge about exhibitions in their mind, they have very limited time to convey their knowledge within the tolerant time duration of audiences. For organizing the interpretation contents in adequate length, exploring the time-satisfaction relationships for interpretation topics is very crucial for the Geopark. Hence, this research aimed to explore the time-satisfaction relationships of interpretation topics in Yehliu Geopark. The empirical analyses showed that, among the 17 standardized interpretation topics, the satisfaction and interpretation time had a significantly positive correlation across 11 topics. Time-satisfaction relationships were not statistically significant for the remaining six topics. The findings of this study suggested that the interpreters should focus on enriching the interpretation contents of the 11 topics with positive time-satisfaction correlation to improve the experience and the effect upon the geological and environmental education for the visitors to Yehliu Geopark.

Keywords: satisfaction, Geopark, interpretation, environmental education

1 INTRODUCTION

Since the Declaration of the United Nations Conference on the Human Environment was issued in 1972, environmental education has already become one of the international topics and responsibilities of the nations around the world.

As a result, Taiwan, in 2010, in order to increase the sense of environmental protection of citizens to achieve the goal of sustainable development, Taiwan has passed ' the educational law of the environment ' in 2010. This legislation shows the attention on and commitment to the environmental protection of the Taiwan Government.

Yehliu Geopark, located in North Coast of Taiwan, has become a popular site for sightseeing and outdoor education due to the landform of the coast and rich natural resources including animals, plant, ecosystem, and a variety of geological landscape. One of the major missions of a Geopark is to convey educational and inspiring information of geology and environment to visitors. Providing a well-designed interpretation service is crucial to Geoparks.

For an interpretation service provider, how to design interpretation contents that cater for the needs of audiences for a guided walk is an important issue. In many cases, interpreters own plenty knowledge about exhibitions in their mind; however they have very limited time to convey their knowledge within the tolerant time duration of audiences.

On the other hand, some studies suggest that the more contents do not guarantee higher interpretation quality (1998). Hence, for organizing the interpretation contents in adequate length, exploring the time-satisfaction relationship for interpretation topics is very crucial for the Geopark.

2 LITERATURE REVIEW

Interpretation is an educational activity which aims to reveal meanings and relationships through the use of original objects, by firsthand experience, and by illustrative media, rather than simply to communicate factual information (Tilden & Heritage, 1977). Mahaffey (1970) described interpretation as a process of activity, which communicates the concepts between the human and the environment.

Interpretation can be a manner of managing the recreational sites. Interpretation provides the functions of increasing recreational benefits, protecting the resources, protecting the visitors, and enforcement (Sharpe, 1982). Providing information and education are two of the major missions of interpretation (Grinder & McCoy, 1985). For the design of a guided tour, Tilden and Heritage (1977) mentions "Nothing in Excess" which implies too much information conveyed may confuse visitors and make them impatient. Beck and Cable (1997) accordingly proposes 15 principles for interpretation and argues that the interpretation contents should be designed in appropriate quantity.

However, the past research about interpretation is more concentrated on the qualitative principle on interpretation. The quantitative research on the optimizing the design of an interpretational tour is still relative rare. For the optimization method, understanding the relationship between resource input and output is important. In interpretation design, the input resource is the interpretation time, and the output is tourist's satisfaction by the tourists. Accordingly, this paper aims at exploring the relationship between interpretation time and satisfaction.

3 METHODOLOGY

3.1 Research design

This study was conducted by using a three-step procedure. First, a list containing 17 topics, which covers subjects of geological landscapes, local cultures, stories, legends, and plants, was identified by interviewing the interpreters. Second, the respondents were asked to answer a questionnaire including the measurement of their perceived satisfaction with each interpretation topic, visitor characteristics (age, gender, place of residence, the purposes of visiting, who are their companion, and how do they know the Geopark) via face-to-face survey. Third, the interviewers listened to the interpretation contents that were recorded by the recording pen attached on the interpreters during the whole process of the guided walks and then investigate the time spent on each topic.

3.2 Data collection

Visitors, who received the Chinese interpretation service, to Yehliu Geopark were surveyed over a 3- month period from July to September in 2011 including weekdays and weekends, as well as national holidays. The respondents were asked to score their perceived satisfaction between 1 and 100 for each topic. A reference spectrum of the scoring (i.e., 1–20 refers to very low satisfaction, 20–40 refers to low satisfaction, 40–60 refers to neutral satisfaction, 60–80 refers to high satisfaction, and 80–100 refers to very high satisfaction) was also provided to ensure that the respondents scored at similar standard. Colored photographs were also attached beside the corresponding topics

to help respondents recalling the interpretation contents.

A total of 550 visitors from 26 guided walks leaded by 21 interpreters were invited participate in the survey, and 500 visitors accepted the invitation. Eventually, 496 valid and usable questionnaires were received.

3.3 Time-satisfaction relationship

In this study, a logistic curve is used to fit the time-satisfaction relationship for each topic. The logistic curve used in this study can be expressed as Equation 1:

$$S_i = \frac{100}{1 + e^{a - rt_i}} \tag{1}$$

where S_i = satisfaction of ith topic; a, r = statistical parameters; and t_i = interpretation time of ith topic.

4 RESULT

4.1 Respondent profile

In total, the respondents included 250 (50.4%) males and 246 (49.6%) females. Based on age group, 31.1 percent of the respondents were 16 and younger, 34.1 percent in the 16–30 age groups which was the largest age group, 26.7 percent aged 31–45, 5.9 percent aged 46–60, and 2.2 percent aged 61 and over. Based on where they live, the respondents from the Northern Taiwan was the largest group (40.4%), with 29.1 percent from Middle Taiwan, 16.8 percent from Southern Taiwan, 11.3 percent from Eastern Taiwan, and 2.4 percent from China and overseas.

Yehliu Geopark is one of the best sites for environmental and geological education in Taiwan. Thus it has been selected by many elementary schools and high schools as an outdoor teaching place. In our survey, 223 (36%) respondents were students who went to the Geopark for outdoor teaching. 32.3 percent of the respondents went to Yehliu Geopark with their family, colleagues and friends.

Regarding the time of the interpretation service, the optimal length for the majority (68%) is 30–90 minutes. Only 25.3 percent of the respondents prefer 90–150 minutes. Only 3.7% prefers 151 minutes or longer.

4.2 Interpretation time of each topic

By listening to the audio files that recorded the whole interpretation contents for each guided walk,

Table 1. The interpretation time of each topic (unit: sec.).

Topic	Max.	Min.	Avg.	Std.
1. Causes for the formation of cuesta	95	40	75	16.13
2. The story of cuesta	75	20	57	17.92
3. The characteristics of Coral Tree	90	20	48	17.92
4. The characteristics of Agar	120	10	43	27.06
5. Causes for the formation of Sea Cave	130	20	73	19.48
6. The story of Sea Cave	95	20	58	20.50
7. Causes for the formation of mushroom rocks	183	60	111	35.18
8. Princess Rock	180	55	89	30.89
9. Japanese Geisha Rock	410	40	89	70.16
10. Causes for the formation of Candle Shaped Rocks	180	60	109	33.58
11. The story of Candle Shaped Rocks	120	10	62	28.08
12. Ice Cream Shaped Rock	110	20	55	21.73
13. Brain Shaped Rock	88	32	63	15.81
14. The statue of Tianzhen Lin	200	45	97	52.68
15. Speed Testing Stand	200	20	74	32.65
16. Lover's Cave	150	20	78	30.47
17. Queen's Head Shaped Rock	140	22	68	30.01

Table 2. The average satisfaction score of each topic.

Topic	Avg	Std.
Causes for the formation of cuesta	76.33	12.76
The story of cuesta	73.34	14.99
The characteristics of Coral Tree	69.27	16.90
The characteristics of Agar	68.08	16.34
Causes for the formation of Sea Cave	76.68	13.52
The story of Sea Cave	74.62	15.17
Causes for the formation of mushroom rocks	81.20	12.66
Princess Rock	78.12	16.70
Japanese Geisha Rock	73.48	15.55
Causes for the formation of Candle Shaped Rocks	78.88	14.25
The story of Candle Shaped Rocks	75.02	17.38
Ice Cream Shaped Rock	75.21	13.78
Brain Shaped Rock	76.96	15.35
The statue of Tianzhen Lin	84.19	13.15
Speed Testing Stand	71.52	15.59
Lover's Cave	73.55	16.76
Queen's Head Shaped Rock	82.22	22.20

Table 3. The parameters calibration results.

Topic	a	r
1. Causes for the formation of cuesta	-0.832[*]	.004[*]
2. The story of cuesta	-0.184[*]	.014[*]
3. The characteristics of Coral Tree	-.040[*]	.015[*]
4. The characteristics of Agar	-0.357[*]	.008[*]
5. Causes for the formation of Sea Cave	-1.123[*]	.000
6. The story of Sea Cave	-0.359[*]	.012[*]
7. Causes for the formation of mushroom rocks	-1.270[*]	.001
8. Princess Rock	-1.316[*]	-.000
9. Japanese Geisha Rock	-1.010[*]	.000
10. Causes for the formation of Candle Shaped Rocks	-0.470[*]	.007[*]
11. The story of Candle Shaped Rocks	-0.144[*]	.015[*]
12. Ice Cream Shaped Rock	-0.869[*]	.004[*]
13. Brain Shaped Rock	-0.471[*]	.011[*]
14. The statue of Tianzhen Lin	-1.441[*]	.002[*]
15. Speed Testing Stand	-0.838[*]	.001
16. Lover's Cave	-1.155[*]	-.001
17. Queen's Head Shaped Rock	-0.773[*]	.011[*]

*Significance level of 5%.

the interpretation time of each topic can be identified as shown in Table 1. The total time spent on a guided walk was around 1 to 1.5 hr. The topics that took the longest average time were "causes for the formation of mushroom rocks" and "causes for the formation of candle shaped rocks" whereas the topic that took the shortest average time was "the characteristics of Agar". This result makes sense because mushroom rocks and candle shaped rocks are the most significant geological landscapes in the park.

4.3 Satisfaction of each topic

For understanding the relative importance and the performance of each topic, respondents are asked to score their perceived importance and satisfaction for the interpretation topics from 1 to 100 under a pre-given scoring reference as mentioned in Section 2.2. The result of the investigation of satisfaction is shown in Table 2 and Table 3.

Based on the observation of the average satisfaction score in Table 2, one may find that "the statue of Tianzhen Lin" receives the highest score. The second highest satisfaction score is obtained by "Queen's Head Shaped Rock". "Causes for formation of mushroom rocks" is the third satisfied topic.

4.4 The time-satisfaction relationships

A tendency analysis illustrated on Figure 1 shows a positive correlation between interpretation time and satisfaction among the 17 interpretation topics of Yehliu Geopark.

Table 3 shows the results of calibrating the parameters in Equation 1. Theoretically, r is positively related to the marginal effect of interpretation time. The higher value of r, the higher marginal contribution brought by interpretation time to satisfaction.

115

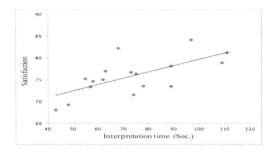

Figure 1. A scatter plot for interpretation time and satisfaction.

As can be seen in Table 3, the time-satisfaction relationship is significantly positive for the majority (11/17 = 0.647) of interpretation topics implying that the higher satisfaction can be brought by providing more contents when interpreting these topics. The satisfaction of the 6 topics including "Causes for the formation of Sea Cave", "Causes for the formation of mushroom rocks", "Princess Rock", "Japanese Geisha Rock", "Speed Testing Stand" and "Lover's Cave" does not significantly affected by interpretation time implying that the current content of these topics is abundant enough to the visitors. The more time spending on interpreting these six topics would not bring higher satisfaction.

5 CONCLUDING REMARKS

In this study, we analyzed the time-satisfaction relationship of the interpretation topics for Yehliu Geopark. According to our analyses, a positive time-satisfaction relationship exists in the majority of interpretation topics. However, 6 topics including "Causes for the formation of Sea Cave", "Causes for the formation of mushroom rocks", "Princess Rock", " Japanese Geisha Rock", "Speed Testing Stand" and "Lover's Cave" have no the significant relationship between interpretation time and satisfaction.

This result implies that the satisfaction at the majority of topics of interpreting Yehliu Geopark can be improved by enriching the information

provided in interpretation service. Although the insignificance time-satisfaction relationship appears in the minority, interpreters are suggested to enrich the contents when interpreting those topics with significant time-satisfaction relationship.

Once the time-satisfaction relationship is confirmed, the further problem is how to optimize the interpretation service by re-allocation the interpretation time among each topic. If the perceived satisfaction can be affected by the interpretation time, the overall all satisfaction with the interpretation service can be enhanced by re-arranging the time allocation of each topic. In principal, interpreters should spend longer time on the topics with higher marginal time effect and spend less time on those topics that the time-satisfaction relationship is insignificant or the marginal time effect is relatively low.

Another thing important is that the topics with insignificant time-satisfaction relationship does not necessary mean they are not important. The insignificant time-satisfaction relationship may exist if the interpretation content or the interpretation style is not attractive enough to the visitors. In this case, the interpretation content or the way the interpreters convey the related information of these topics is subject to revision. Nevertheless, what topics are important for the purposes of geological and environmental education depends on the reality faced by the Geoparks. The authorities in charge of the operation and management of the Geoparks should professionally judge whether a topic should be canceled or revised.

REFERENCES

Beck, L. & Cable, T.T. (1998). *Interpretation for the 21st century: Fifteen guiding principles for interpreting nature and culture*. Champaign, IL: Sagamore Pub.

Grinder, A.L. & McCoy, E.S. (1985). *The good guide: a sourcebook for interpreters, docents, and tour guides*. Scottsdale, AZ: Ironwood Press.

Mahaffey, B.D. (1970). Effectiveness and preference for selected interpretive media. *Environmental Education, 1*(4), 125–128.

Sharpe, G.W. (1982). *An overview of interpretation*. New York, NY: John Wiley & Sons.

Tilden, F. & Heritage, I.O. (1977). *Interpreting our heritage*. Chapel Hill, NC: University of North Carolina Press.

Theory and Practice in Hospitality and Tourism Research – Radzi et al. (Eds)
© 2015 Taylor & Francis Group, London, ISBN 978-1-138-02706-0

Sensory marketing influence on customer lifetime value of the hotel industry

B.S. Hosseini, R. Mohd-Roslin & P. Mihanyar
Arshad Ayub Graduate Business School, Universiti Teknologi MARA, Shah Alam, Malaysia

ABSTRACT: The hospitality sector is a principal participant to the development of the tourism industry and the hotel industry is one of the most prominent contributors to this sector. Nowadays, there is more emphasis for the hotel industry to provide diverse services for specific target customers. The hotels are competing to offer slightly dissimilar services to attain customers' satisfaction. Because of this competitive situation, the hoteliers strived to excel. The notion of Customer Lifetime Value (CLV) has been introduced to emphasize the importance of sustaining customers. CLV has become a preference for many marketers competing in this hypercompetitive business environment (Hosseini & Albadvi, 2009). In the hospitality industry, hotels perceive CLV as a way to balance the challenge of managing additional earnings within specific periods with more long term and strategic value improvisations (McGuire, 2012). The idea of Sensory Marketing (SM) is linked to the creation of CLV in the hotel industry because it could assist in attracting tourists continuously. The sensory marketing postulates that attraction can be developed through the sensory elements experienced by the tourists when they visit specific locations. The question of whether SM incites hotels' occupancy rates and the connection of SM with CLV as a strategic move by the hotels is the direction that this study is proposing. This is a conceptual paper that looks at the relevant literature leading to the linkage between SM and CLV.

Keywords: Customer lifetime value, loyalty, revisits intention, sensory marketing, word of mouth communication

1 INTRODUCTION

Tourism industry is a major economic contributor in many countries in the world and the growth of this industry is an important factor for economic development. The World Tourism Organization (WTO) describes tourists as travellers who are travelling to and staying in places outside their common circumference for not more than one consecutive year for the purpose of resting, business and other specific reasons. Hotels are often regarded as an essential component of the tourism industry. Tourists have various demands while staying in a hotel and they may attain positive or negative experiences depending on what they go through (Lewis & Chambers, 1999). When studies about the hotel industry are linked to the understanding of whether guests are satisfied or not and whether their experience affects their trust when staying in a hotel, the topic of customer life time value will be meaningful. To support this, customers who will return to the hotel or give positive word-of-mouth communication to other tourists are necessary for the hospitality business to succeed (Dominici & Guzzo, 2010).

2 SENSORY MARKETING

Marketing concept is an exclusive function of business and it is a basic issue for an organization (Drucker, 1954). Marketing is a field that pertains to the discovery and comprehension of the demand and requirement of customers. Marketing also helps marketers to realize customers' satisfaction better than competitors. The perceptions of the customers construe the perception about the product or service. How customers perceived the marketers' actions will very much be influenced by the external stimuli generated by the marketers. This can be based on the five senses which are sight, touch, smell, taste, and hearing. These five senses are often characterized as "Sensory Organs or Sensory Receivers". Hult, Broweus, and Van Dijk (2009) termed this as the sensory receivers which are made up of three parts, known as actuate, sentiment and experiments.

The marketing activities that concentrate on such customer's perception are called Sensory Marketing. These activities are used for constructing the marketing strategies based on communication with the customers. If the marketers use stimulus in a proper way, it will be a critical factor that beget cognizance and influence customers' behaviors and it is also capable of affecting decision making and the tendency of customers to spend (Soars, 2009).

According to Goldkuhl and Styvén (2007), sensory marketing activities are more often used in service business nowadays. Services like hotels, restaurants, and department stores use marketing activities via the influence on customers' senses. They also insinuated about the importance of the five senses' in their roles in enhancing the senses relating to the service. McDougall and Snetsinger (1990) believed that the principal trait of service is difficult to influence and that customers do not have the capability to appreciate the quality of the service without relying on their senses. Therefore, the marketers should make the effort to develop an environment where the services are linked to the elements of sounds, color, or scent. Soars (2009) explained that the decisions and positive attitudes of customers is very much linked to them using their sensory tools properly. Each sense could affect customers' behavior differently, and this is described as follows:

Visual dimension: The most prevalent exert on customers is colour because it can influence customers' moods and emotions. The influence varies between one customer and the others. Colours can insinuate different meanings: pink as a romantic colour, green as the symbol of nature, yellow as gaining attention, purple as a luxury colour and red for proactive appetite (Hult et al., 2009).

Aural dimensions (sound): It reflects the physical environment in the service industry and it is used extensively in hotels and restaurants. Music is a considerable factor for everyday living and can influence customers expectations (Dragicevic & Rakidzija, 2012). Bruner (1990) mentioned music is a significant base that can induce customer's demeanour and mood. Hul, Dube, and Chebat (1997) demonstrated that music has positive benefits on customers behaviour in the service surrounding. The marketing manager in a hotel should provide services that affect customers' needs and satisfaction, therefore the hotels should develop a suitable atmosphere with the opportunity for relaxation using music that is relevant for specific age segmentation (Dragicevic & Rakidzija, 2012).

Touch dimensions: It is a perceptible sense where customers can have physical contact with and can feel the element of sharpness, hardness and roundness. It is important for hotels where the comfortable sofas in the lobby and cosy beds in the rooms make customers feel relax (Hultén, Broweus & Van Dijk, 2009).

Taste dimensions: It is the sense where customers can feel in the mouth in relation to the elements of salinity, sweetness, bitterness and sourness. In hotels' the restaurant is the major provider of the taste perception including the foods' name, taste and their scents, also the sound and interior design of the place. For this reason the taste sense relates to the whole senses and not just when putting food into the mouth (Hultén et al., 2009).

Olfactory dimensions (scent): It is the sense which has an effect on the customers brain and mind without any deeper consideration (Davies, Kooijman & Ward, 2003). Guéguen andPetr (2006) asserted that lavender essential oils and lemon essential oils have psychological effect, lavender is creating calmness and lemon is enlivening and energizing when applied.

As such, sensory marketing is seen as having a possible impact on customers' actions which may influence the marketers' performance. Successful implementation of sensory marketing may lead to the creation of loyal customers and therefore lead to the development of Customer Lifetime Value (CLV) for the marketers.

2.1 *Customer Lifetime Value (CLV)*

Hotels are competing with each other to deliver high quality services to satisfy their customers and to be able to sustain them, and the need to sustain customers for long periods of time become important (Choi & Chu, 2001).To know more about the importance of sustaining customers the researcher will elaborate on the concept of customer lifetime value (CLV) (Bohari et al., 2011).CLV is described as an organization's bottom line worth over a specified period of time and it is measured by customer's shares of procurements and relationship continuity (Kim & Cha, 2002).To be able to measure the CLV there is a need to understand three elements known as loyalty, word of mouth and revisit intention (McDonald, 1996).It is noteworthy to mention that retaining loyal customers are important for service sectors such as hotels because losing one customer can relate to losing a stream of revenue and revisits (Kotler & Armstrong, 2003).

Therefore, the measurement of CLV will be based on these elements (loyalty, word of mouth and purchase intention) and were adapted from previous studies of McDonald (1996) and Kim & Cha (2002). Service sectors such as hotels have given more attention to CLV because losing a customer connote much more than a single sale; it relates to losing a stream of cash flows from repeat purchases and revisits (Kotler & Arsmtrong, 2010).

2.2 *Loyalty*

A key factor in CLV is the notion of loyalty. Customer loyalty is nearly linked to corporate performance as a fundamental concept in marketing

(Reichheld, 1992).It has been proven that a 5% increase in customer retention yields 85% more profits in service industry (Reichheld & Sasser, 1990). Moreover, retention and conservation of existing customers costs less than requiring of new customers (Reichheld, 2001).

If customers' experience at a destination is understood as a product (e.g. visiting a hotel), the level of loyalty can be reflected in their behavioural intention to revisit the destination and intention to recommend visiting the hotel to relatives and friends (Oppermann, 2000). It is good to mention that customer loyalty will make the higher costs of services will still be paid by loyal customers; it will bring positive word-of-mouth as the best marketing agent and it will save money for the organization because retaining customers are less expensive than acquiring new customers (Gee, Coates & Nicholson, 2008).

2.3 Word Of Mouth (WOM) communication

Another element of CLV is the tendency of word-of-mouth (WOM) communication among customers. Word of mouth communication serves as a customer to customer (c2c) channel to convey information (Arndt, 1967; Bansal & Voyer, 2000).Word of mouth begins from loyal customers who are committed to a destination or service (Derbaix & Vanhamme, 2003).Word of mouth is the greatest influential communication in the hotel industry and it acts as an alternative source of information to help others (friends and relatives) to make decisions.

2.4 Revisit intention

Customer's repeat purchase in the hotel industry is known as revisit intention and it is also linked to customers' satisfaction with initial purchase (Sirgy & Tyagi, 1986).The qualification of destination effects on tourists' intention to recommend and come back again to the same destination or place (Kozak & Rimmington, 2000) will act as an influencing factor for attracting revisits. Intention of customers is in relation to their satisfaction. If they are satisfied they will repurchase and revisit a destination (Oliver & Swan, 1989). Moreover, when the tourists have enjoyable and unforgettable experiences, they are more likely to plan and return in the future. As a result, satisfaction has a significant effect on planning for the next visit.

3 DISCUSSION

Thus far, the amalgamation of literature appears to point to the realization on the probability of relating SM and CLV in a setting where customers are more likely to react to their experiences. Hotels' guests who stay in a particular hotel and who come back to stay insinuate their direct experiences and it will address elements of physical environment and service substance that indicate their satisfaction or dissatisfaction. As sensory marketing needs the marketers to expand specific strategies that are capable of elevating customer perception, it would be interesting to understand how this translates to the creation of CLV.

This paper illustrates the relevance of sensory marketing in the context of the service industry. The proposed study therefore hopes to address the link between SM and CLV and therefore contributes not only to the body of knowledge relating to the theories of SM and CLV but also to the applied scenario where practitioners, specifically hotels operators can benefit from the understanding of how customers or hotel patrons expressed their experiences during their hotel stay.

4 CONCLUSION

Even though the literature has built on the possibility of associating SM and CLV, there is still a requirement to recognize how these constructs are operationalized in the research setting. Additional assessment is perhaps needed to understand further the procedure of CLV and how this can be interpreted into outstanding information to the academics and the practitioners. Furthermore, understanding of SM also has to be improved if more inferences are to be realized to address related issues of consumer behaviour.

REFERENCES

Arndt, Johan. 1967. Role of Product-Related Conversations in the Diffusion of a New Product. *Journal of Marketing Research (JMR)* 4(3).

Bansal, Harvir S & Voyer, Peter A. 2000. Word-of-mouth processes within a services purchase decision context. *Journal of service research* 3(2): 166–177.

Bohari, Abdul Manaf, Rainis, Ruslan & Marimuthu, Malliga. 2011. Customer Lifetime Value Model in Perspective of Firm and Customer: Practical Issues and Limitation on Prospecting Profitable Customers of Hypermarket Business. *International Journal of Business and Management* 6(8): p161.

Bruner, Gordon C. 1990. Music, mood, and marketing. *The Journal of Marketing* (94–104).

Choi, Tat Y & Chu, Raymond. 2001. Determinants of hotel guests' satisfaction and repeat patronage in the Hong Kong hotel industry. *International Journal of Hospitality Management* 20(3): 277–297.

Davies, Barry J, Kooijman, Dion & Ward, Philippa. 2003. The sweet smell of success: olfaction in retailing. *Journal of Marketing Management* 19(5–6): 611–627.

Derbaix, Christian & Vanhamme, Joelle. 2003. Inducing word-of-mouth by eliciting surprise–a pilot investigation. *Journal of economic psychology* 24(1): 99–116.

Dominici, Gandolfo & Guzzo, Rosa. 2010. Customer satisfaction in the hotel industry: a case study from Sicily. *International Journal of Marketing Studies* 2(2): 3–12.

Dragicevic, Marija & Rakidzija, Ivana. 2012. The Music as an Element of Physical Evidence in Service Organizations. *Procedia Economics and Finance* 3(666–671).

Drucker, Peter. 1954. The principles of management. *New York*.

Gee, Robert, Coates, Graham & Nicholson, Mike. 2008. Understanding and profitably managing customer loyalty. *Marketing Intelligence & Planning* 26(4): 359–374.

Goldkuhl, Lena & Styvén, Maria. 2007. Sensing the scent of service success. *European Journal of Marketing* 41(11/12): 1297–1305.

Guéguen, Nicolas & Petr, Christine. 2006. Odors and consumer behavior in a restaurant. *International Journal of Hospitality Management* 25(2): 335–339.

Hosseini, Monireh & Albadvi, Amir. 2009. Customer value network analysis: improving ways to compute customer life-time value. *International Journal of Electronic Commerce Studies* 1(1): 15–24.

Hul, Michael K, Dube, Laurette & Chebat, Jean-Charles. 1997. The impact of music on consumers' reactions to waiting for services. *Journal of Retailing* 73(1): 87–104.

Hult, B, Broweus, Niklas & Van Dijk, Marcus. 2009. *Sensory marketing*: Palgrave Macmillan.

Hultén, Bertil, Broweus, Niklas & Van Dijk, Marcus. 2009. *Sensory marketing*: Palgrave Macmillan.

Joseph Sirgy, M & Tyagi, Pradeep K. 1986. An attempt toward an integrated theory of consumer psychology and decision-making. *Systems Research* 3(3): 161–175.

Kim, Woo Gon & Cha, Youngmi. 2002. Antecedents and consequences of relationship quality in hotel industry. *International Journal of Hospitality Management* 21(4): 321–338.

Kotler, P & Armstrong, G. 2010. *Principles of Marketing* (13 ed.): Pearson.

Kotler, P Armstrong. G.(2003), Principles of Marketing: Rio de Janeiro: Prentice-hall of Brazil.

Kozak, Metin & Rimmington, Mike. 2000. Tourist satisfaction with Mallorca, Spain, as an off-season holiday destination. *Journal of travel research* 38(3): 260–269.

Lewis, Robert C & Chambers, Richard Everett. 1999. *Marketing leadership in hospitality: foundations and practices*: John Wiley and Sons.

McDonald, Mark Alan. 1996. Service quality and customer lifetime value in professional sport franchises.

McDougall, Gordon HG & Snetsinger, Douglas W. 1990. The intangibility of services: measurement and competitive perspectives. *Journal of Services Marketing* 4(4): 27–40.

McGuire, Kelly. 2012. Customer Lifetime Value—the "Holy Grail" for Hotels. Retrieved 26.9.2013, from http://hotelexecutive.com/business_review/3039/customer-lifetime-value-the-holy-grail-for-hotels.

Oliver, Richard L & Swan, John E. 1989. Consumer perceptions of interpersonal equity and satisfaction in transactions: A field survey approach. *Journal of marketing* 53(2).

Oppermann, Martin. 2000. Tourism destination loyalty. *Journal of travel research* 39(1): 78–84.

Reichheld, Frederick F. 1992. Loyalty-based management. *Harvard business review* 71(2): 64–73.

Reichheld, Frederick F. 2001. *The loyalty effect: The hidden force behind growth, profits, and lasting value*: Harvard Business Press.

Reichheld, Frederick P & Sasser, W Earl. 1990. Zero Defeciions: Quoliiy Comes To Services.

Soars, Brenda. 2009. Driving sales through shoppers' sense of sound, sight, smell and touch. *International Journal of Retail & Distribution Management* 37(3): 286–298.

Theory and Practice in Hospitality and Tourism Research – Radzi et al. (Eds)
© 2015 Taylor & Francis Group, London, ISBN 978-1-138-02706-0

Taxonomical challenges of loyalty program management in Malaysian city hotels

J. Anuar
Faculty of Hotel and Tourism Management, Universiti Teknologi MARA, Terengganu, Malaysia

N. Sumarjan & S.M. Radzi
Faculty of Hotel and Tourism Management, Universiti Teknologi MARA, Shah Alam, Malaysia

ABSTRACT: This study explores the challenges encountered in administering loyalty programs from hotel manager's perspectives. The main aim is to differentiate and analyze the taxonomical challenges encountered. In-depth interviews with six city hotel managers, either Front Office or Marketing department that have loyalty programs were conducted. The interview sessions were recorded using audio visual tapes. The transcripts were then analyzed, coded and summarized into categories. Challenges were then discussed and suggestions were provided. Findings were then structurally compared. Three levels of challenges taxonomy are developed: core challenges, partially common challenges and individual challenges. Findings revealed that these six city hotels shared the same six core challenges namely administering costs of loyalty programs, maintaining, updating and upgrading databases, unattractive reward value, sameness of the rewards by competitors, difficulty to claim reward as well as employee customer knowledge. Findings also revealed that all the participating city hotels encountered three partially common challenges and individual challenges. Future research may focus on guest perceptions of challenges that they encountered with their loyalty program membership. This study furnishes hoteliers with competitor analysis which is helpful and can be a reference for hoteliers in enhancing their loyalty programs.

Keywords: Challenges, city hotels, hotel managers, loyalty programs, taxonomy

1   INTRODUCTION

Loyalty program research has dominated the hospitality realm since its introductory chapters in the 1980s. To date, many huge industry sectors such as rental car companies, cruise lines, spa resorts and hotel chains adopted loyalty programs (Xie & Chen, 2013). Probably, interest in loyalty programs that exploded among many organizations are built mainly on the reason that it is cheaper to market to existing customers than to acquire new ones (Kumar & Reinartz, 2012). Researchers give credence to loyalty programs in enhancing positive behavioral intentions of repeat purchases (Meyer-Waarden, 2008), transmitting customers' share of wallet (Wirt, Mattila & Lwin, 2007) as well as strengthening the customers-service providers bonding (Mattila, 2006). Many hotels' loyalty programs promoted themselves as a passport to excellent value, affordable accommodation, dining, entertainment and other great benefits. However, some advocators claimed that loyalty programs do not perform much (Kim, Lee, Bu & Lee, 2009) and their profitability is uncertain (Wansink, 2003).

Studies also argued on the negative impacts of the program. Researchers' demanded that 'loyalty programs do not create loyalty' (Bellizzi & Bristol, 2004; De Wulf, Odekerken-Schroder & Iacobucci, 2001; Reinartz, 2010). These mixed findings had hindered the proper evaluation of the loyalty programs and suggest a need to understand the programs better, especially from hoteliers' point of views. This study outlined the challenges faced by hotel operators in administering their loyalty programs. Challenges can be regarded as opportunities given the innovative measures taken by hoteliers in addressing the issues. At present, unknown empirical study has been conducted on challenges of loyalty programs comprehensively in Malaysian hotel scenario. This study is unique as it guides hoteliers to refine and execute an advance loyalty programs as well as inspire future research. Additionally, Kumar et al (2012) highlighted that there are only few studies that emphasize on the success or failure of specific loyalty program due to organizations are likely to admit of their poor performance. The remainder of this study is constructed as follows. Section two furnishes the drawbacks of loyalty programs from both management as well as member's perspectives. Section three denotes the methodology followed by findings and discussion in section four. Finally, Section 5 indicates conclusion as well as future research. Thus, this study explores the challenges encountered in administering loyalty programs from hotel manager's perspectives. The main objective is to differentiate and analyze the taxonomical challenges encountered.

2 LITERATURE REVIEWS

2.1 Drawbacks of loyalty programs from management perspectives

Issues on minimizing the undesirable features of loyalty programs have been the prime interest of loyalty program research in hotel industry. Comprehensive reviews of loyalty program studies indicated that the conspicuous drawbacks of loyalty programs evolved from management and members' perspectives. Studies related to costs of managing loyalty programs can be found in Berman (2006) and Wansink (2003) work. Moreover, study by Xie et al (2013) postulated that start-up cost for hotels to acquire new members is higher considering current intense competitive ambience. Thus, a long-term commitment and dedication from hoteliers are needed once customers join loyalty programs. Members' redeemed points should be regarded as hotelier's reliability (Berman, 2006).

2.2 Drawbacks of loyalty programs from members perspectives

Loyalty Program offers both economic and non-economic benefits to customers (Bridson, Evans & Hickman, 2008), however on the other hand several negative perceptions were induced by members on loyalty programs. It includes member frustration, erosion of market saturation and member commitment (Xie et al, 2013). Additionally, though many researchers have demonstrated that loyalty programs provide incentives and reward to its members as part of a marketing strategy (Omar, Abd Aziz & Nazri, 2011; Leenheer, Van Heerde, Bijmolt & Smidth, 2007), Stauss, Schimidt and Schoeler (2005) displayed members' frustrations as struggling to redeem programs rewards and required extensive emotional and material cost to attain benefits. Furthermore, Lacey and Sneath (2006) claimed that revealing members' personal information for loyalty programs might expose members to the probability of personal risk. Moreover, Rowley (2004) highlighted that it's impossible for loyalty programs to lead members purchasing cycles toward commitments to single service pro-

vider. Thus, trying to be unique among competitors probably can entice member's participation and behavior to stay loyal. Hoteliers believed that by providing the best and valuable benefits can managed to drive customer's loyalty through their loyalty programs (Shanshan, Wilco & Eric, 2011).

3 METHODOLOGY

This study used in-depth interviews with six hotel managers to investigate the challenges encountered in administering loyalty programs. Six loyalty programs in Malaysian city hotels were deliberated comprising of Hotel A, B, C, D, E and F. The hotel's names were not revealed for anonymity purposes. Managers uttered their solicitude expressions in revealing themselves as they are exposed to hidden competitive advantage. The characteristics of all hotels were as follows: located in Kuala Lumpur area, five-star rated hotels and listed among the 45 city/business hotels associated and registered with Malaysian Association of Hotels (MAH, 2012). MAH member list (www.hotels.org.my, accessed in October 2012) exhibits 14 out of 45 five-star rated hotels that adopted loyalty programs. Researchers invited all respondents through e-mails and phone calls to take part in this study. Interview appointments were organized for those who responded to the invitation and conformed to involve in this study. Eight hotels declined to take part owing to high occupancy, clashed with staffs training and cluttered with events scheduled at their hotels.

One-on-one in-depth interviews were chosen for this study attributable to richer data could be gathered though it's more time consuming, costly and labor intensive (Creswell, 2008). Purposive sampling was utilized for participants' selection. Front office or marketing department managers were preferred due to their in-depth knowledge in handling loyalty programs at their hotel. Moreover, most hotels appointed these two departments in taking care of all administrative logistics possession to the programs. The interviewees' characteristics for each hotel are exhibited in Table 1.

Table 1. The interviewees' characteristics.

Hotel	Position	Gender
A	Front Office Manager	Male
B	Front Office Manager	Female
C	Front Office Assistant Manager	Female
D	Marketing Assistant Manager	Male
E	Marketing Assistant Manager	Female
F	Front Office Assistant Manager	Female

*Self compiled by researchers.

Each interview session conducted approximately between one to two hours. A set of questions were referred for this semi-structured interviews. Each interview sessions were recorded using audio visual tapes. It was then transcribed verbatim, heavily reviewed and summarized into themes, categories and concepts.

4 FINDINGS AND DISCUSSION

The next stage after the in-depth interview process was generating taxonomical challenges. Table 2 furnished the taxonomy challenges encountered by these six city hotels in Malaysia; where as Table 3 stipulated the summarized challenges for each interviewed hotels. The three levels of challenges taxonomy established were core challenges described as same challenges encountered by all hotels, partially common challenges described as part of challenges that are same among hotels studied and individual challenges described as unparalleled challenges encountered by each hotel. It's interesting to discover that there are six individual challenges in managing the loyalty program.

Researchers clarified six (6) items (no. 1-6) as core challenges. Findings indicated that all six hotels have made ample investments pertaining to administering costs of their programs to assure loyalty of targeted members. It's not about only having a program but it's beyond member's satisfaction and crucial in ensuring the program really

Table 2. The loyalty program's challenges taxonomy.

Core Challenges	Partially Common Challenges	Individual Challenges
1. Administering costs of loyalty programs 2. Maintaining, updating and upgrading databases 3. Unattractive reward value 4. Sameness of the rewards by competitors 5. Difficulty to claim reward 6. Employee customer knowledge	7. High investment on the materials 8. Marketing campaigns 9. Distribution coverage 10. High start-up cost 11. Systems down	12. Member's privacy 13. Yearly activity statements 14. Software applications for ipads and smart phones 15. Member's points expiration 16. Rewards confusion by members 17. Members forget to present card upon check-in

*Source: Self-compiled by researchers.

Table 3. Hotels with summarized challenges.

Hotels	Core challenges	Partially Common Challenges	Individual Challenges
Hotel A	All hotels shared the same challenges (items no. 1-6)	10, 11	12
Hotel B		7, 8, 9, 10	13
Hotel C		8, 9, 11	14
Hotel D		7, 8, 9	15
Hotel E		8, 10	16
Hotel F		7, 8, 9	17

*Source: Self-compiled by researchers.

works. Gable, Fiorito and Topol (2008) corroborated that loyalty programs can be envisioned as an obstacle to entry for competing rivals. Maintaining, updating and upgrading databases also appeared to be among the core challenges. Most of the hotels interviewed have at least an average of 500,000 members worldwide. Thus, hoteliers need to incorporate and update member's profile from time to time. Additionally, unattractive reward value became visible core challenges. This is in line with what had been discussed earlier on in the literature. Members expect products and services of higher grade or quality in retrospect of their commitments. Hence, hoteliers are encouraged to focus on satisfying members without incurring cost. Moreover, the sameness of rewards by competitors contributed as another core challenges. Matilla (2006) claimed that loyalty programs were easy to copy and most loyalty programs look alike. More vigorous efforts need to be put in identifying ways to stay unique and different from competitors (Shanshan *et al,* 2011). Findings also revealed that difficulty to claim reward appeared as core challenges. Hoteliers can restructure back on the reward benefits and redemption for members. Imposing no blackout dates for members reward redemption seems to be an alluring marketing strategy. Item no. 6, employee customer knowledge was exhibited as another core challenges. Hoteliers have to constantly continue their staff training especially in encouraging a positive customer delivery from hotel staffs. In line with Xie *et al* (2013) study, loyalty programs can be used as motivational tools for hotel staffs to render better assistance to members.

For partially common challenges, five (5) items were recognized (no. 7-11). Three (3) hotels claimed they spent high investment on the materials. Moreover, marketing campaigns appeared prominent as it was encountered by five (5) out of six (6) hotels. Though all hotels established the campaigns on the net through their websites, Hotel A seems to fully utilize their online campaigns. Probably other hotels can considered the approach taken by Hotel A that use online forum besides promoting in social media such as face book. Introducing online forum and face book is easy but what matters are usually after the post-introductory stages. More efforts need to be exerted in maintaining members-hotels program rapport. Additionally, distribution coverage tends to contribute as second frequent challenges encountered by all interviewed hotels. Engaging on online marketing can be considered as delighting idea for hoteliers as it will increase more coverage on distribution area. Additionally, three hotels encountered high start-up cost. It's really crucial for hoteliers to really think whether these programs are worth investment for their hotels. It shouldn't be to have a program because everyone else has it. While two (2) hotels experienced systems down and listed them as part of their challenges. These hotels were operated on a worldwide basis. Having systems down can be disaster to their daily operation where they have to contact their hotel chains manually. Thus, maintenance on the hotel systems besides periodic maintenance is extremely crucial.

Six items (no. 12-17) were established as individual challenges and encountered by each hotel. Creative decision-making is needed in tackling the challenges. Hotel A highlighted that member's privacy was their individual challenges. Some members refused to give their particulars such as income statement. This information is vital in assisting hoteliers designing appropriate activities for certain market segmentation in their program. Hotel B revealed that producing yearly activity statements was part of their loyalty programs management challenges. The statements were printed and mailed to all members. However, some statements were returned back due to incorrect, not updated and redundant address. Probably, hard copy yearly activity statements can be converted to soft copy version, sending via members email. No printing process is required in order to reduce hotels loyalty program cost. Findings also revealed that Hotel C had problem with software applications introduced by their hotel program. Members can access hotel information and anything regarding hotel loyalty programs through ipads and smart phones. However, it is a calamitous event especially if it occurs continuously causing inaccessible connections to the hotel. It will lead to member's great dissatisfaction. Xie et al (2013) postulated that it's good effort by the hotel to better communicate, but they need to look at software maintenance as well. Additionally, Hotel D encountered member's point's expiration. It can induce member's frustrations and this indirectly will probably change their attitudes or behaviors towards spending patterns with loyalty programs. Hotel F practiced no expiration point's policy whereas the rests of the hotels imposed certain duration for point's expiration. Perhaps no expiration point's policy can be an attractive strategy where members have more opportunities to redeem their reward. Hotel E encountered rewards confusion by members. Example is their weekend offers. Though it highlight the term week-ends, members misperceived it as their special privileges and it applied on weekdays as well. The miscommunication leads to member's frustration and requires an efficient service recovery from the hotels. Lastly, Hotel F experienced members forget to present card upon check-in for points collection. Though the process of adding points can be done later, but some members expressed their anxieties claiming

hotel staffs were not efficient in reminding them with the program matters. This challenge requires hotel staffs consistency and be persistent in serving members of their hotel loyalty programs.

5 CONCLUSION AND FUTURE RESEARCH

This study suggests that loyalty programs are not a "one-size-fits-all" solution (Xie et al, 2013). Based from different and individual challenges encountered by each hotel, it is therefore strongly recommended that hoteliers review their loyalty programs rather than just designing another copy-cat loyalty program. This taxonomy summary is crucial in guiding hoteliers to improve and accelerate their loyalty programs. Appropriate measures and innovative approaches can be adopted by these hotels especially in dealing with the individual challenges. Challenges can be perceived as opportunities given the unique and competitive measures taken by hoteliers in addressing those issues. Hoteliers need to be focus in conducting their loyalty programs. Failing to do so will lead them to become a master of none.

This study furnishes the loyalty programs literature from the perspectives of management or hoteliers. Future research is required on investigating member's perspectives. Winters and Ha (2012) highlighted that consumer behaviors with regard to loyalty programs are underexplored. Thus, customers' perspectives may be the best approach to explore the value of loyalty programs. Moreover, future research also may concentrate on the lower rated hotels to give different influences and settings.

REFERENCES

Bellizzi, J.A. & Bristol, T. (2004), An assessment of supermarket loyalty cards in one major US market, *Journal of Consumer Marketing,* 21 (2), 144–154.

Berman, B. (2006), Developing an effective customer loyalty program, *California Management Review,* 49 (1), 1–23.

Bridson, K., Evans, J. & Hickman, M. (2008), Assessing the relationship between loyalty program attributes, store satisfaction and store loyalty, *Journal of Retailing and Consumer Services,* 15 (5), 364–374.

Creswell, J.W. (2008), *Educational Research: Planning, Conducting and evaluating quantitative and qualitative research,* Pearson Prentice Hall, NJ.

De Wulf, K., Odekerken-Schroder, G. & Iacobucci, D. (2001), Investment in consumer relationships: a cross-country and cross-industry exploration, *Journal of Marketing Intelligence & Planning,* 65, 33–50.

Gable, M., Fiorito, S.S. & Topol, M.T. (2008), An empirical analysis of the components of retailer customer loyalty programs, *International Journal of Retail and Distribution Management,* 36 (1), 32–49.

Kim, D., Lee, S., Bu, K. & Lee, S. (2009), Do VIP programs always work well? The moderating role of loyalty, *Psychology and Marketing,* 26, 590–609.

Kumar, V. & Reinartz, W. (2012), Loyalty programs: Design and Effectiveness, *Customer Relationship Management,* Springer texts in Business and Econimics, pp 183–206.

Lacey, R. & Sneath, J.Z. (2006), Customer loyalty programs: Are they fair to consumers? *Journal of Consumer Marketing,* 23, 458–464.

Leenheer J., Van Heerde HJ., Bijmolt THA, Smidtsd A. (2007), Do loyalty programs really enhance behavioral loyalty? An empirical analysis accounting for self-selecting members, *International Journal Research Marketing,* 24 (1), 31–47.

Malaysian Association of Hotels (MAH). Available at http://www.hotels.org.my

Mattila, A.S. (2006), How affective commitment boosts guest loyalty and promoted frequent-guest programs, *Cornell Hotel & Restaurant Administration Quarterly,* 47 (2), 174–181.

Meyer-Waarden, L. (2008), The influence of loyalty programme membership on customer purchase behavior, *European Journal of Marketing,* 42 (1/2), 87–114.

Omar, N.A., Abd. Aziz, N. & Nazri, M.A. (2011). Understanding the relationships of program satisfaction, program loyalty and store loyalty among cardholders of loyalty programs. *Asian Academy of Management Journal.* 16 (1), 21–41.

Reinartz, W.J. (2010), Retailing in the 21st century, Understanding customer loyalty programs, Springer, 409–427.

Rowley, J. (2004), Loyalty and reward schemes: How much is your loyalty worth? *The Marketing Review,* 4, 121–138.

Shanshan, N., Wilco, C. & Eric, S. (2011), A study of hotelfrequent-guest programs: Benefits and costs, *Journal of vaccation Marketing,* 17 (4), 315–327.

Stauss, B., Schmidt, M. & Schoeler, A. (2005), Consumer frustration in loyalty programs, *International Journal of Service Industry Management,* 16, 229–252.

Wansink, B. (2003), Developing a cost-effective brand loyalty Program, *Journal of Advertising Research,* 43, 301–309.

Winters, E. & Ha, S. (2012), Consumer evaluation of customer loyalty programs: the role of customization in customer loyalty program involvement, *Journal of Global Scholars of Marketing Science: Bridging Asia and the World,* 22 (4), 370–385.

Wirt, J., Mattila, A.S. & Oo Lwin, M. (2007), How effective are loyalty reward programs in driving share of wallet? *Journal of Service Research,* 9, 327–334.

Xie, K.L. & Chen, Chih-Chien (2013), Progress in loyalty program research: facts, debates and future research, *Journal of Hospitality Marketing & Management,* 22, 463–489.

Theory and Practice in Hospitality and Tourism Research – Radzi et al. (Eds)
© *2015 Taylor & Francis Group, London, ISBN 978-1-138-02706-0*

The mediating effect of superior CRM capability: The impact of organizational wide implementation and training orientation on profitability

H.Y. Liu & T.B. Phung
Department of Business Administration, National Dong Hwa University, Hualien, Taiwan, R.O.C.

ABSTRACT: Customer Relationship Management (CRM) continues to be an important research topic, especially its training orientation and organizational wide adoption because these two aspects seem to have significant effect on profitability. However a careful review of the literature reveals that the effect is directly affected by superior CRM capabilities which are generated by the two aspects. The paper proposes the first unified framework describing the relationships among superior CRM capabilities, CRM training orientation, organizational wide implementation and profitability. The framework was tested through Structural Equation Modeling (SEM) approach and Baron and Kenny method. Our data consisted of 314 questionnaires from 339 hotels in Vietnam. The study revealed that the effect of CRM training orientation on CRM profitability was fully mediated by superior CRM capabilities, whereas the effect of CRM organization-wide was only had partial. Moreover, the study found the size of firm and educational levels also have significant impact on superior CRM capability and CRM profitability.

Keywords: CRM training orientation, CRM organization-wide, superior CRM capability, CRM profitability, mediating effects, hotels, Vietnam

1 INTRODUCTION

CRM is a business strategy that aims to maintain and cultivate relationship with customers. It focuses on the flow of vital, timely, and accurate customer information that is helpful for companies to cultivate customer intimacy to achieve higher profitability (Day, 1994; Kim, Suh & Hwang, 2003). A global CRM study conducted by IBM Business Consulting Service shows that over 50 percent of companies believed CRM can increase performance and 65-75 percent looking to CRM as an important element in delivering revenue growth. However, only 15 percent of companies are fully successful. The companies did not succeed because they fail to consider CRM as a strategy to foster customer relationships (Day & Van den Bulte, 2002) especially they ignored the importance of CRM training orientation and CRM organizational wide adoption (Buttle, 2009; Kim, 2008).

CRM organizational wide adoption is "a higher level process that includes all activities that firms undertake in their quest to build durable, profitable, mutual beneficial customer relationship" to offer customization, simplicity and convenience for employees to keep a good relationship with customers (Gulati & Galino, 2000; Reinartz, Krafft & Hoyer, 2004). The coordination of departments and activities will improve communication among employees (a key indicator of CRM capability) and hence a better understanding of customers. Training orientation is to help new staffs get to adapt the new environment that they will work with. It aims to increase the knowledge, skills and attitude of employees to meet customer demands. CRM training orientation standardizes and simplifies works tasks and consequently achieves higher profitability. The links among CRM organizational wide implementation, training orientation and profitability may seem direct. But a closer review reveals that the increased profitability is actually directly linked to superior CRM capabilities but only indirectly linked to CRM organizational wide implementation and training orientation. Focusing on the CRM capability creates higher profitability because manager and employees try to reach out all potential customers and maximize the value of the customers (Blattberg, Getz & Thomas, 2001). This study aims to examine the mediating effects of superior CRM capability on the CRM training orientation, CRM organization-wide and CRM profitability in hotel industry.

2 REVIEW OF PREVIOUS RESEARCHES

2.1 Customer Relationship Management

Some researchers view CRM as technology, while others see it as data storage and analysis or customer-centric strategy (Bodenberg, 2001). Payne and Frow (2005) pointed that "CRM should be positioned in the broad strategic context because "CRM is not simply an IT solution that is used to acquire and grow a customer base; it involves a profound synthesis of strategic vision; a corporate understanding of the nature of customer value in a multichannel environment; the utilization of the appropriate information management and CRM application; and high-quality operations, fulfillment, and service". CRM strategy consists of the organizational level CRM performances, goals, policies, standard and best practices across department to identify customer behavior trends. CRM helps corporation to increase customer retention, customer loyalty and achieving higher customer profitability by training employees in keeping relationship with customer (Kim, et al., 2003). Hence, providing CRM capabilities to employees is very important that leads to achieving organization performance.

2.2 Superior CRM capability and CRM profitability

Morgan, Slotegraaf, and Vorhies (2009) stated that CRM capability consists of a firm's skill and accumulated knowledge to establish, maintain and enhance sustainable relationships with valuable customers. CRM can help attract new customers and enhance the occupancy rate in hotel industry (Roh, Ahn & Han, 2005). CRM will increase the customer satisfaction and loyalty by understanding their needs. Superior CRM capability could be measured in the activities and processes involving technical, human, and business related capabilities (Barney, 1991; Marchand, Kettinger & Rollins, 2000). Hence the first hypothesis is postulated.

H1: Superior CRM capability is positively related to CRM profitability.

2.3 CRM organization-wide

CRM organization-wide consists of many inter-related sub-process that can be subdivided into a collection of activities to make CRM more comprehensive and successful (Payne & Frow, 2005). Some researchers showed that a fully and successful implementation of CRM with a customer centric cross-functional will have positive effect on customer retention and organization's profitability (Chen & Popovich, 2003; Yim, Anderson & Swaminathan, 2004). Sharing information across the organization will integrate fragmented silos of customer infor-

mation that support for CRM employees to understand customer need and behavior trend. Bohling et al. (2006) suggested CRM organization-wide could increase competitive advantage for employees because it involves the interactions among technology, people, their values and work practices in organization. Thus, we developed H2 as follow:

H2: CRM profitability is positively related to CRM organization-wide. This effect is partly mediated by CRM superior capability

2.4 CRM training orientation

CRM training orientation helps all kinds of employees to understand and serve customers better. Facilitating skills and capabilities will deliver excellent experience in achieving customer value and company alike (Baran, Galka & Strunk, 2008; Buttle, 2009). CRM training orientation helps employees to know how to get things done and how to keep a sustainable relationship with customer in achieving CRM profitability. The more training employees had the better skills and capabilities they could use to provide high quality of products and services to customers. It will lead to increase customer satisfaction via job performance. Employees will be easy to work with CRM application and convert customer data into knowledge by training. Kim (2008) showed that training programs promote employees' specialized skills that are necessary to interact with customers. Therefore, we developed H3 as follows:

H3: CRM profitability is positively related to CRM training orientation. This effect is partly mediated by superior CRM capability.

3 METHODOLOGY

3.1 Sample and procedure

Data was collected in Vietnamese hotel industry. There are 314 usable questionnaires collected from 339 hotels. The questionnaires were handed out by sending through email and by interviewing CRM first-line employees who work in the divisions of marketing, sales, call centers, front office, and reservations. The results showed that 54.5 percent of respondents are female and 30.2 percent of the respondents are from the medium firm size with the total of employees around 51 to 100 employees.

3.2 Measures

To measure CRM training orientation, we used a seven point Likert scale recommended by Kim (2008) with 10 items, while the scale of CRM organization-wide was adopted from Jayachandran, Sharma, Kaufman, and Raman (2005) study. The scales for superior CRM capability were developed by Day and Van den Bulte (2002). To estimate CRM profitability, we adopted the scale of CRM profitability developed by Roh et al. (2005). The five-items with five point likert scale focus on how the implementation of CRM could profit the organization.

3.2.1 Control variables

We controlled five important demographic factors that include: gender, age, level of education, working experience and company size (Amburgey & Rao, 1996).

4 RESULTS

4.1 Confirmatory factor analyses

Cronbachs alpha's result for the scales are 0.91 for CRM training orientation, 0.88 for CRM organization-wide, 0.90 for CRM capability and 0.89 for CRM profitability. All results are greater than 0.70 which is above the acceptable threshold. Composite reliability and average variance extracted are higher than the evaluation criteria, with composite reliabilities larger than 0.70 and average variance extracted are larger than 0.50 with highly

significant factor loading ($p < .01$), providing evidence for convergent validity. Table 1 presents descriptive statistics and variable correlations.

Results of the CFA showed good model-to-data fit ($\chi^2 (224) =$ GFI = 0.86, AGFI = 0.83, CFI = 0.91, IFI = 0.91, RMSEA = .07). The study examined alternative models to confirm discriminant validity as show in Table 2. The fit indices for the four-factor model (baseline model) were superior to those for the three-factor model ($\Delta\chi^2 = 78.44$, $p < .001$), the two-factor model ($\Delta\chi^2 = 89.75$, $p < .001$), and one- factor model ($\Delta\chi^2 = 195.25$, $p > .001$). The baseline model is the best fit among those models. The results provided evidence for the discriminant validity among these core constructs.

4.2 Hypothesis tests

The first, we examined the effects of control variables to dependent variable. Education level and size of firms had a positive impact to CRM profitability. Other control variables were insignificant ($R^2 = .05$, $p < .001$). Model 1 measured the effect of independent variable, CRM training orientation and CRM organization-wide to mediator, superior CRM capabilities ($R^2 = 0.11$, $p < .001$). The model meets the first condition of Baron and Kenny's (1986)

Table 1. Variable descriptive statistical and correlations ($N = 314$).

Variables	Mean	Sd	1	2	3	4	5	6	7	8	9
1. Gender	.45	.50	–								
2. Age	1.90	.89	.05	–							
3. Education level	1.85	.90	.05	(.10)	–						
4. Experience of working	1.69	1.01	(.02)	(.04)	(.03)	–					
5. Size of company	3.24	1.18	(.04)	(.05)	.00	(.09)	–				
6. CRM training orientation	4.82	.97	.03	.07	.06	(.07)	(.04)	–			
7. CRM organization-wide	4.01	.69	.02	.08	.07	(.09)	(.08)	(.06)	–		
8. Superior CRM capability	4.34	.99	.09	.16**	.03	(.01)	.10	.20**	.17**	–	
9. CRM profitability	3.86	.72	.06	.07	.14*	(.04)	.14*	.14*	.15*	.24**	–

Note: *p < .01; *p < .05.

Table 2. Results of the CFA.

Model	χ^2	d.f	$\Delta\chi^2$	GFI	AGFI	CFI	IFI	RMSEA
Four-factor Model (baseline model)	592.25	224		.86	.83	.91	.91	.07
Three-factor Model (Model 1)	1389.40	227	797.15***	.67	.60	.72	.72	.13
Two-factor Model (Model 2)	1916.60	229	1324.30***	.61	.53	.60	.60	.15
One-factor Model (Model 3)	2701.04	230	2108.79***	.50	.40	.40	.40	.19

Note: In the four-factor model, four constructs were treated as four independent factors. In the three-factor model (Model 1), CRM training orientation and CRM organization-wide were combined into one factor. In the two-factor model (Model 2), CRM training orientation and CRM organization-wide were combined into one factor; CRM capabilities and CRM profitability were combined into one factor. In the one-factor model (Model 3), all constructs were combined into one factor; **$p < .01$*$p < .05$.

Table 3. Results of mediating effects.

Variables	Control variables	Model 1: IVs → mediator/superior CRM capability	Model 2: Mediator→ DV Superior CRM profitability	Model 3: IVs → DV Superior CRM profitability	Model 4: IVs & mediator → DV/CRM profitability	Hypothesis Results
Control Variables						
Gender	.08	.16	.05	.08	.06	
Age	.07	.15*	.04	.05	.03	
Education level	.11*	.02	.11*	.09*	.09*	
Experience of working	-.01	.04	-.02	.00	-.01	
Size of company	.09**	.12***	.07*	.10***	.09*	
Hypothesized predictor variables						
Superior CRM capability (H1)			.15***		.12**	Supported
CRM training orientation (H2)		.19***		.09*	.07	Not supported
CRM organization-wide (H3)		.22**		.14*	.17*	Supported
R2	.05***	.11***	.09***	.09***	.11***	

Note: N = 314. IV = Independent Variable; DV = Dependent Variable, H = Hypothesis; ***p < .001;***p < .001.

because the two independent variables were significant predictors of the mediator with CRM training orientation (p < .001) and CRM organization-wide (p < .01). Model 2 measured the effect of mediator, superior CRM capabilities to dependent variable, CRM profitability (R^2 = .09, p < .001). CRM had a positive impact on CRM profitability (p < .001). This suggested that Hypothesis 1 was supported. Model 3 satisfied the third condition of examining the mediating effect by measuring the effect of Independent Variable to Dependent Variable (R^2 = .09, p < .001). There was a significant positive effect between the CRM training orientation (p < .05) and CRM organization (p < .05). Model 4 measured the effect of independent variables and mediator to dependent variable R^2 = 0.11, p < .001). Model 3 fits the last condition of Baran and Kenny's (1986) because superior CRM capabilities had a significant impact to CRM profitable (p < .01). CRM training orientation had no significant effect to CRM profitability in model 4 (p > .05), indicating that the effect of CRM training orientation to CRM profitability is fully mediated by superior CRM capabilities. Hypothesis 2 was rejected. Whereas CRM organization-wide had a significant effect to CRM profitability in model 4 (p < .05), suggesting that the effect of CRM organization-wide to CRM profitability is partly mediated by superior CRM capabilities. Hypothesis 3 was accepted. Size of firm had a positive effect in all models with different p-value. Moreover, educational level also had a positive impact on CRM profitability (p < .05).

5 CONCLUSIONS AND IMPLICATIONS

5.1 Theoretical implications

The results showed that superior CRM capabilities had positive impact on CRM profitability (Hypothesis 1). That finding is consistent with

Reinartz and Kumar (2003). The findings provide further support to the suggestion that superior CRM capability lead to higher CRM profitability. Superior CRM capability provides the necessary capabilities for improving employees and organization skills and capabilities to uncover new market, segments, establishing long term relationship with new and existing customers.

By improving capabilities of employees and organization, hotels are better positioned to set reasonable strategies. Specifically, the paper found that superior CRM capability is a mediator between CRM training orientation, CRM organization-wide and CRM profitability. The effect of CRM training orientation on CRM profitability was fully mediated by superior CRM capabilities (Hypothesis 2). The results suggested that there is no direct effect from CRM training orientation to CRM profitability. This finding is inconsistent to previous research' results that suggested CRM training had a direct effect to customer outcome and CRM profitability (Kim, 2008). They found that training orientation leads to increase customer satisfaction and create a good relationship with customer by reducing operational errors and enhancing attitude and behavior of CRM employees. However, this study indicated that CRM training orientation increases the organization competitive advantage by increasing the superior CRM capability for employees and organizations to reach CRM profitability.

Besides, this study showed that the effect of CRM organization-wide was partly mediated by superior CRM capabilities (Hypothesis 3). It means that there is a direct effect from CRM organization-wide to CRM profitability due to partial mediation by superior CRM capability. This finding is consistent with Wang and Feng (2012). Superior CRM capability will be improved by connecting all departments and functions in an organization.

130

This study revealed that superior CRM capability is a mediator between CRM training orientation, CRM organization-wide and CRM profitability. The findings support and broaden customer relationship management theory. Finally, the results showed evidence on the influence of size of firm and educational level on superior CRM capability and CRM profitability. The findings suggest that the size of hotel has positive effect on profitability because bigger hotel will make better CRM training orientation for their employees and better CRM system.

5.2 Practical implications

CRM training orientation increases CRM capability not only for CRM employees themselves, but also for organization to develop and remain good relationship with customers. Managers can also foster CRM organization-wide by creating working environment with customer-centric culture. CRM organization-wide will connect various functions or departments in organizations that aims to enhance the quality of customer interactions. Second, this study will be very helpful for those who want to build superior CRM capability for organizations.

5.3 Limitations and future research

Future studies may benefit from assessing other constructs and variables to make better customer relationships by improving superior CRM capability. The data collected was from hotel industry of a single country. Future research can use data from other industries and/or countries for comparison.

REFERENCES

Amburgey, T.L. & Rao, H. (1996). Organizational ecology: Past, present, and future directions. *Academy of Management Journal, 39*(5), 1265–1286.

Baran, R.J., Galka, R.J. & Strunk, D.P. (2008). *Principles of customer relationship management*. Mason: Thomson South-Western.

Barney, J. (1991). Firm resources and sustained competitive advantage. *Journal of management, 17*(1), 99–120.

Baron, R.M. & Kenny, D.A. (1986). The moderator–mediator variable distinction in social psychological research: Conceptual, strategic, and statistical considerations. *Journal of personality and social psychology, 51*(6), 1173–1182.

Blattberg, R.C., Getz, G. & Thomas, J.S. (2001). *Customer equity: Building and managing relationships as valuable assets*. Boston: Harvard Business Press.

Bodenberg, T.M. (2001). *Customer Relationship Management: New Ways of Keeping the Customer Satisfied*. Paper presented at the The Conference Board, New York.

Bohling, T., Bowman, D., LaValle, S., Mittal, V., Narayandas, D., Ramani, G. & Varadarajan, R. (2006). CRM Implementation Effectiveness Issues and Insights. *Journal of Service Research, 9*(2), 184–194.

Buttle, F. (2009). *Customer relationship management: concepts and technologies*. Oxford: Elsevier.

Chen, I.J. & Popovich, K. (2003). Understanding customer relationship management (CRM): People, process and technology. *Business Process Management Journal, 9*(5), 672–688.

Day, G.S. (1994). The capabilities of market-driven organizations. *Journal of marketing, 58*(4), 37–51.

Day, G.S. & Van den Bulte, C. (2002). *Superiority in customer relationship management: consequences for competitive advantage and performance*. Cambridge, MA: Marketing Science Institute.

Gulati, R. & Galino, J. (2000). Get the right mix of bricks and clicks. *Harvard Business Review, 78*, 107–114.

Jayachandran, S., Sharma, S., Kaufman, P. & Raman, P. (2005). The role of relational information processes and technology use in customer relationship management. *Journal of marketing, 69*(4), 177–192.

Kim, B.Y. (2008). Mediated effects of customer orientation on customer relationship management performance. *International Journal of Hospitality & Tourism Administration, 9*(2), 192–218.

Kim, J., Suh, E. & Hwang, H. (2003). A model for evaluating the effectiveness of CRM using the balanced scorecard. *Journal of interactive Marketing, 17*(2), 5–19.

Marchand, D.A., Kettinger, W.J. & Rollins, J.D. (2000). Information orientation: people, technology and the bottom line. *Sloan Management Review, 41*(4), 69–80.

Morgan, N.A., Slotegraaf, R.J. & Vorhies, D.W. (2009). Linking marketing capabilities with profit growth. *International journal of research in marketing, 26*(4), 284–293.

Payne, A. & Frow, P. (2005). A strategic framework for customer relationship management. *Journal of marketing, 69*(4), 167–176.

Reinartz, W., Krafft, M. & Hoyer, W.D. (2004). The customer relationship management process: its measurement and impact on performance. *Journal of marketing research, 41*(3), 293–305.

Reinartz, W.J. & Kumar, V. (2003). The impact of customer relationship characteristics on profitable lifetime duration. *Journal of marketing, 67*(1), 77–99.

Roh, T.H., Ahn, C.K. & Han, I. (2005). The priority factor model for customer relationship management system success. *Expert systems with applications, 28*(4), 641–654.

Wang, Y. & Feng, H. (2012). Customer relationship management capabilities: Measurement, antecedents and consequences. *Management Decision, 50*(1), 115–129.

Yim, F.H.-k., Anderson, R.E. & Swaminathan, S. (2004). Customer relationship management: its dimensions and effect on customer outcomes. *Journal of Personal Selling and Sales Management, 24*(4), 263–27.

Theory and Practice in Hospitality and Tourism Research – Radzi et al. (Eds)
© 2015 Taylor & Francis Group, London, ISBN 978-1-138-02706-0

Motivational function and online community participation towards brand commitment: A case of Grand Blue Wave Hotel, Shah Alam

N.S. Shariffuddin & N.A. Ahmad
Faculty of Hotel and Tourism Management, Universiti Teknologi MARA (Shah Alam), Malaysia

ABSTRACT: Due to advance technology, social media has become the latest trend to gather information, communication, knowledge and entertainment. Facebook has become world trend or "King" of Social Media and becoming a popular tool for marketing. In hospitality industry, social media helps marketers to market their business. By building good relationship with online communities, hotel can generate strong commitment to their brands. However, marketers have to consider which motivational functions that influence consumer to participate in online communities and commit toward the hotel brand. This study employs a descriptive research design. A quantitative approach with a structured questionnaire as the research instrument was used among fans of Grand Blue Wave Hotel Facebook page. The findings revealed important insights on issues related to the topic of interest.

Keywords: Brand commitment, Facebook, motivational functions, online community participation, social media

1   INTRODUCTION

Currently, Facebook is the world-leading Social Network (Krahl, 2013) and statistic shows that in April 14, 2013 Facebook had a total number of 978 million active users all over the world (socialtimes. me, 2013) and 680 million monthly active users who used Facebook mobile products as in December 31, 2012 (facebook.com, 2013). With such a huge user base, Facebook is becoming a popular instrument for public relations and advertising professionals to reach mass audiences and Facebook fan pages allow brands to create an online community of brand users on the social networking site (Bushelow, 2012).

Nowadays, owing to an advancement of technology, social media can be considered as an exact mode for the consumer to get the right information at the right time (Naveed, 2012). Accordig to Paris, Lee, and Seery (2010) social media can be defined as a second generation of Web development and design, with the aims to facilitate communication, secure information sharing, interoperability, and collaboration on the World Wide Web. Because social media had reached huge numbers of people far and wide, Hartshorn (2010) testified that it has appeared to be very effective business instrument to involve with consumers and can build a brand name by continuous and prompt correspondence (Seth, 2012).

Lanz et al. (2010) noted that most successful hotels are finding innovative ways to integrate social media with traditional marketing techniques in order to be the frontier. Hotel can generate strong commitment to their brands by constructing good relationship with online communities. Seth (2012) supported that consumers will spend more, get more referrals, and have higher repeat clientele when hotels engage to their customers via social media. In addition, hotels can track the number of conversations that takes place over a new product or service offered and can immediately respond to reviews, comments and feedback (Seth, 2012). Thus, it will enable the hoteliers to overcome the problems fast and put an effort to improve the future brand performance.

2   LITERATURE REVIEW

2.1   *Online community participation*

Participation in online communities can be characterized as passive or active (Kang, 2011). Active community members are those who interact with other members as opposed to those who merely observe information (Madupu, 2006). Passive community members are lurkers or free riders members who browse online communities but rarely become involved in the community activities (Preece, et al., 2004). Moreover, lurkers tend to pursue their own goals and merely take advantage of the benefits of the communities (Ridings, Gefen & Arineze, 2006).

When the community member actively sharing their information and experience at hotel brands in the Facebook fan page it will give good impact to the hotel itself. Muniz and O'Guinn (2001) stated that, members' active participation enhances their knowledge regarding brands and products. Flavián and Guinalíu (2006) noted that active member enables them to offer suggestions to solve problems with product usage and help others in making purchase decisions.

2.2 Motivational functions

Functional benefits: According to Peter, Olson, and Grunert (1999), functional benefit is one that increases the ease and/or efficiency of completing transactions (i.e., purchasing products and services) and exchanging information (i.e., information gathering and sharing). Interaction with other community members in simplifying buying decisions is one of the functional benefits of an online community (Armstrong & Hagel, 1996). Because the information is stored and available within online communities, members can search for and exchange information more efficiently (Wang, et al., 2002). For that reason, social media ease consumer to get information about hotels brand conveniently and efficiently.

Social benefits derived as networking, collaborating and forming relationship with other community members for various product—related, personal or professional reasons (Dholakia, Blazevic, Wiertz & Algsheimer, 2009). In addition Dholakia et al. (2009) stated that retaining interpersonal connectivity denotes social benefits by establishing and maintain contact with other people (e.g., social support, friendship and intimacy). When online community recognizes that they share common interests with other members, they tend to increase the number and length of visits to hotel brand's Facebook fan page, and to actively participate in online community activities, (Hwang & Cho, 2005).

Psychological benefits are derived from feeling connected to community members (include an identity expression through the community), a sense of belonging to the community, and a sense of affiliating with other members (Bressler & Grantham, 2000). As members gain such knowledge about their online communities, they come to understand the community and feel a strong sense of belongings and affiliation, which in turn develops a permanent sense of identification (Wang & Fesenmaier, 2004).

Hedonic benefits Hedonic benefits bring the participant positive reactions and enjoyment (Nambisan & Baron, 2007). In this situation, community's members do not seek for information but seek for entertainment e.g., playing games, and contests. Kang (2011) concluded that if participating in an online community is perceived as fun or entertaining, members are more likely to visit the community and to spend more time visiting it.

In hospitality research, monetary benefits have been considered as part of individualized services that fulfill consumers' specific needs (i.e., special treatment benefits) (Lee, Ahn & Kim, 2008). To attract new members and online community members, these businesses tend to offer special promotions and coupons (Treadaway & Smith, 2010). By promoting special rates of rooms in hotel brands Facebook fan page, the marketers are able to attract community member to participate and increase the room sales.

2.3 Brand commitment of online community members

Brand commitment is defined as a strong and positive feeling of attachment to a brand (Beatty & Kahle, 1988). Online communities are also found to affect customer loyalty and purchase intention (Kim, et al., 2004). In year 2006, Massari and Passiante suggested that satisfaction and commitment are indicators of brand loyalty. Consumers who are highly committed to a specific brand evaluate competing brands less positively or avoid considering competitors' brands when making purchasing decision (Ahluwalia, Burnkrant & Unnava, 2000). They tend to defend their favorable attitudes toward brands when perceiving a threat such as unfavorable information about their preferred brands or favorable information about competing brands (Chaiken, Liberman & Eagly, 1989). Consumers who perceive such threats tend to secure their positive attitudinal position toward their preferred brands by searching for favorable information about their brand (Jain & Maheswaran, 2000). In other words, consumers want to see evidence that their preferred brands are different from other brands (Chaiken, et al., 1989).

3 ISSUE

In June 6, 2013, the Grand Blue Wave Hotel's fan page gained only 760 likes compared to the competitor hotel, Concorde Shah Alam that received 3,693. This shows that their numbers of fans are distinctively different from their competitor, probably due to lack of online community participation from Facebook user.

The manager of Grand Blue Wave Hotel, Mohd Gazali Sayed Ibrahim (2013) pointed out "Grand Blue Wave has no sales goals through social media. But only by promoting the brand and to have the brands loved by people and engage it in a fun way". This issue raised a critical question as to why there is less participation among online community towards Grand Blue Wave Facebook page despite of the hotel using social media to promote, to make people loved their brand and engage in a fun way.

Thus, the factors that would influence customer commitment toward the brand in a social media are worth to be studied further. Specifically, the researcher studied the motivational functions pertaining to Grand Blue Wave Hotel's Facebook page and how the online community participation would have a significant impact on brand commitment.

4 RESEARCH QUESTIONS

The following research questions were formulated in supporting the objectives and direction for this study:

RQ1: What is the relationship between motivational function and brand commitment?

RQ2: What is the relationship between online community levels of participation and brand commitment?

RQ3: To what extent does online community participation mediates the relationship between motivational function and brand commitment?

5 METHODOLOGY

Respondents for this study consist of Grand Blue Wave Hotel Facebook fans that have already participated in Grand Blue Wave Hotel Facebook page. Google Drive was used as the medium to distribute the online questionnaire. Data was gathered through emails. Email invitations were sent to the potential participants, along with a link to the online questionnaire. The email address of the

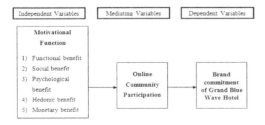

Figure 1. Theoretical framework the role of motivation functions in increasing brand commitment participation adapted from Kang (2011).

potential participants was taken from the personal information from Facebook personal account of fans that participated in the Grand Blue Wave Hotel Facebook page. All data keyed in for analysis using a Statistical Package of Social Science (SPSS), Version 20.0.

6 RESULT

Table 1 shows the result of hierarchical multiple regressions. As can be seen beta (β) value for the first step of analysis (path c) is 0.176 (p = .003). By looking at the beta value there is significantly low contribution to the model. The R value for motivational functions towards brand commitment is .176 while the R^2 value (percentage of variance) for motivational functions towards brand commitment is only 3 percent ($R^2 = .031$, F change = 8.693***).

According to Cohen (1988), when the R value ranges from 0.10 to 0.29 it is predicted as a weak relationship. The second step of analysis (path a) explained that motivational functions do not significantly affect online community participation with standardized coefficient value $\beta = .095$ ($p = 0.1169$)

For the final test, online community participation (mediating variable) contributed minimally

Table 1. Summary of hierarchical multiple regression analysis.

Step and variables	Model 1/ Std. β
Testing step 1 (path c)	
Outcome: Brand commitment	
Predictor: Motivational function	.176**
R	.176
R^2	.031
Adj. R^2	.027
F change	8.693***
Testing Step 2 (path a)	
Outcome: Online Community Participation	.095
Predictor: Motivational function	
R	.095
R^2	.009
Adj. R^2	.005
F change	2.474
Testing Step 3 (path b and c')	Model 2/ Std. β
Outcome: Brand commitment	
Mediator: Online Community Participation (path b)	.173**
Predictor: Motivational function (path c')	-.192**
R	.246
R^2	.060
Adj. R^2	.054
F change	10.579***

Note: *p < .05, **p < .01, ***p < .001.

135

($\beta = 0.173$, p = .004) toward brand commitment. While controlling for online community participation, motivational functions had significant effect on brand commitment. There is some improvement of beta coefficients value which is $\beta = -.192$ (p = .001), but still contributed minimally toward the model. In this case, the R value for online community participation and motivational functions as a predictor towards brand commitment is 0.246 while the percentage of variance was only 6 percent ($R^2 = .060$). The result indicates that there was a significantly low correlation between motivational functions and brand commitment when controlling for mediating effect. It can be said that online community participation significantly influence and mediate the relationship between motivational function and brand commitment.

7 DISCUSSION

RQ1: The results indicated the direction of relationship between independent variable and dependent variable predicted as a very weak relationship. This means low motivational functions tend to lower brand commitment level. When hotel's Facebook page is not fulfilling the benefits of what their fans expected, it will lead to low contribution toward the brand commitment.

RQ2: In this study, online community participation significantly and positively influence brand commitment. When the level of participation in online community is high, it tends to raise the level of brand commitment toward the Grand Blue Wave Hotel. Similarly if the level of participation in online community is low, it tends to lower the level of brand commitment toward the Grand Blue Wave Hotel. The researcher believed that this situation occurs, perhaps, because of mentality of the local that hard to commit to certain brand without having prior experience of the product and services even though they are actively participating in the online community. Interestingly, Kang (2011) reported there is no significant relationship between online community participation and brand commitment.

RQ3: The mediating effects of online community participation on motivational functions and brand commitment increased, but the relationship is still weak. The researcher believed that it is not necessarily for consumers to participate in the online community to commit with the brand of Grand Blue Wave Hotel. Mohd Gazali Sayed Ibrahim (2013) also explained that Grand Blue Wave has their own customer segmentation which is the middle age or baby boomers who less patronized the social media because this type of group preferred a direct conversation.

8 SUGGESTION FOR FUTURE RESEARCH

It is suggested that this study to be conducted using the quantitative approach at other hotels and easily replicated at other hotels so that a larger sample size could be obtained. Future studies also may focus on the enhancement and justification of the scales employed in the present study to help marketers in gaining significant insight into the beneficial aspects of social media communities. The researcher also would like to suggest a study on the level of readiness towards social media in the Malaysian hospitality industry because only limited research has looked at the implications of the 'like' button has for brands.

9 CONCLUSION

The outcomes from this study provide several implications that are essential for the Malaysia hospitality industry. As for the theoretical implication, since little study regarding the impacts of social media toward brand commitment are undertaken, the outcome from this study will add more knowledge and deep comprehension for future study. From practical implication, this study will be an opportunity to marketers to integrate social media as effective business instrument to build brand name and stay forward in the race. Thus, hotel can generate strong relationship and customer commitment towards their brand.

As a conclusion, the outcome from this study will provide additional information for Grand Blue Wave Hotel to enhance its Facebook page because the social media marketing is part of the current trend and able to maintain competitive advantage. The Grand Blue Wave management also should focus to do some improvements in terms of motivational functions that consumer wants and also take proactive actions by implementing necessary strategies to overcome this issue. Due to the importance, uniqueness and rapid growth of social media, marketers and researchers should pay special attention to this phenomenon and further examine the notions and theories in the social media contexts.

REFERENCES

Ahluwalia, R., Burnkrant, R.E. & Unnava, H.R. (2000). Consumer response to negative publicity: The moderating role of commitment. *Journal of Marketing Research, 37,* 203–214.
Armstrong, A.G. & Hagel, J. (1996). The real value of online communities. *Harvard Business Review, 74* (3), 134–141.

Beatty, S.E. & Kahle, L.R. (1988). Alternative hierarchies of the attitude-behavior relationship: The impact of brand commitment and habit. *Journal of the Academy of Marketing Science, 16* (2), 1–10.

Bushlow, E.E. (2012). Facebook Pages and Benefits to Brands. *Strategic Communications Elon University.*

Chaiken, S., Liberman, A. & Eagly, A.H. (1989). Heuristic and systematic information processing within and beyond the persuasion context. In J.S. Uleman & J.A. Bargh (Eds.), *Unintended thought.* New York: Guilford, 212–252.

Cohen, J. (1988). *Statistical power analysis for the behavioral sciences.* Hillsdale, N.J.: L. Erlbaum Associates.

Dholakia, U.M., Blazevic, V., Wiertz, C. & Algsheimer, R. (2009). Communal service delivery: How customers benefit from participation in firm-hosted virtual P3 communities. *Journal of Service Research, 12* (2), 208–226.

Facebook, (2013). Facebook for business. Retrieved February 19, 2013 from: http://www.facebook.com/business

Flavián, C. & Guinalíu, M. (2006). Consumer trust, perceived security and privacy policy. Three basic elements of loyalty to a web site. *Industrial Management & Data Systems, 106* (5), 601–620.

Hartshorn, S. (2010, May 4). *5 Differences Between Social Media and Social Networking.* Retrieved November 18, from: http://www.socialmediatoday.com/SMC/194754

Hwang, Y.H. & Cho, Y.H. (2005). The influence of online community's functions on members' attitude toward the community and off-line meetings. *Korea Tourism and Leisure Research,* 17(4), 141–159.

Jain, S.P. & Maheswaran, D (2000). Motivated reasoning: A depth-of-processing perspective. *Journal of Consumer Research, 27,* 358–371.

Kim, W.G., Lee, C. & Hiemstra, S.J. (2004). Effects of an online virtual community on customer loyalty and travel product purchases. *Tourism Management, 25,* 343–355.

Kang, J. (2011). Social media marketing in the hospitality industry: The role of benefits in increasing brand community participation and the impact of participation on consumer trust and commitment toward hotel and restaurant brands. *Research for Hospitality Management. Iowa State University.* UMI Number: 3494053.

Krahl, A. (2013). Social Media and Internet Tools In Hospitality Marketing. *Bachelor's thesis March 2013Degree Programme in Tourism*

Lanz, L., Fischhof, B. & Lee, R. (2010) *How are Hotels Embracing Social Media in 2010? Examples of How to Start Engaging.* New York: HVS Sales and Marketing Services.

Lee, Y., Ahn, W. & Kim, K. (2008). A study on the moderating role of alternative attractiveness in the relationship between relational benefits and customer loyalty. *International Journal of Hospitality & Tourism Administration, 9* (1), 52–70.

Madupu, V. (2006). Online brand community participation: Antecedents and consequences. *(Doctoral dissertation). Available from ProQuest Dissertations and Theses database.* (UMI No. 3230964).

Mohd Ghazali, S.I. (2013), Interview Social Media towards Brand Commitment.

Muniz, A.M. & O'Guinn, T.C. (2001). Brand Community. *Journal of Consumer Research, 27* (4), 412–432.

Naveed, N. (2012). Role of social media on public relation, brand involvement and brand commitment. Interdisciplinary journal of contemporary research in business, 3 (9).

Paris, C.M., Lee, W. & Seery, P. (2010). The role of social media in promoting special events: acceptance of Facebook Events. *Information and Communication Technologies In Tourism,* 14, 531–541.

Peter, J.P., Olson, J.C. & Grunert, K.G. (1999). *Consumer behavior and marketingstrategy, European edition.* Berkshire: McGraw-Hill

Preece, J., Nonnecke, B. & Andrews, D. (2004). The top five reasons for lurking: Improving community experiences for everyone. *Computers in Human Behavior, 20* (2), 201–223

Ridings, C., Gefen, D. & Arinze, B. (2006). Psychological barriers: Lurker and poster motivation and behavior in online communities. *Communications of the Association for Information Systems, 18,* 329–354.

Seth, G. (2012). The Effects of social media networks in the hospitality industry. *University of Nevada, Las Vegas.* Paper 693

Socialtimes.me, (2013). Social Media Global Info Center. *Facebook Statistics.* Retrived April 14, 2013 from: http://socialtimes.me/

Treadaway. C. & Smith, M. (2010). *Facebook marketing: An hour a day.* Indianapolis, Indiana: Wiley Publishing, Inc.

Wang, Y.C. & Fesenmaier, D. (2004). Towards understanding members' general participation in and active contribution to an online travel community. *Tourism Management, 25* (6), 709–722.

Theory and Practice in Hospitality and Tourism Research – Radzi et al. (Eds)
© *2015 Taylor & Francis Group, London, ISBN 978-1-138-02706-0*

Predictors of guest retention: Investigating the role of hotel's corporate social responsibility activities and brand image

N.Z. Othman
KFCH International College, Malaysia

M.A. Hemdi
Faculty of Hotel and Tourism Management, Universiti Teknologi MARA, Shah Alam, Malaysia

ABSTRACT: The purpose of this study is to investigate the relationship between Corporate Social Responsibility (CSR) activities, hotel brand image and guest retention. Data from 283 respondents from twenty four five-star hotels in Kuala Lumpur were analyzed. Findings from this study showed that CSR activities have a significant positive influence on hotel brand image and consequently influence guest to return. Specially, CSR activities pertaining to sustainable development, environmental practices, involvement in community project and community involvement in quality of life have significant positive influence towards hotel brand image and guest retention. Hotel brand image has also been found to be a significant mediator between CSR activities and guest retention. Theoretical and practical implications of the findings were discussed. Limitations and future research approaches were forwarded.

Keywords: Brand image, Corporate Social Responsibility (CSR), guest retention

1 INTRODUCTION

Tourism industry contributes significantly towards Malaysian economy. By 2020, Malaysia is targeting to receive 36 million tourist arrivals accounted for RM168 billion tourist receipts (*Economic Transformation Programme Annual Report*, 2011). The developments of tourism industry in Malaysia have increased the hotel rooms' supply. As more hotels are built, competitions for guests become stiffer. Thus, hotels need to be innovative and proactive in attracting and retaining guests in order to compete and sustain. Ability to retain loyal guest plays an important aspect towards hotel performance and profitability. Nadiri, Hussain, Ekiz and Erdagon (2008) stated that guest retention occurs when the guest continue purchasing a product or service over an extended period of time.

Besides good services and excellent physical facilities of the hotel, Corporate Social Responsibility (CSR) activities of the hotel towards environment and community are also important in influencing guest retention. Previous studies show that CSR was significant influence towards hotel profits (Mackey, Mackey & Barney, 2007), employee organizational commitment (Ali, Rehman, Ali, Yousaf & Zia, 2010), improved investment (Maigan, Ferrell & Ferrell, 2005) and good organizational image (Cramer, Heijden & Jonker,

2005). Nonetheless, there were limited studies on the effects of CSR activities in hotel setting (Bohdanowicz & Zientara, 2008, Hemdi & Othman, 2013). Hence, the purpose of this study is to investigate whether CSR activities and brand image have significant influence on guest retention.

2 LITERATURE REVIEW

2.1 *Guest retention*

Guests retention is defined as the process when customers continue to buy products and services within a determine time period. Guest retention occurs when a guest is loyal to a company, a brand or a specific product or service, expressing long-term commitment and refusing to purchase from competitors (Narayandas, 1996). The measures of guest retention can be assessed through repurchase intention. Ranaweera and Prabhu (2003) defined future behavioral intentions as the future propensity of a customer to continue or to stay with the service provider. Repurchase intention are usually obtained from surveys of current customers assessing their tendency to purchase the same brand, same product or service from the same company. Cronin, Brandy and Hult (2000) have treated behavioral intentions and repurchase intention as synonymous constructs. This is in line with many previous studies that have used intention to repurchase (Almohammad, 2010; Yen & Lu, 2008) and intention to spread word of mouth (Nadiri et al., 2008) as elements of guest retention.

2.2 Corporate social responsibility toward the environment

According to Thomas and Nowak (2006), CSR involves the corporate sustainability, corporate citizenship, corporate social investment and corporate governance. Mohr, Webb and Harris (2001) defined CSR as company's commitment to minimizing or eliminating any harmful effects and maximizing its long-run beneficial impact on society. Argandona (2010) study on tourism industry proposed sustainability developments as CSR activities. Whooley (2004) classified CSR into four categories: environment, community, workplace and marketplace. For the purpose of this study, CSR dimensions pertaining to the environment and community were chosen since the subjects of the current study were the hotel guests.

The trend in the hospitality and tourism industry was focused on environmental concerns, use of technology and efficient use of energy (Kalisch, 2002). Kassim (2006) argued that hotels of whatever size impacts the environment and society in several ways including energy consumption, water consumption, waste production, waste water management, chemical use and atmospheric contamination and local community initiatives. Energy consumption is the largest environmental impact of every hotel organization and certain hotel actively explore and implement initiatives that could reduce the energy consumption (Kassim, 2006). Hotel's CSR activities pertaining to the environment such as using low energy lighting, regular boiler efficiency audits and installing maximum water temperatures are being practiced (Kalisch, 2002). Nowadays, many guests are becoming increasingly conscious of hotel's CSR activities when making purchasing decision. Study by Ali et al. (2010) found that CSR activities are possible to create a good sense of consumer satisfaction and loyalty. Further, Cooil, Keiningham, Aksoy and Hsu (2007) stated that consumer satisfaction on CSR activities able to retain consumers and sustain the business.

2.3 Corporate social responsibility toward the community

Afiya (2005) suggested that CSR activities with regards to community improvement on quality of life and community events would create positive image in customers' view. To achieve this, Atakan and Eker (2007) proposed that companies need to manage and communicate their CSR activities so that CSR efforts become part of their business identity. Jones et al., (2006) suggested CSR activities such as fundraising for charity programs or sponsoring community public facilities could create a sense of belonging towards the community. Bohdanowicz and Zientara (2008) supported that CSR activities towards stakeholders such as workers, consumers, and society at large would help organization to compete and sustain its business. Within this context, Moir (2001) agreed that one of the reasons for organization to engage in CSR activities is because that is what the society expects and the organization believes that it should take such actions.

2.4 Hotel brand image

Brand image is defined as perceptions about a brand as reflected by the brand associations held in consumer memory (Keller, 2008). Brand image is a feeling that can be translated though brand recognition, brand recall and brand identity. Positive image associated with a brand is found to be a significant purchase influencer. The significant of brand image is based on the proposition that consumers buy not only a product (commodity), but also the image associated to the product, such as power, wealth, sophistication, and most importantly identification and association with other users of the brand. CSR activities toward the environment and community activities undertaken by a company affect the way consumers perceive the company and subsequently create a brand image in the view of consumers. Blumenthal and Bergstrom (2003) highlighted four reasons to adopt CSR: to communicate the brand promise, to maintain customer loyalty, to maximize the effect of investments and to avoid conflicts with stakeholders. Lee and Park (2009) supported the notion that CSR activities will improve company image, increase employee morale and retention, build good relationships with governments and meet the expectations of social groups. In addition, Kayaman and Arasli (2007) found that brand equity components, brand quality, brand image and brand awareness significantly influence hotel's brand loyalty.

Based on the literature review pertaining to the study variables, the framework of this study is proposed as followed:

Figure 1.   Research framework.

3 METHODOLOGY

The survey questionnaire was personally distributed to hotel guests who have had experiences staying at twenty four five-star hotels in Kuala Lumpur. From the 600 questionnaires distributed, a total of 283 questionnaires (47.2%) were completed, coded and analyzed. Regression analysis was used to examine the relationship between variables selected in this study. Guidelines on mediation by Baron and Kenny (1986) were used to test the mediating effect of brand image. The Cronbach's Alpha for the study variables ranged from 0.61 to 0.92.

4 RESULTS

Table 1 shows the results of the regression analysis of CSR activities on guest retention.

Table 1. Summary of hierarchical regression of CSR activities, and Brand Image (BI) on Guest Retention (GR).

Predictors	Model 1 Std. B GR	Model 2 Std. B BI
CSR to Environments:		
• Sustainable development	.03	.13*
• Environmental practices	.24**	.20**
• Energy supply	.02	.60
CSR to Community:		
• Involvement in community project	.15*	.43**
• Involvement in quality of life	.39**	.12*
R^2	.38	.53
Adj. R^2	.37	.53
R^2 Change	.38	.53
F-Change	34.02**	63.24**

Note: $N = 283$, *$p. < 05$; **$p. < 01$, Five-point Likert Scale, GR = Guest Retention, BI = Brand Image.

As illustrated in Table 1, when CSR activities were regressed on guest retention, the five CSR activities were able to explain 38.0 percent ($R^2 = 0.38$, F-Change = 34.02, $p < .01$) of the variance in guest retention (Model 1). From the five CSR activities, community involvement in quality of life ($\beta = 0.39$, $p < .01$) and environmental practices ($\beta = .24$, $p < .01$) significantly contributed to the prediction of guest retention, followed by involvement in community project ($\beta = .15$, $p < .05$).

In Model 2, when CSR activities were regressed on brand image, CSR activities were able to explain 53.0% ($R^2 = .53$, F-Change = 63.24, $p < .01$) of the variance in hotel brand image. Involvement in community project had the most impact on the prediction of hotel brand image ($\beta = 0.43$, $p < .01$), environmental practices ($\beta = 0.20$, $p < .01$), followed by sustainable development ($\beta = 0.13$, $p < .05$) and community involvement in quality of life ($\beta = 0.12$, $p < .05$).

The procedures as suggested by Baron and Kenny (1986) were followed in order to examine the mediating roles of hotel brand image. According to Baron and Kenny (1986), the following conditions must be present for mediation effects: (1) the IV (CSR activities) must significantly affect the mediator (Brand Image), (2) the IV must significantly affect the DV (Guest Retention), (3) the mediator must significantly affect the DV, and (4) the effect of the IV on the DV shrinks upon the addition of the mediator to the model. As shown in Table 1, only environmental practices, involvement in community project, and involvement in quality of life able to meet the conditions for mediation. Table 2 provides the summary of the regression analyses to ascertain the mediation effects of brand image on the relationship between CSR activities and guest retention.

The results from Table 2 showed that hotel brand image was significantly and positively predicted guest retention ($\beta = 0.54$, $p < .01$). On the mediating effect, the effect of environmental practices on guest retention had a decreasing β-value ($\beta = 0.17$, $p < .01$) in the presence of the hotel brand image, implying partial mediation. Similarly, the effect

Table 2. Impact of mediated regression.

Predictors	Model 1 Std. B GR	Model 2 Std. B BI	Model 3 Std. B GR
CSR Activities:			
• Environmental practices	.24**	.20**	.17*
• Involvement in community project	.15*	.43**	.02
• Involvement in quality of life	.39**	.12*	.35**
Brand Image			.54**

Note:. $N = 283$, *$p < .05$; **$p < .01$, Five-point Likert Scale, GR = Guest Retention, BI = Brand Image.

of community involvement in quality of life on guest retention had a decreasing β-value ($\beta = 0.35$, $p < .01$) in the presence of hotel brand image, also implying partial mediation. The effect of involvement in community project became insignificant ($\beta = .02$, $p > .05$) in the presence of hotel brand image, thus, full mediation occurred.

5 DISCUSSIONS

The objective of this study is to determine whether CSR activities and hotel brand image have positive and significant influence on guest retention. Results of this study indicated that CSR activities pertaining to environmental practices, sustainable development, involvement in community projects and community involvement in quality of life have significant and positive influence on hotel brand image and guest retention. Hotel brand image was also found to be a significant mediator on the relationship between CSR activities and guest retention.

Findings from this study concluded that guest who has higher perception of the hotel involvement in improving local quality of life will express higher intentions to return. Specifically, if the hotel were perceived to have employed local workforce and bought products made by the local people, they would expressed higher intentions to return. These findings were in line with Moir (2010) who found that hotel involvements in improving local quality of life have significant influenced on guest retention. Further, this study corroborated Scofidio (2007) whereby community projects such as sponsoring local events, donating for local programs, and volunteering in local activities were significantly influenced guest intentions to return. Respondents in this study also expressed positive brand image and higher intentions to comeback when they believed the hotel have implemented friendly environmental practices such as having water saving devices in bathroom, using recycle packaging for room amenities and using environmental friendly products. Similarly, if the guests perceived that the hotel shows concerns toward conserving the environment through saving the energy, uses pleasant environmental signage in every room, and uses environmental friendly products, they will formed a high positive perception toward the hotel and influence them to spread a positive word of mouth about the hotel. This consequently increase their willingness to return again. These findings were similar with previous researchers (Jones et al., 2006; Lee & Park, 2009).

On theoretical implications this study confirmed the linkage between CSR activities, hotel brand image and guest retention. This study also gave evidence to the importance of hotel brand image as the intervening variable from CSR activities predictor and guest retention. By demonstrating the existence of significant direct and indirect effects of CSR activities, hotel brand image and guest retention, this study provides clear evidence that CSR activities are important in fostering guest retention in the hotel industry. Practically, the results from this study will help hotel managers and operators to plan and implement strategies pertaining to CSR activities. This study suggests that hotel operators should actively involve and organize community projects in order to improve the local community quality of life. Activities such as sponsoring public community facilities and events, hiring more local people to work, and buying products from local community would help to create a strong positive image about the hotel and subsequently could influence guest to spread positive word of mouth about the hotel.

6 LIMITATIONS AND SUGGESTIONS

The CSR activities investigated in this study were limited in scope. Only CSR activities pertaining to environment and community were examined. Future researchers may need to widen the scope of investigations (for example, workplace and marketplace). Additionally, this study was limited to guest who had experienced staying at five-star hotels. Future researchers may want to include guest from three- or four-star hotels from other locality.

REFERENCES

Afiya, A. (2005). Corporate Social Responsibility - Making Business Sense. *Caterer & Hotelkeeper*, 195(5), 4392.

Ali, I., Rehman, K.U., Syed Irshad Ali, S.I., Yousaf, J. & Zia, M. (2010). Corporate Social Responsibility Influences, Employee Commitment and Organizational Performance. *African Journal of Business Management*, 4(12), 2796–2801.

Almohammad, A. (2010). *The Role of Brand Equity in the Effects of Corporate Social Responsibility on Consumer Loyalty*. Master Thesis, Universiti Sains Malaysia.

Argandona, A. (1998). The Stakeholder Theory and the Common Good. *Journal of Business Ethics*, 17, 1093–1102.

Atakan, M.G.S. & Eker, T. (2007). Corporate Identity of a Socially Responsible University: A Case from the Turkish Higher Education Sector. *Journal of Business Ethics*, 76, 55–68.

Baron, R.M. & Kenny, D.A. (1986). The Moderator-Mediator Variable Dinstiction in Social Psychological Research: Conceptual, Strategic amd Statistical Consideration. *Journal of Personality and Social Psychology*, 51, 1173–1182.

Bohdanowicz, P. & Zientara, P. (2008). Hotel Companies' Contribution to Improving the Quality of Life of Local Communities and the Well-Being of Their Employees. *Tourism and Hospitality Research,* 9(2), 147–158.

Blumenthal, D. & Bergstorm, A. (2003). Brand Councils that Care: Towards the Convergence of Branding and Corporate Social Responsibility. *Journal of Brand Management*, 10, 327–341.

Cooil, B., Keiningham, T.L., Aksoy, L. & Hsu, M. (2007). A Longitudinal Analysis of Customer Satisfaction and Share of Wallet: Investigating the Moderating Effects of Customer Characteristics. *Journal of Market,* 71(1), 67–83.

Cramer, J., Heijden, V.D.A. & Jonker, J. (2005). Corporate Social Responsibility: Making Sense through Thinking and Acting. *Business Ethics: A European Review*, 15(4), 9–380.

Cronin, J.J., Brady, M.K. & Hult, T.M. (2000). Assessing the Effects of Quality, Value and Customer Satisfaction on Consumer Behavioral Intentions in Service Environments. *Journal of Retailing, 76*(2), 193–218.

Economy Transformation Programme Annual Report (2011). Tourism Program. Retrived December 23, 2012, from http://www.epu.gov.my/tourism.

Hemdi, M.A. & Othman, N.Z. (2013). Corporate Social Responsibility (CSR) Activities, Brand Image and Hotel Guest Retention. *Proceeding of The International Hospitality & Post Graduate Conference 2013.17–21, UiTM Shah Alam, Malaysia,*

Jones, P., Comfort, D. & Hillier, D. (2006). Reporting and Reflecting on Corporate Social Responsibility in the Hospitality Industry: A Case Study of Pub Operators in the UK. *International Journal of Contemporary Hospitality Management,* 18(4), 329–340.

Kalisch, A. (2002). *Corporate Futures: Social Responsibility in the Tourism Industry,* Tourism Concern, London.

Kasim, A. (2006). The Need for Environmental and Social Responsibility in the Tourism Industry. *International Journal of Hospitality and Environmental Management,* 7(1), 1–22.

Kayaman, R. & Arasli, H. (2007). Customer Based Brand Equity: Evidence from the Hotel Industry. *Managing Service Quality*, 7(1), 92–109.

Keller, K. (2008). *Strategic Brand Management: Building, Measuring and Managing Brand Equity* (3rd *Ed.*): Pearson International Edition.

Lee, S. & Park, S. (2009). Do Socially Responsible Activities Help Hotels and Casinos Achieve their Financial Goals?. *International Journal of Hospitality Management,* 28(1), 105–112.

Mackey, A., Mackey, T.B. & Barney, J.B. (2007). Corporate Social Responsibility and Firm Performance: Investor Preferences and Corporate Strategies. *Academy of Management Review*, 32(3), 817–835.

Maigan, I., Ferrell, O.C. & Ferrell, L. (2005).A Stakeholder Model for Implementing Social Responsibility in Marketing. *European Journal of Marketing*, 39 (9), 77–956.

Mohr, L.A., Webb, D. & Harris, K. (2001). Do Consumers Expect Companies to be Socially Responsible? The Impact of Corporate Social Responsibility on Buying Behaviour. *Journal of Consumer Affairs*, 35(1), 45–71.

Moir, L. (2001). What Do We Mean by Corporate Social Responsibility?.*Corporate Governance*, 1(2), 16–22.

Nadiri, H., Hussain, K., Ekiz, E.H. & Erdagon, S. (2008). An Investigation on the Factors Influencing Passengers' Loyalty in the North Cyprus National Airline. *The TQM Journal*, 20(3), 265–280.

Narayandas, N. (1996). The Link between Customer Satisfaction and Customer Loyalty: An Empirical Investigation. *Harvard Business School*, 97–017.

Ranaweera, C. & Prabhu, J. (2003). The Influence of Satisfaction, Trust and Switching Barriers on Customer Retention in a Continuous Purchasing Setting. *International Journal of Service Industry Management*, 14(3), 374–395.

Scofidio, B. (2007). Incorporate CSR into Your Meeting. *Corporate Meetings and Incentives*, 26(11), 6–6.

Thomas, G. & Nowak, M. (2006). Corporate Social Responsibility: A Definition. *Working Paper Series*, 62, 1–20.

Whooley, N. (2004). Social Responsibility in Europe. Retrieved on August 22, 2012 from www.pwc.com/extweb/newcolth.nsf/0/503508DDA107A61885256F35005C1E35

Yen, C.H. & Lu, H.P. (2008). Factors Influencing Online Auction Repurchase Intention. *Internet Research*, 18(1), 7–10.

Theory and Practice in Hospitality and Tourism Research – Radzi et al. (Eds)
© *2015 Taylor & Francis Group, London, ISBN 978-1-138-02706-0*

Sponsorship leverage and its effects on brand image

A.H. Abdul-Halim, A.R. Mohamad-Mokhtar, N.A. Nordin, A.R. Ghazali,
W.S. Wan-Abdul-Ghani & N.F. Mohd-Sah
Faculty of Business Management, Universiti Teknologi MARA, Shah Alam, Malaysia

ABSTRACT: Although leveraging is not the sponsor's responsibility, substantial numbers of research concluded that sponsorship effectiveness is highly related to the activeness of sponsors to leverage their investment. Because corporation image is perceived by the level of activities it involves in the market, monetary investment by a sponsor in and event is not enough. Sponsoring companies should spend more beyond the sponsorship fee to maximize the Return On Investment (ROI) and Return On Objective (ROO) in the sponsorship. This research discusses the issues of sponsorship, brand image as well as sponsorship leverage strategies through Sponsorship Leveraged Packaging (SLP), association and co-visibility, TV sponsorship, emotional connections and Cause-Related Marketing (CRM). In-depth researches suggested that more studies should be undertaken on the issues of linking leveraging strategies with brand objectives and most importantly, exploring the mechanism to measure the outcomes of both sponsorship and leveraging activities.

Keywords: sponsorship leverage; event management; brand image; sponsorship activation; brand awareness

1 INTRODUCTION

Sponsorship, as distinguished by Woodside and Summers (2009) in support to Olkkonen (2001) involves two principal activities: "an exchange between sponsor and property, where the property receives compensation and the sponsor obtains the right to associate itself with the property"; and "leverage by the sponsor of this association through marketing activities communicates through the sponsorship". In event sponsorship, the amount of money a sponsor invests into an event is merely a start. A good strategist in sponsorship will not hesitate to spend more beyond the sponsorship fee to maximize the return on investment (ROI) and return on objective (ROO) in the sponsorship (Quester & Thompson, 2001; Stokes, 2010). This is called "leverage" and it is a practice encouraged for a sponsor. By spending at least the same value of the sponsorship they purchased in their own promotional activity relating to the sponsored event (Harrison, 2010; Quester & Thompson, 2001), the impact of the sponsoring brand towards the event's audience would be increased. However, it is not unusual for many sponsors to spend several times the property rights fee to support their sponsorship investments (Quester & Thompson, 2001). Most of prior studies on sponsorship activities discussed facets such as sponsoring sportsmen, sporting teams as well as sporting events and other

cultural and special events (Carrillat, Harris & Lafferty, 2010; Smith, 2004; Sylvestre & Moutinho, 2008; Tsiotsou & Alexandris, 2009), but omitting the theory of leveraging (Skildum-Reid, 2008; Uggla, 2006; Woodside & Summers, 2009).

2 AIMS OF THE STUDY

This study is intended to investigate how sponsorship leverage activities could help to increase a corporation's brand image on its target audience. In order to achieve this, literature review on issues relating to public relations, sponsorship, brand association and image transfer as well as sponsorship leverage strategies have been conducted.

3 LITERATURE REVIEW

The function of brands has been described as an approach to identify and differentiate products (Uggla, 2006) and product line decision. However, brand definitions are more complex because "brands can also be defined as products, corporations, persons and places" (De Chernatony, 2001). Uggla (2006) has given an interesting example of how a famous football club in Europe did not only buy a skilful soccer player, but also "reinforced their corporate brand presence and brand portfolio structure" where this particular soccer player "has very strong associations, brand recognition and loyalty among fans". Therefore, understanding the definition of brand and their role in the formation of attitude towards brands is necessary for sponsors to comprehend the dynamics of their brands and how consumers evaluate and make brand choices (James, 2005).

A corporate brand marketing strategists usually believes that the brand itself should be enough to attract it consumers (Schultz & Hatch, 2003). Frustratingly, many corporate brand managers are complacent enough to realize the opportunity of linking the corporate brand and other brand assets in their portfolio with other prospective elements such as associating themselves with events or even other brands. Furthermore, corporate marketing brand managers tend to ignore the power and possibilities in brand association, through sub-brands such as Nescafé by Nestlé or Playstation by Sony and endorsed brands such as Holiday Inn by Intercontinental Hotels Group (Uggla, 2006). Brand associations is how we take advantage of suitable situations, product categories and product attributes to relate with customer which could benefit both parties (James, 2005). Many companies opted to promote the brand as it is, simply because the economics of it. They believed that managing and developing a single corporate brand is a more cost-effective and time-saving rather than managing a portfolio of other sub-brands or involving themselves with events which for them, are of "different nuances" (Aaker, 2004; Melewar & Walker, 2003).

Consumer-derived brand, on the other hand, are communicated in the associations a company make with the consumer of which a strong and positive associations will contributes to the establishment of the brand and this associations act as a leverage depending on the types of associations it has with its consumers (James, 2005). Generally, it is about how likeable a brand is to the consumer and it is the likability that is assisting in the formation of the brand itself. Consumer's associations with corporate brands revolve around three conceptual main streams, associations based on social expectations, associations based on corporate personality traits and associations based on trust (Berens & van Riel, 2004).

3.1 Public relations and sponsorship

Sponsorship, as one of the public relation branches is fast becoming a formidable tool to reach to the core of customers brand awareness and brand associations (Muotka & Mannberg, 2004). One of the primary reasons for undertaking sponsorship is understood as to develop image (Smith, 2004). A report by IEG Network stated that, sponsorship

is regarded as the world's fastest growing form of marketing investment with $24.6 billion worldwide being spent in 2001 alone (Dolphin, 2003). Since a decade ago, sponsorship has also become a conventional marketing communications tool as evidence showed a worldwide sponsorship spending reaching US$33 billion (International Events Group, 2014as cited in Woodside & Summers, 2009). Data published by International Events Group in 2009 showed that "Asia Pacific is the world's fastest-growing sponsorship region, with spending anticipated to increase 7.4 per cent from US$9.5 billion in 2008 to US$10.2 billion in 2009, equivalent to approximately 22 per cent of the global sponsorship market" (Johnston, 2010). Recent report by IEG also shows that Asia Pacific region leads all other geographies with a forecast growth rate of 5.6 per cent (International Events Group, 2014).

Likewise, public relation could also be used to support sponsorship activities in return. Sylvestre and Moutinho (2008) in their study relating to activities involved to leverage sponsorship by four companies in London found that public relations activities were used to support advertising for branding and to promote corporate hospitality and community relations of sponsoring organizations in an event. It could be understood that apart from being one of the major tools in public relations activities itself, sponsorship leveraging activities also include public relations as one of its tool simultaneously.

4 METHODOLOGY

In preparing this paper, the researcher has accessed a number of academic literatures, as well as text books and journal articles to obtain data and information. Existing academic sponsorship discussions form the framework of this study and work as a guideline for the research, which is the most appropriate method considering the research setting. Substantial hour of browsing related websites has also been done. No direct interview or survey has been attempted. Therefore, the outcomes of this research are mainly based on secondary and tertiary sources. Škoda Australia has been selected as the subject of this case study.

5 A BRIEF HISTORY OF ŠKODA AUSTRALIA

The Škoda brand, originated from Czech Republic, first came to Australia in 1965 with the Škoda Octavia model but discontinued its Australian distribution in 1983 (Fallah, 2007). In December 1990, the

Government of the Republic entered into an agreement with the German Volkswagen Group and began to operate on 16 April 1991 under the name Škoda, automobilová a.s. (Škoda Auto, 2011). By the year 2000, the Škoda Auto had been owned 100 percent by Volkswagen Group and after 24 years of absence, in 2007, the Volkswagen Group has announced the return of Škoda Auto to Australia (Fallah, 2007). The models which were re-introduced to Australian were Škoda Fabia and Škoda Roomster (Fallah, 2007; Škoda Auto, 2011).

5.1 *Events sponsored by Škoda in Australia*

Škoda had done tremendously in sponsoring major sporting events in Australia. Two major sporting events sponsored by the brand are Tour Down Under and Melbourne UCI Road World Championships as major suppliers of support vehicles and official car partner (Škoda Auto, 2011).

6 DISCUSSION

There are ranges of activities that could be used to leverage the sponsorship investment. This paper will discuss on five elements namely Sponsorship Leverage Packaging (SLP), association and visibility with other sponsors, TV sponsorship, emotional connections and Cause-Related Marketing (CRM). Stokes (2010) indicated that a sponsor may use many media to meet the objectives and reach out to its audience. These are touch-points, known also as "activation" or "maximization" of sponsorship investment. In other words, these activities are sponsorship leveraging.

6.1 *Sponsorship Leveraged Packaging (SLP)*

SLP involves depicting the sponsored property's image and logos on the sponsoring brand's packaging (Woodside & Summers, 2008). In this case, instead of displaying Škoda's logo on the marketing items for the event, the event's logo is rather displayed on Škoda's products packaging (e.g., Škoda service centre, Škoda car pamphlets and brochures, Škoda cars). It is to associate the event with the car brand, a situation where a customer would be reminded of the event every time they see a Škoda car, passing by Škoda service centre or stumble upon Škoda's promotional items, hence the next time they attend any cycling events for example, they would associate the events with the brand, although the events are not actually sponsored by Škoda. The "association" aspect is highly important in leveraging sponsorship because any association which a sponsor established with the events could also be established in memory of the

consumers with the brand (Woodside & Summers, 2009). As such, when consumers are emotionally involved with a sponsored property and identify with it, it may also lead to a strong sense of attachment with the sponsor (Sirgy, Lee, Johar & Tidwell, 2008).

6.2 *Association and co-visibility*

By actively involving the brand with other types of activities held by different co-sponsors, they could impose the brand on different types of customers. The visibility of the sponsor during the event is conducive of a "halo" effect, where in some competitive environment; this could also be considered as an ambush marketing (Quester & Thompson, 2001). Although some types of ambush marketing is considered against the laws, more elaborated types of ambush marketing might not as long as: (1) they do not contravene any existing law; or (2) enforcement is unlikely, given uncertainty regarding the application of the law, the timing of the event, and/or the costs of litigation (Chan & Hudson, 2007). For example, as many major sponsors and partners of the Tour Down Under or Melbourne UCI Road World Championships comes from different kind of businesses such as hospitality, state government, product manufacturers, clubs to telecommunication companies, different kind of customers would be available to the sponsor. Example of such activities is when a co-sponsor organizes an event; Škoda could send a group of people driving their cars to impose a perspective that most of the event attendees are actually driving their brand.

6.3 *TV sponsorship*

The strength of a sponsorship to persuade its target audience is embedded in "ability to convey the commercial message and influence the consumer in a more voluntary fashion than the standard advertising sell" (Smolianov & Shilbury, 2005). As stated by Pegg and Patterson (2010), television audiences today are more touchable through "sponsored by" advertising rather than television ads that interrupts in the middle of a program. By televising the event, Škoda Australia could be more in touch not only with the audience attending the event, but also those who are watching it on television. It has been proven that airing an event could generate more media exposure and media value, integrates sponsorship messages more closely with audiences and make them stand out more, for example by switching electronic board messages during televised sports events to ensure brands are visible to the right audience at the right time (Comperior Research, 2006).

6.4 Emotional connections with partners

Receiving an exposure is ineffective if a sponsor did not establish emotional connections with their partners and the event organizer. While sponsorship is probably the fastest growing types of marketing it may be under-utilized if it does not provide some kind of added value to the event organizer and co-sponsors. Skildum-Reid, Australian sponsorship guru and author of The Sponsor's Toolkit, in an interview said "too many companies still believe visibility is all". She suggested that if only sponsors opt to use a more "catalyst approach" by "adding value to that sponsorship" can only the sponsorship "add relevance, resonance and real emotional weight to any or all of marketing activities" (Moore, 2008; Skildum-Reid, 2008). Take the example of Wembley construction in London where partners such as Npower and Microsoft are both actively involved in providing the stadium with its energy and IT infrastructure (Comperior Research, 2006). Škoda Australia could practice this by investing in developing the infrastructure of Cancer Council (Tour Down Under charity partner) or providing Škoda cars as the special transportation to the event site.

6.5 Cause-related marketing

The aim of CRM or also could be known as Corporate Social Responsibility (CSR) is to link the brand with the communities; involves a non-profit motivated giving and enables the firm to contribute to a non-profit cause while tying the contributions to sales (Grau & Folse, 2007). It is an increasingly common form of promotion. As reported by IEG, expenditures on this form of communicating with customers are expected to surpass $828 million in North America (Pracejus & Olsen, 2004). Insurance brand Norwich Union for example was quick to spot the potential link between its title sponsorship of UK Athletics and the growing concern about child obesity by responding to integrate a cause-related marketing component into its overall sponsorship strategy based around health and education (Comperior Research, 2006). It was found that CRM had a positive impact on perceptions of the sponsoring firm (Pracejus & Olsen, 2004), particularly when the CRM is targeting consumers with a greater level (Grau & Folse, 2007). Since Australian are among the most wealthiest and steadiest in term of economy (Reserve Bank of Australia, 2010) and majority of them living along the coast (Department of Foreign Affairs and Trade, 2010), the possibility to involve in a non-profit cause related to cycling event is likely to be higher. For example, Škoda could involve in a non-profit cause with Cancer Council. It has also been proven that the "positive impact of CRM was found to be greater when the association was presented as a local, as opposed to a national" (Pracejus & Olsen, 2004). Thus, Škoda could also sponsor a cycling academy for the residents of Adelaide or Melbourne.

7 CONCLUSION

Intelligent marketing strategies, operations, background of the people within the organization and the determination in exploring opportunities are among the vital factors that are required to strengthen a brand. It is indicative that a well devised plan especially in the area of promotion which includes public relations and sponsorship is a requisite. Merely investing in sponsorship is just not enough. A corporation image is highly depending on the level of activities it involves in the market. The most important of sponsorship aspect is the leveraging. Whether it is sports, arts, cultures, entertainment, education or food and beverages, "the type of sponsorship selected may not be as important strategically as how the sponsorship is leveraged" (Sylvestre & Moutinho, 2008). Therefore, active involvement in a sponsored event must become a priority.

With the above mentioned information in mind, there are still very scarce studies on the effect of sponsorship leveraging activities on a corporation's brand image and brand awareness. Most of prior studies on sponsorship activities discussed facets such as sponsoring sportsmen, sporting teams as well as sporting events and other cultural and special events, but omitting the theory of leveraging. Furthermore, as the market for sponsorship itself becomes intensely competitive and challenging, sponsorship investments should be carefully managed and the right measuring method should be used to ensure their effectiveness.

ACKNOWLEDGEMENTS

This study was made possible by the continuous support from Universiti Teknologi MARA through Research Incentive Grant (RIF). To Dr. Wan Edura Wan Rashid, Head of Center for Intergrated Information and Publication, Institue of Business Excellence, Universiti Teknologi MARA, this paper could never have been completed without the motivation and unremitting effort from her.

REFERENCES

Aaker, D.A. (2004). Leveraging the corporate brand. *California management review, 46*(3), 6–18.
Berens, G. & van Riel, C. (2004). Corporate associations in the academic literature: Three main streams

of thought in the reputation measurement literature. *Corporate Reputation Review, 7*(2), 161–178.

Carrillat, F.A., Harris, E.G. & Lafferty, B.A. (2010). Fortuitous brand image transfer: Fortuitous Brand Image Transfer: Investigating the side effect of concurrent sponsorships. *Journal of Advertising, 39*(2), 109–124.

Chan, T. & Hudson, E. (2007). Ambush Marketing Legislation Review. *Melbourne: Frontier Economics Pty Ltd.*

Comperior Research. (2006). Innovation in sponsorship. In D. Smith (Ed.), *Sponsorship works* (3rd ed., Vol. 3, pp. 1–38). London: Sport Business Group.

De Chernatony, L. (2001). A model for strategically building brands. *The Journal of Brand Management, 9*(1), 32–44.

Department of Foreign Affairs and Trade. (2010). About Australia: People, cultures and lifestyles, from http://www.dfat.gov.au/facts/people_culture.html

Dolphin, R.R. (2003). Sponsorship: perspectives on its strategic role. *Corporate Communications: An International Journal, 8*(3), 173–186.

Fallah, A. (2007). Skoda back in Australia Retrieved April 23, 2011, from http://www.caradvice.com.au/1826/skoda-back-in-australia/

Grau, S.L. & Folse, J.A.G. (2007). Cause-related marketing (CRM): The influence of donation proximity and message-framing cues on the less-involved consumer. *Journal of Advertising, 36*(4), 19–33.

Harrison, K. (2010). How you can use leverage to increase your sponsorship value. *Century Consulting Group Cutting Edge PR* Retrieved September 7, 2010, from http://www.cuttingedgepr.com/articles/increase-sponsor-ship-value.asp

International Events Group. (2014). Sponsorship Spending Report: Where the Dollars Are Going and Trends for 2014 Retrieved April 15, 2014, from http://www.sponsorship.com/IEG/files/4a/4aa6a1aa-18a4-46b0-942e-0d4a9cd74de9.pdf

James, D. (2005). Guilty through association: brand association transfer to brand alliances. *Journal of Consumer Marketing, 22*(1), 14–24.

Johnston, M.A. (2010). The impact of sponsorship announcements on shareholder wealth in Australia. *Asia Pacific Journal of Marketing and Logistics, 22*(2), 156–178.

Melewar, T. & Walker, C. (2003). Global corporate brand building: Guidelines and case studies. *The Journal of Brand Management, 11*(2), 157–170.

Moore, P. (2008). Brand leverage through sponsorship: leverage and the dark art. *AdMedia* Retrieved September 8, 2010, from http://www.allbusiness.com/marketing-advertis-ing/marketing-techniques/11467185-1.html

Muotka, D. & Mannberg, M. (2004). *Sport sponsorship: case study of Audi.* (Unpublished Bachelor of Social Science & Business Administration thesis), Lulea University of Technology, Lulea. Retrieved from http://epubl.luth.se/1404-5508/2004/143/LTU-SHU-EX-04143-SE.pdf

Olkkonen, R. (2001). Case study: The network approach to international sport sponsorship arrangement. *Journal of Business & Industrial Marketing, 16*(4), 309–329.

Pegg, S. & Patterson, I. (2010). Rethinking music festivals as a staged event: gaining insights from understanding visitor motivations and the experiences they seek. *Journal of Convention & Event Tourism, 11*, 85–99.

Pracejus, J.W. & Olsen, G.D. (2004). The role of brand/cause fit in the effectiveness of cause-related marketing campaigns. *Journal of Business Research, 57*(6), 635–640.

Quester, P.G. & Thompson, B. (2001). Advertising and promotion leverage on arts sponsorship effectiveness. *Journal of Advertising Research, 41*(1), 33–47.

Reserve Bank of Australia. (2010). A collection of graphs on the Australian Economy and Financial Markets. Reserve Bank of Australia Retrieved August 31, 2010, from http://www.rba.gov.au/chart-pack/graphical-summary.pdf

Schultz, M. & Hatch, M.J. (2003). The cycles of corporate branding. *California management review, 46*(1), 6–26.

Sirgy, M.J., Lee, D.-J., Johar, J. & Tidwell, J. (2008). Effect of self-congruity with sponsorship on brand loyalty. *Journal of Business Research, 61*(10), 1091–1097.

Skildum-Reid, K. (2008). Last generation sponsorship Retrieved September 7, 2010, from www.powersponsorshipdownloads.com/powersponsorship/LastGenerationSponsorship.pdf

Škoda Auto. (2011). Common History. Škoda Auto a.s. Retrieved April 23, 2011, from http://www.skoda-auto.com/com/about/tradition/history/Pages/History.aspx

Smith, G. (2004). Brand image transfer through sponsorship: A consumer learning perspective. *Journal of Marketing Management, 20*(3), 457–474.

Smolianov, P. & Shilbury, D. (2005). Examining Integrated Advertising and Sponsorship in Corporate Marketing Through Televised Sport. *Sport Marketing Quarterly, 14*(4), 239–250.

Stokes, R. (2010). *Measuring and evaluating sponsorship efforts. Event Sponsorship Management (EVNT7002). Lecture Note.* The University of Queensland. Brisbane.

Sylvestre, C. & Moutinho, L. (2008). Leveraging associations: The promotion of cultural sponsorships. *Journal of Promotion Management, 14*(3), 281–303.

Tsiotsou, R. & Alexandris, K. (2009). Delineating the outcomes of sponsorship: sponsor image, word of mouth, and purchase intentions. *International Journal of Retail & Distribution Management, 37*(4), 358–369.

Uggla, H. (2006). The corporate brand association base: a conceptual model for the creation of inclusive brand architecture. *European Journal of Marketing, 40*(7/8), 785–802.

Woodside, F. & Summers, J. (2008). *Packaging exploitation in fast moving consumer goods: consumer processing of sponsorship messages.* Paper presented at the Proceedings of the 2008 Global Marketing Conference, Shanghai.

Woodside, F.M. & Summers, J. (2009). *Consumer awareness of sponsorship: A FMCG context.* Paper presented at the Proceedings of the 2009 Australian and New Zealand Marketing Academy Conference (ANZMAC 2009), Melbourne, Australia.

Theory and Practice in Hospitality and Tourism Research – Radzi et al. (Eds)
© *2015 Taylor & Francis Group, London, ISBN 978-1-138-02706-0*

Price fairness evaluation on hotel distribution channels and emotional response: A concept

W.A.N. Wan-Salman & S.M. Radzi
Faculty of Hotel and Tourism Management, Universiti Teknologi MARA, Shah Alam, Malaysia

ABSTRACT: This paper is intended to conceptualize the price fairness evaluation of the room rate offered by the hotel distribution channels. Generally, the hotel distribution channels consist of two things, namely the direct (hotel websites) and indirect channels (hotels'intermediaries). Apart from that, the existence of the customers' emotional response is expected to contribute to this conceptual paper. It is hoped that this study will contribute to the hotel industry and also the customers in evaluating the room rate through online method.

Keywords: Hotel distribution channels, room rate, price fairness, emotional response

1   INTRODUCTION

Nowadays, the development of technology, especially the internet has helped the hotel industries promoting their facilities in their properties. In addition, the internet has assisted the customers in viewing the room rates and at the same time making the room reservation easier than before. Traditionally, the hotel room rates can be disclosed and reserved by using telephone call, facsimile, and even by mail. According to O'Connor (2003), the traditional method is considered as ineffective and sometimes costly for both parties, which are the customers and the hotel organizations. He further noted that since then, the price has become as an important element in the online pricing method.

Generally, hotels had been practicing the web based pricing in order to survive in the business. In other words, hotels that promote the room rate through their websites are acting as the direct channels of distribution. Apart from that, due to the high degree of perishability in selling the room, hotels have also taken the alternatives promoting their rooms through their intermediaries or also known as indirect channels (e.g.; expedia.com, booking.com, and, agoda.com). In addition, this has created an option for the customers purchasing a hotel room via online method regardless of any means of distribution. As stated by Lim and Hall (2008), online purchase is easier, faster and more convenient to the customers especially when they want to select for the best bargains for a hotel room.

Despite the utilization of the indirect channels in selling the rooms, hotels encountered some challenges, especially with regards to room pricing or rates in the market (Myung, Li & Bai, 2009). According to Lim and Hall (2008), the main issue concerned is involving the inconsistency of the room price offered by both direct and indirect channels. Furthermore, the difference that exists in terms of the room rates between both channels is actually reflecting the same type of rooms of the particular hotel. A few studies that had been conducted on the rate inconsistency (rate disparity) matters and the results indicated that the disparity is still exists (Demirciftci, Cobanoglu, Beldona & Cummings, 2010; Lim & Hall, 2008; Toh, Raven & DeKay, 2011). Not only that, those studies that had been done on the rate (dis) parity were looking at the organizations' perspectives rather than the customers as the purchasers themselves. This has raised another issue, in which Gazzoli, Kim, and Palakurthi (2008) highlighted that the rate disparity will lead to the negative perception by the customers and at the same time will create the matters concerning to the price fairness.

Studies pertaining to the price fairness in hospitality industry have been done by various authors, among them were Choi and Mattila (2009), Taylor and Kimes (2010), and Heo and Lee (2011). However, the above mentioned studies are only reflecting the perception of the price fairness as the outcome. Although there are few studies that had been done beyond the fairness as an outcome, such as customers' loyalty (Hwang & Wen, 2009), customers' response behavior (Ahmat et al., 2011), and behavioral intention (Kim & Mattila, 2011), so far there were no studies pertaining to the customers' evaluation of price fairness and emotional response with regards to the room rate offered by direct and indirect channels. Therefore, this paper seeks to conceptualize the importance of price fairness in relation to the price offered by both channels and the customers' emotional response as the outcome.

2 REVIEW OF RELATED LITERATURE

2.1 *Hotel distribution channels*

O'Connor and Frew (2002) defined distribution channels as a way where customers have been given enough information at the right time and place before purchase decisions take place. Since the development of the electronic marketing, customers can gain more information about the room, rates and their availability (Carroll & Siguaw, 2003) based from the hotel websites as the direct channels. On the other hand, the existence of the indirect channels has also assisted the hotels in selling their unsold rooms to the customers with the same information provided by the direct channels.

Studies with regards to the hotel room rate offered through the direct and indirect channels had portrayed inconsistent results. According to Christidoulidou, Brewer, Feinstein, and Bai (2007), their findings from the panel of experts discovered that the main issues involved were the controlling of the room rates and the distribution channels. Generally, a hotel expects their intermediaries offered the same rate as what had been published in the direct channels' websites (Demirciftci et al., 2010). Unfortunately, findings from the previous studies revealed that the direct channels had quoted the most expensive room rate as compared to the indirect channels (Tso & Law, 2005). In addition, Law, Chan, and Goh (2007) also found that the indirect channels offered the cheapest rate when comparing to both distribution channels.

On another occasion, a study conducted in the United States (US) by Gazzoli et al. (2008) portrayed that the majority of hotel properties offered the lowest rate in all of the available distribution channels. In addition, Toh et al. (2011) further suggested that due to the unsold inventories, hotels are the most aggressive in reducing the room rate. Apart from that, hotels have implemented best rate guaranteed or Best Available Rate (BAR) on their websites in order to influence the customers that they are offering the cheapest price as compared to their indirect channels. Although there are efforts in creating the rate parity between direct and indirect channels (Gazzoli et al., 2008), yet, it is still difficult to solve. Uniquely, even though the rate parity is hard to achieve, but one thing in common were found in both channels is that they tend to reduce the room rate as approaching to the arrival dates (Toh et al., 2011).

Overall, the rate (dis) parity that exists between direct and indirect channels is likely to create confusion for the customers. Moreover, this will also generate another concern, which is the customer's perception of the price fairness.

2.2 *Price fairness perception*

By definition, Xia, Monroe, and Cox (2004) explained fairness as to whether the result is considered as reasonable, acceptable, or just. Studies with regards to price fairness perception have been explored widely in marketing research, which includes hospitality industry as well. Within the context of hotel industry, price fairness perception has created a considerable attention by the researchers and hoteliers due to the rate (dis) parity issues involving direct and indirect channels (Choi, Mattila, Park & Kang, 2009). The price dispersion across distribution channels has made the customers become confused. As a result, this will drive the customers away from making a reservation to any particular hotels (Choi et al., 2009).

Generally, a review of related studies revealed that customers rated price parity as fair when applied to both channels (Choi & Mattila, 2009). It can be said that when the rate parity exists, then the customers will not be involved in searching the best price offered in all channels. In fact, researchers that had studied within this context had suggested the hotels to have an agreement with their indirect channels so as to sustain the rate parity across channels (Gazzoli et al., 2008; Lee, Denizci Guillet & Law, 2012; Myung et al., 2009).

In another perspective, Taylor and Kimes (2010) highlighted that the perception of price fairness may depend on the customers' familiarity with the pricing practices. For instance, a study by Choi and Mattila (2009) found that customers who are not familiar with the differential pricing strategy had rated the perceptions as unfair. In this case, the price variation and fluctuation that exist across channels can be considered as differential pricing practices. Thus, if the customers are familiar with these practices, they will evaluate the fairness perception as fair regardless of any distribution channels, as in the study by Taylor and Kimes (2010). Another issue of price fairness exists when any of the distribution channels published their offerings with certain advantages (such as room with breakfast) at the same price of other channels with the same offering but with certain extra charge. Unfortunately, the fairness evaluation is reflecting the hotel itself rather than their indirect channels. Thus, this will lead to the customers' emotional response on the room rates published by both channels.

2.3 *Emotional response*

The variation in the hotel room rate regardless of any channel will creates the negative effect on fairness. Previous studies have shown that price fairness perception is much related to customers' emotions (Lii & Sy, 2009; O'Neill & Lambert, 2001; Xia et al., 2004). As stated by Demirciftci et al. (2010) that the differences in the price across various channels had created the feeling of disappointment, being cheated and at the same time stir up the unfairness perceptions to the customers. As a result, the unfairness perceptions will initiate various negative emotions (Xia et al., 2004). In addition, Lii and Sy (2009) had suggested that price fairness perceptions influence emotional response in a direct manner. The same authors also further elaborated that when customers rated the price fairness perception as fair, their emotional responses were also portrayed positive reactions (such as joy and satisfaction) and vice versa. These outcomes were also found to be in line with Namkung and Jang (2010) together with Chung and Petrick (2012), in which, when the customers believed that they have paid a fair price, they were likely to feel more positive than negative emotions. Therefore, it can be said that the perceptions of price fairness will exhibit a direct relationship with customers' emotional response.

3 CONCLUSION

As stated in the review of literature, this study intends to enlighten the area, in which is still not being discovered by the previous researchers. Hotels basically relies on the selling of their rooms other than the food and beverage itself. Due to the advancement of technology, hotels have been promoting their room rates for the customers to make a direct reservation through their websites. At the same time, hotels opt to promote their room and rates through the use of their indirect channels. The previous studies have shown that the rate disparity occurs across the channels, thus this will affect the evaluation of price fairness from the customers.

Generally, when the issue of price fairness emerges, it will evoke the emotional response of the customers. To date, as far as this study is concerned, there were no studies that had viewed the price offered by the direct and indirect channels via online method in relation to customers price fairness perceptions and their emotional responses. Therefore, it is hopeful that this conceptual paper will contribute to the body of knowledge in the hotel industry, especially in identifying the fairness aspect and the customers' emotional responses.

REFERENCES

Ahmat, N.H.C., Radzi, S.M., Zahari, M.S.M., Muhammad, R., Aziz, A.A. & Ahmad, N.A. (2011). *The Effect of Factors Influencing The Perception of Price Fairness Towards Customer Response Behaviors.* Paper presented at the 2nd International Conference on Business and Economic Research (2nd ICBER 2011).

Carroll, B. & Siguaw, J. (2003). The evolution of electronic distribution: effects on hotels and intermediaries. *The Cornell Hotel and Restaurant Administration Quarterly, 44*(4), 38–50.

Choi, S. & Mattila, A.S. (2009). The Effect of Cross-Channel Price Dis/parity on Ethicality Evaluations and Purchase Intent: The Moderating Role of Price Frame. *Journal of Marketing Channels, 16*, 131–147.

Choi, S., Mattila, A.S., Park, H. & Kang, S. (2009). The Effect of Cross-Channel Price Dis/parity on Ethicality Evaluations and Purchase Intent: The Moderating Role of Price Frame. *Journal of Marketing Channels, 16*(2), 131–147.

Christidoulidou, N., Brewer, P., Feinstein, A.H. & Bai, B. (2007). Electronic Channels of Distribution: Challenges and Solutions for Hotel Operators. *Hospitality Review, 25*(2).

Chung, J.Y. & Petrick, J.F. (2012). Price Fairness of Airline Ancillary Fees: An Attributional Approach. *Journal of Travel Research, 52*(2), 168–181.

Demirciftci, T., Cobanoglu, C., Beldona, S. & Cummings, P.R. (2010). Room Rate Parity Analysis Across Different Hotel Distribution Channels in the U.S. *Journal of Hospitality Marketing & Management, 19*(4), 295–308.

Gazzoli, G., Kim, W.G. & Palakurthi, R. (2008). Online Distirbution Strategies and Competition: Are The Global Hotel Companies Getting it Right. *International Journal of Contemporary Hospitality Management, 20*(4), 375–387.

Heo, C.Y. & Lee, S. (2011). Influences of consumer characteristics on fairness perceptions of revenue management pricing in the hotel industry. *International Journal of Hospitality Management, 30*(2), 243–251.

Hwang, J. & Wen, L. (2009). The effect of perceived fairness toward hotel overbooking and compensation practices on customer loyalty. *International Journal of Contemporary Hospitality Management, 21*(6), 659–675.

Kim, E.E.K. & Mattila, A.S. (2011). The Effect of Service Duration and Price (Mis) Match on Perceived Price Fairness. Retrieved May 28, 2013, from http://scholarworks.umass.edu/cgi/viewcontent.cgi?article=1302&context=gradconf_hospitality

Law, R., Chan, I. & Goh, C. (2007). Where to find the lowest hotel room rates on the internet? The case of Hong Kong. *International Journal of Contemporary Hospitality Management, 19*(6), 495–506.

Lee, H.A., Denizci Guillet, B. & Law, R. (2012). An Examination of the Relationship between Online Travel Agents and Hotels: A Case Study of Choice Hotels International and Expedia.com. *Cornell Hospitality Quarterly, 54*(1), 95–107.

Lii, Y.-s. & Sy, E. (2009). Internet differential pricing: Effects on consumer price perception, emotions, and behavioral responses. *Computers in Human Behavior, 25*(3), 770–777.

Lim, W.M. & Hall, M.J. (2008). Pricing Consistency Across Direct and Indirect Distribution Channels in South Wesk UK Hotels. *Journal of Vacation Marketing, 14*(4), 331–344.

Myung, E., Li, L. & Bai, B. (2009). Managing the Distribution Channel Relationship With E-Wholesalers: Hotel Operators' Perspective. *Journal of Hospitality Marketing & Management, 18*, 811–828.

Namkung, Y. & Jang, S.C.S. (2010). Effects of perceived service fairness on emotions, and behavioral intentions in restaurants. *European Journal of Marketing, 44*(9/10), 1233–1259.

O'Connor, P. (2003). On-line pricing: an analysis of hotel-company practices. *The Cornell Hotel and Restaurant Administration Quarterly, 44*(1), 88–96.

O'Connor, P. & Frew, A.J. (2002). The Future of Hotel Electronic Distribution: Expert and Industry Perspectives. *Cornell Hotel and Restaurant Administration Quarterly, 43*(3), 33–46.

O'Neill, R.M. & Lambert, D.R. (2001). The emotional side of price. *Psychology & Marketing, 18*(3), 217–237.

Taylor, W.J. & Kimes, S.E. (2010). How hotel guests perceive the fairness of differential room pricing *Cornell Hospitality Report* (Vol. 10): Cornell University.

Toh, R.S., Raven, P. & DeKay, C.F. (2011). Selling Rooms: Hotels vs. Third-Party Websites. *Cornell Hotel and Restaurant Administration Quarterly, 52*(2), 181–189.

Tso, A. & Law, R. (2005). Analysing the online pricing practices of hotels in Hong Kong. *International Journal of Hospitality Management, 24*(2), 301–307.

Xia, L., Monroe, K.B. & Cox, J.L. (2004). The price is unfair! A conceptual framework of price fairness perceptions. *Journal of marketing*, 1–15.

Theory and Practice in Hospitality and Tourism Research – Radzi et al. (Eds)
© 2015 Taylor & Francis Group, London, ISBN 978-1-138-02706-0

Consumer purchase intention towards Sharia Compliant Hotel (SCH)

N.S.N. Muhamad-Yunus, N. Abd-Razak & N.M.A. Ghani
Faculty of Business Management, Universiti Teknologi MARA, Shah Alam, Malaysia

ABSTRACT: The notion of sharia compliant has been widely known to various areas especially in the profit making industries. A progressive Muslim country such as Malaysia has merged the concept to areas such as banking, business, hospitality and tourism. Sharia compliant is defined as an act or activity that follows with the requirements of Islamic law. With the flock of Middle Eastern tourist, the need to implement Sharia compliant hospitality and tourism activities is essential. To no avail, the concept has yet to fully established and in sporadic development. The aim of this study is to discuss on factors that contributed to the consumer purchase intention and its relationship. The awareness on Sharia Compliant Hotel (SCH), brand image and expected service quality are considered as critical factors that possibly provide empirical evidence in relation to consumer purchase intention which measured through the Theory of Planned Behavior (TPB). The study is aligned with Malaysia current tourism agenda that highlights sharia compliant as one of the strategies to attract visitors from Muslim countries to visit Malaysia.

Keywords: Awareness, brand image, purchase intention, service quality, Sharia Compliant Hotel

1 INTRODUCTION

Malaysia has been successful in trying to attract Muslim tourists from all over the world to visit and gain new experience. Due to that, the opportunity for local hotels to provide some facilities and activities in line with the Islamic values indirectly has come into existence. Basically, Sharia Compliant Hotel (SCH) is not a new concept in Malaysia. According to Zakiah and Fadilah (2013), there are few hotels that serve basic facilities to fulfil the Muslim tourists' needs so-called as friendly Muslim hotel. However, the existence of SCH itself is still less being exposed around Malaysia as not all Malaysian knows about it. Obviously, the SCH concept requires further explanation in terms of its definition and attributes that constitutes of what SCH is (Zakiah & Fadilah, 2013).

One of the hotels that practice SCH is De Palma Hotel. De Palma hotel where located around Malaysia have standardized the operation and management based on SCH. De Palma hotel was chosen as the first hotel to offer services that conform to Sharia principles. Among practices are; level 6 is design specifically to locate Islamic room type, large prayer rooms with a full time Imam to lead all prayers, Halal food in their hotel restaurants and the restaurants are certified Halal by Selangor Islamic Religious Department. In addition, all staff especially female is compulsory to wear the dress code that compliant with the Sharia principles.

In reality, there are number of hotel that offers the full SCH in Malaysia but it is still inadequate (Suhaiza, Azizah & Simon, 2011). The number of Muslims tourist to Malaysia increased, but there is lack of services and facilities provided to them as to meet their expectation (Bogas, 2013). Majority of the hoteliers in Malaysia do not see that there are big opportunities in the implementation of Islamic hotel concept as with the growing number of Muslim population and Muslim travellers. This market segment is expected to grow fast especially in attracting travellers from the middle east that have lots of demanding in Islamic hotels (Abuznaid, 2012). Therefore, if Malaysia applies the SCH principle and be able to create a centre of attention to those tourists from the Middle East, it is not impossible Malaysia can produce most of the income from the tourism industry using SCH as an attraction. Based on the given statement, it can be assumed that the SCH in Malaysia are still oblivious and further studies should be carried out (Razali, Abdullah & Hassan, 2012).

Therefore this study discussed the factors that influenced consumer purchase intention towards Islamic hotel in Malaysia. The first discussion is understood the concepts of SCH and theory of planned behavior that discuss on consumer intention to purchase. Next, this study discussed three factors (i) awareness, (ii) brand image, and (iii) services quality that influence the individual intention to purchase. This study highlights that the consumer intention behavior will be strategic to the hotel practitioner to identify the need and fulfill the expectation of them.

2 LITERATURE REVIEW

2.1 *Sharia Compliant Hotel*

Sharia compliant hotel is defined as an act or activity that complies with the requirements of sharia or Islamic law. The whole of hotel practices, activities, management and operations are relying on this concept (Fadil & Zulkifli, 2012). Sharia compliant hotels must follow the overall Islamic values which include from sources of capital to their daily operation. The concept of sharia compliant hotel is very unique and it needs to be promoted not only towards the Muslim market but also non-Muslim.

In essences, sharia compliant hotels are guided by the Islamic law, which are based upon the Qur'an, and the Sunna (the practices and sayings of Prophet Muhammad PBUH). It also includes the Islamic Fiqh and opinion of Muslim legal scholars. According to Henderson (2010), the industry practitioners and the analyst have come out with a set of sharia compliant hotel attributes. The Sharia compliant hotel establishment has highlighted that there are characteristic that is significant in reflecting the practice of Islamic hotel. The importance of this characteristic is to retain and maintain the best service quality to Muslim customer. The characteristics are as follows:

All the elements presented cover the sharia rules as to maintain and attract Muslim tourist to Islamic Hotel. These practices influenced the intention of the consumer to gain experience of SCH. Later the intention to purchase wills discussed the theory and factors that influenced consumer purchase intention towards SCH.

2.2 *Factors influences the intention to purchase*

Purchase intention is the implied promise to one's self to buy the product again whenever one makes next trip to the market (Fandos & Flavian, 2006). The study helps to explain that purchase intention as the number of patrons that has a proposal to buy the products or experience the services in future and make repetition purchases and contact again to the specific product and services.

Besides that, it also portrays the impression of customer retention. Basically, findings from the literature founds that there are certain factors which contributed to the strong influence on the purchase intention of the consumer in service sector such as the consumers' awareness, organization's brand image and the expected service quality.

2.2.1 *Awareness on Sharia Compliant Hotel*

Consumer awareness refers to awareness of a potential or current buyer about a particular product or company which can be as simple as a shopper remembering a television commercial. In the Islamic context literally means experience of something and being well informed of what is happening at the present time on any Halal food, drinks, services and products. Apparently in this context, it is strongly based on the religion belief. Schiffman and Kanuk (2000) assert that members of different religious groups in their decisions to eat, drink or use any product and their awareness are influenced by their religious identity, orientation, knowledge and belief.

Apart from that, Muslim travelers are also looking for the "Halal" label on hotels, restaurants and even airlines when they travel. In fact, 50 per cent of Muslim travelers would use Halal-friendly facilities if they existed and 30 per cent would seek strict Sharia-compliant services (Janmohamed, 2013). Furthermore, there are Muslim travelers that are concerned to be connected with Muslim populations in other countries. They are willing to learn more about the 'ummah', the global Muslim nation and its heritage by visiting other Muslim countries. To sum up the discussion above, consumer awareness are related to individual belief and knowledge on SCH. It is expected that, when Muslim tourist have been acknowledged about the existence and features of SCH, it will directly influence their intention in purchasing SCH without any doubtful.

2.2.2 *Brand image*

According to Berkowitz (2006), branding is an act in which an organization uses a name, phrase,

Table 1. Characteristic of Sharia Compliant Hotel.

Characteristics	Explanation
Food and beverage	Does not serve alcoholic beverages and serve only halal food.
Separate facilities	(i) Separate entrance for women. (ii) Separate recreational facilities such as swimming pool and fitness center for separate genders
Prayer room	(i) Room must be suitable enough to accommodate Muslim for congregation prayer at one time. (ii) Quran prayer mats and arrows indicating the direction of Mecca in every room.
Room	Beds and toilets positioned so as not to be facing the direction of Mecca.
Entertainment	Appropriate entertainment (no nightclubs or adult television channels)
Management practices	(i) Predominantly Muslim staff. Islamic staff dresses code requirements. (ii) Guest dress code must be posted at the hotel entrance or lobby. (iii) Hotel financed through Islamic financial instrument.

design, symbols, or combination of these to identify its products and distinguish them from those of competitors, while brand name is any word, 'device' (design, sound, shape, or colour), or combination of these used to distinguish a seller's good or services. Brand is more than a names, symbols and logos. It is very important to promote and sell product or service. Customers get to know the hotel by their brand images and they view a brand as an important part of the hotel. From the hotel names only, it represents customer's expectation and feeling about the hotel environment and its performance. For example, De Palma Hotel is the first Malaysia's Sharia Compliant Hotel that implement Islamic concept. Based on their names, the customers can predict how good they are in giving a service and maintain their reputation.

The good and Halal brand images cannot be ignored if the concept of Islamic hotel to be successfully implemented. Furthermore, it can be seen that SCH have big opportunities to go further. High levels of brand awareness and positive brand image should increase the probability of brand choice and greater customer loyalty thus decreases vulnerability to compete (Keller, 1993). The significant of brand image towards the influence on consumer purchase intention can be done by enabling the consumers to better visualize and understand the intangible images represented by the company. Hence, this would help in reducing customer's perceived financial, social or safety risk (Simoes & Dibb, 2001).

2.2.3 Expected service quality

In the context of the hotel industry, service quality in terms of both customers' expectation and customers' perception could lead to customer loyalty, enhanced image, reduced costs and increased business performance (Ramanathan & Ramanathan, 2011). Customer expectation is the needs, wants, and preconceived ideas of a customer about a product or service. However, customers' expectations play an important role, because the expectations concerning service significantly differ from those referring to products. Moreover, customers' expectations vary according to the service type. The importance of customers' expectations highlights the fact that product quality represents its ranking according to established standards. When consumers access product or service quality, it is performed according to internal standards, the actual expected quality of service. Therefore, the expectations are internal standards upon which the consumer ranks the quality of delivered service.

Comparing with the manufactured goods, measuring the service quality of hospitality cannot be done objectively. Thus it remains a relatively elusive and abstract construct (Zeithaml, 1990). The evaluation of quality for services is more complex than for products because of their intrinsic nature of heterogeneity, inseparability of production and consumption, perishability and intangibility (Frochot, 2000). These distinguishing characteristics of services make it difficult to define and measure service quality. In the hotel industry, other attributes, such as imprecise standards, short distribution channel, reliability and consistency, face to face interaction and information exchange, and fluctuating demand have been identified and further complicate the task of defining, delivering and measuring service quality.

The SERVQUAL scale is a survey instrument which claims to measure the service quality in any type of service organization on five dimensions which are tangibles, reliability, assurance, responsiveness and empathy (Parasuraman, 1988). Tangibles attribute can be referred to physical facilities, equipment, and appearance of personnel. Kivela, (1996) highlighted that among the service quality for hotel includes; cleanliness, comfortable and well-maintained rooms and convenient location and accessibility, safety and security and room facilities.

Reliability attribute or known as ability to perform the promised service dependably and accurately. Malaysia is a pioneer in introducing a comprehensive standard for Halal products through MS 1500:2004. The standard prescribes practical guidelines for the food industry on the preparation and handling of Halal food, including nutrient supplement, and to serve as a basic requirement for food product and food trade or business in Malaysia. Consumer will have high expectation regarding the Sharia Compliance Hotel in following the basic requirements for the preparation of Halal food.

Another attribute is responsiveness which is willingness to help customers and provide prompt service. The staff is expected willing to serve customers and available when needed. Besides that, assurance is another important attribute which will bring consumer's intention to go SCH. Assurance is knowledge and courtesy of employees and their ability to inspire trust and confidence. Employees must instill a confidence in guests and have in-depth occupational knowledge in serving them. Empathy is the last attribute of SERVQUAL scale and refers to caring, and individualized attention the firm provides its customers. Employees always treat guests in a friendly manner and always willing to serve customers.

2.3 Consumer purchase intention behavior

Intention is a state of a person's willingness to perform the behaviour, and it is considered as an immediate antecedent of behaviour (Ajzen, 1985). The Theory of Planned Behaviour (Ajzen, 1991) deals with the antecedents of attitudes, subjective norms and perceived behavioural control. These elements are used in an attempt to understand people's intention to involve directly or indirectly in a number of activities such as willingness to vote and giving (Hrubes, Ajzen & Daigle, 2001). In a direct measurement of these determinants, attitude refers to an evaluative judgement about the advantages and disadvantages of performing behaviour, while subjective norms refer to a person's perception of social pressure to perform the behaviour and a perceived behavioural control refers to an individual's perceived confidence in the capability of performing the behaviour (Fishbein & Ajzen, 1975).

In measuring the intention to purchase among consumers, TPB theory has been widely used from the day it was developed until today. In a deeper application and usage, TPB theory has also been used by various researchers in measuring people's intention to purchase, consume and accept Halal products and services (Nazahah & Sutina, 2012). By using the Theory of Planned Behavior as a foundation, several researchers agreed that the intention to purchase indicates an individual human belief to buy Halal product and service. For instance, a Muslim consumer has an intention to buy Halal product which presented at the point of purchase displays which arouse their interest onto it (Azis & Vui, 2012). This direct measure provides a guideline to predict human social behavior. In addition, another researcher also concludes that the conceptualizations of TPB presented by Ajzen imply a causal relationship between these variables, namely; attitudes, intentions and behavior".

3 PROPOSED THEORETICAL FRAMEWORK

The conceptual framework explains about the purchase intention of Sharia Compliant Hotel (SCH) that affected by the underlying factors which are applied to guide this study. Based on past literature, there are three factors that affect SCH which is

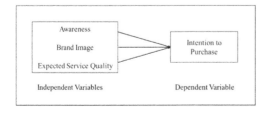

Figure 1. Proposed theoretical framework.

awareness, brand images and expected service quality. These three factors are the most important thing that customers will look first before they make a decision to choose the best hotel especially for Muslim. The independent variables of this framework are awareness, brand images and expected service quality. These independent variables will affect the dependent variable which is the purchase intention of Sharia Compliant Hotels (SCH).

Based on the framework above, several hypotheses are tested as follow:

H1: There is a significant relationship between consumer awareness and intention to purchase a Sharia compliant hotel.

H2: There is a significant relationship between brand image and intention to purchase a Sharia compliant hotel.

H3: There is a significant relationship between consumer expectation on service quality and intention to purchase a Sharia compliant hotel.

4 DISCUSSION AND CONCLUSION

In short, based on the past literature the factors such as consumer awareness on SCH, brand image and the expected service quality are found to have potential influence on consumer purchase intention towards Sharia compliant hotel (SCH). In fact, the factors used are almost consistent with one another. On the other hand, the emergence of consumer awareness, brand image and expected service quality in assessing its relationship between the Theory of Planned Behaviour has captured interesting aspects of the study. Past research then has thoroughly help in developing the hypothesis for this study. In order to prove the hypothesis, further analysis will be done in order to support the reliability and the applicability of the past findings.

As an implication, this study will provide valuable and important information for the hoteliers who are still doubtful and feeling underestimate towards the potential of Sharia Compliant Hotel (SCH) implementation in Malaysia. The findings of this study would provide them with a strong support in deciding to venture in this new and niche concept of hospitality services in Malaysia. This kind of opportunity should not be left behind as its market growth promise bulky revenue that can be earned by the company. A part from that, this study are also expected to facilitate service providers to identify areas of changes and improvements as if they intent to be aligned with this new hotel trend. Therefore, in order to make sure that they could maximize penetration of local market share, those special needs and wants that emerged from Muslim tourist should be given a special attention. The descriptive and relational features of the findings enable company to provide necessary product knowledge, creating right images and put more concentration on service quality aspects through the dedication of the hotel management itself.

REFERENCES

Abuznaid, S., (2012). Islamic Marketing: Adressing the Muslim Market, *An-Najah Univ. J. Res. (Humanities)*, 26(6), 1473–1503.

Ajzen, I., (1985). From Intentions to Actions: A Theory Of Planned Behavior. Action-Control: From Cognition to Behavior, International Journal. Kuhl & J. Beckmann (Eds.), 1–39.

Ajzen, I. (1991). The Theory of Planned Behavior. *Organizational Behavior and Human Decision Processes*, 50, 179–211.

Aziz, Y. A. & Vui, C. N., (2012), The Role Of Halal Awareness And Halal Certification In Influencing Non-Muslim's Purchase Intention, 3rd International Conference On Business And Economic Research, Bandung, Indonesia, Paper 1822.

Bogas, A. (2013, April 27). Thriving Halal Hotels In Malaysia Lack Standardization. *Kuwait News Agency*: Retrieved from http://www.kuna.net.kw/

Eric N. Berkowitz, (2006). *Essentials of Health Care Marketing,* (2nd Ed). Jones And Bartlett.

Fandos, C. & Flavian, C., (2006). Intrinsic and Extrinsic Quality Attributes, Loyalty And Buying Intention: An Analysis For A PDO Product, *British Food Journal*, 108(8), 646–662.

Fishbein, M. & I. Ajzen., (1975). Beliefs, Attitudes, Intentions, and Behavior: An Introduction to Theory and Research. Reading MA: Addison-Wesley.

Frochot, I. H. (2000). Histoqual: The Development Of A Historic Houses Assessment Scale. *Tourism Management*, 21, 157–167.

Fadil, M. Y., & Zulkifli, M., (2012). Introducing Sharia Compliant Hotels as a New Tourism Product. *The Case of Malaysia*, 1142–1146.

Henderson, J.C., (2010). Sharia-compliant hotel. *Tourism and Hospitality Research*, 10(3): p. 246–254.

Hrubes, D., Ajzen, I. & Daigle, J. (2001). Predicting Hunting Intentions And Behavior: An Application Of The Theory Of Planned Behavior. *Leisure Sciences*, 23, 165–178.

Janmohamed, S., (2013, February 4). The Growing Influence Of Muslim Tourists Around The World. Retrieved From Quran Focus: http://www.quranfocus.com/

Keller, K., (1993). Conceptualizing, Measuring And Managing Customer-Based Brand Equity", *Journal Of Marketing*, 1–22.

Kivela, J., (1996). Marketing In The Restaurant Business: A Theoritical Model For Identifying Consumers Determinant Choice Variables And Their Impact On Repeat Purchase In The Restaurant Indstry, *Journal of Hospital Management*, 3.

Nazahah, A. R. & Sutina, J., (2012). The Halal Product Acceptance Model For The Religious Society, *Business Management Quarterly Review*, 3 (1). 17–25.

Parasuraman, A. Z., (1988). SERVQUAL: A Multiple-Item Scale For Measuring Consumer Perceptions Of Service Quality. *Journal Of Retailing*, 1, 12–40.

Ramanathan, U. & Ramanathan R., (2011). Guests' Perceptions Of Factors Influencing Customer Loyalty: An Analysis For UK Hotels. *International Journal of Contemporary Hospitality Management*, 23(1), 7–25.

Razali, R., Abdullah, S. & Hassan, G., (2012). Developing A Model For Islamic Hotels, Paper Presented At Evaluating Opportunities And Challenges, International Conference On Knowledge Culture And Society.

Schiffman, L. G. & Kanuk, L. L., (2000). *Consumer Behavior* (7th Ed.). Upper Saddle River, NJ: Prentice-Hall.

Simoes & Dibb, S., (2001). Rethinking The Brand Concept: New Brand Orientation. *Corporate Communication*, 217–24.

Suhaiza, Z., Azizah, O. & Simon, K., (2011). An Exploratory Study On The Factors Influencing The Non-Compliance To Halal. *Hoteliers In Malaysia*, 1–12.

Zakiah, S., & Fadilah, A. R., (2013). Towards The Formation Of Sharia Compliant Hotel In Malaysia. An Exploratory Study On Its Opportunities And Challenges, 108–124.

Zeithaml, V. P., (1990). Delivering Quality Service: Balancing Customer Perceptions And Expectations, *New York, The Free Press*.

Theory and Practice in Hospitality and Tourism Research – Radzi et al. (Eds)
© 2015 Taylor & Francis Group, London, ISBN 978-1-138-02706-0

Food blog: Examining the blog trustworthiness and purchase intention behavior

W.I. Wan-Ruzanna, A. Zaileen-Elina & A. Roslina
Faculty of Hotel and Tourism Management, Universiti Teknologi MARA, Shah Alam, Malaysia

W.I. Wan-Mahirah
Politeknik Ibrahim Sultan, Malaysia

ABSTRACT: This present study focuses on examining the blog trustworthiness and purchase intention behavior of food bloggers. This study utilized self-administered questionnaire on the reader of food blog www.friedchilies.com. The data was successfully gathered from a sample of 324 respondents, yielding almost 95 percent response rate. The data were analyzed using inferential statistic to confirm the proposed hypothesis, simultaneously answering the research questions. The findings revealed that blog trustworthiness is closely related to the purchase intention behavior. In other word, food blog besides others can be considered as one of the effective marketing tools for restaurants operators to promote, attract their potential customers and retain the existing ones.

Keywords: Food blog, trustworthiness, purchase intention behavior

1   INTRODUCTION

Information technology and foodservice are two different industries which might seem unrelated, but nowadays it almost becomes a backbone of one to another. The modernization of today's era however makes it possible to merge these two different industries. Through the information technology, marketing of the foodservice establishment including restaurants can widely be done widely without any barriers. Out of many, social media is one of the latest means of disseminating information, including food information on any types of restaurant. In line with this, food review through different forms of media such as television, newspaper articles, magazines, websites and blogs are used to help the seekers or customers make a better choice in their food selection. According to (Cox & Blake, 2011) websites or blogs no doubt is one of the important sources of information seekers nowadays. Though the main idea of having a blog is for '*blogger to express their feelings, voice, opinions and share experiences and idea* (Herring, Scheidt, Wright & Bonus, 2005) it is also used as a form of marketing and promotion by the food providers.

Besides is sharing recipes, techniques of cooking food blogs also a review on food and restaurant and inside information on the locations, type of food served, pricing and best-selling menus. The increasing number of food blogs seems to give an impact towards the customer purchase intention behaviour. The customer who reads the food blog usually gets influenced by the pictures and explanation posted which mostly are attractive and capture the attention of the readers. Most positive review posted on a particular restaurant the better chances of readers purchase intention or otherwise.

As far as this study is concerned, there are over 30 popular food blogs are available in Malaysia that are used as a medium for sharing about the food matters and restaurants. The www.friedchillies.com is becoming one of the most famous food websites. The Google Analytics Malaysia revealed that Friedchillies received around 1.8 million viewers per month thus claimed as the most visited food blog in Malaysia. In fact, Friedchillies has expanded their brand by introducing their own food channel through a joint-venture with TM's HyppTV. The food review provided on the Friedchillies blog in fact is used as a guideline to all food lovers not only for local people but also by the international tourists.

Despite this, the question arises whether all the information shared through online blogs can be trusted and whether the information provided would have significant impacts towards customer purchase intention? The objective of this study is thus to analyze the effect of blog trustworthiness on readers purchase intention behavior by the hypothesis that:

H1: Blog trustworthiness affects readers purchase intention behavior.

2 LITERATURE REVIEWS

2.1 *Food blog*

Oxford dictionary refers a blog is a personal website or webpage on which an individual record, opinions or links to another site on a regular basis. One of the important characteristics is that blogs have the frequent updates. (Herring, et al., 2005) affirmed a blog as a mini website which have a reverse chronological order that are updated daily or weekly.

Blogs have existed since the early years of the Internet and have been used widely not only to share an individual thought, but also in sharing knowledge and information with the cyber world. (Schmidt, 2007) mentioning that through blog comments, readers seem to be clustered by the network and simplify the communication among those who have the same interest.

Robinson (2009) states the articles written by food blogger focuses on restaurants, household cooking as well as recipes. As previously highlighted, in addition to sharing recipes, techniques of cooking, food blogs also discuss about food and restaurant review with regard to locations, type of food served, pricing and best-selling menus. The increasing number of food blogs seems to give an impact towards the customer purchase intention behaviour. Bruns (2006) however states that food blogging might turn out to be a failure for a business if there is much critique about the food. On the other hand, attributes that may influence ones in selecting information in the food blogs may be grouped into three broad categories, namely blog trustworthiness, blog involvement and product information highlights.

2.2 *Trustworthiness*

According to Tripp, Jensen, and Carlson (1994) trustworthiness is the degree to which the audience perceives that the communicator intends to convey a valid statement. Hardin (2002) stated that trust involvement is the reason for which the relevant parties consider to be trustworthy. Huang, Shen, Lin, and Chang (2008) affirmed that trust for the blog is considered when the readers rely on the statement in a blog and persuade others to do the same. The trustworthiness of the content in the blogs has a positive impact towards the electronic Word of Mouth (e-WOM) (Chang, Lee & Huang, 2011). Ghazisaeedi, Steyn, and Van Heerden (2012) affirm that the higher trust put on the

blog, the higher the number of visitors will visit the blog.

(Ahn, Jang & Lee, 2012) declared that online reputation can be considered as a main factor for the organization to gain trust. Trust is easier to gain when the companies are well known or have a good reputation in business (Ahuja & Medury, 2010; Castaldo, Premazzi & Zerbini, 2010). Huang et al. (2008), Cyr (2008), and Huffaker (2010) acknowledged that reputation has a significant relationship with trust. On the other hand, Droge, Stanko, and Pollitte (2010) noted that blogs have been used as an important information distribution channel and as a knowledge-sharing tool to gain trust.

According to Ahn et al. (2012) one of the important criteria that might lead to customer trustworthiness is the quality in writing. A good blog comes from those who are expert in their field which might lead to a qualified and informative writing style. Tan, Na, and Theng (2011) affirming that a blogger writing style may gain the readers' trust and influence the customer purchase intention and trust to be one of the factors that lead to online purchases (Teo & Liu, 2007). Ganguly, Dash, and Cyr (2009) affirmed that online purchases were obtained from the trust gained by the readers or customers from the blog they read. The absence of trust might lead to no purchase intentions.

2.3 *Blog involvement and online purchase intention*

Blogs as a new sharing method which allows the blogger and readers to communicate and interact with one another based on their topic of interest (Wu & Lee, 2012). (Doyle, Heslop, Ramirez & Cray, 2012) declared that blog readers mostly get attracted to the annotations that come from individual experience. Involvements have created a communication among blog readers which will lead to the persuasion attempt (Pentina & Taylor, 2010) and positive comment influenced the readers purchase decision making (Chen, 2008).

Chang et al. (2011) contends that the discussion, interaction and participation in conversation in blogs have a positive impact towards online purchase intention compared to the more traditional websites. Shneiderman (2000) pointed out that communication through a suggestion from users shall help to encourage reader trust as well as willingness to purchase.

3 METHODOLOGY

The unit of analysis of this study involved visitors and blog reader of www.friedchilles.com. A total of 384 survey questionnaires were distributed and

324 (over 95%) were completed and met the screening requirement. The instrument which is measuring the variable was adapted from Wu and Lee (2012) and modified to fit the nature of the study. Five items were used to measure trustworthiness which includes the indication on information, writing style and reputation meanwhile five items also used for purchase intention which comprises price, picture and intention to purchase. Five points Likert scale which measuring the level of agreement ranging from 1 = strongly disagree to 5 = strongly agree was applied.

4 RESULTS

The descriptive statistic is used to analyze the respondents' demographic profiles focusing on the visitors and blog reader personal information.

4.1 Respondents profiles

31.3 percent of the respondents were males as opposed to 68.9 percent females. The majority of them is between 26-30 years old and still single (58.8%). 34.6 percent of respondents are students, followed by 32.7 percent as professional and 11.1 percent as an administrative officer. 62.1 percent of respondents were Malay compared to 23.5 percent Chinese, 13.1 percent Indian and 1.3% other ethnics. 40.5% of the respondents are the degree holder followed by 26 percent as Diploma holder, 23.5 percent possessed a Master degree and 6.5 percent obtained a secondary school certificate.

H1: Blog trustworthiness affects readers purchase intention behavior

In examining the relationship between blog trustworthiness and purchase intention behavior, standard regression analysis was undertaken. Result in Table 1 shows that there is a significant relationship between blog trustworthiness and purchase intention behavior. Blog trustworthiness statistically to be as predictor to purchase intention behavior ($\beta = 0.49$, $P < .001$) as 24 percent of the variance of blog trustworthiness influence purchase intention behavior.

5 DISCUSSIONS

Assessing the customer purchase intention behavior through the blog review result revealed that blog trustworthiness is strongly related to readers purchase intention behavior. This result strengthening the previous studies that blog reader mostly get attracted to the products by the clarification come from individual experience (Doyle, et al., 2012) and the communication among the blog

Table 1. Regression analysis of blog trustworthiness and purchase intention behavior.

Predictors	Model 1
Dependent variable: Purchase intention behavior	
Blog Trustworthiness	.49***
R2	.24
Adj. R2	.24
F	48.22***

Note: *p < .05, **p < .01, ***p < .001.

readers will lead to the persuasion and influenced their purchase decision making (Chen, 2008; Pentina & Taylor, 2010). With that, besides others food blog can be considered as one of the effective marketing tools for restaurants operators to promote, attract their potential customers and retain the existing ones. In other word, restaurateurs should not ignore the contribution of the food blog, but take the necessary action of using for the benefit of their business survival as long as developing and maintaining the element of trust in the blog.

6 LIMITATIONS AND RECOMMENDATION

Some limitations pertaining to this study need to be addressed. As this study only limits to foodservice industry specifically in Friedchillies website, the outcomes could not be generalized sub sector of this industry. Due to the time constraints, this study only used a limited number of respondents. Therefore, replication of the study using a broader scope of the food bloggers with more number of restaurants, variables and respondents should be undertaken.

REFERENCES

Ahn, S. H., Jang, K. H. & Lee, C. H. (2012). A Study on the In-fluence of the Characteristics of celebrities on consumers' attitudes and intentions. *Journal of Consumer Respond, 20*(4), 535-547.

Ahuja, V. & Medury, Y. (2010). Corporate blogs as e-CRM tools: Building consumer engagement through content management. *Journal of Database Marketing & Customer Strategy Management, 17*(2), 91-105.

Bruns, A. (2006). The practice of news blogging. In A. Bruns & J. Jacobs (Eds.), *Uses of blogs* (Vol. 38, pp. 11-22). New York: Peter Lang Publishing.

Castaldo, S., Premazzi, K. & Zerbini, F. (2010). The meaning of trust. A content analysis on the diverse conceptualizations of trust in scholarly research on business relationships. *Journal of Business Ethics, 96*(4), 657-668.

Chang, L. Y., Lee, Y. J. & Huang, C. L. (2011). The influence of E-word-of-mouth on the customer's purchase decision: A case of body care products.

Chen, Y.-F. (2008). Herd behavior in purchasing books online. *Computers in Human Behavior, 24*(5), 1977-1992.

Cox, A. M. & Blake, M. K. (2011). *Information and food blogging as serious leisure.* Paper presented at the Aslib Proceedings.

Cyr, D. (2008). Modeling web site design across cultures: relationships to trust, satisfaction, and e-loyalty. *Journal of management information systems, 24*(4), 47-72.

Doyle, J. D., Heslop, L. A., Ramirez, A. & Cray, D. (2012). Trust intentions in readers of blogs. *Management Research Review, 35*(9), 837-856.

Droge, C., Stanko, M. A. & Pollitte, W. A. (2010). Lead Users and Early Adopters on the Web: The Role of New Technology Product Blogs. *Journal of Product Innovation Management, 27*(1), 66-82.

Ganguly, B., Dash, S. B. & Cyr, D. (2009). Website characteristics, Trust and purchase intention in online stores:-An Empirical study in the Indian context. *Journal of Information Science & Technology, 6*(2).

Ghazisaeedi, M., Steyn, P. G. & Van Heerden, G. (2012). Trustworthiness of product review blogs: A source trustworthiness scale validation. *Journal of Business Management, 6*(25), 7498 - 7508.

Hardin, R. (2002). *Trust and trustworthiness* (Vol. 4). New York: Russell: Russell Sage Foundation.

Herring, S. C., Scheidt, L. A., Wright, E. & Bonus, S. (2005). Weblogs as a bridging genre. *Information Technology & People, 18*(2), 142-171.

Huang, C.-Y., Shen, Y.-Z., Lin, H.-X. & Chang, S.-S. (2008). Bloggers' motivations and behaviors: A model. *Journal of Advertising Research, 47*(4), 427-484.

Huffaker, D. (2010). Dimensions of leadership and social influence in online communities. *Human Communication Research, 36*(4), 593-617.

Pentina, I. & Taylor, D. G. (2010). Exploring source effects for online sales outcomes: The role of avatar-buyer similarity. *Journal of Customer Behaviour, 9*(2), 135-150.

Robinson, L. (2009, February 17). The art of food blogging, *Times Online*. Retrieved from 5 October 2013, www.timesonline.co.uk/tol/life_and_style/food_and_drink/real_food/article5753558.ece

Schmidt, J. (2007). Blogging practices: An analytical framework. *Journal of Computer Mediated Communication, 12*(4), 1409-1427.

Shneiderman, B. (2000). Designing trust into online experiences. *Communications of the ACM, 43*(12), 57-59.

Tan, L. K.-W., Na, J.-C. & Theng, Y.-L. (2011). Influence detection between blog posts through blog features, content analysis, and community identity. *Online Information Review, 35*(3), 425-442.

Teo, T. S. & Liu, J. (2007). Consumer trust in e-commerce in the United States, Singapore and China. *Omega, 35*(1), 22-38.

Tripp, C., Jensen, T. D. & Carlson, L. (1994). The effects of multiple product endorsements by celebrities on consumers' attitudes and intentions. *Journal of consumer research, 20*(4), 535-547.

Wu, W.-L. & Lee, Y.-C. (2012). The effect of blog trustworthiness, product attitude, and blog involvement on purchase intention. *International Journal of Management & Information Systems, 16*(3), 265 -275.

Tourism management

Theory and Practice in Hospitality and Tourism Research – Radzi et al. (Eds)
© *2015 Taylor & Francis Group, London, ISBN 978-1-138-02706-0*

Why Malaysian opt out domestic holidays for trips abroad?

M. Hafiz, N. Aminuddin, M.R. Jamaluddin & M.N.I. Ismail
Faculty of Hotel and Tourism Management, Universiti Teknologi MARA, Shah Alam, Malaysia

ABSTRACT: The purpose of this study is to classify the determinants of Malaysian tourist intentions for outbound travels, paying exceptional attention on the motivations aspect. The success of destinations marketing should be supported by strong demand for inbound and domestic tourism demand. The importance of outbound market has drawn scholars' interests in understanding tourists' motivation to travel overseas. Malaysian tourists can be categorized as a quality visitor with ample purchasing power. Even with heavy promotion of 'Cuti-Cuti Malaysia' by the tourism authority, their desire to travel abroad is still soaring. Therefore, it is essential to inspect the divergence between outbound and domestic travel motivation of Malaysian tourists. Data gathered from self-administered questionnaire at Kuala Lumpur International Airports (KLIA) and Low Cost Carrier Terminal (LCCT). Exploratory factor analysis employed to validate the research dimension. The study reveals that there is a combination of pull factors and push factors differing between domestic and outbound groups. The overall findings of this study provide valuable information for the government generally Ministry of Tourism as well as tourism-related organizations to improve their products, facilities and services to be offered to tourist especially among domestic market.

Keywords: Malaysia; motivation; push; pull; domestic; outbound

1   INTRODUCTION

The Malaysian tourism industry started broadening in the late 1980's and early 1990's. Inferior currency, sustainable promotion and political steadiness factors made Malaysia one of the world's most popular tourist destinations. With a market share of 70 percent, West Asia, East Asia and ASEAN countries are Malaysia's most notable tourist market in 2012. Neighboring countries such as Singapore, Indonesia, and Thailand are the three main tourist suppliers in Malaysia, with a total market share of 55 percent. In addition, the inbound western tourist statistic is ever skyrocketing, having Australia, United Kingdom and America in the top ten positions of Malaysia tourist arrivals (UNWTO, 2012).

An exclusive tourism promotional campaign, 'Visit Malaysia Year', was further introduced to foster the inbound demand. The government allocated more funding for advertising, upgrading tourist destinations and infrastructure, as well as on marketing promotions in leading source markets in preparation for the Visit Malaysia Year. As a consequence of the campaign (1990, 1994, 2000 and 2007), total visitor arrivals increased from 7.4 million in 1990 to 20.9 million in 2007 and 25 million in 2012. Tourism receipts also recorded a positive hike from RM4.5 billion 1990 to RM60.6

billion in 2012. With the latest 'Malaysia is Truly Asia' tagline promoted by Tourism Malaysia and the expected 30 million tourist arrivals in 2014 based on the Visit Malaysia Year 2014 campaign, the promotional effort is forecasted to be a success (UNWTO, 2012).

However, in spite of exceptional growth in inbound demand, diminutive amount of research has been undertaken on the outbound tourism development. Lately, Malaysian had been in a better place to travel abroad due to the improving economy, elevated purchasing power and weakening foreign currencies (British Pound and USD). In addition, the offering of a growing number of both medium and long haul flights, mostly by low cost and premium carriers (e.g., AirAsia, MAS and FireFly), encouraging Malaysians to travel overseas (Hanafiah & Harun, 2010).

UNWTO (2012) reported outbound tourism has a favourable position in the global market, with the number of trips anticipated growing at the rate of 3 percent per year. Furthermore, the ever growing number of aircraft and routes introduced will enhance the potential for outbound tourism. Based on the table below, outbound tourism expenditure registered the outflows of RM25.67 billion in 2012 as compared to RM21.34 billion in 2011 and RM9.09 billion in 2005. The outbound tourism expenditure propelled by country-specific

Table 1. Outbound tourism expenditure by products.

Products	2010^r (RM mil)	2011^r (RM mil)	2012^p (RM mil)
Accommodation	3,686.0	4,094.5	4,900.0
Food and beverage	1,545.7	2,047.2	2,807.3
Transport	4,076.7	4,141.4	5,763.9
Travel agencies	730.4	949.8	1,276.0
Cultural, sports and recreational	815.3	1,013.1	995.3
Country-specific tourism goods	5,826.3	8,653.3	9,187.5
Country-specific tourism services	305.8	443.2	740.1
Total (RM Billion)	16,98	21,34	25,67

Source: Tourism Satellite Account (2012).

tourism characteristic goods (shopping) followed by passenger transport services and accommodation services.

In basic economic, outbound tourism is viewed as an import. It is considered as leakage in the national economy and affect negatively on the national account (Smeral & Witt, 1996). Abundant studies have been conducted to assess the factors that drive people to travel (Bieger & Laesser, 2002; Kozak, 2002; Prebensen, 2006; Winter, 2009). However, studies on outbound tourism motives are scarce. Thus, this paper explores the determinants for outbound tourism with the intention to construct a structural model on travel motivation dimension.

2 LITERATURE

2.1 *Tourism motivation*

Motivation theory can be classified into two forces suggesting that people travel because they are pushed and pulled to do so by "some forces" or factors (Dann, 1981). According to Uysal and Hagan (1993), these forces describe how individuals are pushed by motivation variables into making travel decisions and how they are pulled or attracted by destination attributes. Push motivations are akin to psychological or emotional aspects. Contrarily, pull motivations connect to external, situational, or cognitive aspects (Yoon & Uysal, 2005, Hanafiah, et al., 2010). Furthermore, Josiam, Kinley and Kim (2005) analyses the concept of 'push' and 'pull' motivation. 'Push' motivations are the socio-psychological needs that motivate a person to travel.

Based from Kim and Lee (2002), the influence of attractions in a destination area considered as exerting a pull response on the individual. Resources considered pulling factors include natural attractions, cultural resources, recreational activities, special events or festivals, and other entertainment opportunities. Some destinations feature an assortment of these various resources to meet a variety of motives while others represent one unique resource and focus on a specific market segment.

Hanqin and Lam (1999) identified several dimensions of the travel motivations such as reputation, novelty, and service attitude/quality. Chinese repeat tourists perceived that there would be nothing new for them to explore on their returns, this is consistent with the study of Beerli and Martín (2004). However, the study of Li, Cheng, Kim, and Petrick (2008) stated that repeat tourists had higher satisfaction levels than first-time tourists. Consequently, Rittichainuwat, Qu and Mongkhonvanit (2008) found significant differences in travel motivations between first-time and repeat tourists to Thailand. Further, Liu (2006) argued that destinations possess different stocks of resources that offer a different appeal to people, such that a tourist will likely travel with different motivation on different occasions. This theory, therefore, fails to capture the diverse and changing motives for travel to different destinations by individuals or groups of people.

A literature review about motivation reveals that people travel because they are pushed into making travel decisions by internal psychological forces and pulled by the external forces of the destination attributes (Uysal & Hagan, 1993). The push-pull framework provides a practical technique for examining the motivations underlying tourist and visitation behavior. In this framework, push factors refer to the specific forces that affect a person's decision to take a trip (i.e., to travel outside of one's everyday environment), while pull factors refer to the forces that affect the individual's choice of which specific destination should be selected.

Additionally, Beerli and Martín (2004) found that motivations had influenced the affective components of the image (pleasant/unpleasant, exciting/boring); for example, first-time tourists who identified relaxation as the motivation found the sun and beach destination as attractive, whereas the repeat tourists went there to increase knowledge of the destination. The study identified that the more experiences with the destination the tourists had, the preferable destination image they would have because they were more familiar with the destinations.

Mansfeld (1992) suggested that an analysis of the motivational stage can show the way in which people set goals for their destination choice and how these goals are then reflected in both their choice and travel behavior. Additionally, this may provide tour operators, tourism planners, and other tourist-related institutions with a better understanding of the real expectations, needs and goals of tourists. Such an understanding is necessary to create travel products designed to meet these needs and expectations.

In line with that, Jang and Cai (2002) suggested that destination marketers need to build a secure fit between their destination attributes and the motivations of their target markets through effective marketing and promotional programs. In addition, in terms of the role of push and pull factors in the destination choice of tourists, the tourism literature emphasizes the importance of both push and pull factors in shaping tourist motivations and therefore in choosing holiday destinations. According to Crompton (1979), travel motivations, including push and pull factors have a significant impact on the decision to choose a destination.

Further, Yuan and McDonald (1990) identified five push factors which are escape, novelty, prestige, enhancement of family relationships, and relaxation/hobbies. Pull items included budget, culture and history, nature, ease of travel, cosmopolitan environment, facilities, and hunting. Differences found among the four countries were culturally defined. They concluded that, although individuals may travel for similar reasons, reasons for choosing certain destinations and the level of importance attached to each factor might differ.

Table 2. Factor analysis on travel motivation (domestic).

	Factor Loading	
	1	2
Push Factors		
Escape from the routine of work or life	.775	
Do something exciting	.721	
Reduce stress	.702	
Away from home	.683	
Relax and doing nothing at all	.657	
Pull Factors		
Safe destination		.781
Low budget		.765
Reliable weather		.743
Available attraction		.684
Eigenvalues	4.8	4.2
% of variance	26.41	21.14
Cumulative variance (%)	26.41	43.28
Cronbach's alpha score	.87	.85
Kaiser-Meyer-Olkin	0.89	

3 METHOD

3.1 Sampling and data collection

There were 1,435,091 domestic outbound movement recorded by Malaysia in 2012 (WTO, 2012). Based from Sekaran (2003), it is suggested that the sample size for more than 1 million populations is 384 units. Convenience sampling was used to acquire a large number of completed questionnaires swiftly and economically (Zikmund, 2003).

A self-administered questionnaire was used to collect data at the Kuala Lumpur International Airport (KLIA) and Low Cost Carrier Terminal (LCCT) in Kuala Lumpur. The information was collected during different departure flights, days, and times to reduce bias. The tourists were approached at the international departure halls with the assistance of KLIA personnel.

The questionnaire was validated by a pilot study and adjusted in content and structure based on the content validity and reliability. Research items validated through exploratory factor analysis technique (EFA) to ensure the instrument used is valid. The EFA was widely used to gather information on the variables interrelationship. In this process, all items were gathered and captured in a different pattern of correlation.

3.2 Exploratory factor analysis

The table below shows the result of EFA practice on travel motivation (domestic) items. The Barlett's

Test of Sphericity shows statistical significance with the Kaiser-Meyer-Olkin value of 0.89, exceeding the recommended value (Hair, Anderson, Tatham and Black, 1998). From the Varimax-rotated factor matrix, two factors representing 47.55 percent of the explained variance were extracted from 18 variables. The results showed the alpha coefficient for all three factors ranged from 0.85 to 0.87.

Table 2 reports the 18 attributes factor analysis on travel motivation (domestic) resulted in 8 attributes in two component groupings. For the first factor grouping named Push Factors, they are "escape from the routine of work or life", "do something exciting", "reduce stress", "away from home" and "relax and doing nothing at all". For the second factor grouping named Pull Factors, they are "safe destination", "low budget", "reliable weather" and "available attraction". Table 2 shows the result of the EFA procedure on travel motivation (outbound) items. The Barlett's Test of Sphericity shows statistical significance with the Kaiser-Meyer-Olkin value of 0.91, exceeding the recommended value (Hair, et al., 1998). From the Varimax-rotated factor matrix, two factors representing 53.89 percent of the explained variance were extracted from 18 variables. The results showed the alpha coefficient for two factors ranged from 0.85 to 0.89.

Table 3 reports the 18 attributes factor analysis on travel motivation (outbound) resulted in

Table 3. Factor analysis on travel motivation (outbound).

	Factor Loading	
	1	2
Push Factors		
Social status	.812	
Experience new/different lifestyles	.735	
Impress my friends and family	.711	
Be together with my family	.695	
Enjoy shows and entertainments	.687	
Beautiful environment, scenery, beaches	.654	
Pull Factors		
Modern atmospheres		.821
Available activities		.803
Outstanding scenery		.784
Interesting attraction		.654
Nightlife and entertainment		.636
Eigenvalues	4.9	4.6
% of variance	29.11	24.78
Cumulative variance (%)	28.16	46.17
Cronbach's alpha score	.89	.85
Kaiser-Meyer-Olkin	0.91	

Table 4. Descriptive findings on travel motivation (domestic).

	Mean	S.D
Push Factors		
Escape from the routine of work or life	3.84	0.743
Do something exciting	3.83	0.817
Reduce stress	3.81	0.854
Away from home	3.97	0.746
Relax and doing nothing at all	3.65	0.695
Pull Factors		
Safe destination	4.12	0.759
Low budget	3.78	0.767
Reliable weather	4.04	0.694
Available attraction	4.17	0.718

11 attributes in two component groupings. For the first factor grouping named Push Factors, they are "social status", "experience new/different lifestyles", "impress my friends and family", "be together with my family", "and enjoy shows and entertainments", "beautiful environment, scenery, beaches". For the second factor grouping named Pull Factors, they are "modern atmospheres", "available activities", "outstanding scenery", "interesting attraction" and "night life and entertainment".

4 RESULTS

The study offers an attempt to understand outbound/domestic visitor motivation and extend the theoretical and empirical evidence on the relationships among push and pull motivations, although the relationship is not unusual. Nevertheless, the study suggests that the conceptual framework of push and pull factors in the literature was generally supported. The study reconfirms that tourists' travel behavior is driven by internal and external factors. In other words, they decide to go on a vacation because they need to fulfill their intrinsic desires, and at the same time, their decisions on where to go are based on destination attributes.

Based on the Table 4, looking into factors identified as push attributes, the study claimed that the needs to escape from the routine of work or life, to do something exciting and to reduce stress are among important motives which trigger the need to travel domestically. Crandall (1980) argued that people travel with specific motives to explore themselves and to improve the kinship relationship. The pull attributes, on the other hand, indicate that Malaysia is a safe destination, lower budget and understandably have attraction available. This clearly shows that Malaysia has a variety of offerings which could potentially improve the demand for domestic tourism by capitalizing on its heritage, natural attractions, food and culture. The abundance and diversity of tourism resources are widely recognized as an essential tourism asset for the country to develop its tourism industry. Besides, the country is also seen as an affordable, safe destination with friendly image and hospitable local people.

Based on the items identified as push factors in Table 5, the social status, impress friends and family and new/different lifestyles are among notable motives which induce the need to travel. This substantiation is consistent with Crandall's evidence that people travel with specific motives to gain prestige (Crandall, 1980). On the other hand, the pull attributes show that Malaysian travel outbound because these destinations offer modern atmospheres & activities, outstanding scenery, night life and entertainment and interesting attraction. The Malaysian government should re-assess their product availability and improve their information dissemination domestically. The government authorities and tourism operators should make the most out of this by enhancing the service sector, particularly in tourist facilities and infrastructure at the destination as the findings reveal that attraction availability is the most significant pull factor.

Table 5. Descriptive findings on travel motivation (outbound).

Push Factors	Mean	S.D.
Social status	4.14	0.695
Experience new/different lifestyles	3.77	0.737
Impress my friends and family	3.83	0.855
Be together with my family	3.79	0.816
Enjoy shows and entertainments	3.05	0.995
Beautiful environment, scenery, beaches	3.89	0.764

Pull Factors		
Modern atmospheres	4.27	0.776
Available activities	4.13	0.822
Outstanding scenery	3.81	0.759
Interesting attraction	3.62	0.669
Nightlife and entertainment	3.74	0.711

5 CONCLUSION

To adequately manage the domestic tourist flow to other secondary locations, Malaysia must utilize in developing a new thrust of tourism product such as nature-based or adventure type of tourism to attract specific niche markets. Mass media, at the same time, should play a vital role in forming a desired destination image for Malaysia. The strategic challenge for destination is not only on how to acquire images that influence travel, but also within the country.

A successful matching of push and pull motives is ideal for a marketing strategy in destination areas and proficient in segmenting markets, designing promotional programs, and decision-making about destination development. Finally, there is an impending need to develop comprehensive tourism policies to boost domestic tourism and reduce the outflow of foreign exchange from outbound tourism.

ACKNOWLEDGEMENT

The work described in this study was funded by the Universiti Teknologi MARA Malaysia (UiTM) under Research Managament Institute (RMI).

REFERENCES

Beerli, A. & Martin, J.D. (2004). Tourists' characteristics and the perceived image of tourist destinations: a quantitative analysis—a case study of Lanzarote, Spain. Tourism Management, 25(5), 623–636.

Bieger, T., Laesser, C. (2002). Travel segmentation by motivation—The Case of Switzerland. Journal of Travel Research 41(3), 68–76.

Crandall, R. (1980). Motivations for leisure. Journal of Leisure Research, 12(1), 45–54.

Crompton, J.L. (1979). Motivation for pleasure vacation. Annals of Tourism Research, 6(4), 408–424.

Dann, G.M. (1981) Tourism motivations: An appraisal. Annals of Tourism Research, 8 (2), 189–219.

Hair Jr JF, Anderson RE, Tatham RL, Black WC. Multivariate data analysis. New Jersey: Prentice-Hall; 1998.

Hanafiah, M.H.M., Othman, Z., Zulkifly, M.I., Ismail, H. & Jamaluddin, M.R. (2010). Malaysian tourists' motivation towards outbound tourism. International Journal of Tourism, Hospitality and Culinary Arts, 4–1:17–21.

Hanafiah, M.H.M., Harun, M.F.M., (2010), "Tourism demand in Malaysia: A cross-sectional pool time-series analysis", International Journal of Trade, Economics, and Finance, 1–1:80–83.

Hanqin, Z. & Lam, T. (1999). An analysis of Mainland Chinese visitors' motivations to visit Hong Kong, Tourism Management, 20(5), 587–594.

Jang, S. & Cai, L.A. (2002). Travel motivations and destination choice: A study of British outbound market. Journal of Travel & Tourism Marketing, 13(3), 111–133.

Josiam, B.M., Kinley, T.R. & Kim, Y.K. (2005). Involvement and the tourist shopper: using the involvement construct to segment the American tourist shopper at the mall. Journal of Vacation Marketing, 11(2), 135–154.

Kim, S. & Lee, C. (2002). Push and Pull relationships. Annals of Tourism Research, 29(1), 257–260.

Kozak, M., 2002. Comparative analysis of tourist motivations by nationality and destinations. Tourism Management 23, 221–32.

Li, X., Cheng, C., Kim, H. & Petrick, J.F. 2008. 'A systematic comparison of first-time and repeat visitors via a two-phase online survey', Tourism Management, 29: 278–293.

Liu, A. (2006). Tourim in rural area: Kehad, Malaysia. Tourism Management, 27, 878–889.

Mansfeld, Y. 1992 From motivation to actual marketing, Annals of Tourism Research, 19, 399–419.

Pearce, P.L. (1993): Fundamentals of tourist motivation. In D.G. Pearce and R.W. Butler (Eds.)

Prebensen, N.K. (2006), "A grammar of motives for understanding individual tourist behavior", Thesis submitted to the Department of Strategy and Management at the Norwegian School of Economics and Business Administration.

Rittichainuwat Ngamsom Bongkosh, Qu Hailin., and Brown Tom. The image of Thailand as an international travel destination. Cornell Hotel and Restaurant Quarterly 2001; 42 (2) April: 82–95.

Sekaran, U. (2003). Research methods for business (4th ed.). Hoboken, NJ: John Wiley & Sons.

Smeral and Witt, 1996, Smeral E., Witt S., Econometric Forecasts of Tourism Demand to 2005, Annals of Tourism Research 23, 1996, 891–907.

Tourism Research: Critique and Challenges (pp. 113–134). London: Routledge.

UNWTO (2012): World Tourism Organization Report

Uysal, M. and Hagan, L. (1993): Motivations of pleasure travel and tourism. In M. Khan, M. Olsen & T. Var

(Eds.), Encyclopedia of Hospitality and Tourism (pp. 798–810), New York: Van Nostrand Reinhold.

Winter, C. (2009). Tourism, social memory and the great war.Annals of Tourism Research,36(4), 607–626.

Yoon, Y., and M. Uysal (2005): An examination of the effects of motivation and satisfaction on destination loyalty: a structural model. Tourism Management 26:45–56.

Yuan, S. & McDonald, C. (1990). Motivational Determinants of International Pleasure Time. Journal of Travel Research, 29(1), pp. 42–44.

Zikmund, W.G. (2003) Business Research Methods, (7th edn), Thompson South-Western: Ohio.

Theory and Practice in Hospitality and Tourism Research – Radzi et al. (Eds)
© 2015 Taylor & Francis Group, London, ISBN 978-1-138-02706-0

Exploring responsible tourism development facet

M. Hafiz, M.R. Jamaluddin, M.I. Zulkifly & N. Othman
Faculty of Hotel and Tourism, Universiti Teknologi MARA, Shah Alam, Malaysia

ABSTRACT: Island tourism sector is regarded as an essential economic generator, creating employment, job and revenue. However, this conventional economic development mechanism affects the economy, socio-cultural and environment, especially in the aspect of quality of life negatively. Thus, responsible tourism management strategy had been widely adapted to embrace the planning, management, product development and marketing in order to churn out the positive impact towards the destination in terms of economic, social, cultural, and environmental. The responsible tourism framework has been identified as an ideal framework to sustain the development and minimize the constructive effects. This research focused on developing the responsible tourism practices dimension by exploring the community perceives value. Self-administered questionnaires were distributed to the Pangkor Islands' local community using quota sampling method. This study utilized quantitative approaches to analyze 214 local residents' feedback. Through a series of descriptive and factor analyses, various useful understandings on the issues of interest revealed. The study results contribute to the existing literature and recommend future research to explore the concept of responsible tourism, particularly in Malaysia.

Keywords: Responsible tourism; development, framework, local community, Pangkor Island

1 STUDY BACKGROUND

1.1 Introduction

Tourism products are extremely diverse, and they contribute to the overall development towards the potential range of business, cultural exchange, and other earnings (Cottrell & Vaske, 2006). Despite, tourism developments offer opposite effect towards a destination. This implication may arise from the visiting tourist behavior, improper planning, and development by the government, which neglected the importance of local residence viewpoint. Therefore, it is due to the government role as the main stakeholders in planning the tourism development, boosting tourists' satisfaction and also promoting the destination (Nunkoo & Gursoy, 2012). Furthermore, it is essential for the individuals, particularly residents, to incorporate in environmentally responsible behavior as a step in preserving natural resources and the environment.

According to Badaruddin (1996), Malaysia is extremely delighted with the economic benefits gained from nature tourism, but also cried for the harmful effects left. This focused on uncontrolled visitation, overuse of beaches, degrading the beach conditions and destroying the flora and fauna (Priskin, 2003). Furthermore, Honey (1999) argued that one of the consequences of such tightly populated areas is human-induced marine and coastal pollution. This comes through waste waters released from human settlements and poor waste disposal practices.

Pangkor Island is one of the island resorts that require further research and exhaustive feasibility studies due to the increasing number of population. 1.7 million tourists have visited the Perak State and total numbers of arrival to Pangkor Island were more than 400,000 tourists in 2012 and this number is predicted to grow every year. The increase in tourist visits was due to the ongoing tourism development from 1995 until 2015 under Draf Pembangunan Pulau Pangkor (Majlis Perbandaran Manjung, 2000). However, it has led to some issues specifically in ecological, social-cultural and economic such as natural degradation, cleanliness, marine pollution, over coastal development as well as improper waste management.

Toriman and Shukor (2007) stated that due to the increasing population in Pangkor Island, a feasibility study must be done to support the island especially in the water conservation. Furthermore, if coastal resources are to be used sustainably, coastal residents are also responsible for conserving natural resources and the environment. Authorities should put the residents' quality of life in the utmost matter where, if the quality of life perceived is lesser, residents may be reluctant to endorse tourism activities in their community (Kim, 2002). In line with prior notion, it is clear that countries embracing tourism must also take responsibility for ensuring that it is responsible and sustainable (Goodwin, 2007).

Consequently, many researchers argued a fundamental question of what is the finest solution to minimize the impact of tourism. Are these issues reflecting the local residents' quality of life? What are the Pangkor Island residences' perceptions towards this development? In other words, does responsible tourism concept may be the solution and contribute to quality of life? With this notion, in line with the absence of such study, an empirical research needed to be undertaken, particularly looking at the responsible tourism development practices from the local residence point of view.

2 LITERATURE REVIEW

2.1 *Responsible tourism*

Responsible tourism is a management strategy embraces planning, management, product development and marketing to bring about positive economic, social, cultural, and environmental impacts. The term 'responsible tourism' has emerged as a new gallows for developing and managing tourism development. Responsible tourism is still in the hypothetical phase. However, there were growing inclinations of public awareness of the harmful effects from negligent tourism developments. Husbands and Harrison (1996) define responsible tourism as "a framework and a set of practices that chart a sensible course between the fuzziness of ecotourism and the well-known negative externalities associated with conservative mass tourism".

The essential goal of responsible tourism is to diminish and reduce the negative impacts. Conceptually, responsible tourism includes governments, communities, investors, and visiting tourists knowing taking responsibility on their action to maximize the benefits of tourism while controlling the negative ones. The main argument for responsible tourism is the multi-level collaboration and stakeholder involvement in policy making and development planning. The involvement of stakeholders may yield future benefits to a tourism destination such as avoiding the cost of conflicts, enhancing political realism, generating enhanced, effective and innovative plans.

Responsible tourism development is expected to reach the three following fundamental goals (Scheyvens, 1999). The responsible practices promote equitable distribution of tourism benefits among communities, the business sector, and tourists themselves. Local economic benefits can be maximized through increasing linkages, reducing

leakages, and ensuring that communities were involved in tourism planning. Responsible tourism attempts to provide a comprehensive social experience while ensuring that socio-cultural diversity maintained. Local cultures need to be presented in an authentic way by allowing host communities to determine the manner of their presentation. Responsible tourism focuses on the management of natural diversity, sustainability, and appropriate systems for minimizing waste and over-consumption, integrate environmental considerations into all economic considerations, and verify any development is environmentally just.

While these three key goals of responsible tourism development are equally indispensable, little research efforts being given to exploring the social-cultural facet of tourism (Brunt & Courtney, 1999). Therefore, this study particular concentration is devoted to developing a framework on responsible tourism practices.

2.2 *Responsible tourism facet*

Moving forward, few researchers urged the responsible tourism concept to be another alternative tourism (Brunt & Courtney, 1999); (Carasuk, 2011). Settachai (2009) suggested future consideration to investigate responsible tourism as a new robust paradigm for developing and managing tourism. Focusing on a narrow definition, responsible tourism offers a comprehensive perspective on the practical implication of tourism that cut across many aspects of the community and involves multiple stakeholders. It is believed that the concept may become an effective guidelines to support and protect tourism planner from overwhelmed by mass tourism.

This notion is supported by the World Tourism Organization (2009) where responsible tourism is not only aiming to the improvement of the destination damaged from mass tourism activities, but it aims to provide benefits to the local residents in terms of economy, environment and socio-cultural. To ensure the sustainability of the resources (economy, environment and socio-cultural), responsible tourism practices may help the industry by preserving the future through natural and cultural preservation (Carasuk & Fisher, 2008). Adapting the unique setting of responsible tourism practices in another country, (Stanford, 2006) stated that the behavioral standard of practice in responsible tourism consists of recycling, water conservation, crime prevention and promoting local culture. Gauteng Tourism Authority (2003) guidelines for responsible tourism covered all aspects of the development; stakeholder involvement, environment, practices and also benefit received by the local residents. This responsibility merged together with the tourism activities through stability and sustainable design that includes environment, economy and socio-cultural (Ramchander, 2004). The concept also enables local residents to have a better quality of life where socio-economic and environment targeted to increase.

Responsible tourism practices need to be clearly understood avoiding ignorance towards other ideas that broadly shares a similarity among each other. As a new development concept, responsible tourism is still in the conceptual building stages. It reflects growing trends of rising public awareness of the harmful effects and irresponsibility of numerous tourism activities (Settachai, 2009). (Spenceley et al., 2002) added that responsible tourism practices may come from social responsibility and ethical dimension based on the local perspectives. Therefore, the responsible tourism practices assessment was narrowed down to the local's insight on the practices.

3 METHOD

3.1 Approach and sampling

(Leedy & Ormrod, 2005) suggested that in order to determine the variables with the purpose to explain, predict and control the phenomena, the quantitative data approach is the most appropriate procedure in achieving the goal of the study. It is supported by Abeyasekera and Savitri (2000) where quantitative data analysis may provide an exceptional value to the researcher to produce significant findings from a large amount of information.

In line with the above concept, variables were adapted from few researchers, combined and EFA was used to determine the variable dimensional structure (Spenceley et al., 2002); (Ramchander, 2004); Settachai, 2009). Information was obtained through a self-administered questionnaire towards the local residents. Furthermore, a cross-sectional approach also was used in this study and variables were measured at the same period of time. Quota sampling technique was used to obtain representative from the whole group of islands for a broad range of areas. 246 questionnaires were completed from 300 respondents approached, representing 82 percent response rate for this study. Descriptive statistic was used to analyze the average score for each item tested.

3.2 Factor analysis

Exploratory factor analysis (EFA) conducted to determine the links between the observed indicators for unknown or irregular construct. Since the measurement scale in this study was recently measured and quite exploratory in nature, the determination of how and to what extent the observed indicators are linked to the construct of the destination competitive strategy was necessary. Typically, the principal factors derived from EFA are represented as correlations between sets of various interrelated variables.

4 RESULT

4.1 Exploratory factor analysis

The results of varimax rotation were reported so that the extracted factors were independent and not correlated with each other. The sample size (N = 246) was study enough to run the factor analysis because it is generally recommended that a ten-to one fraction of the sample size is acceptable (Hair, et al., 1998). Consequently, a total of 16 items of support for responsible tourism practices was utilized for EFA. As an initial analysis, the Anti-image matrix indicated that most of the values were negative or had a little value of partial correlation. The Kaiser-Meyer-Olkin value was 0.77 conforming the partial correlations among variables are adequate. Basically, these examinations confirmed that since the initial analysis was sufficient, further factor analysis was possible.

4.2 Descriptive analysis

Mean score obtained from descriptive statistic pertaining to the locals' perceived value towards responsible tourism practice were reported in this section. Looking at the mean score in (Table 2), most of the respondents agreed towards all items in this dimension. The means of 3.0 to below 4.0 suggest the viability of responsible tourism practice in Pangkor Island.

5 DISCUSSION

As presented in Table 1, two factors were derived from the 16 responsible tourism practices items. All of the factor loadings were over 0.60 and had an eigenvalue >1.0. The first factor explained 25.17 percent of the variance. This factor was termed "Responsible Destination Planning" on the basis of the interpretation of the overall item context. The item with the highest loading was "Development strategies and action programs are specifically set up and adapted to achieve goals". The second factor explained 23.79 percent of the variance. This factor was termed "Responsible Environmental Practice" since all of the variables loading on this factor were related to environmental items.

Table 1. Factor analysis on responsible tourism practice.

	Factor Loading	
	1	2
Responsible Destination Planning		
Development strategies and action programs are specifically set up and adapted to achieve goals.	.824	
Tourism development enhanced the character of a place, promote the preservation of historic buildings, and improve the quality of the physical environment of the host community.	.798	
Locally-oriented images and identities been incorporated into the tourism promotion.	.747	
Tourism development and promotion, enhanced neighborhood bonds, community connections, and people attachments to a place.	.691	
Government, planners and decision makers communicate with and involve the local community in the planning process.	.675	
Host community has been explained on how a certain tourism development project may affect them.	.651	
Host community was given the opportunity to participate in and influence decision-making about tourism activities	.624	
Responsible Environmental Practice		
Restoration programs have been carried out effectively in areas that have been damaged or degraded by past activities		.794
Local residence recycles the reusable product		.776
Sports and outdoor activities, particularly in ecologically sensitive areas are managed in a way that they fulfil the requirements of nature and biological diversity conservation		.739
Local residence limits their activities related to pollution and did not exceed the ecological carrying capacity		.681
	4.7	4.5
% of variance	25.17	23.79
Cumulative variance (%)	25.17	48.96
Cronbach's alpha score	.885	.861
Kaiser-Meyer-Olkin	0.77	

Table 2. Descriptive analysis of responsible tourism practice items.

	Mean	S.D.
Responsible Destination Planning		
Development strategies and action programs are specifically set up and adapted to achieve goals.	3.95	.816
Tourism development enhanced the character of a place, promote the preservation of historic buildings, and improve the quality of the physical environment of the host community.	3.78	.907
Locally-oriented images and identities been incorporated into the tourism promotion.	3.97	.847
Tourism development and promotion, enhanced neighborhood bonds, community connections, and people attachments to a place.	3.87	.846
Government, planners and decision makers communicate with and involve the local community in the planning process.	4.02	.773
Host community has been explained on how a certain tourism development project may affect them.	4.06	.882
Host community was given the opportunity to participate in and influence decision-making about tourism activities	4.07	.811
Responsible Environmental Practice		
Restoration programs have been carried out effectively in areas that have been damaged or degraded by past activities	3.56	.964
Local residence recycles the reusable product	3.76	.858
Sports and outdoor activities, particularly in ecologically sensitive areas are managed in a way that they fulfil the requirements of nature and biological diversity conservation	4.02	.773
Local residence limits their activities related to pollution and did not exceed the ecological carrying capacity	4.12	.783

Additionally, the highest loading item was "Restoration programs have been carried out effectively in areas that have been damaged or degraded by past activities". Overall, 48.96 percent of the variance was explained by two factors for responsible tourism practices. "Responsible Destination Planning" and "Responsible Environmental Practice" were identified as appropriate strategies supported by the community.

The study outcome is in line with Carasuk and Fisher (2008), Davina (2006), and Ramchander

(2004). The result shows that community proposition on responsible tourism was based on environmental practice and development planning. Tourism stakeholders therefore should be responsive and the application of responsible tourism practice might help magnify community support towards future tourism development. For further planning policy, tourism policy maker should ensure that the residents are involved in the planning process that assist practitioners to better understand these residents' concerns.

Reflecting on the responsible tourism practice mean score (Table 2), the residents perceived responsible tourism practice provided the appropriate solution in combating the inappropriate development. It can be seen through their success in conducting restoration program towards the destination that damaged from the previous activities. Furthermore, the destination management practices recycle and reuse to protect the environment from pollution. In increasing the awareness towards this practice, cooperation between NGO's and residents association also have been done to educate the people from the grass root level.

6 CONCLUSION

The local residents, officials, and other tourism stakeholder should take a proactive effort by providing a vital support to the residents in Pangkor Island. Responsible tourism activities should be highly organized in the beginning stages such as school, kinder garden, religious center and social responsibility. Authority and tourism supplier must assess the level of tourism development and responsible tourism practices in Pangkor Island and may increase the destination image of Pulau Pangkor in the future. In the future, one could focus on the comparison between different stakeholder such as profit establishment and non-profit organization might also benefit. In addition, diverse perspectives on responsible tourism practices from different point of views may be investigated. Finally, future exploration pertaining to the exact responsible practices from other stakeholder may provide a clear depiction in providing responsible tourism development guidelines for Malaysian context.

ACKNOWLEDGEMENT

The work described in this study was funded by the Universiti Teknologi MARA, Malaysia (UiTM) under Research Management Institute (RMI).

REFERENCES

Badaruddin, M. (1996). *Local perception and involvement in tourism-keys to sustainable tourism.* Unpublished Phd Thesis. Rikkyo University, Tokyo.

Brunt, P. & Courtney, P. (1999). Host perceptions of sociocultural impacts. *Annals of tourism Research, 26*(3), 493–515.

Carasuk, R. (2011). *Responsible tourism Qualmark accreditation: a comparative evaluation of tourism businesses and tourists' perceptions*: Lincoln University. LEaP.

Carasuk, R. & Fisher, D. (2008). *Staycation: how global warming become a tourism constraint.* Paper presented at the New Zealand Tourism and Hospitality Research Conference 2008: Re-creating tourism.

Cottrell, S.P. & Vaske, J.J. (2006). A framework for monitoring and modeling sustainable tourism. *e-Review of Tourism Research, 4*(4), 74–84.

Goodwin, H. (2007). *Responsible tourism in destinations.* London, UK: CMRB.

Honey, M. (1999). *Ecotourism and sustainable development: Who owns paradise?* (Vol. 2nd). USA: Island Press.

Kim, K. (2002). *The effects of tourism impacts upon quality of life of residents in the community.* Virginia Polytechnic Institute and State University, Blacksburg, Virginia.

Leedy, P.D. & Ormrod, J.E. (2005). *Practical research: Planning and design* (Vol. 8th ed). Upper Saddle River, NJ: Prentice Hall.

Nunkoo, R. & Gursoy, D. (2012). Residents' support for tourism: An identity perspective. *Annals of tourism Research, 39*(1), 243–268.

Priskin, J. (2003). Issues and Opportunities in Planning and Managing Nature-based Tourism in the Central Coast Region of Western Australia. *Australian Geographical Studies, 41*(3), 270–286.

Ramchander, P. (2004). Towards the responsible management of the socio-cultural impact of township tourism. *Tourism and politics: Global frameworks and local realities*, 149–173.

Scheyvens, R. (1999). Ecotourism and the empowerment of local communities. *Tourism Management, 20*(2), 245–249.

Spenceley, A., Relly, P., Keyser, H., Warmeant, P., McKenzie, M., Mataboge, A., Seif, J. (2002). Responsible Tourism Manual for South Africa, Department for Environmental Affairs and Tourism, July 2002. *Responsible Tourism Manual for South Africa, 2*, 3.

Stanford, D. (2006). RESPONSIBLE TOURISM, RESPONSIBLE TOURISTS: What makes a responsible tourist in New Zealand?

Toriman, M.E. & Shukor, M.N. (2007). An Analysis of Rainfall Interception on the selection Experimental Plot of Pangkor Hill Reserved Forest. *Journal of Wildlife and Parks, 22*(2), 169–178.

Theory and Practice in Hospitality and Tourism Research – Radzi et al. (Eds)
© 2015 Taylor & Francis Group, London, ISBN 978-1-138-02706-0

The development of sports tourism towards residents of Langkawi Island, Malaysia

P.H. Khor & N.F.H. Ibrahim-Rasdi
Faculty of Sport Science and Recreation, Universiti Teknologi MARA, Perlis, Malaysia

K.C. Lim
College of Law, Government and International Studies, Universiti Utara Malaysia, Malaysia

ABSTRACT: Sports tourism has been documented as a platform for various developments towards sports events destination. Understanding the impact of sports tourism development in Langkawi Island from the aspects of economic, environmental, and socio-cultural can facilitate the planning of significant shared development strategies between the local government and residents in promoting sports tourism in Langkawi Island. The objectives of this study were answered based on survey research conducted among local communities. The results revealed that in general, the local residents have a high perception of economic and environmental impact on the development of sports tourism in Langkawi Island. Additionally, development of sports tourism brings positive effects towards the residents of Langkawi Island. Significant differences were also identified in terms of gender and marital status. Further specific study could be conducted on the attributes of sports events which contribute to positive development impact in promoting Langkawi Island as an international sports tourism destination.

Keywords: Sports tourism, economic impact, environmental impact, socio-cultural impact

1   INTRODUCTION

Langkawi Island is located in the Andaman Sea, some 30 km off the mainland coast of northwestern Malaysia. It is a duty-free-island, with a population of about 65,000 people and administered by Langkawi Development Authority (LADA). Equipped with outstanding services on the transportation, communication, accommodation and sporting infrastructures, Langkawi Island is in a good position to promote itself as a choice for international sporting events. Le Tour De Langkawi, Ironman Langkawi, Langkawi International Mountain Bike Challenge, and Royal Langkawi International Regatta are a few examples of annual sporting events that attract global tourist (Ministry of Human Resources Malaysia, 2012).

Sports tourism is currently adding five billion Ringgits annually to the Malaysia's revenue. Hence, it is rated as one of the best growing sectors in the tourism industry, as related by Malaysia Sports Tourism Council. With a yearly growth of eight to ten percent, it is gaining major attention for its impact on the economic, environmental, and socio-cultural towards the related destination. Previous research on tourism has focused wholly on

the development of tourism as a whole and does not distinguish among the impacts on different aspects. This study examined the impact of sports tourism development towards Langkawi residents in relation to the economic, environmental and socio-cultural aspect, as Hritz and Ross (2010) reported that sports tourism is strongly associated to economic, environmental, and socio-cultural impacts.

The findings of this study provided additional data in understanding the theoretical perspective of effects on sports tourism development towards the islanders. The information could also increase the likelihood for residents' support in promoting Langkawi Island as a sports tourism destination. Perceptions involving different age and marital status groups *display the* demographic *of the populations in Langkawi Island*. The hypotheses generated were:

(i)   There are differences in sports tourism impacts observed by local community in Langkawi Island, Malaysia in term of age group.
(ii) There are differences in sports tourism impacts observed by local community in Langkawi Island, Malaysia in term of marital status.

2 LITERATURE REVIEW

Sports tourism, as identified by Gibson and Fairley (2011), is a leisure-based travel that makes individuals temporarily being away from their home communities to involve in physical activities, to watch physical activities, or to venerate attractions associated with physical activities. Being a hospitality and service business, the host communities must be acknowledged as they are directly impacted (Bob & Swart, 2009).

The increase in the importance of sports tourism has created great impacts for the host communities. Generally, the impacts can be classified as either positive or negative, and often can have concurrent effect on the communities (McCartney et al., 2010). Hritz & Toss (2010) differentiated these impacts into four factors, namely social benefits, environmental benefits, economic benefits, and general negative impacts.

Previous studies have indicated clear positive and negative impacts of major multi-sport events on host populations. The positive impacts include development of the community facility, job opportunity, and increased cultural identity and social interaction (Hritz & Ross 2010; McCartney, et al. 2010). However, the economic impact on the host destinations is only significant in terms of certain parameters such as gross domestic product performance and unemployment in the short term (McCartney, et al. 2010; Tien, Low & Lin (2011). Nevertheless, a systematic organised environmental programme linked to a sporting event and multicultural investment could possibly contribute a handsome return to a host destination and its surrounding region (Dodouras & James, 2004).

It has been reported that local residents have suffered from living in a sports tourism area (Hritz & Ross 2010). The negative impacts include destruction caused by construction of facilities and restoration of transport systems which degrade the physical and biological environment through the air, water, soil, and visual pollution (Tatoglu, Erdal, Ozgur & Azakli, 2002), more traffic problems (Zhou & Ap, 2009), increased noise (Konstantaki & Wickens, 2010) and forced evictions and relocation of local housing and business (Porter, Jaconelli, Cheyne, Eby & Wagenaar, 2009). Waste, travel, and food and drink consumption have also created large ecological footprints (Collins & Flynn, 2008). Other negative impacts include lack of unity among the locals as their standard of living had improved, and the increase in prices of goods and services (Hritz & Ross, 2010).

3 RESEARCH METHODS

Self-administered questionnaires were distributed to participants selected from the attendees of Ironmen Langkawi, Langkawi International Mountain Bike Challenge, and Royal Langkawi International Regatta. The items in the questionnaire were adopted from the measurement scale developed by (Brida, Osti & Faccioli, 2011).

4 DATA ANALYSIS

The statistical significance of both hypotheses was tested using MANOVA. A significance level $p < .05$ was set to decide the significance level of each research hypothesis.

5 RESULTS

5.1 Factor analysis and reliability of measurement scale

The value of item loading for the measurement scale was greater than 0.40, with eigenvalues greater-than-one for the three subscales, while the value of item-total correlation for each subscale was more than 0.45. The coefficient alpha for the scale of economic impact, environmental impact and socio-cultural impact was 0.60, 0.67 and 0.78 respectively.

5.2 The economic impacts of sports tourism development towards residents of Langkawi Island

The main component identified was economic and environmental impact. The lists are as displayed in Table 1. Specifically, both economic impacts reading "Attracts more investments and spending" and "Brings more positive than negative impacts" were identified as main overall impacts of sports tourism development towards local communities of Langkawi Island.

5.3 The socio-cultural impacts of sports tourism development towards residents of Langkawi Island

Table 2 identified the significant socio-cultural impacts as "Learns tourist's culture through interaction", "Tourists are keen to learn local culture", "Gains positive experience", "Presents culture to tourists in an authentic way", "Provides incentives for historical restoration", and "Changes in local traditions and culture".

Table 1. Mean and percentage of frequencies of economic impacts of sports tourism development towards residents of Langkawi Island (N = 300).

Economic Impacts	Mean	*SD*	Percentage of frequency Strongly Disagree	Disagree	Agree	Strongly Agree
Attracts more investments and spending.	3.34	.753	2.0	11.0	38.0	49.0
Brings more positive than negative impacts.	3.34	.746	0.5	15.0	34.5	50.0
Creates job opportunities for outsiders than the locals.	3.21	.813	12.0	18.5	35.5	44.0
Increases standard of living.	2.75	.953	12.0	24.5	39.5	24.0
Benefits only certain groups of individuals.	2.67	.826	13.5	15.5	62.0	9.5
Overall	3.06					

Table 2. Mean and percentage of frequencies of socio-cultural impacts of sports tourism development towards residents of Langkawi Island (N = 300).

Socio-Cultural Impacta	Mean	*SD*	Percentage of frequency Strongly Disagree	Disagree	Agree	Strongly Agree
Learn tourist culture through interaction.	3.29	.705	3.5	0.4	52.5	40.0
Tourists are keen to learn local culture.	3.29	.748	0.5	16.0	37.0	46.5
Gains positive experience .	3.24	.588	0.5	6.5	61.0	32.0
Presents culture to tourists in an authentic way.	3.16	.798	2.0	19.0	40.0	39.0
Provides incentives for historical restoration.	2.58	.998	18.5	23.5	39.0	19.0
Changes in local traditions and culture.	2.52	.783	1.5	61.5	20.5	16.5
Lowers quality of life culturally.	1.91	.591	19.5	72.0	0.6	2.5
Overall	2.86					

5.4 *The environmental impacts of sports tourism development towards residents of Langkawi Island*

Table 3 identified the significant environmental impacts as "Provides incentive for the conservation", "Unpleasant crowd and inaccessible places during peak season", and "Changes local traditions and culture".

5.5 *The impacts of sports tourism development towards residents of Langkawi Island in term of age*

The MANOVA's results for economic, socio-cultural and environmental impacts tested on the five age groups was significant. Follow-up ANOVA tests extracted significant results for all the five economic impacts, reading "Attracts more investments and spending", $F(4, 195) = 5.017, p = .001, \eta^2 = 0.8$, "Brings more positive than negative impacts", $F(4, 195) = 3.597$, $p = .007$, $\eta^2 = 0.5$, "Creates job opportunities for outsiders than the locals", $F(4, 195) = 7.056$, $p = .000$, $\eta^2 = 1.1$, "Increases standard of living", $F(4, 195) = 4.555, p = .002, \eta^2 = 0.7$, and "Benefits only certain groups of individuals", $F(4, 195) = 16.493$, $p = .000$, $\eta^2 = 2.4$. The group of resident aged 50 years and above highly agreed that sports tourism attracts more investments and spending, increases standard of living although it benefits only certain groups of individuals. Residents of group aged 30–39 years observed that sports tourism creates job opportunities for outsiders than the locals.

ANOVA's results were found significant for socio-cultural impacts reading "Tourists are keen to learn local culture", $F(4, 195) = 6.290, p = .000, \eta^2 = 1.0$, "Provides incentives for historical restoration", $F(4, 195) = 5.094$, $p = .001$, $\eta^2 = 0.8$, "Changes in local traditions and culture", $F(4, 195) = 8.042, p = .000$, $\eta^2 = 1.2$, and "Lowers quality of life culturally", $F(4, 195) = 3.988$, $p = .004$, $\eta^2 = 0.6$. Residents of group aged 20–29 years old believed that tourists are keen to learn local cultures, while the residents aged below 20 years observed that sports tourism provides incentives for historical restoration and lowers quality of life culturally. Residents of group aged 30–39 years old complained that sports tourism changes the local traditions and culture.

Report on ANOVA's tests observed significant results for two environmental impacts namely

Table 3. Mean and percentage of frequencies of environmental impacts of sports tourism development towards residents of Langkawi Island ($N = 300$).

Evironmental Impacts	Mean	SD	Percentage of frequency			
			Strongly Disagree	Disagree	Agree	Strongly Agree
Provides incentive for conservation of natural resources.	2.76	.886	10.0	24.0	46.0	20.0
Unpleasant crowd and inaccessible places during peak season.	2.63	.803	3.5	46.5	62.0	9.5
Changes local traditions and culture.	2.52	.782	1.5	61.5	20.5	16.5
Adds to traffic congestion.	2.42	.887	9.5	56.0	17.0	17.5
Projects of hotels and other tourists' facilities destroy natural environment.	1.89	.690	26.5	60.5	10.0	3.0
Overall	3.06					

"Provides incentive for conservation of natural resources", $F(4, 195) = 6.613$, $p = .000$, $\eta^2 = 1.0$, and "Projects of hotels and other tourists' facilities destroy natural environment", $F(4, 195) = 3.213$, $p = .014$, $\eta^2 = 0.4$, as observed by the residents of group aged 30–39 years old.

5.6 The impacts of sports tourism development towards residents of Langkawi Island in term of marital status

MANOVA's score was reported significant for the economic, socio-cultural and environmental impacts tested on the three marital status groups. ANOVA's results were found significant for economic impacts reading "Brings more positive than negative impacts", $F(2, 197) = 4.881$, $p = .009$, $\eta^2 = 0.4$, "Creates job opportunities for outsiders than the locals", $F(4, 195) = 11.449$, $p = .000$, $\eta^2 = 1.0$, "Increases standard of living", $F(4, 195) = 10.464$, $p = .002$, $\eta^2 = 0.9$, and "Benefits only certain groups of individuals", $F(4, 195) = 11.262$, $p = .000$, $\eta^2 = 1.0$. Residents of divorced/widow status group highly believed that development of sports tourism brought more positive than negative impacts. It increases standard of living and

creates job opportunities, more for outsiders than the locals. The single status group felt that the development of sports tourism benefits only certain groups of individuals.

Significant results were noted for socio-cultural impacts, such as "Learns tourist's culture through interaction", $F(2, 197) = 3.884$, $p = .022$, $\eta^2 = 0.3$, "Tourists are keen to learn local culture", $F(2, 197) = 3.905$, $p = .022$, $\eta^2 = 0.3$, "Presents culture to tourists in an authentic way", $F(2, 197) = 3.073$, $p = .048$, $\eta^2 = 0.2$, "Provides incentives for historical restoration", $F(2, 197) = 4.512$, $p = .012$, $\eta^2 = 0.4$, "Changes in local traditions and culture", $F(2, 197) = 8.897$, $p = .000$, $\eta^2 = 0.8$, and "Lowers quality of life culturally", $F(2, 197) = 6.440$, $p = .002$, $\eta^2 = 0.5$. The residents from the married status group revealed that development in sports tourism changes in local traditions and culture. According to the observation of the single status group, tourists are keen to learn local culture and sports tourism provides incentives for historical restoration.

Significant ANOVA's results were identified for environmental impacts reading "Provides incentive for conservation of natural resources", $F(2, 197) = 3.511$, $p = .032$, $\eta^2 = 0.3$, and "Adds to traffic congestion", $F(2, 197) = 16.071$, $p = .000$, $\eta^2 = 1.3$, as observed by the residents of divorced/widow status group.

6 DISCUSSION

The findings in this study confirm previous report by Hritz and Ross (2010) that sports tourism is closely related to economic, environmental, and socio-cultural impacts. Resident's great experience of economic impacts clearly support studies by Hritz and Ross (2010), McCartney et al. (2010), and Tien et al. (2011) that major sporting events do fetch positive economic development. Hence, sports tourism are highly valued as according to Kurtzman (2005), "Tourism is a trillion dollar business and sport is a multi-billion dollar industry and has become a leading force in the lives of millions people globally".

The elderly residents (30 years and above) claimed that development of sports tourism attracted investments, besides increasing the spending consumption of tourists on Langkawi Island, a finding in line with Chappelet and Junod (2006). The younger generation (below 30 years) observed that sports tourism gave the tourists opportunities to learn the local culture, and it also act as a medium for historical restoration, a finding in line with Hritz and Ross (2010) and McCartney et al. (2010). The group aged between 30–39 years has mixed perceptions about the environmental impact.

Although development in sports tourism provides incentive for conservation of natural resources, it destroyed natural environment caused by constructions of hotels and other tourists' facilities.

Significantly, residents of divorced/widow status perceived positive economic development in sports tourism as it helped to present culture to tourists in an authentic way, and also helped the locals to learn tourists' cultures. Conversely, they felt that the development of sports tourism benefited the non-islanders as more job opportunities are available in Langkawi Island. Although it increased traffic congestion as identified by Zhou and Ap (2009), it provides incentives for historical restoration, and also exposed tourists to local cultures. Nevertheless, the married residents complained that developments in sports tourism change the local traditions and culture. These findings complement Brida et al.'s (2011) revelation that tourism gave the residents opportunities to meet people and learn their culture.

7 CONCLUSION

Langkawi Island is listed among the ten best islands in Asia ("Langkawi among top islands in Asia", 2014). LADA was established to inspire, implement, expedite, and execute socio-economic development in Langkawi. The identification of the lists of economic, socio-cultural, and environmental impacts could facilitate LADA in planning the strategies for future developments of the island. These considerations are important as LADA needs the local communities' supports for sport tourism development which could enhance the quality of their life. Future studies could focus on the long-term sports tourism impacts. Specific study on the attributes of international sports events could contribute to positive development in promoting Langkawi Island as an international sports tourism destination.

REFERENCES

Bob, U. & Swart, K. (2009). *Resident perceptions of the 2010 FIFA Soccer World Cup stadia development in Cape Town.* Paper presented at the Urban Forum.

Brida, J.G., Osti, L. & Faccioli, M. (2011). Residents' perception and attitudes towards tourism impacts: A case study of the small rural community of Folgaria (Trentino–Italy). *Benchmarking: an international journal, 18*(3), 359–385.

Chappelet, J. & Junod, T. (2006). A tale of 3 Olympic cities: What can Turin learn from the Olympic legacy of other Alpine cities. *Major Sport Events as Opportunity for Development. Valencia: Valencia Summit proceedings.*

Collins, A. & Flynn, A. (2008). Measuring the environmental sustainability of a major sporting event: a case study of the FA Cup Final. *Tourism Economics, 14*(4), 751–768.

Dodouras, S. & James, P. (2004). Examining the sustainability impacts of mega-sports events: Fuzzy Mapping as a new integrated appraisal system.

Konstantaki, M. & Wickens, E. (2010). Residents' perceptions of environmental and security issues at the 2012 London Olympic Games. *Journal of Sport & Tourism, 15*(4), 337–357.

Kurtzman, J. (2005). Economic impact: sport tourism and the city. *Journal of Sport Tourism, 10*(1), 47–71.

McCartney, G., Thomas, S., Thomson, H., Scott, J., Hamilton, V., Hanlon, P., . . . Bond, L. (2010). The health and socioeconomic impacts of major multi-sport events: systematic review (1978–2008). *BMJ: British Medical Journal, 340.*

Porter, L., Jaconelli, M., Cheyne, J., Eby, D. & Wagenaar, H. (2009). Planning Displacement: The Real Legacy of Major Sporting Events "Just a person in a wee flat": Being Displaced by the Commonwealth Games in Glasgow's East End Olympian Masterplanning in London Closing Ceremonies: How Law, Policy and the Winter Olympics are Displacing an Inconveniently Located Low-Income Community in Vancouver Commentary: Recovering Public Ethos: Critical Analysis for Policy and Planning. *Planning Theory & Practice, 10*(3), 395–418.

Tatoglu, E., Erdal, F., Ozgur, H. & Azakli, S. (2002). Resident attitudes toward tourism impacts: The case of Kusadasi in Turkey. *International journal of hospitality & tourism administration, 3*(3), 79–100.

Tien, C., Lo, H.-C. & Lin, H.-W. (2011). The Economic Benefits of Mega Events: A Myth or a Reality? A Longitudinal Study on the Olympic Games. *Journal of Sport Management, 25*(1), 11–23.

Zhou, Y. & Ap, J. (2009). Residents' perceptions towards the impacts of the Beijing 2008 Olympic Games. *Journal of Travel Research, 48*(1), 78–91.

Theory and Practice in Hospitality and Tourism Research – Radzi et al. (Eds)
© 2015 Taylor & Francis Group, London, ISBN 978-1-138-02706-0

Travel behavioral intention of choosing Malaysia as destination for medical tourism

A. Aziz, R. Md-Yusof, N.T. Abu-Bakar, S.N.H. Taib & M. Ayob
Universiti Teknologi MARA, Melaka, Malaysia

ABSTRACT: This paper aims to explore how the image of destination could allure the medical tourists in choosing Malaysia as their chosen destination. This conceptual paper offers an in-depth literature review regarding destination image and behavioral intention of the potential tourists and medical tourists. These tourists and medical tourists from many countries will be tapped in order to get information and to test the hypotheses. Understanding the relationships between behavioral intention and destination image, the destination country would have a better idea how to build up and attractive image and improve their marketing efforts in attracting as many medical tourists as possible. This study is perhaps will contribute to an understanding what factors that will influence the intentions of medical tourists through investigation by targeting the customers' perception and satisfaction.

Keywords: Perceived, destination image, behavioral intention

1 INTRODUCTION

The growth of medical tourism manage to attract many countries to venture in this field and at the same time compete for new customers and maintaining existing customers. Malaysia is one of the top five destinations and it's targeting patients is from developed and less developed countries in the medical tourism sector by providing good facilities and competitive rates compared to other parts of the world. How do we ensure that Malaysia will be the leading choices of destination? International accreditation alone is not good enough to attract the foreigner; marketing do play an important role in the maintenance of sustainable competitiveness. In this new era of globalization, it is important to understand the attitudes and behaviors of the medical tourists in order to support the government agencies and stakeholders to facilitate and formulate appropriate tourism policies.

Research reported that international demand for medical interventions from developed countries has grown dramatically due to lower cost health care services which is provided in the respective countries(Crooks, Kingsbury, Snyder & Johnston, 2010). For example around 750,000 Americans travel to the developing countries in the year 2007 (Connel, 2006). Other reasons are lack of medical insurance or underinsured, long waiting list and low exchange rates (Andaleeb, 2001; Opperman, 1999). For example in the year of 2011, 52 percent of health care consumers in France, 45 percent from

Germany, and 36 percent form United Kingdom has expressed their frustrations with long waiting list of medical treatment in their countries ("2011 Survey of Health Care Consumers Global Report Key Findings, Strategic Implications," 2011).

Nowadays, medical travel has been a popular scenario since many people realize the significant of travelling and benefits that they are getting. Malaysia has been known as one of Southeast Asia's topmost travel destinations, providing many interesting place for recuperation, attractive place for shopping and wonders place of natural habitat. Healthcare expenditure in Malaysia is determined by augmented privatization within the healthcare service provision. Moreover it was found that the (The Malaysian Tourism Promotion Board (MTPB, 2009) has been supporting the healthcare services. Malaysia has been considered as heaven for healthcare facilities and hospitals. In fact, Malaysia has been receiving approximately around 75 percent of medical tourists from the ASEAN region, Europe and Japan at 3 percent each, India 2 percent and others at 17 percent in 2008 (Wood, 2009). Since the Malaysian government has a significant budget in these activities, this research will be very helpful because it can justify the promotional activities such as creating value, suitable strategies to improve the medical tourism promotion. Furthermore, Malaysia Prime Minister Datuk Seri Najib Tun Razak has announced in the Malaysian budget 2012, saying that The Malaysian Healthcare Travel Council will be privatized to promote and develop Malaysia as a health care destination.

Table 1. Medical revenue in Malaysia, 2000–2011.

Year	Value (RM Million)	Growth (%)
2000	33	48.4
2001	44	35.7
2002	36	−18.7
2003	59	63.6
2004	105	78.2
2005	151	43.0
2006	204	35.0
2007	254	24.6
2008	299	17.8
2009	288	−3.7
2010	378	31.5
2011	511	34.9

Source: Medical Tourism Revenue (2012).

Table 2. Total number of foreign patients in Malaysia.

Year	Number of Foreign Patients	Growth Rate (%)
2001	75,210	33.99
2002	84,585	12.47
2003	102,946	21.71
2004	174,189	69.20
2005	232,161	33.28
2006	296,687	27.79
2007	341,288	15.03
2008	374,063	9.60
2009	336,000	−10.18
2010	392,956	16.95
2011	583,296	48.44

Source: Medical Tourism Revenue (2012).

From the above table, it was found that even though the number of revenue is increasing throughout the years, unfortunately the growth rate is very unstable and it does not exhibit a trend. The growth rate from 2001 to 2011 is around 25.37 percent per annum ("Medical Tourism Revenue," 2012). Table 2 shows that number of foreign patients in Malaysia from 2001 until 2011. The total numbers of foreign patients in Malaysia keep increasing from 2001 until 2011 except in 2009 the numbers decline.

2 LITERATURE REVIEW

2.1 Behavioral intention

Many researchers have found various definitions of behavioural intention (Caruana, 2002; Jacoby & Chestnut, 1978). Zeithmal, Berry, and Parasuraman (1996) described that behavioural intention

act as a signal whether the customers will remain or exit the relationship with the service provider that is the private hospital. In addition, Zeithmal et al. (1996) mentioned that there are two dimensions to measure behavioral intention—favorable and unfavorable. Favorable intentions means that the customers will convey a positive word of mouth, repurchase intention and loyalty (Ladhari, 2009; Zeithaml, Berry & Parasuraman, 1996). On the other hand, unfavorable behavioral intention tends to spread negative word of mouth due to their negative experiences to other customers (Caruana, 2002). This will make them directly switch their intention to competitors (Athanassopoulos, Gounaris & Stathakopoulos, 2001).

In this situation, the relationship focuses on the number of customer who comes back to buy and continues to buy until it creates a positive attitude towards the company products and services. This will definitely will creates a customer loyalty and repurchase intention (Zeithaml et al., 1996). Loyalty is defined as "a deeply held commitment to re-buy or re-patronize a preferred product/service consistently in the future" (Kesar & Rimac, 2011). Repurchase-intention can be defined as customer will maintain a relationship with the services provider (Zeithaml et al., 1996). Whereas word of mouth can be defined as a customer will inform positive experience of relationship with friends, relatives or others (Woodside & Moore, 1987). As in the hospital context, patients satisfied with the hospital are obviously will recommend their treatment to other patient (Finkelstein, Harper & Rosenthal, 1999). For example Kessler and Mylod (2011) reported that patients satisfaction will significantly influenced the patient's intention to return to the same hospital. In addition, if the patient's are highly satisfied with the admissions, facilities provided, services and other processes, all of this will definitely lead to patient's returning to the same hospital (Kessler & Mylod, 2011). In the Asian culture, friends, relatives, colleagues and neighbors have great influence on the customers when it comes to making decisions regarding an institutions and they are really depends on the personal recommendation from family and friends (Owusu-Frimpong, Nwankwo & Dason, 2010).

2.2 Perceived destination image

Generally, it has been accepted in the literature that image has influenced the tourist behavioral intentions in choosing destination (Bigne, Sanchez & Sanchez, 2001; Fakeye & Crompton, 1991; Lee, Lee & Lee, 2005). Image is not only applicable to brand but apply to company, service, person or place. This is consistent with the concept of the product, which can be defined as physical goods, service, place, person and even ideas (Kotler & Armstrong, 2001). Image of a destination is definitely an ongoing process and establishing it in the minds of the potential visitors (Gartner, 1989). Furthermore, a number of the researchers have studied the destination image which has influence the behavioral of the international tourists in choosing the selective place for medical treatment and travelling satisfaction (Bigne, Sanchez & Sanchez, 2001; Fakeye & Crompton, 1991; Lee, Lee & Lee, 2005).

Places need to differentiate themselves among their competitors in order to attract tourists, and investors. This image includes the reason why the consumer chooses the destination for certain purposes, evaluation and also for future behavioral intention. Destination image derived from organic image to induced image and to a complex image. Fortunately these image were connected to the information, persuasion and the remaining function of promotion (Fakeye & Crompton, 1991). Therefore, destination image comprises of perceptions of individual attributes such as climate, facilities and friendliness of the people as well as the overall impressions of the place (Echtner & Ritchie, 2003). In addition, the respective places need to use a creative method and identify a unique destination that will enable potential tourists to choose the destination. Due to the increasing of competition, the respective places such as countries, cities or even town have been using all the appropriate tools to promote and improve their image in order to reach their objectives and creating a positive experience for the medical tourist (Milman & Pizam, 1995).

Jayawardena (2002), points out that the prospect of tourism markets rely on the ability of the service provider to deliver a high and good quality of the products that relates to the changing needs, wants and demands of the tourists. One of the most marketing challengers which arise from this situation is the need to have an effective and strong destination positioning strategy. Since many countries nowadays are focusing on medical tourism, the choices of destination chosen by the medical tourists continue to grow. In order to be successfully promoted in the destination choices, the chosen country must be favorably different from its competitor. Hankinson (2005) had identified eight clusters of brand image attributes. They are physical environment, economic activity, business tourism facilities, accessibility, social facilities, and strength of reputation, people characteristics and destination size. According to (Echtner & Ritchie, 2003), the important image of a destination, understanding the tourist behavior, designing an effective marketing strategies, plays an important role in order to develop more specific and more complicated

frameworks and methodologies in order to reliably and validly measure destination image.

It was observed that the attractiveness of a destination, the quality of services, facilities provided, attractive location and accessibility of centres has been considered as tourist destination choice (Ali & Howaidee, 2012). In addition, image of a destination has been one of the important factors in determining the purchase decision by the customer. If the consumer feels that they are being influenced by a different stages of risk, the consumer's tend to change their decision, modify or worst avoid making a purchase decision(Kotler, 2003). Furthermore the amount of risk varies such as the time spending for a treatment, psychological effect, and social effect, functional and physical effect. Due to that the marketers are supposed to reassures customers and aware the factors that will provoke positive or negative feelings in consumers and at the same time the marketers are supposed to provide information to support reducing the perceived risk.

3 FUTURE OF MALAYSIAN HEALTH TOURISM SECTOR

According to previous Health Minister Liow Tiong Lai, Malaysia is estimated to rake in RM342 million (US$110 million) and it showed that the country still has room to grow and be at the forefront of health tourism in the region. He also added that the healthcare industry has been identified as one of the key areas in achieving high income status. One of the way is by joint venture with an establish medical centre for example signing an agreement between AriyanDana Equities and Narayana Hrudayalaya India to establish the Narayana International Medical Centre (NIMC) in Nilai located in the state of Negeri Sembilan. Furthermore, Liow also said that this joint venture was moving the right direction in planning to promote health tourism, creating a hub to attract foreigners while being beneficial to Malaysians. Furthermore he said that the merger will also become a good option for patients from other parts of the regions and a source of foreign revenue which will help to speed up Malaysia's transformation into high income nation.

Malaysia has many advantages in the health tourism sector. Among the factors of advantages are its cost-competitiveness compared to the regional and international markets, the good infrastructure and the fact that English is widely spoken here. In addition, the overall performance of Malaysia's healthcare system is considered remarkably well by the standards of the World Health Organization (WHO).

4 DESIGN/METHODOLOGY/APPROACH

This conceptual paper offers an in-depth literature review regarding destination image and behavioral intention of the potential medical tourists and tourists. Based on this, contents of the behavioral intention and destination image are discussed.

REFERENCES

2011 Survey of Health Care Consumers Global Report Key Findings, Strategic Implications. 2011. Retrieved from www.deloitte.com

Ali, J.A. & Howaidee, M. 2012. The Impact of Service Quality on Tourist Satisfaction in Jerash. *Interdisciplinary Journal of Contemporary Research in Business* 3(12): 164–187.

Andaleeb, S.S. 2001. Service Quality perceptions and patient satisfaction: a study of hospitals in developing country. *Social Science & Medicine 52*: 1359–1370.

Athanassopoulos, A., Gounaris, S. & Stathakopoulos, Behavioural responses to customer satisfaction: an empirical study. *European Journal of Marketing* V. 2001.*35,*(5/6): 687–707.

Bigne, J.E., Sanchez, M.I. & Sanchez, J. 2001. Tourism image, evaluation variables and after purchase behaviour: inter-relationship. *Tourism Management 22*(6): 607–616.

Caruana, A. 2002. Service loyalty: the effects of service quality and the mediating role of customer satisfaction. *European Journal of Marketing, 36*(7/8): 811–828.

Connel, J. 2006. Medical Tourism: Sea, Sun, sand and... surgery. *Tourism Management,* (6). Retrieved from http://www.sciencedirect.com

Crooks, V.A., Kingsbury, P., Snyder, J. & Johnston, R.2010. What is known about the patient's experience of medical tourism? A scoping review. *BMC Health Services Research 10*(1): 266.

Echtner, C.M. & Ritchie, J.R.B. 2003. The Meaning and Measurement of Destination Image. *The Journal of Tourism studies 14*(1).

Fakeye, P.C. & Crompton, J.L. 1991. Image differences between prospective, first-time, and repeat visitors to the Lower Rio Grande Valley. *Journal of Travel Research 30*(2): 10–16.

Finkelstein, B.S., Harper, D.L. & Rosenthal, G.E. 1999. Patient assessments of hospital maternity care: a useful tool for consumers? *Health Services Research 34*(2): 623.

Gartner, W.C. 1989. Tourism image: attribute measurement of state tourism products using multidimensional scaling techniques. *Journal of Travel Research 28*(2): 16–20.

Hankinson, G. (2005). Destination brand images: a business tourism perspective. *Journal of Services Marketing, 19*(1), 24–32.

Jacoby, J. & Chestnut, R.W. 1978. *Brand loyalty: Measurement and management*: Wiley New York.

Jayawardena, C. 2002. Mastering Caribbean tourism. *International Journal of Contemporary Hospitality Management 14*(2): 88–93.

Kesar, O. & Rimac, K. 2011. Medical Tourism Development in Croatia. *Zagreb International Review of Economic & Business 14*(2): 107–134.

Kessler, D.P. & Mylod, D. 2011. Does patient satisfaction affect patient loyalty? *International Journal of Health Care Quality Assurance 24*(4): 266–273.

Kotler, P. (2003). *Marketing Management* (11 ed.): Prentice Hall.

Kotler, P. & Armstrong, G. 2001. *Principles of Marketing* (9 ed.): Prentice Hall.

Ladhari, R. 2009. Service quality, emotional satisfaction, and behavioural intentions: A study in the hotel industry. *Managing Service Quality 19*(3): 308–331.

Lee, C.-K., Lee, Y.-K. & Lee, B. 2005. Korea's destination image formed by the 2002 World Cup. *Annals of Tourism Research 32*(4): 839–858.

The Malaysian Tourism Promotion Board (MTPB), T.M. 2009. MALAYSIA:Tracking Malaysian medical tourist statistics. *International Medical Travel Journal:News.*

Medical Tourism Revenue. 2012. *Penang Monthly.*

Milman, A. & Pizam, A. 1995. The role of awareness and familiarity with a destination: The central Florida case. *Journal of Travel Research 33*(3): 21–27.

Opperman, M. 1999. Predicting Destination Choice—A Discussion of Destination Loyalty. *Journal of Vacation Marketing 5*(1): 78–84.

Owusu-Frimpong, N., Nwankwo, S. & Dason, B. 2010. Measuring service quality and patient satisfaction with access to public and private healthcare delivery. *International Journal of Public Sector Management 23*(3): 203–220.

Wood, L. 2009. Researh and Markets: Malaysia Medical Tourism Outlook 2012- Despite the Global Economic Slowdown, Malaysia Received Around 22 Million International Tourists in 2008. *Business and Economics.*

Woodside, A.G. & Moore, E.M. 1987. Competing resort hotels Word-of-mouth communication and guest retention. *Tourism Management 8*(4): 323–328.

Zeithaml, V.A., Berry, L.L. & Parasuraman, A. 1996. The Behavioral consequences of service quality. *Journal of Marketing 60*: 31–46.

Theory and Practice in Hospitality and Tourism Research – Radzi et al. (Eds)
© 2015 Taylor & Francis Group, London, ISBN 978-1-138-02706-0

Cruise tourism in Malaysia: A SWOT analysis

K.L. Chong

Centre for Tourism, Hospitality and Culinary Management, Sunway University, Selangor, Malaysia

ABSTRACT: According to Tourism Malaysia, the Asian and international cruise is growing at the average of 14 percent annually over the past 10 years while number of cruise passengers in Malaysia also expected to exceed half a million in 2013. Despite its potential, the cruise industry in Malaysia is facing ever-growing of competitions from its neighbouring countries. An SWOT analysis was used to discuss the potential of improving strategic decision and planning for cruise tourism in Malaysia. According to findings, notable strengths include a growing market and the government's push to upgrade ports' infrastructure. Weaknesses include oligopolistic competition and negative perception of Malaysian cruises being overly gambling-oriented rather than holiday making. Opportunities are seen in three areas: 1) increase of spending in travel and cruising; 2) exemption from the Malaysia Cabotage Policy was given to all international cruises by the Malaysian government, hence, attracting more international cruises to stopover to more destinations in Malaysia; 3) strong supports from Malaysian government, as well as their initiatives to concern, and develop this business now and future. Piracy and safety have been identified as the main threats that create risks for the cruise business in Malaysia.

Keywords: Cruise, tourism, SWOT analysis

1 INTRODUCTION

According to Tourism Malaysia, the Asian and international cruise in growing at the average of 14 percent annually over the past ten years while number of cruise passengers in Malaysia also expected to exceed half a million in 2013 (The Star, 2013a). In fact, this is considered as part of the result for Tourism Malaysia to promote cruise tourism of Malaysia actively in several international cruise conferences and exhibitions since 2009 (Tourism Malaysia, 2013).

As the cruise industry in Malaysia is still growing now, there are many challenges that are facing in this sector. Although this business is growing worldwide yet the Asia Pacific market share is minute and it remains in the product-introduction phase. This trend is unlikely to change anytime soon, as the barriers to entry to the market are increasingly high and many Asian destinations adequate port infrastructure (WTO, 2012).

In order to sustain and obtain benefits from this business, cruise liners need to work together with stakeholders such as governments, port operators, communities and more. Meanwhile, they also should take into consideration on various internal and external factors which might influence the operations or the whole industry. Hence, the purpose of this study aims to discover the internal (strengths and weaknesses) and external factors (opportunities and threats) that are potential in influencing the development and operational decision of cruise tourism or industry in Malaysia.

2 METHODOLOGY

SWOT analysis was adopted as a model in assessing the cruise industry of Malaysia. SWOT analysis is not only approach that widely used for reviewing the strategy, position and direction of a particular organization and product yet it is also applicable for any industry (Morrison, 2014). Nevertheless, PESTLE analysis (Political, Economy, Social, Technology, Legal and Environment) was used as an analytical tool in addressing the external environment (Opportunities and Threats) while capability, cultural and organizational analysis was used in assessing the internal environment of the industry (Strengths and Weakness). Latest news, industrial annual reports, statistical reports and reviews pertaining to cruise tourism in Malaysia were the main sources used in the analysis.

3 FINDINGS

3.1 *Strengths*

3.1.1 *Adequate of ports and cruise infrastructures*
Ports and cruise infrastructures are playing a very significant role in the cruise industry as their

relationship is like a complementary product for each other (Manning, 2006; World Cruise Industry Review, 2013). One of the strengths for a cruise industry in Malaysia is that the government has significantly invested in the development of ports and cruise infrastructures for embarkations and disembarkations in the. According to Kevin Leong (2013), general manager for Asian Cruise Association (ACA), developments on this area have significantly attracted more and more cruise liners to venture into a particular country or region. For instance, there are a total of six main ports that are strategically and adequately located in Malaysia for cruise's stopover which including Port Klang, Penang, Malacca, Kuching, Kota Kinabalu, and Langkawi (ETP, 2012). Indeed, Malaysian government also eventually increased the cruise ship capacity of Port Klang in order to cater to larger ships and more passengers as part of the Entry Points Project (EPP) six which focusing in cruise tourism (The Star, 2013b; ETP, 2012).

3.1.2 *Strategic location of malaysia*
Location of Malaysia is strategically considered as a cruise destination. Indeed, this is due to the reason that South-East Asia is having more than 25,000 islands within the region that provide a natural strength for this area in the international cruise industry. Eventually, Malaysia will be able to gain advantage from this region's growing cruise market as it is geographically located in South-East Asia (The Star, 2013a). Furthermore, this area also allowed the industry to access to markets that comprised of 60–70 percent of the world population.

3.1.3 *Award winning ports and cruise operators*
Reorganization towards the cruise industry in Malaysia is gained through the world class awards that received by both the cruise terminals and local cruise operators such as Star Cruises. For instance, Star Cruises Terminal in Port Klang and Jetty Terminal in Langkawi Island Malaysia were proudly awarded the Statements of Compliance under the International Ship and Port Facility Security (ISPS) Code 2002 (Genting Group, 2014; UC Cruises, 2014). This code is primarily a security threat preventive framework which involves several parties including cooperation and contracting governments, government agencies, local authorities and members of the shipping and port fraternity.

In terms of the local cruise operators, Star Cruises is also well recognized for its world class service standards of hospitality not only in the cruise industry of Malaysia but the whole Asia Pacific (UC Cruises, 2014). These international recognized awards would then play a significant role and act as the strength of the cruise industry in Malaysia as it is not only a testimonial towards the cruise facilities and cruise service standards but also strengthen the positive impression and perception towards Malaysia's cruise industry.

3.2 *Weaknesses*

3.2.1 *Market structure of cruise industry— oligopoly*
The seven major cruise operators who dominated the whole cruise industry now are owned by only three corporations who overall controlled nearly 80 percent of the whole cruise market (Wood, 2004; Lekakou, Pallis & Vaggelas, 2009). Carnival Cruise Lines, Costa Cruises, Holland America, Aida Cruises, Cunard Line, Ocean Village, P&O Cruises, Seabourn Cruise Line, Windstar Cruises and Princess Cruises are all owned by Carnival Corporation (World's Leading Cruise Lines, 2014); Celebrity Cruises, Pullmantur, Azamara Cruises, Croisieres de France (CDF) and Royal Caribbean International are owned by Royal Caribbean (Royal Caribbean International, 2014); while Star Cruises Group of Malaysia also owns NCL America, Norwegian Cruise Line, Orient Lines as well as Star Cruises (Star Cruises, 2014; NCL, 2014).

The barrier of new entry into this cruise business market is high as it is dominated just by these three conglomerates. Besides, the products and services provided by the cruise liners in this market would be very homogenous as the competition is not high. The development of the cruise industry either in Malaysia or internationally might be constrained by having a very homogenous products and services in the market.

3.2.2 *Gambling versus Muslim*
In an Islamic country like Malaysia, gambling is an illegal activity for the Muslims and even strict rules and regulations are imposed to the very limited number of licensed casinos that available in the country. However, Clarke (1999) found that casinos or gambling facilities are not only an on-board activity that enlightens the cruise passengers but it also plays a significant role in the cruise industry as a major profit contributor in their annual profit and loss reports. Furthermore, it is also not uncommon for cruise operators to set up a casino or provide gambling facilities on board as casino that is operating offshore or on international sea is not regulated by legislation or government (Ng & Kwortnik, 2007).

Hence, it can be seen as a weakness for the cruise industry in Malaysia when the cruise liners start to concern and focus on the gambling activities and forgetting the importance of the Muslims market in Malaysia, as they are the biggest population in Malaysia yet are illegal to gamble according to the law.

3.2.3 *Labour shortage*

Although the cruise industry is well-known for its globalised workforces on-board yet labour shortage has become one of their weaknesses that might harm the cruise liners or even the cruise industry itself (Gibson, 2008; Terry, 2011; Larsen et al., 2012). Burke (2009) had identified that mistreat of cruise employees is part of the labour issues that lead to labour shortages in this industry as cruise liners will have a chance to escape from labour laws when they registered their cruises and company in foreign countries.

Additionally, Larsen et al. (2012) also reveal that jobs on-board is considered as very isolated as compared to others whereby the employees have no choice, but to be cut off from any recreational activities, their own families and friends. Indeed, low work engagement, job autonomy and departmental resources will lead to low employee retention.

3.3 *Opportunities*

3.3.1 *Increase of Spending in Travel and Cruising*

As reported in the study done by Travel Weekly (2012), 43 percent of consumers are expected to spend significantly more while 31 percent spend somewhat more for travelling. This economy factor is considered as one of the opportunities for cruise tourism in Malaysia to boost as this situation will enable more people around the world to travel with cruises as their spending on travel is increasing. Despite international market, the cruise industry of Malaysia also might be benefits from its domestic market as 73 percent of Malaysian indicated that they would increase their budget for travelling in 2013 in the research done by Trip Advisor (The Star, 2013b).

According to Travel Weekly (2012), the number of global cruise passengers increased approximately two million (10%) as the figures increased from 18.7 million in 2010 to 20.6 million for 2011. In terms of Asia and Pacific region, its total number of cruise passengers goes up to almost 800,000 which contributed a 6 percent market share with a combined growth rate of 90 perecent from year 2001 to 2004 (WTO, 2012). Hence, these evidences strongly shown that people were not only spent more in travelling but there is an increasing trend for cruising.

3.3.2 *Relieved from Cabotage Policies*

In Malaysia, Cabotage Policy was initially implemented with the main objective to improve Malaysian ownership and local shipping as well as to control the dependence of Malaysia on foreign vessels that lead to the outflow of foreign exchange (Malaysiakini, 2009).

However, it is an opportunity for the cruise industry of Malaysia whereby an exemption from the Malaysia Cabotage Policy was given to all international cruises by the Malaysian government (ETP, 2012). By having this opportunity to be exempted from this government policy, international cruises are now allowed to disembark and re-embark cruise passengers at more than one Malaysian port in any of its stopover destinations throughout the itinerary (ETP, 2012; Malaysiakini, 2009). In return, this will helps to develop the cruise industry in Malaysia by attracting more international cruises to stopover to more destinations in Malaysia.

3.3.3 *Government supports and initiatives*

Cruise industry is part of the high impact project of Malaysia, which is the Economy Transformation Programme (ETP) that introduced and launched in year 2010 (ETP, 2012). Thus, cruise industry in Malaysia is being considered as one of the national key economic areas by the government. Furthermore, a policy-making public-private stakeholders' advisory committee were also formed under this programme with the function of providing direction for policies, developments and frameworks required by the cruise industry in Malaysia (ETP, 2012). In fact, this advisory committee is known as The Malaysia Cruise Council (MCC) whereby another six sub-task forces were also formed underneath to address and streamline the specific issues of those six identified ports in Malaysia (ETP, 2012).

3.3.4 *Free from severe natural disaster*

The fact that Malaysia is free from any severe natural disaster such as Tsunami, earthquakes, typhoons and volcanic eruptions since it is strategically located out of the "Pacific Rim of Fire" (Disaster Management Division of Malaysia, 2011). This is an opportunity to promote and develop Malaysia as an international cruise destination because the safety issues of a particular country or destination is basically a main concern and consideration when choosing a cruise destination either by the cruise operators or cruise passengers (Manning, 2006; London, 2011). This is also justified by the research done Travel Canada (2014) who found that 87 percent of the target respondent (cruise passengers) revealed that the safety issue is definitely their main concern for cruising.

3.4 *Threats*

3.4.1 *Negative perceptions towards cruising*

One of the most common negative perceptions towards cruise tourism would be the safety issues of the cruises due to the emergent of cruise incidents recently. In fact, these cruise incidents have negatively impacted on the demand for cruise tourism as well as the prices for the cruise tour. For instance, the demand for cruise tourism was found approximately 15 percent to 20 percent in the weeks after the cruise incident of Costa Concordia that stranded off the coast of Isola del Giglio, Italy on 13th of January 2014 (CBC News, 2012).

Another negative perception towards cruising will be misperception of the customer whom portrait cruises as a cheap and gambling focus excursion, especially for the Asian market (Golden, 2013). Asian always have the conservative thought that assumed on-board casino as the main attraction that commonly used to promote a particular cruise or cruise liner yet this is not the case.

3.4.2 Piracy

Externally, piracy also has been identified as one of the threats that create risks for a cruise industry (Dowling, 2006; Global Travel Industry News, 2009; Bundhun, 2011). According to a report presented by Statista (2013), the statistic of pirate attacks was the highest in year 2010 recorded 445 cases. Although the number of pirate attacks reduced to 297 in year 2012 yet it does create a threat towards the safety of the cruise industry (Dowling, 2006).

Indeed, the primary routes of international cruising as well as the geographical location of piracy include the Indian Ocean, Straits of Malacca, Red Sea or Horn of Africa, and Indonesian and Malaysian waters which contributed to the areas of concern in the cruise industry (Global Travel Industry News, 2009; Dowling, 2006). This factor will then take into consideration of the cruise liners when designing on the itinerary of the cruises as well as when consider a cruise destination due to the reason that passenger safety and security are always the primary concerns of them, says Sasso as the chairman of the Cruise Lines International Association (CLIA) in the Global Travel Industry News (2009).

4 CONCLUSION

Overall, there are several great opportunities that can be used on using the strengths that are currently available in the industry or market. On the other hand, the government and cruise operators should also take into consideration of overcoming the threats that posed by the environment and the industry itself. It is very crucial for all the stakeholders and shareholders of the cruise industry in Malaysia to work together for future development and sustainability.

In terms of threats such as the negative perception, threats of substitute tourism products and piracy, the government might look into the reinforcement of government policy regarding the safety and security of Malaysian waters in order to minimise or avoid from severe piracy. Moreover, the cruise liners may try to improve on the perceptions of people towards cruising in Malaysia or other regions. Perhaps, this can be done through educating the public or market with more specific knowledge about cruising or through a properly plan media and social network as the societies now are much influenced by the media and social networks in their daily life.

All in all, there are rooms for the cruise industry in Malaysia to further develop and sustain in the industry. However, this is only provided when the cruise operators, port authorities, communities, and the government are to work together in planning and strategy developments.

REFERENCES

AIPA. (2011). *Brief note on the roles of the National Security Council, Prime Minister's Department as National Disaster Management Organization (NDMO)*. 3rd AIPA CAUCUS REPORT.

Bundhun, R. (2011). *Gulf cruises under threat from pirates | The National.*

Burke, C. (2009). A qualitative study of victimization and legal issues relevant to cruise ships.

CBC News. (2012). *Cruise ship incidents drive down demand, prices—World—CBC News.*

Clarke, J. (1999). *Cruise lines just say no.*

Dowling, R. (2014). *Cruise ship tourism*. India: CABI Publisher.

Genting. (2014). *Group Profile—Star Cruises Limited.*

Gibson, P. (2008). Cruising in the 21st century: Who works while others play? *International Journal of Hospitality Management, 27* (1), 42–52.

Global Travel Industry News. (2009). *Cruise industry vs. piracy cruise lines weigh tougher response to pirate threat near Somalia.*

Golden, F. (2013). Cracking challenging Asia cruise market is no slam-dunk.

Larsen, S., Marnburg, E. & Øgaard, T. (2012) Working onboard—Job perception, organizational commitment and job satisfaction in the cruise sector. *Tourism Management, 33*(3), 592–597.

Lekakou, M., Pallis, A. & Vaggelas, G. (2009, June). *Is this a home-port? An analysis of the cruise industry's selection*. Paper presented at *International Association of Maritime Economists (IAME) Conference*, Copenhagen, Denmark.

London, W.R. (2011). *Economic risk in the cruise sector.*

Malaysiakini. (2009). Cabotage policy will be reviewed.

Manning, T. (2006). Managing cruise ship impacts: Guidelines for current and potential destination communities.

Morrison, M. (2014). *SWOT Analysis—History, Definition, Templates & Worksheets, RAPIDBI.*

Ng, I. & Kwortnik, R.J. (2007). Balancing cruise revenue sources: The case of empress cruise lines. *Case Research Journal, 27*(2), 105–127.

NLC. (2014). *Our Family & Star Cruises | NCL Cruise Family | Norwegian Cruise Line*.

Royal Caribbean International. (2014). Investor relation.

Statista. (2013). Cruise passenger share source market worldwide 2013 | Statistic.

Terry, W.C. (2011). Geographic limits to global labor market flexibility: The human resources paradox of the cruise industry". *Geoforum, 42*, 660–670.

The Star. (2013a). Smooth sailing for cruise industry—Malaysia" | *The Star Online*.

The Star. (2013b). Malaysians to spend more on holidays—Nation "| *The Star Online*.

Tourism Malaysia. (2012). *ETP Annual Report 2012. NKEA: Tourism* (Tourism Malaysia Report, pp. 118–139).

Travel Canada. (2014). *Most popular cruise destinations*.

Travel Weekly. (2012). *Global cruise passengers top 20 million*.

UC cruises. (2014). UCCRUISES.COM—*Powered By Unique Choice*.

WTO. (2012). *Asia-Pacific Newsletter*.

Wood, R.E. (2004). *Neoliberal globalization: The cruise ship industry as a paradigmatic case*.

World Cruise Industry Review. (2013). *Untapped potential—World Cruise Industry Review*.

World's Leading Cruise Lines. (2014). *Find a Cruise & Compare Cruises*.

Theory and Practice in Hospitality and Tourism Research – Radzi et al. (Eds)
© 2015 Taylor & Francis Group, London, ISBN 978-1-138-02706-0

The way youth enjoying KILIM Geopark

A. Marzuki & D. Mohamad
School of Housing, Building and Planning, Universiti Sains Malaysia, Pulau Pinang, Malaysia

ABSTRACT: This paper presents the result of a research that centers on the impact of demographic impact towards visitation to a nature-based tourism attraction. More specifically, the main objective of this paper is to provide insights on youth visitation to KILIM Geopark in relation to tourism activities, hospitality services expectation and three visitation characteristics (visit companion, length of stay and visit purpose). Within a two-month data gathering and mining, data of total 142 youth respondents aged 18 to 30 with various nationality backgrounds was collected. As geopark tourism activity is the new trend of nature-based tourism industry, it is assumed that the populace will take pleasure in a distinctive tourism experience. Additionally, the populace is set as the targeted group given their publicly known indirect pronounced tradition of road trip engagement. In light of the results presented, practical and sensible solutions are discussed in relation to how youth perceived KILIM Geopark.

Keywords: Youth visitors, KILIM Geopark, tourism activities, hospitality services

1 INTRODUCING KILIM: FROM THE THIRD CATEGORY TO A WORLD CLASS ATTRACTION

Heritage-based tourism as well as other tourism's segments proposes the potential for numerous economic benefits for a geopark, thereby motivating the local economy both directly and indirectly through the multiplier or trickle-down effect (Herbert, 2001). A geopark is a nationally protected area containing a number of geological heritage sites being part of an integrated concept of protection, education and sustainable development (Farsani, Coelho & Costa, 2011). KILIM Geopark, located in Langkawi Island of the State of Kedah, is well established for its valuable nature-based tourism industry and famous for its duty free zone status effective since 1987 (Othman, Rosli & Harun, 2011). The area is promoted as a unique archipelago comprises of ninety nine scenic small islands. KILIM Geopark's limestone heritage value is worth 450 million years and more importantly, the unique and outstanding characteristics have crowned KILIM Geopark as the complete archipelago, to date (Leman, Komoo, Mohamed, Ali & Unjah, 2007). The idea of promoting this unique archipelago is to educate and deepen the understanding on the importance of sustainability development (Huang, 2010). This directly suggests the significance of KILIM Geopark as an '...*outstanding universal value for humanity*' (Elliott & Schmutz, 2012, p. 256). In 2007, KILIM Geopark has been gazette as the 52nd Global Geopark by

UNESCO where this crowns KILIM Geopark as the first global geopark both in Malaysia and in Southeast Asia (Azman, Halim, Liu & Komoo, 2011). Prior to tourism movement, KILIM Geopark generally and Langkawi Island specifically, primarily operated on the agricultural- and fisheries-based economy (Abdul Halim, Komoo, Salleh & Omar, 2011). At present, geopark would be considered as a new movement helping travellers to increase their knowledge about natural resources, cultural identity of the hosts and ways of preserving them. Emphasizing on the conventional wisdom that seldom relates youth tourism to ideas of education/culture exchange.

2 LITERATURE REVIEW

2.1 *Youth tourism*

The first youth tourism conference dated back to 1991 where the importance of youth tourism was analyzed for the purpose of constructing pragmatic strategies, policies, facilities and service (Abdel-Ghaffar, Handy, Jafari, Kreul & Stivala, 1992). The first to show a promising development progress and potential profits compared to international tourism (Abdel-Ghaffar, et al., 1992) and seldom referred to as a trend setter (Wilkening, 2010), conventional wisdom seldom relates youth tourism to ideas of, but not limited to, budget travel, backpacking, summer camp, winter holiday and education/culture exchange. Schnhammer (1992) views youth tourism as a transition process to adulthood or more specifically, it is '...*an expression of the adolescent drive to expand the region of free movement*' (Schnhammer, 1992, p. 19). Youth tourism was initially motivated by the idea of global peace and cultural exchanges (Mohammad Taiyab, 2005). In addition, the fast-pace growing progress is also said to be contributed by the affordable transportation cost (Moisă, 2010b) and the accessibility to and opportunity of financial supports (Mohammad Taiyab, 2005). Richards and Wilson (2003) conceptualise youth travelers are highly educated students aged below 26 and according to Wilkening (2010), youth travelers are defined as a person who aged between 18 to 30 years old. Youth tourists are sometimes viewed as people who dare themselves to experience things that are culturally, socially and psychologically differ from their root (Maoz, 2007).

2.2 Satisfaction and acceptance

When it comes to analyzing the tourism destinations' successfulness, conventional wisdom relates this to visitors' satisfaction in relation to services provided at a tourism spot. In the context of nature-based tourism, which operates on the limited natural resources, visitors' satisfaction is an indicator of '...*the performance of attraction providers in terms of providing service to their visitors*' (Nowacki, 2013, p. 18). Satisfaction is viewed as '...*expectation, performance, expectancy discomfirmation (expectation minus performance), attribution, emotion and equity*' (Bowen & Clarke, 2002, p. 297). Chen, Goodman, and Li (2013) envisage satisfaction as '...*a construct with experience dependency because its formation requires a consumer to have actual experience with the consumption*'. This interlinks the visitors' satisfaction with their acceptance towards a particular tourism spot, which directly influences its popularity (Yüksel & Yüksel, 2003). Echoing Yuksel and Yuksel's (2003) finding, Su (2004, p. 397) demonstrates the visitors' satisfaction (and acceptance) fundamental role in dealing with '...*competitive differentiation and customer retention*'. Here, acceptance in relation to tourism experiences is guided by visitors' preferences and perceptions (Dorwart, Moore & Leung, 2009), where central to this perspective is the belief that visitors interpret their environment in terms of their needs, and prefer settings in which they are likely to function more effectively (Kaplan & Kaplan, 1989).

2.3 Summary

Youth travelers were observed to be spending more than the mainstream tourists (Khoshpa-

kyants & Vidishcheva, 2010) and factors differentiating youth travelers and other travelers' travel include urbanization process, the population's dynamic and socio-economic of the family (Moisă, 2010b). Moisa (Moisă, 2010a) argued that youth tourism assessment propensity was more towards leisure activities undertaken rather than focusing on specific tourism activities experienced. At a deeper level, youth tourism is usually studied in relation to the quality of tourism facilities provided and accessibility to hospitality services (Richards & Wilson, 2004). Against this background, this paper aims to provide insights on youth travelers' satisfaction and acceptance towards KILIM Geopark. As the youth definition is changeable as emphasized by United Nations (2001, p. 2): "...*the definitions of youth had changed continuously in response to fluctuating political, economic and socio-cultural circumstances*", for the purpose of this paper, Kale. Mcintyre, and Weir (1987) criteria (individuals that fit in the 18 to 35 age group) is employed in defining youths.

3 RESEARCH METHOD

This paper is a part of KILIM Geopark research where it specifically focuses on youth travelers' satisfaction and acceptance towards the quality of activities and hospitality offered by KILIM Geopark. Given the nature of exploratory aspect, quantitative research method was employed in addition to revisiting the body of knowledge. Self-administered questionnaire survey, which utilized both close-ended questions and likert type scale, was constructed by working closely with the tourism professionals (hotel managers, travel agents, and tourist attraction's representative). Within a two-month data collection timeframe, 142 questionnaires survey were retrieved from the youth travelers who visited KILIM Geopark. Three hypotheses were tested, namely: [1] each dependent variable (a tourism activitity) is well connected to independent variables (tourism activities), [2] each dependent variable (a tourism activity) is well connected to independent variable (services and environment variables) and [3] KILIM Geopark's nature attractions encourage people to participate in non-water-based activities.

4 THE YOUTHS PERSPECTIVE

For the purpose of this research, the tourism activities were categorized into groups of water-based (fish farm, open sea, fishing trip, kayaking, island tour and fish feeding) and non-water-based (mangrove sightseeing, crocodile cave, limestone cave and eagle feeding) activities. One-Way ANOVA and T-test analyses were run to evaluate the signification relationship between [1] nationality and visitation characteristics and [2] genders and visitation characteristics. Both nationality and genders were found significant with visit companion [$F(4, 137) = 6.461$, $p = .000$]. The results suggest that higher number of female visitors were accompanied during visitation to KILIM Geopark. Interestingly, nationality and genders were insignificant with visitation purpose for ANOVA [$F(4, 137) = 0.601$, $p = 0.663$] and significant for t-test [$(M = 2.09, SD = 1.085)$ conditions; $t(-3.582) = 140$, $p = .000$)]. Meanwhile, ANOVA revealed significant result [$F(4, 137) = 3.022$, $p = .019$] and T-test showed insignificant result [$M = 2.53$, $SD = 1.276$) conditions; $t(-.015) = 140$, $p = 0.913$)] with length of stay.

4.1 Non water-based tourism activities

This section examines the non-water-based tourism activities with [1] services variables (adequate safety facilities, cheap recreational activities and good condition) and [2] environment variables (experience beautiful nature, experience unspoiled nature and unique). Correlation analysis was done in order to understand the ability of non-water-based tourism activities to satisfy the youth respondents' expectation. Analysis between non water-based tourism activities and services exhibits the following: [1] significant results except for crocodile cave and adequate safety facilities ($r = 0.129$, $n = 142$, $p = 0.127$) as well as crocodile cave and good condition ($r = 0.134$, $n = 142$, $p = 0.113$) and [2] weakest correlation is recorded by crocodile and cheap recreational activities ($r = 0.318$, $n = 142$, $p = .000$), meanwhile strongest correlation is observed for adequate safety facilities and good condition ($r = 0.674$, $n = 142$, $p = .000$). Further observation (excluding crocodile cave activity) shows weak correlation results for limestone cave and adequate safety facilities ($r = 0.409$, $n = 142$, $p = .000$) and limestone cave and cheap recreational activities ($r = 0.389$, $n = 142$, $p = .000$). The high significant results present the ability of limestone cave to be further improved.

On the other hand, while environment variables are found significant (experience beautiful nature and experience unspoiled nature: $r = 0.622$, $n = 142$, $p = .000$; experience beautiful nature and unique: $r = 0.655$, $n = 142$, $p = .000$; experience unspoiled nature and unique: $r = 0.609$, $n = 142$, $p = .000$), correlation analysis shows none significant relationship between non water-based tourism activities and environment except for

mangrove sightseeing and experience unspoiled nature ($r = -0.230$, $n = 142$, $p = .006$) as well as eagle feeding and experience unspoiled nature ($r = -0.234$, $n = 142$, $p = .005$). Of importance, the correlation results were not influenced by distance between tourism activities location, which indirectly explained respondents' accessibility level to each tourism activity. Despite the importance of sustaining and maintaining KILIM Geopark, an alarming result was observed in relation to youth respondents' sustainable tourism development supportiveness. The fact of only one significant model (limestone cave and emphasis on limits: $r = 0.175$, $n = 142$, $p = .037$) observed shows KILIM Geopark inefficient campaign and/or respondents' indifferent attitude towards nature.

Additionally, results obtained from Regression analysis state that an activity is well explained by the remaining activities and services variables: mangrove sightseeing [$R^2 = 0.599$, $F(6, 135) = 33.570$, $p = .000$], limestone cave [$R^2 = .621$, $F(6, 135) = .36.862$, $p = 000$], crocodile cave [$R^2 = 0.557$, $F(6, 135) = 28.286$, $p = .000$] and eagle feeding [$R^2 = 0.554$, $F(6, 135) = 27.902$, $p = .000$]. Further observation reveals the following: [1] mangrove sightseeing is well explained by eagle feeding ($b = 0.355$, $t(4.695) = p = .000$), [2] limestone cave is well explained by crocodile cave ($b = 0.492$, $t(7.301) = p = .000$), [3] crocodile cave is well explained by limestone cave ($b = 0.575$, $t(7.301) = p = .000$) and [4] eagle feeding is well explained by mangrove sightseeing ($b = 0.395$, $t(4.695) = p = .000$). Similarly, an acitivity is well described by the remaning activities and environment variables: mangrove sightseeing [$R^2 = 0.604$, $F(6, 135) = 34.271$, $p = .000$], limestone cave [$R^2 = 0.593$, $F(6, 135) = 32.791$, $p = .000$], crocodile cave [$R^2 = 0.499$, $F(6, 135) = 22.399$, $p = .000$] and eagle feeding [$R^2 = 0.486$, $F(6, 135) = 21.288$, $p = .000$]. At a deeper level, it shows that: [1] mangrove sightseeing is well explained by eagle feeding ($b = 0.392$, $t(5.804) = p = .000$), [2] limestone cave is well explained by crocodile cave ($b = 0.426$, $t(6.241) = p = .000$), [3] crocodile cave is well explained by limestone cave ($b = 0.525$, $t(6.241) = p = .000$) and [4] eagle feeding is well explained by mangrove sightseeing ($b = 0.509$, $t(5.804) = p = .000$).

4.2 Water-based tourism activities

This section examines the water-based tourism activities with [1] services variables (adequate safety facilities, cheap recreational activities and good condition) and [2] environment variables (experience beautiful nature, experience unspoiled nature and lakes). Correlation results present the following: [1] assessments between water-based activities indicate strongest correlation for open sea and island tour ($r = 0.795$, $n = 142$, $p = .000$) and weakest correlation for kayaking and fish farm ($r = 0.380$, $n = 142$, $p = .000$), [2] assessments between water-based activities and services variable point towards strongest correlation between open sea and cheap recreational activities ($r = 0.633$, $n = 142$, $p = .000$) and weakest correlation between kayaking and good condition ($r = 0.245$, $n = 142$, $p = .000$), [3] none significant relationship between water-based activities and environment variables except for kayaking and lakes ($r = -.191$, $n = 142$, $p = .023$) and [4] positive correlations were observed between the environment variables (experience beautiful nature and lakes: $r = 0.647$, $n = 142$, $p = .000$; experience unspoiled nature and lakes: $r = 0.528$, $n = 142$, $p = .000$). Of importance, the correlation results were not influenced by distance between tourism activities location, which indirectly explained respondents' accessibility level to each tourism activity. In summary, correlation analysis signifies the youth respondents' propensity towards open sea activity and additional attention should be given to kayaking activity in terms of upgrading the facilities and improvising the activity's attraction.

Findings from regression analysis points out the dependency between water-based activities as follows: fish farm [$R^2 = 0.508$, $F(5, 136) = 28.037$, $p = .000$], fish feeding [$R^2 = 0.680$, $F(5, 136) = 57.716$, $p = .000$], open sea [$R^2 = 0.755$, $F(5, 136) = 83.721$, $p = .000$], island tour [$R^2 = 0.699$, $F(5, 136) = 63.060$, $p = .000$], fishing trip [$R^2 = 0.660$, $F(5, 136) = 52.800$, $p = .000$] and kayaking [$R^2 = 0.723$, $F(5, 136) = 71.066$, $p = .000$]. Further, it is learned that: fish farm is well predicted by fish feeding ($b = 0.549$, $t(5.763) = p = .000$), fish feed is well predicted by open sea ($b = 0.487$, $t(5.487) = p = .000$), open sea is well predicted by island tour ($b = 0.452$, $t(6.757) = p = .000$), island tour is well predicted by open sea ($b = 0.556$, $t(6.757) = p = .000$), fishing trip is well predicted by kayaking ($b = 0.781$, $t(11.575) = p = .000$) and kayaking is well predicted by fishing trip ($b = 0.636$, $t(11.575) = p = .000$). In the context of services and environment variables, it is interesting to find that all tourism activities are well predicted by cheap recreational activities variable where: [1] fish farm [$R^2 = 0.255$, $F(6, 135) = 7.699$, $p = .000$; $b = 0.311$, $t(3.081) = p = .003$], [2] fish feeding [$R^2 = 0.402$, $F(6, 135) = 15.148$, $p = .000$; $b = 0.381$, $t(4.215) = p = .000$], [3] [$R^2 = 0.466$, $F(6, 135) = 19.650$, $p = .000$; $b = 0.408$, $t(4.769) = p = .000$], [4] island tour [$R^2 = 0.396$, $F(6, 135) = 14.725$, $p = .000$; $b = 0.471$, $t(5.182) = p = .000$], [5] fishing trip [$R^2 = 0.235$, $F(6, 135) = 6.909$, $p = .000$; $b = 0.415$, $t(4.056) = p = .000$] and [6] kayaking [$R^2 = 0.334$, $F(6, 135) = 11.266$, $p = .000$; $b = 0.510$, $t(5.342) = p = .000$).

5 CONCLUSION AND DISCUSSION

Based on the presented results, it is observed that hypotheses one and three are proven. As for the second hypothesis, the water-based tourism activities are only well predicted by cheap recreational activities (environment variable). Meanwhile, for non-water-based tourism activities, findings indicate positive result only for mangrove sightseeing, eagle feeding and experience unspoiled nature. Despite the insignificant results observed for nationality, it is within this research interest for KILIM Geopark management specifically and future studies generally to further evaluate on nationality from the perspectives of demographic background and nature-based tourism regulations practiced by respondents' countries.

In addition, it is also important to study on respondents' understanding on Malaysia's nature-based tourism regulations and respondents' expectations with regards to Malaysia's nature-based tourism industry specifically and KILIM Geopark tourism activities generally. In terms of genders, despite focusing on similarities and differences in relation to participation in tourism activities, attention should be given towards determining the time spend on tourism activities where this will provide insights on differences and similarities of types of non-tourism activities-related actions (for example eating, taking pictures and exploring) and level of expectations. By doing so, this will deepen the understanding on the psychological aspect pertaining behavior-related issues, which described by Theory of Reasoned Action: '…[a person will] evaluate the possible implication of [his or her] action before [deciding whether] to engage or not in particular decision' (Bestard & Nadal, 2007, p. 194). On top of this, this research comes to conclude the significance of upgrading and enhancing KILIM Geopark's public relations activities in order to empowering the people on the importance of proper planning management (as documented by Bohdanowicz, Zanki-Alujevic & Martinac, 2004) in addition to justify the value of KILIM Geopark. Against this background, KILIM Geopark specifically and future studies generally could employed Kytzia et al's (2011) evaluation model where in this model, development and/or planning proposal will be assessed for input-output structure, tourists' behavior and industries, space required for activities implementation and productivity functions for land and labor. In doing so, within this research interest, issue of encouraging tourists' active participation in all tourism activities offered could be addressed.

ACKNOWLEDGEMENTS

The funding for this project is made possible through the research grant obtained from the Ministry of Higher Education, Malaysia under the Long Term Research Grant Scheme 2011 [LRGS grantNo.: JPT.S (BPKI)2000/09/01/015Jld.4(67)].

REFERENCES

Abdel-Ghaffar, A., Handy, M., Jafari, J., Kreul, L. & Stivala, F. (1992). Conference reports: Youth tourism. *Annals of Tourism Research, 19*(4), 792–795.

Abdul Halim, S., Komoo, I., Salleh, H. & Omar, M. (2011). The Geopark as a potential toll for alleviating community marginality. *Shima: The International Journal of Research into Island Cultures, 5*(1), 94–113.

Azman, N., Halim, S.A., Liu, O.P. & Komoo, I. (2011). The Langkawi Global Geopark: Local community's perspectives on public education. *International Journal of Heritage Studies, 17*(3), 261–279.

Bestard, A.B. & Nadal, J.R. (2007). Attitudes toward tourism and tourism congestion. *Region et Developpement, 25*, 193–207.

Bohdanowicz, P., Zanki-Alujevic, V. & Martinac, I. (2004). Attitudes towards environmental responsibility among Swedish, Polish and Croatian hoteliers. *Proceedings of the BEST Sustainable Tourism Think Tank IV: "Sustainability and Mass Destinations: Challenges and Possibilities", Esbjerg, Denmark.*

Bowen, D. & Clarke, J. (2002). Reflections on tourist satisfaction research: past, present and future. *Journal of Vacation Marketing, 8*(4), 297–308.

Chen, X., Goodman, S. & Li, E. (2013). Modelling the impact of the cellar door experience on visitor satisfaction and loyalty intentions, from http://anzmac.org/conference/2013/papers/anzmac2013–304.pdf

Dorwart, C.E., Moore, R.L. & Leung, Y.-F. (2009). Visitors' perceptions of a trail environment and effects on experiences: A model for nature-based recreation experiences. *Leisure Sciences, 32*(1), 33–54.

Elliott, M.A. & Schmutz, V. (2012). World heritage: Constructing a universal cultural order. *Poetics, 40*(3), 256–277.

Farsani, N.T., Coelho, C. & Costa, C. (2011). Geotourism and geoparks as novel strategies for socio-economic development in rural areas. *International Journal of Tourism Research, 13*(1), 68–81.

Herbert, D. (2001). Literary places, tourism and the heritage experience. *Annals of Tourism Research, 28*(2), 312–333.

Huang, S. (2010). The geological heritages in Xinjiang, China: Its features and protection. *Journal of Geographical Sciences, 20*(3), 357–374.

Kale, S.H., Mcintyre, R.P. & Weir, K.M. (1987). Marketing overseas tour packages to the youth segment: an empirical analysis. *Journal of travel research, 25*(4), 20–24.

Kaplan, R. & Kaplan, S. (1989). *The experience of nature: A psychological perspective.* New York: Cambridge University Press.

Khoshpakyants, A.V. & Vidishcheva, E.V. (2010). Challenges of youth tourism. *European researcher*(1), 101–103.

Leman, M.S., Komoo, I., Mohamed, K.R., Ali, C.A. & Unjah, T. (2007). Geopark as an answer to geoheritage conservation in Malaysia: The Langkawi Geopark case study. *Geolog Soc Malaysia, 53*, 95–102.

Maoz, D. (2007). Backpackers' motivations the role of culture and nationality. *Annals of Tourism Research, 34*(1), 122–140.

Mohammad Taiyab, M. (2005). *The role and importance of youth tourism in the Malaysian tourism industry.* Paper presented at the Youth Tourism Conference: Perspectives and Prospects, Putrajaya, Malaysia.

Moisă, C. (2010a). Factors Influencing The Evolution Of Youth Travel. *Management and Marketing Journal, 8*(2), 308–316.

Moisă, C. (2010b). Main destinations and tourist flows on the youth travel market. *Annals of Faculty of Economics, 1*(2), 418–424.

Nowacki, M. (2013). The determinants of satisfaction of tourist attractions' visitors, from http://otworzksiazke.pl/images/ksiazki/the_determinants/the_determinants.pdf

Othman, P., Rosli, M.M. & Harun, A. (2011). The impact of tourism on small business performance: Empirical evidence from Malaysian islands. *International Journal of Business and Social Science, 2*(1), 11–21.

Richards, G. & Wilson, J. (2003). *Today's Youth Travellers: Tomorrow's Global Nomads. New Horizons in Independent Youth and Student Travel*: International Student Travel Confederation (ISTC).

Richards, G. & Wilson, J. (2004). *The global nomad: Backpacker travel in theory and practice* (Vol. 3). Great Britain: Channel View Publications.

Schnhammer, R. (1992). Youth tourism as appropriation of the world: a psychological perspective. *Phenomenology & Pedagogy, 10*, 19–27.

Su, A.Y.-L. (2004). Customer satisfaction measurement practice in Taiwan hotels. *International Journal of Hospitality Management, 23*(4), 397–408.

United Nations. (2001). Implementation of the worl programme of action for the youth to the year 2000 and beyond: report of the Secretary General

Wilkening, D. (2010). Youth matters: the most neglected travel market, from http://www.travelmole.com/news_feature.php?id=1145165

Yüksel, A. & Yüksel, F. (2003). Measurement of tourist satisfaction with restaurant services: A segment-based approach. *Journal of Vacation Marketing, 9*(1), 52–68.

Theory and Practice in Hospitality and Tourism Research – Radzi et al. (Eds)
© *2015 Taylor & Francis Group, London, ISBN 978-1-138-02706-0*

Tourism signatures and moderating effect of by-products in building Sarawak state destination image

A. Emaria, M.S.M. Zahari & M.Z. Nur-Adilah
Faculty of Hotel and Tourism Management, Universiti Teknologi MARA, Shah Alam, Malaysia

ABSTRACT: Scholars classified the attributes that influence tourism destination image based two major signature products namely primary and secondary products. The primary includes natural features and natural resources while secondary includes the manmade features that are built either by the government, tourism authorities, individuals and private sectors. Besides these two major signature or core tourism products, local by-products like traditional and ethnic crafts, local souvenirs, miniatures, food products that synonym and portraying a particular destination could also contribute in increasing and strengthening the tourism image of a destination. This paper discusses and probing questions on the role of the local by products in addition to signature tourism products in strengthening the Sarawak state destination image.

Keywords: Signature, tourism, by-products, destination, image

1 INTRODUCTION

Image or the overall perceptions or impressions toward a destination is one of the important determinant influence tourists' decision to choose a specific holiday destination (Jenkins, 1999). This has encouraged many countries, tourism authorities and tourism destination to continuously develop or sustain their image among the local and international tourists. Out of many examples, the French Riviera which is referring to the Mediterranean coastline of the southeast corner of France including the sovereign state of Monaco is able to sustain and having a strong and positive destination image which continuously attracting tourists around the world. Most of the tourists keep in their mind about the image of French Riviera based on their famous casinos namely Monaco-Monte Carlo and its hotel "Promenade des Anglais" in Nice. Tourists describe the image of the French Riviera with the three "S" which are Sea, Sun and Sand (Di Marino, 2008).

Despite the importance and the relationship between tourists and destination images, both have been widely discussed from the scope of understanding the formation processes of image that can help in improving a tourist destination needs to be further explored (Yoon & Kim, 2000). Tourism scholars postulated that the destination image is formed through cognitive and affective means (Lin, 2000; Walmsley & Young, 1998). From cognitive means, the image is evaluated through the ideas,

knowledge and beliefs of individual has toward the all attributes about a particular destination namely resources and attractions Affective element on the other hand refers to feelings or emotional dimension that a person associates toward the tourism destination (Beerli & Martín, 2004).

Kim and Agrusa (2005) classified the attributes that influence tourism destination image based two major signature or core products namely primary and secondary products. The primary includes natural features such as natural resources, scenery, climate, culture, ecology, historical, architecture and many others. For the secondary, it includes the manmade features that are built either by the government, tourism authorities, individuals and private sectors such as resorts, transportation, hotels, entertainment and catering outlets. These two signature tourism products strongly influenced the image of a tourism destination and in fact, the mixture of these two, in addition to those characteristics plus services and intangible products may also linked to a destination image (Pechlaner, 2000).

Besides these two signature products, local by-products like traditional and ethnic crafts, souvenirs, miniatures, food products, merchandise that represents local culture and other items synonym and portraying a particular destination could also contribute in increasing and strengthening the tourism image of a destination (Tzuhui, David & Ching-Cheng, 2009; Wicks et al., 2004). Vladimir, a city in Russia not only attract a substantial number of tourists because of its culture and architectural sights, but the unique local crafts without exception contribute the image of the city. Yüksel and Akgül (2007) noted local by-products also create the image of Turkey; Ali-Knight (2011) in fact posited that other local byproducts in addition to the core tourism products could moderate the image of a particular country or destination.

2 LITERATURE REVIEW

2.1 Signature tourism products

There are numerous ways to define signature or tourism core products. Out of many, Kotler (2001) postulate that tourism signature products as an object that consist of one or a mix of components including physical goods, services, information, ideas, places, experiences, events, persons, properties and organizations in which these components can create the whole satisfaction of tourists during their visit to a destination. Brass (1997) divided the components of signature tourism products into two main attributes namely attractions and facilities. Attractions are including natural and human-made features while facilities closely related to accommodation, roads, airports, hospitals, railway, parking areas, water, power services and many others. Both components (attractions and facilities) together create a set of intangible "subjective experiences" for tourists.

Other scholars categorized tourism products into four levels namely the signature or core products, the facilitating products, the supporting products and augmented products (Kotler, Bowen & Makens, 1996; Swarbrooke & Page, 1995). The core or signature products are major products or main attraction being offered by a destination to tourists. This type of products includes natural resources, scenery, climate, culture, ecology and historical architecture or manmade features. Facilitating products refer to related services and goods that must be together with the core product. The supporting product, or known as extra products or by-products are the added value products to the core products being offered by a destination to help distinguish it from other destination. These products are stand-alone including traditional and ethnic crafts, local souvenirs, miniature, food products and merchandise. The last is augmented products which refer to individual or tourists' perception, attitudes, participation and interaction with the service organizations.

Xu (2009) posited that tourism products can be in the form of tangible and intangible elements or combination of both. Tangible elements are the physical plant that includes cultural resources, adventure resources and natural resources, while intangible elements refers to related hospitality, services, local peoples as well as local by-products. The following sections look at specific attributes of the core or signature tourism products like culture, adventure and nature.

2.2 Culture

Culture is the set of attitudes, beliefs, ideas, behaviors, values and a way of life shared by a group of peoples which is passed from one generation to another generation. The importance of culture as one of the major assets in tourism development has been highlighted in the literature (Beeton, 2005). In fact, culture and tourism have a strong relationship which can strengthen the attractiveness of the destinations and this is popularly known as cultural tourism.

Anholt (2006) postulated that cultural tourism as a tourist travelling to a destination for experiencing and learns about the local traditions, arts, folk lore, the pilgrimage, historic sites and monuments without violating the respecting of the local community and the surrounding environment. Beeton (2005) noted that cultural tourism besides other products of tourism is the most powerful factor to draw tourists to visit a particular destination. MacKay and Fesenmaier (2000) on the hand, emphasized that most of the tourism destination image is represented by its culture and the culture in a particular destination like Bali clearly attracting a substantial number of international tourists.

2.3 Adventure

Millington, Locke and Locke (2001) refers adventure as risky outdoor activities that usually takes place in a remote, unusual or wilderness area and those activities cannot be separated from tourism industry thus it is called an adventure tour. It is undeniable that adventure tourism needed a guide, experience and specialized equipment's to perform. The activities involve elements of danger, risk, emotion, exploration and discovery, challenge, escape from reality, exciting and stimulating (Page, Bentley & Walker, 2005). The adventure activities includes caving, diving/snorkeling, cycle tours/mountain, kayaking, touring, rock climbing, guided walks, paragliding, safaris, horse riding, whitewater rafting, off-road driving, mountain biking and expeditions. As such, Romania is popular with adventure tourism particularly on mountain biking activities.

2.4 Nature

According to Jafari (2000) nature consists of plants, the landscapes, animals, and other products and features in the earth that are being conserved and protected for future generations and directly related to the tourism industry as tourists nowadays are seeking for natural attraction.

Nature tourism is denotes as travelling to a destination to experience natural places that involve activities such as sightseeing, wildlife viewing, beaches, lakes, waterfalls, national parks, fishing and other nature activities (Silvennoinen & Tyrväinen, 2001). Weiler and Hall (1992) described the largest components of nature tourism goes to national parks and conservation reserves.

Owing to the large ethnic groups and home of adventure and nature, Sarawak state is having the most exotic flora and fauna in Malaysia (Hamzah, 2004). Out of many natural attractions in the state, Semenggoh Wildlife Center, Bako National Park and Matang Wildlife Centre are among the highest tourists' receipts and this directly boosts Sarawak as nature tourism destination (Sarawak Tourism Board, 2014).

2.5 *Relationship between by-products and tourism*

Wicks et al. (2004) postulated that local by-products are a traditional crafts, local souvenirs, miniature, food products, merchandise and other items that represent the local culture and purchased by tourists at a particular destination. Yüksel and Akgül (2007) argue local by-products in addition to signature products are part of numerous reasons tourists travel to a destination. It is becoming common practices that tourists when traveling to any particular destination will purchase local by-products as memories or gift for family members or friends (Reisinger & Turner, 2000). They further posited that when traveling domestically, tourists are normally spent 2/3 from their total cost purchase of the local by-product of a particular destination while 1/5 was spent when traveling internationally. Yüksel and Akgül (2007) stated that local by-products such as souvenirs or holiday postcard in addition to other tourism products create and building the image of Turkey; Ali-Knight (2011) therefore deduce the local by-products in addition to the core tourism products could moderate the image of a particular country or destination.

2.6 *Destination image*

Beerli and Martin (2004) named three key factors that influence image formation: 1) stimulus factors (information sources); 2) previous experience and distribution and 3) personal factors (psychological and social). Information sources are the main stimulus factors that have an effect on the forming of cognitive perceptions and evaluations. Um and Crompton (1990) asserted that individuals form perceptual/cognitive evaluation of destination attributes subsequent to an exposure to various information sources such as symbolic stimuli (promotional efforts through media), social stimuli (WOM and recommendations) and information acquired from previous visitation.

Essentially, a primary image is formed through either personal experience or actual visit to the destination. The destination image formed by the latter has the propensity to be more realistic and complex as opposed to the one formed through secondary sources of information (Gartnerand & Hunt, 1987; Pearce, 1982; Phelps, 1986). Phelps (1986) makes a proposition that the image formed by organic, induced and autonomous sources of information is dubbed as secondary image. Another assertion by Beerli and Martin (2004) has it that the secondary sources of information play a pertinent and crucial role in forming cognitive dimension of image. More importantly, Baloglu and McCleary (1999) stressed that the variety, the amount and type of information sources are proxies for cognitive evaluation of images.

The process of image formation can be conceived as a continuum of separate agents or forces that generate a specific result either independently or in combination to form a destination image unique to an individual (Gartner, 1994). It is further affirmed that the detailed breakdown of agents is vital because destination selection itself is a process of narrowing alternatives from an initial opportunity set of all possible destinations to an evoked set of approximately three destinations. This contention is indeed, in support of the general consumer behavior theory (Howard & Sheth, 1969). Moreover, destination images are imperative to the process of narrowing alternatives and they act as pull factors toward the final selection.

3 SIGNATURE TOURISM PRODUCTS, BY-PRODUCTS AND SARAWAK DESTINATION IMAGE

Out of many states in Malaysia, Sarawak also known as Bumi Kenyalang (Lands of the Hornbills) is particularly rich in diversification of primary tourism products such as culture, heritage, caves, national parks, mountain, forest, exotic foods, longhouses, nature and wildlife. Owing to diversification of primary tourism products, the Sarawak Tourism Board (2014) is continuously promoting its tourism signature products like culture, adventure and nature or simply known as CAN to the local and international tourists (Sanggin, 2009). These core or Sarawak signature tourism products are claimed to have strong influenced in attracting local and international tourist to the state. Report from Ministry of Tourism Sarawak (MOT) (Sarawak Tourism Board, 2014) revealed that there is a positive growth of tourist receipts both domestic and international from 3,280656 in 2009 to 4,069023 in 2012 and the figure is expected to flourish yearly. In conjunction with Visit Malaysia Year (VMY) 2014, Sarawak is also aggressively promoting the state through the launched of the Visit Sarawak Year (VSY) 2014 to draw more international tourists to visit the so called Lands of the Hornbills (Sarawak Tourism Board, 2014).

According to Sarawak Minister of Tourism, Datuk Amar Abang Haji Abdul Rahman Zohari, an excellent range of tourism products together with the promotion and advertisements directly establishing Sarawak as one of the international recognized culture, adventure and nature tourism destination. A total of 431,505 tourists to have visited the sixteen Sarawak national parks in 2012 meanwhile the Sarawak Cultural Village (SCV) has remained as a popular tourist attraction for more than two decades (Sarawak Tourism Board, 2014). It is not too overwhelmed to say that this positive scenario indicates that the three core or signature tourism products could continue building or strengthening the Sarawak state destination image.

Apart from this, tourists when travelling to any destination in addition to the core tourism activities will also take the opportunity to get local by-products that synonymous with the destination (Reisinger & Turner, 2000). In this sense, it is argued and based on to what have been highlighted by Wicks et al (2004), Tzuhui et al. (2009), Yüksel and Akgül (2007), and Ali-Knight (2011) by-products of a particular destination could also contribute in increasing and strengthening the tourism image of the destination.

In line with this statement and in the case of Sarawak, the local and international tourists who visited this state in addition of experiencing the culture, adventure and nature as a signature tourism products are assumed to have purchased the popular local Sarawak by-products such as traditional and ethnic crafts (the rattan products, e.g. mats, hats and basket, the 'sumpit', the 'shape' and others) local souvenirs, merchandises, food products (ikan terubuk and kek lapis,) and others. Based on this, some questions related to local Sarawak by-products could be raised. To what extent the local Sarawak by-products in addition to its signature tourism products (Culture, Adventure, and Nature or CAN), contribute to increasing its destination image? In other words, what are the moderating effects of the local Sarawak by-products in building its tourism destination image?

To date, as compared to culture, adventure and nature (CAN), there is still unclear understanding of how and to what extent the local by-products

helps in strengthening the Sarawak tourism destination image (Sanggin, 2009) thus this issue is still under investigation.

4 CONCLUSION

Understanding the role of Sarawak signature tourism products and the moderating effect of its by-products in strengthening state destination image will create awareness on the importance of by-products to the state tourism development. In addition, besides maintaining its signature tourism products, by-products which represent all ethnics could aggressively be promoted to the local and international tourists in particular thus the by-products together with signature tourism products is gradually recognizing internationally and at the same time creating its image and boost up the state economy.

REFERENCES

Ali-Knight, J. (2011). The role of niche tourism products in destination development. *International Journal of Wine Marketing, 12*(3), 70–80.

Anholt, S. (2006). Competitive identity: The new brand management for nations, cities and regions. *Journal of Brand Management, 14*(6), 474–475.

Baloglu, S. & McCleary, K.W. (1999). A model of destination image formation. *Annals of Tourism Research, 26*(4), 868–897.

Beerli, A. & Martín, J.D. (2004). Tourists' characteristics and the perceived image of tourist destinations: A quantitative analysis a case study of Lanzarote, Spain. *Tourism management, 25*(5), 623–636.

Beeton, S. (2005). *Film-induced tourism.* Clevedon: Channel View Publications.

Brass, J.L. (1997). *Community Tourism Assessment Handbook* Logan: Western Rural Development Center.

Di Marino, E. (2008). The strategic dimension of destination image. An analysis of the French Riviera image from the Italian tourists' perceptions. *Journal of Travel and Tourist Marketing, 9*(4), 47–67.

Gartner, W.C. (1994). Image formation process. *Journal of Travel & Tourism Marketing, 2*(2–3), 191–216.

Gartnerand, W.C. & Hunt, J.D. (1987). An Analysis of State Image Change Over a Twelve-Year Period (1971–1983. *Journal of travel research, 26*(2), 15–19.

Hamzah, A. (2004). *Policy and planning of the tourism industry in Malaysia.* Paper presented at the 6th ADRF General Meeting, Bangkok, Thailand.

Howard, J.A. & Sheth, J.N. (1969). *The theory of buyer behaviour.* London: John Wiley and Sons, Inc.

Jafari, J. (2000). *Introduction.* London: Routledge.

Jenkins, O.H. (1999). Understanding and measuring tourist destination images. *International Journal of Tourism Research, 1*(1), 1–15.

Kim, S.S. & Agrusa, J. (2005). The positioning of overseas honeymoon destinations. *Annals of Tourism Research, 32*(4), 887–904.

Kotler, P. (2001). *A framework for marketing management*. Upper Saddle River, NJ: Prentice-Hall.

Kotler, P., Bowen, J.T. & Makens, J.C. (1996). *Marketing for hospitality and tourism*. Upper Saddle River, NJ: Prentice Hall.

Lin, W. (2000). *Study on influence of compensation system and organizational climate on job satisfaction and job performance*. Master's thesis, Graduate Institute of Industrial Engineering, Da Yeh University.

MacKay, K.J. & Fesenmaier, D.R. (2000). An exploration of cross-cultural destination image assessment. *Journal of travel research, 38*(4), 417–423.

Millington, K., Locke, T. & Locke, A. (2001). Occasional studies: Adventure travel. *Journal of Travel and Tourism, 4*(1), 65–97.

Page, S.J., Bentley, T.A. & Walker, L. (2005). Scoping the nature and extent of adventure tourism operations in Scotland: How safe are they? *Tourism management, 26*(3), 381–397.

Pearce, P.L. (1982). *The social psychology of tourist behaviour*. Oxford: Pergamon Press.

Pechlaner, H. (2000). Cultural heritage and destination management in the Mediterranean. *Thunderbird International Business Review, 42*(4), 409–426.

Phelps, A. (1986). Holiday destination image—the problem of assessment: An example developed in Menorca. *Tourism management, 7*(3), 168–180.

Reisinger, Y. & Turner, L. (2000). Japanese tourism satisfaction: Gold coast versus Hawaii. *Journal of Vacation Marketing, 6*(4), 299–317.

Sanggin, S.E. (2009). Community involvement in culture and nature tourism in Sarawak. *Journal of Southeast Asia Social Sciences and Humanities, 77*, 149–165.

Sarawak Tourism Board. (2014). Sarawak tourism products. Sarawak Ministry of Tourism, from http://www.mot.sarawak.gov.my/

Silvennoinen, H. & Tyrväinen, L. (2001). Demand for nature tourism services. In T. Sievänen (Ed.), *Outdoor recreation 2000* (pp. 112–127). Finland.

Swarbrooke, J. & Page, S.J. (1995). *Development and management of visitor attractions*. Oxford: Butterworth-Heinamann.

Tzuhui, A.T., David, Y.C. & Ching-Cheng, S. (2009). *The use of souvenir purchase as an important medium for sustainable development in rural tourism*. Paper presented at the National Extension Tourism (NET) Conference, Dahu, Taiwan.

Um, S. & Crompton, J.L. (1990). Attitude determinants in tourism destination choice. *Annals of Tourism Research, 17*(3), 432–448.

Walmsley, D.J. & Young, M. (1998). Evaluative images and tourism: The use of personal constructs to describe the structure of destination images. *Journal of travel research, 36*(3), 65–69.

Weiler, B. & Hall, C.M. (1992). *Special interest tourism*. London: Belhaven Press.

Wicks, B., Do, K., Hsieh, P.-C., Komorowski, A., Martin, K., Qiu, X., Rimdzius, M., Strzelecka, M., Wade, K. & Yu, G. (2004). Direct marketing of crafts and souvenirs to Vladmir Visitors. *Vladimir tourism development project. University of illinois at Urbana-champaign college of applied life studies, department of recreation, sport and tourism*.

Xu, J.B. (2009). Perceptions of tourism products. *Journal of Tourism Management, 21*(2), 53–62.

Yoon, S.-J. & Kim, J.-H. (2000). An empirical validation of a loyalty model based on expectation disconfirmation. *Journal of Consumer Marketing, 17*(2), 120–136.

Yüksel, A. & Akgül, O. (2007). Postcards as affective image makers: An idle agent in destination marketing. *Tourism management, 28*(3), 714–725.

Theory and Practice in Hospitality and Tourism Research – Radzi et al. (Eds)
© 2015 Taylor & Francis Group, London, ISBN 978-1-138-02706-0

International tourists revisit intention: Has it prevalence in United Arab Emirates?

J.M. Abdul-Rahim
Fujairah Tourism and Antiquities Authority, Fujairah, UAE

M.S.M. Zahari, S.A. Talib & M.Z. Suhaimi
Faculty of Hotel and Tourism Management, Universiti Teknologi MARA, Shah Alam, Malaysia

ABSTRACT: The connection between past travel experience and tourists revisits behavioral intentions has not been widely explored but the existing studies suggest a close relationship between them. Tourism-related products of a country can equally be construed to have effects on the attitudes of the tourists at the end of their actual visitation and satisfaction of a tourist with his or her travel experiences contribute to a loyalty to a particular destination. It has also been widely acknowledged that the destination image affects tourists' subjective perception, consequent behavior and destination choice. This paper is reviewing the constructs and dimensions which associated with international tourists revisit intention using United Arab Emirates as contextual study setting and proposed the conceptual study framework.

Keywords: Tourist, revisit intention, United Arab Emirates, food, tourism core products, destination image

1   INTRODUCTION

Consumer behavioral studies revealed that past experiences influence the satisfaction (Licata, Mills & Suran, 2001; Mittal, Kumar & Tsiros, 1999). The link between past experiences, customer satisfaction and company success has historically been a matter of faith and numerous satisfaction studies has also supported the case (Hill & Alexander, 2000). Customer experiences and satisfaction has always been considered an essential business goal because it was assumed that satisfied customers would buy more. However, many companies have started to notice a high customer defection despite high satisfaction ratings (Oliver, 1999; Taylor, 1998). This phenomenon has prompted a number of scholars (Jones & Sasser, 1995; Oliver, 1999) to criticize the mere satisfaction studies and call for a paradigm shift to the quest of repurchase as a strategic business goal. As a result, customer experiences and satisfaction measurement have recently been displaced by the concept of customer loyalty; primarily because loyalty is seen as a better predictor of repurchase behavior.

In tourism perspectives, previous experience is found to have given significant impact on future tourists' visitation behavior (Oppermann, 2000) and an actual visitation which is construed as a past travel experience to be a primary source of influence on revisiting behavioral intentions (Sönmez & Graefe, 1998). Tourism-related products of a country can equally be construed to have effects on the attitudes of the tourists at the end of their actual visitation and satisfaction of a tourist with his or her travel experiences contribute to a loyalty to a particular destination (Bramwell, 1998; Pritchard & Howard, 1997). (Oppermann, 2000) noted that tourist loyalty towards a destination is reciprocated by their intention to revisit the destination apart from willingness to recommend it to others.

This paper is reviewing the constructs and dimensions which associated with the international tourist revisit intention using United Arab Emirates as contextual study setting and proposed the study conceptual framework.

2   LITERATURE REVIEW

2.1   *Tourist total experience and revisit intention*

The connection between past travel experience and future travel behaviour has not been explored widely but the existing studies suggest a close relationship between them (Dolnicar & Huybers, 2010; Hosany & Witham, 2010; Oppermann, 2000; Sönmez & Graefe, 1998).

Oppermann (2000) states tourists' loyalty towards a destination is reciprocated by his or her intention to revisit the destination apart from his or her willingness to recommend it to others. Past travel experience appears to be a powerful influence on behavioural intentions and individuals with past travel experiences to various destinations may become more confident as a result of their experience and thus be more likely to travel back to those places of interests (He & Song, 2009; Sönmez & Graefe, 1998) and repeat tourists are expected to be more likely than first-timers to choose the same destination in the future (Sampol, 1996). The first-timers cannot be totally relied upon as they may visit other destinations. In other words, repeat visitation can be developed and it is contingent upon their total experience with promotion, price, core products, facilities and front employees while they were on vacation for the first time.

2.2 Destination images

It has also been widely acknowledged that the destination image affects tourists' subjective perception, consequent behaviour and destination choice (Assaker & Hallak, 2013; Chi, 2012; Prayag & Ryan, 2012). Researchers modeled image as a function of marketing information or other external stimuli (Fakeye & Crompton, 1991; Gartner, 1989). For example, in one of the first conceptualizations of destination image, (Gunn, 1972) proposed that destination images are formed from the types of information that tourists received. It is argued that destination images are divided into two levels: 1) organic and 2) induced images. An organic image is formed as a result of exposure to stimuli from non-tourism market oriented information whereas an induced image derives from conscious efforts of marketers and advertisers to develop promote and advertise a destination. Fakeye & Crompton (1991) states that information as an important determinant which distinguished an organic image from an induced image.

In tourism, tourists develop a more complex and differentiated image from induced image throughout the actual visitation experience. Tasci et al (2007) contend that the composition of tourism images as advanced by place image has made inroads into the gambit of tourism marketing. Product-country image or simply country image and tourism destination image are both focusing on place image (Gallarza, Saura & García, 2002). In this sense, place image relates on the buyer attitudes towards products from various origins. In other word, destination image represents the effects of beliefs, ideas and impressions that a person has of a destination.

In line with above notion, tourists' behavioral intention is expected to be partly conditioned by the image that they have of destinations. Image will influence tourists in the process of choosing a destination, the subsequent evaluation of the trip and in their future intentions. Destination image exercises a positive influence on perceived quality and satisfaction. A positive image deriving from positive travel experiences would result in a positive evaluation of a destination (Gallarza et al., 2002; Tasci, Gartner & Cavusgil, 2007). Tourist satisfaction would improve if the destination has a positive image and destination image also affects tourists' behavioural intentions. (Pakaleva, 1998) on the other hand contended that more favorable image will lead to higher likelihood to return to the same destination as long it is free from any destruction or environmental turbulences.

2.3 Tourists responses

It is also evident that satisfaction and commitment with travel experience contributes to destination loyalty (Bramwell, 1998; Hosany & Witham, 2010; Oppermann, 2000; Pritchard & Howard, 1997). The degree of tourists' satisfaction and commitment to a destination is reflected in their intentions to revisit the destination and in their willingness to recommend it to others (Oppermann 2000). Tourists' positive experiences of service, products and other resources provided by tourism destinations could produce repeat visits as well as positive word-of-mouth effects to friends and/or relatives. Recommendations by previous visits can be taken as the most reliable information sources for potential tourists. Recommendations to other people (word-of-mouth) are also one of the most often sought types of information for people interested in travelling.

On different note, tourist commitment which is preceded by an act of planning to visit is an important aspect of response as planning in itself is a commitment to the intended behaviour (Warshaw & Davis, 1985). Essentially, most scholars view commitment as being represented by the following dimensions: 1) expectation of continuity, 2) desire of continuity (Geyskens, Steenkamp, Scheer & Kumar, 1996), and 3) willingness to revisit (Siguaw, Simpson & Baker, 1998).

3 ISSUES RELEVANT TO THE UNITED ARAB EMIRATES

It is without doubt that UAE is blessed with a location that is strategic and provides a bridge to the connecting link between Europe, the Indian subcontinent but both the Far East and Africa (Henderson, 2006). It is still somewhat a melting of culture as the population originates from different cultures but somehow the dominant culture remains Arabic despite the fact that even the Arabs themselves are from different corners of the Arab world. Against such a backdrop, UAE appears to be a beneficiary to the convenience to air travel which has contributed towards the significantly increased the number of visitors (DTCM, 2011).

In line with the above notion, tourism industry has contributed a steadily increasing percentage of the United Arab Emirates GDP (Gross domestic product) which according to some estimates at 20 per cent (United Arab Emirates Tourism Report, 2012). Some views are mind-boggling when they claim tourism is expected to similarly important as oil exports as a major source of revenue in the near future for UAE. Although this only a prediction, trends showing that despite recession the arrivals of tourists are still increase with unrelenting campaign by UAE Department of Tourism and Commerce Marketing (DTCM, 2011) and hotels are willing to slash their rate.

Since January 1997, subsequent to the Department of Tourism and Commerce Marketing (DTCM, 2009) taking over from the Tourism and Trade Promotion Council, there has been renewed focus on worldwide promotion of UAE as an ideal tourist destination apart from being a thriving commercial and business centre and very attractive for Dubai property investors. What followed after the takeover was the setting up of the DTCM representative offices in many countries across the globe as well as participation in numerous international tourism fairs to promote the country.

UAE as a country is also rich with culture and history and exciting place to visit with numerous events held throughout the year. Dubai Shopping festival and Dubai Desert Classic are unique to the city and UAE with other national festivals are vibrant for the tourists. The Dubai Desert Classic is normally takes place at the Emirates Golf Club and this is one of the main golf tournaments not only attracting golfers but visitors around the world (Sharpley, 2002).

Emirates also is renowned as one of the largest horse races in the entire world with US6 million of the winning price. Besides these, catchphrase of the event is also "shop, save and celebrate" and most of the city's malls and other outlets offer massive discounts on their products while the activities held during this time are divided into categories such as arts, food, nature and adventure. In addition, apart from being a thriving commercial and business centre and very attractive for property investors, Dubai, Abu Dhabi and Fujairah are three major cities and district are influx with local and international tourists, ideal tourist destination and creating its own images.

According to (DTCM, 2009), 3.95 million visitors visited UEA in the first 6 months of 2009 compared to 1 million visitors annually in the last 10 years and only 600 thousands during 80's. Similarly, the demand for hotel room tremendously increased with 255 international hotels in the city of Dubai alone with a total of 17,253 rooms compared to less than 100 in the last decade. Not to exaggerate with a strong marketing efforts, UAE not only considered successfully attract the foreign investors but becoming the most popular tourism destination in the Arab region and the developed it tourism image.

Despite the above developments, to what extent tourist experience on promotion, price, core products, facilities and frontline employees provide by the UEA tourism organizations, hotels, destinations and others and the impact of their level of satisfaction, commitment toward revisit intention are not known. In other words, can the visit of the international tourists' to UEA, Dubai, Abu Dhabi and Fujairah in particular and their experience with promotion, price, core products, facilities and frontline employees) together with country, destination and hotel image along with their satisfaction, commitment be translated into revisit intention?. In short, four questions are posted which also related to the problem statement of this proposed study.

1. Does total tourist experience whose underlying dimensions are promotion, price, core products, facilities and frontline employees have a positive relationship with the tourism images which comprises of country image, destination image and hotel image?
2. How do tourism images affect tourist satisfaction, commitment most and importantly their revisit intention in the face of the unfavorable occupancy rate and declining revenue per available hotel?
3. To what extent does the total tourist experience is able to create tourist satisfaction and commitment?
4. How can the hotel operators, the Department of Tourism and Commerce Marketing (DTMC) of UAE and travel agencies ensure that the country image, destination image and hotel image are sustained to generate tourist responses which will develop into positive revisit intention?

All those questions warrant an empirical investigation. In addition, the influence of the individual underlying dimensions of the total tourist experience (promotion, price, core products, facilities and frontline employees) on tourism images (country, destination and hotel) and tourists responses (satisfaction and commitment) and the impact of these attributes on tourist revisit intention need holistically be investigated.

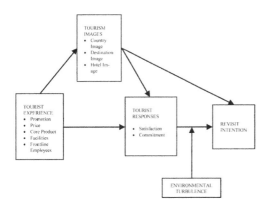

Figure 1. Conceptual framework.

4 CONCEPTUAL FRAMEWORK/MODEL

Based on the literatures and the issue pertaining to the United Arab Emirates tourism perspectives, the conceptual framework is proposed in Figure 1.

This conceptual study framework which is also referred to hypotheses diagrammed depicts the role of total tourist experience in creating country image, destination image, hotel image and the consequences it has on tourist responses and his or her intention to revisit. The total tourist experience is operationalized by: 1) promotion, 2) price, 3) core products, 4) facilities and 5) frontline employees whereas the underlying dimensions of tourist responses are: 1) tourist satisfaction, 2) tourist commitment and 4) revisit intention.

5 CONCLUSION

As the study is still under investigation, significant academic contributions to the existing body of knowledge will be accomplished by way of testing the hypotheses and confirming whether they are supported or rejected and relating the findings to empirical evidence drawn from the literature. From the practical perspective, the Department of Tourism and Commerce Marketing (DTMC) of UAE and its representative offices and travel agencies may take heed to the findings. For instance, by adopting human resource strategies the frontline employees are productive and motivated enough to help the tourism industry to create its tourism images. Its aggressive promotion should be able to induce positive image among potential visitors by providing adequate information on UAE as a tourist destination which is a prerequisite to destination selection and visitation.

As tourist satisfaction is found to be insufficient to bring about loyalty and revisit intention, then it is incumbent upon the policy-makers to develop a loyalty programme so that both tourist commitment and satisfaction are enhanced to ensure that their revisit intention are more significant. The major interest of the stakeholders is to improve the tourism images as a marketing tool to bring more and more visitors to UAE and to inculcate or implant their revisit intention.

REFERENCES

Assaker, G. & Hallak, R. (2013). Moderating effects of tourists' novelty-seeking tendencies on destination image, visitor satisfaction, and short-and long-term revisit intentions. *Journal of Travel Research, 52*(5), 600–613.

Bramwell, B. (1998). User satisfaction and product development in urban tourism. *Tourism Management, 19*(1), 35–47.

Chi, C.G.-q. (2012). An examination of destination loyalty differences between first-time and repeat visitors. *Journal of Hospitality & Tourism Research, 36*(1), 3–24.

Dolnicar, S. & Huybers, T. (2010). Different Tourists—Different Perceptions of Different Cities Consequences for Destination Image Measurement and Strategic Destination Marketing. In J.A. Mazanec & K. Wober (Eds.), *Analyzing International City Tourism* (pp. 127–146). Vienna/New York: Springer.

DTCM. (2009). Developments in Selected Major Government and Private Sector. Dubai: Department of Tourism and Commerce Marketing.

DTCM. (2011). Developments in Selected Major Government and Private Sector. Dubai: Department of Tourism and Commerce Marketing.

Fakeye, P.C. & Crompton, J.L. (1991). Image differences between prospective, first-time, and repeat visitors to the Lower Rio Grande Valley. *Journal of Travel Research, 30*(2), 10–16.

Gallarza, M.G., Saura, I.G. & García, H.C. (2002). Destination image: towards a conceptual framework. *Annals of Tourism Research, 29*(1), 56–78.

Gartner, W.C. (1989). Tourism image: attribute measurement of state tourism products using multidimensional scaling techniques. *Journal of Travel Research, 28*(2), 16–20.

Geyskens, I., Steenkamp, J.-B.E., Scheer, L.K. & Kumar, N. (1996). The effects of trust and interdependence on relationship commitment: a trans-Atlantic study. *International Journal of research in marketing, 13*(4), 303–317.

Gunn, C. (1972). *Vacation scape. Austin Bureau of Business Research*: University of Texas.

He, Y. & Song, H. (2009). A mediation model of tourists' repurchase intentions for packaged tour services. *Journal of Travel Research, 47*(3), 317–331.

Henderson, J.C. (2006). Tourism in Dubai: overcoming barriers to destination development. *International Journal of Tourism Research, 8*(2), 87–99.

Hill, N. & Alexander, J. (2000). *Handbook of customer satisfaction and loyalty measurement*: Gower Publishing, Ltd.

Hosany, S. & Witham, M. (2010). Dimensions of cruisers' experiences, satisfaction, and intention to recommend. *Journal of Travel Research, 49*(3), 351–364.

Jones, T.O. & Sasser, W.E. (1995). Why satisfied customers defect. *Harvard business review, 73*(6), 88-&.

Licata, J.W., Mills, G.N. & Suran, V. (2001). Value and satisfaction evaluations during a service relationship. *Services Marketing Quarterly, 22*(3), 19–42.

Mittal, V., Kumar, P. & Tsiros, M. (1999). Attribute-level performance, satisfaction, and behavioral intentions over time: a consumption-system approach. *The Journal of Marketing*, 88–101.

Oliver, R.L. (1999). Whence consumer loyalty? *Journal of marketing, 63*(4).

Oppermann, M. (2000). Tourism destination loyalty. *Journal of Travel Research, 39*(1), 78–84.

Pakaleva, K. (1998). *Transition: Environmental Consequences of Political and Economic Transformation.* U.S.A: Ashgate.

Prayag, G. & Ryan, C. (2012). Antecedents of Tourists' Loyalty to Mauritius The Role and Influence of Destination Image, Place Attachment, Personal Involvement, and Satisfaction. *Journal of Travel Research, 51*(3), 342–356.

Pritchard, M.P. & Howard, D.R. (1997). The loyal traveler: examining a typology of service patronage. *Journal of Travel Research, 35*(4), 2–10.

Sampol, C.J. (1996). Estimating the probability of return visits using a survey of tourist expenditure in the Balearic Islands. *Tourism Economics, 2*(4), 339–352.

Sharpley, R. (2002). Barriers to Sustainable Development:Jordan's Sustainable Tourism Strategy. *Journal of Asian and African Studies, 41*, 439–457.

Siguaw, J.A., Simpson, P.M. & Baker, T.L. (1998). Effects of supplier market orientation on distributor market orientation and the channel relationship: The distributor perspective. *Journal of marketing, 62*(3).

Sönmez, S.F. & Graefe, A.R. (1998). Determining future travel behavior from past travel experience and perceptions of risk and safety. *Journal of Travel Research, 37*(2), 171–177.

Tasci, A.D., Gartner, W.C. & Cavusgil, S.T. (2007). Conceptualization and operationalization of destination image. *Journal of Hospitality & Tourism Research, 31*(2), 194–223.

Taylor, T.B. (1998). Better loyalty measurement leads to business solutions. *Marketing News, 32*(22), 41–42.

United Arab Emirates Tourism Report. (2012). Business Monitor International (pp. 61).

Warshaw, P.R. & Davis, F.D. (1985). Disentangling behavioral intention and behavioral expectation. *Journal of experimental social psychology, 21*(3), 213–228.

Theory and Practice in Hospitality and Tourism Research – Radzi et al. (Eds)
© 2015 Taylor & Francis Group, London, ISBN 978-1-138-02706-0

A survey of hikers' characteristics at prominent mountains in Malaysia

S.H. Taher, S.A. Jamal & N. Sumarjan
Faculty of Hotel and Tourism Management, Universiti Teknologi MARA, Shah Alam, Malaysia

ABSTRACT: This paper presents findings of a study of hikers at the seven most prominent mountains in Malaysia. An increasing number of hikers visit these mountains each year; however, they are not well-understood. The purpose of this study was to identify hiker characteristics in terms of socio-demographic characteristics, group characteristics, hiking experience, hiking behaviours, and motivations. To achieve this purpose, the methodology includes site observation, literature reviews and questionnaire surveys. The findings of the study suggest that there is a wide range of mountain hikers, extending from beginners to experienced enthusiasts with different hiking motivations. For many hikers, nature appreciation and solitude is just as important as finding a different experience along the hiking trails. The findings of this study have important implications for mountain tourism policy and practice.

Keywords: Mountain tourism, hikers' characteristics, hikers' motivation, hiking experience

1 INTRODUCTION

Malaysia's support for mountain tourism was part of the 10th Malaysia Plan and Tourism National Key Economic Area (NKEA), and the Tourism Transforming Plan aims to attract high-yield tourist markets and contribute significantly to the country's GDP. The Plan will chart Malaysia's tourism progress with the goal of welcoming RM36 million foreign tourists and bringing in foreign revenue of RM168 billion (approximately USD 54.3 billion) by the year 2020.

Mountain tourism has been continuously promoted as one of the sectors targeting a niche segment of the market. The effort to promote greater local and international participation in mountain tourism was further emphasised through the launch of the "5-Mountain Motorcycle Tour" package by the Prime Minister in March 2012, which was part of the agenda to promote the scenic mountains of the country and other attractive tourist destinations (Wonderful Malaysia, 2012).

The country's efforts to promote mountain tourism have continued through the hosting of the 26th Annual International Climbathon in 2012. The event took place on 13 to 14 October 2012, when hundreds of hikers attempted to complete what is widely known as "The World's Toughest Mountain Race". Among the most popular and most challenging mountains selected for the event are the G7, which includes Mount Tahan (Pahang), Mount Korbu (Perak), Mount Yong Belar (Perak), Mount Gayong (Perak), Mount Chamah Kelantan), Mount Yong Yap (Perak), and Mount Ulu Sepat (Perak-Kelantan border).

The objectives of this study are to add to existing knowledge of G7 hikers and to provide a current assessment of hikers' characteristics, experiences, and behaviours. Specifically, the objectives are to identify hiker characteristics in terms of socio-demographic characteristics, group characteristics, hiking experience, hiking behaviours, and motivations.

2 LITERATURE REVIEW

2.1 *An overview of mountains in Malaysia*

Malaysia is one of the most popular emerging mountain tourism destinations, as the region's tallest peak Mount Kinabalu (4,095 m) is located there. Situated in the Kinabalu National Park in Sabah, the mountain is a popular destination among locals and international tourists. The eight-kilometre climb usually starts from Timpohon Gate, near the park headquarters (1,800 m). Hikers usually opt to stay the night at Laban Rata (3,273 m) before continuing to climb the summit the next day. For an exciting experience, hikers can also choose to descend Mount Kinabalu via Ferrata, a route equipped with fixed cables, stemples, ladders, and bridges (Malaysia Traveller, 2014).

Other popular climbing sites among hikers in Malaysia include the Lost World of Tambun, Nipah Village on Tioman Island, Taman Etnobotani in

Gua Musang, Kelantan, Gunung Jerai in Kedah, Gunung Angsi in Negeri Sembilan and Gunung Ledang in Johor. One of the most challenging mountains in the country is Gunung Tahan, the Peninsula's highest peak, which crosses over the rainforest, rivers and mountain ridges. Hikers can also choose to climb the Batu Caves, a limestone outcrop with eight crags and approximately 170 routes. This formation is closer than others to the capital, Kuala Lumpur.

2.2 *Defining mountain tourism*

Pomfret (2011) categorised mountain tourism, mountain hiking, mountain trekking and ice and rock hiking in mountain regions anywhere in the world as adventure tourism. These adventure activities can take on a variety of sights and routes, including ice, snow, glaciers and even tropical rainforests. Mountaineering can be a subset of other tourism categories: nature based tourism (Whitlock, Romer & Becker, 1991), eco-tourism (Johnston & Edwards, 1994), and adventure tourism (Carroll, 1999; Pomfret, 2006). Currently, mountaineering activities have evolved globally, creating the opportunity for industry players to generate creative packages and inspired marketing strategies for a vast growth in revenue.

A review of the literature reveals that there is a lack of commonly accepted term use for mountain climbers who recreationally scale or hike up mountains with the specific goal of reaching the summit (Lasco, 2009). On the other hand, a person who performs an outdoor activity that consists of walking in natural environments, often on hiking trails, is termed a hiker (Heer, Rusterholz & Baur, 2003). In some countries, the mountaineer is also a term used by climbers to characterise the types and landscapes of the mountains (Pomfret, 2006), thus suggesting differences in terminology that are inevitable across cultures and countries.

2.3 *Hikers motivations*

In mountaineering, the goal is usually to arrive at the peak. It is a phenomenon known as a peak experience, which leads to feelings of joy and self-fulfilment upon reaching the summit and is bounded in time rather than enduring over time (Privette & Bundrick, 1991). A peak experience is characterised as a transformational experience and one that surpasses the usual level of intensity, meaningfulness, and richness (Privette, 1983).

Historically, years ago, individuals hiked mountains due to spiritual motives. They built altars and houses to offer human sacrifices on the peaks. Pfister (2000) described mountainous areas as pure and being connected with God, consecrated

images, spirits and legends. Godde, Price and Zimmermann (1999) reported that a scientific investigation has later become a reason for philosophers and scientists to scale the Alps.

On the aspect of risk, there are generally two categories of hikers: a soft adventure group, who scale peaks less than 6000 metres, and a hard adventure group, who scale peaks greater than 6000 metres (MacLellan, Dieke & Thapa, 2000; Pomfret, 2006). Soft adventure expeditions are mountaineering activities with a perceived risk that requires only basic skills and the accompaniment of skilled guides; the hikers may be either experienced or inexperienced. Most of the soft adventure hikers spend fewer days on the mountains (Snowdon, Slee, Farr & Godde, 2000). Soft adventure expeditions comprise guided activities that cover an introductory mountain climbing course for easy trails, mainly to attract children and beginners.

Hard adventure expeditions, on the other hand, involve a high level of risk and require serious commitment and a high level of skill on the part of the hikers. These expeditions involve climbing along rough trails, reading maps and compasses as well as other advanced equipment. Much of the previous research has focused on the supply aspects of mountain tourism, exploring mountain ecology for instance, the sustainability of mountains as protected areas (Ranger & Turpault, 1999) and tourism development in mountain regions (Beedie & Hudson, 2003; Godde, et al., 1999). There is limited research investigating the demand phase of mountain tourism (Muhar, Schauppenlehner, Brandenburg & Arnberger, 2007).

3 METHODOLOGY

The methodology includes site observation, literature reviews and structured questionnaire surveys. The data for this study were obtained through a questionnaire-based survey that was conducted over a 6-month period, distributed to the mountain hikers who agreed to participate in a survey regarding their hiking experiences. Hikers who had hiked the seven most prominent mountains, known as the G7 (Malaysia Traveller, 2014) Gunung Tahan in Pahang, Gunung Korbu, Gunung Gayong, Gunung Yong Belar and Gunung Yong Yap in Perak, Gunung Camah and Gunung Ulu Sepat in Kelantan were surveyed.

The questionnaire was provided after they completed their mountain hiking activities. The questionnaire was designed to collect a wide range of information, including demographics (e.g., gender, country of origin, source of promotional information, hiking trip frequency, organisation mode and duration of trip hiking experience), group

Table 1. Hiker demographic characteristics ($n = 378$).

Characteristics	Percentage (%)
Gender,	
Male	57.94
Female	42.06
Origin,	
Malaysian	85.00
Singapore	3.17
Indonesia	2.38
Hong Kong	1.32
Australia	8.20
Age,	
18–20 years old	29.40
21–25 years old	31.00
26–30 years old	26.10
Above 30 years old	13.50

Table 2. Hiker socio-economic characteristics.

Characteristics	Percentage (%)
Level of education, $n = 351$	
High/Secondary School	29.40
Diploma	34.50
Bachelor's Degree	26.45
Master's Degree	9.50
PhD	0.15
Annual household income, $n = 320$	
Below RM50,000	53.50
Above RM50,000	46.50

characteristics, hiking experience, hiking behaviours, and motivations. All data were analysed through Statistical Package for the Social Sciences software (SPSS) version 21.0.

4 RESULTS AND DISCUSSION

From the survey, 396 responses were collected and 18 responses were discarded due to incompleteness. Overall, 57.94 percent of the hikers were male (see Table 1). The proportion of male and females is con\sistent across the seven mountains. 85 percent were reported as Malaysian, and 15 percent were from Singapore, Indonesia, Hong Kong, and Australia.

The largest group of respondents were those in the age range of 21–25 at 31.0% ($n = 117$), followed by respondents under 20 years old at 29.4 percent ($n = 111$). A majority of the hikers (53.5%) reported a total annual household income below RM50,000 (refer Table 2). This finding is in line with reported majority who are below 25 years old.

For the majority of the hikers, their experience in the field was less than 5 years and their

Table 3. Importance of motivation categories.

| Categories | Level of Importance | | |
	Unimportant %	Neutral %	Important %
Nature appreciation	13.5	1.3	85.2
Solitude	7.4	0.5	92.1
Family togetherness	46.5	51.0	2.5
Testing skills	29.5	18.0	52.5
Being with others	32.4	62.2	5.4

participation in hiking trips was less than 5 times per year. A total of 63 percent of the respondents ($n = 240$) indicated that they were part of an organised group and hired mountain guides for their hiking trip. Just 37 percent of the respondents organised their trip on their own. A large majority of the day hiking groups were composed of four or fewer people (84.5%). A pair of hikers was the most frequent hiking group size (46.3%), and the rest were respondents hiking alone (14.2%). Most respondents (62.7%) visited the G7 for two or three days. Most hikers (55.9%) have previously made at least one trip to one of the mountains of the G7. While the rest reported that their current hike has been their first hike at the G7.

Several questions were asked to understand the motivation of the hikers. Respondents rated the importance of twenty items that were grouped into five categories (Backlund, Stewart, Schwartz & McDonald, 2006). The items asked indicate the importance of various types of experiences in motivating a hike at the G7. The hikers were asked to rate the items unimportant, neutral or important.

Similar to what has been reported by Backlund et al. (2006), the majority of the respondents rated nature appreciation and solitude as important to them, 85.2 percent and 92.1 percent, respectively. More than half of the respondents also rated testing skills as important to them (52.5%). Finally, family togetherness and being with others was rated as either unimportant or neutral by almost all respondents (see Table 3).

5 CONCLUSIONS

These findings highlight that there is a broad diversity among mountain hikers, ranging from beginners to experienced enthusiasts. In spite of their differences, many are looking for and finding similar experiences in similar environments and social surroundings. For many hikers, finding unique values and agendas is just as important as findings a different experience along the hiking trails. Future management should seek to further define and refine the variety of hiking opportunities to allow hikers to find trails that fit with their preferences and abilities.

Mountain tourism thus, should be promoted to the appropriate target segment. This focus could prevent excessive promotional costs, scattered media marketing, and off-target messages that result from targeting an overly broad market. These findings provide some interesting insights into the indicators and specific motivations of hikers in the context of the G7 mountains. These insights will help clarify some specific actions that mountain tourism organisers should undertake to improve hikers' satisfaction of specific hiking motivations highlighted by them for future visitations.

ACKNOWLEDGMENT

This study was made possible by the continuous support from Universiti Teknologi MARA, UiTM—Grant No: 600-RMI/DANA 5/3/ RIF(234/2012).

REFERENCES

Backlund, E.A., Stewart, W., Schwartz, Z. & McDonald, C. (2006). *Backcountry Day Hikers at Grand Canyon National Park*. University of Illinois.

Beedie, P. & Hudson, S. (2003). Emergence of mountain-based adventure tourism. *Annals of Tourism Research, 30*(3), 625–643.

Carroll, A.B. (1999). Corporate social responsibility evolution of a definitional construct. *Business & society, 38*(3), 268–295.

Godde, P.M., Price, M.F. & Zimmermann, F.M. (1999). Tourism and development in mountain regions: moving forward into the new millennium. *Tourism and development in mountain regions.*, 1–25.

Heer, C., Rusterholz, H.P. & Baur, B. (2003). Forest perception and knowledge of hiker and mountain bikers in two different areas in North-Western Switzerland. *Environmental Management, 31*, 709–723.

Johnston, B.R. & Edwards, T. (1994). The commodification of mountaineering. *Annals of Tourism Research, 21*(3), 459–478.

Lasco, G. (2009). Mountaineering vs. hiking vs. trekking Retrieved October 10, 2012, from http://www.pinoy-mountaineer.com/2009/07/essay-mountaineering-vs-hiking-vs.html

MacLellan, L.R., Dieke, P.U. & Thapa, B.K. (2000). Mountain tourism and public policy in Nepal. *Tourism and development in mountain regions*, 173–197.

Malaysia Traveller. (2014). Climb Malaysia mountains: There are hundreds to choose from Retrieved April 16, 2014, from http://www.malaysia-traveller.com/malaysia-mountains.html

Muhar, A., Schauppenlehner, T., Brandenburg, C. & Arnberger, A. (2007). *Alpine summer tourism: The mountaineers' perspective and consequences for tourism strategies in Austria.* Paper presented at the Forest Snow and Landscape Research.

Pfister, R.E. (2000). Mountain culture as a tourism resource: Aboriginal views on the privileges of storytelling. *Tourism and development in mountain regions*, 115–136.

Pomfret, G. (2006). Mountaineering adventure tourists: a conceptual framework for research. *Tourism management, 27*(1), 113–123.

Pomfret, G. (2011). Package mountaineer tourists holidaying in the French Alps: An evaluation of key influences encouraging their participation. *Tourism management, 32*(3), 501–510.

Privette, G. (1983). Peak experience, peak performance, and flow: A comparative analysis of positive human experiences. *Journal of personality and social psychology, 45*(6), 1361–1368.

Privette, G. & Bundrick, C.M. (1991). Peak experience, peak performance, and flow: Correspondence of personal descriptions and theoretical constructs. *Journal of Social Behavior & Personality, 6*, 169–188.

Ranger, J. & Turpault, M.-P. (1999). Input–output nutrient budgets as a diagnostic tool for sustainable forest management. *Forest Ecology and Management, 122*(1), 139–154.

Snowdon, P., Slee, B., Farr, H. & Godde, P. (2000). The economic impacts of different types of tourism in upland and mountain areas of Europe. *Tourism and development in mountain regions*, 137–145.

Whitlock, W., Romer, K.v. & Becker, R.H. (1991). *Nature based tourism: an annotated bibliography*. Clemson, SC: Strom Thurmond Institute of Government and Republic Affairs, Clemson University.

Wonderful Malaysia. (2012). Tourism Malaysia 5-Mountain motorcycle tour, from http://www.wonderfulmalaysia.com.

Theory and Practice in Hospitality and Tourism Research – Radzi et al. (Eds)
© 2015 Taylor & Francis Group, London, ISBN 978-1-138-02706-0

Tourism Industry Compensation Fund (TICF) in Malaysia: Some legal and policy considerations

A.A. Hasan & N.C. Abdullah
Faculty of Law, Universiti Teknologi MARA, Shah Alam, Malaysia

ABSTRACT: Tourism Industry Compensation Fund (TICF) is a new ministerial agenda for the protection of package tour purchased by travelers. This paper endeavors to encourage further discussion and aims to assist the policy drafter in attending at some conceptual implementation concerns of TICF. By applying qualitative and comparative research method, the paper examines the regulatory framework establishing TICF and make comparative assessment with equivalence model adopted by selected foreign jurisdictions. The paper concludes that the operating mechanism of TICF needs to align with the principles of fairness and equity to secure the mutual interest of travelers and travel agencies in Malaysia.

Keywords: compensation fund, insurance, package tour, tourism industry

1 INTRODUCTION

Tourism industry is an important foreign exchange earner under the National Key Economic Areas (NKEA). In temporal parallel to the National Transformation Programme, TICF is viewed as an innovative approach to consumer grievance management through regulative means. The present paper analyses the practical feasibility of TICF as the effective way to promote a positive balance between consumers' rights and sustainable tour and travel industry. Since TICF is still at the consultation stage, the questions and issues raised in this paper may appear to be speculative in nature. However, the issues raised are grounded on some analysis on the present state of the industry, existing regulatory framework currently in force as well as comparative study with the existing features of equivalence model in other countries.

2 THE GENESIS OF TICF

For many years, there are many complaints against travel and tour agencies. The complaints centered on travel agencies' failure to perform tour packages as contracted, failure to adhere the tour itinerary, last-minute cancellation and failure to account money in respect of tour cancellation by the company, and improper performance of service quality (Tourism Licensing Division, Ministry of Tourism and Culture Malaysia, 2012). In furtherance of the complaints lodged, several travel agencies were found to have committed various offences against

the Tourism Industry Act 1992 and its accompanying regulations. Actions taken against travel agencies include revocation of business licenses and issuance of compounds in respect of fraud, operating business without license, operating business outside the permitted scope of the license, non-compliance with the terms and conditions of tour brochures, and engagement of unlicensed tour guide in package tour (Tourism Licensing Division, Ministry of Tourism and Culture Malaysia, 2012). In response to travelers' complaints against unethical misconduct of some travel agencies in the industry, the Malaysian Ministry of Tourism and Culture (MOTAC) has recently proposed the Tourism Industry Compensation Fund (TICF) TICF with the ultimate aim to protect customers who purchase package tour from travel agencies licensed by the Ministry (Tourism Licensing Division, Ministry of Tourism and Culture Malaysia, 2012).

According to the proposal, TICF is modeled from the legal framework currently adopted by the United Kingdom and Queensland, Australia. No justification, however, was given in support of such preferences. Hence, in order to get a broader perspective on the subject matter, comparative study with other selected jurisdictions is made. Essentially, any merit derived from the comparative jurisdictional assessment needs to be molded with the local trading atmosphere. The implementation of the TICF has been agreed by the Drafting Division of Attorney General Chamber and Ministry of Finance (Tourism Licensing Division, Ministry of Tourism and Culture Malaysia, 2012). The proposal for the introduction of TICF is available to the travel agency associations for their review and feedback. It has been contemplated that TICF would be implemented, at latest, by end of 2012 (BERNAMA, 2012a; Bernama 2012b; Dato' Dr. James Dawos Mamit, 2012). However, to date, TICF has not yet been tabled at the Parliament.

3 OVERVIEW OF THE SCOPE OF TICF

The principal purposes of the proposed TICF in Malaysia are threefold; (i) to protect the legitimate interest of travellers who purchased package tour, (ii) to serve as assurance to travellers on the package tour purchased and (iii) to protect domestic and foreign travellers from cases arising from fraud, negligence and insolvency. The general concept of TICF to protect outbound travellers against travel agencies' insolvency and failure to carry out contractual obligation has been the underlying idea behind the establishment of similar compensation fund in Brunei (section 26(1)(i), Travel Agents Act 1982), British Columbia (regulation 18, Travel Industry Regulations 2004), California (section 17550.36(a), California Business and Professions Code), Ontario, Canada (regulation 57(1), Ontario Regulation 26/05), People's Republic of China (article 15, Regulation on Travel Agencies 2009), Quebec, Canada (section 36, Travel Agents Act; regulation 37, Regulation Respecting Travel Agents), and Queensland, Australia (section 40(1), Travel Agents Act 1988).

Another way of looking at the function of compensation fund is to protect travelers' pre-payment of tour fare rather than the fulfillment of package tour as a whole. Protection of pre-payment in this context, conceptually, covers situations in respect of abscondment and travel agency's insolvency. This has been the view adopted by Hong Kong (rules 4(1) and (2), Travel Industry Compensation Fund (Amount of Ex Gratia Payments and Financial Penalty) Rules), Australian Capital Territory (section 149(1), Agents Act 2003), New South Wales, Australia (regulation 15.1, Travel Agents Regulations 2006), South Australia (regulation 15.1, Travel Agents Regulations 1996), and Western Australia (regulation 15.1, Travel Agents Regulation 1986). Some countries have their own unique system of protection for travellers. Travel agencies in Florida (section 559.929(2), Florida Statute (Seller of Travel)), and Jamaica (Travel Agencies Regulation Act, section 10(a)(iii) (1956)), for instance, are required to secure surety bond to compensate unethical misconduct of travel agencies. In United Kingdom (regulation 16(1), Package Travel, Package Holidays and Package Tours Regulations 1992) and Republic of Ireland (section 22(1), Package Holidays and Travel Trade Act 1995), travel agencies are required to provide adequate security to guarantee the travellers refund of tour fare paid and repatriation in the event travel agencies went into insolvency. Such arrangements for security may either be made by entering into a bond with a body approved by the relevant authority or by being insured under appropriate insurance or by holding the travellers' pre-payment in a trust account until the completion of the tour packages (regulation 17, 18, 19, 20 and 21, UK Package Travel, Package Holidays and Package Tours Regulations 1992; sections 23, 24 and 25, Irish Package Holidays and Travel Trade Act 1995).

Ontario, Canada (regulation 25, Regulation 26/05), Quebec, Canada (regulation 28, Regulation Respecting Travel Agents Regulation) and California (sections 17550.11 and 17660.37, California Business and Professions Code) practice hybrid system of protection whereby travel agencies are required to be a contributor to compensation fund on top of the requirement to provide a form of financial security or surety bond. In Hawaii (section 468L-5, Hawaii Revised Statutes (Chapter 468L)), Illinois (section 6, Travel Promotion Consumer Protection Act), Virginia (section 59.1-447.1, Code of Virginia (Travel Club Act) 2010) and Washington (section 19.138.040, Revised Code of Washington), protection of pre-payment is ensured through the requirement for every travel agency to deposit such payment into a trust account maintained in a federal insured financial institution. In Hong Kong, the Travel Industry Compensation Fund provides payment of compensation for loss sustained by outbound travelers due to an accident which resulted in the death or injury during outbound travel service organized by licensed travel agency (rule 5A(1), Travel Industry Compensation Fund (Amount of Ex Gratia Payments and Financial Penalty) Rules,). However, for the purpose of the application, no provision stating that proof of negligence in respect of the accident is required.

4 THE MODUS OPERANDI

4.1 *Composition of the board of trustee or committee*

In general, the Board of Trustee or Committee may be divided into two administrative components which are the management and the evaluation panels. The TICF will be managed by a Board of Trustees or Committee presided by the Commissioner of Tourism with the Licensing Division acting as the secretariat. It would appear that the role of the Commissioner is administrative in nature. As the law currently stands, the Commissioner plays a prominent role as the adjudicator in determining the quantum of compensation to be awarded to travelers in cases involving breach of the terms and conditions stated in the tour brochure (regulation 6(1)(m)(iii), Tourism Industry (Tour Operating and Travel Agency Business) Regulations 1992). The award of compensation is made out of the insurance policy, cash deposit or bank guarantee which the travel agencies are required to secure before they may commence travel agency business.

Every compensation claim made through the TICF will be decided by a panel which comprises of the following members: (i) the Controlling Officer of the Fund which is the Chief Secretary to the Ministry of Tourism who will be acting as the Chairman; (ii) the Director General of Malaysia Tourism Promotion Board or his representative; (iii) four senior public officers responsible in tourism sector, and (iv) representatives of the travel and tour associations recognized by the MOTAC. Given the present sittings, it appears that the voice of travel agencies associations remains to be of practical importance in dispute resolution regime. At this juncture, it remains to be observed as to the mode of selection of the senior public officers and the representatives of the travel and tour associations. It is foreseeable, that the panels will be deciding cases filed by travelers from other jurisdictions. In this respect, proper mechanism should be set in place in order to accommodate to the claims made by foreign travelers.

Although all members of the panel are individuals with savvy travel and tour background, they may not possess the necessary legal knowledge in granting the award. The Board of Trustee or Committee would be able to serve its function much better if the chairs for legal experts could be allocated within the sittings to assist in deciding claims involving pure economic loss, issue on mitigation of loss, and the legal formula in computing money arising from frustrated package tour contract.

4.2 *Mandatory contribution*

Under the proposed TICF framework, all travel agencies licensed under the Ministry must become participants to the TICF. The compulsory contribution carries with it legal sanction in the form of suspension of license and cancellation of license, in the case of continuous default, i.e. a measure to ensure compliance on the part of the tour and travel agencies. The amount of contribution payable is determined by the scope of license issued to the travel agencies and the amount differs from license issued for company's headquarters and its branches. All licensed travel agencies will be required to re-contribute to the TICF once the remaining balance reach below 25% of the total reserves. Despite of the underlying virtue sought by TICF, the scheme of contribution give rise to some

implementation concerns. The flat rate of contribution does not take into consideration the annual total revenue of small and medium-scale travel agencies in comparison with the large-scale travel agencies. Such an inequitable scheme of contribution fly in the face of the initial virtue sought by the Ministry in removing the regulatory burden which would otherwise impacts the small and medium-scale travel agencies. A more proportionate scheme of contribution should be adopted by the TICF by taking into account the extent of annual total revenue made by the respective travel agencies.

The TICF is, essentially, a pool of common contribution of licensed travel agencies. Where claims are made against a travel agency for a value more than its contribution, the deficit is supplemented by the contribution made by other travel agencies. Although it would seem that justice is appropriately serve to the aggrieved travelers, the same does not equally apply to travel agencies which have clean record of business and have been operating the business with utmost integrity. The present state may be distinguished with the operating mechanism of some professional indemnity insurance in other jurisdictions where the amount of indemnity corresponds to the annual premium calculated on the amount of coverage and the number of travelers indemnified (Dautch, 1941). However, the loss resulting from claims made under fraudulent, malicious, dishonest and criminal acts are generally not covered under the policy contract, where the proposed TICF is intended as a mechanism to resolve these unethical actions of the travel agencies (Hooker & Pryor, 1990).

It is interesting to note that in Quebec, Canada, it is the travelers of licensed travel agencies who are required to contribute to the fund, the amount of which is to be collected by the travel agencies dealing directly with the travelers (regulation 39, Regulation Respecting Travel Agents Regulation). The contribution payable is calculated by multiplying the total cost of tour fare by a designated percentage varying according to the surplus accumulated in the fund. In Hong Kong, the compensation fund consist of, among others, contribution from travel agencies in the form of levies computed at 0.15% of each outbound tour fare received (section 32H, Travel Agent Ordinance), financial penalty for the late payment or failure to make due payment of the levy payable (section 32G(2)(c), Travel Agent Ordinance), and income from investment made out of the fund (section 32F(2)(d), Travel Agent Ordinance). Similarly, in Ontario, Canada, the amount of contribution is calculated at $0.15 for every $1000 or part of $1,000 of sales in Ontario including $25 being the applicable taxes (Payment Schedule, Compensation Fund Contribution Rates, Travel Industry Council of Ontario).

In British Columbia, travel agencies are required to contribute to the compensation fund for a value of $100 or an amount equal to 0.05 out of the gross sale, whichever is greater (regulation 17(3), Travel Industry Regulation 2004). A travel agencies may not need to make further contribution if the book value of the compensation fund is at least $2,000,000 and the travel agencies has paid for successive contribution periods totaling 3 years (regulation 17(4), Travel Industry Regulation 2004). It is most unfortunate that in Australia, every travel agency's contribution is fixed at $7,430.00 for principal location and $5,000.00 for branch location irrespective of the capacity and size of the travel agency's business scale (Fee Schedule, Travel Compensation Fund). At present, TCF will no longer be in operation in Australia commencing from 30 June 2014 with claims paid until March 2015.

4.3 *Procedures for claiming compensation*

4.3.1 *Type of claims*
Under the proposed TICF framework, any aggrieved 'individual' or traveller seeking for redress may resort to the TICF. Eligibility to claim is, however, circumscribed to cases in respect of (i) fraud, negligence, or breach of terms and conditions of package tour; and (ii) insolvency of the travel agent company, subject to the settlement of all claims through legal procedure or insurance claims. It would seem to be the case that individual other than travelers who purchase package tour from licensed travel agencies may also claim under the TICF. In the absence of specific interpretation, the term 'individual' is ambiguous. The term may refer to traveler's next of kin or where the traveler is incapacitated or a minor, the traveler's guardian *ad litem*. In the context of claim made following the travel agency' insolvency, it is questionable as to whether the term 'individual' may also refer to the travel agency's creditors. If the answer is affirmative, it would appear that the compensation claimed may be of an excessive sum, causing undue burden to the rest of travel agencies.

4.3.2 *Conditions and amount of claim*
In relation to claims made by traveler, several prerequisites must be complied with. Travelers, who purchased package tour from non-contributor of the TICF, will be barred from claiming. Such prerequisite penalizes imprudent travelers in that by purchasing tour packages from unlicensed travel agencies, the payment made by the travelers is not protected. The maximum amount of claim which can be made under the TICF is 80% of the amount of loss claimed subject to the total amount approved by the Board of Trustee or Committee. In the light of type of claims which may be made

to the TICF, such an amount of compensation is by far a large amount. By comparison, the Hong Kong Travel Industry Compensation Fund covers loss for a maximum of 90 percent out of the outbound fare or HK$10,000, whichever is lower (rules 4(1) and (2), Travel Industry Compensation Fund (Amount of Ex Gratia Payments and Financial Penalty) Rules). In relation to loss sustained by outbound travelers in respect of an accident, the maximum amount claimable is HK$100,000 for every medical expenses incurred in the destination country, expenses incurred for funeral outside Hong Kong or for the return of the dead body to Hong Kong, and visitation expenses incurred by the relatives of the outbound traveler to the destination country (rule 5C, Travel Industry Compensation Fund (Amount of Ex Gratia Payments and Financial Penalty) Rules). In British Columbia, the maximum amount claimable is $5,000 for each person in respect of a claim (regulation 22(1), Travel Industry Regulation 2004). In Australian Capital Territory, traveler is entitled to claim the amount of the actual financial loss suffered less any amount that he has recovered or can recover from a source other than the compensation fund (section 149(2), Agents Act 2003).

Apart from the maximum limit of amount which is claimable, the proposed TICF further requires that the loss sustained by the traveler must be such a loss that is not covered by his travel insurance or that he had failed to pursue his claims through the Tribunal for Consumer Claims or other legal procedure. These two requirements merit further discussion. The Tourism Industry (Tour Operating and Travel Agency Business) Regulations 1992 neither requires outbound traveler to purchase travel insurance nor requires travel agencies to advice travelers on the benefit of being insured (clauses 12 and 5.2, Fourth Schedule, Tourism Industry (Tour Operating Business and Travel) Regulation 1992). The existing travel insurance coverage that is currently available in the market protects travelers against trip cancellation, physical injury, medical evacuation or repatriation, loss of baggage, loss of travel deposit as well as against travel agency's insolvency.

It is important to note that the entitlement to compensation under TICF is strictly circumscribed to loss not covered by travel insurance. It is questionable as to the status of travelers who travelled without any travel insurance coverage. Would the travelers be able to claim under the TICF if they do not purchase any travel insurance? If the answer is in the affirmative, it would seem that the TICF favourly treats imprudent travelers who travelled without travel insurance against those who are insured. Aside for practical and uniformity purposes, the TICF needs to be reconciled with established principles of law to do fairness to travelers who are insured. It is interesting to note that travel insurance and the associated risks of travelling uninsured are amongst the mandatory information to be delivered to travelers in United Kingdom (regulation 8(2)(d), Package Travel, Package Holidays and Package Tours Regulations 1992), the Republic of Ireland (section 12(1)(c), Package Holidays and Travel Trade Act 1995), Ontario, Canada (regulations 34(1) and 36(d), Regulations 26/05) and other European countries (Article 4(1)(b)(iv), European Union Council Directive 90/314/EEC on package travel, package holidays and package tours). From the provision, the practice in the United Kingdom and the Republic of Ireland seem to imply that travel insurance may be amongst the inclusive components of package tour sold by travel agencies.

4.3.3 *Evidentiary concern*

Every claim must be supported with sufficient evidence for evaluation by the Board of Trustee or Committee. It is yet to be determined as to the application of rules on evidence and it may appear to be a challenging task for both travelers and the Board of Trustee or Committee to establish a case founded on contractual negligence and fraud. In fact, to avoid complexity, claim for personal injury and death is expressly ousted from the jurisdiction of the Tribunal for Consumer Claims (section 99(3), Consumer Protection Act 1999). It is important to note that the ceiling-amount of compensation claimable is 80% of the amount of loss claimed subject to the total amount approved by the Board of Trustee or Committee. Through such percentage-based scheme of compensation, the total amount claimable may exceed RM25,000 being the maximum amount of claim which may be awarded by the Tribunal for Consumer Claims (section 98(1), Consumer Protection Act 1999). Therefore, it would appear that claims involving complex factual and legal issues are inevitable.

Another evidentiary concern is on the onus of proof applicable to the travelers in establishing their claims. It is settled that proof of fraud in civil cases is on balance of probabilities (*Lau Hee Teah v. Hargill Engineering Sdn Bhd & Anor* [1980] 1 MLJ 145 as per Abdoolcader J.). However, the applicable degree of probability very much depends on the gravity of the allegation (*Lau Hee Teah v. Hargill Engineering Sdn Bhd & Anor* [1980] 1 MLJ 145 as per Abdoolcader J.). The degree of probability in cases involving fraud as required in civil court is higher than that which it would require in consideration of whether negligence was established (*Bater v. Bater* (1950) 2 All ER 458 as per Lord Denning). In the absence of clear guideline on the existing proposed TICF, the burden of proof in claims from the TICF remains questionable.

5 CONCLUSION

A consensus has almost always existed among jurisdictions under review on the prominent role of travel compensation fund as an instrument for travelers' protection through a much equitable scheme of contribution. The regulatory reform currently taking place in Australia is an indicative lesson to be learnt by Malaysia. It remains to be observed whether the TICF would be a pragmatic mode of alternative dispute resolution involving package tour. In the meantime, for the mutual interest of travelers and travel agencies in Malaysia, the implementation concern as highlighted demand utmost vigilance. The appropriate path to approach the implementation concerns is always the middle path between travelers' interest and travel agencies interest.

REFERENCES

Ahmad Afiq Hasan & Nuraisyah Chua Abdullah (2013). Doctrine of Frustration as the Second Level of Protection: The Case of Frustrated Travellers. *Proceedings of the International Conference on Consumerism*, Perpustakaan Negara Malaysia, 312–320. ISBN 978-967-5920-07-3.

Ahmad Afiq Hasan & Nuraisyah Chua Abdullah (2013). Force Majeure in Holiday Package Contract: Issues & Challenges for the Travelers. In Ng Siew Imm, Hamimah Hassan and Lee Shin Ying (ed.) 2013). *Readings on Hospitality and Tourism Issues*. Mc Graw Hill, 6–15. ISBN: 9789675771880).

BERNAMA. (2012a, May 3). 8 agensi pengendalian pelan-congan dibatalkan lesen kerana menipu, *Berita Harian online*. Retrieved from: http://www.bharian.com.my/bharian/articles/8agensipengendalianpelancongandibatalkanlesenkeranameni-pu/Article/

BERNAMA. (2012b, May 4). 12 kes penipuan libatkan pakej umrah, pelancongan, *Utusan online*. Retrieved from: http://www.utusan.com.my/Utusan/info.asp?y=2012&dt=0504&pub=Utusan_Malaysia&sec=Parlimen&pg=pa_03.htm

Dato' Dr. James Dawos Mamit, *Parliament official statement (Dewan Negara)*, 3 May 2012, pp. 3–4. Available: http://www.parlimen.gov.my/files/hindex/pdf/DN-03052012.pdf

Dautch, C. (1941). Lawyers' Indemnity Insurance, *Commercial Law Journal*, 46 (11), 412–414.

Hooker, N.D. & Pryor, L.M. (1990). Professional Indemnity Insurance, *Journal of the Staple Inn Actuarial Society*, 32, 37–69.

Tourism Licensing Division, Ministry of Tourism and Culture Malaysia. (2012). Proposal for the establishment of the Tourism Industry Compensation Fund. Available: http://www.matta.org.my/index.php/news2/announcements/56-tourism-industry-compensation-fund-ticf.

Theory and Practice in Hospitality and Tourism Research – Radzi et al. (Eds)
© 2015 Taylor & Francis Group, London, ISBN 978-1-138-02706-0

Community-based rural tourism as a sustainable development alternative: An analysis with special reference to the community-based rural homestay programmes in Malaysia

K. Kayat, R. Ramli, M. Mat-Kasim & R. Abdul-Razak
Universiti Utara Malaysia, Kedah, Malaysia

ABSTRACT: For Community-Based Tourism (CBT) to be accepted as an alternative to sustainable development, it should mirror the characteristics that signify a community in a rural area working together to plan, operate and manage tourism businesses that both satisfy the visitors and develop the people and the place where the community live in. CBT that achieve these two goals will succeed in becoming a sustainable development alternative. This paper explored the required criteria for a successful Community-Based Rural Homestay (CBRH) programmmes in Malaysia from the perspectives of different individuals who are familiar with the CBRH programmes through a case study. As a CBT product, the CBRH programmes have been well imaged, commoditized, and packaged to tap the rural tourism potential of Malaysia. The identification of the CBRH programmes success factors is of critical importance for the notion of sustainable development alternatives through tourism.

Keywords: Community-based rural tourism, sustainable development, homestay programmes, success factor

1 INTRODUCTION

1.1 *Tourism and sustainable development*

The importance of tourism lies in its contribution to the global citizens' economic and social standards. Two broad groups of beneficiaries from tourism are the tourists and the hosts. Tourists receive psychological benefits as an exchange for the price they pay to travel and visit the destinations of their choice. Tourism generates important benefits to the hosts as well; it produces earning opportunities and enables them to increase their income sources (Samimi, Sadeghi & Sadeghi, 2011). Employment created by tourism also contributes to poverty alleviation which leads to the hosts to a better welfare and quality of life (Manyara & Jones, 2007). Additionally, there are empirical evidences that tourism expansion is relevant and significant to the nations' growth and economic development. Fayissa, Nsiah, and Tadasse (2008) for instance, found that a 10 percent increase in the spending of international tourists leads to a 0.4 percent increase in the GDP per capita income in African countries. Finally, tourism develops a nation further through modernization—for example, tourism brings about higher levels of employment and literacy, improvement in medical services, as well as wider access to other elements related to comfort and security.

Thus, tourism is beneficial to both the tourists and the local community as it allows them to experience other cultures, which broadens understanding among citizens of the world.

However, mere statistics reflecting the growth of income and wealth spurred by tourism is not enough to qualify tourism as an important contributor to the development of a host community. Mundt (2011) argues that too much concern with numbers, which is the basic tenet of the concept of growth, can actually lead to destructions. A development agenda must also cultivate the minds and increase the capacity and capability of the people within the community to continue thriving. Tourism development must 'develop' the economic capability and knowledge of the receiving residents in order to create economic sustainability. Income generated from tourism should ideally be used at the national and local level to support education, improve infrastructure, finance conservation efforts, and to foster more responsible tourism. In that way, tourism is allowed to become a crucial strategy for sustainable development. Sustainable tourism strives to reduce economic leakages and increase economic linkages as well as to conserve the environment through effective energy use. In addition, sustainable tourism requires respect for local culture and involvement of local community in tourism development, planning and monitoring (Scheyvens, 1999). Thus, the concept of sustainable tourism can be interpreted as a process of tourism development as well as the outcome of tourism development.

In addition, sustainable tourism development defies the need to gratify each and every demand made by tourists' as some of the needs may not correspond to the principles of sustainable development. Instead, the importance is put into educating the tourists to respect the principles of sustainable tourism. A sustainable approach in tourism development should aim to satisfy the visitors as well as to develop the people and the place where the community lives in. Striving only to satisfy the tourists will lead to unfairness as a large number of tourists can generate social costs as they impinge upon the lifestyles of the host community as well as deteriorate the community's ecological and sociocultural resources (Tasci, Gurbuz & Gartner, 2006).

1.2 *CBT as the tourism that produce sustainable development*

Community-based tourism (CBT) is the type of tourism which is strongly linked to the community as it is the community who is the main actor of its planning and execution as well as the primary party that benefit from it. This is in accord with Russell (2000) who argues that a tourism project cannot be termed as a CBT unless it has the support and participation of local people, economically benefits the people living at or near the destination and protects local people's cultural identity and the natural environment. These noticeable similarities with the sustainable development concept lead to an easy conclusion that CBT is a sub-category of sustainable tourism. CBT is often linked to rural tourism as communities in rural areas are usually the ones that have the propensity to work together on community projects such as tourism.

CBT presents an opportunity to empower local communities, particularly in developing countries, to develop a more apt 'grass-roots' form of sustainable tourism than mass tourism and to contribute to local economic development and poverty reduction. It is argued that through developing CBT enterprises, communities can be empowered by raising pride, self-esteem and status, improving cohesion and community development and creating an equitable community political and democratic structure. Thus, the success of a CBT is based on both competitive and sustainability criteria. CBTs must generate individual and collective benefits for community members (Simpson, 2008), which must exceed costs to all involved and counterbalance to tourism impacts produced (Novelli & Gebhardt, 2007). The benefits generated must accrue both to

individuals and the whole community, and exceed costs to those involved. Benefits may be financial and/or non-financial, e.g. to include for example social, cultural, environmental and educational opportunities.

However, due to its 'saint-like' nature, several parties use the term 'CBT' like others use the term 'ecotourism' as a marketing gimmick to attract consumers who are then made to believe that they are supporting a good cause—which is to travel responsibly. It is argued that there is little tangible evidence of the benefits produced (Goodwin, 2007). Previous research has found that unfortunately many CBT enterprises do not succeed or do not produce intended benefits, or do not sustain. For example, Goodwin & Santilli (2009) surveyed 116 CBT initiatives identified by experts as successful: of 28 responses secured, 15 qualified as CBT enterprises, and only six were economically sustainable. A research done by Dixey (2008) in Zambia brought similar findings when only three of 25 CBT enterprises surveyed were "generate enough net income per year for tangible development and social welfare in the wider community", all of which had a private sector backing. If the initiative fails, investments and efforts made by the community will make an already vulnerable community worse off (Mitchell & Muckosy, 2008).

1.3 *CBRH programmes in Malaysia*

Community-based rural homestay (CBRH) programmes which are located at the village communities throughout Malaysia are promoted by the Malaysian Government as the Malaysian version of CBT as they claim that the CBRH programmes which are closely related to the nature of certain rural areas strive for culture preservation and environment protection. A CBRH programme offers visitors accommodation, food, cultural and/or nature activities as well as education. Although the programmes are managed by the local people, usually through a committee formed by the communities, the committees are not legally allowed to sell packages unless if they have are licensed tour operators. Most of them do not own tour operating license due to the high start-up capital imposed by the government. Just as there is an uncertainty about the actual benefits brought by CBTs, CBRH programmes' actual benefits that can qualify them as tools for sustainable development are still vague due to lack of research. Thus, this study is to explore the required criteria for a successful CBRH programmes from the perspectives of different experts who are familiar with the CBRH programmes. In addition, the extent that the criteria mirrored the criteria of a CBT is examined.

2 METHODOLOGY

2.1 *Methodology*

The study employed a qualitative method, involving an open-ended email interview that aims to explore perceptions and ideas from those who have high familiarized with CBRH. Forty-four participants were contacted and invited to respond to a set of open-ended interview questions using convenience and snowball sampling method, with the qualifying criteria that they understand the nature and operation of CBRH. There was an attempt to construct a sample, which represented a good range of groups with different interests in the programmes. The profile of interviewees who participated in the Email interviews is presented in Table 1.

Through the Emails, the researcher introduced herself and her study. The participants were asked to provide feedbacks to a list of questions attached to the email. The list of questions is as follows:

a. From your observation, please name (up to five) successful homestay programmes that are promoted by the government in the rural areas of Malaysia.
b. Can you explain why you consider the homestay programmes your named above as 'successful'?
c. In your opinion, what are the critical factors that lead to the success of the homestay programmes?

Question b and question care given to purposely cross check participants' responses in order for the study to conclude if the participants would equate 'success of the Homestay programmes with 'the critical factors that lead to the success of the Homestay programs'. The questions used in the interview uses the term 'successful' as opposed to 'sustainable' as to understand whether the terms carry the same meaning to the interviewees.

The feedbacks received from the interviewees were compiled, tabulated and categorized in order to conclude the interviewees' perceptions with regards to CBRH programmes' success factors.

3 RESULTS

At the time of this article is written, there are 166 homestay programmes or clusters throughout the rural areas in Malaysia. Among these homestay programmes, Banghuris Homestay in Selangor, was named as the most successful by the majority of the participants. The Banghuris Homestay has won numerous awards and recognitions such as the winner of the Ilham Desa Competition in 2003 (Central Zone) and 2005, well as the Malaysia's Best Homestay Award in 2004 and 2013. This

Table 1. Participant profiles.

No.	Occupation	Gender	Age
1.	Researcher	Male	45
2.	Researcher	Male	50
3.	Researcher	Male	35
4.	Researcher	Male	54
5.	Researcher	Female	52
6.	Researcher	Female	49
7.	Researcher	Female	48
8.	Researcher	Female	50
9.	Researcher	Male	54
10.	Researcher	Male	47
11.	Researcher	Male	36
12.	Researcher	Male	45
13.	Researcher	Male	38
14.	Researcher	Male	49
15.	Researcher	Male	45
16.	Researcher	Male	53
17.	Researcher	Male	60
18.	Researcher	Male	52
19.	Researcher	Female	45
20.	Ministry Tourism Officer (Homestay Programme)	Female	28
21.	Ministry Tourism Officer (Homestay Programme)	Male	31
22.	Ministry Tourism Officer (Homestay Programme)	Female	26
23.	Member, Malaysia Homestay Association	Male	56
24.	Tourist Guide	Male	40
25.	Tourist Guide	Female	35
26.	Tourist Guide	Male	26
27.	Tourist Guide	Male	45
28.	Tourist Guide	Male	34
29.	Tourist Guide	Male	28
30.	Tourist Guide	Male	30
31.	Tourist Guide	Female	28
32.	Tourist Guide	Male	35
33.	Tourist Guide	Male	42
34.	State Director, Malaysia Tourism Promotion Board	Male	48
35.	State Director, Malaysia Tourism Promotion Board	Male	36
36.	State Director, Malaysia Tourism Promotion Board	Male	53
37.	State Officer, Malaysia Tourism Promotion Board	Male	45
38.	State Officer, Malaysia Tourism Promotion Board	Female	35
39.	State Officer, Malaysia Tourism Promotion Board	Female	29
40.	State Director, Malaysia Tourism Promotion Board	Male	43
41.	State Director, Malaysia Tourism Promotion Board	Male	48
42.	State Director, Malaysia Tourism Promotion Board	Female	35
43.	State Director, Malaysia Tourism Promotion Board	Male	45
44.	State Director, Malaysia Tourism Promotion Board	Male	48

homestay programme has attracted a large number of tourists to visit this rural area to enjoy the beauty of nature and at the same time, learn and be aware of the resident cultures. A previous study found that rural tourism in Banghuris has provided numerous economic benefits to the villagers (Hamzah & Ismail, 2003). Other homestay programmes considered as successful by the participants in this study are Miso Walai Homestay in Sabah, Teluk Ketapang Homestay in Terengganu, Sungai Haji Dorani Homestay in Selangor, Kampung Parit Bugis Homestay in Johor and Santubong Homestay in Sarawak.

The analysis on the feedbacks indicates that most of the experts interviewed in this study associate success of a homestay programme with the following criteria:

1. *The number of visitors they receive and to its ability to pull tourists.* The pull factors mentioned by the interviewees are comfortable and clean accommodation facilities, hospitable hosts, creative packaging/programming, and the uniqueness of activities, local resources and attractions included in the homestay package.
2. *The effectiveness of the management especially with regards to the operations of the homestay programmes.* The factors that lead to management effectiveness hence the success of the homestay programmes are strong support system, leadership, creativity, promotional strategies (including through media coverage) and participation by the villagers where the programmes operate.
3. *Benefits produced by the programmes to the villagers.* These include contribution to their income, enhancement of their entrepreneurial skill and ability, contribution to their socio-economic development and strengthening ties among members of the communities surrounding the programmes. Only one of the participants chose to name a successful homestay programme based on the programme's contribution to the conservation of local culture.

The findings indicate that the success of a homestay programme perceived to be based on its commercial success. Thus, the criteria of a successful homestay programmes perceived by the participants is linked to the amount of revenue generated by the programmes. However, revenue generation from a CBT is only one of the criteria that qualify CBT as a tourism alternative that aims for sustainable development. As mentioned earlier, a CBT that focuses on sustainable development must also be supported by resident participation and must also stress on the protection of local people's cultural identity and natural environment (Russell, 2000).

The feedbacks given by the participants further indicate that they indeed equate 'success of the homestay programmes' with 'the critical factors that lead to the success of the homestay programmes' as the feedbacks from questions b and c are similar. The critical factors are perceived by the interviewees to be the factors that lead to the success of the Homestay programmes i.e. the

amount of revenue generated by the programmes. Furthermore, they perceive that the success of the homestay programmes relate more to the profit sustainability of the programme than to the conservation of the resources in the programme. There is an indication that the survey participants are more concerned with the success of the homestay programmes as business entities than their success as an alternative to sustainable development. However, a successful Homestay program that attracts high number of visitors and that create much profit does not always focus on sustainable development. It can be proposed here that the perceptions that the survey participants have with regard to success is separated from their perceptions toward sustainable development.

4 CONCLUSION

In recent years, increasing attention has been paid to the effects of tourism and related developments upon the culture and environment resources of destination areas. The increasing popularity of the concept of sustainable development has resulted in tourism being viewed as a means to achieve sustainable development (Butler, 1991). Findings from this exploratory study concludes that the view is too simplistic and naïve. A CBT product such as the CBRH programme is quick to be claimed as a responsible and sustainable development alternative as it portrays the saint-like elements such as 'grassroot project' (Ghai & Vivian, 2014) and 'people-centered development' (Sharpley, 2009). However, the present study concludes that the success factors of these CBTs perceived by the participants who are supposed to understand the operations of the CBRH programme do not mirror the criteria of sustainable development.

REFERENCES

Butler, R.W. (1991). Tourism, environment, and sustainable development. *Environmental conservation, 18*(03), 201–209.
Dixey, L. (2008). The unsustainability of community tourism donor projects: Lessons from Zambia. *Responsible tourism: Critical issues for conservation and development*, 323–341.
Fayissa, B., Nsiah, C. & Tadasse, B. (2008). Impact of tourism on economic growth and development in Africa. *Tourism Economics, 14*(4), 807–818.
Ghai, D. & Vivian, J.M. (2014). *Grassroots environmental action: people's participation in sustainable development*: Routledge.
Goodwin, H. (2007). *Responsible tourism in destinations*. London, UK: CMRB.
Goodwin, H. & Santilli, R. (2009). Community-based tourism: A success. *ICRT Occasional Paper, 11*(1), 37.

Manyara, G. & Jones, E. (2007). Community-based tourism enterprises development in Kenya: An exploration of their potential as avenues of poverty reduction. *Journal of Sustainable Tourism, 15*(6), 628–644.

Mitchell, J. & Muckosy, P. (2008). *A misguided quest: Community-based tourism in Latin America*. London, England: Overseas development institute (ODI).

Mundt, J.W. (2011). *Tourism and sustainable development: Reconsidering a concept of vague policies*: Erich Schmidt Verlag.

Novelli, M. & Gebhardt, K. (2007). Community based tourism in Namibia:'Reality show'or 'window dressing'? *Current Issues in Tourism, 10*(5), 443–479.

Russell, P. (2000). Community-based tourism. *Travel & Tourism Analyst*(5), 89–116.

Samimi, A.J., Sadeghi, S. & Sadeghi, S. (2011). Tourism and economic growth in developing countries: P-VAR approach. *Middle-East Journal of Scientific Research, 10*(1), 28–32.

Scheyvens, R. (1999). Ecotourism and the empowerment of local communities. *Tourism Management, 20*(2), 245–249.

Sharpley, R. (2009). *Tourism development and the environment: Beyond sustainability?* Sterling VA: Earthscan.

Simpson, M.C. (2008). Community benefit tourism initiatives—A conceptual oxymoron? *Tourism Management, 29*(1), 1–18.

Tasci, A.D., Gurbuz, A.K. & Gartner, W.C. (2006). Segmented (differential or discriminatory) pricing and its consequences. *Progress in tourism marketing*, 171.

Theory and Practice in Hospitality and Tourism Research – Radzi et al. (Eds)
© 2015 Taylor & Francis Group, London, ISBN 978-1-138-02706-0

Image formation, tourism images and tourists visit intention: Developing constructs for empirical investigation

Z. Saleki, M.S.M. Zahari & S.A. Talib
Faculty of Hotel and Tourism Management, Universiti Teknologi MARA, Shah Alam, Malaysia

ABSTRACT: Image is a dynamic concept that may be altered or changed due to earlier experiences or when the individual is exposed to various sources. Induced (promotional materials), organic (word of mouth) and autonomous sources of information (news articles, educational materials) are the image formation factors which perceived by individual before experiencing a destination. Perceiver characteristics (socio—demographic) who normally filter the information could also contribute in shaping the image. These constructs are considered as secondary image that create individual intent than that primary image which is formed by individual actual visitation. This paper reviews the pre—image formation factors, tourism images and tourists visit intention constructs subsequently developing conceptual study framework showing the logical relationship between them.

Keywords: Image formation, tourism images, tourist visit intention

1 INTRODUCTION

Recognizing travelers' behavioral intentions and revisit intention is a vital objective of both host destinations and destination marketing organizations. In general terms, behavioral intention as an immediate antecedent of behavior and to direct behavior in an organized and thoughtful manner (Fishbein & Ajzen, 1975) and behavioral intentions means an individual's predicted or planned future behavior (Swan & Trawick, 1981). Behavioral intentions in the tourism industry context denote a potential traveler's expectancy of their future trip to the destination of interest (Chen & Tsai, 2007).

As a general rule, the stronger the intention to involve in a behavior, the more to be expected the behavior will be completed. In this sense, when an opportunity to act arises, the intention will lead to actual behavior, thus if the intention is measured correctly, it would provide a noble predictor of behavior (Fishbein & Ajzen, 1975).

(Martineau, 1958) construed that human behavior also is relying on image than objective reality and following that many scholars in different fields have been attracted to the concept of images. The adoption of the concept of the image has directed to "image theory" which suggests that the world is a psychological or inaccurate picture of objective reality existing in the mind of the individuals (Myers, 1968). When initial credibility varies from the public's perception of merchandise, the perception of the image will probably define that

product's success or failure. Images therefore can give a pre-test for the travelers and transposed picture of an area into the mind of tourists. Therefore, it has become one of the most popular topics in the tourism literatures (Pike, 2002).

A positive correlation between image and intention was found in numerous studies and the findings were consistent with consumer and tourist behavior. Country image has been investigated largely from various viewpoints including the international marketing and tourism. The tourism literatures have mainly focused on the country image as a tourism destination image which centrally relate to loyalty, satisfaction and perceptions of service quality (Bigne, Sanchez & Sanchez, 2001). Though, these two constructs (country image and tourism destination image) have been previously examined under one construct, scholars lately have called for the need to explore the impact of tourism activities on both country and destination image as two different related constructs (Martínez & Alvarez, 2010). Scholars even believed that hotel image should be considered as third and independent construct together with country image and destination image when it comes to the exploration of tourism activity impacts in a particular country.

Image in actual fact is a dynamic concept that may change due to earlier experiences or when the individual is exposed to various sources of information (Dann, 1996). The image shaped by induced, organic, and autonomous sources of information that essentially perceived before experiencing a destination, which Phelps (1986) named them as the secondary image than that primary image which is formed by actual visitation. Induced agents which are controllable by marketers are the promotional materials. Organic agents are word of mouth from friends and relatives based on their own knowledge or experience about the places which are uncontrollable by the destination marketers (Gartner, 1993). Autonomous agents on the other hand, consist of things such as news articles, educational materials, movies and popular culture which are independent and also uncontrollable by marketers. Besides induced, organic and autonomous agents and perceiver characteristics who normal filter the information could also contribute to shaping the image which cannot be controlled by marketers (Dann, 1996; Gartner, 1993).

Several studies have been conducted in determining how these sources of information impact the image formation process and scholars have distinguished between those sources that initiate from marketing activities and which are managed, and those that cannot be controlled for such as news, movies and other media (Beerli & Martin, 2004; Gartner, 1993; McCartney, Butler & Bennett, 2008) However, there is insufficient and lack of research looking at the effect of uncontrollable sources of information on the image of a country as well as hotel image. Moreover, latest studies explored the difference between country image, destination image and hotel image (Martínez & Alvarez, 2010) shows that there is a need to separate third concepts when analyzing the effect of communication strategies on the image of a country, destination and hotel. It is therefore, important to establish and understand the extent to which information sources which are not under control of marketers and cannot be managed have a varied influence on the general country image, destination image and hotel image, tourists visit and revisit intention to a particular country.

2 LITERATURE REVIEW

2.1 Pre-image formation factors

The foundation for the image kinds was established by Gunn (1972) and expanded by Gartner (1993) which also named these stimuli as image formation agents. Gartner (1993) denoted that image is a dynamic concept that may be altered or change due to earlier experiences or when the individual is exposed to various sources of information (Dann, 1996). To address the image, determinants need to reflect on the organic (word of mouth, past visit) induced (marketing tools) and autonomous image formation agents (news, documentaries) as

well as the outcome of it which is shaped the complex images (Gartner, 1993). The image shaped by induced, organic, and autonomous sources of information are essentially what have been perceived before experiencing a destination, which Phelps (1986) named them as the secondary image than that primary image which is formed by actual visitation. In addition to preceding determinants, perceiver characteristics such as socio—demographic and culture are another source of destination image determinants. It is purported to be the consumers (perceivers) who filter the information from information sources and form images about travel destinations (Dann, 1996; Gartner, 1993). With that, the subsequent sub-section reviews all the determinants.

2.1.1 Induced

Induced agents or promotional materials which Gartner (1993) classified as (a) overt induced and found in conventional advertising in the mass media, from information delivered by the relevant institutions in the destination or by tour operators and wholesalers; and (b) covert induced is using celebrities in the destination's promotional activities or destination reports or articles. Most of the conducted studies well recognized the relationship between supply side (induced) image formation agents with the destination image (Gartner, 1993; Iwashita, 2003). The image hold by visitors is differing from non-visitors also on some dimensions the expectation of non-visitors exceed from the actual performance mention by visitors. As a result, there must be some disconnection between what the destination marketers presents in their marketing and promotional efforts and the genuine delivery of products and services.

2.1.2 Organic

Organic agents are word of mouth (WOM) from friends and relatives based on their own knowledge or experience about the places whether the information was requested or volunteered and a visit to the destination which both are not controlled by destination marketers (Gartner, 1993). Literatures indicated that receiving WOM has an impact on the receiver's awareness, attention, consideration, brand attitudes, intentions and expectation (Laczniak, DeCarlo & Ramaswami, 2001; Mikkelsen, Van Durme & Carrie, 2003). The significance of WOM on businesses have been broadly argued and examined particularly since the worldwide implementation of Internet technologies, which have revolutionized the spreading and influenced by WOM (Stokes & Lomax, 2002). Online user-generated reviews about travel destinations, hotels, and tourism services have become important sources of information for travelers and positive online reviews advance the perception of hotels among the potential travelers (Vermeulen & Seegers, 2009).

2.1.3 *Autonomous*

The name "autonomous agents" given by Gartner (1993) consist of things such as news articles, educational materials, movies and popular culture. Autonomous agents are assumed to have more influence on image formation since they have higher authority and capability to reach mass crowds than the destination-originated information (Gartner, 1993; Hanefors & Mossberg, 2002). Autonomous agents generate a general understanding about a destination, and are independent and out of a destination's direct control. Image can change in a short period of time when the information is shown to large audiences through the media. Also in the case of a distant country which lack of knowledge exists, the autonomous agents are able to cause a more effective change in image due to their high credibility and ability to reach mass crowds.

2.1.4 *Perceivers characteristics*

Another source of destination image determinants is the consumers (perceivers) who filter the information from these sources and form images about travel destinations (Dann, 1996; Gartner, 1993). Perceivers' socio-demographics are assumed to play a role in this image formation process, and, therefore, have been investigated in terms of its relationships with the destination image. Many scholars have argued and examined the influence of perceivers' socio-demographic characteristics including age, gender, household status, education, income, residence/geographic distance on destination image (Baloglu, 2001; Rittichainuwat, Qu & Brown, 2001). A few academics have investigated the impact of respondents 'residence or distance from the study destination (Hunt, 1975; Walmsley & Young, 1998).

2.2 *Tourism images*

Kotler, Haider, and Rein (1997) state that the image of a place refers to "the sum of beliefs and impressions that people hold about places. The images represent a simplification of the large number of relations and information linked to the place. They are a product of the mind trying to evaluate and pick out critical information from massive amounts of data about a place. Thus, the image is seen as a multidimensional impression that includes an individual's valuation of different aspects of a place (Baloglu, 1997; Gallarza, Saura & Garcióa, 2002).

Tourism literatures have mainly focused on the country image as a tourism destination image

which centrally relate to loyalty, satisfaction and perceptions of service quality (Bigne, et al., 2001). The country image and tourism destination image have been previously examined under one dimension. However, scholars lately have called for the need to explore the impact of tourism activities in countries, destination image and hotel image as different related dimension (Martínez & Alvarez, 2010).

2.3 *Travel constraints*

In general terms, constraints are the factors which prohibit people from participating and enjoying leisure offered three groups of constraint; structural constraints (e.g., family life-cycle stage, financial resources, season, climate, available time such as work schedule, availability of opportunity, etc.), intrapersonal constraints (individual psychological states such as stress, depression, anxiety, perceived self-skill, etc.) and interpersonal constraints (constraints resulting from interrelationships such as finding a suitable traveling partner) (Jackson, 1988). Earlier studies discovered that "critical constraints" play a significant role in the travel decision-making procedure. Critical constraints range from social, political, physical, financial, time, health, family stage, lack of interest, fear and safety, transportation, lack of partners, overcrowding, distance, limited information (e.g., images), to previous visitation (Jackson, 1988; Lee & Tideswell, 2005). The most studied constraints are time, cost, and space (e.g., destination attributes such as distance) and are considered to be the most critical constraints in leisure and tourism research (Hong, Fan, Palmer & Bhargava, 2005; White, 2008; Wong & Yeh, 2009).

Leisure constraints may also be considered in two categories; real and perceived. Real constraints are those physical, financial, and practical related constraints such as sickness, costs, and time. Perceived constraints categorized as internal constraints (e.g., knowledge about destination, anxiety, stress, or personal health) and external constraints (e.g., work, time availability, or family obligation) (Jackson, 1988). Literature also has connected numerous external or situational factors or constraints on travel behavior, affecting pre-trip decisions including social, political, physical, financial, time, and distance (Gartner, 1989, 1993; Hunt, 1975).

2.4 *Behavioral intention*

As in the theory of Planned Behavior, intentions are expected to capture the motivational factors that impact a behavior and they arc pointers of how much of an effort individuals are planning to apply or how hard they are willing to try in order

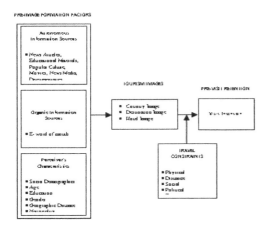

Figure 1. Conceptual framework.

to achieve a behavior (Ajzen, 1991). Behavioral intention even can be defined as revisit intention and recommend to others as well (Bigne, et al., 2001). The Theory of Planned Behavior (TPB) was widely used in many fields and disciplines and start getting attention among the hospitality and tourism scholars and many of them were looking travelers revisit intention and recommendation behavior (Baker & Crompton, 2000; Castro, Martín Armario & Martín Ruiz, 2007).

3 CONCEPTUAL FRAMEWORK

Based on the literatures and related issue highlighted, the conceptual framework is proposed as in Figure 1 which is still under investigation.

The framework decomposed to facilitate first-order investigation of the relationships between the underlying dimensions of the pre-image formation and tourism images. The second order constructs are to examine the relation between tourism images and behavior intention while the third order construct is to test the moderating effect of travel constraint attributes and behavior intention. The last or four order construct is to observe the relationship between pre-image formation factors and behavior intention.

4 CONCLUSION

Recognizing the relationship between pre—image formation factors, tourism images and tourists visit intention will help industry practitioners to understand which image should be promoted to which market segments and help them to identify their target market. In addition, understanding the

effect of different sources of information on tourism images (country, hotel and destination) will also aid the hoteliers, travel agencies and government organization in attracting either local and international tourists even without through mainstream media coverage.

REFERENCES

Ajzen, I. (1991). The theory of planned behavior. *Organizational behavior and human decision processes, 50*(2), 179–211.

Baker, D.A. & Crompton, J.L. (2000). Quality, satisfaction and behavioral intentions. *Annals of tourism research, 27*(3), 785–804.

Baloglu, S. (1997). The relationship between destination images and sociodemographic and trip characteristics of international travellers. *Journal of Vacation Marketing, 3*(3), 221–233.

Baloglu, S. (2001). Image variations of Turkey by familiarity index: Informational and experiential dimensions. *Tourism Management, 22*(2), 127–133.

Beerli, A. & Martin, J.D. (2004). Factors influencing destination image. *Annals of tourism research, 31*(3), 657–681.

Bigne, J.E., Sanchez, M.I. & Sanchez, J. (2001). Tourism image, evaluation variables and after purchase behaviour: Inter-relationship. *Tourism Management, 22*(6), 607–616.

Castro, C.B., Martín Armario, E. & Martín Ruiz, D. (2007). The influence of market heterogeneity on the relationship between a destination's image and tourists' future behaviour. *Tourism Management, 28*(1), 175–187.

Chen, C.-F. & Tsai, D. (2007). How destination image and evaluative factors affect behavioral intentions? *Tourism Management, 28*(4), 1115–1122.

Dann, G.M.S. (1996). Tourists' images of a destination-an alternative analysis. *Journal of Travel & Tourism Marketing, 5*(1/2), 41–55.

Fishbein, M. & Ajzen, I. (1975). *Belief, attitude, intention and behavior: An introduction to theory and research.* Boston, MA: Addison-Wesley.

Gallarza, M.G., Saura, I.G. & Garcióa, H.C. (2002). Destination image: Towards a conceptual framework. *Annals of tourism research, 29*(1), 56–78.

Gartner, W.C. (1989). Tourism image: attribute measurement of state tourism products using multidimensional scaling techniques. *Journal of Travel Research, 28*(2), 16–20.

Gartner, W.C. (1993). Image formation process. In M. Ulysal & D.R. Fesenmaier (Eds.), *Communication and Channel Systems in Tourism Marketing, edited by* (pp. 191–215). New York: Haworth Press.

Gunn, C. (1972). Vacationscape. *Bureau of Business Research, University of Texas, Austin, TX.*

Hanefors, M. & Mossberg, L. (2002). TV travel shows: A pre-taste of the destination. *Journal of Vacation Marketing, 8*(3), 235–246.

Hong, G.-S., Fan, J.X., Palmer, L. & Bhargava, V. (2005). Leisure travel expenditure patterns by family life cycle stages. *Journal of Travel & Tourism Marketing, 18*(2), 15–30.

Hunt, J.D. (1975). Image as a factor in tourism development. *Journal of Travel Research, 13*(3), 1–7.

Iwashita, C. (2003). Media construction of Britain as a destination for Japanese tourists: social constructionism and tourism. *Tourism and Hospitality Research, 4*(4), 331–340.

Jackson, E.L. (1988). Leisure constraints: A survey of past research. *Leisure sciences, 10*(3), 203–215.

Kotler, P., Haider, D.H. & Rein, I. (1997). *Marketing places: Attracting investment, industry and tourism to cities, states and nations.* Nova Iorque: Free Press.

Laczniak, R.N., DeCarlo, T.E. & Ramaswami, S.N. (2001). Consumers' responses to negative word-of-mouth communication: An attribution theory perspective. *Journal of Consumer Psychology, 11*(1), 57–73.

Lee, S.H. & Tideswell, C. (2005). Understanding attitudes towards leisure travel and the constraints faced by senior Koreans. *Journal of Vacation Marketing, 11*(3), 249–263.

Martineau, P. (1958). The personality of the retail store. *Journal of Retailing 52*, 37–46.

Martínez, S.C. & Alvarez, M.D. (2010). Country versus destination image in a developing country. *Journal of Travel & Tourism Marketing, 27*(7), 748–764.

McCartney, G., Butler, R. & Bennett, M. (2008). A strategic use of the communication mix in the destination image-formation process. *Journal of Travel Research, 47*(2), 183–196.

Mikkelsen, M., Van Durme, J. & Carrie, D. (2003). *Viewers talking about television advertising: A supplimentary measure of advertising effectiveness.* Paper presented at the 32nd EMAC, Glasgow, May, 20–23.

Myers, J.G. (1968). *Consumer image and attitude.* Berkeley: Institute of Business and Economic Research, University of California.

Phelps, A. (1986). Holiday destination image: The problem of assessment: An example developed in Menorca. *Tourism Management, 7*(3), 168–180.

Pike, S. (2002). Destination image analysis: A review of 142 papers from 1973 to 2000. *Tourism Management, 23*(5), 541–549.

Rittichainuwat, B., Qu, H. & Brown, T. (2001). Thailand's international travel image: mostly favorable. *The Cornell Hotel and Restaurant Administration Quarterly, 42*(2), 82–95.

Stokes, D. & Lomax, W. (2002). Taking control of word of mouth marketing: the case of an entrepreneurial hotelier. *Journal of Small Business and Enterprise Development, 9*(4), 349–357.

Swan, J.E. & Trawick, I.F. (1981). Disconfirmation of expectations and satisfaction with a retail service. *Journal of Retailing, 57*(3), 49–67.

Vermeulen, I.E. & Seegers, D. (2009). Tried and tested: The impact of online hotel reviews on consumer consideration. *Tourism Management, 30*(1), 123–127.

Walmsley, D.J. & Young, M. (1998). Evaluative images and tourism: The use of personal constructs to describe the structure of destination images. *Journal of Travel Research, 36*(3), 65–69.

White, D.D. (2008). A structural model of leisure constraints negotiation in outdoor recreation. *Leisure sciences, 30*(4), 342–359.

Wong, J.-Y. & Yeh, C. (2009). Tourist hesitation in destination decision making. *Annals of tourism research, 36*(1), 6–23.

Theory and Practice in Hospitality and Tourism Research – Radzi et al. (Eds)
© *2015 Taylor & Francis Group, London, ISBN 978-1-138-02706-0*

Economic values of Ulu Chepor Recreational Park, Perak: A travel cost approach

M.A.R. Abdul-Samad
Faculty of Economics and Management, Universiti Putra Malaysia, Malaysia

S. Naim
School of Business and Economics, Universiti Malaysia Sabah, Malaysia

ABSTRACT: The estimation of economic values of Ulu Chepor Recreational Park (UCRP) is needed because no study has been done so far in valuing this area. By applying the travel cost model, the travel information is then used for the estimation of the recreational values in UCRP. The recreational value of Willingness To Pay (WTP) per person derived is estimated at RM2.30 per visit. The total recreational net benefit is estimated at RM33,120 for the year 2012. This implies that the significant economic values of eco-tourism will be lost from any large scale development by degrading natural environment. On top of that, the management of UCRP could generate more income and use it for maintenance purposes. For the time being, visitors will be charged RM2 per car per entry. In fact, there is no charge for visitors that come by motorcycle. Therefore, this finding is useful for them and can be used as a basis to change the current practice for the future benefit of Ulu Chepor Recreational Park.

Keywords: Recreational values, eco-tourism, willingness to pay, economic valuation

1 INTRODUCTION

Ulu Chepor Recreational Park (UCRP) is located in Chemor, Perak, Malaysia approximately about 10 km from the city of Ipoh. This area is on the outskirts of Ipoh City and it is the upstream areas being developed into tourist centres. This is evident when UCRP area is included in Perak tourism map as one of the tourist attractions in Perak. UCRP has been commercialized and maintained by the Ipoh City Council. On weekends, visitors arriving by car are required to pay RM2 per car and RM 0.50 per motorcycle to the management of Village Development and Security Committee (*JKKK*). There is no fee charged on weekdays. There are various facilities available in the UCRP such as toilets, prayer room, food stalls and small huts for shelter visitors. The path to this area is also in good condition.

UCRP is one of the popular places for picnic and other recreational activities. Besides the locals, this area is also becoming more popular among foreign tourists. Conditions of clean river water, waterfalls and the tranquil ambience of UCRP make it suitable for family recreation activities.

According to Smith (1993), travel cost method (TCM) is an indirect market based approach and claimed to be the oldest of the non-market valuation techniques. TCM is predominantly used in outdoor recreation modeling, with fishing, hunting, boating and forest visits among the most popular applications.

The objectives of this study are to i) examine the determinant of tourist arrival at UCRP, ii) to analyze the pricing policy made by the UCRP, and iii) to investigate the consumer surplus value per visit. This paper is divided into five sections; where section one is the introduction, followed by section two which describes the location of the study. Section three the methodology and source of data used in study. Empirical results are presented in section four while the last section offers some discussion and concluding with regard to consumers' willingness to pay assessment of the net benefit of recreational resource in UCRP.

2 METHODOLOGY

In this study, the data were gathered through personally administered questionnaire. The primary data for the study were gathered directly from respondents in a face-to-face setting. This approach is deemed appropriate as other methods, such as mail or telephone interviews, could not allow the interviewers to explain the actual issue in detail to the respondents. This approach was anticipated to obtain more accurate and complete responses and the information given was meaningful and able to produce high quality data.

A total of 250 respondents from diverse demographic and socio-economic backgrounds were surveyed. Data were collected at UCRP. Random sampling of the visitors was done throughout the day to capture variations in the origin of visitors at different times of the day. The survey instrument was pre-tested and question wording was refined based on the results of a pilot test. The respondent's socio-economic characteristics obtained include state of origin, age, gender, race, marital status, education, occupation and income.

2.1 The travel cost method

This study adopts the TCM because it is consistent with the consumer demand theory. TCM is the practical method that estimates actual consumption of the environment, rather than a hypothetical one.

Three functional forms were used to estimate the economic model of UCRP visitor demand, which are the linear model (LM), semi-log (SL) and double-log (DL). SL and DL models were transformed by taking the natural logarithm of both the dependent and continuous independent variables. Advantages of using the semi-log functional form include minimizing the problem of heteroscedasticity, as well as eliminating the potential problem of negative trip prediction, which may happen in linear functional form (Loomis & Cooper, 1990). The econometric models for the three models are defined as follows:

Linear Model (LM)

$$V_i = \beta_0 + \beta_1 TC_i + \beta_2 TimeC_i$$
$$+ \beta_3 INC_i + \beta_4 AGE_i + \varepsilon \qquad (1)$$

Semi-Log Model (SL)

$$\ln V_i = \beta_0 + \beta_1 TC_i + \beta_2 TimeC_i$$
$$+ \beta_3 INC_i + \beta_4 AGE_i + \varepsilon \qquad (2)$$

Double-Log Model (DL)

$$\ln V_i = \beta_0 + \beta_1 \ln TC_i + \beta_2 \ln TimeC_i$$
$$+ \beta_3 \ln INC_i + \beta_4 \ln AGE_i + \varepsilon \qquad (3)$$

Where V_i is the number of visit for the past three year of the i-th visitor, TC_i is the round-trip travel cost of the i-th visitor, $TimeC_i$ is the opportunity cost for the i-th visitor, INC_i is the monthly household income for the i-th visitor, AGE_i is the age in years of the i-th visitor. Travel cost consisted of travelling cost and time cost.

Table 1. Visitors' profile ($N = 250$).

Visitors' profile	Frequency	Percentage (%)
Age		
Below 21	10	4.0
22 – 25	43	17.2
26 – 30	49	19.6
31 – 35	43	17.2
36 – 40	52	20.8
41 – 45	24	9.6
46 – 50	20	8.0
Above 51	9	3.6
Gender		
Male	161	64.4
Female	89	35.6
Race		
Malay	235	94.0
Chinese	6	2.4
Indian	9	3.6
Marital status		
Single	68	27.2
Married	182	72.8
Education level		
Primary school	2	0.8
Secondary school	68	27.2
College/Institute	92	36.8
University	88	35.2
Occupation		
Government sector	69	27.6
Private sector	102	41.2
Self-employed	28	11.2
Student	27	10.8
Housewife	21	8.4
Others	2	0.8
Monthly household income		
Below RM1,001	5	2.0
RM1,001 - RM2,000	88	35.2
RM2,001 - RM3,000	67	26.8
RM3,001 - RM4,000	31	12.4
RM4,001 - RM5,000	10	4.0
Others	49	19.6

Source: Survey (2012).

3 RESULT AND DISCUSSION

3.1 Socio-economic profile

This section presents the results of the descriptive analysis of visitors' socio-economic variables such as age, gender, race, marital status, educational level, occupation and monthly household income. Table 1 depicts the summary of the visitors' profile. Most of the visitors (20.8%) were between 36 to 40 years old, followed by the age group 26 to 30 years old (19.6%).

By gender, the majority of visitors (64.4%) were male. This finding is in line with Chin Moore, Wallington, and Dowling (2000) that the visitors to Bako national Park were aged between 16 and 40 (76%). This is because the younger generation has

Table 2. Characteristics of visits.

Definitions and coding	Percentage (%)
State of origin	
Perak	92.4
Kedah	0.8
Pulau Pinang	0.8
Selangor	4.4
Negeri Sembilan	0.4
Kelantan	1.2
Will visit in future	
Will visit again	100
Will not visit again	0

Source: Survey (2012)

Table 3. Estimated travel cost method (RM) for Ulu Chepor Recreational Park.

	LM	SL	DL
Constant	15.679	19.694	19.024
TC	-0.555***	-0.890***	-0.612***
	(0.000)	(0.000)	(0.000)
TimeC	0.111	0.021	0.113
	(0.191)	(0.947)	(0.278)
INC	-0.132	-0.103	-0.239**
	(0.199)	(0.463)	(0.028)
AGE	-0.108	-0.063	-0.436**
	(0.513)	(0.781)	(0.020)

***Significant at 1% level, **Significant at 5% level, *Significant at 10% level.

Table 4. Comparison of selected functional forms.

	LM[a]	SL[b]	DL[c]
R^2	0.371	0.487	0.501
Adj. R^2	0.329	0.455	0.469
F-value	8.883	14.836	15.661

Note: [a,b,c]Selected functional forms, refer to Eq. 1, 2 and 3.

Table 5. WTP and expected net benefit (RM) for Ulu Chepor Recreational Park.

Variable	WTP per visit	Expected Total Benefit
Value	RM2.30	RM33,120

a greater intention and desire to venture of visiting the tropical rain forests and recreational parks. The visitors interviewed were 94 percent Malay, 2.4 percent Chinese and 3.6 percent Indian. In terms of marital status, 27.2 percent of the visitors were single and 72.8 percent were married. It is worth mentioning that the conditions of clean river, waterfalls and huge rocks in UCRP contribute to the fact that this location makes a perfect place for family recreation activity. Furthermore, most parents want their children to enjoy the natural heritage such as the scenic beauty of national park and want to take part in conservation of the rain forest for future generation. In terms of education level, most visitors (36.8%) have gone through college or institute, followed by 35.2 percent university and 27.2 percent secondary school.

Table 2 outlines the characteristics of UCRP visits. Approximately 38.7 percent of the visitors were first time visitors, while 61.63 percent were repeat visitors. In terms of the state of origin, majority of visitors were from Perak, which contributed, 92.4 percent while the rest (7.6%) were from other states in Malaysia such as Selangor, Kelantan, Kedah, Pulau Pinang and Negeri Sembilan. The results show that 100 percent of visitors stated that they would visit the UCRP again in future.

This shows that there are elements in the UCRP that attracts and make them intend to visit again in the future. The main reasons for the visitors to return among others are ideal picnic and relaxation spot, green environment surrounded by green forest that makes this place more serene. This finding is very important for the ecotourism operators and ecotourism planners in marketing their products and developing marketing strategies.

3.2 Multiple-regression results

All OLS regressions were conducted using Statistical Package for Social Sciences (SPSS) version 20.

The estimated results of travel cost (TC) method for UCRP are shown in Table 3. All of the estimated coefficients for TC were highly significant at 1 percent level under the three functional forms namely LM, SL and DL models.

The elasticity of demand can be estimated with respect to expenditure (travel cost). In a DL model, the coefficient of the travel cost estimated can be considered as elasticity. It explains that for a 10 percent increase in expenditure, the number of visit decreases by 6.12 percent which means that it is inelastic.

The goodness of fit model is shown in Table 4. In term of R^2 value amongst the three functional forms, DL model (50.1%) outperforms the SL (48.7%) and LM (37.1%). This is supported by the F-value where DL model shows the highest value (15.661).

3.3 Willingness To Pay (WTP) and expected net benefit

Based on the estimation results, equivalent WTP were calculated using regression analysis at expenditure and level of visit (see Table 5). The calculated mean of WTP ranged from RM2.30 per visit, and expected net benefit of RM33,120. Calculating the area below the cumulative distribution function yields the mean WTP estimate for each respondent.

From these values of consumers' surplus or the willingness to pay to UCRP, one can compute the additional net benefit of the park for the estimated year by the number of visitors to this park.

4 CONCLUSION

The recreational value of willingness to pay (WTP) per person derived is estimated at RM2.30 per person. The total recreational net benefit is estimated at RM33,120 for the year 2012. We could conclude that the significant economic values of eco-tourism will be lost from any large scale development by degrading natural environment. On top of that, the management of UCRP could generate more income and use it for maintenance purposes. It is because in the current practice, they just charge RM2 per car per entry. In fact, there is no charge for a visitor that comes by motorcycle. Therefore, this finding is useful for them and can be used as a basis to change the current practice to the best management practice.

REFERENCES

Chin, C.L., Moore, S.A., Wallington, T.J. & Dowling, R.K. (2000). Ecotourism in Bako National Park, Borneo: Visitors' perspectives on environmental impacts and their management. *Journal of Sustainable Tourism, 8*(1), 20–35.

Loomis, J. & Cooper, J. (1990). Comparison of environmental quality-induced demand shifts using time-series and cross-section data. *Western Journal of Agricultural Economics, 15*(1), 83–90.

Smith, V.K. (1993). Nonmarket valuation of environmental resources: An interpretive appraisal. *Land Economics, 69*(1), 1–26.

Theory and Practice in Hospitality and Tourism Research – Radzi et al. (Eds)
© *2015 Taylor & Francis Group, London, ISBN 978-1-138-02706-0*

Homestay accommodation for tourism development in Kelantan, Malaysia

M.H. Bhuiyan
Institute for Environment and Development (LESTARI), Universiti Kebangsaan Malaysia, Malaysia

A. Aman
Center for Enterpreneurship and SMEs Development, Universiti Kebangsaan Malaysia, Malaysia
Center for Advancement of Social Business (CASB), Universiti Kebangsaan Malaysia, Malaysia

C. Siwar & S.M. Ismail
Institute for Environment and Development (LESTARI), Universiti Kebangsaan Malaysia, Malaysia

M.F. Mohd-Jani
Center for Enterpreneurship and SMEs Development, Universiti Kebangsaan Malaysia, Malaysia

ABSTRACT: Homestay accommodation can create a scope to the local communities for active participation in tourism activities. The Malaysian homestay program has announced officially as a tourism product in 1995. This accommodation provides multi ethnic life condition with cultural experiences to tourists and economic well beings for the local people. Kelantan state is full of natural resources, unspools beauties, beaches, recreational forests, waterfalls and cultural attractions. Homestay accommodation may be a potential tourism and economic activities in Kelantan. The present study analyzes the potentialities and examines the overall situations of homestay accommodation for tourism development in Kelantan. The study used secondary data like documents, acts, regulations and policies which were collected from reliable sources. The homestays of eight villages in Kelantan are officially registered by Ministry of Tourism. The participation of local communities in homestay accommodations are increasing year by year. Kelantan is showing backward position in guest arrivals and incomes compare to other states. Kelantan has suitable conditions to develop homestay accommodation such as low cost, hospitality, entrepreneurship development, small scale investment, treasure of tourism resources and cultural and social benefits. The study recommended several requirements such as easy access, basic facilities, hygiene, safety and security, proper management and promotion and lucrative packages for homestay development in Kelantan. Finally, proper planning, regulations, financial allocation and managerial efficiency will be ensuring homestay development in this state as well as well beings for local communities.

Keywords: Homestay, tourism, development, Kelantan, Malaysia

1 INTRODUCTION

Homestay is contributing in tourism development in a country or an area. Tourism segmentations such as rural tourism, ecotourism and cultural tourism are facing accommodation problems at near the tourism destination. Homestay can solve the accommodation problem for this type's tourism. Local communities have active participation in rural and cultural tourism destinations. Moreover, eco-tourism is ensuring strong community participation in tourism activities with decrease the environmental degradation. Homestay accommodation can create a scope to the local communities for

active participation in tourism activities. It is developing relationship between local government and communities to understand and adjust knowledge regarding tourism activities (Saeng-Ngam, Chantachon & Ritthidet, 2009). The homestay owners have improved and developed their living standard and economic condition through various programs of government. Again, they are working with the village's Development and Security Committee for the rural development. It can attract potential visitors, who can gather their knowledge and experience by their visits (Din & Mapjabil, 2010).

The Malaysian homestay program has announced officially as a tourism product in 1995.

Homestay can be highlighted the rural lifestyle and culture of local communities. It has recognized a potential tourism product in the rural tourism master plan in 2001 (UNDP, 2003). In Malaysia, homestay accommodation mainly operates and organized by the Kampung (village) people. The homestay owners are keeping Malay culture and activities in their accommodations. This accommodation provides multi ethnic life condition with cultural experiences to tourists and economic well beings for the local people (Liu, 2006). According to the Malaysian Homestay Association (MHA), students and foreign visitors from Japan, Australia and Korea is the main client for homestay accommodation (Kayat, 2007). The Malaysian homestay program differs from the other commercial home stay in the world. Here guests are living with the family members of homestay owner during their stay period. The operators' involves with the guests eating, cooking and other activities which exchange their culture and can learn from each other (Peterson, 2004). The homestay programs in Malaysia get huge support from government. The government has allocated required capital to the homestay operator for capacity building and empowerment. This allocation is largely provided to the lower socioeconomic groups (Din & Mapjabil, 2010). Bhuiyan, Siwar and Mohamad Ismail (2013) identified through a study on Terengganu that homestay operations have positive socio-economic impact on local communities. The operators have earned a major portion of their monthly expenses from the homestay accommodations. Furthermore, home stay is potential business operation for the local entrepreneurs. This accommodation has ensured employment opportunities and economic advancement for the local people (Bhuiyan, Siwar & Ismail, 2006).

There are some economic and social backwardness situations are remaining in Kelantan rather than other states of Malaysia. These are low incomes, unemployment, poverty, low urbanization, insufficient infrastructure development and low level of investment. This state is full of natural resources, unspools beauties, beaches, recreational forests, waterfalls and cultural attractions. Tourism activities may be developed in this state on the basis of these attractions and natural resources. Tourism has been an important economic tool in this state, attracting both domestic and foreign tourists to the state's charming natural and cultural attractions. Most of the tourism attractions of this state are situated in the rural and non-urban areas (ECER Master Plan, 2007).

There are eight Kampungs (village) are involving mainly in homestay accommodation in Kelantan. These villages are Seterpa, Pantai Suri, Kubang Telaga, Kemunchup, Batu Papan, Bukit Jering, Renok Baru and Jelawang. Homestay accommodation may be a potential tourism and economic activities in Kelantan. This accommodation program should emphasize on local custom and cultural practices, to be attract the tourists as well as to create authentic interest and commitment among the local youth in this state. The traditional food, local life style, fruits, cultural heritage and local musical performances are attracting to the guests of home stay accommodations in this state. Local people of this state can be participating in home stay operation for their economic advancement which ensures the economic development in this state. The present study analyzes the potentialities of homestay accommodation for tourism development in Kelantan. The aim of the study is examining the overall situations of homestay accommodation in Kelantan.

2 METHODS

The study used secondary data like documents, acts, regulations and policies which were collected from reliable sources. This study reviewed published materials such as relevant research reports, articles and books in order to accumulate secondary data and justify arguments.

3 ANALYSIS AND DISCUSSION

The homestays of eight villages in Kelantan are officially registered by Ministry of Tourism. Kelantan office of this Ministry is responsible for operating and supporting homestays in these villages. These villages are famous for natural attractions, waterfalls, activities, caves, limestone hills, jungle tracking, sightseeing, traditional custom and culture, tropical rainforest, handicraft, traditional lifestyle, fishing and traditional games. These tourism resources are attracting guest in homestays from home and abroad. The participation of local communities in homestay accommodations are increasing year by year. In 2006, 123 operators in six villages were participating in homestay while it reached in 152 operators from eight villages in 2013 (Table 1). This is indicating that homestay is economically and socially potential for the rural communities in Kelantan.

Table 2 highlights the guest arrivals and incomes from homestay accommodation for the three states of East Coast Economic Region in 2013. Kelantan is showing backward position in guest arrivals and incomes compare to other two states of ECER. In this circumstance, more attentions are needed to homestay accommodations for income generation as well as communities' well-beings.

Table 1. Homestay accommodations in villages and operators in Kelantan.

Year	Villages	Operators
2006	6	123
2008	8	125
2013	8	152

Source: Modified from Malaysia, 2013.

Table 2. Guest arrivals and income from homestay accommodation in ECER, 2013.

State	Guest Arrivals		Income (RM)
	Local	Foreign	
Pahang	122,584	2,838	7,263,919
Terengganu	4,457	521	327,601
Kelantan	3,061	140	234,959

Source: Malaysia, 2013a.

Table 3. Projected guests and accommodation rooms for Kelantan and Kota Bharu.

	Guests		Accommodation Rooms	
Year	Kelantan	Kota Bharu	Kelantan	Kota Bharu
2005	689,520	619,884	3,462	–
2020	2,511,565	2,260,409	22,937	20,643

Source: Malaysia, 2010 and ECER, 2007.

Table 3 shows the projected guests and accommodation rooms in Kelantan and its' capital Kota Bharu for the year 2020. Based on 2005, the guest arrivals will increase more than three times both in Kelantan and Kota Bharu. Moreover, accommodation rooms' demand will increase more than six times in Kelantan as well as Kota Bharu. Homestay can meet the demand of accommodation rooms both in Kelantan and Kota Bharu.

Kelantan has suitable conditions to develop homestay accommodation based on tourism sites of this state. There are some potentialities are remaining in this state for establishing homestay which should be helpful in tourism development. These are:

Low cost: The accommodation, meal and other charges of homestay are cheaper than other tourist accommodation. The guests are normally spent RM60 to RM120 in a day for their all activities in homestay.

Hospitality: The rural people of Kelantan are famous for their traditional hospitality. The guests of homestay get warm hospitality from the operators and become actually guests for the whole village areas.

Entrepreneurship development: Homestay program is helpful for the entrepreneurs to develop small and medium enterprises (SME) in the rural areas. It provides employment and business opportunities to the rural communities using their local resources.

Small scale investment: Homestay owners can operate their business with a small scale investment. They have withdrawn their investment within a short period. Moreover, this accommodation gives profitable income to the operator.

Treasure of tourism resources: Kelantan is full of natural resources, rich biodiversity, unspools beauties, beaches, recreational forests and waterfalls. These are suitable for homestay development in this area. Homestay gives opportunities to the tourists for staying closely with the tourism resources.

Cultural and social benefit: Kelantan is rich in traditional Malay culture, custom and life-style. This cultural performance is an integral component of the homestay program. Moreover, Homestay gives opportunities to the guests to stay with local communities and gather experiences form their social value, belief and life-style.

Ministry of Tourism has developed some requirements for homestay accommodation development in Malaysia. The homestay accommodations of Kelantan must be followed these standards for successful operation. According to the MT (2009), the necessary requirements for homestay operations are as follows.

Easy access: Homestay accommodations must be situated in a suitable location for easy accessibilities and near the tourism attractions. If the visitors are staying in the homestay, they can enjoy easy access in the tourism attractions.

Ensure basic facilities: Proper and separate spaces are needed for home stay accommodation. These accommodations have given emphasize to the comfort of tourists, including facilities such as dining space, living rooms and toilets.

Hygiene: Most of the homestays of Kelantan are not in suitable conditions in terms of hygicnc. These accommodations should be clean and free from pollution for comfortable stay of guests. Good toilets and drainage systems are needed to be in homestay for avoiding bad smell.

Safety and security: Security of houses is important elements for the homestay operation in Malaysia. Moreover, safety steps like fire protection facilities to be arranged by the homestay operators.

Packages offered: The guests are interested for attractive tourism activities in any destinations. Homestay accommodations can be arranged and offered lucrative service packages for the guests.

Homestay management and promotion: Most of the homestay owners of Kelantan have lack of managerial efficiencies for operating the business. They can enhance their managerial capacity through the training programs arranged by Ministry of Tourism. Moreover, promotional activities such as advertisement, website development, face book group initiatives, attending in carnival and participate in exhibition can expand the business opportunities of homestay.

4 CONCLUSION

The Malaysian government has taken several initiatives for economic enhancement of this state by tourism development. Homestay accommodation can be one of the major activities for tourism development in this state. The local communities of this state will be benefited economically, socially and culturally through the development of this accommodation. Joint efforts between State office of Tourism Ministry, tour operators, homestay operators and investors are necessary for the development of this accommodation in this state. Proper marketing and promotional activities, various services for customer satisfaction, effective policies of government and research initiatives are necessary for sustainable homestay development in Kelantan. Finally, proper planning, regulations, financial allocation and managerial efficiency will be ensuring homestay development in this state as well as well beings for local communities.

ACKNOWLEDGEMENTS

Financial assistance provided from the Exploratory Research Grant Scheme (ERGS/1/2013/SS05/UKM/02/7) on 'Exploring Islamic Perspective on Transformative Service for Societal Well-being' and Fundamental Research Grant Scheme (FRGS/1/2013/STWN01/UKM/03/1) on 'A Study on Green Tourism Development for Rural Community Transformation in Malaysia' is gratefully acknowledged. The authors are grateful to the Ministry of Tourism, Kelantan office to provide necessary information and assistances.

REFERENCES

Bhuiyan, M.A.H., Siwar, C. & Ismail, S.M. (2006). The role of homestay in community based tourism (CBT) development in Malaysia. In J. Hummel, H. Jong & K. Dhiradityakul (Eds.), *Innovating CBT in ASEAN: Current Directions and New Horizons* (pp. 89–99).

Bhuiyan, M.H., Siwar, C. & Mohamad Ismail, S. (2013). Socio-economic impacts of home stay accommodations in Malaysia: A study on home stay operators in Terengganu State. *Asian Social Science, 9*(3), 42–49.

Din, A.K.H. & Mapjabil, J. (2010). *Tourism research in Malaysia: What, which way and so what?* (1st ed.). Sintok: Universiti Utara Malaysia Press.

ECER Master Plan. (2007). *East Coast Economic Region Master Plan.* Kuala Lumpur: East Coast Economic Region Development Council Retrieved from http://www.ecerdc.com/ecerdc/dc.htm.

Kayat, K. (2007). Customer orientation among rural home stay operators in Malaysia. *ASEAN J. Hospitality Tourism, 6*, 65–78.

Liu, A. (2006). Tourism in rural areas: Kedah, Malaysia. *Tourism management, 27*(5), 878–889.

Peterson, M. (2004). Home stays in Malaysia. *Transitions Abroad Mag, 28*(3), 56–57.

Saeng-Ngam, A., Chantachon, S. & Ritthidet, P. (2009). The Organization of Cultural Tourism by the Community People in the Region of Toong Kula Rong Hai. *Journal of Social Sciences, 5*(4), 342–347.

UNDP. (2003). *Rural tourism master plan for Malaysia.*

Theory and Practice in Hospitality and Tourism Research – Radzi et al. (Eds)
© *2015 Taylor & Francis Group, London, ISBN 978-1-138-02706-0*

The influence of generic and specific features of destination attractiveness on behavioural intentions: Taman Negara National Park

P. Mihanyar, S.A. Rahman & N. Aminudin
Faculty of Hotel and Tourism Management, Universiti Teknologi MARA, Shah Alam, Malaysia

ABSTRACT: This article explores the relationship of generic and specific features of tourism destination with tourists' behavioral intentions by measuring the importance of generic and specific features of tourism destination. Generic and specific features are conceptualized as attributes of destination attractiveness which are both important to tourists. This study refers to the fact that on the demand side those tourists visiting national parks exert more demand on specific attributes rather than generic attributes, therefore the results suggests that evaluation efforts should include assessment of both generic and specific attributes (demand and supply side) while marketing attractiveness of tourism destination.

Keywords: Destination attractiveness, pull factor, demands and supply, generic and specific attributes

1 INTRODUCTION

Many tourists travelling to natural areas with the desire to see, touch and in overall experience new things and be inspired by nature. The key tourist attractions in many tourist destinations are natural areas, as it has been estimated that nature tourism has risen approximately from 2 percent in the late 1980s to approximately of 20 percent today (Newsome & Moore, 2012). In the context of current study, destination attractiveness refers to the attributes of the host destination national parks that attract or pull international tourists. The attraction of a tourist destination extensively depends on the climatic conditions and the natural resources of the geographical place in which it is located (Amelung, Nicholls & Viner, 2007; Gössling, Scott, Hall, Ceron & Dubois, 2012). According to Nyberg (1995) the tourism system which encompasses tourists, destination and linkage among the two is dependent on destination attractiveness. Destination attractiveness can be viewed as generic and specific attributes. According to Dann (1977) specific attributes are those that motivate a person to go for holiday to a specific destination (pull factor). However, generic attributes are shopping, accommodation, food and transportation (Klenosky, 2002; Pesonen, Komppula, Kronenberg & Peters, 2011), but how generic and specific features of national parks will influence tourists' behavioural intentions has not been investigated.

2 LITERATURE REVIEWS

Environmental management in hotel is typically categorized into the following segments; energy and water conservation as well as waste reduction and recycling management (Molina-Azorín, Claver-Cortés, Pereira-Moliner & Tarí, 2009). Meanwhile, the span of environmental management is built around both front and back-of-house operations, organizational system and culture as well as external business relationship (Park, 2009). Environmental efforts in hotels must be planned as exercised with care; hence hoteliers need to strike a balance between creating unobtrusive environmental strategies to all guests while keeping it visible to satisfy environmentally-conscious clientele. For such reason, understanding both guests' perception towards existing environmental practices as well as its influence towards their future behavior intention are considered to be among the indicators to determine one's business success.

2.1 *Pull factor*

Destination attractiveness is a pull factor, whether the attraction is to the country, a region or just a single feature of the destination. It represents the demand and supply side of tourism, where the tourist creates demand and the destination elements are the supply (Formica, 2000; Prayag & Ryan, 2011).

2.2 Demand side

On the demand side, attractivenss can be evaluated through number of tourist arrivals and their expenditures, as well as through tourist's prefrences, which are the most reliable indicators (Formica & Uysal, 2006; Kim & Perdue, 2011). Tourists' perceptions of attractivenss influence the success or failure of a destination. These perceptions were based on personal and cultural beliefs and were influenced by promotional activities and previous experince (Milman & Pizam, 1995; Prayag & Ryan, 2012).

2.3 Supply side

On the supply side, destination attractivenss (generic and specific attributes) can be evaluated through the measurment of tourism resources to create an inventory of existing tourism resources (Formica & Uysal, 2006). For instance, Jafari (1982) measured three elements on the supply side; tourism-oriented products, resident-oriented products and background tourism elements. Accommodations, food service, transportation, travel agencies and tour operators, recreation and entertainment and other travel-trade services are considered as tourism-oriented products. As tourists stay longer at destination sites, they may increase their use of resident-oriented products, which include clinics and banks. While visiting the host destination, tourists also are exposed or experienced the background tourism elements, such as natural, sociocultural and manmade attractions, that frequently constitute tourists' main reasons for travel. These elements collectively shape tourists experience and their perception of the destination's quality of performance, thus leading their behavioral intentions towards revisit and willingness to recommend (Frechtling, 2012; Pyo, Uysal & McLellan, 1991).

2.4 Destination attractiveness evaluation

The investigation and evaluation of attractiveness in the literatures has moved from the seller's market to the buyer's market over time, and more recently investigations have used a combination of both markets (Formica & Uysal, 2006; Prayag & Ryan, 2011). Since attractivenss can be examined from different perspectives, definitions of the term differ slighlty. From the buyer's side, Hu and Ritchie (1993) defined destination attractivenss as an individual's feelings, beliefs, and opinions about a destination's capability to provide satisfaction in relation to one's special vacation needs. Meanwhile, Mayo and Jarvis (1981) summarised destination attractivenss as "specific benefits that are desired by travellers and with the capability of the destination to deliver them". More generally, the supply approach defines destination attractivenss as the attractive force generated by the attrbites of a given place at a certain time (Kaur, 1981).

From behavioral point of view, the nature of interaction among demand and supply implies the fact that people travel or involve in leisure activities because they are pushed or pulled by tourists' motivations and destination attributes. The interaction between demand and supply is essential for the vacation and leisure experience to take place (Formica & Uysal, 2006). The review of previous research (Formica, 2000; Moore, 2010) has exposed that destination attractiveness is a function of the resource base (attraction) and of demand (those who are attracted). The attraction of tourism destinations extensively depends on the natural environment and climatic conditions (Alqurneh, Md Isa & Othman, 2010; Amelung, et al., 2007). Thus, for a destination to respond significantly to demand or to reinforce push factors, it must be perceived and valued (Brayley, 1990; Deery, Jago & Fredline, 2012).

2.5 Hypotheses development

Based on the above literature, the hypotheses are as follows:

H1: There is a significant relationship between generic attributes and behavioral intentions.

H2: There is a significant relationship between specific attributes and behavioral intentions.

3 METHODOLOGY

This study examined the relationship of generic and specific features of tourism destination with tourists' behavioral intentions in Taman Negara National Park during February 2014 to April 2014. This study is based on quantitative methodology to investigate the influence of generic and specific features of destination attractiveness on tourists' behavioral intentions. The research study use survey questionnaires. The sample of this study consisted of international tourists who are visiting Taman Negara National Park. The study instrument is a self-administrated questionnaire encompassing generic and specific attributes of tourism destination. The items measured using 7-point Likert scale, ranging from 1 (strongly disagree) to 7 (strongly agree). The Likert scale was developed after reviewing previous studies (Hu & Ritchie, 1993; Lew, 1987).

4 RESULTS AND DISCUSSION

This section discusses the results of the research gained from the tourists perspectives in Taman Negara National Park. The total of

Table 1. Destination attractiveness attributes.

Dimensions	Mean score
Generic attributes (shopping, accommodation, food and transportation)	3.02
Specific attributes (beautiful scenery and natural attractions)	3.07

Table 2. Items indicators for specific& generic attributes.

Dimensions and items	Mean	Std. D
Specific attributes	3.07	0.61
natural environment	3.76	0.59
fauna and flora	3.54	0.81
peacefulness	3.16	0.51
climate	3.04	0.65
nature activities	3.00	0.57
beautiful trails	2.73	0.59
historical attractions	2.72	0.58
cultural attractions	2.68	0.58
Generic attributes	3.02	0.49
Staff actively and aggressively provide services	3.62	0.72
Staff are friendly and courteous	3.09	0.54
Accommodation price is reasonable	2.95	0.57
Accommodation are clean	2.78	0.57
Public transportation	3.84	0.54

Table 3. Behavioural intentions.

Items	Mean score
Recommend to family and friends	4.25
Revisit	4.20

500 questionnaires was distributed to international tourists and the total of 384 valid responses was gathered and analyzed using convenience sampling.

The results presented in Table I support hypotheses H1, H2. However, specific attributes considered as the most important factor when visiting Taman Negara National Park. If the park management desire to attract more international tourists, they must focus more on specific attributes.

In Table 2 indicators items of destination attractiveness has been shown.

In Table 3, most of the respondents are willing to recommend the Taman Negara National Park to family and friends.

5 CONCLUSION

The results gained from this research can be used by Taman Negara National Park authorities to improve the specific attributes in order to respond effectively to international tourists. In general, this research provided support for a strong relationship between destination attributes (generic and specific attributes) and behavioral intentions. More importantly by concentration on fulfilling the demand side, the behavioral intentions may significantly increase to the desire stage.

REFERENCES

Alqurneh, M., Md Isa, F. & Othman, A.R. (2010). Tourism destination image, satisfaction and loyalty: A study of the Dead Sea in Jordanian curative tourism.

Amelung, B., Nicholls, S. & Viner, D. (2007). Implications of global climate change for tourism flows and seasonality. *Journal of Travel Research, 45*(3), 285–296.

Brayley, R.E. (1990). An analysis of destination attractiveness and the use of psychographics and demographics in segmentation of the within-state tourism market. *Dissertation Abstracts International. A, Humanities and Social Sciences, 51*(5).

Dann, G. (1977). Anomie, ego-enhancement and tourism. *Annals of tourism research, 4*(4), 184–194.

Deery, M., Jago, L. & Fredline, L. (2012). Rethinking social impacts of tourism research: A new research agenda. *Tourism Management, 33*(1), 64–73.

Formica, S. (2000). *Destination attractiveness as a function of supply and demand interaction.* Virginia Polytechnic Institute and State University.

Formica, S. & Uysal, M. (2006). Destination attractiveness based on supply and demand evaluations: An analytical framework. *Journal of Travel Research, 44*(4), 418–430.

Frechtling, D. (2012). *Forecasting tourism demand*: Routledge.

Gössling, S., Scott, D., Hall, C.M., Ceron, J.-P. & Dubois, G. (2012). Consumer behaviour and demand response of tourists to climate change. *Annals of tourism research, 39*(1), 36–58.

Hu, Y. & Ritchie, J.B. (1993). Measuring destination attractiveness: A contextual approach. *Journal of Travel Research, 32*(2), 25–34.

Jafari, J. (1982). The tourism market basket of goods and services: The components and nature of tourism. In T. Singh, J. Kaur & D. Singh (Eds.), *Studies in tourism, wildlife, parks, conservation* (pp. 1–12).

Kaur, J. (1981). Methodological approach to scenic resource assessment. *Tourism Recreation Research, 6*(1), 19–22.

Kim, D. & Perdue, R.R. (2011). The influence of image on destination attractiveness. *Journal of Travel & Tourism Marketing, 28*(3), 225–239.

Klenosky, D.B. (2002). The "pull" of tourism destinations: A means-end investigation. *Journal of Travel Research, 40*(4), 396–403.

Lew, A.A. (1987). A framework of tourist attraction research. *Annals of tourism research, 14*(4), 553–575.

Mayo, E.J. & Jarvis, L.P. (1981). *The psychology of leisure travel. Effective marketing and selling of travel services*: CBI Publishing Company, Inc.

Milman, A. & Pizam, A. (1995). The role of awareness and familiarity with a destination: The central Florida case. *Journal of Travel Research, 33*(3), 21–27.

Molina-Azorín, J.F., Claver-Cortés, E., Pereira-Moliner, J. & Tarí, J.J. (2009). Environmental practices and firm performance: an empirical analysis in the Spanish hotel industry. *Journal of Cleaner Production, 17*(5), 516–524.

Moore, W.R. (2010). The impact of climate change on Caribbean tourism demand. *Current Issues in Tourism, 13*(5), 495–505.

Newsome, D. & Moore, S.A. (2012). *Natural area tourism: Ecology, impacts and management* (Vol. 58): Channel View Publications.

Nyberg, L. (1995). Determinants of the attractiveness of a tourism region. *Tourism marketing and management handbook, 2*.

Park, J. (2009). *The Realationship between Top Manager's environmental attitudes and ENvironmental managment in hotel companies.* Virginia Polytechnic Institute and State University.

Pesonen, J., Komppula, R., Kronenberg, C. & Peters, M. (2011). Understanding the relationship between push and pull motivations in rural tourism. *Tourism Review, 66*(3), 32–49.

Prayag, G. & Ryan, C. (2011). The relationship between the 'push'and 'pull'factors of a tourist destination: The role of nationality–an analytical qualitative research approach. *Current Issues in Tourism, 14*(2), 121–143.

Prayag, G. & Ryan, C. (2012). Antecedents of Tourists' Loyalty to Mauritius The Role and Influence of Destination Image, Place Attachment, Personal Involvement, and Satisfaction. *Journal of Travel Research, 51*(3), 342–356.

Pyo, S.S., Uysal, M. & McLellan, R.W. (1991). A linear expenditure model for tourism demand. *Annals of tourism research, 18*(3), 443–454.

Theory and Practice in Hospitality and Tourism Research – Radzi et al. (Eds)
© 2015 Taylor & Francis Group, London, ISBN 978-1-138-02706-0

Supply analysis in the continuity of community-based tourism

N. Aminudin, W.S.Z. Yahya & N. Sumarjan
Faculty of Hotel and Tourism Management, Universiti Teknologi MARA, Shah Alam, Malaysia

ABSTRACT: Community-based tourism is well known for its novelty in diversifying the economy, alleviating poverty, and improving life-quality of the community. Push factors for community-based tourism usually comes from government support and coordination and cooperation among the community itself. Although the financial reward is not that great, this supplementary income is not to be belittled. Little is known however of the continuity capability of this homestay programme after a few years of its operation. Currently there is an indication that the programme is facing a declining growth in terms of tourism receipts. Plotting this against Butler's *Tourist Area Life Cycle* model would mean it is in the *stagnation* stage before it eventually reaches the *decline* stage. Nonetheless this should not be the case because it has not reached the necessary carrying capacity limit to reach the stagnation stage. A continuation of this digression would be a loss of income at a premature level. Therefore this study aims to identify the causes of the declining growth. A qualitative research method with case study was adopted. One to one interview with five informants which were homestay providers, personal observation and analysis of related documents are the approaches the researchers adopted. A purposive sampling criterion was used in selecting the informants. The findings are manually transcribed and analyzed. Results show that high dependency on the authority and lack in the interest of the second generation in the programme contributed to the situation. The identifying of factors contributing to the discontinuity of the homestay from the suppliers' perspective is hoped to be a pointer for the homestay programme to be rejuvenated and continue its competitiveness.

Keywords: Community-based tourism, homestay programme, tourist area life cycle

1 RESEARCH OVERVIEW

Community-based tourism is well known for its novelty in diversifying the economy, alleviating poverty, and improving life-quality of the community (Aminudin & A Jamal, 2006) and in some countries major investments were involved (Holladay & Powell, 2013). It is common that community-based tourism (CBT) requires support from local authority and the co-ordination and cooperation of the local community (Sriprasert, Chanin & Suttara, 2011). These become push factors that enable a CBT to be successful. It usually does not have problem in attracting tourists since most CBT continues to be a niche market within the hospitality industry (Holland, Martin & Shakur, 1998) since it portrays the lifestyle, culture and heritage of the rural host community (Aminudin & A Jamal, 2006). Although the financial reward is not that great, this supplementary income to the host is not to be belittled. It is an economic alternative which optimize available resources. Other than the economic benefit, the chance of running own business catering to the needs of clients from diverse nationalities instills a sense of pride and personal

satisfaction (Holland, et al., 1998). The Malaysian scenario of CBT, homestay programme has been officially in existence about 20 years ago. It is fully supported by the government and was shaped to be the platform to introduce the uniqueness of the rural area and its communities (Ministry of Tourism, 2009). It was recognized through the UNWTO Ulysses Award in 2012 in the Innovation in Public Policy and Governance category.

This CBT registered revenue of only RM15.74 million in 2011, compared to revenue of RM58.3 billion for the whole tourism sector, merely 0.26 percentage (Statstic Department of Malaysia, 2012). Its revenue is small but in general it is on a rising trend especially when more households are joining the programme. However, tourism receipts from this programme are showing that in some states, there is a declining growth (Statstic Department of Malaysia, 2012). What is worrying is if the declining situation is on the *stagnation* stage of Butler's (1980) *Tourist Area Life Cycle* (TALC) model. If no effort taken, it will reach the *decline* stage soon, a loss of income at a premature level and could cause the programme to be shot-lived. According to TALC, destination areas will undergo a fairly uniform transformation where increases in visitation to a destination can be followed by a decrease in visitation as carrying capacity to the destination is reached. Nonetheless in the case of this CBT, it should not be the case because it has not reached the necessary carrying capacity limit to reach the stagnation stage (personal comm. President of Negeri Sembilan Homestay Association). Therefore the main aim of this study is to identify the causes of the declining growth of the homestay programme. The research is based on a case study basis where a survey of Negeri Sembilan homestay providers was undertaken to gain a better understanding of the reasons for the declining situation. The state was chosen because it showed the highest decline, from RM1.24 million to RM0.69 million in 2009 to 2011 respectively, despite the increase in numbers of participating villages

2 LITERATURE REVIEWS

2.1 *Homestay and community*

Homestay is described as activities receiving tourists to stay with the host during the holiday began as early as in the 70's where there are villages along the coast became a destination for tourists. The tourist experience greatly enhanced when the host provides food such as breakfast and lunch. He pointed out that the concept of homestay is different between countries, particularly in South-east Asia. Although there is different interpretation, the basic concept of homestay remains the same, "living with host families". The staying with the host families resulted in an indirect exchange of culture through eating together, participating in rural daily activities such as fishing, rubber tapping and picking coconuts palm and attend gatherings, provide a different experience for tourists (Kayat., 2009). Homestay tourists are those who want to know the ins and outs of the culture of human societies in Malaysia (Peterson, 2004), it is a combination of the relationship between local culture, the cultural and natural environment in local areas (A Jamal, Othman & Nik Muhammad, 2011). On the supplier's side, CBT is focused on the concept of allowing local communities to preserve their community and identity, especially in the era of globalization Youl (1997).

2.2 *Tourist area life cycle by Butler (1980)*

Butler (1980) has introduced Tourist Area Life Cycle (TALC) in his study for the management of tourism resources. According to Butler, a tourist destination will experience the evolution either of the physical environment or on its ability to meet

demand. These evolutions are due to many factors, including changes in priorities and meet the needs of visitors, the gradual deterioration and possible replacement of physical plant and facilities, and changes or possible loss of natural and cultural attractions of origin is responsible for the initial popularity of the area. Evolution experienced by a destination will have an impact on whether the tourist experience will visit again or not. This is likely because the evolution occurring will give effect to the tourists to make their choice of destination (Ritchie & Crouch, 2003).

In TALC, there are seven phases of development transformation of a destination. The cycle utilizes number of tours (visitation) against time and goes through a few steps of transformation from *exploration, involvement, development, consolidation* to *stagnation*. From *stagnation*, the cycle has options, either to *rejuvenate* or *decline*. These steps however relate to the concept of carrying capacity, which means increases in visitation to a destination can be followed by a decrease in visitation as carrying capacity to the destination is reached. Butler (1980) proposed that the change in visitor arrivals to a tourist destination follows the S-shaped curve of the product life cycle. It refers to the dynamic nature of the destination and it has proposed a general process of development and the potential decline can be prevented with appropriate intervention interested parties at the destination (Butler, 2000).

Plotting the Negeri Sembilan homestay in TALC, an S-shaped appears as in Figure 1. The curve indicates that it has surpassed the point to rejuvenate its destination but it is still not too late to revive. Nonetheless, the argument here is that the carrying capacity on the supply side has not been reached yet. Resources in terms of houses are still available. Therefore, the main aim of this research is to identify what are the factors that could have contributed to the declining situation.

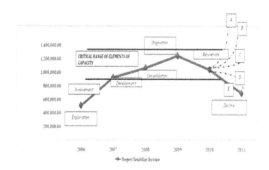

Figure 1. Negeri Sembilan Homestay in Butler's TALC Graph.

3 RESEARCH METHODOLOGY

3.1 Research design: Exploratory approach with case study

A qualitative method with a case study approach is used as the research design to explore the phenomenon of Negeri Sembilan homestay development. Exploratory research method is defined as a research that has not previously been studied and attempt to identify new knowledge, new insight, new understanding; new meanings and to explore new factors related to the topic (Qu & Dumay, 2011). The researchers are using semi-structured interview with open-ended questions. The interviews were carried out at informants' residence, which were convenient them. With permissions granted by informants, interviews were audio recorded, while the interview notes taken by researchers in writing during interviews. The keynote is active listening, allowing the informant to talk freely and ascribe meanings, while keeping in mind the broader aim of the research as suggested by Silverman (2006). In some interviews researcher used a more standardized technique, yet most of the questions occurred randomly and freely. Researcher tried to interact with each informant and understand their experiences, opinions and ideas in order to get their perceptions of the matter. This method is particularly useful when accessing individuals' attitudes and values. Interviews offer a rich source of data, which provides access to how people account for their troubles as well as joy. The duration of each interview conducted is varied and unexpected. Nevertheless, the overall level, every interview held between in the hour and not more than 1 hour and 30 minutes. The time is adequate, given the issues raised required a verbal explanation together with the help of documents provided to support the informant of the statement. The interview questions are directed on the informant experiences and knowledge about the phenomenon with the direct involvement of the homestay management.

In order to not solely rely on deep-interviews of the respondents' self-report, the research consisted of observations too. This method is an important resource in understanding the character of informants and the communities. Observation approach gives a better understanding and clear picture on what is going on, rather than presume researcher own opinion and perceived mind to interpret the information from the interview.

3.2 The case study: Homestay in Negeri Sembilan

Within the qualitative research approach, researchers has chosen to study a case in order to emphasise *"...the embeddedness of a phenomenon in its real-life context."* (Blumberg, Cooper & Schindler, 2008, p. 375). Homestays in Negeri Sembilan were chosen due to its declining performance and personal connections that researchers have with some of the providers. The fieldwork in five of nine homestay in Negeri Sembilan was carried out and observed over a period of two months. Researchers chose a smaller sample in order to get closer to the respondents perspective. As Silverman (2006, p. 9) indicates: *"...qualitative researchers are prepared to sacrifice scope for detail"*, and here detail means finding out precise data about people's lives, perceptions and interactions. The selected sample in this research was mostly chosen through the purposive sampling. This is because, the selection of the informant is based on the informants' direct experience with the issues and yet, it was ensured that the resulting sample was relevant for the study. The diversity of views and opinions helped researchers to relate to why certain themes or issues occur and have impacts on the researched situation.

4 FINDINGS AND DISCUSSION

The research finding has led to a better understanding of the S-shaped curve destination development in Negeri Sembilan Homestay. The results of this study indicated that a high dependency on the government agencies in marketing the homestay program is one of the reasons the declining growth in tourist arrivals to the homestay in Negeri Sembilan. Another significant finding is that the fading interest in participating in the homestay program because of its insignificant economic return to the homestay providers and their successors. Additionally, the absent of a leader in leading the CBT. These two major problems, government dependency and lack of motivation have been as contributors to the slacking in performance of the homestay program in Negeri Sembilan. This is in line with the Social Development Theory as argued by Andriotis (2005) that the community offset the costs and benefits of tourism development, and support for tourism is dependent on the outcome of cost-benefit equation gained by them in the process of development.

Initial efforts undertaken by the government agencies in raising awareness of the opportunities created from participation in tourism in the early days of homestay villages, have made it as if it is their responsibility to keep on marketing the CBT. While the provider of homestay should act as a business entity and carry out independently marketing efforts by themselves, not all are successful in doing so. By failing to do so, the high dependence on the government's efforts has resulted in some homestay facing a stalemate and has a difficulty to maintain its sustainability as a tourist destination. Thus making some homestay less competitiveness compared to others. However, no doubt homestays have vision and passion; they successfully act as a team and focus on marketing and operating interesting CBT activities. This is an example of why it is necessary for the homestay to have the planning and development of an action plan after the government has helped to develop the village as a tourist destination. However, there is no denying that the effort by the government, particularly in the development of much-needed infrastructure and facilities. The role undertaken by Tourism Malaysia to promote the homestay, it appears indispensable, especially in a space and platform for homestay to build relationships with industry players outside Malaysia.

Associated with community support, it can be concluded that there is a relationship between community residents' support for the development of tourism and community perceptions of tourism impacts. Negative perceptions are likely to cause a loss of interest among homestay providers. The negative perceptions could come from unexpected yield which caused lack of motivation. Findings from this study suggest that people will act to maximize the benefits and reduce costs in different situations. They also weigh the benefits against the cost of implementing their decision to participate in decision-making tourism and tourism development planning (Kayat, 2002; Lawler, 2001; Yoon, Gursoy & Chen, 2001). In order to have a CBT supported by all members of the community of life, the benefits of tourism must be evenly distributed. Obviously, it is appropriate for the community to unite in making the form and function of tourism development and have sustainable management of any tourism scheme at their location. In reality, however, communities suffer from a lack of experience, human resources, leadership and therefore even if interest is required to create a successful tourism venture. All of these factors, particularly the last one, clearly the case in empirical research for this study.

5 CONCLUSION

In the early stages of this study, researchers were using and adapting Butler's (1980) TALC for homestay development as a tourist destination, however upon conducting the field survey it is found that the homestay program in Negeri Sembilan has yet to reach maturity in the cycle since its supply capacity has not been fully utilized. Although the earlier S-shaped curve showed that the income of homestay providers have reached

the point of stagnation, in real situations homestay program has not yet fulfill it full carry capacity. Results show that high dependency on the authority and lack in the interest of the second generation in the programme contributed to the situation. The identifying of factors contributing to the discontinuity of the homestay from the suppliers' perspective is hoped to be a pointer for the homestay programme to be rejuvenated and continue its competitiveness.

ACKNOWLEDGEMENTS

This research is funded by Universiti Teknologi MARA through Research intensive Faculty Grant (RIF) 600-RMI/DANA 5/3/RIF (885/2012).

REFERENCES

A Jamal, S., Othman, N. & Nik Muhammad, N.M. (2011). Tourist perceived value in a community-based homestay visit: An investigation into the functional and experiential aspect of value. *Journal of Vacation Marketing, 17*(1), 5–15.

Aminudin, N. & A Jamal, S. (2006). *Homestay Selangor: Keunikan dan pengalaman pengusaha*: Pusat Penerbitan Universiti (UPENA), Universiti Teknologi MARA.

Andriotis, K. (2005). Community groups' perceptions of and preferences for tourism development: Evidence from Crete. *Journal of Hospitality & Tourism Research, 29*(1), 67–90.

Blumberg, B., Cooper, D.R. & Schindler, P.S. (2008). *Business research methods* (7th ed.). London, UK: McGraw-Hill Higher Education.

Butler, R.W. (1980). The concept of the tourist area life-cycle of evolution: implications for management of resources. *Canadian Geographer, 24*(1), 5–12.

Butler, R.W. (2000). The resort cycle two decades on. In B. Faulkner, E. Laws & G. Moscardo (Eds.), *Tourism in the 21st Century: Reflections on experience* (pp. 284–299). London: Cassell.

Holladay, P.J. & Powell, R.B. (2013). Resident perceptions of social-ecological resilience and the sustainability of community-based tourism development in the Commonwealth of Dominica. *Journal of Sustainable Tourism, 21*(8), 1188–1211.

Holland, J.D., Martin, K. & Shakur, S. (1998). *A survey of farmstay tourism in New Zealand.* Paper presented at the Third International Conference on Tourism in Indi-China & Southeast Asia: Development, marketing and sustainability.

Kayat, K. (2002). Power, social exchanges and tourism in Langkawi: Rethinking resident perceptions. *International Journal of Tourism Research, 4*(3), 171–191.

Kayat., K. (2009). *Community based toursim in developing countries.* Paper presented at the International Seminar On Community Based Tourism.

Lawler, E.J. (2001). An affect theory of social exchange. *American Journal of Sociology, 107*(2), 321–352.

Ministry of Tourism. (2009). *Homestay Statistic 2009. Putrajaya, Malaysia.*

Peterson, M. (2004). Home stays in Malaysia. *Transitions Abroad Mag, 28*(3), 56–57.

Qu, S.Q. & Dumay, J. (2011). The qualitative research interview. *Qualitative Research in Accounting & Management, 8*(3), 238–264.

Ritchie, J.B. & Crouch, G.I. (2003). *The competitive destination: A sustainable tourism perspective*: Cabi.

Silverman, D. (2006). *Interpreting qualitative data: Methods for analysing talk, Text and interaction* (3rd ed.). London: Sage.

Sriprasert, P., Chanin, O. & Suttara, R. (2011). *Exploring the relationship between managerial functions and the success of home stay community based tourism in Thailand: A case study of Phomlok, Nakhon Si Thammarat, Thailand.* Paper presented at the 2nd International Conference of Business & Economics Research proceeding.

Statstic Department of Malaysia. (2012). Malaysia tourism satelite account Retrieved Jan 10th, 2014, from http://www.statistics.gov.my

Yoon, Y., Gursoy, D. & Chen, J.S. (2001). Validating a tourism development theory with structural equation modeling. *Tourism management, 22*(4), 363–372.

Youl, R. (1997). Landcare: Positive and proven force for rural landscapes (pp. 190): Parkwatch.

Theory and Practice in Hospitality and Tourism Research – Radzi et al. (Eds)
© *2015 Taylor & Francis Group, London, ISBN 978-1-138-02706-0*

The perspective and expectation among Singaporean tourists toward destination image of Perak

A.M.F. Wahab, M.A.A. Bashir, M.D. Darson & M.H. Zamri
Universiti Teknologi MARA, Penang, Malaysia

ABSTRACT: The importance of tourist destination's image is universally acknowledged since it affects individual's subjective perception and consequent behavior as well as destination choice. Images significantly affected tourists' behavior, starting from their mental constructions about destination attributes to travel decision making process. In line with that notion, this study aims to investigate Singaporean tourists' perspective as well as expectation towards destination image of Perak. Using Perak as contextual setting, 150 Singaporean tourists were successfully surveyed. Findings showed that Singaporean tourists were slightly disagreed with certain quality attributes of destination image in Perak with shopping facilities and safety condition image recorded the lowest and highest score respectively. Apart from that, destination image was found significantly affecting expectation formation in tourists mind. Finally, implications and recommendations for the state government and the local tourism service providers were emphasized.

Keywords: Destination image, expectation, shopping facilities, safety

1   INTRODUCTION

The effort to develop a positive image of the tourist destination in target markets' mind is crucial as it will help the destination to reach its real competitive advantage (Gartner, 1994). In line with that, Baloglu and Mangaloglu (2001) have contended that destinations primarily compete based on their perceived images relative to other competitors in the marketplace. Identifying images of tourist destination will help tourism marketers to promote their destination efficiently in the marketplace (Leisen, 2001) and it will significantly assist identification of destination strengths and weaknesses (Chen & Uysal, 2002). In line with that statement, it is argued that tourism destination image is one of the key challenges in tourism study nowadays, since it is an essential component of tourism destination marketing that influences tourists' behavior by stimulating multiple creative activities and experiences (Nicoletta & Servidio, 2012).

A review of tourist arrivals revealed that Singaporean remains the leading group of international tourist to Malaysia with 13.37 million arrivals in 2011 (Tourism Malaysia, 2012). In year 2012, tourists from this country made up 52 percent from 25.03 million total international tourist arrivals in Malaysia. With this positive indication, many states in Malaysia including Perak are competing each other in attract the neighboring country tourist by strengthening their destination images.

With a mission to promote tourism and foreign investment to the state especially from Singapore, State Government of Perak has signed a five-year strategic collaboration with Firefly, and one of the focuses is to enable international air connection between Sultan Azlan Shah Airport in Ipoh, Perak and Changi International Airport in Singapore (Firefly, 2010). This action were driven by the fact that Singapore is the biggest tourism spender from South East Asia region, sitting in the 11th place in the world ranking of international tourism spenders in the year 2012, jumped two places from 2011 (United Nation World Tourism Organization, 2013). According to the same source, an astonishing RM69.4 billion total expenditure worldwide has been recorded by tourists from this island country in the year 2012. It indicated that this targeted group of tourists has strong capability in spending during travelling and some consideration should be taken to cater this valuable market.

In addition, Perak state in collaboration with Tourism Malaysia Singapore has geared an effort to tap the growing number of Singaporean tourists to Malaysia (Sgtravellers.com, 2010). The effort showed an impressive result when Singaporean tourists arrival to Perak rose 700 percent to 72,000 between January and August 2012 (The Borneo Post, 2012). However, the percentage of Singaporean tourists to Perak is only 0.553 percent from 13.01 million of their total arrivals to Malaysia in 2012.

Although Perak state offers many invaluable attractions and facilities, it is worth understanding whether such attractions have developed a significant sense of image among Singaporean. Besides, it raises another query regarding their expectation toward destination images of this silver heritage land. To date, there is still lack of studies per se looking at this issue. Understanding the perspective of Singaporean tourists towards Perak as their destination choice is therefore considered important to strategize the marketing approach and maximizing the use of available resources among this highly profitable tourist market. Following such contemplation, a hypothesis was therefore tested.

H1: There is a significant relationship between destination image and expectation towards that image

2 LITERATURE REVIEW

2.1 Destination image

The study about destination image started in early 1970s and among the first group of pioneered scholars were, Hunt (1971, 1975), and Gunn (1971) and Mayo (1973). Several efforts have been identified in the last two decades to provide an overview of the previous destination image studies in order to assist future researchers to better navigate the field. Chon (1990) had initiated the earliest reviews when the researcher explored the role of image in tourism destination before Echtner and Ritchie (1991) redefined the meaning as well as measurement of destination image.

Many definitions of destination image have been discussed by previous scholars. For some tourism researchers, destination image is simply described as an overall impression about the destination. Hunt (1971) mentioned that state tourism image is the impression that an individual has about a state in which they do not stay. Crompton (1979) further broadens the views when the researcher defined destination image as the summation of beliefs and ideas as well as impressions that a person has about the destination. While for Fridgen (1987), this researcher defined destination image as mental representation of an object or place which is not physically appeared in front of the observer.

2.1 Expectation

Expectation is the desire of customer about what they believe a service or product should or will perform in the future (Zeithaml & Bitner, 1996). They further noted that the specific wants, needs, and preferences as well as past experiences also contribute to the expectations of a prospective customer.

Gnoth (1997) contended that since expectations can significantly influence tourist choice processes as well as perceptions of experiences, managing tourists' expectations is a very crucial task. While Hung, Huang, and Chen (2003) suggested that in order to achieve high customer satisfaction, it is crucial to comprehend customer expectations.

Additionally, Peter, Olson, and Grunert (1999) have noted that personal communication from friends and relevant others have a powerful strength in persuading people while advertising, publicity, and sponsorships as external communication can play a vital role in influencing people's expectations. In line with that, Rodríguez del Bosque, San Martín and Collado (2009) argued that past experience, external communication, word of mouth communication and destination image can be considered as expectations-generating factors of a future destination experience.

2.2 Relationship between destination image and expectation

In the tourism industry, a mental representation of a destination is crucial as it will assist individuals to expect their experiences (Jenkins, 1999). In line with this, Bigné Sánchez and Sánchez (2001) noted that before people visit a destination, the preconceived image will drive their expectation toward the destination. Image will act as a generating factor of expectation for a future encounter with the tourist service (Rodríguez del Bosque, San Martín & Collado, 2006). Previous studies by academicians have established a significant relationship between image and expectations in several service industries such as catering and travel agencies (Clow, Kurtz, Ozment & Ong, 1997; Rodríguez del Bosque, et al., 2006).

Driving by intention to explore more expectation formation in destination marketing, Rodri'guez del Bosque et al. (Rodríguez del Bosque, et al., 2009) have examined the factors contributing to the expectations toward tourist destination and theoretical as well as empirical evidence about the role of different factors in generating tourists' expectation. It is proven in the study that preconceived image of tourist destination is the most influential factor in generating tourists' expectation of a future destination experience besides past experience, external communication and word-of-mouth communication. Image is proven play a significant role as it represents the true capabilities of a tourist destination, at least in the eyes of tourists. As a result, it will gear individuals' confidence to form their expectations of a future destination experience.

3 METHODOLOGY

Singaporean tourists that visited Perak and stay at least one night at the state have been selected as samples for this study. A convenience sampling

Table 1. Reported mean scores for destination image (*N = 150*).

	Variables	Mean (M)	Std. Deviation (SD)
	Destination Image		
1.	Great variety of flora and fauna	3.95	1.965
2.	Beautiful scenery	3.21	1.615
3.	Beautiful parks	3.30	1.678
4.	Variety of cultural attraction	3.26	1.681
5.	Interesting cultural activities	3.37	1.628
6.	Appealing local custom	3.38	1.771
7.	Shopping facilities	3.01	1.614
8.	Quality accommodation	3.45	1.709
9.	Good value for money	3.33	1.645
10.	Safety	4.33	1.665
11.	A relaxing destination	3.49	1.666
12.	Pleasant environment	3.43	1.607
13.	Exciting places	3.29	1.688
14.	Different cuisine	3.87	1.792
15.	Quality food	3.56	1.603

Note: 1 = Strongly Disagree, 2 = Disagree, 3 = Slightly Disagree, 4 = Neither Disagree nor Agree, 5 = Slightly Agree, 6 = Agree, 7 = Strongly Agree.

Table 2. Reported mean scores for expectation (*N = 150*).

	Variables	Mean (M)	Std. Deviation (SD)
	Destination Image		
1.	Great variety of flora and fauna	3.95	1.965
2.	Beautiful scenery	3.21	1.615
3.	Beautiful parks	3.30	1.678
4.	Variety of cultural attraction	3.26	1.681
5.	Interesting cultural activities	3.37	1.628
6.	Appealing local custom	3.38	1.771
7.	Shopping facilities	3.01	1.614
8.	Quality accommodation	3.45	1.709
9.	Good value for money	3.33	1.645
10.	Safety	4.33	1.665
11.	A relaxing destination	3.49	1.666
12.	Pleasant environment	3.43	1.607
13.	Exciting places	3.29	1.688
14.	Different cuisine	3.87	1.792
15.	Quality food	3.56	1.603

Note: 1 = Very Low, 2 = Low, 3 = Moderately Low, 4 = Neutral, 5 = Moderately High, 6 = High, 7 = Very High.

method was used. The data was collected at Sultan Azlan Shah Airport in Ipoh, Perak in addition to four selected hotels around Ipoh and Lumut that frequently stayed by Singaporean while visiting Perak. Since the study was undertaken during low seasonal period, only 150 questionnaires were successfully obtained.

4 RESULT AND ANALYSES

4.1 *Demographic profiles*

Majority of the respondents were male with 64 percent (*n* = 96). Most of the respondents' ages between 25–44 years old (52.7 percent as opposed to 32 percent which over 44 years old). Moving into education background, 78 percent graduated from university while the remaining obtained high school certificate.

4.2 *Destination image*

The magnitudes of mean score between 2.5 to below 3.5 indicated that majority were slightly disagreed with most of the Perak destination image components. Slightly disagree feeling was expressed to the beautifulness of scenery and parks, varieties of cultural attraction, cultural activities and local custom. Similar feelings also goes to facilities for shopping which recorded the lowest mean value, accommodations' quality, destination's value for money and Perak's image as relaxing destination, pleasant environment as well as possess exciting places.

Fall in the scales of mean score between 3.5 to below 4.5, the inclination toward neither disagree nor agree were identified which included several images such as safety condition that recorded the highest score, diversity of flora and fauna, different cuisine and quality food.

4.3 *Expectation*

This descriptive analysis described Singaporean tourists' expectation toward destination image of Perak. From the mean scores listed in Table 2, most images recorded mean score between 3.5 and below 4.5 indicating neutral expectation from tourists while some images are recorded fall in the magnitude of mean between 2.5 to below 3.5 that indicate the inclination toward moderately low expectation.

In general, tourists portrayed neutral expectation towards most images of destination in Perak. It can be seen via flora and fauna diversity, the beautifulness of scenery and parks, quality of accommodations, destination's value for money and safety condition which again recorded the highest mean score among all images. The inclination toward neutral expectation was also portrayed by tourists through the Perak's image as relaxing destination, pleasant environment and possesses exciting places as well as different cuisine and quality food provided.

Four images have been identified getting a moderately low expectation from tourists. The images are Perak's cultural attractions, cultural activities and local custom as well as shopping facilities that

Table 3. Regression model for destination image vs expectation.

Model Summary[b]				
Model	R	R Square	Adjusted R Square	Std. Error of the Estimate
1	.744[a]	.554	.551	.72542

Model Summary[b]				
Change Statistics				
R Square Change	F Change	df1	df2	Sig. F Change
.554	184.037	1	148	.000

Coefficients[a]				
	Unstandardized Coefficients		Standardized Coefficients	t
	B	Std. Error	Beta	
1 (Constant)	-.900	.208		4.323
D. Image	.777	.057	.744	13.556

Coefficients				
Sig.	Correlations			Collinearity Statistics
	Zero-order	Partial	Part	Tolerance	VIF
.000					
.000	.744	.744	.744	1.000	1.000

a. Predictors: (Constant), Destination Image
b. Dependent Variable: Expectation

once again recorded the lowest score among all images listed. This repeated figure might be driven by the fact that many tourist destination in Perak is built on the basis of eco-tourism which consequently had contributed to the minimum extensive facilities efforts.

4.4 Destination image vs expectation

A simple linear regression analysis was used to approximate the coefficient of linear equation between destination image and their expectation towards those images. Result in Table 3 revealed that the destination image affecting 55.4 percent ($R^2 = 0.554$, F-Change = 184.037, $p < .01**$) of the variance in the expectation formation. It showed that the destination image attributes significantly contributed to the expectation formation in tourist mind. It is proven when the destination image ($\beta = 0.744$, $p < .01**$) was found to positively and significantly affect the expectation formation which can be claimed that the hypothesis is supported.

The analysis result for the hypothesis was consistent with Bigné et al. (2001) and Rodríguez del Bosque and Martín (2008) studies as they noted that before people visit their desired place, image will influence the expectation formation in their mind toward the destination. This was supported by previous studies by Clow et al. (1997) and Rodrígue del Bosque et al. (2006) when both group

of researchers found an existence of a significant relationship between image and expectations in catering and travel agencies respectively.

5 DISCUSSION AND CONCLUSION

By looking at the perspective of Singaporean tourists toward destination image of Perak, the analysis discovered that tourists were slightly disagreed quality of certain destination images listed with shopping facilities recorded the lowest score. Viewing the relationship between destination image and expectation, analysis indicates that image appears to have significant effect on expectation formation in tourists mind where most of them having a neutral expectation toward the images of destination in Perak.

The finding is consistent with past studies that confirmed the relationships of image and expectation (Bigné, et al., 2001; Clow, et al., 1997; Rodríguez del Bosque & Martín, 2008; Rodríguez del Bosque, et al., 2006). It is suggested that serious efforts should be taken to further strengthen the destination image of Perak in destination marketing strategies in order to facilitate the state effort to grab more Singaporean tourists market.

Theoretically, the findings will become a base reference to further clarify more deeply regarding which aspects of destination segment that should be investigate more in the future research. It could also assist the state government of Perak and other related authorities to reshaping their strategy in a promotional campaign to tap the Singaporean tourists market. All the input will help those authorities to plan in advance the suitable approach so that the potentially negative image in tourists' mind about the destination before can be reduced.

In order to develop a robust and clear image of tourism destination in Perak, it is suggested that the priority should be given to policy planning as it play a crucial role in rejuvenating the image. Emphasize should be focused on elements that received poor rating among tourists. As such, shopping facilities scored the lowest rating among Singaporeans. Therefore, related authorities should undertake a comprehensive effort to improve this image. Improvement in service quality and physical appearance could be one way to revitalize this image apart from injecting more unique elements that depict the culture of Perak in the shopping premises.

In conclusion, comprehensive action and collective involvement from all related parties are needed to rejuvenate and refurbish the image of destination in Perak. Hence, the mission of state government to tap the Singaporean tourist market can be achieved successfully.

REFERENCES

Baloglu, S. & Mangaloglu, M. (2001). Tourism destination images of Turkey, Egypt, Greece, and Italy as perceived by US-based tour operators and travel agents. *Tourism management, 22*(1), 1–9.

Bigné, J.E., Sanchez, M.I. & Sanchez, J. (2001). Tourism image, evaluation variables and after purchase behaviour: inter-relationship. *Tourism management, 22*(6), 607–616.

Chen, J.S. & Uysal, M. (2002). Market positioning analysis: A hybrid approach. *Annals of Tourism Research, 29*(4), 987–1003.

Chon, K.-S. (1990). The role of destination image in tourism: A review and discussion. *Tourism Review, 45*(2), 2–9.

Clow, K.E., Kurtz, D.L., Ozment, J. & Ong, B.S. (1997). The antecedents of consumer expectations of services: an empirical study across four industries. *Journal of Services Marketing, 11*(4), 230–248.

Crompton, J.L. (1979). An assessment of the image of Mexico as a vacation destination and the influence of geographical location upon that image. *Journal of travel research, 17*(4), 18–23.

Echtner, C.M. & Ritchie, J.B. (1991). The meaning and measurement of destination image. *Journal of Tourism Studies, 2*(2), 2–12.

Firefly. (2010). Strategic Collaboration between Firefly and Perak Retrieved December 10, 2012, from http://firefly.com.my/news-releases/strategic-collaboration-between-firefly-and-perak

Fridgen, J.D. (1987). Use of cognitive maps to determine perceived tourism regions. *Leisure Sciences, 9*(2), 101–117.

Gartner, W.C. (1994). Image formation process. *Journal of Travel & Tourism Marketing, 2*(2–3), 191–216.

Gnoth, J. (1997). Tourism motivation and expectation formation. *Annals of Tourism Research, 24*(2), 283–304.

Gunn, C.A. (1971). *Vacationscape: Designing tourist regions.* University of Texas, Van Nostrand Reinhold.

Hung, Y., Huang, M. & Chen, K. (2003). Service quality evaluation by service quality performance matrix. *Total Quality Management and Business Excellence, 14*(1), 79–89.

Hunt, J.D. (1971). *Image: A factor in Tourism.* Unpublished Ph.D. dissertation, Colorado State University, Fort Collins.

Hunt, J.D. (1975). Image as a factor in tourism development. *Journal of travel research, 13*(3), 1–7.

Jenkins, O.H. (1999). Understanding and measuring tourist destination images. *International Journal of Tourism Research, 1*(1), 1–15.

Leisen, B. (2001). Image segmentation: the case of a tourism destination. *Journal of Services Marketing, 15*(1), 49–66.

Mayo, E.J. (1973). *Regional images and regional travel behavior. Research for Changing Travel Patterns: Interpretation and Utilization.* Paper presented at the Travel Research Association, Fourth Annual Conference.

Nicoletta, R. & Servidio, R. (2012). Tourists' opinions and their selection of tourism destination images: An affective and motivational evaluation. *Tourism Management Perspectives, 4*, 19–27.

Peter, J.P., Olson, J.C. & Grunert, K.G. (1999). *Consumer behavior and marketing strategy.* Berkshire, London: McGraw-Hill

Rodríguez del Bosque, I.A. & Martín, H.S. (2008). Tourist satisfaction a cognitive-affective model. *Annals of Tourism Research, 35*(2), 551–573.

Rodríguez del Bosque, I.A., San Martín, H. & Collado, J. (2006). The role of expectations in the consumer satisfaction formation process: Empirical evidence in the travel agency sector. *Tourism management, 27*(3), 410–419.

Rodríguez del Bosque, I.A., San Martín, H. & Collado, J. (2009). A framework for tourist expectations. *International Journal of Culture, Tourism and Hospitality Research, 3*(2), 139–147.

Sgtravellers.com. (2010). Perak seeks bigger share of Singapore tourist dollar Retrieved December 11, 2012, from http://sgtravellers.com/travel-article/perak-seeks-bigger-share-ofsingapore-tourist-dollar/368/2

The Borneo Post. (2012). Singaporean tourist arrivals to Perak soar 700 percent between Jan to Aug, says State Exco Retrieved December 7, 2012, from http://www.theborneopost.com/2012/11/05/sporean-tourist-arrivals-to-perak-soar-700-pct-between-jan-to-aug-says-state-exco/

Tourism Malaysia. (2012). *Tourist arrivals to Malaysia for Jan-Dec 2012* Retrieved from http://corporate.tourism.gov.my/images/research/pdf/2012/TouristArrivals_JanDec_2012.pdf.

United Nation World Tourism Organization. (2013). Worlds' top tourism spenders Retrieved May 2o, 2013, from http://dtxtq4w60xqpw.cloudfront.net/sites/all/files/pdf/tsen_0.pdf

Zeithaml, V.A. & Bitner, M.J. (1996). *Service Marketing.* Singapore: McGraw-Hill.

Theory and Practice in Hospitality and Tourism Research – Radzi et al. (Eds)
© *2015 Taylor & Francis Group, London, ISBN 978-1-138-02706-0*

Understanding the impact of cultural tourism in small town sustainable development

K.W. Awang, M.F. Ong, Y.A. Aziz & G. Jeahnichen
Universiti Putra Malaysia, Selangor, Malaysia

ABSTRACT: By recognizing the nature and culture in small towns as an asset that could become tourist attraction, this research is aimed at exploring the potential of small towns in Malaysia as the sustainable tourism product. In any tourism planning and destination impact on the surrounding especially to the surrounding community or local resident can be negative or positive. Nevertheless, local resident support is essential to ensure long-term success in tourism development. This is particularly important because local community is the main player of the small town development to support the overall small town sustainability. The overall purpose of the study is to be a model of small town development and become a pioneer research of a well self-sustained tourism destination that enhances the economy, quality of life, and the community pride of small towns.

Keywords: Culture, development, small town, sustainable, tourism

1   INTRODUCTION

As the centuries move on, small towns are being abandoned. Deterioration of small towns are especially significant in developed countries. The population of small towns has migrated to great commercial and industrial areas and cities. Not only have the communities in small towns lost population, businesses have closed or relocated. The physical environment is deteriorating, the community's spirit is low, and the agricultural base had challenged by world markets and technology.

Many researchers relate their study of small towns with globalization challenges (e.g. Courvisanos and Martin, 2005), urban sprawl (e.g. Lambe 2008; Leinberger, 2005), aging population (e.g. Nicholls, 2005; Lambe, 2008), tourism development (e.g. Altinay and Hussain, 2005), industrial development (e.g. Leinberger, 2005), and entrepreneurship.

Though, many small town planners have identified the small town potential assets both tangible and intangible before the place became completely silent. They repackage the resources of small town into tourism products that attract millions to come.

Small town research studies have been a popular subject in western countries. According to Lambe (2008), towns are divided into few categories:

i. Small towns that are recreation or retirement destinations
ii. Small towns that have abundance of natural assets

iii. Small towns with historical, cultural or heritage assets
iv. Small towns with college campuses

By recognizing the social and economic benefits of tourism, many town planners even the country planners have had goals that involve tourism development.

2   LITERATURE REVIEWS

2.1   *Malaysia tourism industry*

The tourism sector has been Malaysia's second-largest foreign exchange-earner. In 2009, although the world was hit by the global financial crisis, Malaysia's tourism arrivals continued to increase 7.2%; 23.65 million compared to 22.05 million the previous year (Malaysia Economic Report 2010/2011). In 2010, Malaysia ranked 16th in terms of global inbound tourist receipts, capturing approximately 2% of the global market share (Tourism's New Drive, 2010).

Malaysia's tourism industry has been significantly important to the country as one of the 12 National Key Economic Area (NKEA) where Malaysia has the potential to excel. We can see the Ministry of Tourism and Culture has developed many new tourism products such as the Homestay program, rail tourism, and art tourism, in-line with the government target to improve Malaysia's position within the top 10 in terms of global tourism receipts in 2015. That is to reach RM115 billion and provide 2.7 million jobs (Tourism's New Drive, 2010).

Malaysia is expecting to achieve 36 million tourist arrivals and RM 168 billion in tourism receipts by 2020. The main attraction for the tourists is Mother Nature, culture and heritage. The concept of 'peace and quiet', 'slower pace of life', 'fresh air', 'gentle' and relaxing are used to describe the rural/countryside tourism (Tourism's New Drive, 2010). The Ministry of Tourism and Culture has allocated a promotional budget that covers domestic and international promotions, which include participation in international tourism fairs, advertising campaigns, sales missions and mega familiarization programs every year. The reliance on tourism as a tool for development is used by the Malaysian government as the Ministry of Tourism and Culture promotes the Homestay program which involves mostly the rural, sub-urban community as well as small towns to offer their accommodation for the tourists.

2.2 Small town issues in Malaysia

The impact of tourism planning and destination definitely will go to the surrounding especially to the surrounding community or local resident whether negatively or positively. Nevertheless, local resident support is essential to ensure long-term success in tourism development. This is particularly important because local community is the main player of small town development and support the overall small town. As Chandralal (2010) noted that it is impossible to sustain tourism ata destination that is not supported by the local people.

According to Tatoglu (2000), there were negative perception invoke of the community due to the unstructured tourism planning, uncontrolled constructions, increase of noise level, pollution, and congestion. Mason (2000) supported this and stated that the negative environment impacts which are frequently highlighted include littering, overcrowding, traffic congestion as well as pollution of water and soil. Chandralal (2010) however mentioned that the most important benefit that the residents felt had flowed from tourism were increased employment opportunities, property values, image of the city, appearance and infrastructure of the city and improved pride as the residents.

Humans are the major force that changes the condition of land on Earth. Consequently, land transformation effects many of the planet' physical, chemical, and biological systems that impact directly on humans. Thus, a critical challenge for land use and management are to be study to overcome the problems of haphazard uncontrolled development, deteriorating environment quality, loss of flora and fauna habitat and to make sure the sustainability of the environment.

Besides that, small towns are often shown as a place with many interpersonal relationship and high in social capital. Social capital is a resource or force that can influence the quality of life of a community living environment. As explained by Putnam (2000), social capital defined as the relationship between people characterized by trust, norm, reciprocity that facilitate coordination and cooperation for mutual goal. Courvisanos and Martin (2005) said that by understanding the forces that operate on small country towns, local communities and governments will be better able to develop actions and policies which can make these towns be resilient, more viable and sustainable.

This study of small town in Malaysia contributes to the industry especially to the tourism planner as a comprehensive master plan in developing small town as a tourism product. This research also benefits the local authority in future planning for the local town as a tourism destination, hence enhance the local economy and improve the quality of community life.

Besides, this paper explores and adds-on the knowledge of tourism sector in Malaysia in creating a new direction or a new aspect on promoting our country locally as well as internationally. On the other hand, the insight resulting from this research is valuable as a beginning for other academic scholars who wish to further investigate into small towns.

3 RESEARCH PROBLEMS

This research is aimed at exploring the potential of small towns in Malaysia as a sustainable tourism product. The research issue concern includes the host-community residents' attitudes toward tourism. The sustainability of tourism development is related to the host-community response and co-operation. Therefore the strategies that need to be included in the small towns tourism planning not only comprise of the internal and external environment of the town but also the deep understanding of the host-community.

i. The level of the availability of community resources influences the quality of community life.
ii. The social patterns in small town in-terms of community bonding, bridging and linking give impact on the sense of belonging of the place.
iii. Identifying the community values of the place and determine the local intangible cultural assets and uniqueness.
iv. The level of development of central business area will influence the liveliness of the central business area.

4 RESEARCH DESIGN

According to Brewer (2000), ethnography is the study of people in naturally occurring settings or 'fields' by methods of data collection which capture their social meanings and ordinary activities, involving the researcher participating directly in the setting, if not also the activities, in order to collect data in a systematic manner but without meaning being imposed on them externally. This research method is used because the information is new and unfamiliar, besides the information requested is too subtle to be interpreted by quantitative techniques. The data collection is done through in-depth interviewing, observation and document review. This research is using ethnography method base on a case study of a small town.

There are four types of data collection methods in this study: checklist, interviews, direct observation and participant observation. In the field-work composition: a researcher's drafts (log-book), transcripts and recordings of interviews with respondents, videotapes and notes from direct field observations, and checklists.

The secondary data is gathered from the documents from the government. The report of overall economic performance also can be found on the website. The secondary information was gathered from previous research and journals, articles, blogs, newspaper, films, statistic data, documents found in the website and magazines.

As the centuries move on, small towns are being abandoned. Deterioration of small towns is especially significant in developed countries. The population of small towns has migrated to great commercial and industrial areas and cities. Not only have the communities in small towns lost population, businesses have closed or relocated. The physical environment is deteriorating, the community's spirit is low, and the agricultural base had challenged by world markets and technology.

Many researchers relate their study of small towns with globalization challenges (e.g. Courvisanos and Martin, 2005), urban sprawl (e.g. Lambe 2008; Leinberger, 2005), aging population (e.g. Nicholls, 2005; Lambe, 2008), tourism development (e.g. Altinay and Hussain, 2005), industrial development (e.g. Leinberger, 2005), and entrepreneurship.

Though, many small town planners have identified the small town potential assets both tangible and intangible before the place became completely silent. They repackage the resources of small town into tourism products that attract millions to come.

Small town research studies have been a popular subject in western countries. According to Lambe (2008), towns are divided into few categories:

- Small towns that are recreation or retirement destinations
- Small towns that have abundance of natural assets
- Small towns with historical, cultural or heritage assets
- Small towns with college campuses

By recognizing the social and economic benefits of tourism, many town planners even the country planners have had goals that involve tourism development.

5 STUDY FRAMEWORK

Please refer to Figure 1 Study Framework. This research framework shows the flow of the research

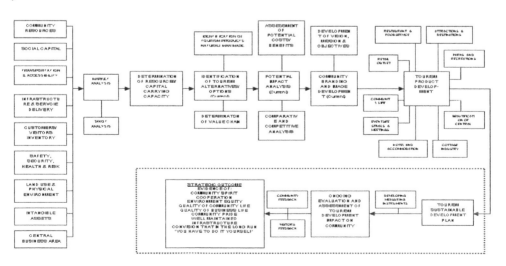

Figure 1. Study Framework.

study to be conducted in order to develop a small town research project. The diagram provides as a master guideline to the researcher to fully cover the area with the depth insight of each element in the boxes. The major study starts with nine elements that include: community resources, social capital, transportation and accessibility, infrastructure and service delivery, customers and visitors inventory, safety and security, land use and physical environment, intangible assets, and central business area as the market analysis on strengths, weaknesses, opportunity and threats. The tourism product development would be recognized and the strategic outcome of the research would show the evidence of community spirit, cooperation, environment equity, quality of community life, quality of business life, community pride, well-maintained infrastructure, and conviction that in the long run the community can be self-sustained.

6 OUTLOOK

Malaysia is a country with its richness of different cultures and races which attracted thousands of tourists to visit this colorful country. The international tourists have shown more preference on natural or ecotourism in the current decade. Small towns offer resources and culture that have potential to be repackaged into tourism products. An in-depth study needed to get the mutual understanding and support from the local community of the small town. Local community support is the main reason for the overall small town sustainability. Critical challenges for this study involve interpersonal relationship between internal and external players; to facilitate coordination and co-operation for mutual goal. The strategies include strengthening the "kampong spirit" among the local community and much can be adapted from the previous researchers on small town tourism planning. This research is to discover the nature and culture in small town as a potential destination for tourist. The importance of this study is to be a pioneer study of small town in Malaysia contributing to the industry especially to the tourism planner as a comprehensive master plan in developing small town as a tourism product. This research also benefits the local authority in future planning for the local town as a tourism destination, hence enhance the local economy and improve the quality of community life.

REFERENCES

Abdul Aziz, A., Bakhtiar, M.F.S., Che Ahmat, N.'H., Balakrishnan, M.S. (2009). Commentary Strategic Branding Of Destination: A Framework. European Journal Of Marketing Vol.43 No.5/6, pp. 611–629.

Brewer J. D. (2000) Ethnography.Open University Press, pp. 10. P1.

Chandralal, K.P.L. (2010). Impacts of tourism and community attitude towards tourism: A case study in Sri Lanka. South Asian Journal of Tourism & Heritage, 3(2), 41–49.

Courvisanos, J. and Martin, J (2005). Developing Policy for Australia's Small Town: from Anthropology to Sustainability. 2nd Country Towns Conference. paper.

Kotler, P. (1991). Marketing Management (7th Ed.). New Delhi: Prentice-Hall.

Destination Brand Identity—The Case Of The Alps, Tourism Destination Development And Branding Eilat 2009 Conference Proceedings, Vol 1, pp. 10–26.

Laksiri, W.M.R., And Falkenburg, A. (2009). Marketing Sri Lanka As An International Tourist Destination. Saarbrucken: Vdm Verlag.

Leinberger, C.B. (2005). Turning Around Downtown: Twelve Steps to Revitalization. Metropolitan Policy Program. The Brookings Institution.

Lambe W. (2008) Small Towns Big Ideas Case Study in Small Town Community Ecomonic Development. N.C. Rural Economic Development Center.

Malaysia Economic Report 2010/2011. Ministry of Finance Malaysia www.treasury.gov.my/pdf/economy/er/1011/chap3.pdf viewed on 15th July 2013

Mason, P. (2000), "Residents' attitudes to proposed tourism development", Annalsof Tourism Research, 27(2), 391–411.

Mihalis, K. (2005). Branding The City Through Culture And Entertainment. Journal Aesop 05 Vienna, pp. 1–7.

Nicholls, S., Vogt, C., Soo, H.J. (2004) as cited in Fariborz Aref, Ma'rof B Redzuan and Zahid Emby (2009) Heeding the Call for Heritage Tourism: More Visitors Want an "Experience" in Their Vacations—Something a Historical Park can provide.

Putnam, R.D. (2000) Bowling Alone: The Collapse and Revival of American Community. Simon & Schuster, 2001.

Tatoglu, E. (2010). Resident perceptions of the impact of tourism in a Turkish Resort Town.Beykent University, Department of Management, para 745–755.

Tourism's New Drive. 2010. Malaysian Business magazine. September 2010 issue.

Wagner, O. And Peters, M. (2009). The Development And Communication Of.

Theory and Practice in Hospitality and Tourism Research – Radzi et al. (Eds)
© *2015 Taylor & Francis Group, London, ISBN 978-1-138-02706-0*

STEP analysis and health tourism development in Malaysia

M.R. Dzulkipli & N.A. Mohamed-Yunus
Faculty of Business Management, Universiti Teknologi MARA, Puncak Alam, Malaysia

I. Zakaria
Faculty of Business Management, Universiti Teknologi MARA, Kelantan, Malaysia

ABSTRACT: Health tourism can be defined as the pursuit of other countries' patient to seek treatment abroad. It was noted that the utmost reason why medical tourists seek medical attention abroad was due to the economy and cost saving factors. In Malaysia, the notion of health tourism has started to emerge in the turbulence recession time in the late 90s. A dawdling economic development had forced the government to diverse its country income, hence resulting the born of health tourism activity in healthcare. This study aims to define the relationship of STEP analysis to measure the development of health tourism in Malaysia. This analysis is useful to measure the external environment that may affect healthcare organization. Apart from that, it will be useful to look the healthcare organization's potential of health tourism development from various factors as listed by STEP analysis. Few recommendations were put forward to expedite the development of health tourism to the desired state.

Keywords: Economy, health tourism development, politic, social, technology

1 INTRODUCTION

1.1 *Background*

Health tourism has started to boom in the health care industry of Southeast Asia since the recession in 1997. To maintain the survival of their business, most private hospitals have opened its doors to cater the need of the wealthy patients of the neighboring countries. As a result, most of private hospitals were resorted to promote health care services to the foreigners to compensate the "refusal" of the local population to seek treatment in the countries private hospitals (Lunt, Mannion & Exworthy, 2013).

Health tourism is defined as a vacation that require the patient to travel abroad to obtain medical services offered or offered health tourism as part of their health services (Alvarez, Chanda & Smith, 2011; Heung, Kucukusta & Song, 2010). According to Horowitz and Rosensweig (2007), health tourism can be further expanded as a mean to get appropriate and affordable medical services in which the patient seeks in developing country (which is relatively low in cost) compared to their own countries. On the other hand, health tourism activity can be redefined as exporting the medical services to the other nations particularly among developed countries. However, it must be noted that the "exportation" will take place in the place where the services is offered.

Malaysia is one of the Southeast Asia countries that are pursuing to promote health services to foreigners. Malaysia through various government efforts has put the medical expertise and services into profitable business by launching campaign and promotional activities to sell the services to the potential consumer abroad (Lunt et al., 2013). This step is taken as one of the method to vary the national income with twofold activities such as health tourism. Health tourism is believed to be able to provide appropriate medical care to the patient and in the same time allow the patient to enjoy the vacation after they recover.

Among the contributing factors that lead to the potential growth of health tourism are long waiting list that is faced by the patient in developed countries, high cost of medical services and lacking of insurance coverage to cover the higher cost of medical in certain developed countries (Garcia-Altes, 2005). Previously, it was noted that most of the patient who go overseas to obtain medical services were due to better medical technology and high skilled expertise at lower cost. Though the price to get medical services overseas was considered high, in recent development, the trend has changed whereby the opportunity to get the treatment abroad is now can be bought by the middle income countries (Leng, 2007). As a result, more involvement can be expected from developing countries that offer affordable medical cost to the potential foreigner's patient. This is further supported by Garcia-Altes (2005) who stated that tourist patients were seeking medical treatment in other countries as they are facing affordability problem to get the same treatment in their home country. Apart from that, lack of insurance coverage to certain diseases has led to the mushrooming of health tourism development to the countries that offered health tourism services such Malaysia.

In addition to that, the notion of health tourism is receiving its support due to the high medical cost in the developed country that forced their patient to find alternatives to treat their medical predicaments (Garcia-Altes, 2005). Lower health care cost which can be found in most of developing countries such as Malaysia, Thailand and India has lure most of this patient's group to get treatment in the respective countries that offered health tourism to the foreigners (Chinai & Goswami, 2007; Gupta, 2008). However, more efforts need to be done by the host countries, to combat the skeptic view in convincing the potential visitors about the quality of medical services offered are of the same level with the patient's home country.

1.2 *Gaps in health tourism development*

Medical tourism is one of the potential generating income sectors contributing to the growth of Malaysia's economy. Even though the medical tourism has boomed since late 1990s, fewer efforts were seen from the government to further commercialize the potential of health tourism. Due to this, Malaysia potential of being the leader of the health care provider in the Southeast Asia might be surpassed by the neighboring countries such as Thailand and Singapore. Venturing in medical tourism can be thought as having the twofold objectives. Medical tourism will simultaneously provide two sources of income to the country through healthcare and tourism services. This study is sought to examine the utilization of STEP analysis to measure the possible factors that influence the development of medical tourism in Malaysia. The outcome of this research is hoped to provide the policy maker pertinent information and ways to enhance the medical tourism activities in the country thus pave the formulation of parallel policy by the government to promote medical tourism to the highest extent.

2 LITERATURE REVIEW

2.1 *Social*

The change in people preferences does influence the changes and values about the services delivery in healthcare industry, which indirectly impacts medical tourism development (Piggot, 2000).

Therefore, patient preferences and need should be highlighted in delivering the healthcare. This is supported by Zhang, Dixit, and Friedmann (2010) who emphasized the need to attain customer satisfaction is vital in ensuring high profitability to the organization.

Demographic trends, population shift to the major cities, life expectancy, lifestyle, consumer preferences and expectation are among the factors representing social factors that might influence the healthcare organization (Harris & Associates, 2006; Piggot, 2000). Consumer preference is assumed as one of the important criteria to be considered in offering medical tourism services to the potential foreign patients. The quality of medical services offered need to be on par with what has been offered by the developed countries. It was found that the booster of medical tourism especially in lower and middle countries is due to the increasing awareness of the level of quality possessed by the healthcare providers (Crooks, Turner, Snyder, Johnston & Kingsbury, 2011). More healthcare providers are keen to acquire appropriate quality certification as one of the assurance to the potential patient of their ability in the services offered. This has led to the major shifting in the perception of the good and services offered by the lower and middle income countries are of inferior in term of quality (Johnston, Crooks, Snyder & Kingsbury, 2010).

2.2 *Technology*

The modern technology helped healthcare to get through of its evolution. The two sectors in medical tourism, which are medicine and tourism, rely heavily on the Internet to spread information (Bookman & Bookman, 2007). When potential tourist patients hit the web, they are posted with plethora of hospitals to choose from. It is also noted that rapid advancement in technology and lower cost has attracted thousands of health visitors to receive medical treatment abroad (Carruth & Carruth, 2010). An immense number of medical tourists come to seek medical care to developing countries that not only offering vast medical specialties and skills, but also the state of the art medical technology that were offered. However, several essentials issues need to be addressed such as training and development as this will acculturate the utilization of modern technology among medical personnel (McNish, 2002). In addition, it was suggested that tax relief should be initiated and given to the hospitals that offer medical tourism to support the need for the latest medical equipment technology (Aniza, Aidalina, Nirmalini, Inggit & Ajeng, 2009). Consumer expectation regarding technological usage and advances as well as trend and development in medical research and primary care efforts are among the factors representing technological factors that might influence the healthcare organization (Harris & Associates, 2006; Piggot, 2000).

2.3 *Economic*

From the economic perspective, Chanda (2002) stated that medical tourism can generate an investment through foreign exchange and additional resources derived from international companies who are interested to boost their healthcare services arm abroad. This is consistent with the government aim, which is to diverse national income through the introduction of medical tourism in Malaysia healthcare. Thailand and India are among the leading countries in Asia that enjoy the economic growth due to medical tourism (Hakonsen, Horn & Toverud, 2009). A supportive government roles in providing assistance such as tax reduction, land subsidies and special tariff for imported medical equipment's helps as a catalyst to the development of medical tourism and enhance the tourism rate in the country (Johnston et al., 2010).

Consumer expectation regarding technological usage and advances as well as trend and development in medical research and primary care efforts are among the factors representing economic factors that might influence the healthcare organization (Harris & Associates, 2006; Piggot, 2000).

Furthermore, the exchange rate that is promising especially to the high income country visitors might attract more potential patient to seek for medical tourism services abroad. This is noted especially when the healthcare services in the home countries offer higher prices of medical services compared to the same treatment offered in the lower or middle income countries (Johnston et al., 2010). The lower charge and cost for medical treatment in lower and middle income countries are owing to the cheap labor and medical facilities cost without compromising the credential of the medical professionals (Horowitz & Rosensweig, 2007).

Apart from that, the offering of medical tourism services in the healthcare system will boost the economic development of the country. It was found that medical tourist spend double of their money as compared to those that travel without medical seeking intention (Crooks et al., 2011; Johnston et al., 2010). That explained the eagerness of certain Asia countries India and Thailand to assist the development of their medical tourism industry specifically in the formulation of related policies (Alvarez et al., 2011). Nevertheless, there are consequences that need to be acknowledged in return of the lucrative economic's yield from

medical tourism activity. The flow of medical tourist will increase the price of medical services in local hospital. This condition will deprive the local especially underserved population from getting medical treatment in the local hospital participating in medical tourism.

2.4 *Politic*

Politic cannot be dichotomized with any activities that are happening in the country. The political factor affect largely to how the population and the apparatus involved in the decision making react and response to certain issues.

According to Piggot (2000), the trends and priorities of the state and national level will contribute on how politician will voice their views and react to health tourism. This is supported by (Gwatkin, Bhuiya & Victoria, 2004) who explained that politician behavior that prefer to support the trends will support their survival in politic. Therefore, the influence of political factor is assumed to have an impact towards medical tourism development. Thus, this research will be a milestone to the nation to advance the notion of medical tourism and placing Malaysia as an affordable and quality assured medical hub center in the South East Asia. Furthermore, it was noted that the support from the government in terms of policy formulation and regulations to govern the medical tourism activity plays a role in affecting the tourism industry (Lee, 2010).

Current state of political status in the country, health policy trends, change in arrangement involving national health system and insurance, legislative and current regulation enforced by the authority as well the influence from non-government organization and interest groups to the government are among the factors representing political factors that might influence the healthcare organization environment (Harris & Associates, 2006; Piggot, 2000).

3 DISCUSSION & CONCLUSION

3.1 *STEP analysis and health tourism development*

Malaysia has a viable and sound industry of healthcare. With a promising quality of healthcare services and lower labor cost, Malaysia can be a good hub in offering the healthcare tourism services to the potential foreigners.

Malaysia has achieved its independence from the British Crown since 1957 and inherited pretty much the same system with the Englishman. Since then, the healthcare industry has emerged and expand to offer a dual healthcare system which allow the formation of both public and private healthcare sector to operate simultaneously (Barraclough, 1999; Kananatu, 2002). To further widen the profit sources of private hospital, the notion of medical tourism has started to boom to cater the rich and wealthy potential patients from abroad. The phenomenon is believed to start during late 1990s in which the recession was at it peaks. The recession has then forced the private hospitals to find another sources of income as the local patient shift their demand to the public hospitals. As a result, the private hospitals have changed their target customers to the wealthy tourist from the neighboring countries in Asia region (Leng, 2007; Yeoh, Othman & Ahmad, 2013). The problem occurred in the patient's home countries such as waiting time, uninsured patient, availability of medical procedures could be the factors for the potential hospitals or entrepreneur to create a business in the medical tourism industry (Yeoh et al., 2013). With all these strengths, it will be a waste of opportunity if the medical tourism area is left uncultivated. For comparison, Malaysia has been left behind from the level of achievement the neighboring countries have enjoyed from medical tourism activity. Many countries such as Singapore, Thailand, Hungary, India and Korea managed to gain beneficial return from this wellness industry (Chinai & Goswami, 2007; Kiss, 2012; Leng, 2010). Therefore, it is vital for the government to strategize and realign our healthcare plan to gain fruitful vintage from this opportunity.

There are several factors that suggest the need for Malaysia to equip aggressively of medical tourism activity in healthcare industry. A good return of income from medical tourism industry for both government and the hospitals, advancement of medical technology and retention of medical professionals are among the reason why the effort to promote medical tourism should be emphasized. The STEP analysis proposed by this study is expected to conceal the holes of health tourism industry thus equip the betterment for areas of improvement in the future.

3.2 Conclusion

Health tourism industry is a promising area that needs more attention by the policy makers. The ability to provide return to the government in the form of tax yielded from the health tourism activity as well as to flourish the business activities is eminent. However, with the recent development, Malaysia is seen as in slow pace of developing the industry like the neighboring countries such as Thailand and Singapore in becoming a leading nation in health tourism. More efforts need to be done by the relevant authorities, interest groups and the industry players to boost this activity to its ultimatum. Policies related to the medical regulations need to be introduced, more promotion activities need to be undertaken to attract potential visitors to the health tourism activities. Apart from that, government support in term of tax exemption for high technology medical equipment should be given to furnish the available facilities and indirectly provide sort of niche or advantages to our healthcare providers in the eye of potential health tourism patient. These recommendations are important as to allow us to expand our grasp to the healthcare industry in Malaysia as well as to be the best leading healthcare providers in Asia.

REFERENCES

Alvarez, M.M., Chanda, R. & Smith, R.D. (2011). The potential for bi-lateral agreements in medical tourism: A qualitative study of stakeholder perspectives from the UK and India. *Globalization and Health, 7*(11), 1–9.

Aniza, I., Aidalina, M., Nirmalini, R., Inggit, M. & Ajeng, T. (2009). Health tourism in Malaysia: The strength and weaknessess. *Journal of Community Health, 15*(1), 7–15.

Barraclough, S. (1999). Constraints on the retreat from a welfare-oriented approach to public health care in Malaysia. *Health Policy, 47*, 53–67.

Carruth, P.J. & Carruth, A.K. (2010). The Financial and Cost Accounting Implications of Medical Tourism. *International Business & Economics Research Journal, 9*(8), 135–140.

Chanda, R. (2002). Trade in health services. *Bulettin of the World Health Organization, 80*(22), 158–163.

Chinai, R. & Goswami, R. (2007). Medical visas mark growth of Indian medical tourism. *Bulleting of the World Health Organization, 85*(3), 164–165.

Crooks, V.A., Turner, L., Snyder, J., Johnston, R. & Kingsbury, P. (2011). Promoting Medical Tourism to India: Messages, Images, and the Marketing of International Patient Travel. *Social Science & Medicine*, 726–732.

Garcia-Altes, A. (2005). The development of health tourism services. *Annals of Tourism Research, 32*(1), 262–266.

Gupta, A.S. (2008). Medical tourism in India: winners and losers. *Indian Journal of Medical Ethics, 5*(1), 4–5.

Gwatkin, D.R., Bhuiya, A. & Victoria, C.G. (2004). Making health systems more equitable. *The Lancet, 364*, 1273–1280.

Hakonsen, H., Horn, A.M. & Toverud, E.-L. (2009). Price control as a strategy for pharmaceutical cost containment—What has been achieved in Norway in the period 1994–2004. *Health Policy, 90*, 277–285.

Harris, M.G. & Associates. (2006). Strategy and organisational design in healthcare. In J. Madern, M. Courtney, J. Montgomery & R. Nash (Eds.), *Managing Health Services: Concepts and practice* (pp. 273–276). Sydney: Elsevier.

Heung, V.C.S., Kucukusta, D. & Song, H. (2010). A conceptual model of medical tourism: Implications for future research. *Journal of Travel & Tourism Marketing, 27*, 236–251.

Horowitz, M.D. & Rosensweig, J.A. (2007). Medical Tourism—Health Care in the Global Economy. *Trends*, 24–30.

Johnston, R., Crooks, V.A., Snyder, J. & Kingsbury, P. (2010). What is known about the effects of medical tourism in destination and departure countries? A scoping review. *International Journal for Equity in Health, 9*(24), 1–13.

Kananatu, K. (2002). Healthcare financing in Malaysia. *Asia Pac J Public Health, 14*(23), 23–28.

Kiss, K. (2012). Analysis of demand for wellness and medical tourism in Hungary. *Applied Studies in Agribusiness and Commerce, 6*(5).

Lee, C.G. (2010). Health care and tourism: Evidence from Singapore. *Tourism Management, 31*, 486–488.

Leng, C.H. (2007) Medical Tourism in Malaysia: International Movement of Healthcare Consumers and the Commodification of Healthcare. *ARI Working Paper No. 83*. Asia Research Instititute of the National University of Singapore: National University of Singapore.

Leng, C.H. (2010). Medical tourism and the state in Malaysia and Singapore. *Global Social Policy, 10*(3), 336–357.

Lunt, N.T., Mannion, R. & Exworthy, M. (2013). A Framework for Exploring the Policy Implications of UK Medical Tourism and International Patient Flows. *Social Policy & Administration, 47*(1), 1–25.

McNish, M. (2002). Guidelines for managing change: a study of their effects on the implementation of new information technology projects in organisations. *Journal of Change Management, 2*(3), 201–211.

Piggot, C.S. (2000). *Business planning for healthcare management*: Open University Press.

Yeoh, E., Othman, K. & Ahmad, H. (2013). Understanding medical tourists: Word-of-mouth and viral marketing as potent marketing tools. *Tourism Management, 34*, 196–201.

Theory and Practice in Hospitality and Tourism Research – Radzi et al. (Eds)
© 2015 Taylor & Francis Group, London, ISBN 978-1-138-02706-0

Determinants of tourist perception towards responsible tourism: A study at Malacca world heritage site

H.S. Fatin
Universiti Putra Malaysia, Serdang, Malaysia

A.S. Amirah & N.O. Khairani
Universiti Teknologi Mara, Shah Alam, Malaysia

ABSTRACT: The study aims to explore the perceptions of tourist both foreigner and local in term of cultural, experience and environment towards responsible tourism. The role of the government in implementing responsible tourism in Malacca as endorsed by UNESCO for world heritage site will be explored as well. The theoretical framework developed in this study comprised of cultural, experience and environment as independent variables, responsible tourism as dependent variable and government as the moderator. An analysis was conducted based on the questionnaire developed for 200 local and foreigner tourist who visit Malacca world heritage site. A hierarchical regression was used to test the hypotheses. The results revealed that the government did not play an important role as a moderator in developing responsible tourism among the tourist. Tourists also perceived that cultural and environments were two important variables in responsible tourism while experience was not the indicator in describing responsible tourism. In conclusion, in order to improve the perception of tourist towards responsible tourism, the government needs to highlight their roles in terms of managing responsible tourism among tourist. Interpretations of results, implications and future research were also discussed.

Keywords: Malacca world heritage site, perception, responsible tourism, tourist attitude

1 INTRODUCTION

Responsible tourism has become an important aspect with the emergence of sustainable tourism and as consumers shift towards lifestyle marketing and ethical marketing and ethical consumption (Goodwin & Francis, 2003). Responsible tourist behavior is complex, dynamic and multi-faceted and their expectations are varying (Bramwell & Lane, 2008). In addition, the lack of responsibility and careless packaging of tourism destinations as mass tourism can be damaging to the tourism industry in the long run. Therefore, it is crucial to investigate the tourists' perception towards responsible tourism in terms of their understanding of the concept of responsible tourism, culture, experience and environment.

Past studies showed that most of the studies were done on heritage sites benefit the Western countries. Since Malacca was awarded by UNESCO as world heritage site recently, a study is needed to investigate on the responsible tourism within the Malaysian context. This study was also been done since no research has been done lately regarding this topic.

Therefore, the main objective of this study was to investigate the perception of tourists both local and foreigner's perception towards responsible tourism at the Malacca Heritage Site and also to identify the role of government of Malacca towards responsible tourism.

2 LITERATURE REVIEW

2.1 Responsible tourism

European commission (2003) had stated that sustainable tourism is a tourism that is economically and socially viable without detracting from the environment and local culture. Responsible tourism not only attempts to not harm, but it is also strives to contribute the general wellbeing of local people, the tourist and the places visited.

Langen (2006) mentioned that there are an increasing number of publications and emergence of tour operators and travel agents in the responsible tourism segment. Goodwin et al. (2001) mentioned that government and the private sector are committed to work in partnership with the people of South Africa to develop and market tourism experiences that demonstrate our social, economic, environmental, technical, institutional and financial responsibility.

2.2 Cultural

Responsible tourism recognises and celebrates the diversity of the world's cultures and environments (Goodwin, 2010).

Responsible tourism is a concept of responsibility to the local community, nature and tourists. In order to fulfill the responsibilities to the local community, it should focus both on the local economy and local culture.

Culture represents an important tourism attraction of destinations. The development of tourism products based on cultural heritage requires a framework—tourism policy that provides the necessary conditions. Tourism development should accordingly protect cultural values, preserve environmental assets and maximize economic results. (Datzira, 2006). Tourist wanted information about local customs and appropriate dress and behavior for them (Goodwin, 2003). Therefore the concept of responsible tourism is concern on the culture of destination include the custom, people, place and language of the originality.

Tourists visit Malacca due to its full of cultural heritage element including historic buildings, sites, cultures and other invaluable assets that encapsulate a nation's soul and spirit. Thus, tourists that come to Malacca perceive that respecting the culture is part of their responsibility during their visit to Malacca.

2.3 Experience

Zepple and Hall (1991) see heritage tourism as a broad field of specialty travel, 'based on nostalgia for the past and the desire to experience diverse cultural landscapes and forms'. Responsible tourists have specific motivation toward and need of tourism experiences. Responsible tourists want to ensure that their tourism experience impacts positively on local host communities and environment.

In order to analyse the tourism experience literature, key themes are developed to explicate the concept. The following themes are discussed; emotional elements & social inclusion, environment to experience, involvement, social science and marketing management approach, experience embedded in long term memory (Murray, 2010).

Malacca as a Historical City has a lot to offer a great experience to tourists in terms of visiting historical sites. Buyong and Rajiani (2011) proved that Malacca has fulfilled the requirements to differentiate the city as a unique tourist destination. The positioning of Malacca as Malaysia truly Asia and World Heritage Site should be translated into the rational benefit of encountering unspoiled historical side and multi-racial living cultures. Positive unique image creation leads to intention to revisit and recommend others experiencing the world heritage and history of Malacca. When tourist has a great experience, they will have the sense of responsible on their vacation at the Malacca World Heritage Site.

2.4 Environment

It was crucial for tourist to have their holiday that did not damage the environment. The tourists are also willing to pay more to assist in preserving the local environment and reversing some negative environmental effects of tourism (Goodwin, 2003). In term of environmental impacts, tourism has infected serious "architectural pollution" as the historical buildings and churches of once quaint sleepy towns and village either have been pulled down (Loannides, 1995).

According to Stankovic (1991), the protection of nature is not a task of individuals and of specialized agencies. Nature has to be protected by all, for thereby they protect themselves, they preserve the man in it. The modern idea of the problems and processes concerning on protection of nature for tourism need to be based on a concept of determining the relationship between human and society to it. In connection with this, it stress on the substance to the concept in active protection of nature. According to this concept, the protection of nature must not be restricted to the protection of individual natural rarities and isolated reservations. Nature should be protected in it's entirely.

In national environment policy, Malaysia's overall environmental policy will take the factors of the need to maintain a healthy environment for human habitation and the need to preserve the country's unique and diverse natural heritage, all of which contribute to the quality of life. As Malacca has become the tourist attraction, so there is the need of responsible tourism in persevering and maintaining the environment.

2.5 Government

In implementing the responsible tourism in the tourism destination, government play an important role due to its management of historic properties and culture and consequences of their action, whether their concerns result in responsible behavior and did not yet forego business opportunities from promoting heritage tourism. Implementing the guidelines improves relations with the community and protects the environment (Merwe, 2007). Therefore, the government should implement the guidelines to hotel, tourists' operators and tourist provider in order to protect the environment.

Government is the moderating variable because that they are middleman between tourist provider, government agencies and industry association as well as local community towards maintaining the listing status and preserving cultural heritage, cultures, languages and food in the long run. Government challenges are to manage the cultural heritage and capitalizing on Malacca's tourism potential at the same time.

3 METHODOLOGY

Based on the above past studies, the following model is developed to be tested in this study.

The independent variables recognized in this study are cultural, experience, and environment while dependent variable is the responsible tourism and this study also use the role of government as moderating variable. The sampling size constituted of 200 tourists that visit Malacca World Heritage Site.

The sampling design used in this research was the simple random sampling. The study was administered personally by distributing questionnaires to the respondents. The questionnaire item were extracted and adapted from Harold Goodwin (2003). The respondents were required to rate their level of agreement with statements given using five point scales ranking from strongly disagree (1) to strongly agree (5). In this study, SPSS version 17 was utilized to analyze information gathered. In order to interpret the data analysis for this study, this study used the quantitative analyses which are frequency distribution, reliability of measure using Cronbach's Alpha, multiple regressions, and descriptive statistics and coefficient correlations.

There were 4 hypotheses have been developed for this study:

Hypothesis 1: There is a significant relationship between culture and responsible tourism

Hypothesis 2: There is a significant relationship between experience and responsible tourism

Hypothesis 3: There is a significant relationship between environment and responsible tourism

Hypothesis 4: There is a significant relationship between government and responsible tourism

4 FINDINGS AND DISCUSSION

4.1 Discussion

Summary of the hierarchical regression on government as a moderator in relationship between cultural, experience and environment with responsible tourism is explained in Table 1.

The model shows that 71.7 percent of the dependent variable is explained by the changes of independent variables. It indicates that this model is useful with adjusted $r^2 = 0.676$, means of 67.6 percent indicating that the factors may influence responsible tourism that can be explained by all the independent variables (experience, cultural and environment) and moderating variable (government).

H1: There is a significant relationship between culture and responsible tourism

The study also found out that there is a strong linkage between cultural and responsible tourism ($B = 0.577$, $p = 0.000$), therefore cultural is significant at 0.05 level with responsible tourism. Based on the findings, there is a significant relationship between cultural and responsible tourism. Thus,

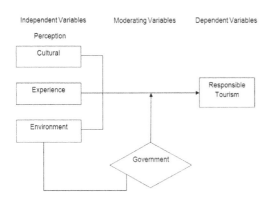

Figure 1. Research model (Sources adapted from Goodwin, H., and Francis, J., (2003)).

Table 1. Summary of result on regression on government as a moderator in relationship between cultural, experience and environment with responsible tourism.

Model (Independent Variable)	Standardized Beta	t-value	t-value sig.
Cultural	0.577	3.894	.000*
Experience	-0.126	-0.805	0.423
Environment	0.401	2.532	.013*
Government	-.038	-0.297	0.767
Cul_Gov	.015	0.147	0.884
Exp_Gov	-.029	-0.277	0.783
Env_Gov	-0.108	-1.136	0.259
Model Summary			
R-Sq		0.717	
Adjusted R-Sq		0.676	
F Value		1.185	
F value Sig		0.320	
Durbin Watson		1.407	

hypothesis 1 is supported. The characteristic of responsible tourism are culturally sensitive, engenders respect between tourists and hosts, and builds local pride and confidence. It shows the important of cultural in order to educate tourist to be more responsible (Goodwin, 2003).

H2: There is an insignificant relationship between experience and responsible tourism

For the second independent variable which is experience, the study revealed that there is a weak linkage between experience and responsible tourism ($B = -0.126$, $p = 0.423$), therefore experience is insignificant with responsible tourism. Hypothesis 2 was rejected. It was supported by Goodwin and Francis (2003) and stated that tourist will engaged in responsible tourism if they have real holiday by having a better experience in their vacation. In this study, tourist does not really engaged themselves in their holiday experience as they only experience with the place and their visit at Malacca World Heritage Site did not emotionally attached with real holiday.

H3: There is a significant relationship between environment and responsible tourism

For environment, this study found out that there is also a strong linkage between environment and responsible tourism ($B = 0.401$, $p = 0.013$), therefore environment is significant at 0.05 level with responsible tourism. Hypothesis 3 was accepted. This was also supported by Zoe (2005) stated that many travellers seek out pristine environments to visit, and it is important to the vast majority of them that their trip not damage local ecosystems. They are interested in patronizing hotels that are committed to protect the local environment, and increasingly view local environmental and social stewardship as a responsibility for the businesses they support thus; it shows that responsible tourism has an important impact on the environment.

H4: There is an insignificant relationship between government and responsible tourism

For moderating variable which is government, this study found out that there are several weak and insignificant linkages between government and responsible tourism ($B = -.038$, $p = 0.767$); cultural and responsible tourism with government ($B = .015$, $p = 0.884$); experience and responsible tourism with government as moderator ($B = -.029$, $p = 0.783$); and environment and responsible tourism ($B = -0.108$, $p = 0.259$) with government. It can be concluded that tourist perceived that government does not play an important role in responsible tourism. Therefore, hypothesis 4 was rejected.

This was supported by Goodwin (2003) as tourist perceive do not see the role of government in responsible tourism as they only focus on hoteliers and tour operators in development of tourism. As the study has been done in Malacca World Heritage Site, the role of Non-Government Organization (NGO) such as Malacca Tourism Association has been seen as a major player in developing responsible tourism as compared to the government role.

4.2 Summary of hypothesis testing

Based on the findings of the regression analysis shown, there are significant relationships between independent variables (cultural and environment) and dependent variable (responsible tourism) whereby hypothesis 1, and 3 are supported. Hypothesis 2 and 4 are rejected because there are insignificant relationships between perceptions on experience towards responsible tourism and government and responsible tourism. In conclusion, based on the hierarchical multiple regression models, the researcher found that cultural and environment has a significant influence towards perception of responsible tourism at the Malacca World Heritage Site. Table 2 shows the summarization on the hypothesis testing.

5 CONCLUSION

The result of this study showed that only two significant variables relationship existed i.e. cultural and environment towards responsible tourism. This research will provide new findings pertaining to tourists' perception towards responsible tourism from Malaysian perspective since there are little study has been devoted to responsible tourism practices in Malaysia. Future researchers will find this study as a significant reference in order to help their government to have constant advancement in a country's tourism industry.

This research also will give the opportunity for tourists to understand their responsibility towards their holiday destination by preserving the natural

Table 2. Summarization of hypothesis testing result.

Research objectives	Hypothesis testing	Result
To investigate perception of tourists towards responsible tourism at Malacca Heritage Site.	There is a significant relationship between cultural and responsible tourism.	H1 Accepted
	There is an insignificant relationship between experience and responsible tourism.	H2 Rejected
	There is a significant relationship between environment and responsible tourism.	H3 Accepted
To identify the role of government of Malacca towards responsible tourism at Malacca Heritage Site.	There is an insignificant relationship between government and responsible tourism.	H4 Rejected

aspect of building and cultural environments of Malacca as the heritage site.

The government should be aware of the responsible tourism practice at Malacca as a World Heritage Site. In order for them to practice, they need to allocate some fund accordingly based on the market segment that has been recognized within this research. They also can focus on the advertising campaign to foreign and local tourist in order to increase the number of arrivals.

The methodology used can be further enhanced by using structural equation modeling (SEM) that can establish a model as depicted in the theoretical framework which can be viewed holistically. In conclusion, based on the hierarchical regression models, the study found that the cultural and environment factors has a significant impact towards responsible tourism at Malacca World Heritage Site while experience and role of government factors are not significantly associated with responsible tourism.

REFERENCES

Bramwell, B. & Lane, B. (2008), A. 2005. 'Priorities in sustainable tourism, *Journal of Sustainable Tourism.*

Buyong E, Rajiani I (2011),7th Global brand conference of the Academy of Marketing's Brand, Brand, Identity and Reputation: Exploring, creating new realities and fresh perspectives on multi-sensory experiences, corporate identity and reputation SIG April 5th–7th 2011.

Datzira, J., and Masip., (2006). Cultural Heritage Tourism—Opportunities for Product Development: The Barcelona Case, *Journal Tourism Review*, Vol 61, No 1.

Duncan, I. (1995). Tourism development: The role of government. Wellington: New Zealand Institute of Economic Research.

Goodwin, H., (2005). Responsible Tourism and the Market, Occasional Paper No. 4.

Goodwin, H., and Francis, J., (2003). Ethical and responsible tourism: Consumer trends in the UK, *Journal of Vacation Marketing*, Vol. 9, No. 3, pp. 271–284.

Goodwin, H, Stuart Robson and Sam Higton (2004).The International Centre for Responsible Tourism.

Langen, K.C., (2006). Responsible tourism management at world heritage site.

Loannides, D., (1995). Planning for International Tourism in Less Developed Countries: Toward Sustainability, *Journal of Planning Literature*, Vol 9, pp. 235.

Merwe, M.V.D., and Wocke, A., (2007). An investigation into responsible tourism practices in the South African hotel industry, *Journal of S.A fr. J. Business Management*, pp. 38(2).

Murray, N., Foley, A., and Lynch, P., (2010). Understanding the tourist experience concept The RIKON Group, School of Business, Waterford Institute of Technology.

Stankovic, S.M., (1991). The Protection of Life Environment and Modern Tourism.

Zeppel, H., and Hall, C.M., (1992).Arts and Heritage Tourism. Special Interest Tourism., *Belhaven Press*: London.

Zoë Chafe (2005) Consumer Demand and Operator Support for Socially and Environmentally Responsible Tourism The International Ecotourism Society CESD/TIES Working Paper No. 104.

Theory and Practice in Hospitality and Tourism Research – Radzi et al. (Eds)
© *2015 Taylor & Francis Group, London, ISBN 978-1-138-02706-0*

Entrepreneurial knowledge and hospitality awareness of agro-tourism

Z. Siti-Hajar, M.S. Fadzilah, M. Muzzamir & A.H. Norhaslin
Faculty of Hotel and Tourism, Universiti Teknologi MARA, Shah Alam, Malaysia

ABSTRACT: This article deals with how farm diversification and farm based tourism influence the agro-tourism of farmers. Some growing loses on entrepreneurial knowledge and lack of hospitality awareness by farmers and this issue starts to receive significant attention among the academic scholars. Understanding the process of sharing entrepreneurial knowledge are important same goes to hospitality awareness in ensuring the survival and continuing involvement of farmers to tourism, thus this paper conceptually discusses the entrepreneurial knowledge and hospitality awareness and relates it into the contextual teaching and learning of these knowledge for Malaysia scenario.

Keywords: Agro-tourism, farm diversification, farm based tourism, entrepreneurial knowledge, contextual learning

1   INTRODUCTION

1.1   *Background of study*

Tourism industry is the second largest contributor that gives effect to Malaysia. In 2008 alone, Malaysia recorded 22.05 million tourist arrivals, and the tourism industry contributed RM 49.6 billion (USD 13.4 billion) of revenue (Fahmi, Hamzah, Muhammad, Yassin, Samah, D Silva & Shaffril., 2013). This indicates that the tourism industry helps in enhancing local economic growth as well as the economic multiplier effects (Jaafar, Kayat, Tangit & Yacob, 2013).

Tourism, particularly ecotourism in developing countries reflects richness of natural and cultural capital, and is considered a sustainable source of revenue for indigenous and rural communities (Jaafar et al., 2013). Besides, desire between rural and urban family for relaxation and recreation increase the market potential for agro-tourism (Gao, Barbieri & Valdivia, 2013). In many areas, rural tourism has been associated with agro-tourism or farm tourism and national park and wilderness (Jaafar et al., 2013). Agro tourism is among rural tourism that allows travelers to experience the life of a farmer as well as visit the farms (Kunasekaran, Ramachandran & Shuib, 2011).

From the economic views, tourism industry gives benefits like foreign-exchange inflows, employment creation and diversification of individual income, improve economic structure and increase standard of living (Jaafar et al., 2013). However, there are disagreements on the subject because from Malaysians' views based on previous study at Kedah shown most of the local residents are not getting the benefits from agro-tourism activities. It is because they are less involved and they unable to respond to the new job opportunities created. This situation shows that the authorities do not take care of promoting the local population to participate and help in decision-making programs that suit with their live environment (Fahmi et al., 2013). Many studies looking at the effects of tourism on the community, however the main character in the tourism sector itself are seldom discussed (Kunasekaran et al., 2011). Although many studies have been done on the perception and impact of agro-tourism, study on the perception of farmers on agro-tourism is less which result of lack of knowledge and theoretical framework (Kunasekaran et al., 2011).

As mention early, the study found that the majority of the local community keenly to participate in agro-tourism activities either directly or indirectly. However, these opportunities are dependent on the government on how to improve the level of confidence of the participants to ensure the successful agro-tourism projects besides increase their participation (Fahmi et al., 2013). Additionally, it is agreed by (Kunasekaran et al., 2011), that the utilization of rural resources will shrink when there is lack of involvement and inclusion from the local community.

According to Kunasekaran et al. (2011), most study already focused on the impacts of tourism.

However, there is a gap like awareness and entrepreneurial knowledge were not found in literature. In line with the above notion, there are great concerns among nations about the potential loss both entrepreneurial knowledge and awareness about farm based tourism (Kunasekaran et al., 2011); (Muhammad, Hamzah, Mohamed Shaffril, Lawrence, D Silva, Yassin, Abu Samah & Tiraieyari, 2012). Thus, this paper conceptually discusses the entrepreneurial knowledge and hospitality awareness of agro-tourism in Malaysia scenario.

2 LITERATURE REVIEW

2.1 *Agro-tourism*

In essence, agro-tourism is tourism activities involving agriculture carried out in rural areas in groups or individually. It is an alternative tourism product where tourists will pay a sum of money as resident guest or day trippers to experience recreational activities and know-how life with the farming community (Fahmi et al., 2013). The benefits of agro-tourism in term of economics are unemployment rate and rural migration can be reduced because youngster no needs to travel to urban because they can create income from agro-tourism (Fahmi et al., 2013).

In the European Union (EU) farmers from that countries can access the LEADER program that offers grants for the promotion of rural development. Whereas in lesser government support, agritourism has also emerged as an alternative economic activity among farmers in the United States of America (US) (Gil Arroyo, Barbieri & Rozier Rich, 2013).

Agro-tourism also been known as agritourism in certain place. Normally definition of agritourism is visiting a working farm or any other agricultural situation for pleasure, education or vigorous involvement in operation's activities. However, there is much debate over the meaning of agritourism scientifically, as there is a geopolitical context associated with the government. For example, (Gil Arroyo et al., 2013) sum the meaning of agritourism as a set of policies establishing specific guidelines, obligations and incentives to assist and encourage farmers to diversify their entrepreneurial portfolio through tourism and hospitality services fostered the development of agritourism.

2.2 *Farm diversification*

The farm diversification into tourism demands a major changes because it requires new abilities and capabilities that persuade mentality and identity (Brandth & Haugen, 2011). According to (Di &

Miller, 2012), data from UK Government Department for the Environment, Food and Rural Affairs (DEFRA), total income from farming per head and total income from farming has fallen. Therefore, in order to continue in business, farmers are encouraged to venture into different areas such as farm tourism other than family-based business.

Farm diversification previously is seen as a challenge to farmers and their identity against this change because they want to maintain their status and identity. However, recently farm diversification has been seen as a diversion to find other income rather than committed to conventional agriculture only (Own & Me, 2010). It is agreed by (Brandth & Haugen, 2011), the diversification of farming into tourism in many ways a fundamental change since it demands new skills and competencies and may influence mentality and identity.

However, some studies found that farmer's identity are not unaffected by diversification. It is because for example, entrepreneurial identity will fit well with how farmers conceive of themselves. The diversified that farmers see are as both entrepreneurial and farmers. In addition, the farms that diversify into tourism are making several sources of income not transition wholly (Brandth & Haugen, 2011).

According to (Di & Miller, 2012), the challenges that farm families have to face about diversification are issues on their identities, ways of life and preferences. Other dilemmas they have to encounter are perceptions and experiences of authenticity and their relation with farming.

2.3 *Farm based tourism*

The meaning of farm-tourism has changed over the years. Early definition stated that visitation tended to emphasis the farm stay and romanticism of the countryside. However, with the multifaceted phenomenon today, farm tourism is still a form of rural tourism even it is more varied and become more difficult to describe (Own & Me, 2010). (Own & Me, 2010) described the transition from 'tourism on farms' to 'farm tourism' and they suggest on various factors such as the level of marketing, competition, entrepreneurship and investment and the level of tourism versus agricultural income.

Farm-based recreation and tourism is a plan under European Union's (EU) Common Agricultural Policy (CAP) to help the farm families who are facing farm incoming problems and for those who want to stay behind the land operations (Phelan & Sharpley, 2012).

The findings from (Brandth & Haugen, 2011) stated that the goodness of combining farming with tourism was that great until they have to sold their sheep and leased their land to their neighbor farmer to help them cater the business during the peak periods.

2.4 Entrepreneurship in tourism

Definition of entrepreneurship in general is the initiative to change through new creation and innovation that bear by risk (Zhao, Ritchie & Echtner, 2011). Scholars also mentioned that entrepreneurship is about discovering and exploiting opportunities (Zhao et al., 2011). It is also suggested that entrepreneurship research should deal with the phenomena of emergence, such as how opportunities are detected and acted upon, or how new organizations come into being.

Entrepreneurial skills have been dispute either it can be learned and experientially acquired or it can be manipulate from a certain strategic measures like business education (Ismail & Ahmad, 2013). Some scholars' mentions recommend that entrepreneurial elements can be positively persuading by educational programmes. Then, a group of scholars from USA thought that someone cannot be teach to be an entrepreneur but still can be teach about the entrepreneurial skills needed in order to be a successful entrepreneur (Ismail & Ahmad, 2013).

It is said that tourism industry will catch the attention of incomers to start businesses because it will initiate new skills and access to network beyond the local economy. However, it leads to questions because not all tourism businesses are good in entrepreneurial or innovative (Bosworth & Farrell, 2011).

2.5 Contextual teaching and learning

Contextual teaching and learning (CTL) is one of the hot topics in education today. The contextual approach is a learning philosophy that emphasizes students' interest and experiences. The CTL was developed by the Washington State Consortium, which involve 11 universities, 20 schools and some education organizations in the United States (Satriani, Emilia & Gunawan, 2012).

Based on Piagetian constructivism, cognitive constructivist see the students construct their own meaning of reality (Porcaro, 2011). It is similar with Johnson (2002), CTL is a learning concept that helps teacher connects the learning material to the real condition of the student and encourages students to use their own knowledge in their daily life. This method will help students to be a more independent and natural learners in their effort to develop their knowledge. CTL has the potential to be more than just another blip on the screen of ephemeral classroom practices. CTL offers a pathway to academic excellence all students can follow

According to Johnson (2002), CTL has the potential to be more than just another blip on the screen of ephemeral classroom practices. CTL offers a pathway to academic excellence all students can follow. CTL is a learning concept that helps teacher connects the learning material to the real condition of the student and encourages students to use their own knowledge in their daily life. This method will help students to be a more independent and natural learners in their effort to develop their knowledge.

Therefore, CTL is one of the innovative approaches in teaching and learning. According to Tural (2012), contextual learning theory occurs only when students (learners) process new information or knowledge in such a way that it makes sense to them in their own frames of reference (their own inner worlds of memory, experience and response). This approach to learning and teaching assumes that the mind naturally seeks meaning in context that is, in relation to the person's current environment and that it does so by searching for relationships that make sense and appear useful.

Thus, CTL approach will be apply in transferring entrepreneurial knowledge and hospitality awareness to the farmers to achieve the research objectives later. The entrepreneurial knowledge and hospitality awareness will be transferred to the farmers on class organized for them.

3 CONCLUSION

From the studies, there is a great concern on entrepreneurial knowledge and hospitality awareness among the person that get involved in agro tourism in Malaysia. Kunasekaran et al., (2011), found factors like awareness and entrepreneurial knowledge would be beneficial for an investigation of general perception of the local community on tourism. However, these factors were not found in literature because most studies concentrated only on the impacts of tourism. The involvement of rural communities in the agro tourism programmes is not well integrated into the rural development strategies and participation in entrepreneurial activities related to agro tourism is still low (Muhammad et al., 2012).

To date, entrepreneurial knowledge among the farmers and the hospitality awareness have not been widely researched. In fact, there have been limited comprehensive studies on the entrepreneurial knowledge and hospitality awareness among the farmers in Malaysia. To answer such questions besides creating literature and lay the groundwork, some empirical evidences on the highlighted issues needs to be explored.

REFERENCES

Bosworth, G. & Farrell, H. (2011). Tourism entrepreneurs in Northumberland. *Annals of Tourism Research*, *38*(4), 1474–1494.

Brandth, B. & Haugen, M.S. (2011). Farm diversification into tourism—Implications for social identity? *Journal of Rural Studies*, *27*(1), 35–44.

Di, M. & Miller, G. (2012). Farming and tourism enterprise : Experiential authenticity in the diversi fi cation of independent small-scale family farming. *Tourism Management*, *33*(2), 285–294.

Fahmi, Z., Hamzah, A., Muhammad, M., Yassin, S.M., Samah, B.A., D Silva, J.L. & Shaffril, H.A.M. (2013). Involvement in Agro-Tourism Activities among Communities in Desa Wawasan Nelayan Villages on the East Coast of Malaysia. *Asian Social Science*, *9*(2), 203–207.

Gao, J., Barbieri, C. & Valdivia, C. (2013). Agricultural Landscape Preferences: Implications for Agritourism Development. *Journal of Travel Research*, *53*(3), 366–379.

Gil Arroyo, C., Barbieri, C. & Rozier Rich, S. (2013). Defining agritourism: A comparative study of stakeholders' perceptions in Missouri and North Carolina. *Tourism Management*, *37*, 39–47.

Ismail, M.Z. & Ahmad, S.Z. (2013). Entrepreneurship education: an insight from Malaysian polytechnics. *Journal of Chinese Entrepreneurship*, *5*(2), 144–160.

Jaafar, M., Kayat, K., Tangit, T.M. & Yacob, M.F. (2013). Nature-based rural tourism and its economic benefits: a case study of Kinabalu National Park. *Worldwide Hospitality and Tourism Themes*, *5*(4), 342–352.

Johnson, E.B. (2002). *Contextual Teaching and Learning*. California: A Sage Publication Company.

Kunasekaran, P., Ramachandran, S. & Shuib, A. (2011). Development of Farmers ' Perception Scale on Agro Tourism in Cameron Highlands, Malaysia. *World Applied Sciences Journal*, *12*, 10–18.

Muhammad, M., Hamzah, A., Mohamed Shaffril, H.A., Lawrence D Silva, J., Md. Yassin, S., Abu Samah, B. & Tiraieyari, N. (2012). Involvement in Agro-tourism Activities among Fishermen Community in Two Selected Desa Wawasan Nelayan Villages in Malaysia. *Asian Social Science*.

Own, S.Y. & Me, N. (2010). Agritourism and the Farmer as Rural Entrepreneur : A UK Analysis, (April).

Phelan, C. & Sharpley, R. (2012). Exploring entrepreneurial skills and competencies in farm tourism. *Local Economy*, *27*(2), 103–118.

Porcaro, D. (2011). Applying constructivism in instructivist learning cultures. *Multicultural Education & Technology Journal*, *5*(1), 39–54.

Satriani, I., Emilia, E. & Gunawan, M.H. (2012). Contextual Teaching And Learning Approach To Teaching Writing. *Indonesian Journal of Applied Linguistics*, *2*(1), 10–22.

Tural, G. (2012). The Process of Creating Context Based Problems by Teacher Candidates. *Procedia—Social and Behavioral Sciences*, *46*, 3609–3613.

Zhao, W., Ritchie, J.R.B. & Echtner, C.M. (2011). Social capital and tourism entrepreneurship. *Annals of Tourism Research*, *38*(4), 1570–1593.

Theory and Practice in Hospitality and Tourism Research – Radzi et al. (Eds)
© *2015 Taylor & Francis Group, London, ISBN 978-1-138-02706-0*

HR issues of Gen Y in tourism: Anticipating the future challenges

N. Aqilah-Ahmad
School of Graduate Studies, Universiti Putra Malaysia, Malaysia

A.H. Jantan & D. Zawawi
Faculty of Economics and Management, Universiti Putra Malaysia, Malaysia

M. Othman
Faculty of Food Science and Technology, Universiti Putra Malaysia, Malaysia

ABSTRACT: Tourism industry has become one of the largest contributors to Malaysia's economy. In 2012, there were 25.03 million tourist arrivals contributing 60.6 billion ringgits to Malaysian economy. This resulted in the tourism industry becoming one of the top three contributors of foreign exchange to the economy. However, employees' turnover is reported to be high in this industry and it is increasing over the years. Ministry of Human Resource reported that in 2009, the tourism employee turnover was 16 percent and turnover of hoteliers contributed half of the number. This paper conceptually discusses the human resource challenges in the tourism industry and at the same time provides recommendation for future research.

Keywords: Tourism, turnover, Gen Y, retention

1 INTRODUCTION

International tourism is one of the important sources of revenue in directing Malaysian economy towards a higher growth (Mazumder & Ahmed, 2009). The industry has become the second largest foreign exchange earner after the manufacturing industry in Malaysia. Specifically, there were 25.03 million tourist arrivals contributing to 60.6 billion ringgits to Malaysian economy in 20012. This resulted in the tourism industry becoming one of the top three contributors in term of foreign exchange to the economy (Economic Transformation Programme Annual Report, 2012). Despite the increase in tourist visits, Ministry of Human Resource reported a 16 percent turnover of tourism employees in 2009, and half of the turnovers were the hoteliers (Saad, Yahya & Pangil, 2012). For instance, Abdullah et al. (2010) found that the employees in hotel industry who worked for more than two years but less than five years changed job constantly compared to those who worked for less than two years or more than five years. Despite the growth, the tourism and hospitality industry is facing numerous challenges in retaining their employees and this issue has severely affected the Malaysian tourism industry (Alonso & Neill, 2009; Chiang & Birtch, 2008; Karatepe & Uludag, 2012; Roney & Öztin, 2007; Poulston, 2008).

There are four key themes discussed in this paper. Firstly, the study evaluated the literature on the nature of work in tourism. The second theme addresses the entrance of the new generation employees in the workforce known as generation Y. The characteristics and expectations of work environment and industry of Generation Y could be relevant to the intention to leave many organizations (Doherty, Guerrier, Jamieson, Lashley & Lockwood, 2001; Jenkins, 2001; Pavesic & Bymer, 1990; Zacerelli, 1985). This study also looked at the emerging issues from the literature focusing on the role that work-life balance (WLB) play towards the intention to leave. Lee and Shin (2005) found that the psychological dimensions specifically job burnout, and cynicism were significant predictors of turnover intentions. The discussion on the retention issue in the tourism work industry is then followed.

2 LITERATURE

2.1 *Nature of work in tourism industry*

Numerous tourism and hospitality management graduates distance themselves from the industry, and fail to enter the industry due to low job satisfaction, poor employment conditions, and absence of motivating factors. These also result in high turnover and wastage of many well-trained and experience personnel (Doherty, Guerrier, Jamieson, Lashley & Lockwood, 2001). Kulsuvan and Kulsuvan (2002) found that, in Turkey, workers or interns in the hospitality industry claimed that their negative attitude towards tourism careers and intention to leave were impacted by the nature of work in the industry.

2.2 *Generation-Y*

Over the past decade, new generation of employee has entered the workforce. This generation is known as generation Y who were born in year 1981–1999. Generation Y has dramatically different expectations of the work environment and industry compared to their predecessors (Chen & Choi, 2008; Gursoy, Maier & Chi, 2008).

Generally, it can be said that generation Y has a strong sense of morality, is willing to fight for freedom, is sociable, and values home and family. Besides that, generation Y are seen to have high expectation of job compare to previous generation, including high expectation of pay, conditions, promotion, and advancement. In addition, generation Y also tends to want an intellectual challenge, needs to succeed, strives to make a difference, and seeks employers who will further their professional development. The Generation Y cohort is accustomed to being active in family decisions. Thus, they are more likely to expect a similar amount of authority or ability to contribute to decisions in employer organizations (Johns, 2003). In the workplace, Generation Y favors an inclusive style of management; they tend to dislike slowness, and desire immediate feedback about their performance (Francis-Smith, 2004).

2.3 *Work life balance*

The issues relating to obtaining a work life balance (WLB) have received substantial attention over recent years, especially in the area of hospitality and tourism. Work in long hours is consistently associated with worse work-life outcomes and all our work-life measures (Pocock, Skinner & Williams, 2008). Thus, the authors suggest that there is unhealthy acceptance of long working hours, especially in service industries. In addition, for young workers, as they are seeking for work life balance and flexibility in work, these longer working hours that offered by hospitality and tourism industry is unsocial in the way that such workers conduct their social or family lives. Likewise, the most significant predictor of turnover intentions was the job burnout dimension of cynicism (Lee & Shin, 2005). Mulvaney, O' Neill, Cleveland, and Crouter (2006) suggested that the levels of conflict between work and family will be impacted

or moderated by the levels of support employees. Namasivayam and Zhao (2007) and Karatepe and Uludag (2007), together with Rowley and Purcell (2001) argue work and family conflict does influence employees' organizational commitment and ultimately lead to employee intention to leave.

2.4 *Retention*

Attracting and retaining well-educated, skilled, enthusiastic and committed workers is a chronic problem in the hospitality and tourism industries (Lucas & Johnson, 2003). With the lack of experienced and knowledgeable employees, the tourism and hospitality industry runs a risk of having problems in generating profits, and hindered growth. Due to this, the issue of retention is considered vital for any player in the industry. Therefore, the tourism and hospitality organizations are being forced to rethink their human resource policies and strategies in variety of areas, including training, recruitment, and compensation and scheduling.

Previous works have focused on the important roles that appropriate recruitment plays in retaining good staff (Collins, 2007; Dermody, Young & Taylor, 2004; Martin, Mactaggart & Bowden, 2006; Reynolds, Merritt & Gladstein, 2004). Improving the quality and quantity of tourism and hotel employees leads to improving the image of the industry, together with more strategic ways of managing work rosters and workloads.

3 HUMAN RESOURCE CHALLENGES IN TOURISM

Human resource management plays a major role in assisting organization to attract and retain employees. It is important to understand the needs of Generation Y as they are the generation who are going to take over the workforce in the future. As stated by Morton (2002), generation Y employees show a tendency towards valuing equality in the workplace and they seek for positions that offer reasonable wages and good opportunities for career advancement. In addition, the Y generation respects managers who empower workers and who are open and honest with employees. New employees or fresh graduates are eager to start work in hospitality and tourism, however it seems the hospitality industry work environment is not well prepared for the new arrival (Richardson, 2010a, 2010b). Thus these would lead to shortage of skilled workers in the industry (Ferris, Berkson & Harris, 2002; Freeland, 2000; Hinkin & Tracey, 2000) and cause organizations having difficulties in several aspects related to economic and performance of the business organization, including delaying in developing new product and service, increasing the operating cost and having barriers to meeting the required quality standards (Marchante, Ortega & Pagan, 2003).

Besides, the changing nature of employee demographic and attitude require new ways of understanding values, attitude and behaviors needed (Solnet & Hood, 2008). The main argument by the authors is that the current nat ure of work offered by hospitality and tourism organizations do not appear to motivate generation Y employee. Thus, employers must have the understanding of the needs and aspiration of the new generation of employee to ensure the industry can offer the opportunities seek by this group. With a proper human resource management practice by organizations, such as recruitment, intrinsic and extrinsic benefits and performance management, it will generate a positive organizational outcomes leading to higher profitability, lower turnover and better customer retention.

In Malaysia, the major challenge in upcoming years for tourism industry is to ensure that Malaysia's tourism industry's service is at level of excellent. Malaysia received about 22 million tourists and they have a different kind of expectation. Thus, it is the challenge for Ministry of Tourism and Tourism Malaysia to ensure that the tour operators and tour guides are well monitored. In order to ensure that the supply of labor is continuous, the tourism industry will need to rebrand itself. This forum is to create a favorable image of the industry so that future candidates feel interested to join the tourism workforce, and later reside their careers with the tourism industry.

4 FUTURE RESEARCH

Based on previous studies, a niche industry like tourism will usually require a group of well-educated, skilled, enthusiastic and committed to work. Since this industry is unique in nature, it is necessary to conduct extensive research on its human resource requirements and challenges before continuing to develop a typology of skill set and values that would fit into it. The intention of developing a typology is to make better projections of the future workforce needs. At the same time, the study also intends to look at the variation in term of skills, values and expectations of Gen Y pertinent to the current recruitment practices.

REFERENCES

Abdullah, R., Mohd Alias, M.A., Zahari, H., Karim, N.A., Abdullah, S.N., Salleh, H. & Musa, M.F. (2010). The study of factors contributing to chef turnover in hotels in Klang Valley, Malaysia. *Asian Social Science, 6*(1), 80–85.

Alonso, A.D. & O'Neill, M.A. (2009). Staffing issues among small hospitality businesses: A college town case. *International Journal of Hospitality Management, 28*, 573–578.

Chen, P. & Choi, Y. (2008). Generational differences in work values: A study of hospitality management. *International Journal of Contemporary Hospitality Management, 20*, 595–615.

Chiang, F.F.T. & Birch, T.A. (2008). Achieving task and extra-task-related behaviors: A case of gender and position differences in the perceived role of rewards in the hotel industry. *International Journal of Hospitality Management, 27*, 491–503.

Collins, A. (2007). Human resources: A hidden advantage? *International Journal of Contemporary Hospitality Management, 19* (1), 78–84.

Doherty, L., Guerrier, Y., Jamieson, S., Lashley, C. & Lockwood, A. (2001). *Getting ahead: Graduate careers in hospitality management.* London, England: Council for Hospitality Management Education/Higher Education Funding Council for England.

Dermody, M., Young, M. & Taylor, S. (2004). Identifying job motivation factors of restaurant servers: Insight for the development of effective recruitment and retention strategies. *International Journal of Hospitality and Tourism Administration, 5* (3), 1–13.

Francis-Smith, J., (2004). Surviving and thriving in the multigenerational workplace. *Journal Record, 1.*

Ferris, G.R., Berkson, H.M. & Harris, M.M. (2002). The recruitment interview process persuasion and organization promotion in competitive labour markets. *Human Resource Management Review, 12*, 359–375.

Gursoy, D., Maier, T.A. & Chi, C.G. (2008). Generational differences: An examination of work values and generational gaps in the hospitality workforce. *International Journal of Hospitality Management, 27*, 448–458.

Hinkin, T.R. & Tracey, J.B. (2000). The cost of turnover. *Cornell Hotel and Restaurant Administration Quarterly, 43*(1), 14–21.

Jenkins, A.K. (2001). Making career if it? Hospitality students' future perspectives: An Anglo-Dutch Study. *International Journal of Contemporary Hospitality Management, 13*(1), 13–20.

Karatepe, O. & Uludag, O. (2007). Conflict, exhaustion and motivation: A study of frontline employees in Northern Cyprus hotels. *International Journal of Hospitality Management, 26*, 645–65.

Karatepe, O.M. & Karadas, G. (2012). The effect of management commitment to service quality on job embeddedness and performance outcomes. *Journal of Business Economics & Management, 13*, 614–636.

Kulsuvan, S. & Kulsuvan, Z. (2000). Perceptions and attitudes of undergraduate tourism students towards working in the tourism industry in Turkey. *Tourism Management, 21*, 251–269.

Lee, K.E. & Shin, K.H. (2005). Job burnout, engagement and turnover intention of dieticians and chefs at a contract foodservice management company. *Journal of Community Nutrition, 7* (2), 100–6.

Lucas, R. & Johnson, K. (2003). Managing students as a flexible labour resource in hospitality and tourism in Central and Eastern Europe and the UK. In S. Kuslavan (Ed.), *Managing employee attitudes and behaviors in the tourism and hospitality* (pp. 153–170). New York, NY: Nova.

Marchante, A.J., Ortega, B. & Pagan, R., (2003). *Determinants of skills´ shortages and hard-to-fill vacancies in the hospitality sector*. Retrieved from http://www.fweb.vu.nl/ersa2005/final_papers/21.pdf

Martin, A., Mactaggart, D. & Bowden, J. (2006). The barriers to the recruitment and retention of supervisors/managers in the Scottish tourism industry. *International Journal of Contemporary Hospitality Management, 18*, 380–97.

Mazumder, M.N.H. & Ahmed, E.M. (2009). Does tourism contribute significantly to the Malaysian Economy? Multiplier Analysis Using I-O Technique. *International Journal of Business and Management, 4*(7), 146–159.

Morton, D.L. (2002). Targeting generation Y. *Public Relations Quarterly, 47*(2), 46–48.

Mulvaney, R., O'Neill, J., Cleverland, J. & Crouter, A. (2006). A model of work-family dynamics of hotel managers. *Annals of Tourism Research, 34*, 66–87.

Namasivayam, K. & Zhao, X. (2007). An investigation of the moderating effects of organisational commitment on the relationships between work-family conflict and job satisfaction among hospitality employees in India. *Tourism Management, 28*, 12–23.

Pavesic, D.V. & Brymer, R.A. (1990) Job satisfaction: What's happening to the young managers. *Cornell Hotel and Restaurant Administration Quarterly, 31*(1), 90–96.

Performance Management & Delivery Unit (PEMANDU), Prime Minister's Department of Malaysia. (2012, February). *Economic Transformation Programme Annual Report* [Data file]. Retrieved from http://etp.pemandu.gov.my/annualreport/upload/Eng_ETP2012_Full.pdf

Poulston, J. (2008). Hospitality workplace problems and poor training: a close relationship. *International Journal of Contemporary Hospitality Management, 20*, 412–427.

Pocock, B., Williams, P. & Skinner, N. (2008). Work-life outcomes in Australia: Concepts, outcomes, and policy. In C. Warhurst, D. Eikhof & A. Haunschild (Eds.), *Work Less Live More?* (pp. 22–43). New York, NY: Palgrave Macmillan.

Reynolds, D., Merritt, E. & Gladstein, A. (2004). Retention strategies for seasonal employers: An exploratory study of US based restaurants. *Journal of Hospitality & Tourism Research, 28*, 230–43.

Richardson, S.A. (2010a). Generation Y's perceptions and attitudes towards a career in tourism and hospitality. *Journal of Human Resources in Hospitality & Tourism, 9*, 179–199.

Richardson, S.A. (2010b). Understanding generation Y's attitudes towards a career in the industry. In P. Benckendorff, G. Moscardo & D. Pendergast (Eds.), *Tourism and generation Y* (pp. 131–142). Wallingford, UK: CABI Publishing.

Roney, S.A. & Öztin, P. (2007). Career perceptions of undergraduate tourism students: A case study in Turkey. *Journal of Hospitality, Leisure, Sports and Tourism Education*, 6 (1), 4–18.

Rowley, G. & Purcell, K. (2001). As cooks go, she went: Is labour churn inevitable?. *International Journal of Hospitality Management, 20*, 163–85.

Saad, S., Yahya, K.K. & Pangil, F. (2011). Integrated business strategy and its constructs: Pilot study at hotels in Malaysia. Journal of Global Management, 3(1), 34–42.

Solnet, D. & Hood, A. (2008). Generation Y as hospitality employees: Framing a research agenda. *Journal of Hospitality and Tourism Management, 15*(1), 59–68.

Zacerelli, H.E. (1985). Is the hospitality/food service industry turning its employees on or off? International *Journal of Hospitality Management, 4*, 123–124.

Theory and Practice in Hospitality and Tourism Research – Radzi et al. (Eds)
© 2015 Taylor & Francis Group, London, ISBN 978-1-138-02706-0

Community-Based Tourism (CBT) industry: Operators awareness in the revenue management practices

N.A. Ahmad, A.F. Amir, S.M. Radzi, A.A. Azdel & M.S.Y. Kamaruddin
Faculty of Hotel and Tourism Management, Universiti Teknologi MARA, Shah Alam, Malaysia

ABSTRACT: This study gauge the level of awareness on the revenue management practices in Community-Based Tourism (CBT) by highlighting practices or strategy of revenue management from the perspectives of homestay operators. The homestay business was strongly supported by the government over the years. As most of homestay operators are from the village areas, their knowledge on revenue management is still inaccessible. As the characteristics of homestay suggest the ability to practice revenue management, it provides an opportunity to look into this gray area as limited research had been conducted. The study is set to identify the level of knowledge regarding revenue management among the homestay operators. A qualitative approach was opted and interviews were done with several homestay operators. The study provides significant insights on revenue management knowledge as well as revenue management strategies, tactics and practices used by the homestay operators.

Keywords: Awareness, Community-Based Tourism (CBT), homestay, revenue management

1 INTRODUCTION

The tourism industry in Malaysia has witnessed a robust growth in recent years. In fact, this industry has become an important source of revenue and contributes to sustainable development for Malaysia's economy (Tenth Malaysia Plan, 2011–2015, 2011). As one of the world's major service industries, tourism has contributed significantly to the global economy in recent decades (Law, Rong Vu, Li and Lee 2011). World Travel and Tourism Council (WTTC) (2011) reported that Travel and Tourism makes a contribution of 5% or MYR124.7 billion of GDP in 2011 to the Malaysian economy and supports 1.6 million jobs or 13.8% of total employment. Tourism Malaysia (2012) reported in their news release that Malaysia recorded an overall growth of +1.2% tourist arrivals with 9,438,592 tourists for the first five months of 2012 as compared to 9,323,827 for the same period last year of 2011. A growing sector in rural tourism also known as community based tourism (CBT) which focuses on culture and experiential tourism provides a significant impact to the community and the tourism in general. Homestay industries fall under the culture and heritage tourism in which they contribute towards culture preservations and community economic gains. Kayat (2010) stated that, apart from

its potential contribution to sustainable development, community-based cultural rural tourism is said to be able to bring immediate benefits to both the hosts and guests. Su (2011) stressed that many governments aggressively promote the development of this so called "chimney-free industry".

Tourist receipts from homestay program for the first five months of 2012 also proliferated to RM 7,376,446.50 (+53.1%) compared to RM4, 817,158.30 in January to May 2011. It shows a significant indicator of the growth and sustainability of CBT in Malaysia. Xuan and Bowie (2009) showed that many hospitality and service industries have been reaping benefits by employing the concepts of yield/revenue management (RM) in their businesses. On the other hand, due to the background of the operator's, revenue management strategies and practices are still vague due to lack of research done in this specific field. Thus, with the existence of this gap, it provides an opportunity to look into this gray area in providing a better understanding and contribute some insights in revenue management practices from the homestay operators' perspective. This paper embarks to identify the level of homestay operator's awareness on revenue management practices that they have been applying while running the homestay operations.

2 COMMUNITY BASED TOURISM, HOMESTAY AND REVENUE MANAGEMENT

2.1 *Community based tourism*

Community based tourism starts during 1950s and 1960s; as community development was introduced as an approach to rural development. It was popularized by the United Nations during the same period as many countries in the less developed world gained independence and were decolonized (Catley, 1999). Since then, it has evolved into revenue generating tourism industry in which it promotes cultural experience as well as community based development and economic gains. Community-based homestay tourism is a form of tourism that is closely related to nature, culture and local custom and is intended to attract a certain segment of the tourist market that desires authentic experiences (Jamal, Othman and Nik, 2012).

The origin of homestay programme was first detected as early as the 1970's where it began in Kampung Cherating Lama, Pahang when a local lady by the name of Mak Long has provided all the accommodation including breakfast and dinner to the 'drifter enclave' at home (Amran, 2008). Aminudin and Jamal, (2006) added that Malaysian homestay is a type of community based tourism that employs a somewhat different concept than homestay tourism in other region. Homestay is an alternative tourism product that has the potential to attract tourists due to a marked increase in international demand for tourism that enhances tourist knowledge by allowing them to observe experience and learn about the way of life of the local residents of their destinations (Tourism Malaysia, 2012).

2.2 *Revenue management and homestay*

The nature of homestay business is similar to the operation of a hotel as managing bookings, customer segmentation, promotions, pricing and etc. In line with these statements is a valid ground to test the revenue management in this particular setting and looking into the criteria of revenue managements set by Kimes (2003) who discussed revenue management as technique to optimize the revenue earned from a fixed, perishable resource. The main challenge was to sell at the right resources to the right customer at the right time without neglecting the basic principles of supply and demand economics in a tactical way to generate incremental revenues. Haddad, Roper and Jones (2008) added revenue management as follows: Revenue management is an important tool for matching supply and demand by segmenting customers into different segments based on their willingness to pay and

allocating limited capacity to the different segments in a way that maximizes company's revenues. Cross (1997) highlighted the seven principle of revenue management as dynamic pricing, demand management, market segmentations, capacity management, forecasting, product management and profitability management. Thus, the literature review will focus on these seven variables in establishing the theoretical aspect of revenue management.

2.2.1 *Dynamic pricing*

Pricing strategies are more flexible than other marketing strategies and more easily adjusted to a changing environment. They are closely related to seasonality, price regimes and different facilities (Espinet, Saez, Coenders and Fluiva, 2003). Kimes (2003) supported that pricing and demand based on pricing strategies would be useful for revenue managers as they are really popular in revenue management strategies.

2.2.2 *Demand management*

Demand management refers to the ability of the operators to understand demand pattern. As for tourism product it is usually affected by various reasons. According to Nadal, Font and Rosello (2004) the first element relates to temporal variations in natural phenomena, particularly those associated with climate and season of the year. Seasonality is one of the unique, indeed a major problem, that tourism industry has to face (Jang, 2004). Gurbuz (2011) discussed that demand generation is the art and science of creating, nurturing, and managing purchase interest in your products and services through campaign management, lead management, marketing analysis, and data management. Thus, it revolves in the issue of generating demand when the right seasonality or time comes.

2.2.3 *Market segmentations*

Most literature defines market segmentation as to select the best market for the business. It can be accomplished using different types of data, including geographic, demographic, psychographic, and behavioral variables (Reid and Bojanic, 2006). Different segments demand different prices. To maximize revenue and stay competitive, prices must vary to meet the price sensitivity of each market segment. A simple understanding of market segmentation strategy by Kozak and Martin (2011) focused on identifying and nurturing the most profitable customers is an alternative to maximizing visitor counts. It focused on selecting the right target market in business.

2.2.4 *Capacity management*

Fixed capacity and perishable inventory provides a credible threat heightening acquisition risk, resulting in the existence of advanced demand for the practice of revenue management (Gurbuz, 2011). Irene (2005) discussed that positive fixed capacity may encourage buyers to buy in advance because of acquisiion risk, but threat of non-acquisitions (from the consumer's perspective) may also be generated through consumers" inability to afford higher spot prices.

2.2.5 *Forecasting*

Espinet et al (2003) stressed that forecasting is one of the cornerstones of RM. Forecasting refers to the ability to predict future demand based on historical records and current demand pattern. Kimberley, Stuart and Parker (2010) added that forecasting is an act of estimating, calculating, or predicting conditions in the future.

2.2.6 *Product management*

Amran & Hairul (2003) asserted that homestay is a form of accommodation whereby tourists will get the chance to stay with the chosen house-owner or host, communicate with them as well as go through the family's daily routine which in a way let the tourists have a live experience of Malaysian cultures. Fauziah and Mohd, (2012) proposed that homestay need to be viewed as a complete tourism product. They also added that homestay are dormant and are less involved in development focusing on way of life which is unique to attract visitors to try the homestay experience.

2.2.7 *Profitability management*

Jonathan (2011) stated that meeting budgeted and forecasted targets in terms of internal measures is challenging particularly for hospitality managers to maximize profits when a hotel is not meeting budgeted and forecasted revenue. Keeping track of revenue and profit is vital to ensure the community gets returns from their contributions. Gurbuz (2011) pointed that profitability is the indicator of return on investment for performance in the hospitality industry nowadays.

2.3 *Marketing effort in homestay*

The marketing of the homestay program is currently controlled by Ministry of Tourism Malaysia (MOTOUR, 2013). The main platform used is the corporate website, homestay e-marketing and mobile apps. Other than that, the Ministry of Tourism use social media platform such as Facebook, Twitter and YouTube to promote the homestay program. Apart from marketing support the ministries also provide financial support, infrastructure and development, as well as training and technical operation on how to manage the homestay (Rojulai 2014).

3 METHODOLOGY

3.1 *Preliminary findings*

This qualitative case study was conducted focuses on homestay in Selangor. Selangor recorded the highest generated income in year 2011 – RM 2,243,926 and the second highest homestay in Malaysia with a total of 15 villages, 458 operators with 660 rooms available (MOTOUR 2013). Data were gathered mainly from in-depth interviews with the homestay operators.

3.2 *Population and sampling*

This study began with intensive literature review to understand the relationship between homestay and revenue management principles and practices. In addition, interviews with homestay operators were conducted to probe their awareness of revenue management practices in their current operations. The conversations have been audio taped. This study managed to capture 6 informants as the data had been saturated on the 6th informant. A purposive sampling technique was used to determine the sample.

4 DISCUSSION

Based on the interview with the homestay operators, several themes of revenue management practices have been developed and categorized as follows: revenue management, pricing, demand generation, customer segmentation, capacity management, forecasting product and profitability management.

In regards to revenue management concept in general the interview highlighted that homestay has a brief understanding and most responded as follows.

"*….We a have a slight understanding what is revenue management in which it is ensuring there are customers for the business, the knowledge and experience are from the effort of the ministry of tourism by providing courses, campaign and workshop, it exposed us to the prospect of the business, revenue management and revenue generation …*" (Respondent 1, 2 and 5)

Second identifiable theme identified is pricing in which it is crucial in giving the value perspectives to the tourist and it must be viewed as a whole packages. The respondent feedback are as follows"*…our prices are being dictates by the ministry…for a 2 day 1 night package at RM150 to RM250 in which it is inclusive of tour and performance.*"..(Respondent 2,

3, and 6). And when they were asked about charging difference prices at difference season most of them replied "that *is not possible. Rates are monitored as best that we stick to as homestay coordinator ensure that we do so.. any rate revises will be informed* " (respondent 1,2,3, and 6)

Other theme that is significant in line with revenue management is demand generation. The interview provided insight that the demand for the homestay program lays in the facilities as well as activities and performance prepared by each village. Tourists are attracted to homestay as it is based on culture and nature beside the warm hospitability as the host. "*.......we try our best to ensure we provide genuine experience making tourist eager to know our culture, languages, food and custom such as wedding....among others and activities such as rubber tapping, padi planting and catching eel* "... "*Making it unique experience...thus demand to our villages is maintained*". (Respondent 1, 3, and 6)

The forth theme identifiable is customer segmentation which requires operators to separate customer into categories. Respond of the interview show segmentation practices done. Operators respondent are as follows...*"Currently most of our customer are student and foreign tourist, this are the segment we focus on we operate to show our hospitality and culture, to show tourist our way of life... for sure we would not practice treating our customer differently.*" (Respondent 2,4,5 and 6).

The next theme that come into light is capacity management relies on having to separate room in to different set of prices to be sold at different channel. From the interview homestay does not practice setting up different rates for their packages "*...we rely most of the marketing channel via ministry website and the agent will call and set it up with us...the minimum room that need to be provided is 10 then we can be consider as a homestay...some customer would come directly and most of the time call to make a booking and most of the time same rate is being charged*" (Respondent 2,4,5 and 6).

The interview also highlighted forecasting practices which enable homestay operators to be prepared for busy business period and most operators agreed that tourist arrival data provides them with a lead start. "*...knowing when more tourist come to our homestay really help us in preparing and maintaining our homestay while coordinating activities and cultural performance and this is usually during the holiday season...*"(Respondent 1, 3, 4 and 6)

Other emerging theme was managing homestay as a product relies on maintaining the nature and culture element that signifies them as a homestay operator. At the same time ensuring tourist experience is important rather than tourist staying in the room."*...we believe our homestay program rely on our culture and they way we are living now...it is*

the real experience and to ensure our product sell we must portray the best in our culture ...if I'm not mistaken we have won an award by UMWTO in 2012" (Respondent 1, 2, 3, 6)

The last significant theme identified from the interview was profitability management. It is important to maintain the profitability to ensure homestay business. The findings showed that operators gained enough profit to compensate their effort and involvement. "*...right now we are able to maintain our business profitability.. With the training done by the ministry enable us to understand that this business is very good, provided that we manage our profit wisely.. Treating each tourist fairly.. Ensure that our activities excite them.. Thus spreading the good word of mouth to their peers...making other to come to our homestay.. And currently we are making an average of RM600 – RM800 per month, a good additional income.* (Respondent 1, 2, 3, and 4)"

5 CONCLUSION

The interview provides a reasonable ground to gauge the awareness of homestay operators on the practices of revenue management. Operators have a brief understanding and awareness of revenue management practices in their daily operations. Themes raised from the interview are in line with revenue management concept and practices. Despite of having less knowledge on operating a lodging operation, training provided have been successful as most respondent clearly understand their roles in making this program successful. Although many still rely on the Ministry of Tourism effort in marketing and generating demand for the business. Among others they have to comply strictly with the operating guidelines set by the Ministry of tourism, cultural program, lodging condition and setting rates. Thus, with this understanding homestay operators are able to sustain their business over the years.

ACKNOWLEDGEMENTS

This research is funded by Universiti Teknologi MARA through Research Intensive Grant 600-RMI/DANA 5/3/RIF (879/2012).

REFERENCES

Aminudin, N. and Jamal, S.A. (2006). Homestay Selangor: Keunikan dan Pengalaman Pengusaha. Selangor: Pusat Penerbitan Univesiti (UPENA).
Amran, H. and Hairul, N.I. (2003). Kajian penilaian kesan sosio-ekonomi program homestay di Kampung

Banghuris, Sepang, Selangor (Report FRGS Vot 71538). Skudai, Johor: UTM.

Amran, H. (2008). Malaysia Homestay from the Perspective of Young Japanese Tourist: the quest for furusato in Janet Cochrone (Ed). Asian tourism, growth and change (pp. 193–207). Amsterdam: Elsevier.

Catley, A. (1999). Methods on the move: A review of veterinary uses of participatory approaches and methods focusing on experiences in dry land Africa. London: International Institute for Environment and Development.

Cross, R.G. (1997) Revenue Management: Hard—Core Tactics for Market Domination, Broadway Books, Bantam Doubleday, Dell.

Espinet, J.M., Saez, M., Coenders, G., and Fluiva, M., (2003). Effect on prices of the attributes of holiday hotels: a hedonic price approach. Tourism Economics 9, pp. 165–177.

Fauziah, C.L. and Mohd, R.H.(2012) Homestay tourism and pro-poor tourism strategy in banghuris selangor, Malaysia. Elixir Geoscience 45, pp. 7602–7610.

Gurbuz, E.G. (2011). Revenue Management Operations in Hotel Chains in Finland. Unpublished Bachelor Thesis. Saimaa University of Applied Sciences Tourism and Hospitality, Imatra.

Haddad, R.E. Roper, A. and Jones, P. (2008), The Impact of Revenue Management Decisions on Customer Attitudes and Behaviors: A Case Study of Leading UK Budget Hotel Chain.

Irene, C.L.NG (2005), Differentiation, Self-Selection and Revenue Management: Manuscript submitted to The Journal of Revenue and Pricing Management.

Jamal, S.A. Othman, N. And Nik, M.N.M. (2012). Tourist Perceived Value in a Community-Based Homestay Visit: An investigation into the functional and experiential aspect of value. *Journal of Vacation Marketing*. 17(1), pp. 5–15.

Jang, S., 2004. Mitigating tourism seasonality. A quantitative approach. *Annals of Tourism Research* 31 (4), pp. 819–836.

Jonathan, A.H. (2009) Accounting and Financial Analysis in the Hospitality Industry.

Kayat, k. (2010). The Nature of Cultural Contribution of A Community-based Homestay Programme. *Tourismos: An International Multidisciplinary Journal of Tourism.* Volume 5, Number 2, Autumn 2010, pp. 145–159.

Kim, H.-G., and Ko, S.-T. (2008). Development of social capital in rural tourism: perspectives of community leaders. *Journal of Tourism and Leisure Research*, 20(2), pp. 29–49.

Kimberley, A.T., Stuart, H.T. and Parker., J. (2010). An introduction to Revenue Management For The Hospitality Industry: Principles and Practices For The Real World.

Kimes, S.E. (2003), "Revenue Management: A Retrospective", Cornell Hotel and Restaurant Administration Quarterly, Vol. 30, No. 3, pp. 14–19.

Kimes, S.E., and Wirtz, J. (2003) Has revenue management become acceptable? Findings from an international study on the perceived fairness of rate fences, Journal of Service Research, 6(2), pp. 125–135.

Kozak., M. and Martin., D. (2011). Tourism life cycle and sustainability analysis: Profit-focused strategies for mature destinations. Tourism Management 33 (2012) pp188–194.

Law, R., Rong, J., Vu, H.Q., Li, G., and Lee, H.A., (2011). Identifying changes and trends in Hong Kong outbound tourism. Tourism Management 32 (5), pp. 1106–1114.

Nadal, J.R., Font, A.R., and Rosselo, A.S., 2004. The economic determinants of seasonal patterns. *Annals of Tourism Research.* 31 (3), pp. 697–711.

Reid, R.D., and Bojanic, D.C., (2006). Hospitality Marketing Management. John Wiley & Sons, New Jersey.

Rojulai, N. (2014). Report in Community Based Tourism "Malaysia Homestay Experience Program" Ministry of Tourism Malaysia.

Su, B., (2011). Rural tourism in China. Tourism Management 32 (6), pp.1438–1441.

Tourism Malaysia (2012). Ministry of Tourism Malaysia News Release, ETP: TRANSFORMING TOURISM TO THE NEW HEIGHTS Tourist Arrivals in Malaysia Rise 1.2% in First 5 Months. Retrieved online from www.tourism.gov.my on 10 October 2012.

Tourism Malayis (2013) from http://www.tourism Malaysia.gov.my/facts_figures/retrived on 18 February 2013.

Vernon, R., and Wells, L.T., Jr. (1976). Economic environment of international business (2nd ed.). Englewood Cliffs, NJ: Prentice Hall.

WTTC (2011). Prime Minister of Malaysia: tourism key to country's economic transformation. Retrieved online from http://www.wttc.org/news-media/news-archive/2011/prime-minister-Malaysia-tourism-key-countrys-economic-transforma/ on 10 October 2012.

Xuan L.W. and Bowie, D. (2009), Revenue Management: The Impact on Business to Business Relationships.

Theory and Practice in Hospitality and Tourism Research – Radzi et al. (Eds)
© 2015 Taylor & Francis Group, London, ISBN 978-1-138-02706-0

Tourists perceived destination competitiveness: A case of Langkawi Island, Malaysia

Z. Zainuddin
Tourism Malaysia, Malaysia

S.M. Radzi & M.S.M. Zahari
Faculty of Hotel and Tourism Management, Universiti Teknologi MARA, Shah Alam, Malaysia

ABSTRACT: Destination competitiveness has become a critical issue and creates increasingly challenging in tourism market. A successful tourism destination should embrace an integrated approach towards the many components of the tourism system and competitive advantage of a destination is closely relates to the quality of the product offered, which means the quality of tourist experience provided by the destination. This research note discussed the tourism destination competitiveness and highlight the Langkawi Island as one of competitive tourism destination in Malaysia as a case study which is still under investigation.

Keywords: Tourist, perceived destination, competitiveness, behavioral intention

1   INTRODUCTION

Tourism destination around the world competing each other's owing to increasing global mobility of the tourists. Every tourism destination is trying hard to be more competitive. Meng (2006) noted that in the current competitive tourism market, competitiveness has increasingly been seen as a critical influence on the performance of tourism destinations. He further argued destination competitiveness has become a critical issue and creates increasingly challenging in tourism market. Crouch and Ritchie (2003), Jones and Haven (2005) postulated that a successful tourism destination must embrace an integrated approach towards the many components of the tourism system. However, the tourism industry players like the government, tourism enterprises, tourists, and local communities may have very different approaches to destination competitiveness.

Many studies have indicated that tourists and their needs stand as the ultimate driving force which influences competition and competitiveness in the tourism destination. Today, destinations eventually compete on the quality of tourism experience offered to visitors. In this note, competitiveness in tourism denote a destination is compatible, attract visitors, increase tourism expenditure and providing them with satisfying memorable experiences. It also enhancing the well-being of destination resident's and preserving the natural capital of the destination for future generation (Crouch & Ritchie, 2003).

Most destination competitiveness research suggests that the competitive advantage of a destination closely relates to the quality of the product offered, which means the quality of tourist experience provided by the destination (Chon & Mayer, 1995; Crouch & Ritchie, 1999; d'Hauteserre, 2000; Faulkner, Oppermann & Fredline, 1999; Hassan, 2000).

2   LITERATURE REVIEW

2.1   *Destination competitiveness*

D'Hauteserre (2000) defined destination competitiveness as the ability of a destination to sustain its market position and share and/or to improve them through time. Destination competitiveness is similarly defined by Hassan (2000) as the ability of one destination to create and integrate value-added goods that maintain its resources while also carry on its own market position concerning those of competitors. The most detailed work on overall tourism competitiveness was undertaken by Crouch and Ritchie (1999, 2003). They contended that to be competitive, a destinations development of tourism must be sustainable, not just economically, ecologically, but socially, culturally and politically. They focus on long-term economic prosperity as the yardstick by which destinations can be assessed competitively. Thus the most competitive destination is that which most effectively creates sustainable well-being for its residents.

2.2 Destination resources and attributes

There are several studies have been conducted on the attributes or characteristics of destination competitiveness (Buhalis, 2000; Crouch & Ritchie, 2003; Dwyer & Kim, 2003; Dwyer, Mistilis, Forsyth & Rao, 2001; Hassan, 2000; Kozak & Rimmington, 1999; Mihalič, 2000; Wilde & Cox, 2008). Dwyer (2001) posited that to achieve competitive advantage for its tourism industry, any destination must ensure that its overall attractiveness and the tourism experience must be superior to that of the many alternative destinations open to potential visitors. According to Dwyer et al. (2001), the key success factors in determining destination competitiveness can be classified under eight main headings: Endowed Resources (natural/heritage); Created Resources; Supporting Resources; Destination Management (Government/Industry); Situational Conditions and Demand. In an earlier model, Crouch and Ritchie (2003) developed similar factors, but categorized them into five general industry levels as well as mainstream tourism destination attractiveness attributes including: Supporting Factors and Resources; Core Resources and Attractors; Destination

It is interesting to see that both research studies such as Crouch and Ritchie (2003) and Dwyer et al. (2001) revealed similar competitiveness factors. The only difference is the descriptive terms which are used by these researchers and other studies such as those carried out by Kozak and Rimmington (1999), Wilde and Cox (2008). In addition to this, scholars (such as Buhalis, 2000; Dwyer & Kim, 2003; Hassan, 2000; Kozak & Rimmington, 1999; Mihalič, 2000; Wilde & Cox, 2008) also identified a destination's resources as universally important factors in determining its competitiveness.

2.3 Destination competitiveness measurement approaches

The measurement of tourism competitiveness has attracted many researchers indentation as they are vital factors for the success of tourism destinations (Crouch & Ritchie, 1999; Dwyer, et al., 2001; Go & Zhang, 1997; Kozak & Rimmington, 1998, 1999; Mihalič, 2000). A model of competitiveness particularly on the tourism sector is based on product and service offered (Murphy, Pritchard & Smith, 2000). According to Dwyer and Kim (2003), there are no single or unique unit indicators that can exploit and apply to all destinations at all times. However, there are two kinds of variables used

which are objectively measured variables such as visitor numbers, market share and subjectively measured variables such as image, climate and so on. Kozak and Remmington (1998, 1999) used both measures in surveying perceptions and opinions of visitors such as friendliness of local citizens, shopping facilities and others in measuring the competitiveness of one destination. Poon (1993) suggested four main principles for the destinations to be competitive namely; strongly sustaining environment, making tourism a leading sector, strengthening the distribution channels in the market and building a dynamic private sector.

3 LANGKAWI ISLAND AS TOURIST DESTINATION

Many scholars claim that competitive destination is one which brings about the greatest success not only in developing a particular destination but increase in tourists' arrival or receipt, market share, and that is, the greatest well-being for its residents on a sustainable basis (Crouch & Ritchie, 2003).

In line with this notion, Malaysia government through Ministry of Tourism continually is proactive in promoting the nation by using all the available tourism resources in making this country as competitive destination. Attractions like shopping with duty-free prices, conventional tourism like diving, flora, fauna, cultural, heritage and now step—up promotion for gastronomic, eco-business tourism, sports, Meeting, Incentive, Convention and Exhibition (MICE) tourism and many others while committed in preserving Mother Earth for future generations well-being are some of the initiative undertaking.

In 2010, tourism has been allocated RM899 million (approximately US$ 267.4 million) and this funding has increased the revenue for local in capitalizing the economy. RM36 millions of tourist arrival and RM168 billion in revenue are set by the government in with the 2020 Vision for tourism industry (MTPB, 2012).

In making competitive destinations, some of the popular islands are also included. Langkawi besides Penang, Tioman since the inclusion of it as a prospective competitive tourist destination in 1975 is one of the popular destinations aggressively developed. It was first declared as a tax-free island in order to draw more visitors to shop and spend besides appreciating the exotic beauty of the island.

The rapid investments by the federal government and the private sector can still be seen in making Langkawi at least compatible with Phuket and Bali Island. Not only that, in positioning Langkawi as an international tourist destination Langkawi

Development Authority (LADA) is local govern agency were formed responsible for expanding and supporting tourism development in Langkawi by encouraging and carry out the economic restructuring of lower-productivity to higher productivity sectors. LADA is also responsible to create attractive opportunities for foreign investors using the existing available tourism products and resources. This government body is operated based on social, economic and physical development of Langkawi in line with the Malaysian government policies as well as preserving the natural resources and establish conducive environment.

The above mentioned efforts are evidences when Langkawi is experiencing significant changes on the supply side of tourism, as well as in demand trends or significantly turned this Island into a popular destination and a shopping heaven for local and foreign tourists. In early month of January 2014, the Prime Minister, Dato Seri Najib Tun Abdul Razak has launched the Visit Malaysia Year with the target of 28 million tourist arrivals and Langkawi Island is one of the popular spots promoted among the international tourists.

The rapid expansion of both international and domestic tourism to Langkawi Island has increased the need for a comprehensive view of the social, economic, cultural, environmental problems and processes related to tourism development and understand its dynamics and impacts. As earlier mentioned, the competitive advantage of a destination, closely relates to the quality of the product offered, which means the quality of tourist experience provided by the destination (Crouch & Ritchie, 1999) (Chon & Mayer, 1995; d'Hauteserre, 2000; Faulkner, et al., 1999; Hassan, 2000). Dwyer and Kim (2003) and Enright and Newton (2005) in addition, argues that the principal factors contributing to competitiveness are varied amongst destinations. They suggested that destination must take a more tailored approach to enhancing and developing tourism competitiveness, rather than adopting a single, universal policy or strategy. In fact, Meng (2006) noted in a highly competitive tourism destination market, tourists' experiences and their opinions and attitudes should be understood in order to enhance the performance of destination products and services and promote destination development strategies.

In this sense, the relationship between Langkawi Island tourism performance on the local economic and socio-cultural impacts is rather vague. Further, the international tourist perceived Langkawi Island as a competitive tourism destination is hardly discussed or investigated.

4 EMPIRICAL INVESTIGATION

With the preceding notion, the empirical investigation of the highlighted issue pertaining to Langkawi Island particularly on its competitiveness from international tourist's perspective is still under investigation. The study employs a quantitative approach with structured questionnaires as the research instrument. A descriptive research design, correlation approach and a cross sectional study is used where data gathered just once at the same point of time for each participating individual international tourist. The popular spots among the international tourists like Kuah, Pantai Cenang, Padang Matsirat and many others are the data collection contextual setting.

5 CONCLUSION

It is expected that the insightful from this study would help the local authority and agencies, particularly LADA, tour operators, travel agencies the hotel operators and the federal government, particularly the Ministry of Tourism in improving, continuously promoting and maintaining, the competitiveness of Langkawi Island as a tourist destination without degrading the environment, economic and social aspects of it.

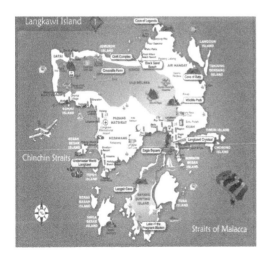

Figure 1.   Map of Langkawi Island, Malaysia.

REFERENCES

Buhalis, D. (2000). Marketing the competitive destination of the future. *Tourism management, 21*(1), 97–116.
Chon, K.-S. & Mayer, K.J. (1995). Destination competitiveness models in tourism and their application to Las Vegas. *Journal of Tourism Systems and Quality Management, 1*(2), 227–246.

Crouch, G.I. & Ritchie, J.B. (1999). Tourism, competitiveness, and societal prosperity. *Journal of Business Research, 44*(3), 137–152.

Crouch, G.I. & Ritchie, J.B. (2003). *The competitive destination: A sustainable tourism perspective.* Wallingford, Oxon, UK: Cabi Publishers.

d'Hauteserre, A.-M. (2000). Lessons in managed destination competitiveness: the case of Foxwoods Casino Resort. *Tourism management, 21*(1), 23–32.

Dwyer, L. & Kim, C. (2003). Destination competitiveness: determinants and indicators. *Current Issues in Tourism, 6*(5), 369–414.

Dwyer, L., Mistilis, N., Forsyth, P. & Rao, P. (2001). International price competitiveness of Australia's MICE industry. *International Journal of Tourism Research, 3*(2), 123–139.

Enright, M.J. & Newton, J. (2005). Determinants of tourism destination competitiveness in Asia Pacific: Comprehensiveness and universality. *Journal of travel research, 43*(4), 339–350.

Faulkner, B., Oppermann, M. & Fredline, E. (1999). Destination competitiveness: An exploratory examination of South Australia's core attractions. *Journal of Vacation Marketing, 5*(2), 125–139.

Go, F. & Zhang, W. (1997). Applying importance-performance analysis to Beijing as an international meeting destination. *Journal of travel research, 35*(4), 42–49.

Hassan, S.S. (2000). Determinants of market competitiveness in an environmentally sustainable tourism industry. *Journal of travel research, 38*(3), 239–245.

Jones, E. & Haven, T.C. (2005). *Tourism SMEs, service quality, and destination competitiveness.* Wallingford, Oxon,: CABI Publishers.

Kozak, M. & Rimmington, M. (1998). Benchmarking: Destination attractiveness and small hospitality business performance. *International Journal of Contemporary Hospitality Management, 10*(5), 184–188.

Kozak, M. & Rimmington, M. (1999). Measuring tourist destination competitiveness: conceptual considerations and empirical findings. *International Journal of Hospitality Management, 18*(3), 273–283.

Meng, F. (2006). *An Examination of Destination Competitiveness from the Tourists' Perspective: The Relationship between Quality of Tourism Experience and Perceived Destination Competitiveness.* Virginia Polytechnic Institute and State University.

Mihalič, T. (2000). Environmental management of a tourist destination: A factor of tourism competitiveness. *Tourism management, 21*(1), 65–78.

MTPB. (2012). *Malaysia Tourism Promotion Board. Annual Report 2012.*

Murphy, P., Pritchard, M.P. & Smith, B. (2000). The destination product and its impact on traveller perceptions. *Tourism management, 21*(1), 43–52.

Poon, A. (1993). *Tourism, technology and competitive strategies.* Wallingford, Oxon, UK: CAB international.

Wilde, S.J. & Cox, C. (2008). *Linking destination competitiveness and destination development: Findings from a mature Australian tourism destination.* Paper presented at the Travel and Tourism Research Association (TTRA) European Chapter Conference—Competition in tourism: Business and destination perspectives, Helsinki, Finland.

Theory and Practice in Hospitality and Tourism Research – Radzi et al. (Eds)
© 2015 Taylor & Francis Group, London, ISBN 978-1-138-02706-0

The role of perceived authenticity as the determinant to revisit heritage tourism destination in Penang

Z.M. Rani, N. Othman & K.N. Ahmad
Faculty of Hotel and Tourism Management, Universiti Teknologi MARA, Shah Alam, Malaysia

ABSTRACT: The recent growth in the heritage tourism has led to a renewed interest in the study of revisit intention. Heritage tourism is one of the fastest-growing segments that can be classified as the peculiar fraction in tourism industry. The popularity of heritage tourism escalates as the segment continues to provide diversity and variation of cultural and heritage elements that offer more authentic experience to the visitors. Cultural and heritage elements and activities often portrayed the authenticity of a destination. As a nation, Malaysia is often cited as among the main destinations in Asia that flourishes with culture and heritage elements. However, it is afraid that the heritage tourism product in Malaysia is losing the touch of authenticity and originality due to mass tourism. This has led to major concern on the sustainability of heritage tourism market. Significantly, the issue of authenticity and originality are the foremost to determine the magnitude of the success of heritage tourism. Therefore, the study has two main objectives: (1) to examine the relationship of perceived authenticity and satisfaction to revisit Penang, (2) to analyze the role of satisfaction as mediation in the relationship of perceived authenticity in revisiting Penang.

Keywords: Tourist behavior, perceived authenticity, heritage tourism, and heritage destination

1 INTRODUCTION

Tourism industry has become a more important component to the country's economy and further grant a number of advantages which in turn benefits the country and residents in terms of monetary and recognition (Waligo, Clarke & Hawkins, 2013). In addition, tourism is known as one of the major industry that contributes to the proliferation and growth of many developed and developing countries.

In the interim, the acceleration of the industry globalization and increasing in demand from mass tourist are among the major reasons of many countries strive in establishing modern infrastructures such as hotels, tourist attractions, recreation areas and other related infrastructures and facilities in the direction to remain competitive within the industry. Without doubt, the globalization and development of tourism industry have acted as catalysts for the growth of heritage tourism market.

According to Naef (2011), globalization helps to safeguard the heritage from dilution or disappearance of the identity as it has become a product that attracts people to come to cherish and learn (Mohamed, 2005). According to Mohamed (2005), the post-modernism has brought up the value of heritage. However, he added that the post-modernism in some cases, could bring the downfall of heritage sites worldwide.

A conflict arises as heritage tourism is concerning more on preservation while tourism focuses on development. In consequence, the issue has grown in importance in light of the need for preservation of authenticity, which is essential because heritage tourism is about experience, the core product of which is identity. Hence, preceding literatures have uncovered that the issue of authenticity and originality are the foremost notion to determine the magnitude and success of heritage tourism.

Consequently, the study has two main objectives: (1) to examine the relationship of perceived authenticity and satisfaction to revisit Penang, (2) to analyze the role of satisfaction as mediation in the relationship of perceived authenticity in revisiting Penang.

The findings of this paper indicate a distinct relationship between tourist's perception of authenticity and intention to revisit.

2 LITERATURE REVIEW

2.1 *Perceived authenticity*

Authenticity is among the severe issue in tourism studies especially heritage (Zhou, Zhang &

Edelheim, 2013) and is the main focused on within the heritage tourism (Wang & Wu, 2013).

The rapid growth of the tourism industry has strengthened the competition among tourist destinations (Lee, Cho & Hwang, 2014). Prideaux and Timothy (2008) argue that the tourism industry works best if there is competition between destinations, all of whom are marketing their uniqueness and different experiences to the tourist. As reported by George and Anandkumar (2014), due to high levels of industry competitiveness, numerous destinations are striving to ensure that their presence is firmly established. Therefore, a major problem is that many heritage destinations are struggling to establish their destinations' uniqueness in attracting more tourist vis-à-vis their rivals. For that reason, to be different from others requires a lot of changes in order to attract tourists' attention.

Xianger, Jianming, Zhenshan, and Webster (2014) have recognized that authenticity has become relevant to the growth of heritage resources. The issue of authenticity then becomes of major concern, as it is one of the underlying factors, which determines the success of heritage tourism, and also acts as a magnet that influences the tourist's decision to travel. Furthermore, a major challenge of this industry is to highlight and maintain the authenticity and the natural experience (Engeset & Elvekrok, 2014) and to take action to tourists' emotion (Lin, Kerstetter, Nawijn & Mitas, 2014).

Tourists' perception towards the authenticity of a particular heritage sites is one of the underlying factors which determine their decision to travel to a destination (Chhabra, 2010; Kolar & Zabkar, 2010).

2.2 *Satisfaction*

Tourist satisfaction is the tourist overall assessment with the travel experience (Lee, Lee & Lee, 2013). Sun and Kim (2013) mentioned that tourist satisfaction differs from customer satisfaction in other industries including manufacturing, retail, and business. Beard and Ragheb (1980) defined tourist satisfactions as a positive notion and experience obtain from the particular leisure activity. As reported by Zalatan (1994), tourist satisfaction is a foreseeable final result from the tourist experience and expectation at a destination. Wang, Zhang, Gu and Zhen (2009) posit that tourist satisfaction is a feeling produced by means of cognitive and emotional facets of tourism activities, including the accrued assessment of varied elements and destination features. Equally important, Pawitra and Tan (2003) posit that tourist satisfaction is acquired by means of examining the gap involving the expected and perceived service.

It is vital to monitor tourist satisfaction, as it is a significant steps in applying techniques to strengthening the performance of the tourism destination (Moital, Dias & Machado, 2013) and also influences choice destination, the consumption of products and services, and the decision to revisit to the same destination (Kozak & Rimmington, 2000). Del Bosque and Martín (2008) mentioned tourist satisfaction as one of the crucial issues in tourism discipline; it is also important to the academics and researchers, but also for the host communities and society.

Generally, tourist's satisfaction can be examined after the occurrence of the purchasing process and beyond doubt after the tourist experiences it by his or herself. The questions about satisfaction need to be understood as it has the compelling influence towards tourist behavior intention positively or negatively.

2.3 *Revisit intention*

Revisit intention continues to be considered a crucial topic in tourism research (Huang, Cai, Yu & Li, 2013; Um, Chon & Ro, 2006) academically and practically. Monitoring tourist revisit intentions is crucial, and according Assaker, Vinzi and O'Connor (2011), behavior intention frequently changes over time. As reported by Um et al. (2006) revisit intentions are derived from tourist satisfaction as opposed to being an initiator of revisit decision making process. On the other hand, Han, Back and Barrett, (2009) described revisit intention as an established likelihood to revisit derived from positive attitudes and perspectives regarding the service provider.

In the marketing literature, revisit intention is similar to the notion of repurchase that has grown to be important in the primary body of modern-day marketing approaches (Luoand & Hsieh, 2013). Revisit intention is defined as likelihood and willingness to return to visit the same destination as well as to recommend to others. As reported by several researchers (Cole & Chancellor, 2009; Kim, Kim & Kim, 2009) revisit intention derives from overall satisfaction towards the destination and positively contributes to the destination loyalty. Revisit intention is a significant concern in management of tourism organization (Mao & Zhang, 2014). This is due to revisit intention being able to attract more tourist arrival and revenue generated from tourists (Jayaraman, Lin, Guat & Ong, 2010; Mao & Zhang, 2014).

Revisit intention in the tourism industry is important to the growth of the industry as it represents a stable source of income and also acts as a medium of information to the potential tourist (Reid & Reid, 1994), and as an attractive and cost effective market segment (Lau & McKercher, 2004).

Furthermore, Mohamad et al. (2013) add that according to a Bucket Theory of Marketing, "bringing in a new tourist and increasing the revenue from existing customers are said to "fill the bucket". They add that tourists stop purchasing or a decreased arrival for a period of time is portrayed as "a hole in the bucket". They suggest an understanding of the post-behavioral intention of the tourist will help to stop money from leaking.

So far, however, insufficient attention and limited empirical investigation has been given to the relationship between intention to revisit a heritage destination, and the crucial factors of perceived authenticity.

3 RESEARCH METHOD

The data was collected through a personally administered questionnaire using 10-point Likert scale, to tourist traveled heritage destination in Penang. Convenience sampling was use to obtain participants. Both international and domestic tourists have an equal chance of being chosen as respondents; however a screening interview was conducted to domestic respondents and they were asked a few questions to determine their origin. This procedure was compulsory in order to differentiate tourists from local people and to avoid biases.

Well-established and popular heritage destinations in Penang were chosen as the appropriate location for data collection, these include, Town Hall, The Queen Victoria Memorial Clock Tower, Kapitan Keling Mosque, For Cornwallis, Penang State Museum, War Museum and Cheong Fatt Tze Museum.

4 RESULTS

A total of 255 useable questionnaires were collected from the respondents. Analysis of the data revealed that majority of the respondents was male (52.9%) and female (47.1%). The age group were fairly evenly distributed with the exception of "21–30" group consisting of 51 percent, and the group of "31–39" consisting of 40 percent. 54.1 percent of the respondents were married and 25.1 percent were single and the rest were divorced (20.8%). The respondents travel to heritage tourism destination had a high level of education 85.1 percent from tertiary education background, with a moderately even split of those were manager or executive, 53.7 percent. Eighty percent of the respondents were from international tourist while 20 percent were from domestic tourist. Majority of the respondents (80.8%) indicated that this was their first trip to their heritage tourism destinations.

Confirmatory factor analysis (CFA) was conducted prior to Structural Equation Modeling as proposed by (Gerbing & Anderson, 1988). Findings from the assessment of fitness for the measurement model shows that all fitness indexes required in evaluating the fitness of a model are achieved. It was indicated that the result of (CFA) was excellent. Most of the fitness indexes have achieved the required level for measurement model. The measurement model fit was great with p-value = .000, GFI = 0.927, CFI = 0.953, RMSEA = .080 and NORMEDCHISQ = 2.656 indicating a strong model fit.

The result of Cronbach alpha, average variance extracted (AVE) and composite reliability (CR) for perceived authenticity is (α = 0.829, AVE = 0.648, CR = 0.901), for satisfaction (α = 0.880, AVE = 0.726, CR = 0.888), and for revisit intention (α = 0.863, AVE = 0.614, CR = 0.864).

To validate the assessments of reliability, convergent validity and discriminant validity, a tool developed by Gaskin (2012) was utilized, namely the validity master test. Gaskin (2012) proposed several measures which are suitable to gauge validity and reliability. The results from the test show (Table 1) that there were no issues concerning reliability, convergent validity and discriminant validity. Therefore, the final measurement model can proceed to the structural model, as the model has achieved the requirements for a structural model.

Table 2 shows that the result of the structural model fit was great with p-value = .000, GFI = 0.915, CFI = 0.944, RMSEA = .080 and NORMEDCHISQ = 2.941 indicating a strong predictive validity.

4.1 Hypothesis testing

The purpose of this study is to study the effect of perceived authenticity on revisit intention (H[1]), the mediation effect of satisfaction between perceived authenticities and revisit intention (H[2]). H[1] was supported with a significant level of (p < .000). H[1] is supported as the unstandardized beta = 0.408, standard beta = .065, t-value = 6.254.

As for the mediation effect (H[2]), the direct effect of *PA* on *RIN* is significant, as β = .065 and p < .000. As *SATIS* entered the model as mediator, the direct effects of *PA* on *RIN* is reduced from .065 to .064 due to the moderation effect of *SATIS*. However, it is still significant since the probability

Table 1. Validity test.

	CR	AVE	MSV	ASV	PA	SATIS	RIN
PA	0.921	0.854	0.217	0.187	0.924		
SATIS	0.888	0.726	0.356	0.257	0.397	0.852	
RIN	0.864	0.614	0.356	0.287	0.466	0.597	0.784

Table 2. Goodness-of fit measures for the structural equation model ($N = 255$).

Name of Category	Name of Index	Index Value	Comments
1. Absolute fit	RMSEA	0.080	The required level is achieved
Absolute Fit	GFI	0.915	The required level is achieved
2. Incremental Fit	CFI	0.944	The required level is achieved
3. Parsimonious Fit	Chisq/df	2.626	The required level is achieved

of getting a critical ratio as large as 2.702 in absolute value are .007. In other words, the regression weight for *PA* in the prediction of *RIN* is significantly different from zero at the .01 level. In this case, the study concludes that *SATIS* has a partial mediation effect in the relationship between *PA* and *RIN* (Zainudin, 2013).

The type of mediation for H^2 is called partial mediation since the direct effect of *PA* on *RIN* is still significant after *SATIS* entered the model even though the effect is reduced from .065 to .064. In this case, *PA* has a significant direct effect on *RIN* as well as a significant indirect effect on *RIN* through the mediator variable, namely *SATIS*.

5 CONCLUSION

The results of the present study indicate that perceived authenticity clearly influence tourist to revisit the same destination. Regardless of numerous studies tourist behavior intention in tourism industry, the implication of the study will be valuable to the heritage tourism destination especially to the heritage site.

The findings also indicate that satisfaction has a significant mediation effect between the relationship of perceived authenticity and revisit intention. The result proved that regardless of the perception of authenticity that tourist have towards the heritage destinations, satisfaction strengthen the tendency to revisit to that particular destination.

ACKNOWLEDGEMENT

This research paper would not have been possible without the support of many people. Above all, the authors would also like to convey thanks to the Research Management Institute (RMI), Universiti Teknologi MARA, Shah Alam for providing the financial means and guidance to complete this research.

REFERENCES

Assaker, G., Vinzi, V.E. & O'Connor, P. (2011). Examining the effect of novelty seeking, satisfaction, and destination image on tourists' return pattern: A two factor, non-linear latent growth model. *Tourism Management*, *32*(4), 890–901.

Beard, J.G. & Ragheb, M.G. (1980). Measuring leisure satisfaction. *Journal of Leisure Research*, *12*(1), 20–33.

Chhabra, D. (2010). Back to the past: A sub-segment of generation Y's perceptions of authenticity. *Journal of Sustainable Tourism*, *18*(6), 793–809.

Cole, S.T. & Chancellor, H.C. (2009). Examining the festival attributes that impact visitor experience, satisfaction and re-visit intention. *Journal of Vacation Marketing*, *15*(4), 323–333.

Del Bosque, I.R. & Martín, H.S. (2008). Tourist satisfaction a cognitive-affective model. *Annals of Tourism Research*, *35*(2), 551–573.

Engeset, M.G. & Elvekrok, I. (2014). Authentic concepts: Effects on tourist satisfaction. *Journal of Travel Research*. doi:10.1177/0047287514522876.

Gaskin, J. (2012). Validity Master, Stats Tool Package. Retrieved from http://statwiki.kolobkreations.com

George, J. & Anandkumar, S.V. (2014). Portrayed and perceived online destination personality of select island destinations. *Anatolia*, 1–10.

Gerbing, D. & Anderson, J. (1988). An updated paradigm for scale development incorporating unidimensionality and its assessmet. *Journal of Marketing Research*, *25*(2), 186–192.

Han, H., Back, K.-J. & Barrett, B. (2009). Influencing factors on restaurant customers' revisit intention: The roles of emotions and switching barriers. *International Journal of Hospitality Management*, *28*(4), 563–572.

Huang, Z. (Joy), Cai, L.A., Yu, X. & Li, M. (2013). A further investigation of revisit intention: A multigroup analysis. *Journal of Hospitality Marketing & Management*, *16*(1).

Jayaraman, K., Lin, S., Guat, C. & Ong, W. (2010). Does Malaysian tourism attract Singaporeans to revisit Malaysia? An empirical study. *Journal of Business and Policy Research*, *5*, 159–179.

Kim, T. (Terry), Kim, W.G. & Kim, H.-B. (2009). The effects of perceived justice on recovery satisfaction, trust, word-of-mouth, and revisit intention in upscale hotels. *Tourism Management*, *30*(1), 51–62.

Kolar, T. & Zabkar, V. (2010). A consumer-based model of authenticity: An oxymoron or the foundation of cultural heritage marketing? *Tourism Management*, *31*(5), 652–664.

Kozak, M. & Rimmington, M. (2000). Tourist Satisfaction with Mallorca, Spain, as an off-season holiday destination. *Journal of Travel Research*, *38*(3), 260–269. 8

Lau, A.L.S. & McKercher, B. (2004). Exploration versus acquisition: A comparison of first-time and repeat visitors. *Journal of Travel Research*, *42*(3), 279–285.

Lee, B.C., Cho, J. & Hwang, D. (2014). An integration of social capital and tourism technology adoption-A case of convention and visitors bureaus. *Tourism and Hospitality Research*, Ahead–of–Print.

Lee, B., Lee, C.-K. & Lee, J. (2013). Dynamic nature of destination image and influence of tourist overall satisfaction on image modification. *Journal of Travel Research*, 53(2), 239–251.

Lin, Y., Kerstetter, D., Nawijn, J. & Mitas, O. (2014). Changes in emotions and their interactions with personality in a vacation context. *Tourism Management*, 40, 416–424.

Luoand, S.J. & Hsieh, L.Y. (2013). Reconstructing revisit intention scale in tourism. *Journal of Applied Sciences*, 13(18), 3638–3648. 48

Mao, I.Y. & Zhang, H.Q. (2014). Structural relationships among destination preference, satisfaction and loyalty in Chinese tourists to Australia. *International Journal of Tourism Research*, 16(2), 201–208.

Mohamad, M., Manan Ali, A., Ghani, A., Izzati, N., Rusdi Abdullah, A. & Mokhlis, S. (2013). Positioning Malaysia as a Tourist Destination Based on Destination Loyalty. *Asia Social Science*, 9(1), 286–292.

Mohamed, B. (2005). *Heritage tourism management in Japan*. Penang, Malaysia: Penerbit Universiti Sains Malaysia.

Moital, M., Dias, N.R. & Machado, D.F.C. (2013). A cross national study of golf tourists' satisfaction. *Journal of Destination Marketing & Management*, 2(1), 39–45.

Naef, P. (2011). Reinventing Kotor and the Risan Bay, a study of tourism and heritage conservation in the New Republic of Montenegro. *European Countryside*, 3(1), 46–65.

Pawitra, T.A. & Tan, K.C. (2003). Tourist satisfaction in Singapore—A perspective from Indonesian tourists. *Managing Service Quality*, 13(5), 399–411.

Prideaux, B.R. & Timothy, D.J. (2008). Themes in cultural and heritage tourism in the Asia Pacific region. In B. Prideaux, D.J. Timothy & K. Chon (Eds.), . New York, NY: Taylor & Francis.

Reid, L.J. & Reid, S.D. (1994). Communicating tourism supplier Services: Building repeat visitor relationships. *Journal of Travel & Tourism Marketing*, 2(2–3), 3–19.

Sun, K.-A. & Kim, D.-Y. (2013). Does customer satisfaction increase firm performance? An application of American Customer Satisfaction Index (ACSI). *International Journal of Hospitality Management*, 35, 68–77.

Um, S., Chon, K. & Ro, Y. (2006). Antecedents of revisit intention. *Annals of Tourism Research*, 33(4), 1141–1158.

Waligo, V.M., Clarke, J. & Hawkins, R. (2013). Implementing sustainable tourism: A multi-stakeholder involvement management framework. *Tourism Management*, 36, 342–353.

Wang, J. & Wu, C. (2013). A process-focused model of perceived authenticity in cultural heritage tourism. *Journal of China Tourism Research*, 9(4), 452–466.

Wang, X., Zhang, J., Gu, C. & Zhen, F. (2009). Examining antecedents and consequences of tourist satisfaction: A structural modeling approach. *Tsinghua Science & Technology*, 14(3), 397–406.

Xianger, C., Jianming, C., Zhenshan, Y. & Webster, N. (2014). Set relationships between tourists' authentic perceptions and authenticity of world heritage resources. *Journal of Resources and Ecology*, 5(1), 20–31.

Zalatan, A. (1994). Tourist satisfaction: A predetermined model. *Tourism Review*, 49(1), 9–13.

Zhou, Q. (Bill), Zhang, J. & Edelheim, J.R. (2013). Rethinking traditional Chinese culture: A consumer-based model regarding the authenticity of Chinese calligraphic landscape. *Tourism Management*, 36, 99–112. do.

Theory and Practice in Hospitality and Tourism Research – Radzi et al. (Eds)
© *2015 Taylor & Francis Group, London, ISBN 978-1-138-02706-0*

The prospect of Malay cultural heritage in Melaka: A preliminary evaluation

J. Jusoh & N.F.A. Hamid
Universiti Sains Malaysia, Penang, Malaysia

ABSTRACT: The aim of this paper is to highlight the preliminary evaluation of Malay cultural heritage that is located within the Melaka state. The history of Melaka began with the Malay Sultanate Empire thousands years ago. As such, this legacy has left many cultural heritages that belong to the Malay community. Nevertheless, only a few attractions of Malay cultural heritages were exposed to the tourists. Hence, this research has been carried out under the Exploratory Research Grant Scheme (ERGS) to explore the hidden Malay cultural heritages that are still being preserved and practiced in Melaka. A qualitative method through observation and interview were used as a medium to evaluate the prospect of Malay cultural heritage in Melaka. The preliminary observation is made to identify the existing scenario of Malay cultural heritage in Melaka either still being practiced or otherwise. Besides, the interview sessions were conducted with the National Heritage Department and a group of Malay cultural heritage product practitioners. The findings were extracted by using SWOT analysis. The findings indicate that the Malay cultural heritage product in Melaka represents the identity of the Malay community in Malaysia. However, this product is lacking in documentation system to conserve and safeguard the cultural heritage. However, the Malay cultural heritage product has an opportunity to become new tourist attraction in Melaka if the aspect of marketing is being taken into consideration. In conclusion, the Malay cultural heritage product will be able to compete if all stakeholders participate and sit together to market this product globally.

Keywords: Malay, tangible heritage, intangible heritage, hidden attraction, Melaka

1 INTRODUCTION

1.1 *The cultural heritage*

The cultural and heritage presents the local identity of a place. It is about the way of life, the beliefs and the achievement of a nation and its citizens (Abd Aziz & Abdullah, 2011). It is a combination of tangible and intangible heritage that are being inherited from one generation to another. The tangible heritage is defined as cultural properties including monuments, groups of buildings and site (UNESCO, 1999). On the other hand, the intangible heritage is tradition or living expression that is inherited from our ancestors and passed on to our descendants. It includes oral traditions, performing arts, social practices, rituals, festive events, knowledge and practices concerning nature and the universe or the knowledge and skills to produce traditional crafts (UNESCO, 1999). In Malaysia, the cultural heritage is varied according to the ethnics. There are three major ethnics that are living in harmony in Malaysia. They are Malay, Chinese and Indian. The cultural heritage from these ethnic groups can be identified through the religious buildings such as the mosque and temple, food, attire and festival (National Heritage Department, 2013). This paper is focusing on the Malay cultural heritage in Melaka.

1.2 *The cultural heritage tourism*

Department of Tourism, Republic of South Africa (2010) stated that cultural and heritage tourism is one of the oldest types of tourism. Tourism is one of the sectors that brings cultural heritage to the world. Through the conservation of tangible heritage such as buildings and monuments and by safeguarding the intangible heritage, millions of tourists were able to re-capture the moment of the earliest times. According to Robinson and Picard (2006), the difference among the cultures will strongly motivate the tourists to visit the heritage tourism destination. The influence of cultural heritage tourism development can be seen through the world heritage sites that are inscribed by UNESCO. The Coliseum in Rome, pyramid in Egypt, Taj Mahal in India and Porte De Santiago in Malaysia is tangible heritage that is still

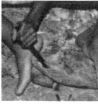

Figure 1. Example of the tangible and intangible heritage (Source: http://whc.unesco.org).

being preserved until today. Meanwhile, Tombs of the Buganda Kings at Kasubi, Uganda, Royal Ancestral Rite and Ritual Music of the Jongmyo Shrine at Republic of Korea and water puppetry in Vietnam are examples of intangible cultural heritage that remained until today (Park, 2011). Thus, tourism sector creates an opportunity to the international and domestic tourists to understand and appreciate the cultural heritage that are still being preserved and practiced until today.

1.3 The Malay cultural heritage in Melaka

In Malaysia, UNESCO inscribed Pulau Pinang and Melaka as World Heritage Site in the year 2008, among the reason because of their cultural heritage. Among these heritage sites, Melaka is significant with Malay cultural heritage because the Empire of Melaka Sultanate begins in the year 1261 (Yaakob, 2013). Since then, Melaka has become the Malay Kingdom with Melaka River plays a significant role as the core of trading and administrative center of the 14th century. The Malay cultural heritage in Melaka is classified into tangible and intangible heritage. The Malay tangible heritage includes monuments, traditional villages and buildings. On the other hand, Malay intangible cultural heritage is batik printing, rattan weaving, wood carving, traditional food, martial art weapons, song and dance.

Preliminary investigation reveals that, some original location of palace, a mosque and fortress related to the Malay heritage in Melaka could not be found. During that time timber is used for building construction and it is easily broken. Furthermore, the destruction of historical buildings by the colonialist namely Portuguese, Dutch and British caused the buildings of the Malay Sultanate Empire could not be discovered (Yaakob, 2013). Besides, there are not much of the intangible cultural heritage of the Malay community in Melaka have been exposed to the tourists. The reason being, the Melaka world heritage site mainly stands with its architectural and historical buildings that represent the colonialist history. In contrast, the Malay cultural heritages are located scattered within Melaka especially in Jasin and Alor Gajah district. This distinction caused the tourists lack of awareness about the Malay cultural heritage in Melaka. Therefore, this research is conducted to identify the hidden Malay cultural heritage attraction in Melaka. This paper highlighted the preliminary evaluation towards the prospect of Malay cultural heritage that is located in Melaka.

2 METHODOLOGY

The objective of this study is to have a preliminary evaluation towards the prospects of Malay cultural heritage that is located in Melaka, a state that is rich with multicultural heritages. Thus, a qualitative method is being used as to achieve the objective of the study. In order to have general information regarding to the current trends of Malay cultural heritage in Melaka, interview sessions were conducted with two stakeholders from the National Heritage Department South Zone (Malaysia) and three practitioners of Malay cultural heritage in Melaka. The sample chosen are based on previous research conducted (see: Chen, 2013; Rahman, 2012). They used representative from the local authority and individual or group that is involved in the heritage industry. Besides that, the information about Malay entrepreneurs was obtained from the National Department for Culture and Arts, Melaka (2013) and a source for a research by Khairul (2012, December 3).

An appointment was then made by the interviewees through email and phone calls. The in-depth interview was applied to access the current information and scenario regarding the Malay cultural heritage in Melaka. The researchers used an informal approach during the interview because it is more flexible and comfortable for both interviewers and interviewees (Chen, 2013). In addition, observation method is also being used to explore the Malay cultural heritages that are available in the three districts of Melaka namely Melaka Tengah, Jasin and Alor Gajah. Finally, to evaluate the findings from the secondary and primary data collected from the Malay cultural heritage expertise, a SWOT analysis have been used. This method has been used widely in numbers of fields, namely urban planning, counselling psychology, business and tourism (Kam, 2013). This tool is the most frequently used for the purpose of analysing the business environment (Glaister & Falshaw, 1999) and it helps in verdict the crucial factors of a future plan (Nikolaou, Ierapetritis & Tsagarakis, 2011). A SWOT analysis is a combination of strengths, weaknesses, opportunities and threats. Therefore, this study applied a SWOT analysis in order to have a preliminary understanding of the Malay

Table 1. Summary of total stakeholders as respondents.

Category	Identity	Code	Total Respondent
A	National Heritage Department South Zone (Malaysia)	A1	1
B	Malay Cultural Heritage Practitioners	B1, B2,B3	3

Table 2. The question have been asked to the respondents.

Question 1 (Strengths)	What is the strength and element of cultural heritage in Melaka that is significant to the Malay community?
Question2 (Weaknesses)	What are the weaknesses of Malay cultural heritage attraction in Melaka that the stakeholders should take into account in order to ensure the preservation of this heritage?
Question 3 (Opportunities)	What are the opportunity factors of Malay cultural heritage attraction in Melaka that the stakeholders should take into account in order to ensure the preservation of this heritage?
Question 4 (Threats)	What are the threat factors of Malay cultural heritage attraction in Melaka that the stakeholders should take into account in order to ensure the preservation of this heritage?

cultural heritage in Melaka. The responses from respondents were coded as A1, B2, B3 and B4. Table 1 indicates the summary of total stakeholders as respondents.

As this study applied the SWOT analysis, the question have been asked to the respondents are derived according to the strengths, weaknesses, opportunities and threats (Table 2).

3 FINDINGS AND DISCUSSIONS

The data acquired from the researchers' observation and interview sessions were critically analysed using a SWOT analysis.

3.1 The strength of Malay cultural heritage in Melaka

The strength of Malay cultural heritage in Melaka is a symbol and identity of Malay community that are inherited from one generation to another. According to A1, *"The attractions that symbolize Malay cultural heritage attractions within the world heritage site of Melaka include Melaka Sultanate Palace Museum, Malay and Islamic World Museum, and Morten Village"*. Based on secondary data, the most significant Malay cultural heritage attractions are the Malay traditional house, *Asam Pedas Melaka, Hang Tuah* Village, Malay warrior's mausoleum and Bullock cart (Melaka State Government, 2013, September 5). Most of the Malay heritages that can be identified easily are tangible heritage such as traditional Malay house and mosque. Such examples are the House of Munshi Abdullah and traditional houses in Morten Village and Duyong village (Melaka State Government, 2013, September 5). These Malay cultural heritage attractions portray the tradition and practices of the Malay community during the Melaka Sultanate Empire. According to B1, *"Malay cultural heritage represents the way of life of Malay people from the past generation and still being practiced until today. Malay cultural heritage in Melaka is strong. It is proven through the scenario of people wearing Tanjak in Umbai, Melaka"*. Those who are wearing *Tanjak* (one of the Malay attires) are among a group of Malays that is involved in a martial art which is called *Silat*. According to B3, *"The Malay cultural heritage in Melaka is unique. For example, the model of the Malay traditional house can only be found in Melaka"*. Therefore, based on the feedback from interviewees, the strength of Malay cultural heritage in Melaka is where the local people are still practicing the traditions and keeping the valuable legacy till today. However, it is still new and the people who are practicing the traditions are based on group and availability of events like wedding and cultural ceremony.

3.2 The weaknesses of Malay cultural heritage in Melaka

The Malay cultural heritage may be divided into different types of heritage. It includes performing arts, customs and culture, language, literature, fine arts and living person (National Heritage Department, 2013). However, not many Malay cultural heritages in Melaka are easy to be identified and have vast information. According to A1, *"Malaysia is lacking in term of literature and weak in documentation system. Consequently, the history becomes a myth. The validity of the story is doubtful. As a result, many Malay cultural heritages cannot be appointed as world heritage"*. On the other hand, B1 has a different perspective about the weakness of Malay cultural heritage in Melaka, *"More attention have been given to the traditional Malay dances and music compared to the other type of cultural heritage like Silat, Malay weapons and attires"*. This scenario caused inequality in term of promoting the Malay cultural heritage in Melaka to the outsiders. Another weakness about Malay cultural heritage in Melaka is regarding the inheritor or the next generation practitioner.

According to B3, he is having a problem to continue his business by declaring that," *my sons do not have any interest in building the traditional Malay house"*. In this case, it caused the skill in making a traditional Malay house to be moved out from the industry. Lastly, B2 has expressed his view towards the mentality of the public and the Malay communities about a Malay cultural heritage product that is related with mystical elements. Hence, not so many people do believe in the culture and will not continue the tradition. This scenario happened to one of the Malay weapons known as *Keris* where this product was only used as 'furniture' during traditional Malay ceremony. Overall, the weaknesses of Malay cultural heritage in Melaka are lacking and inefficient in terms of the record system, exposure, marketing and literature, the inheritor and mystic issue.

3.3 *The opportunities of Malay cultural heritage in Melaka*

Based on secondary data and on site observation, the tangible heritages such as Malay traditional house, museums and a traditional village are among the attractions that are always being visited by many tourists in Melaka. However, only a few attractions related to the intangible Malay heritages were exposed to the tourists. This study has found that the Malay cultural heritage in Melaka specifically the intangible heritage has the potential to be the new tourist attractions in Melaka. According to A1, *"The local authorities are now focusing on the intangible cultural attraction that is located outside of Bandar Hilir (Core Heritage Zone) of Melaka"*. In this case, Malay cultural heritage in the district of Jasin and Alor Gajah has a chance to be promoted and commercialized to the tourists. The other opportunity is creating awareness about the existence of Malay cultural heritage to the local people as well as domestic and international tourists. B2 mentioned that *"... not all Malays noticed and know about Malay cultural heritage. You can ask Malay people, especially the younger generation a simple question like the name of traditional food; I believed that not all of them are able to identify these elements"*. Thus, by emphasizing on the Malay cultural heritage, it would give opportunities to enhance the people's knowledge and responsiveness to safeguard the Malay identity. In summary, the Malay cultural heritages have the opportunities to become a new attraction in Melaka for the tourism related authorities of Melaka have decided to focus on the attraction that are located outside of Bandar Hilir. Furthermore, the awareness towards Malay cultural heritage could be created among local public of Melaka and the tourists.

3.4 *The threat of Malay cultural heritage in Melaka*

Malay cultural heritage is related and inherited by the Malays in Melaka, Malaysia. Besides Malaysia, the Malays can also be found in Thailand, Indonesia, Filipina, Cambodia, Brunei and South Africa. Malay people are categorized differently according to their ethnics which include Javanese, Bugis, Minangkabau, Acehnese, aboriginal and *Peranakan*. Thus, A1 has stressed, *"The multi-ethnics have caused difficulties in creating patent for the heritages"*. For example, Batik Painting can be found in Indonesia as well as Malaysia. However, which one represents the Malay heritage? On the other hand, B2 raised the competitive issue about Malay cultural heritage. He said that, *"Malay handicraft products such as Tanjak and Keris faced the competition from neighboring countries. It is because Tanjak and Keris made by neighboring countries are more competitive as these countries offered lower prices and produced an accep*table *quality"*. B1 said, *"Bangladesh also has factories that produced Tanjak"*. B2 added that most of the buyers preferred to buy *Tanjak* and *Keris* from the outsiders because of their cheaper price. B2 emphasized, *"Only activists in Malay cultural heritage can identify the difference in quality and the originality of this heritage product"*. Besides that, B3 has similar views about competitiveness. He said, *"I have tried to sell my handmade crafts to one of the retailer in Bandar Hilir at a price RM 18 per each. Unfortunately, the retailer said the price is too expensive and decided to buy my handmade crafts at RM 4 for each crafts"*. This scenario shows that the handicraft offered by other countries like Thailand, China and India can be purchased at a cheaper price. Therefore, retailer in Malaysia can sell the craft at a competitive price. In conclusion, the threat of Malay cultural heritage is a difficulty to patent the product as the Malays consist of different ethnics and the competition from other neighbouring countries like Thailand and Indonesia are extremely high.

4 CONCLUSIONS

The Malay cultural heritage is about the Malay traditions that are still being practiced and well preserved until today. This study reveals that the Malay community in Melaka is practicing the traditions; however, it involves only a small percent of the Malay community. These heritages are unique that could attract tourists to see, learn and amaze with the products. The Malay cultural heritages have many different types. However, only few Malay cultural heritages can be identified because of lack of awareness and poor in documentation and record system. Furthermore, only Malay dances and music were exposed to the tourists compared to the other type of heritage. These problems lead to the loss of Malay cultural heritage in Melaka. Yet, as Melaka is receiving the highest number of tourists' arrival, this legacy should be safeguarded by introducing it as a new attraction in Melaka. However, the Malay cultural heritages have many ethnics that caused difficulties to patent and to be pointed out as a world heritage. In addition, the competition from neighbouring countries discouraged the local practitioner to continue the legacy of conserving the original Malay heritage culture.

ACKNOWLEDGEMENTS

The researchers would like to extend their deepest gratitude to the Ministry of Higher Education, Malaysia (MOHE) for financing this research under the Exploratory Research Grant Scheme (ERGS) (203/ PPBGN/ 6730137).

REFERENCES

Abd Aziz, K. & Abdullah, F.Z. (2011). Cultural heritage tourism development in Kota Lama Kanan, Kuala Kangsar, Perak. *Universiti Tun Abdul Razak E – Journal 7*(2), 1–10.

Chen, G.H. (2013). Stakeholder's perception on authenticity of cultural products at Jonker Street, Melaka: Universiti Sains Malaysia.

Department of Tourism Republic of South Africa. (2010). The workshop report heritage and cultural tourism strategy. South Africa: Department: Tourism Republic of South Africa.

Glaister, K.W. & Falshaw, J.R. (1999). Strategic planning: still going strong? *Long Range Planning, 32*(1), 107–116.

Kam, H. (2013). Understanding China's hotel industry: a SWOT analysis. *Journal of China Tourism Research, 9*(1), 81–93.

Khairul. (2012, December 3) Retrieved January 9, 2014, from http://kronikeltanjakmelayu.blogspot.com

Melaka State Government. (2013, September 5) Retrieved March 10, 2014, from http://www.melaka.gov.my

National Department for Culture and Arts Melaka. (2013) Retrieved March 10, 2014, from http://www.jkkn.gov.my/en/national-department-culture-and-arts-jkkn-melaka

National Heritage Department. (2013) Retrieved December 16, 2013, from http://www.heritage.gov.my/

Nikolaou, E., Ierapetritis, D. & Tsagarakis, K. (2011). An evaluation of the prospects of green entrepreneurship development using a SWOT analysis. *International Journal of Sustainable Development & World Ecology, 18*(1), 1–16.

Park, S.-Y. (2011). The intangible cultural heritage courier of Asia and the Pacific. Republic of Korea: Intangible Cultural Heritage Centre for Asia and the Pacific (ICHCAP).

Rahman, S. (2012). *Heritage tourism and the built environment.* University of Birmingham, United Kingdom.

Robinsin, M. & Picard, D. (2006). *Cross-cultural behaviour in Tourism: Concepts and analysis.* Oxford: Butterworth-Hinemann.

UNESCO. (1999). Operational guidelines for the implementation of the world heritage convention. Paris: World Heritage Committee.

Yaakob, M.K. (2013). Melaka sebagai bandar warisan dunia: Pandangan mengenai warisan dan pelancongan Melaka. Melaka: National Heritage Department.

Theory and Practice in Hospitality and Tourism Research – Radzi et al. (Eds)
© 2015 Taylor & Francis Group, London, ISBN 978-1-138-02706-0

Re-visit alternative dispute resolution in resolving disputes amongst travel operators and holidaymakers in Malaysia

N.C. Abdullah
Faculty of Law, Universiti Teknologi MARA, Shah Alam, Malaysia

ABSTRACT: The commercial travel agencies and travel agencies which offer *umrah* and *hajj* recorded the highest disputed cases in the Tribunal for Consumer Claims since 2010–2013. This is attributed to the fact that the Consumer Protection Act 1999 allows the President of the Tribunal for Consumer Claims (TCC) to assist holidaymakers and travel operators in negotiating an agreed settlement in relation to their disputes. Progressing on the same tone, the courts in Malaysia have embarked on the practice of mediation as one of the forms of alternative dispute resolution (ADR) through two choices of mediation; judge-led mediation or mediation to be conducted by the Malaysian Mediation Centre (MMC) under the auspices of the Malaysian Bar Council. In judge-led mediation, if mediation is successful, the judge mediating shall record a consent judgment on the terms agreed by holidaymakers and holiday providers. In mediation by MMC, a successful mediation will be reduced into writing in a settlement agreement signed by the holiday provider and holiday maker and similar to the judge-led mediation, the holiday provider and holidaymakers are then mandated to record the terms of settlement as a consent judgment. Whereas, in the case where the holiday provider and holidaymaker fail to reach a settlement in the judge-led mediation, the case will revert to the original judge to hear and complete the case. It is also argued that ADR can be used to utilize administrative influence, formally or informally, to make the ADR process efficient. In view of the limited literature pertaining to disputes between holidaymakers and holiday provides, using the qualitative approach, this paper explores the likelihood of the judge-led mediation in court and administrative ADR to overtake the popularity of the TCC in view of the regulatory limitation of claims by holidaymakers and the prohibition of legal representation by the holiday operators and holidaymakers in the TCC.

Keywords: Alternative dispute resolution, tribunal for consumer claims, court, holidaymakers, holiday providers

1 INTRODUCTION

Section 107(1) of the Consumer Protection Act 1999 allows the President of the Tribunal for Consumer Claims to assist parties in negotiating an agreed settlement (not exceeding RM 25,000) in relation to their claim. The provision implies the application of alternative dispute resolution (ADR) in the Tribunal for Consumer Claims (herein after refers to TCC). Since its introduction, the commercial travel agencies and travel agencies which offer *umrah* and *hajj* recorded the highest disputed cases since 2010–2013, where the total number of cases, within the three years are, 947 cases relating to travel agencies and 1156 cases concerning *umrah* and *hajj* (TCC, 2013). The application of ADR in the court system is also significant nowadays since in many jurisdictions, courts are moving towards active case management which moves the traditional parties-activated adversarial system

to the new philosophy of court-managed system of litigation. Active case management includes, inter alia, encouraging holiday makers and holiday providers to use an alternative dispute resolution (ADR) such as mediation by a third party. A mediation which results in a consent judgment between holidaymakers and holiday providers is final and certainly avoid prolonged arguments and reduce tension between both parties. Progressing on the same tone, the courts in Malaysia has embarked on the practice of mediation as a one of the forms of ADR through the issuance of a Practice Direction No 5 of 2010, Practice Direction on Mediation (herein after referred to as the PD) by the Chief Registrar of the Federal Court of Malaysia and came into effect on 16 August 2010. The PD covers all on-going civil litigation pending in the High Court and the Subordinate Courts. The Judge to whom the action has been assigned shall make such orders and give such directions as to the future conduct of the action to ensure its just, expeditious and economical disposal. The PD is in complete consonance with the procedural reforms in the mechanics of modern civil litigation. Nevertheless, the National Consumer Complaints Centre (NCCC) is also playing an increasing important role in the dispute between the holidaymakers and holiday providers.

2 LEGAL REPRESENTATION

Since TCC aims at providing justice to holidaymakers at a very limited cost, i.e. RM 5 as the application cost and in a speedy manner, TCC does not permit any party to be represented by a lawyer (section 108 Consumer Protection Act 1999 and Instructions to claimant and respondent in Form 1, TCC Regulations). Consequently, principles of law are not put forward by parties to TCC. Therefore, it depends on the presiding President of TCC whether to highlight the relevant principles of law to the travel operator and holidaymakers or otherwise. It is observed that if the President of TCC is not well-versed in the area of claim governing the dispute before him (Safei & Chua Abdullah, 2013), such area of law will be left uncovered in the hearing.

On legal representation, it may appear that the travel operator may be on a better position compared to the holidaymakers because even though section 108 (2) of the Consumer Protection Act 1999 expressly prohibits legal representation in the hearing of TCC, the travel operator can have its full-time paid employee who has a legal background to represent his case by virtue of section 108 (3)(a) of the Consumer Protection Act 1999 (Amin & Abu Bakar, 2010). Due to the fact that in TCC, the holidaymakers and holiday providers cannot be represented, it is seen that settlement through ADR in TCC sometimes do not give force to the values embodied in authoritative texts such as the statutes, since the parties often are not able to argue the legal principles and the spirit of the law. However, in court-led mediation, the disputants' lawyers are allowed to participate in the mediation process, as will be discussed further in this paper.

3 LEGAL STATUS OF ADR FORUM

In the Malaysian courts, where the mediation Judge is unable to bring about an amicable settlement, the case is reverted to the hearing Judge for disposal. This is not the case for the TCC. In TCC, The claim of the travel dispute can only be initiated by the holidaymaker and not the travel operator. Once the petition is being initiated by the holidaymaker, the travel operator has to go through the hearing at the TCC. Though ADR is statutorily permissible in the TCC, its level of implementation can be questioned as there are occasions where judgments are made by the president of tribunal rather than by a discourse between the holidaymakers and travel operators. The travel operator cannot deny the right of the holidaymaker to claim for compensation at the TCC and is not allowed to challenge the TCC as the rightful forum and is not allowed to opt for the court as the forum to hear the case. The finality of the president's award in occasions where the disputants have not reached a settlement is one of the aspects which may be of concern; both to the parties in dispute, i.e. the travel operator and consumer. Alternatively, administrative ADR under the purview of National Consumer Complaints Centre (NCCC) does not have the power to compel the disputants to reach to any settlement.

4 APPLICATION OF FACILITATIVE AND EVALUATIVE MEDIATION

Generally, the president in TCC indirectly adopts the facilitative mediation approach when in practice, many presidents during the introduction of the hearing stage, enquires the holidaymakers and travel operators as to whether they had discussed the disputed matter and whether they had tried attempted to resolve the disputes. Facilitative mediation is obvious as if the answer is in negative, the president would persuade and sometimes request the holidaymakers and travel operators to try to resolve the matter at the waiting corner. The travel operators and holiday makers are given assistance to arrive at their own conclusions. The President helps the holidaymakers and travel operators to explore the options. The disputants may find one of the options appealing to them. Since the presidents of TCC had opportunity to analyses cases before the hearing of the cases in TCC, it is reasonable to assume that presidents of TCC tend to adopt facilitative mediation in travel disputes which are complicated. However, the difference between ADR in TCC and court is that while the court applies facilitative mediation, the travel operators and holidaymakers and their lawyers would participate actively in the discussion and negotiation.

In court-led mediation, lawyers and travel disputants do not make submissions before the mediator. Unlike the TCC, in court-led mediation, the disputants, with the assistance of their lawyers would offer options or counter-proposals to persuade each other, while the facilitative mediator offers his expertise and experience in the discussion and negotiation. The involvement of lawyers is significant in the ADR process as they are able to give their advice to their clients. In an evaluative mediation which is also applied in the TCC and court-led mediation, the mediator evaluates the travel disputants' position and point of view, the strength and weakness of the respective parties' case and offer suggestions to them. The difference is that in TCC, the President listens to the disputants' arguments and on the other hand, in court-led mediation, evaluative mediators would even hear the counsel's arguments presented for the disputants. However, in TCC, the President, after hearing the arguments from both the holiday providers and holidaymakers, makes a binding award, which is final; which can only be revised in a court. Unlike the TCC, in court-led mediation, there is no cohesive or coercive power to make any binding decision between a holiday provider and holidaymaker. The disputants do not have to accept the mediator's evaluation of the relative merits of their case. Evaluative mediation gives the holiday providers and holidaymakers the opportunity to obtain an objective and unbiased evaluation of the case in the Court of first instance or at the appellate stage, without the expense of a hearing of the case or appeal proper.

5 CONFIDENTIALITY OF INFORMATION

The hearings in TCC are opened to public. Hence, sometimes trade secrets and business strategies of travel agencies are being exposed in the TCC. Unlike the TCC, the court-led mediation assists in maintaining the status quo of confidentiality where the disputants who have confidential information may wish to disclose it to the mediator but not to the other side. In maintain the confidentiality of the information; the mediator may enter into a caucus meeting, in the absence of the other disputants and legal representative. The mediator can use the confidential information to structure discussion in order to explore or broaden the scope of the options for settlement. All disclosures, admissions and communications made in and during a mediation session are strictly "without prejudice", by virtue of section 23 of the Evidence Act 1950. Such communications do not form part of any record. The mediator shall not be compelled to divulge such records or testify as a witness or consultant in any judicial proceeding, unless all parties to the Court proceedings and the mediation proceedings consent to their inclusion in the record or for any other use. Hence, the concept of confidentiality in the court-led mediation is a unique feature which would reduce tension between the travel operators and holidaymakers as sometimes information is legally irrelevant and yet exceptionally

important to understanding the conflict and facilitating resolution. If such information is revealed to the public, this may be detriment to the good name of either party or both parties.

6 ADMINISTRATIVE ADR

In some jurisdictions, for example, Japan and Korea, administrative ADR is used widely to facilitate disputants and this is an option which is possible in the travel industry. One of the advantages of the administrative ADR is that they can utilize administrative influence, formally or informally, to make the ADR process efficient. For example, a travel operator will appear at the ADR session and negotiate with the claimant (holidaymaker) faithfully, because it wishes to avoid the risk of negative evaluation by the administrative agency. Under the umbrella of the administrative agency whose influence over the companies is strong, ADR may give the claimant better remedies than those he/she would be awarded in court. It is also expected that the agency monitors the implementation of the promised remedies. For the holidaymakers, the administrative ADR can be more effective than court proceedings. For the travel operator, because of the confidentiality of the process administrative ADR is more desirable than court proceedings.

In Malaysia, administrative ADR can be illustrated by the National Consumer Complaints Centre (NCCC) which is officially established on 13 July, 2004. The NCCC was initiated jointly by the Education and Research Association for Consumers (ERA Consumers), Selangor and Wilayah Persekutuan Consumer Association and the Ministry of Domestic Trade and Consumer Affairs. NCCC functions as a one stop centre to help consumers with their problems and complaints by ensuring that complaints are forwarded to relevant authorities and solutions are obtained while also acting as a go-between consumers and enterprises in settling disputes. At present, the NCCC is an option that is used by the holidaymakers to express their dissatisfaction towards travel agencies. Officers from NCCC does take the initiative to call up travel agencies and make visits to travel agencies to make further discussions with the travel agencies as regards to the complaints received by the holidaymakers. There are travel agencies who felt obliged to conform to the recommendations made by the officers from NCCC as the travel agencies wish to avoid the risk of negative evaluation by the administrative agency.

At the same time, administrative ADR can be beneficial to the Ministry of Domestic Trade and Consumer Affairs and Ministry of Tourism because a 'dispute' is full of useful and specific information about the travel industry which the agency must regulate. It alerts the Ministry of Domestic Trade and Consumer Affairs in general and specifically Ministry of Tourism as to the necessity of regulation not only of the accused travel operator but also the travel industry as a whole. In the absence of comprehensive precautionary regulations in the travel industry, in view of the laissez-faire economy approach taken by the government, the administrative ADR come to be a good means for the Ministry of Tourism to find out where a problem is and to regulate the travel industry.

7 CONCLUSION

In the light of the distinct features of the TCC as discussed above, holidaymakers may turn to the practice of mediation in courts, or alternatively rely on administrative ADR. However, the application of ADR in TCC can be upgraded with the modification of the existing features of the TCC. It is suggested that legal representation in exceptional circumstances in TCC should be allowed for example, where the issue of law arises and where both, the travel operator and holidaymaker consented to it. This would facilitate to the application of ADR in TCC. However, at this juncture, the high statistics on cases pertaining to travel agencies indicate that TCC is still much favored by holidaymakers, most probably due to the cheaper and speedier method of award as compared to the court-led mediation and the holidaymakers' ignorance of the concept of administrative ADR applied in the National Consumer Complaints Centre (NCCC).

ACKNOWLEDGMENT

The author wishes to thank the Research Management Institute (RMI) UiTM Shah Alam for funding this project entitled "The Suitability of Hybrid Alternative Dispute Resolution (ADR) Methods in Courts and Tribunals (Phase 2)", under the Principal Investigator Support Initiative (PSI) grant.

REFERENCES

Amin, N. & Abu Bakar, E. (2010). ADR for consumers: An appraisal of the tribunal for consumer claims in Malaysia. In M.N. Ishan Jan & A.A. Ali Mohamed (Eds.), *Mediation in Malaysia: The law and practice* (pp. 171–185). Kuala Lumpur: Lexis Nexis.
Safei, S. & Chua Abdullah, N. (2013). *Mediating claims in the tribunal for consumer claims: Some considerations.* Paper presented at the International Conference on Consumerism, Equatorial Hotel, Bangi, Selangor, Malaysia.

Theory and Practice in Hospitality and Tourism Research – Radzi et al. (Eds)
© *2015 Taylor & Francis Group, London, ISBN 978-1-138-02706-0*

Exploring conceptual framework of tourism SMEs performance in heritage sites

M.M. Rashid, M. Jaafar, N. Dahalan & M. Khoshkam
School of Housing, Building and Planning, Universiti Sains Malaysia, Penang, Malaysia

ABSTRACT: In order for tourism to be beneficial in terms of economic development, income earning, poverty diminution, and improving rural livelihoods, it should be linked with local economic activities of micro and small scale enterprises. Tourism is composed of conservative Small and Medium-sized Enterprises (SMEs) as indicated by the fact that most tourist facilities are run by small and medium-sized businesses. The main purpose of this study is to identify the gap in existing literatures to comprehensively understand tourism SMEs performance in an archaeological heritage of the Lenggong Valley located in a developing country of Malaysia. This study uncovered factors associated with the tourism SMEs performance. While developing this part of tourism SMEs performance concept, tourism approach is particularly useful in relation to three aspects: SMEs owner manager characteristics, SME business practice characteristics and networking. The implication of this study is that it is an attempt to provide a general overview of SMEs business and tourism performance concept that can be used by stakeholders and managers in the heritage sites.

Keywords: Small and Medium Enterprises (SMEs), tourism SMEs performance, heritage sites

1 INTRODUCTION

The Tourism has been recognized as one of the most popular industries in terms of socio-economic activity in the entire world. World Tourism Organization's (2012) (UNWTO) reported that international tourism would continue to grow in 2013, regardless of natural catastrophes around the world, political revolution issues in the Middle East and North Africa and the doubtful global economy. The number of international tourist arrivals rose to 4.6%, with a record 982 million tourists travelling worldwide. World Travel and Tourism Council (2012) (WTTC) reported that: *"travel and tourism's total contribution in 2011 was US$6.3 trillion in GDP, US$743 billion in investment, US$1.2 trillion in exports, and created 255 million jobs. This contribution signified 9% of GDP, 5% of investment, 5% of exports and 1 in 12 jobs"*.

As reported in World Travel & Tourism Council (2012) (WTTC) in their Travel and Tourism: Economic Impact, 14.8% of Malaysia's GDP in 2011 was directly related to the contribution of tourism and travel, an economic value of MYR125.4 billion.

Jenkins (2007) claimed that business activity generates jobs, develops inter-firm linkages, allows technology transfer, creates human capital and physical infrastructure, generates tax revenues for government, and evidently tenders a variety of products and services to consumers and other businesses. Philip (2010) stated that SMEs induce private ownership and entrepreneurial skills, adjust quickly to shifting market situations, create employment, vary economic activity, and make a large contribution to exports and trade.

Previous studies on the relationship between tourism and the SMEs focus more on hospitality or accommodation service (Morrison & Teixeira, 2004; Mshenga & Owuor, 2009; Seppälä-Esser, Airey & Szivas, 2009; Thomas, Shaw & Page, 2011), and very less focus on other types of SMEs in tourism.

Consequently, this study is a conceptual article with the aim to develop a SMEs tourism concept through business and management perspectives. The main purpose of this study is to identify the gap in existing literature and obstruction for Archaeological Heritage of the Lenggong Valley, a UNESCO World Heritage Site in developing country of Malaysia to comprehensively understand tourism SMEs performance from three aspects: SMEs owner manager, SME business practice and networking.

2 FACTORS INFLUENCING FORMATION OF TOURISM SMES PERFORMANCE CONCEPT

2.1 *Owner manager characteristics*

Within the extensive category of owner-manager characteristics, researchers had classified the variables into two categories which are general background and entrepreneurial characteristics.

2.1.1 *General background*

Previous literature had proved that banks may impose more difficult requirements on female operated businesses in concern for loans; hence it might hamper their ability to grow (Riding & Swift, 1990). On the other hand, Ahmad (2005) presented results to indicate that while the majority of small firms are owned and managed by men, more women owner-managers are involved in tourism and hospitality industry. He added that 81.1% male owner-managers rule the small industry in Malaysia, and this is as a result of cultural influence, which recognizes the male as the decision maker in the family.

The limited empirical evidence suggests that the entrepreneurs' age was negatively related to firm survival (Davidsson, 1991). Alasadi and Abdelrahim (2007) proved that those firms in Syria whose owner-managers were young and had previous training are more successful.

Bates (1990) declared that the successful ownership of a business required at least minimal level of educational accomplishment. Cooper, Gimeno-Gascon, and Woo (1994) proved that in ten out of seventeen empirical studies, there was a positive relationship between level of education and business performance.

Course and training particularly for the small business owner manager allow them to gain the necessary skills to ensure the survival and success of their businesses. Past researchers had argued that course and training management for human resource management (HRM) is a key factor in small business performance (Blackwood & Mowl, 2000; Huang & Brown, 1999).

Several studies (e.g., Carlsen, Getz & Ali-Knight, 2001; Getz & Carlsen, 2000; Getz & Nilsson, 2004) agreed that the majority of SMEs in the tourism sector, along with most family businesses, fit into lifestyle entrepreneurs' category. Naturally, lifestyle entrepreneurs allow lifestyle choices to affect their business. According to Getz and Carlsen (2000), the majority of the small business owners in Australia agreed that two main goals in getting their businesses started is to live in the right environment and to enjoy a good lifestyle. However, this is in contrast to the findings of research on small and

medium island chalet operators in selected islands of Terengganu and Pahang conducted by Maideen (2010) whereby money was apparently the important goal.

2.1.2 *Entrepreneurial characteristics*

Entrepreneurial characteristics have been one of the important features in understanding the owner-manager characteristics. Previous researches had study the relationship of the entrepreneurial characteristics and the business performance. Among the important values in entrepreneurial characteristics are (1) need of achievement motivation (Nieman & Nieuwenhuizen, 1997; Perren, 1999); (2) active risk taker (Papadaki & Chami, 2002; Perren, 2000; Rody & Stearns, 2013); (3) innovativeness (Papadaki & Chami, 2002; Perren, 2000); (4) self-confidence, (Ho & Koh, 1992); (5) ability to learn from failure (Morrison, Breen & Ali, 2003); (6) independence (Hankinson, Bartlett & Ducheneaut, 1997; Nieman & Nieuwenhuizen, 1997) and (7) pro-activeness (Rody & Stearns, 2013).

2.2 *Business practise characteristics*

Understanding the business practice characteristics of SMEs is crucial because the on-going business itself is the main cause in determining the SMEs performance. However, identifying the business practices in SME might seem challenging because their activities are commonly not documented, and are not part of any drafted strategy (Grayson & Dodd, 2007). In this study, researchers focused on four important business characteristics which are financial resources, marketing, technology adoption and product innovation.

2.2.1 *Financial resources*

Study by Glancey and Pettigrew (1997) showed that own funds (34%), bank lending (20%) and a combination of both (46%), were the major sources of finance. However, empirical evidence by Ahmad (2005) added that owner-managers depended on strong ties with family members and close friends during the early years of business. The study by Sharma and Upneja (2005) also proves that bank loans appeared as the third most likely source of start-up finance, while the most preferable source for raising working capital finance was through internal funds.

2.2.2 *Marketing*

According to Manan and Jan (2010), marketing capability is the ability of a firm to utilize its resources in executing marketing activities and thus the customers' needs can be fulfilled. Manan and Jan (2010) added that marketing capability is unique as it requires the interaction of many resources, and it is relatively hard to recognize any other single resource that could substitute for marketing capability that could create a similar outcome.

2.2.3 *Technology adoption*

The development of technology led to the introduction of new products, changes in techniques and organization of production, changes in the quality of resources and products, novel ways of allocating products and new ways of storing and broadcasting information. Papadaki and Chami (2002) found that the significant measurements in business practice characteristics are business innovation, technology adoption of e-business and focusing on the local market, all were prove to have a significant influence on the growth of business.

2.2.4 *Product innovation*

Product innovation can play an important role in determining the survival and performance of new entrance and existing businesses. New and small businesses are more exposed to risk particularly in the first few years of entry (Caves, 1998; Geroski, 1995). Chittithaworn, Islam, Keawchana, Yusuf, and Hasliza (2011) and Philip (2010) has found that innovative product is among the keys to small business success.

2.3 *Networking*

Networking is a key strategic dimension in external environment in SME performance. Prior researchers had identified the significance of networking to business performance (Chittithaworn, et al., 2011; Philip, 2010). According to Morrison et al. (2003), networking is defined as "Joiners" included multi-participation across an extensive range of social and business fields, placing high value on learning through networks. Based on this definition, researchers describe networking as the participation of all external parties to the business, either directly or indirectly. This may include customers, suppliers, peers, government, financial banking, business associations, politicians and the local community. Naudé (1998) states that access to government tender and policies are also a contributing factor in micro businesses' performance in order to help them grow.

3 DEVELOPMENTS OF TOURISM SME PERFORMANCE IN HERITAGE SITES CONCEPT

3.1 *Rural tourism and its heritage sites*

Rural areas have unique potential to attract tourists through building a connection between rural areas and their ethnic, cultural, historic, and geographical roots (Dimitrovski, Todorović & Valjarević, 2012). Strategic planning in developing the rural tourism will encourage creating new source of income and occupation; also at the same time can eliminate the social isolation as well as becoming a crucial aspect in the country's development. Nevertheless heritage or cultural attraction is more than just museums, monuments, and archaeological treasures. They also include showplaces for natural wonders such as botanical gardens and aquariums as well as parks and preserves of natural resources that are dedicated to public enjoyment (Mansor, Ahmad & Mat, 2011).

UNESCO has declared Lenggong Valley, Perak as a world heritage site on 30 June 2012 for its archaeological discovery. Excavation activities revealed the finding of a Paleolithic culture and the oldest human skeleton named as "Perak Man" dates to about 11,000 years before present. Lenggong has been awarded the "Pre-historic Heritage Town," signifying its significance as a rich source of natural success, heritage, and culture that has attracted different segments of tourists. Heritage attractions include the natural attraction such as the caves, the waterfall cascade, the rainforest, also the richness of flora and fauna. The beautiful scenery of Lenggong Valley which is surrounded by the rainforest of Titiwangsa Ranges and Bintang Ranges, also having more than 20 caves to be explored are among the attraction. Based on the uniqueness and place attributes, Lenggong Valley has a potential to be a competitive tourism destination.

3.2 *Significance between tourism and SMEs*

The mutual significance between tourism and SMEs activities are (a) provide business opportunities, (b) employment chances (c) encourage local economic development.

Archaeological Institute of America (2007) (AIA) stated that tourism supports local retail businesses such as restaurants, hotels, local crafts and souvenir shops, while simultaneously creating various employment chances, includes the recruiting and training of guides and interpreters. Furthermore, UNESCAP (2005) also contends that tourism industry creates opportunities in local communities, enabling the not so rich to start a small business because capital costs for and barriers to entry in this sector are relatively low and can even be accessible.

A study done by Mbaiwa (2003) proved that local people in Okavango Delta, Bostwana was employed in tourism-related businesses, such as retail, wholesale, handicraft and souvenir industry, transportation, airlines, and also in the provision of accommodation. Consequently this indicates that

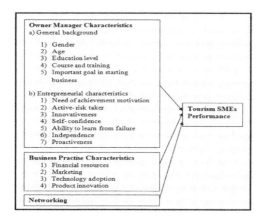

Figure 1. Proposed conceptual framework of the study.

tourism contributes employment opportunities and a chance for the community to engage in an entrepreneurial activity such as manufacturing, agricultural, service and construction-related businesses. However, the ability of the tourism to generate employment creation and entrepreneurship development is essentially reliant on a number of variables included the nature and location of the tourism project, the size and source of investment, the policy intentions (if any) associated to the investment and the level of support accessible to entrepreneurs (Kirsten & Rogerson, 2002). In developing a nations, tourism's economic purposes of increasing earnings, investment, foreign exchange, creating job opportunities, and also reducing adverse social and cultural consequences are best promoted through enhancing micro and small enterprises (Wanhill, 2000). Besides, in order for tourism to be advantageous in terms of economic development, income earning, poverty diminution, and improving rural livelihoods, it should be associated with local economies activities such as agriculture and micro and small scale enterprises (Mshenga & Owuor, 2009).

The mutual significance between tourism and SMEs activities explained that tourism and SMEs activities are at the same time benefit each other. Consequently, it is necessary to measure the performance of tourism SMEs in Lenggong Valley, in the attempt to promote the heritage site as a competitive world tourism destination.

Therefore, the proposed conceptual framework of tourism SMEs performance will be measured by three important variables of owner manager characteristics, business characteristics and networking. Figure 1 shows the relation between these variables to establish the conceptual framework of this study.

4 CONCLUSION

This study exhibits the relationship between factors associated with tourism SMEs performance. While developing this part of tourism SMEs concept, tourism approach is particularly useful in relation to three aspects: owner manager characteristics, business practice characteristics and networking, as well as an effort to determine the factors that may contribute to the performance. The concept of tourism SMEs performance is quite limited, because less research has been done specifically in literature, theory, and framework, leaving a gap in the knowledge of researchers. Therefore, through this study was a preliminary study, the factors which are discussed in this study contributed to the formation of conceptual framework of tourism SMEs performance with the relationship between three mentioned aspects. Thus, discussion with regard to this study will encourage more study on tourism SMEs performance concept in heritage sites as well as enable business managers and entrepreneurs to manage heritage sites in developing countries such as Malaysia, in particular in Lenggong Valley. The implications are to provide a general overview of SMEs business and tourism performance concept that can be used by SMEs owner-managers, policymakers, government agencies and other stakeholders that are closely assisting the tourism SMEs in a heritage site.

ACKNOWLEDGEMENT

The authors would like to extend their appreciation to the Universiti Sains Malaysia for the research grant entitled (Heritage Awareness and Interpretation) [Grant No 1001/PTS/8660012] that makes this presentation possible.

REFERENCES

Ahmad, G. (2005). Small firm owners-managers networks in tourism and hospitality. *International Journal of Business and Society, 6*(2), 37–54.

Alasadi, R. & Abdelrahim, A. (2007). Critical analysis and modeling of small business performance (case study: Syria). *Journal of Asia Entrepreneurship and Sustainability, 3*(2).

Archaeological Institute of America. (2007). *A guide to best practices for archaeological tourism. Los Angeles: The Getty Conservation Institute.* Retrieved January 12, 2012, from http://www.archaeological.org/pdfs/AIATourismGuidelines.pdf

Bates, T. (1990). Entrepreneur human capital inputs and small business longevity. *The review of Economics and Statistics*, 551–559.

Blackwood, T. & Mowl, G. (2000). Expatriate-owned small businesses: measuring and accounting for success. *International Small Business Journal, 18*(3), 60–73.

Carlsen, J., Getz, D. & Ali-Knight, J. (2001). The environmental attitudes and practices of family businesses in the rural tourism and hospitality sectors. *Journal of Sustainable Tourism, 9*(4), 281–297.

Caves, R.E. (1998). Industrial organization and new findings on the turnover and mobility of firms. *Journal of economic literature*, 1947–1982.

Chittithaworn, C., Islam, M., Keawchana, T., Yusuf, M. & Hasliza, D. (2011). Factors affecting business success of small and medium enterprises (SMEs) in Thailand. *Asian Social Science, 7*(5), 180.

Cooper, A.C., Gimeno-Gascon, F.J. & Woo, C.Y. (1994). Initial human and financial capital as predictors of new venture performance. *Journal of business venturing, 9*(5), 371–395.

Davidsson, P. (1991). Continued entrepreneurship: Ability, need, and opportunity as determinants of small firm growth. *Journal of business venturing, 6*(6), 405–429.

Dimitrovski, D.D., Todorović, A.T. & Valjarević, A.D. (2012). Rural tourism and regional development: Case study of development of rural tourism in the region of Gruža, Serbia. *Procedia Environmental Sciences, 14*, 288–297.

Geroski, P. (1995). Markets for technology: Knowledge, innovation and appropriability. In P. Stoneman (Ed.), *Handbook of the economics of innovation and technological change* (pp. 90–131). Oxford and Cambridge: Blackwell Publishers.

Getz, D. & Carlsen, J. (2000). Characteristics and goals of family and owner-operated businesses in the rural tourism and hospitality sectors. *Tourism management, 21*(6), 547–560.

Getz, D. & Nilsson, P.A. (2004). Responses of family businesses to extreme seasonality in demand: The case of Bornholm, Denmark. *Tourism management, 25*(1), 17–30.

Glancey, K. & Pettigrew, M. (1997). Entrepreneurship in the small hotel sector. *International Journal of Contemporary Hospitality Management, 9*(1), 21–24.

Grayson, D. & Dodd, T. (2007). Small is sustainable (and beautiful!): Encouraging European smaller enterprises to be sustainable.

Hankinson, A., Bartlett, D. & Ducheneaut, B. (1997). The key factors in the small profiles of small-medium enterprise owner-managers that influence business performance: The UK (Rennes) SME survey 1995–1997 An international research project UK survey. *International Journal of Entrepreneurial Behaviour & Research, 3*(3), 168–175.

Ho, T. & Koh, H. (1992). Differences in psychological characteristics between entrepreneurially inclined and non-entrepreneurially inclined accounting graduates in Singapore. *Entrepreneurship, Innovation and Change: An International Journal, 1*(11), 243–254.

Huang, X. & Brown, A. (1999). An analysis and classification of problems in small business. *International Small Business Journal, 18*(1), 73–85.

Jenkins, B. (2007). Expanding economic opportunity: The role of large firms. *Journal of Economic Opportunity Series, 17*.

Kirsten, M. & Rogerson, C.M. (2002). Tourism, business linkages and small enterprise development in South Africa. *Development Southern Africa, 19*(1), 29–59.

Maideen, S.A. (2010). *Small and Medium Island Chalet Operators in the selected Islands in Terengganu and Pahang*. Unpublished Master thesis, Universiti Sains Malaysia, Pulau Pinang.

Manan, D.I.A. & Jan, N.M. (2010). Do resources contribute to firms' performances? Exploring Batik Industry in Malaysia. *International Review of Business Research Papers, 6*(3), 189–204.

Mansor, N., Ahmad, W.A.W. & Mat, A.C. (2011). Tourism Challenges among the SMEs in State of Terengganu. *International Journal of Business and Social Science, 2*(1), 101–112.

Mbaiwa, J.E. (2003). The socio-economic and environmental impacts of tourism development on the Okavango Delta, north-western Botswana. *Journal of Arid Environments, 54*(2), 447–467.

Morrison, A., Breen, J. & Ali, S. (2003). Small business growth: intention, ability, and opportunity. *Journal of small business management, 41*(4), 417–425.

Morrison, A. & Teixeira, R. (2004). Small business performance: A tourism sector focus. *Journal of Small Business and Enterprise Development, 11*(2), 166–173.

Mshenga, P.M. & Owuor, G. (2009). Opportunities for micro and small scale businesses in the tourism sector: The case of the Kenya Coast. *KCA Journal of Business Management, 2*(2), 52–56.

Naudé, W.A. (1998). On the platinum road. *The South Africa Report, 1*, 82–87.

Nieman, G. & Nieuwenhuizen, C. (1997). *Female entrepreneurs in the hospitality Trade: A case study in South Africa*. University of Pretoria, South Africa.

Papadaki, E. & Chami, B. (2002). *Growth determinants of micro-businesses in Canada*. Canada: Citeseer.

Perren, L. (1999). Factors in the growth of micro-enterprises (part 1): developing a framework. *Journal of Small Business and Enterprise Development, 6*(4), 366–385.

Perren, L. (2000). Factors in the growth of micro-enterprises (part 2): exploring the implications. *Journal of Small Business and Enterprise Development, 7*(1), 58–68.

Philip, M. (2010). Factors affecting business success of Small & Medium enterprises. *Asia Pacific Journal of Research in Business Management, 1*(2).

Riding, A.L. & Swift, C.S. (1990). Women business owners and terms of credit: Some empirical findings of the Canadian experience. *Journal of Business Venturing, 5*(5), 327–340.

Rody, R.C. & Stearns, T.M. (2013). Impact of entrepreneurial style and managerial characteristics on SME performance in Macao SAR, China. *Journal of Multidisciplinary Research, 5*(1), 27–44.

Seppälä-Esser, R., Airey, D. & Szivas, E. (2009). The dependence of tourism SMEs on NTOs: The case of Finland. *Journal of travel research, 48*(2), 177–190.

Sharma, A. & Upneja, A. (2005). Factors influencing financial performance of small hotels in Tanzania. *International Journal of Contemporary Hospitality Management, 17*(6), 504–515.

Thomas, R., Shaw, G. & Page, S.J. (2011). Understanding small firms in tourism: A perspective on research trends and challenges. *Tourism management, 32*(5), 963–976.

UNESCAP. (2005). Major issues in tourism development in the Asian and Pacific region: Enhancing the role of tourism in socio-economic development and poverty reduction Retrieved 29 November, 2007, from http://www.unescap.org

Wanhill, S. (2000). Small and medium tourism enterprises. *Annals of Tourism Research, 27*(1), 132–147.

World Tourism Organization. (2012). Tourism Highlighted. Madrid: World Tourism Organization.

World Travel & Tourism Council. (2012). *Travel and Tourism Economic Impact: Malaysia*. London.

Technology and innovation in hospitality and tourism

Theory and Practice in Hospitality and Tourism Research – Radzi et al. (Eds)
© *2015 Taylor & Francis Group, London, ISBN 978-1-138-02706-0*

Social media: Credibility, popularity and its benefits towards events' awareness

M.H. Zamri, M.D. Darson & A.M.F. Wahab
Universiti Teknologi MARA, Shah Alam, Malaysia

ABSTRACT: This study investigates the credibility, popularity and benefits of social media towards event industry. It is also analyzing the interrelation between popularity of social media in events marketing and the public awareness of the event. 101 respondents were successfully surveyed using convenience sampling. Findings revealed that there is a partial positive relationship exists between popularity of social media in events marketing and the public awareness of the event. Finally, the implications toward the event marketing effort and recommendations for event organizer were suggested and potential future research was also emphasized.

Keywords: Social media, credibility, popularity, benefits, awareness

1 INTRODUCTION

A social media is a recent and complex phenomenon. Nowadays, companies are striving to understand how to capture effectively and utilize social media as part of their business portfolio and service offering. The 21st century is witnessing an explosion of Internet-based messages transmitted through these media. They have become a vast apparatus in influencing various aspects of consumer behavior including awareness, information acquisition, opinions, attitudes, purchase behavior, and post-purchase communication and evaluation.

As claimed by Benson. Haghighi, and Barzilay (2011), social media messages are short, often use a simple language and require situational context for interpretation. Not all details of an event can be expressed in a single message and the relation between messages and event context is not clear. These characteristics of social media streams make extracting techniques significantly are not effective. While such resources are widely available online, they are usually high precision, but low recall. Social media is a right place for users to discover new events and it's provide a pool of information such as date, time and attendance confirmation if someone planning to attend the occasion.

Getting involved in social media will helps a lot of new company and increased their company transparency to public. Each company should consider it before involve and active in social media of how the company wants to be, which type of information the company wants to share and the utilization reason of this medium. It is also important to create understanding and acceptance internally for the use of social media. It is necessary to find the balance between what feels good for the company and what is interesting for the reader. A too strict information policy probably leads to the contents that are not interesting and that nobody reads them (Johansson, 2010).

According to Young (2012), the social media elements that generate business opportunity for companies to extend their brands are also have the same elements that created IT-related risk. Like the borderless nature of social media itself, the various risks surrounding social media can be faced by multiple enterprise functions at the same time, challenged companies to understand how, when and where to engage their IT functions or plug risk coverage gaps.

In line with that notion, although social media have a huge impact to various aspects, question raise whether its credibility highly influenced the event management, the benefits of social media usage in event promotions and the interrelation between popularity of social media and events' awareness of people towards event. To date, there is still lack of studies looking at these issues as the studies before are being concentrated more on other angles such as the influence of social media in sociocultural and the impact being brought by these technology to the business and politics. Understanding all these issues are therefore considered important to strategize the marketing approach and maximizing the capability of this unique communication tools. With that, two research objectives are formulated to be further investigated and one hypothesis is aim to be tested.

1. To investigate the credibility of social media in event management.
2. To distinguish the benefits of social media usage in event promotions.
3. To study the interrelation between popularity of social media in events marketing and the public awareness of the event.

2 LITERATURE REVIEW

2.1 Social media

Social media is a term that is used to describe web services that receive most of the content from their users or that aggregate the content from other sites as feeds. The sites build on social networks and on the creativity of the participants of one or more communities. In social media anyone can become a producer, but many of the people see themselves as participants who engage in the community rather than producers (Lietsala & Sirkkunen, 2008).

At its best, social media builds the foundations for a participatory economy where participants gain use-value as the result of community action. People collaborate on social media, and as a return, the action of the individuals produces something new, even unexpected results. The emergence may be profitable business and provide income for firms, but it also has an impact on the social relations and the wellbeing of individuals.

2.2 Credibility of social media

According to Kiousis (2001), it has been suggested that the credibility of the channel/medium of communication influences the selective involvement of the audience with the medium. Furthermore, individual audiences are paying closer attention to the media that they perceive to be credible. When individual audiences rely more on a certain communication medium for information seeking, they are likely to rate the medium more credible than other media.

Individual audiences are paying closer attention to the media that they perceive to be credible. It is widely accepted, when individual audiences rely more on a certain communication medium for information seeking, they are likely to rate the medium more credible than other media.

2.2 Benefits of social media

Social media opens up new channels of communication that give marketers direct access to customers and opinion formers. Moreover, social media allows marketers to listen to customers more easily and more cost-effectively. Social media also relies on a 'sharing' culture, which means sharing information and being helpful (Smith & Zook, 2011).

Besides, it is widely accepted that social media allow users to interact, to share content and to create content collectively. Social media connect a large set of people around the world, thereby increasing the number of potential participants in prediction markets. Moreover, social media might increase the diversity of participants, thereby potentially improving the quality of predictions.

By looking at social media as a tool for marketing, Arca (2012) has stated that social media can be used as an unconventional means to achieve conventional goals, through the use of creativity, community, and relationships instead of big budgets to achieve marketing objectives. These methods are powerful guerrilla marketing strategies. It could be said that, the use of social media is a great change in the world of marketing. Moreover, with millions of users, social media sites make a convenient target base for people who are trying to market products or services to people worldwide.

2.3 Popularity of social media

It is widely accepted that the growing popularity of social media leads advertisers to invest more effort into communicating with consumers through online social networking. Social media, enable users to present themselves, establish and maintain social connections with others, and articulate their own social networks. In this setting, brand managers often seek two types of marketing communication: interactive digital advertising and virtual brand communities. However, such social media marketing trends also raise concerns about how to optimize the effects of marketing communication in the context of online social networking.

Social media pertains essentially to social networking among users, so marketing communication approaches in such user-centered social networking contexts reflect the views of marketing practitioners (Mulhern, 2009). The most prevailing communication approach focuses on the media features of social networking, such that social connections transform into personal channels for brand communication (Russell, 2009).

According to Becker, Naaman, and Gravano (2010), the ease of publishing content on social media sites brings to the Web an ever-increasing amount of content captured during and associated with real world events. Sites like Flickr, YouTube, Facebook and others host user-contributed content for a wide variety of events. These range from widely known events, such as presidential inaugurations, to smaller, community-specific events, such as annual conventions and local gatherings. By automatically identifying these events and their associated user-contributed social media documents, which is the focus of this paper, we can enable powerful local event browsing and search, to complement and improve the local search tools that Web search engines provide. In this paper, we address the problem of how to identify events and their associated user-contributed documents over social media sites.

2.4 Relationship between popularity of social media and public awareness about the event

The popularity of contributed items also shows this extreme diversity. Relatively few of the four billion images on the social photo-sharing site Flickr, for example, are viewed thousands of times, while most of the rest are rarely viewed. The number of viewers are commonly been determined by the strength of the images and the actual messages that the sender intended to share.

Looking at other example, with more than 900 million users worldwide (Cohen, 2012), Facebook has become one of the most successive social network in world history. Undoubtedly, the capability of this famous social media in connecting people worldwide with its unique feature and smart approach via incorporating its function in various operating system and mobile devices make Facebook are above their competitors in the same field. Looking at this phenomenon, it is widely accepted that to be survive in this media line, it is critically important for social media providers to offer their users with unique and update tools to help them enjoy and aware about any latest news and changes around them.

According to Bruhn, Schoenmueller, and Schäfer (2012), social media offer companies numerous opportunities to listen to their consumers, to engage with them, and to even influence their conversations. Companies providing social network platforms bring like-minded consumers together and give them the opportunity to talk about event-based topics.

3 METHODOLOGY

This study was focused on how social media as promotion aid may influence participants' awareness towards the event. Therefore, targeted audiences were form AVON Kiss Goodbye Breast Cancer event launching on December 2014 which was held in Mydin Supermarket, Subang Jaya, Selangor, Malaysia. According to Krejecie and Morgan (1970), a hundred and three respondents were sufficient to represent 140 populations Subsequently,

the samples obtained for this study were 101 respondents which represented 72 percent of total population. 98.5 percent return rate was recorded and analyzed for further discussion. Moreover, the convenience sampling method was used to select the potential respondents. Five points Likert scale also was used in measuring the tendency of answer given by the respondents. The data sets were coded and keyed for analysis using a Statistical Package of Social Science (SPSS), Version 20.0.

4 RESULT AND ANALYSIS

4.1 Credibility of social media towards event awareness

The most outstanding result in the sub-section analysis is related to the social media as an active and fast moving medium (item 5, $m = 4.62$) and information to influence platform (item 3, $m = 4.38$). This could be true as social media has strong power in influencing the end user. As a result, social media become a tool to seek out information (item 1, $m = 4.10$). In addition, social media become a regular place for expressing thoughts and opinions (item 2, $m = 4.07$), potential registrants turn to social channel (item 6, $m = 3.98$) and different platform has its own focus (item $= 4$, $m = 3.87$). This could be supported by saying that the credibility of social media may influence the event awareness through the functionality of internet.

4.2 Benefit of social media influence public in event promotion

Most of the respondents agreed that the benefits obtain from social media may drive public towards event promotion which indirectly related to their awareness. It can be shown through the statement of social media sites are inexpensive (item 1, $m = 4.32$), social media is a communication tools (item 2, $m = 4.56$), the use of social media is readily available (item 3, $m = 4.25$), Attendees are empowered to become reviewers (item 4, $m = 4.09$), Expose an event to thousands of potential registrants (item 5, $m = 4.25$). Positive agreement by the respondents have shown that the social media may bring benefits and influenced public in event promotion which eventually broaden up the people awareness.

4.3 Popularity vs public awareness

The relationship between these two variables was diagnosing using Pearson Correlation method. Result in Table 3 below shows that the interrelation between popularity of social media in events marketing and the public awareness of the event is (0.252). Thus, it shows that there is a relationship between the popularity of social media and its influence to the public awareness towards the event. However, the relationship was a weak but a definite relationship (Salkind, 2012).

5 DISCUSSION AND CONCLUSION

Referring to Table 1, the mean is 4.17 which showed that respondents agree on how social media can influence their thoughts. Besides result proved that the social media is very active and fast moving medium. For an instance, when using Facebook to promote an event, all the comments that are written in the event page could sway possible attendees to actually come and attend the event only by reading the good feedbacks. Considering this, it seems that the event organizer in this case study has the possibilities of really reaching and maintaining the strongest event awareness by the use of social media.

By looking at Table 2, it is proven that social media is beneficial when the mean was recorded at 4.29. It could be said that, social media has provide a fast track of communication as people can send, receive or share the information almost immediately with each other's. From the event organizers context, they can just send or post notices about news or activities to groups of friends or members of an

Table 1. Credibility of social media towards event awareness *(n = 101)*.

No.	Item	Mean	S.D
1.	Seek out information	4.10	.728
2.	Express thoughts and opinions	4.07	.697
3.	Information platform to influence platform	4.38	.719
4.	Different platform has its own focus	3.87	.770
5.	Active and fast moving medium	4.62	.563
6.	Potential registrants turn to social channel	3.98	.761
	Total Mean Score	25.02	
	Population Mean	4.17	

Table 2. Benefit of social media influence public in event promotion *(n = 101)*.

No.	Item	Mean	S.D
1.	Social media sites are inexpensive	4.32	.761
2.	Social media is a communication tools	4.56	.623
3.	The use of social media is readily available	4.25	.754
4.	Attendees are empowered to become reviewers	4.09	.694
5.	Expose an event to thousands of potential registrants	4.25	.805
	Total Mean Score	21.47	
	Population Mean	4.294	

event by just clicking and sharing the info faster—in social media like Facebook and Twitter rather than using conventional tools of communication such as letter. This medium helped the organizer to efficiently spreading the information and assists them in controlling the total cost of the event.

By referring to Pearson Correlation in Table 3, result showed that the mean recorded is 0.252 which showed that the partial positive relationship exists between popularity of social media and public awareness towards the event. In line with that result, it is widely accepted that majorities of people around the world own a cell phone with easy internet access and mankind now see social media as a more trustworthy source of information about any kind of products and services. Social networking is ranked as the most popular content category for worldwide engagement. Literally meaning that one in every five minutes, people spent on social networking sites globally. It's not just young people using social networking anymore, it's everyone. Mobile devices are fueling the social addiction and they represent the future of social networking from a technology point of view. With that, it could be said that the result in this study is parallel with the phenomena and undoubtedly, the social media with its strong popularity, can still give an influence to the public awareness even the antecedent between variable was not assertively reacted.

Practically, this study suggests that event organizer should acknowledge more on the differences of each social media platform in order to use them as a marketing tool. For example, generally, Facebook page is known as a platform that connect people around the world, as technology mature, Facebook can now also be a medium to market services and products, thus event organizer should take digital marketing into consideration finding all the functions that be can be used in order to execute as one of the instrument for marketing.

In order to manage successful event marketing via social media, it is best that event organizer could create a buzz in each of social media platform so that it could be visible to all Internet users globally. Creating hype such as hash tags and retweet also played a big part in getting the success of marketing an event.

Table 3. Correlation model between popularity of social media and event awareness *(n = 101)*.

		Mean Popularity
Mean Awareness	Pearson Correlation	.252[*]
	Sig. (2-tailed)	.011
** Correlation is significant at the .01 level (2-tailed).		
**.Correlation is significant at the .05 level (2-tailed).		
**.Correlation is significant at the .01 level (1-tailed).		

Encourage users in participating to give feedback on an event, so that event organizer could take in consideration the points and remarks of event attendees. This is a way of showing that the event organizer wants to communicate with their participants other than making profits.

REFERENCES

Arca, C. (2012). Social Media Marketing Benefits for Business. Why and how should every business create and develop its Social Media Sites?, 90.

Becker, H., Naaman, M. & Gravano, L. (2010). *Learning similarity metrics for event identification in social media.* Paper presented at the Proceedings of the third ACM international conference on Web search and data mining.

Benson, E., Haghighi, A. & Barzilay, R. (2011). *Event discovery in social media feeds.* Paper presented at the Proceedings of the 49th Annual Meeting of the Association for Computational Linguistics: Human Language Technologies-Volume 1.

Bruhn, M., Schoenmueller, V. & Schäfer, D.B. (2012). Are social media replacing traditional media in terms of brand equity creation? *Management Research Review, 35*(9), 770–790.

Cohen, N. (2012). The breakfast meeting: Grilling for James Murdoch, and Facebook tops 900 million users Retrieved 22 Feb, Retrieved on 2014 from, from http://mediadecoder.blogs.nytimes.com/2012/04/24/the-breakfast-meeting-grilling-for-james-murdoch-and-facebook-tops-900-million-users/?_php=true&_type=blogs&_r=0

Johansson, M. (2010). Social media and brand awareness. *Bachelor thesis, Luleå University of Technology, Sweden.*

Kiousis, S. (2001). Public trust or mistrust? Perceptions of media credibility in the information age. *Mass Communication & Society, 4*(4), 381–403.

Krejcie, R.V. & Morgan, D.W. (1970). Determining sample size for research activities. *Educational and psychological measurement, 30*(3), 607–610.

Lietsala, K. & Sirkkunen, E. (2008). Social media. Introduction to the tools and processes of participatory economy. *Social Media.*

Mulhern, F. (2009). Integrated marketing communications: From media channels to digital connectivity. *Journal of marketing communications, 15*(2–3), 85–101.

Russell, M.G. (2009). A call for creativity in new metrics for liquid media. *Journal of Interactive Advertising, 9*(2), 44–61.

Salkind, N.J. (2012). *Exploring research* (8th ed.). University of Kansas: Pearson Education Inc.

Smith, P.R. & Zook, Z. (2011). *Marketing communications: Integrating offline and online with social media* (5th ed.). London: Kogan Page Ltd.

Young, E. (2012). Protecting and strengthening your brand. Insights on governance, risk and compliance.

Theory and Practice in Hospitality and Tourism Research – Radzi et al. (Eds)
© 2015 Taylor & Francis Group, London, ISBN 978-1-138-02706-0

Travelers' sharing behaviors on the internet: Case of Taiwan

P.S. Hsieh
National Dong Hwa University, Hualien, Taiwan

ABSTRACT: Travelers integrate portable computers and the internet, and more recently mobile phones and social media into their travel and tourism practices. The present study intended to explore a deeper understanding of travelers' (the blog-owners or users of social media) online sharing behaviors. A qualitative approach using netnography and in-depth interviews was employed to examine travelers' online sharing behaviors. Recruitment of participant was through snowballing sampling technique and resulted in 32 participants. An analysis of interviewing participants' online sharing behaviors emerged three themes: (a) social connections, (b) conspicuous consumption, and(c) identity formation. The implications for future research by considering the popular use of the internet are discussed.

Keywords: Travelers, online sharing behaviors, Taiwan

1 INTRODUCTION

The increasingly widespread use of information and communication technologies, and social media, has triggered major changes in tourist behaviors in terms of the organization of travel, the travel actually undertaken, and post-travel behavior (Parra-López, Gutiérrez-Taño, Díaz-Armas & Bulchand-Gidumal, 2012). Molz (2012) indicates that travelers are integrating portable computers and the Internet, and, more recently, mobile phones and social media, into their travel and tourism practices. It has become popular for travelers to share their travel experiences, including diaries, photos, and itineraries, online with the public. Among those who travel with cameras, 89 percent take photos during their journeys and 41.4 percent post travel photos online (Lo, McKercher, Lo, Cheung & Law, 2011). It has become popular for travelers to share their travel photos online with the public. While an increasing number of blog-owners post travel photos during or after travel, little research has examined the behaviors of tourists with respect to their online sharing. While an increasing number of travelers share travel experience during or after travel, there is little research examining the behaviors of tourists with respect to their online sharing. To date, Lo et al. (2011) have compared users across different media platforms and found that age and income significantly impact online photo-sharing behaviors. Most studies of tourism and the Internet focus on consumer behaviors and overlook the diversity and inherent social and cultural meanings of the virtual space. Systematic observation and analysis of travelers'

online sharing behaviors through their personal blogs and social media can help us understand the construction and communication of an individual's travel experiences. Thus, this study aims at exploring, and arriving at a deeper understanding of, travelers' online sharing behaviors, including the posting of photos, diaries, experiences, and itineraries related to traveling.

2 LITERATURE REVIEWS

Urry (1990) develops the paradigm of the tourist gaze as a means of understanding how tourism is produced and reproduced as a socially constructed phenomenon. The tourist gaze is elaborated as a particular "way of seeing", and the fundamental motivation of tourists presented as a yearning to gaze on the panoramas, landscapes, buildings, people, and other manifestations of places they have already viewed in tourism advertisements, movies, brochures, travel guidebooks. Many studies have been done to try and understand tourism through the lens of the tourist gaze. Some critiques of the tourist gaze suggest that the visual experience is over-emphasized and the travelers' actual, physical experience is overlooked. Therefore, in *Tourist Gaze 3.0*, Urry and Larsen (2011) extend the concept of the tourist to include a "performance turn" in order to examine the production to tourism. The performance turn employs Goffmanian performative metaphors to conceptualize the thematic and staged nature of tourist locales, the scripted and theatrical corporealities, and the embodied actions of tourist bodies. Tourism/travel is as a site of stage, and travelers take up the roles of actors, cast members, guides as directors, stage managers, and so forth (Edensor, 2000). The performance turn highlights how tourists experience places in multi-sensuous ways that involve bodily sensations and affect. Through the lens of the performance turn, tourism is an activity accomplished through performance. However, some literature examines how tourists are not only audience members, but also performers. Edensor (1998) explores how tourists at the Taj Mahal perform walking, gazing, photographing and remembering, while Bærenholdt. Haldrup, Larsen and Urry (2004) examined the performances involved in strolling, beach life and photography.

The Internet can serve as a platform for displaying a variety of individuals' activities including tourism and tourists, and many travelers post their travel diaries and photos online. Molz (2012) indicates that there are no clear-cut statistics on the number of travel blogs published online. A website, i.e., Travelblog.org, one of the most popular travel-related web 2.0 sites, reports that it has over 200,000 members and hosts over five million photos, as well as more than 600,000 blog entries. Another similar travel blog hosting site (Travelpod.com) claims that more than 75,000 'travel experiences' from 181 countries are shared on its site in any given week. These statistical snapshots suggest that travel blogging has become a popular practice (Molz, 2012). As performance turn, travelers are performers in how they choose to walk, photograph, sing, and so on. If the tourism stage is transferred on to the Internet, how do these travelers perform their experiences via blogs and social media?

3 METHODOLOGY

The present study employed a qualitative approach using netnography and in-depth interviews to explore travelers' online sharing behaviors. Taiwanese bloggers and users of social media (Facebook, Twitter, Line etc.) who focused on documenting their travel experiences over a two-year period were potential participants. Recruitment of participants was conducted through a snowballing sampling technique and resulted in 32 participants who consenting to participate. The participants were twenty-one females and eleven males with an average age of 36 years (range: 20 to 51); their educational backgrounds ranged from college to doctoral degree. Their travel experiences occurred over three years, and most took digital devices such as laptops, cameras, smartphone or iPad on the road to record their travel experiences. The researcher contacted each participant and scheduled a time that was convenient for the interviewee that lasted between 20 minutes and 2.5 hours. The researcher also reviewed participants' travel blogs online by reading the stories published, looking at their photographs, paying careful attention to the back-and-forth comments posted by other readers, and joining the participants' networks on social media. The study was conducted from May 2012 to November 2013. The Atlas.ti 6.0 was used to data analysis.

4 RESULTS AND DISCUSSION

Based on the data collection and analysis, the following discussion addresses three key sets of findings: (a) social connections, (b) conspicuous consumption, (c) and identity formation.

4.1 Social connections

These travelers are standing for significant shifts in how they use mobile technologies every moment to engage with each other and the world while on the move. In the present study, all participants took digital devices with them while traveling, and they uploaded their travel photos and stories of their journeys to their blogs and social media. Hookway (2008) also points out that blogs capture a "tight union between everyday experience and the record of that experience". For example, a female participant who has travel to 10 countries over eight years wrote:

I took a laptop and posted travel experiences and photos on my blog or facebook almost every day while I was on the road. I wanna tell my friends and family what did I do while traveling (Informant C)

Travel blogs or social media do not merely document daily activities, but rather serve as nodes of travel connection and social interaction between travelers and their family members and friends in a virtual space. A male participant who has been travelling for over five years interprets them thus:

I think the fee of international phone call is very expensive. I can keep in touch with my family and friends on the blog. I can share my travel photos online immediately after I return to the hostel. My family can confirm that I am safe and secure when they see the photos on my blog and Facebook (Informant E)

As a network of friends, families, and colleagues extends across geographical space, social life now involves multiple forms of co-presence established through physical travel, online interactions, and mobile communications (Larsen, Urry & Axhausen, 2006). Another female informant regarded travel photos on her Facebook as a way to keep in touch with her friends:

When I post travel photos online, I record what I see, eat, and experience. I feel my friends are joining me in the trip and we are traveling together when they follow my blog (Informant K)

Indeed, the online sharing behaviors of travelers tighten the social relations with friends. Travelers experiment with technology to engage in new mobile lifestyles, establish new ties to certain spaces and places, and navigate new modes of co-presence with friends and family members while on the road and far away (Vertovec, 2004). During physical travel, mediated communication serves as "social glue".

Travel photos on the Internet are not only online diaries, but also virtual sites that mediate and mobile and great social interaction (Molz, 2012). When travelers post their travel experiences on blogs and social media while traveling and further develop social relations and networks through online sharing, they constitute a kind of it can be interactive travel. According to Urry (2007), such interactions produce and shape the contours of mobile social life: the corporeal travel of people; the physical movement of objects; imaginative travel enabled by various media; virtual travel; and communicative travel. Travelers keep constructing their traveling memories and documenting the places they visited in a way that conveys the essence of tourist experiences.

4.2 Conspicuous consumptions

Travel is a very conspicuous behavior when traveling experiences are shared online. As one informant said:

I want to share experiences with others about where I went and what I ate. Actually, I just want to show off (Informant D)

Veblen (1912) argues that conspicuous consumption is a quality of slaves and livestock, while leisure is social signal and means of displaying one's elevated status. Leisure is a strategy for obtaining respect from others, and a form of psychological displacement. As one informant explained:

My traveling experience is unique. I go to a lot of destinations around the world. I post photos taken from different angles on my blog although I went to the same destinations as my friends did (Informant I)

Another informant pointed out:

One of my friends really show off. She stayed at a five-star hotel when she visited Hong Kong. When I visited Hong Kong, I chose a six-star hotel, and posted photos of the interior view, facilities, as well

as everything of the hotel room on my blog. But I spent a lot of money (Informant B)

Urry and Larsen (2011) indicated people consumed for a short period because they are supposedly generating pleasurable experiences which differ from those typically encountered in everyday life. The other informant said much the same:

Normally, I have to work. When I am on vacation or holiday, I absolutely enjoyed the moments. I shared my travel story and photos online to provide evidence that I was on vacation and stayed in a perfect hotel, ate exotic cuisine. I was on vacation, and you had to work (Informant X)

Some of the informants in this study post travel photos of luxurious and expensive forms of consumption, such as eating in a highly-rated Michelin restaurant, collecting special souvenirs, or availing themselves of remarkable forms of transportation, etc. A female informant explained:

I like to post what I bought during or after traveling, such as souvenirs, post cards, local special snack and so forth. One thought is show off, the other thought is to check what I bought.

4.3 Identity formation

Travel experiences are not simply a description of self, but a construction of one's identity. According to Nardi, Schiano, and Gumbrecht (2004, p. 223), bloggers reach out "to connect with and insert themselves into the virtual space of others in their personal social networks". Nardi et al. (2004, p. 224) further explain that" blogs create the audience, but the audience also create the blogs". One interviewee's statement echoed this perspective and stated:

I like to reply comments or messages on my blog. When my friends or others leave message for me such as praising my photographic skills (e.g., your photography skills is very good, your photos compositions are professional), these messages inspire me to take lots of photos and post them on my blog more often (Informant M)

And the other female informant pointed:

I like to check message from my Facebook, and I reply the questions that left on my Facebook. When I answered or solved friends' questions, I thought I am a wonderful person (Informant Y)

By replying to messages on the blog, the travelers' social identity is converted into self-identity and a form of circulation. The Internet affords a communication space for the expression of both social mobility and privatization that means individuals can upload private files but they also can interact with others cross the boundary in this virtual space. This virtual space creates a stage on which travelers can perform their travel experiences, and construct themselves. Studies (i.e., Bortree, 2005; Boyd & Ellison, 2008) show that blogs are effective in producing self-presentation because an individual can display various types of online activities such as. In addition to self-presentation, the blogs and social media are also a platform for identity formation through social construction.

5 CONCLUSION

This present study was aimed at in-depth understanding of travelers' online sharing behaviors. The present study addressed three key sets of findings: (a) social connections-building up social tie and relations with family members and friends who were far away; sense of togetherness with friends during journeys; keeping connections with friends; (b) conspicuous consumption-showing off unique or luxurious travel experiences; and (c) identity formation—constructing self-identity through social interaction on personal blogs and social media.

Studies (e.g., Strano, 2008; Trammell & Keshelashvili, 2005) suggest that young people, in particular, use online media to present and manipulate a desirable self-image. The demographic profiles of participants in the present study were similar to those who participated in other research on tourists, backpackers and interactive travelers (O'Reilly, 2006; White & White, 2007). The travelers in this study were young, professionals, middle class or students who hold passports to travel around the world. Race, class, and nationality present the desire to travel the world, enable travelers' reach to mobility, and shape their communication practices (Molz, 2012). In other words, these travelers in this study had high cultural and economic capitals because they were capable to interact with far-reaching people using the technology. Furthermore, sharing travel experiences on the Internet may play a significant role in identity construction or re-construction.

Traditionally, studies on tourism tend to investigate travel experiences that are subject to time and space. The increasing number of people sharing and posting their travel experiences on the Internet charts possible directions for researchers to extend study by considering the popular use of the Internet. Systematic observation and analysis of travelers' online sharing behaviors through their personal blogs and social media can help us understand the construction and communication of an individual's travel experiences.

REFERENCES

Bærenholdt, J.O., Haldrup, M., Larsen, J. & Urry, J. (2004). *Performing tourist places.* Aldershot, England: Ashgate Publishing, Ltd.

Bortree, D.S. (2005). Presentation of self on the web: An ethnographic study of teenage girls' weblogs. *Education, Communication & Information, 5*(1), 25–39.

Boyd, D.M. & Ellison, N.B. (2008). Why youth social net-work sites: Definition, history, and scholarship. *Journal of Computer-Mediated Communication, 13*(1), 210–230.

Edensor, T. (1998). *Tourists at the Taj: Performance and meaning at a symbolic site.* London: Routledge.

Edensor, T. (2000). Staging tourism: Tourists as performers. *Annals of Tourism Research, 27*(2), 322–344.

Hookway, N. (2008). Entering the blogosphere': some strategies for using blogs in social research. *Qualitative research, 8*(1), 91–113.

Larsen, J., Urry, J. & Axhausen, K.W. (2006). *Mobilities, Networks, Geographies.* Aldershot: Ashgate Publishing Ltd.

Lo, I.S., McKercher, B., Lo, A., Cheung, C. & Law, R. (2011). Tourism and online photography. *Tourism management, 32*(4), 725–731.

Molz, J.G. (2012). *Travel connections: Tourism, technology and togetherness in a mobile world.* Oxon, England: Routledge.

Nardi, B.A., Schiano, D.J. & Gumbrecht, M. (2004). *Blogging as social activity, or, would you let 900 million people read your diary?* Paper presented at the Proceedings of the 2004 ACM conference on Computer supported cooperative work.

O'Reilly, C.C. (2006). From drifter to gap year tourist: Mainstreaming backpacker travel. *Annals of Tourism Research, 33*(4), 998–1017.

Parra-López, E., Gutiérrez-Taño, D., Díaz-Armas, R.J. & Bulchand-Gidumal, J. (2012). Travellers 2.0: Motivation, Opportunity and Ability to Use Social Media. In M. Sigala, E. Christou & U. Gretzel (Eds.), *Social media in travel, tourism and hospitality: Theory, practice and cases* (pp. 171–187). Burlington, VT: Ashgate Pub.

Strano, M.M. (2008). User descriptions and interpretations of self-presentation through Facebook profile images. *Cyberpsychology, 2*(2), 1–11.

Trammell, K.D. & Keshelashvili, A. (2005). Examining the new influencers: A self-presentation study of A-list blogs. *Journalism & Mass Communication Quarterly, 82*(4), 968–982.

Urry, J. (1990). *The tourist gaze: Leisure and travel in contemporary societies.* London: Sage Publications.

Urry, J. (2007). *Mobilities.* Cambridge: Polity Press.

Urry, J. & Larsen, J. (2011). *The tourist gaze 3.0.* Los Angeles: Sage.

Veblen, T. (1912). *The theory of the leisure class.* New York: Macmillan.

Vertovec, S. (2004). Cheap calls: The social glue of migrant transnationalism. *Global networks, 4*(2), 219–224.

White, N.R. & White, P. B. (2007). Home and away: Tourists in a connected world. *Annals of Tourism Research, 34*(1), 88–104.

Theory and Practice in Hospitality and Tourism Research – Radzi et al. (Eds)
© *2015 Taylor & Francis Group, London, ISBN 978-1-138-02706-0*

Exploring the accessibility and content quality of the Go2homestay website

S.A. Sabaruddin, N.H. Abdullah, S.A. Jamal & S. Tarmudi
Faculty of Hotel and Tourism Management, Universiti Teknologi MARA, Shah Alam, Malaysia

ABSTRACT: This study addresses a two-part research question: Is the Go2homestay website easy to access and use, and does it provides visitors with sufficient information to make informed choices? A variety of standard methodologies were employed, including content evaluation, a focus group and a questionnaire survey. Demographic information was recorded for each participant. A review of the Go2homestay website by visitors revealed that the visitors were likely to evaluate this homestay guide website as acceptable. In general, visitors rate the accessibility and content quality of Go2homestay website as above average.

Keywords: Website evaluation, website, accessibility, content quality, Go2homestay

1 INTRODUCTION

Due to its emerging market, Malaysia has become one of many countries that are rapidly embracing online travel planning (Vinod, 2011). More than two-thirds of Malaysians use online booking systems for travel and tourism products and services, including airlines, hotels and private car rental (Abd Aziz, Tap, Osman & Mahmud, 2013). Despite the substantial number of consumers opting to book travel online, previous Malaysian studies on the accessibility and content quality of destination websites are remarkably limited in quantity and scope. A recent study by Pourabedin and Nourizadeh (2013) focussed exclusively on the influence of visual website design features on the intention to travel in Malaysia. Selamat and Ismail (2008) primarily focussed on Content Based Image Retrieval for the Malaysia Tourism Website to examine user perceptions and experiences.

Another study focussed mainly on how tourism websites reflect the role of Islam in promoting destinations to Muslim and non-Muslim tourists (Hashim & Murphy, 2008). Despite the numerous studies examining website characteristics and effectiveness, especially in hotel organisation (Musante, Bojanic & Zhang, 2009; Schmidt, Cantallops & dos Santos, 2008), travel agencies (Chiou, Lin & Perng, 2011; Ruiz-Molina, Gil-Saura & Moliner-Velázquez, 2010), rural tourist accommodations (Gregory, Wang & DiPietro, 2010; Herrero & San Martín, 2012) and official tourism (Lepp, Gibson & Lane, 2011), little is known about the accessibility and content quality of homestay tourism websites.

Many previous studies have focussed on the relationship between website effectiveness and purchase intention (Bai, Law & Wen, 2008; Ranaweera, Bansal & McDougall, 2008; Shi, 2011; Thongpapanl & Ashraf, 2011; Woodside, Vicente & Duque, 2011) and the relationship between effectiveness and revisit intention (Aziz, Musa & Sulaiman, 2010; Faullant, Matzler & Füller, 2008; Kabadayi & Gupta, 2011). Nevertheless, little analysis has been conducted of the Go2homestay website's accessibility and content quality and their influences on visitors' intentions to make initial visits to the site's destinations. Thus, this study aims to explore the Go2homestay website, an official website for homestay tourism that serves as a platform for tourists or visitors to make enquiries or reservations, from the visitors' perspective to determine its effectiveness in encouraging more visitors to homestay in Malaysia.

2 LITERATURE REVIEW

2.1 *Go2homestay website*

Web-based technology has played an important role and become a necessity in the travel and tourism industry (Lu, Lu & Zhang, 2002). Due to the spread of Internet access and an increase in the demand for online services, most tourism sectors, such as accommodations, transportation and tourist destinations, now offer their products and services online. The Malaysian Ministry of Tourism has endorsed the Go2homestay website (http://www.go2homestay.com) as the official homestay website in the country. The website promotes all registered homestays in the country through narratives and multimedia presentations on each homestay provider. The website's objective is to provide users with information on Malaysian homestays and act as a gateway for information on all homestays in Malaysia.

As a government-endorsed website, Go2homestay is an example of the efforts that many governments around the world are making to embrace Internet technology. However, the introduction of official websites has encountered numerous problems, especially in developing countries (Prins, 2001). Go2homestay, despite being an official website, does not have an interactive feature to allow customers to make online reservations.

In addition to Go2homestay, several other websites are owned by the state governments, including the Selangor Tourism Website (http://tourismselangor.my/homestay), Perak Tourism Website (www.peraktourism.com.my), Sabah Tourism Board Official Website (www.sabahtourism.com/sabah-malaysian-borneo/en/homestay), and Sarawak Homestay Website (www.right.sarawak.gov.my/homestay). Moreover, several homestay operators in Malaysia have created their own websites and blogs to promote their products and services.

2.2 *Web accessibility*

The Internet is now one of the primary tools for searching for information and facilitating communication between consumers and businesses. Consumers increasingly use the Internet when making travel decisions and reservations. Therefore, the use of established criteria to evaluate tourism destination websites is important for the future of the tourism industry (Zulfakar & Buchanan, 1999). The factors influencing the success of a website include accessibility, content quality and web design. A website's success can also be explained by its effectiveness as a communication tool (López & Ruiz, 2011).

Web accessibility refers to the degree to which online information is accessible to users. The goal of web accessibility is to allow anyone using any web browser to visit the site, access all its information and use all its interactive features if necessary (W3C, 2003). The variety of resources and guidelines can capture whether a particular website achieves its accessibility potential. Therefore, it is important to evaluate the accessibility and content quality of the recently launched official Malaysian homestay tourism website. The scope of this evaluation includes not only accessibility but also the quality of the site's content from the perspective of its intended purpose and audience.

2.3 *Web content/information quality*

The basic goal of a website is to provide information to prospective and existing customers and stakeholders (Huang & Shyu, 2008). Gregory et al. (2010) noted that information is the main reason that individuals initially visit a website, and information has been argued to be the most fundamental component of a website. Xiang and Gretzel (2010) stated that providing travellers with the ability to search for travel and tourism information when they are planning their trips is important in facilitating travellers' access to tourism-related information online.

A number of elements are crucial to encouraging website users to make purchases, including providing related, relevant, comprehensive and helpful information (Kozak, Bigne & Andreu, 2005). As Kozak et al. (2005) and Kaynama and Black (2000) note, the design and presentation of a website are critical in meeting users' expectations. Additionally, providing comprehensive information that includes packages and pricing helps capture the attention of users browsing a website (Han & Mills, 2006).

3 METHODOLOGY

3.1 *Population and participants*

The aim of this study, through a combination of relevant accessibility and content quality testing methodologies, is to answer the following questions: (i) Is the Go2homestay website easily accessible? (ii) Are visitors readily able to use the site to locate the sources of homestay tourism information they require (adequate and appropriate) to make informed choices? Overall, the aim is to generate a set of findings than can be used to assess the usefulness, value and appropriateness of the site's content with respect to visitors' needs.

The target population of this study is online travel customers in Malaysia, specifically individuals who have visited the Go2homestay website. To achieve its objectives, this study employed a quantitative research approach. Thirty participants were recruited to take part in the preliminary and main phases by completing self-administered questionnaires. This process followed the guidance on sample size published by Billingham, Whitehead and Julious (2013), which stated that the median sample size per arm across all types of pilot or preliminary study is 30. The surveys were conducted in person and by email through personal contacts with the homestay providers.

3.2 *Instruments and procedures*

A pilot study of 20 respondents was first conducted to clarify the overall structure and research approach and validate the measurement instrument. All minor problems or misinterpretations of the questions that occurred in the pilot study were addressed to improve the questionnaire in both form and content (e.g., the layout of the questions and answer choices and the wording of the questionnaire and instructions).

A sample of 406 web-based survey questionnaires was collected in Klang Valley, Malaysia in March 2013. It is believed that Klang Valley residents tend to be more Internet-connected and—savvy, better educated, and more English-speaking than residents of other parts of Malaysia (Malaysiakini, 2003). Intercept and snowball sampling methods were used as informal methods to reach the target population. Because the aim of this study is primarily explorative and descriptive, snowball sampling offered practical advantages (Hendriks, Blanken, Adriaans & Hartnoll, 1992). Six questionnaires were eliminated due to excessive missing data, leaving a total of 400 users included in this study. Accessibility and content quality were examined using instruments adapted from the scale derived by Li and Wang (2011) and focus group interviews.

The survey instrument consisted of four sections. The aim of the first section was to filter out irrelevant respondents and collect general information from the relevant respondents. The aim of the second section was to measure the respondents' perceptions of the website's accessibility and content quality. The respondents were asked to indicate their perceptions on a 7-point Likert scale that ranged from 1 (strongly disagree) to 7 (strongly agree). The third section was intended to examine the respondents' overall level of satisfaction with the Go2homestay website, and they were asked at the end to provide an overall evaluation of their intention to actually undertake the homestay of their choice. The final section collected socio-demographic information (e.g., age, gender, income, educational background, marital status and race).

4 RESULTS AND ANALYSIS

4.1 Demographic analysis

Approximately 69.5 percent of respondents in this study were female, and the remaining 30.5 percent were male. Among the respondents, 56 percent earned less than RM 1,500 per month, 20 percent earned between RM 1,500 and RM 2,500 and 24 percent earned more than RM 2,500 a month. 61.5 percent of respondents were between 18 and 25 years of age, while the remaining 38.5 percent were 26 or above. Many of the respondents (53.5%) had become aware of homestay tourism through the Internet.

4.2 Information and design evaluation

The information assessment was not intended to be all-inclusive but was only designed to assess the information available on and performance of the Go2hometay website. According to the mean score table, the most notable result in this subsection of the analysis concerns the website's photo gallery, which visitors considered vital to finding relevant information (Mean=5.68, Standard Deviation=1.02). This high mean score was due to the use of graphics and pictures on the site encouraging respondents to view the experiences of those who have visited homestays.

Typically, email addresses, phone numbers, mailing addresses and other contact information are listed together, and the visitors ranked the provision of such information as the second most vital feature (Mean=5.56, Standard Deviation=0.96). After contact information came information on accommodations (Mean =5.31, Standard Deviation=1.09), entertainment (Mean=5.12, Standard Deviation=1.08), and nearby attractions (Mean=5.22, Standard Deviation=1.13).

Beyond these components, respondents did not believe that the Go2homestay website provided any unique content. The keyword search, maps and directions, and selected links to external websites were considered useful resources but ones that could be found on other related websites. One exception was the event calendar, which enables visitors to plan visits in conjunction with events they find interesting.

4.3 Accessibility and content quality evaluation

This study's first research question concerned how visitors rated the accessibility of the Go2homestay website. The results show that the mean scores of six features among the accessibility criteria are above 4.0. These criteria also present low standard deviations, thereby supporting the consistency of the distribution. Therefore, the majority of the features were likely to be considered satisfactory. It is clear that the Ministry of Tourism Malaysia is endeavouring to meet the expectations of visitors regarding the promotion of homestay tourism in the country.

The notable results in this subsection of the analysis include the following: (i) the ease of sub-menu use; (ii) ease of navigation; and (iii) multi-lingual capabilities, with Mean=5.18, Mean=5.18 and Mean=5.03, respectively. These results may indicate that the Go2homestay website met visitors' expectations. Visitors expected to access information without experiencing any technical or linguistic difficulties. These results are consistent with previous findings that highlighted the critical importance of accessibility in determining website ratings; in the absence of accessibility, the issues of navigation, visual attractiveness, and informational content generally become irrelevant (Fu Tsang, Lai & Law, 2010; Tanrisevdi & Duran, 2011).

The second criterion identified in this study, content quality, consisted of eleven features regarding the accuracy of the information on the Go2homestay website and the degree to which the information was up-to-date and complete when customers used the site. According to the mean score table, the most notable result in this subsection of the analysis concerns the website's photo gallery, which visitors considered the most important feature in searching for information and which was awarded the highest mean score (Mean=5.68, Standard Deviation=1.01).

This result may be attributable to the aesthetically pleasing manner in which the relevant graphics and photographs were presented on the website. Moreover, this result indicates that potential tourists were interested in viewing the experiences of previous visitors. However, the respondents assigned the information on the event calendar one of the lowest rankings, which is surprising given the emphasis placed on this information by the focus group participants. Generally, the respondents did not consider information provided on the event calendar sufficient. It is possible that the events listed do not meaningfully affect visitors' homestay decisions.

5 DISCUSSION AND CONCLUSION

By gathering primary data from a website's users, the results of this study indicate that the respondents were likely to evaluate the official website for homestay tourism as acceptable. In general, visitors rated the accessibility and content quality of the Go2homestay website as above average. This finding is not consistent with those of some previous studies, such as Oertel, Haße, Scheermesser, Thio and Feil (2004) reported that the accessibility guidelines were only partially followed on the European Union's tourism websites, and the United States' official tourism websites were not able to benefit from the functionality (Lee, Cai & O'Leary, 2006).

Practically, the findings of this study should assist the tourism ministry and the Go2homestay website creator by demonstrating that website users are interested in having access to comprehensive information on homestay destinations, such as culture, activities, nearby attractions, pricing and contact information. The respondents considered the information in featured travel videos on Go2homestay's website to be relatively acceptable.

This feature enables visitors to better understand the experiences offered at a particular homestay destination. As found by Kim and Lee (2005), pleasant aesthetics were reported to be an important feature in website usability.

ACKNOWLEDGMENT

This study was made possible by the continuous support from Universiti Teknologi MARA, UiTM—Grant No: 600-RMI/DANA 5/3/RIF(645/2012).

REFERENCES

Abd Aziz, P.N.R., Tap, M., Osman, A. & Mahmud, M. (2013). *Computer-supported cooperative work in Malaysian homestay industry.* Paper presented at the International Conference on Information and Communication Technology for the Muslim World (ICT4M).

Aziz, N.A., Musa, G. & Sulaiman, A. (2010). The Influence of Travel Motivations and Social Factors on Travel Web Site Usage among Malaysian Travellers. *Asian Journal of Business and Accounting, 3*(1), 89–116.

Bai, B., Law, R. & Wen, I. (2008). The impact of website quality on customer satisfaction and purchase intentions: Evidence from Chinese online visitors. *International Journal of Hospitality Management, 27*(3), 391–402.

Billingham, S.A., Whitehead, A.L. & Julious, S.A. (2013). An audit of sample sizes for pilot and feasibility trials being undertaken in the United Kingdom registered in the United Kingdom Clinical Research Network database. *BMC medical research methodology, 13*(1), 1–6.

Chiou, W.-C., Lin, C.-C. & Perng, C. (2011). A strategic website evaluation of online travel agencies. *Tourism Management, 32*(6), 1463–1473.

Faullant, R., Matzler, K. & Füller, J. (2008). The impact of satisfaction and image on loyalty: the case of Alpine ski resorts. *Managing Service Quality, 18*(2), 163–178.

Fu Tsang, N.K., Lai, M.T. & Law, R. (2010). Measuring e-service quality for online travel agencies. *Journal of Travel & Tourism Marketing, 27*(3), 306–323.

Gregory, A., Wang, Y.R. & DiPietro, R.B. (2010). Towards a functional model of website evaluation: a case study of casual dining restaurants. *Worldwide Hospitality and Tourism Themes, 2*(1), 68–85.

Han, J.H. & Mills, J.E. (2006). Zero acquaintance benchmarking at travel destination websites: What is the first impression that national tourism organizations try to make? *International Journal of Tourism Research, 8*(6), 405–430.

Hashim, N.H. & Murphy, J. (2008). *Online destination marketing: A case study of Malaysia.* Johor, Malaysia: Penerbit UTM.

Hendriks, V.M., Blanken, P., Adriaans, N.F. & Hartnoll, R. (1992). *Snowball sampling: A pilot study on cocaine use:* IVO, Instituut voor Verslavingsonderzoek, Erasmus Universiteit Rotterdam.

Herrero, Á. & San Martín, H. (2012). Developing and testing a global model to explain the adoption of websites by users in rural tourism accommodations. *International Journal of Hospitality Management, 31*(4), 1178–1186.

Huang, J.-H. & Shyu, S.H.-P. (2008). E-government web site enhancement opportunities: a learning perspective. *The Electronic Library, 26*(4), 545–560.

Kabadayi, S. & Gupta, R. (2011). Managing motives and design to influence web site revisits. *Journal of Research in Interactive Marketing, 5*(2/3), 153–169.

Kaynama, S.A. & Black, C.I. (2000). A proposal to assess the service quality of online travel agencies: an exploratory study. *Journal of Professional Services Marketing, 21*(1), 63–88.

Kim, W.G. & Lee, H.Y. (2005). Comparison of web service quality between online travel agencies and online travel suppliers. *Journal of Travel & Tourism Marketing, 17*(2–3), 105–116.

Kozak, M., Bigne, E. & Andreu, L. (2005). Web-based national tourism promotion in the Mediterranean area. *Tourism Review, 60*(1), 6–11.

Lee, G., Cai, L.A. & O'Leary, J.T. (2006). WWW. Branding. States. US: An analysis of brand-building elements in the US state tourism websites. *Tourism Management, 27*(5), 815–828.

Lepp, A., Gibson, H. & Lane, C. (2011). Image and perceived risk: A study of Uganda and its official tourism website. *Tourism Management, 32*(3), 675–684.

Li, X. & Wang, Y. (2011). Measuring the effectiveness of US official state tourism websites. *Journal of Vacation Marketing, 17*(4), 287–302.

López, I. & Ruiz, S. (2011). Explaining website effectiveness: The hedonic-utilitarian dual mediation hypothesis. *Electronic Commerce Research and Applications, 10*(1), 49–58.

Lu, Z., Lu, J. & Zhang, C. (2002). Website development and evaluation in the Chinese tourism industry. *Networks and Communication Studies, 16*(3–4), 191–208.

Malaysiakini. (2003). Survey: Malaysiakini readers mostly professionals, Klang Valley-based, from http://www.malaysiakini.com/news/15112

Musante, M.D., Bojanic, D.C. & Zhang, J. (2009). An evaluation of hotel website attribute utilization and effectiveness by hotel class. *Journal of Vacation Marketing, 15*(3), 203–215.

Oertel, B., Haße, C., Scheermesser, M., Thio, S.L. & Feil, T. (2004). *Accessibility of Tourism Web sites within the European Union.* Paper presented at the 11th International Conference on Information and Communication Technologies in Tourism (ENTER 2004), ISBN.

Pourabedin, Z. & Nourizadeh, A. (2013). Designing persuasive destination website: The role of visual aesthetic. *Journal of Basic and Applied Scientific Research, 3*(2), 675–680.

Prins, C. (2001). *Designing e-government: On the crossroads of technological innovation and institutional change* (Vol. 12): Kluwer Law International.

Ranaweera, C., Bansal, H. & McDougall, G. (2008). Web site satisfaction and purchase intentions: impact of personality characteristics during initial web site visit. *Managing Service Quality, 18*(4), 329–348.

Ruiz-Molina, M.-E., Gil-Saura, I. & Moliner-Velázquez, B. (2010). The role of information technology in relationships between travel agencies and their suppliers. *Journal of Hospitality and Tourism Technology, 1*(2), 144–162.

Schmidt, S., Cantallops, A.S. & dos Santos, C.P. (2008). The characteristics of hotel websites and their implications for website effectiveness. *International Journal of Hospitality Management, 27*(4), 504–516.

Selamat, A. & Ismail, M.K. (2008). *Gustafson-Kessel algorithm in content based image retrieval for Malaysia tourism website.* Paper presented at the Information Technology, 2008. ITSim 2008. International Symposium on, Kuala Lumpur on August 26–28.

Shi, M. (2011). Website characteristics and their influences: A review on web design *Association of Business Information System* (pp. 127–135). Houstan, Texas.

Tanrisevdi, A. & Duran, N. (2011). Comparative evaluation of the official destination websites from the perspective of customers. *Journal of Hospitality Marketing & Management, 20*(7), 740–765.

Thongpapanl, N. & Ashraf, A.R. (2011). Enhancing online performance through website content and personalization. *Journal of Computer Information Systems, 52*(1), 1–13.

Vinod, B. (2011). The future of online travel. *Journal of Revenue & Pricing Management, 10*(1), 56–61.

W3C. (2003). Evaluation, repair, and transformation tools for web content accessibility, from http://www.w3.org/WAI/ER/existingtools.html

Woodside, A.G., Vicente, R.M. & Duque, M. (2011). Tourism's destination dominance and marketing website usefulness. *International Journal of Contemporary Hospitality Management, 23*(4), 552–564.

Xiang, Z. & Gretzel, U. (2010). Role of social media in online travel information search. *Tourism Management, 31*(2), 179–188.

Zulfakar, M.R. & Buchanan, J. (1999). Effective tourism web-sites, Part 1: Literature review and features survey, from http://www.mngt.waikato.ac.nz/depts/mnss/john.

Theory and Practice in Hospitality and Tourism Research – Radzi et al. (Eds)
© 2015 Taylor & Francis Group, London, ISBN 978-1-138-02706-0

SoLoMo and online trust towards generation Y's intention to visit boutique hotels

A.A. Ahmad, M.N.I. Ismail & S.M. Radzi
Faculty of Hotel and Tourism Management, Universiti Teknologi MARA, Shah Alam, Malaysia

ABSTRACT: The expansion of information technology is leading the online marketplace, thus affecting the marketing industry. Information technology has also contributed to the progression of Malaysia as a tourist destination. Henceforth, there are greater increments in revenue and occupancy rate in the hotel sector, particularly boutique hotels. Generation Y represents the digital natives that is concerned on information technology. Thus, generation Y favors comprehensive application that meets their needs as information seekers. The accessibility of the internet has led the SoLoMo as a feature in marketing strategies. SoLoMo trends have been changing the way of generation Y behaves as consumers. Moreover, SoLoMo offers a unique opportunity for companies to gain a competitive edge. Therefore, this paper will explore on awareness of SoLoMo trends, whilst focusing on each element in SoLoMo, which includes social, local and mobile as having online trust as moderating factors that result in intention of visiting a boutique hotel.

Keywords: SoLoMo, intention of visiting, boutique hotel

1 INTRODUCTION

Information technology is defined as a process of obtaining and gathering data by using software and hardware in order to generate the information required. Nowadays, information technology and tourism are some of the major contributor in today's economy (Buhalis & Law, 2008). In Malaysia, this connection has given strategic opportunities and powerful components in the economic growth. There are 17,723,000 internet users in Malaysia.

The expansions in the hospitality industry are leading to fierce competition in targeting the market sector. Recently, boutique hotel is in trend and also is one of the growth contributors to the hotel industry. As stated by Horner and Swarbrooke (2005), boutique hotel rapidly grows in the industry because its targeted audience is generation Y. It was discovered that generation Y met and suited the characteristics as the target market for boutique hotel (Olga, 2009). However, the peak competition in the industry has led the marketers to find different marketing channel in order to introduce and convey their products and excellent services to the customers. Thus, Thakran and Verma (2013) suggested that digital marketing is the best medium for reaching potential consumers and retain customer relationships.

According to a statistic, generation Y is the highest internet user's ranking which represents 64 percent between ages of 15–34, followed by 3 percent for those aged 35–44 and the least about 13 percent of aged 45–55. Furthermore, generation Y has greater disposable income rather than their past generation. Nowadays, patrons can retrieve information anywhere and every time through information platform which is the Internet. Internet has influenced the generation Y in making a purchase decision. Besides, Maswera, Edwards and Dawson (2009) mentioned that online trust also has influenced generation Y's behavior and lead as a contributor to the growth of e-commerce business for the hotel industry.

2 RESEARCH GAP

According to Buhalis and Law (2008), the research on SoLoMo trends is still at infancy and needs to be discovered thoroughly for the greater opportunities in return. SoLoMo is not only a growing trend but it is giving the momentum on a global basis for each industry. As mentioned by Dimitrovski, Todorović, and Valjarević (2012), the intersection of SoLoMo should be explored more since there is little exposure about these. This is because SoLoMo helps the industry to deal with engagements with digital natives and rivals. By this time, SoLoMo grants the advantages for retailer, marketers, campaigners, ambassadors and consumer that suits with every business in e-commerce (Buhalis & Law, 2008).

Simultaneously, the advancements of generation Y in obtaining information on the Internet are giving challenges and opportunities to marketers. Hence, the new effective marketing channel should be explored in order to meet the expectations from the consumer particularly generation Y as information seekers. Thus, it is important to discover on awareness of each element in SoLoMo from both marketers and consumers perspectives. Since SoLoMo is changing the generation Y purchasing behavior, it is important to measure online trust as moderator. The perceived usefulness, enjoyment and ease of use of SoLoMo will indicate the acceptance in having decision making for both parties.

3 LITERATURE REVIEW

3.1 SoLoMo

The vast attention has been received regarding SoLoMo terms for many years. Recently, the buzzword that is derived from SoLoMo has been accepted in the marketing world. SoLoMo was invented by the famous American venture capital investor, John and the French Blogger and entrepreneur, Loïc Le Meur. They have contributed the popularity towards the term of SoLoMo in the LeWeb 11 conference. Revolution of the SoLoMo era is resulted from the rapid growth of smartphones and tablet users. Sequentially, the consumers demand for high pace of information searching via mobile devices in order to meet their life desires on the go.

The definition of SoLoMo merges from Social as So, Location as Lo and Mobile as Mo (Phil, 2012). Social can be referred to as social media networking that acts as a medium to connect with others whereby local based application refer to identification of place or location. Mobile devices can be referred to the devices that enable sharing, delivering information and engaging with others. The application of the SoLoMo terms can be explained whereby individuals are using the social media sites along with a location based application that interacts and transacts through mobile devices. However, there are some uncertainties on acceptance of SoLoMo for people who are unfamiliar with the term, thus tend to ignore it. Many people, especially marketers identified that SoLoMo is nothing more than a transient fad. In addition, this sweeping trend is upsetting the entire industry in terms of distribution channels for conveying product and services, thus engaging the customers' via media marketing.

3.2 Online trust

As regards to the development of tourism industry, Fam, Foscht and Collins (2004) identified that online trust is becoming the determinant factor in e-commerce marketing growth. This is because trust will be a measurement tool of purchase intention (Sichtmann, 2007). As a result, the greater uncertainty of trust by consumers will lead to a decline on intention of buying decision and tarnishing the relationship between both parties. Ninety-two percent of consumers rely and are concerned on recommendations by their peers or family. This will affect the uncertainty before deciding on buying decision. McCole (2002) clarified that trust becomes a basis word in every industry. However, hospitality industry must concern more on trust due to its nature as providing services that are categorized as intangible products. Furthermore, the fluctuating demands and inseparability of production and consumption will rely on trust because it is a priority factor for retaining the relationship with consumers.

There is a study by Corritore, Kracher and Wiedenbeck (2003), about online trusts that focuses on people towards informational websites. According to the researcher, online trust consists of several elements which are credibility, ease of use and risk. Each of these elements significantly affects the consumers' attitude towards the brands. All those elements are derived from external factors as an agent that enables the shifting of person's actions.

3.3 Gen Y's intention to visit boutique hotel

Generation Y can be defined as people that are born between 1977 and 1994 which are currently aged 19–34 (Herbig, Koehler & Day, 1993). Generation Y were born after generation X and grows up with technology advancements. Their characteristics are knowledgeable, technology-savvy, social-savvy and exposed to new experiences. Therefore, generation Y is digital natives that are always connected with technology and actively seek information. The advancements of the Internet have empowered them as information finders. Having a greater disposable income than previous generations, generation Y is more concerned on branding and marketing. They take reviews from online websites or considering their peer and family's review about branding as earnestly. Generation Y also utilizes the Internet to maximize to obtain information and build trust through reviews or comments in internet as reliable source before making any intention to purchase or visit the product.

Intention to visit can be described as an act for choices decisions or behavior (Mohsin, 2005).

Intention to visit was influenced by social media platforms and search engines as a dimension of decision making. The result of intention to visit was influenced by several factors, depending on the situation or other indicators. There is insufficient literature in expanding and discovering the indicators that lead to the intention as result.

The boutique hotel was described as a small scale lodging industry that has less than 100 rooms and delivering the experience that conveys status and lifestyle through difference décor by targeting the focus niche market. As cited by Olga (2009), the target market of boutique hotel ranges the age from early twenty to fifty, thus suited with the definition of generation Y. Another definition by Aggett (2007), boutique hotel delivers the uniqueness of lifestyles through the design elements for building, décor, furniture as their competitive advantage. The different strategies that are implemented by boutique hotel were driving it to massive growth in popularity in hospitality and tourism industry. Despite of boutique hotels growth in the tourism industry, Aggett (2007) highlighted that there is still an insufficient effort in exposing the potential on organization sides. Therefore, the academicians should stress on a potential position of boutique hotel in industry since this segment have generated high growth in industry, Freund de Klumbis and Munsters (2005) suggested exploring the boutique hotels based on consumer's perspective.

4 DISCUSSION

SoLoMo still remains as an unfamiliar term even though the SoLoMo rapidly grows its trend in marketing industry recently. In addition, SoLoMo marketing is fresh and innovative in the research area to be explored due to huge opportunities as a return. Therefore, this study will help to contribute more on literature in the related field as well. This is the opportunity study; where more outcomes can be explored due to the richness in this topic area and it will add more on the body of knowledge. For the matter referred above, the expected findings can be used as key explanations apart from adding to the existing literature in the field information technology that cover on social, location and mobile that will benefit to all sectors because information technologies are becoming the foundation and essential for each of us. Furthermore, the adoption of smartphones and tablet devices as a necessity in the digital lives currently aligns with the growth of online services, e-commerce, and e-marketing in information technology industry knowledge.

It is expected to result in a new approach to market because SoLoMo becomes a new effective Omni channel that acts as communication, marketing and distribution channel in the industry, thus it will lead to the growth in revenue by offering great return on investment. Hence, SoLoMo is valuable and suits well with the characteristic of boutique hotel itself that offering the uniqueness of lifestyle to the customer. Furthermore, SoLoMo will help the boutique hotel industry to keep up with the massive demand due to the popularity and growth of boutique hotel itself by being the Omni channel distribution that giving the cohesive experience.

Affected by the Internet development, SoLoMo grasps the benefits in digital marketing to build closer relationship with consumers by conveying the information about products in social media sites by tagging the location via mobile devices. It is expected to result that SoLoMo provides the database of consumers and the potential consumers as well. Thus, it will help the marketers in each industry to enhance their relationship with customers in both worlds—reality or digital world, next reaching a broader market segment in each industry. As a result, this consumer engagement will lead to the consumer and brand loyalty that leads to a greater return on investment further. In the meantime, the awareness of customers as SoLoMo's consumer will be projected.

5 CONCLUSION

The SoLoMo era can be clarified as the third wave of tsunami in the digital marketing industry because it generates the storm impact in marketing segment. Also, SoLoMo is ready as a game changer in whole new economy of marketing by itself, renovating the people on dealing with ideas in the commerce world through internet. SoLoMo can be proven to be more than just growing trends that consists of opportunities to the marketers and consumers. Henceforth, both parties should acknowledge this trend as the basic movement in the hospitality industry.

With the preceding notion, the empirical investigation of the highlighted issue pertaining to SoLoMo and online trust is still under investigation. The quantitative research design with the close ended questionnaire will be used as the research instrument. The close ended questionnaires will be distributed through online surveys and will be self-administered. The screening question will be asked before the questionnaires are distributed to indicate generation Y as SoLoMo consumers. The reason is to ensure that the information gathered is reliable and valid. The respondents will be generation Y that ranges between 19 to 36 years old that had experiences as SoLoMo consumers. With that, each boutique hotel in the four states (Penang, Kuala Lumpur, Melaka and Johor) will be chosen as there is a steady growth of boutique hotels within this region. According to Tourism Malaysia, occupancy percentages of boutique hotels are 64 percent for Penang, 69 percent for Kuala Lumpur, 62 percent for Melaka and 56.1 percent for Johor. In addition, these states are popular as tourist destinations in Malaysia due to the culture and heritage in Penang and Melaka and fast growing development in Kuala Lumpur and Johor.

ACKNOWLEDGEMENTS

This study was made possible by the continuous support from Universiti Teknologi Mara, Shah Alam (UiTM).

REFERENCES

Aggett, M. (2007). What has influenced growth in the UK's boutique hotel sector? *International Journal of Contemporary Hospitality Management, 19*(2), 169–177.

Buhalis, D. & Law, R. (2008). Progress in information technology and tourism management: 20 years on and 10 years after the Internet—The state of eTourism research. *Tourism management, 29*(4), 609–623.

Corritore, C. L., Kracher, B. & Wiedenbeck, S. (2003). On-line trust: Concepts, evolving themes, a model. *International Journal of Human-Computer Studies, 58*(6), 737–758.

Dimitrovski, D. D., Todorović, A. T. & Valjarević, A. D. (2012). Rural tourism and regional development: Case study of development of rural tourism in the region of Gruža, Serbia. *Procedia Environmental Sciences, 14*, 288–297.

Fam, K. S., Foscht, T. & Collins, R. D. (2004). Trust and the online relationship: An exploratory study from New Zealand. *Tourism management, 25*(2), 195–207.

Freund de Klumbis, D. & Munsters, W. (2005). Developments in the hotel industry: design meets historic properties. *International Cultural Tourism*, 162–184.

Herbig, P., Koehler, W. & Day, K. (1993). Marketing to the baby bust generation. *Journal of Consumer Marketing, 10*(1), 4–9.

Horner, S. & Swarbrooke, J. (2005). *Leisure marketing: A global perspective*. Oxford: Butterworth-Heinemann.

Maswera, A. G., Edwards, N. & Dawson, E. (2009). E-commerce adoption and practice in sub saharan Africa. *Tourism Business*(5), 89–95.

McCole, P. (2002). The role of trust for electronic commerce in services. *International Journal of Contemporary Hospitality Management, 14*(2), 81–87.

Mohsin, A. (2005). Tourist attitudes and destination marketing: The case of Australia's Northern Territory and Malaysia. *Tourism management, 26*(5), 723–732.

Olga, A. (2009). *The alternative hotel market.* Paper presented at the International Conference on Management Science and Engineering, 2009 (ICMSE 2009).

Phil, H. (2012). How Solomo is connecting consumers and brands to almost everything. *Marketing Research.*

Sichtmann, C. (2007). An analysis of antecedents and consequences of trust in a corporate brand. *European Journal of Marketing, 41*(9/10), 999–1015.

Thakran, K. & Verma, R. (2013). The emergence of hybrid online distribution channels in travel, tourism and hospitality. *Cornell Hospitality Quarterly, 54*(3), 240–247.

Theory and Practice in Hospitality and Tourism Research – Radzi et al. (Eds)
© 2015 Taylor & Francis Group, London, ISBN 978-1-138-02706-0

The integration of Technology Readiness (TR) and Customer Perceived Value (CPV) in tablet-based menu ordering experience

M.I. Zulkifly, M.S.M. Zahari, M. Hafiz & M.R. Jamaluddin
Faculty of Hotel and Tourism Management, Universiti Teknologi MARA, Shah Alam, Malaysia

ABSTRACT: Traditional paper-based menu ordering has long been considered as the most appropriate way to communicate the items in the menu with the customer. Many argue about the ability of the menu card to generate more sales and profit through the information given but not all restaurant operators fully utilize them to meet those purposes. In fact, it is impossible to put all the information regarding the items in a single menu card. Due to the problem, the use of technology namely the tablet-based menu ordering seems to be the best option to replace the old practice especially in time when even customers demand the foodservice industry to keep pace with the technology. It is not a very straightforward solution since the Technology Readiness (TR) and the Customer Perceived Value (CPV) are yet to be examined. TR is believed to have a significant impact on CPV which later translates into the behavior and actual use of this technology among customers.

Keywords: Menu, tablet, technology readiness, and customer perceived value

1 INTRODUCTION

Menu is considered as the 'heart' of any foodservice operation (Payne-Palacio & Theis, 2004; Walker, 2014; Warner, 1994). The word 'heart' is used to symbolize the capability of menu in catalyzing the other functions in an operation. It is practically true since the functions of purchasing, receiving, storing, preparation and serving will only start once the menu is decided. Thus, the phrase "Everything starts with menu" could not get any better looking at the role menu plays in a foodservice operation (Payne-Palacio & Theis, 2004). Those were just the role of menu in running the operation. It is also proven that the right choice of menu will bring in crowds which in turn help generating the revenue if marketing, pricing and quality are done well. The togetherness of these elements will ensure the continuity of the business in a longer run.

It is known that the choice of menu based on the demographic and psychographic factors should be the top priority of a foodservice establishment. Getting it right is just the beginning of many. Once the menu is decided after careful consideration on the above factors, the strategies of getting it sold take another measure. In this circumstance, it is important to create an effective communication between the restaurant and the customers about the menu offered and most of the times menu cards or books hold the aces (Payne-Palacio & Theis, 2004) (McVety, Ware & Ware, 2008). This is particularly true since foodservice establishments have been using the menu card to speak with the customers and sometimes they tend to use it in an intolerable and abusive level. There were studies on developing a good menu card in which the psychology of reading and the eye-gaze motion of customers had been taken into consideration but it is really up to the restaurateurs to take the opportunities to fully utilize it. In addition, a good menu card is the one which uses 'descriptive wording' wisely especially to the so-called 'killer' items and other items with a good profit margin.

The menu, when used properly is a great marketing tool that can influence customer's behavior and decision-making process. There are number of factors in relation to menu which influences customer's decision (Panitz, 2000). Factors such as the physical menu's background, texture, color, photos, item's positioning, fonts, size and pricing come into the frame of one's mind during the menu ordering process. All these factors shape the menu design and displays of a particular restaurant. Mills and Thomas (2008) argued that menu design and displays therefore can be said as the central part in merchandising strategy in the restaurant business. Menu ordering experience among customers can be enhanced through the use of descriptions and images. Customer nowadays would like to know more about what they are getting on their table simply from reading the menu. On top of that, they are now more health conscious and demanding for nutritional information and food claims from the restaurants.

It is however impossible to stuff all the information about items in the menu in a single menu card. Otherwise, it will greatly shift the burden to customers who will be trying so hard to be decisive in choosing their meals. Considering this limitation, it is said that technology application is the only solution to the above mentioned problem as agreed by Oronsky and Chathoth (2007) who mentioned that technology can enhance or bring about change in the dining experience of customers as a whole including the menu ordering. Leung and Law (2007) added that is more relevant than ever to use technology and innovative ideas to provide additional support for the restaurants' business processes and decision making although the use of 'traditional' paper-based menu ordering is still considered by many as one of the most important moment-of-truth prior to the other processes later on.

Nonetheless, along with the increment in the importance of technology there has been an increase in the use of tabletop and handheld ordering devices (Rousseau, 2011). Significantly, the use of these devices, especially tablets like iPad and Galaxy Tab has enables customers to review menu items, nutritional information and preparation of menu items. Rousseau (2011) further added that tablet menus are capable to enhance customer's informational satisfaction by reducing customer indecision and uncertainty in ordering decision. On top of that, instead of going through number of pages or folds of traditional menu cards, customers will find it easy to go through innovative displays of information developed through intuitive interfaces of those tablets. The author in his study also demonstrated that the customers were fascinated by the new technology where he explained one example where wine consumption in Chicago Cut Steakhouse had increased since the implementation of tablet-based menu ordering by using iPad. Perhaps, customer satisfaction survey can also be done by using the same device.

Many restaurants throughout the world are using tablet-based menu ordering (iPad, Samsung Galaxy Tab) lately. The most interesting part here is that these restaurants allocate one tablet-based gadget on every table for customers to place their orders thus triggering a service called Self-Service Technologies (SSTs) in table service menu ordering applying the same concept of SSTs used in Airline and hotel industry to name a few. For example, in China, the number of restaurants providing this kind of experience is increasing especially among seafood restaurants and experts are anticipating that it will be a new phenomenon that will shake the foodservice industry in the country. A study has been conducted by Dixon and Kimes (2012) from Cornell University regarding the consumer preferences for restaurant technology innovation and the study revealed that the technology customers value the most currently is 'table virtual menus' where customers love the fact that menu is presented in more meaningful ways with additional information on nutritional value, origin of ingredients. Not only that, customers can place orders, play games, pay their bills and even watch movie trailers with the gadgets on their table.

It sounds nice to have something like this in the restaurant but the menu ordering experience on the customers' behalf is yet to be examined. For long, restaurant industry relies heavily on the traditional paper-based menu ordering where face-to-face interaction is at the utmost important. The tablet-based menu ordering however is said to be the most appropriate approach considering the changes among customers and issues regarding human errors and service failures exposed by the traditional method of taking and delivering the order. In fact, Yieh, Chen and Wei (2012) claimed that customers in this age demand businesses to keep pace with them in terms of technological use. However, it is also depends on the technology readiness among customers and how they value technologies prior to their use. Individuals will have different reactions to computer interaction therefore customers' technological readiness plays an important role in influencing the overall customers' information satisfaction. Hence the assessment on Technology Readiness (TR) among customers on this technology to focus on their reactions and level of comfort with technological tools is very much needed (Dixon, Kimes & Verma, 2009). On top of that, the capability of businesses to create a good Customer Perceived Value (CPV) based on this system could also influence the behavior and actual use of this technology among customers.

2 LITERATURE REVIEW

2.1 *Technology Readiness (TR)*

According to Parasuraman and Colby (2001), Technology Readiness (TR) is defined as the people's propensity or inclination to embrace and use new technologies for accomplishing goals in life and at work. The TR as proposed by the above authors is categorized into four distinct categories namely optimism, innovativeness, discomfort and insecurity. The first two dimensions are actually the motivator that increase a customer's TR while the other two are identified as inhibitors that restrain TR. In short, TR will indicate a person's openness to technology in general.

It is important to elaborate on all four dimensions of TR to better comprehend the concept in this research. Optimism in TR is referred to a customer's constructive view of technology in which control, flexibility, convenience and efficiency are among the concerns. Walczuch, Lemmink and Streukens (2007) defined optimism as the tendency to believe that one will generally experience good versus bad outcomes in life. They added that optimists often look for strategies than pessimists and these strategies will produce more effective and favorable outcomes. These types of customers will be less likely to bother on the negative side thus they are more likely to accept and confront the situation rather than hide away from it.

The other driver is innovativeness which is defined as the tendency to be a thought leader and technological pioneer (Parasuraman, 2000). This dimensions specifically dig out the level individuals assume themselves as being among the forefront adopters of technology. Karahanna, Straub and Chervany (1999) added that there are individuals who are among the first adopters who have less complex belief sets about new technology hence increase their willingness to try out any latest technology emerge in the market.

One of the inhibitors is the dimension of discomfort which refers to the situation in which the individual is unable to manipulate technology and gets intimidated with his or her incompetence. Insecurity on the other hands is defined as the level of distrust among individuals towards technology and its potential to function properly. Kwon and Chidambaram (2000) talked about the apprehensiveness among individuals who have fears of using technology probably due to their skepticism on the matter.

All in all, the four TR dimensions represent different characteristics and psychological processes underlying technology adoption. Both motivators and inhibitors will dominate every individual in shaping up their behavior towards technology adoption (Lam, Chiang & Parasuraman, 2008).

2.2 Customer Perceived Value (CPV)

Customer Perceived Value (CPV) is one of the most important elements in this research. According to Yieh et al. (2012) CPV has been considered as one of the critical mean to analyze service quality, customer satisfaction and even consumer behavior. The dimensions under CPV could be tailored to the context of study since there is no universally accepted construct and conceptualization to this day (Sánchez-Fernández, Angeles Iniesta-Bonillo & Holbrook, 2009). CPV construct is centered on the trade-off between the benefits and sacrifices when using technology. It is noteworthy to understand that people buy products or services not only for what they do but also for what they mean to do.

Therefore, a multidimensional approach has been used to assess CPV as what has been used by Yieh et al. (2012) study who measured CPV through four specific dimensions; in-use value, social value, emotional value and security value. The in-use value scale was actually originated from Keeney (1999) and it is one of the measurements of the rational aspect of CPV. This measurement is beneficial in assessing one's opinion on items like convenience, time and flexibility.

Meanwhile, for the social value developed by Sweeney and Soutar (2001), it would measure individuals' feelings and their impression towards the use of technology and further understand how they perceive the usage in enhancing their sense of belongings. The authors also came out with emotional value scale to measure the experiential aspect of CPV which is more complicated and pragmatic at the same time. It is as simple as measuring one's happiness, contentment and frustration towards technology.

Realizing the relationship with technology and the usage of internet for the entire system in tablet-based menu ordering, it is obvious that the researcher must also explore CPV from e-business perspective where security could be one of the most critical issues to be tackled (Chen & Dubinsky, 2003; Keeney, 1999). As mentioned in the previous chapter, customer can also use the tablet-based menu ordering for different activities other than just placing the order hence some of their information might get exposed during the experience. In measuring this dimension the security value constructed by Torkzadeh and Dhillon (2002) will be referred to in relation to the CPV.

2.3 The relationship between TR and CPV

Many studies have looked at the relationship between the TR and technology acceptance model (TAM) in this related field. There is only one study found which addressed the relationship between TR and CPV which has been done by Yieh et al. (2012) in which the perceived ease-of-use and perceived usefulness often used in TAM was modified and further expanded into a more comprehensive way as explained in CPV. The researcher has decided to look at the similar dimension as proposed by Yieh et al. (2012) who clearly demonstrated that TR has a significant impact on CPV. A similar concept with Theory of Reasoned Action (TRA) and Theory of Planned Behavior (TPB) can be seen here to say the least where TR act as the main factor in determining consumer attitude, intention and behavior while CPV is the antecedent of attitude, intention and behavior.

3 ISSUE FROM MALAYSIAN PERSPECTIVE

Malaysia, as with the other countries, is also experiencing the technological advancement from many perspectives including the foodservice industry. Foodservice industry in this country is perhaps one of the largest industries in terms of the economic contribution and has been identified as one of the major sources of employment among any other industries. The fact that Malaysian spent 40 to 50 percent of their incomes might have contributed to the growth of this industry in general (Selamat, Shamsudin & Dulatti, 2002). The use of technological advancement devices has been popular in the country for quite some time particularly in the back-of-the-house operation. There are number of restaurants which use cards and PDAs to take their order but the number of restaurant who use tablet-based device such as iPad are still lacking. Assuming that the anticipation of scholars who believe that the adoption of this technology will be a big hit in the near future will hold true, the researcher believe that the same phenomenon will be the major talking point in the industry in just a matter of time. The number or statistic of restaurants with this technology adoption is not available at the moment but the researcher has already identified a few based on self-observation at shopping complexes and the Internet.

This technology is a relatively new concept in this country, but it will be booming once restaurants realized the potential of this system for menu ordering process. For the time being, it is appropriate to say that this paper will be looking at the idea of integrating TR and the impact of it towards their perceived value (CPV) prior to their experience and consequently examine their information satisfaction based on their actual usage (ordering experience).The idea behind this integration was inspired by the Yieh et al. (2012) who argued that TR has a significant impact on CPV which will further influence the behavior and actual use of this technology among customers. Prior to their study, most of the research integrated the TR with the well-known Technology Acceptance Model (TAM). TAM worked well in that integration but the idea of replacing it with CPV will not disregard the element of TAM. CPV by any means is trying to comprehend and expand the existing dimension under TAM especially on the emotional and security side of using technologies.

There were studies in Malaysia which looked into almost identical concept but the number is not something to be proud of and those studies were actually conducted in the other fields or industries. It is therefore important to start looking at the opportunity to examine Malaysian's TR and CPV on tablet-based which is practically limited where very few studies have been undertaken on the topic. Furthermore, overall customer information satisfaction on the usage of the system is almost non-existent thus it is hard to tell whether the system is fulfilling the goals.

4 CONCLUSION

This paper is proposing the idea of integrating TR and CPV on the tablet-based menu ordering and surely the academic contributions to the existing body of knowledge will be accomplished through a proper research probably looking at the conceptual model and testing all the hypotheses drawn from it.

Consequently, more empirical analysis is needed to consolidate this knowledge. That information can be beneficial to analyze or foresee the Malaysian readiness to accept this kind of system. The data on usability and overall experience on the other hand would be beneficial to help restaurants with the system to tailor their system with more 'value' for customers. In addition, the result from this study could be a 'look-to' reference for foodservice operations which are yet to invest on the system.

REFERENCES

Chen, Z. & Dubinsky, A.J. (2003). A conceptual model of perceived customer value in e-commerce: A preliminary investigation. *Psychology & Marketing, 20*(4), 323–347.

Dixon, M. & Kimes, S.E. (2012). Technology customers value most Retrieved October 3, 2014, from http://restaurantbriefing.com/2010/02/technology-customers-value-most/

Dixon, M., Kimes, S.E. & Verma, R. (2009). Customer preferences for restaurant technology innovations. *Cornell Hospitality Report, 9*(7), 4–16.

Karahanna, E., Straub, D.W. & Chervany, N.L. (1999). Information technology adoption across time: A cross-sectional comparison of pre-adoption and post-adoption beliefs. *MIS quarterly, 23*(2), 183–213.

Keeney, R.L. (1999). The value of Internet commerce to the customer. *Management science, 45*(4), 533–542.

Kwon, H.S. & Chidambaram, L. (2000). *A test of the technology acceptance model: The case of cellular telephone adoption.* Paper presented at the International Conference 33rd Annual System Sciences, 2000, Hawaii, January 3–6.

Lam, S.Y., Chiang, J. & Parasuraman, A. (2008). The effects of the dimensions of technology readiness on technology acceptance: An empirical analysis. *Journal of interactive marketing, 22*(4), 19–39.

Leung, R. & Law, R. (2007). Information technology publications in leading tourism journals: a study of 1985 to 2004. *Information Technology & Tourism, 9*(2), 133–144.

McVety, P.J., Ware, B.J. & Ware, C.L. (2008). *Fundamentals of menu planning*: John Wiley & Sons.

Mills, J.E. & Thomas, L. (2008). Assessing customer expectations of information provided on restaurant menus: a confirmatory factor analysis approach. *Journal of Hospitality & Tourism Research, 32*(1), 62–88.

Oronsky, C.R. & Chathoth, P.K. (2007). An exploratory study examining information technology adoption and implementation in full-service restaurant firms. *International Journal of Hospitality Management, 26*(4), 941–956.

Panitz, B. (2000). A promising future. *Restaurant USA: The Magazine of the National Restaurant Association, 20*(2), 13–18.

Parasuraman, A. (2000). Technology Readiness Index (TRI) a multiple-item scale to measure readiness to embrace new technologies. *Journal of Service Research, 2*(4), 307–320.

Parasuraman, A. & Colby, C.L. (2001). *Techno-ready marketing: How and why customers adopt technology*. Avenue of the Americas, New York: Free Press.

Payne-Palacio, J. & Theis, M. (2004). *Introduction to foodservice* (10th ed.): Prentice Hall.

Rousseau, C. (2011). Restaurants uploading menus onto iPads for diners. *Chicago Sun-Times, available at: www. suntimes. com/lifestyles/food/3143444–423/ipads-chicago-wine-ipadmenu. html.*

Sánchez-Fernández, R., Angeles Iniesta-Bonillo, M. & Holbrook, M.B. (2009). The conceptualisation and measurement of consumer value in services. *International Journal of Market Research, 51*(1), 93–113.

Selamat, J., Shamsudin, M. & Dulatti, M. (2002). Malaysia food safety. *Pacific Food System Outlook, 2003.*

Sweeney, J.C. & Soutar, G.N. (2001). Consumer perceived value: the development of a multiple item scale. *Journal of Retailing, 77*(2), 203–220.

Torkzadeh, G. & Dhillon, G. (2002). Measuring factors that influence the success of Internet commerce. *Information Systems Research, 13*(2), 187–204.

Walczuch, R., Lemmink, J. & Streukens, S. (2007). The effect of service employees' technology readiness on technology acceptance. *Information & Management, 44*(2), 206–215.

Walker, J.R. (2014). *The restaurant: From concept to operation* (7th ed.): John Wiley& Sons.

Warner, M. (1994). Noncommercial, institutional, and contract foodservice management.

Yieh, K., Chen, J.-s. & Wei, M.B. (2012). The Effects of Technology Readiness on Customer Perceived Value: An Empirical Analysis. *Journal of family and economic issues, 33*(2), 177–183.

Theory and Practice in Hospitality and Tourism Research – Radzi et al. (Eds)
© *2015 Taylor & Francis Group, London, ISBN 978-1-138-02706-0*

Perception of environmental strategies in hotels and the influence towards future behavior intention: Locals' perspective in Malaysia

M.F.S. Bakhtiar, A.A. Azdel, M.S.Y. Kamaruddin & N.A. Ahmad
Faculty of Hotel and Tourism Management, Universiti Teknologi MARA, Shah Alam, Malaysia

ABSTRACT: An ongoing concern for environmental protection among society is much apparent today; the same is evident in the hotel industry. Despite continuing pressure to sustain with such demand, hoteliers need to balance between creating unobtrusive environmental strategies to guests while keeping it visible to satisfy environmentally-conscious clientele. Hence, it is noteworthy to evaluate guests' perception about environmental strategies applied in hotels and how such implementation may influence one's future behavior intention. Through self-reported questionnaire, 288 valid responds were gathered using convenience sampling method to unravel both matters. From the analysis, respondents were inclined to refuse strategies that compromise personal privacy, preferences and hygiene. Strategies with much inconveniences, cutback and discrimination in service delivery as well as cost saving approach toward the organization solely are advised to be avoided. Interestingly, respondents were unaffected by eco-friendly guestroom design. They were willing to tolerate reuse, recycle and environmental campaigns together with green gadgets implementation. Overall, there is a strong association between environmental strategies applied in hotels and guests' future behavioral intention. It was found that respondents were willing to spread positive Word Of Mouth (WOM) about hotels with environmental initiatives and recommend the establishment to others. Contrarily, few respondents were willing to spend extra to support environmental initiatives in hotels.

Keywords: Environmental strategies, future behavior intention, hotel, perception

1   INTRODUCTION

The ongoing global concern for environmental protection has led to inevitable changes in the way people live. Demand for environmental products and services are becoming more noticeable with no exception in hotel industry. There were studies attempting to capture hotel guests' changing attitude towards protection of environment. Swanger, Benson and Paxson (2009) study conducted in the US emphasized that majority of hotel guests prefer to stay in an environmental strategies practiced hotel. Hence, it is essential for hoteliers to understand the role environmental issues play to encourage guests' purchasing decision.

Creating a balance environmental atmosphere in hotels is crucial; hoteliers must be able to simultaneously keep environmental strategies unobtrusive so that guests do not feel inconvenienced, but make them visible enough to satisfy environmentally-conscious clientele (White, 2010). Past research has proven that unwarranted environmental strategies have caused guests to respond negatively when such policies impacted their experiences (El Dief & Font, 2012). Thus, environmental strategies should not at any time compromise personal

privacy, preferences and hygiene as well as other fundamental individual needs.

According to the conceptual model of service quality, return intention will not occur if poor service is experienced by customers (Parasuraman, Zeithaml & Berry, 1985). Interestingly, similar evidence is seen in various studies within service and hospitality field (Abdul Aziz, Saiful Bakhtiar, Che Ahmat, Kamaruddin & Ahmad, 2012; Saiful Bakhtiar, Mohd Zahari, Azhar & Kamaruddin, 2012).

Based on the aforementioned contemplation, it is crucial for hoteliers to understand how each environmental strategy applied within their establishments is perceived by current or potential guests. Albeit the fact many that researchers have addressed this issue in the U.S., Europe and several Asian countries, minimal is established in Malaysia and countries with similar geographical, climate and cultural characteristics. Findings from a recent study by Saiful Bakhtiar, Mohd Zahari, Azhar, dan Kamaruddin (2014) have revealed potential guests' perception with several environmental strategies in hotel. In addition, this study extends the earlier research by evaluating the influence of such strategies towards one's future behavior intention.

2 LITERATURE REVIEW

Environmental management in hotel is categorized into the following segments; energy and water conservation, waste reduction, and recycling management (Molina-Azorín, Claver-Cortés, Pereira-Moliner & Tarí, 2009). Meanwhile, the span of environmental management is built around both front and back-of-house operations, organizational system and culture as well as external business relationship (Park, 2009). Environmental efforts in hotels must be planned and exercised with care; hence hoteliers need to strike a balance between creating unobtrusive environmental strategies to all guests while keeping it visible to satisfy environmentally-conscious clientele. Thus, understanding both guests' perception toward existing environmental practices as well as its influence towards their future behavior intention are considered to be part of the indicators that could determine one's business success.

2.1 Environmental strategies in hotel

The hotel industry in general consumes considerable amount of electricity and fossil fuel energy in various operational areas (Park, 2009). The usage of energy efficient bulbs can save about 25 percent of energy costs, depending on the scale of the hotel. In addition other countries also implement numerous energy management initiatives including application of renewable energy program such as the use of wind, solar, and run-of river power; adopting automated energy control system through key-card; replacing incandescent bulbs with fluorescent lighting; installing energy-efficient equipment in major operational areas such as laundry; using digital thermostats to control guestroom energy consumption; and installing occupancy sensor to automatically turn off lights when guests leave their room (Kasimu, Zaiton & Hassan, 2012).

An average hotel consumes about 209 gallons of water per occupied room daily (Brodsky, 2005). The high cost consideration has driven many hotels to implement water conservation program. For example, Marriott International introduced 'linen reuse program' by encouraging guests to reuse linens and towels during their stay in the hotel (Marriott International, 2007) and this program has been implemented by other hotel brands worldwide (Kasimu, et al., 2012).

Waste reduction and recycling are part of environmental initiatives in hotel. The main purpose of waste management is to reduce quantity and toxicity of waste produced that mostly derived from papers and food waste (Wastecare Corporation, 2013). The following are other commonly applied waste and recycling management strategies; placing recycling bins in all front and back-of-house areas; purchasing used or recycled-content products; adopting a donation program for leftover guest amenities, old furniture and appliances; and composting organic kitchen waste (Park, 2009). Other strategies include using refillable amenity dispensers; providing reusable items such as cloth napkins, glass cups, ceramic dished with food and beverage service; grinding guest soaps to use as laundry detergent for hotel uniforms; purchasing food items and cleaning chemicals in bulk containers; and recovering used cooking oil and food waste are also practiced in hotels.

Besides the three major areas of environmental management mentioned earlier, there are several other practices designed to encourage eco-friendly hotels. Kasimu et al. (2012) suggested utilizing eco-friendly cleaners and detergent, purchasing locally produced supplies, buying from environmentally responsible suppliers, establishing a formal channel to cooperate NGOs, establishing guests education programs, supporting local communities to enhance local environment, and incorporating environmental reporting in corporate control systems.

2.2 Future behavior intention

A future behavioral intention is a person's subjective possibility to perform possible behavior in the future and key to profitability to the most of the service provider (Rise, Sheeran & Hukkelberg, 2010). Customers tend to demonstrate positive behavior intention when they perceive high levels of value from consumption experiences (Ha & Jang, 2010). Willingness to recommend to others and stronger intentions to revisit a place are another result from positive perceived service quality (Chen & Peng, 2012). Other studies within the spectrum of hospitality and tourism for example, Godes (2013) reported similar outcome. Another manifestation of behavioral intention can be seen through word of mouth or WOM (Buttle, 1998). WOM is essential in hospitality marketing and considered the most consistent and powerful tool deriving through personal expression (Ferguson, Paulin & Bergeron, 2010), while the impact is long-term (Keller, 2007). Valls Andrade and Arribas (2011) demystified that demand becomes greater when a product or service is rare; consequently user is willing to spend more to obtain it and becoming less sensitive to the price.

3 METHODOLOGY

In accessing respondents' perception and future behavioral intention toward environmental practices in hotels, a quantitative approach was employed through self-administered questionnaire. Following simple screening questions, local travelers at public airport terminals in central region of Malaysia were chosen for data collection using convenience sampling technique. A total of 288 valid responds were gathered and analyzed using descriptive and correlation analysis.

4 FINDINGS AND DISCUSSION

4.1 *Demographic profile of the respondents*

Prior to commencing for further statistical analysis, all data were checked for irregularities, missing item, or unrealistic responses. Frequency test was used against all demographic variables (such as gender, age, and education background). Of the 288 respondents, 51.1 percent were male as opposed to 48.9 percent female. Moving to education profile, over 90 percent of respondents own at least undergraduate academic background; hence it is supposed that they may have some basic knowledge about environmental strategies and able to respond well to the subject matter. Meanwhile, a review of respondents' occupation revealed that majority (45.1 percent) were corporate professionals, followed by 24.7 percent working in the government sector, while the remaining were either self-employed or students. Because majority of the respondents were classified as professionals, it is assumed that those individuals could be business travelers and have various hotel stay experiences.

4.2 *Perception on energy management program in hotels*

In response to the objective of the study, mean score was used to indicate respondents' perception on three categories of environmental practices: (1) energy, (2) water management as well as (3) waste reduction and environmental protection campaign presented in Table 1. From the table, majority of respondents were inclined towards agreeing with energy management program in hotels. The use of key card to enable electricity in guestroom received the highest score ($M = 4.27$, $SD = 0.740$). Contrarily, installation of automated (motion sensor) air-conditioning system along hallway got the lowest score. Being at a country with warm tropical climate all year round; the idea of pursuing automated air-conditioning system may be perceived as operationally ineffective. Although other energy management strategies were viewed positively, standard deviation value for all items was high; between 0.740 and 0.938. This shows that not all of them were willing to accept suggested strategies

Table 1. Mean score for perception of environmental strategies in hotels.

Energy Management			
Variables	n	(M)	SD
Air-conditioning activated by motion sensor along hallways	288	3.92	0.938
Use key-card to enable electricity usage in guestrooms	288	4.27	0.740
Adopt "smoke-free policy" within hotel to save energy	288	4.13	0.905
Installation of reflective coating at guestroom's window	288	4.24	0.770
Tubular skylight room design	288	4.15	0.768
Water Management			
Not changing bed sheet	288	3.53	1.162
Not changing bath linen	288	3.38	1.189
Water efficient devices in all toilets/bathrooms	288	3.98	0.877
Reuse waste-water in hotel's toilet flushing system	288	3.91	1.126
Waste Reduction and Environmental Campaign			
Recycle bins in guestroom and public areas	288	4.09	0.803
Using refillable amenities (i.e. soap & shampoo) dispenser	288	3.93	0.819
Heavy environmental awareness promotion program	288	4.09	0.815
Reusable items in food & beverage service	288	4.01	0.855
Complimentary newspapers in public areas only	288	3.97	0.966
Special car park entitlement to hybrid vehicles driver	288	3.77	0.992
Replace hand-towel tissue paper with automated hand dryer	288	4.03	0.867
Brochure via email; printed copy upon request	288	4.05	0.797

Scale: 1 = Strongly Disagree, 2 = Disagree, 3 = Moderate, 4 = Slightly Agree, 5 = Totally Agree.

evenly. As such, smokers might be less enthusiastic with the idea of "smoke free zone" if it runs at the entire guestrooms and all public areas. Interestingly, respondents were not affected by eco-friendly guestroom design; installation of reflective coating and tubular skylight as those strategies were perceived positively.

As compared with energy management, majority gave a moderate score for water management program in hotels. Although the range of mean score was only between 3.38 and 3.98; an in-depth analysis reveals that they were still inclined towards favoring 3 out of 4 strategies listed. Again, standard deviation for all items was high between 0.877 and 1.189 indicating varies feedback among respondents. Installation of water efficient devices in toilet was perceived most acceptable ($M = 3.98$) compared to others. This may be because the strategy has no significant changes to the way such devices being used, thus there is none credible reason to refuse such strategy. Similar argument may be applied to the application of waste water in hotel's flushing system. On the other hand, fewer

Table 2. Descriptive analysis for future behavior intention.

Future Behavior Intention			
Variables	*n*	*(M)*	*SD*
Recommend to others	288	4.06	0.763
Positive word of mouth	288	4.07	0.822
Willingness to purchase/repurchase	288	3.80	0.737
Willingness to spend more	288	3.66	0.771

Scale: 1 = Strongly Disagree, 2 = Disagree, 3 = Moderate, 4 = Slightly Agree, 5 = Totally Agree.

Table 3. Pearson correlation test.

Environmental Strategies to Future Behavior Intention		
Variables	*r*	Sig.(2-tailed)
Energy management program and future behavioural intention	0.537	.000
Water management program and future behavioural intention	0.264	.000
Waste reduction management and environmental protection campaign and future behavioural intention	0.559	.000
All hotels' environmental strategies and future behavioural intention	0.543	.000

**Correlation is significant at the .01 level (2-tailed).

respondents agreeing to not changing bed sheet during a guest's stay ($M = 3.53$). Even fewer agreed to not changing bath linens during their stay in hotel. This may indicate that respondents were unlikely to compromise cleanliness and personal hygiene for meeting environmental goals.

A glance through the figures about respondents' perception on waste reduction management and environmental protection campaign in hotels revealed that majority respondents were inclined towards agreeing with all strategies listed. They favored the idea of having recycle bin in guestroom and public areas ($M = 4.09$); heavy promotion about environmental awareness does not seem to bother them ($M = 4.09$); willing to accept hotel's brochure and promotion via email while printed is made upon request ($M = 4.05$); replacing hand-towel tissue paper with automated hand dryer was perceived positive ($M = 4.03$), even using reusable items in food and beverage service was deemed acceptable ($M = 4.01$). From another standpoint, the strategy to offer complimentary newspapers only at public areas as opposed to delivering it to each guestroom were minimally favorable ($M = 3.97$); maybe respondents perceived the ideas as a cost saving approach benefiting the organization rather than the guests or they were make inconvenienced as a result for supporting such strategy. In similar gesture, using refillable amenities dispenser ($M = 3.93$); as well as offering special parking entitlement to guest arriving with hybrid vehicles ($M = 3.77$) were less favorable strategies. Based on the three final items described, maybe respondents were unlikely to accept a cutback and discrimination of service delivery to uphold hotel's environmental program.

4.3 *Future behavior intention*

Descriptive result for future behavior intention is detailed in Table 2; overall score between 3.66 and 4.07 indicating respondents' minimal support with all statements.

In general, majority respondents were willing to spread positive word of mouth and recommend hotels implementing environmental initiatives to others. Conversely, they were minimally attracted to repurchase environmental friendly services ($M = 3.80$) and even to spend extra ($M = 3.66$) to support such practices.

4.4 *Relationship between environmental strategies towards future behavior intention*

Table 3 revealed a significant correlation result between energy management program and future behavioral intention (sig. value = .001 < .05) while $r = 0.537$ indicates strong positive relationship coefficient of correlation value. Comparable result was achieved for the relationship between waste reduction/environmental protection campaign and future behavioral intention; with slight differences in values. Interestingly, the correlation between water management program and future behavioral intention was significant; yet coefficient correlation value was minimal with $r = 0.264$, $p < .001$ indicating a moderate relationship. In general, there was a significant correlation between all hotels' environmental strategies and future behavioral intention was achieved with strong relationship of coefficient correlation, $r = 0.543$, $p < .001$.

5 CONCLUSION

Overall, this study revealed that not all participating respondents were willing to accept suggested strategies evenly. Therefore, any environmental management strategy implemented in hotel should be planned with care especially if it involves limitation in individual's activity or preference. Based on the findings, it was clear that respondents were inclined to refuse strategies that compromise personal privacy, preferences and hygiene as well as other fundamental individual needs. Respondents were unlikely to support if the strategy is perceived causing much inconvenience to them. Furthermore, any cutback and discrimination in service delivery made in order to uphold hotel's environmental program should be avoided; hence it is crucial for hoteliers to understand that the strategy implemented must not be viewed as a cost saving approach benefiting the organization. Interestingly, respondents were unaffected by eco-friendly guestroom design. They were willing to tolerate reuse, recycle and other environmental campaigns in addition to green gadgets implementation positively.

From the correlation perspective, there was an overall significant relationship between hotel environmental strategies and guests' future behavior intention. This result is identical with finding from previous study by Manaktola and Jauhari (2007) that analyzed guests' attitude and behavioral intention towards environmental practices in hotels. Within the same research, respondents also revealed that they were less willing to spend more for environmental initiatives despite their positive attitude manifestation. Again, this is similar with current findings; low mean score was achieved for respondents' willingness to spend for environmental initiatives in hotel. It is also worth to mention about finding from similar study by Lee Hsu, Han, and Kim (2010) that test the correlation of identical variables. Their study revealed that the value and quality of environmental attributes will enhance the overall image of a hotel thus contribute to the positive guest future behavior. Based from the above contemplation, it is vital for hoteliers to understand and assess guests' perception toward existing environmental strategies in use; so unconvincing strategies may be improved to encourage guests' positive behavioral intention.

ACKNOWLEDGEMENTS

This research is funded by Universiti Teknologi MARA through Research Intensive Faculty Grant 600-RMI DANA 5/3/RIF (105/2012).

REFERENCES

Abdul Aziz, A., Saiful Bakhtiar, M.F., Che Ahmat, N.H., Kamaruddin, M.S.Y. & Ahmad, N.A. (2012). The usage of ICT applications in 5 star hotels in Kuala Lumpur, Malaysia. In A. Zainal, S.M. Radzi, R. Hashim, C.Thamby Chik & R. Abu (Eds.), *Current Issues in Hospitality and Tourism Research and Innovations: Proceedings Conference of the International Hospitality and Tourism Conference (IHTC 2012), Kuala Lumpur, Malaysia, 3–5 Sept 2012 (pp 231–235)*. London: Taylor & Francis Group.

Brodsky, S. (2005). Water conservation crucial to energy savings. *Hotel & Motel Management, 220*(13), 12.

Buttle, F.A. (1998). Word of mouth: Understanding and managing referral marketing. *Journal of strategic marketing, 6*(3), 241–254.

Chen, A. & Peng, N. (2012). Green hotel knowledge and tourists' staying behavior. *Annals of Tourism Research, 39*(4), 2211–2216.

El Dief, M. & Font, X. (2012). Determinants of Environmental Management in the Red Sea Hotels Personal and Organizational Values and Contextual Variables. *Journal of Hospitality & Tourism Research, 36*(1), 115–137.

Ferguson, R.J., Paulin, M. & Bergeron, J. (2010). Customer sociability and the total service experience: Antecedents of positive word-of-mouth intentions. *Journal of Service Management, 21*(1), 25–44.

Godes, D. (2013). Product policy in markets with word-of-mouth communication. *Available at SSRN 2275617*.

Ha, J. & Jang, S. (2010). Perceived values, satisfaction, and behavioral intentions: The role of familiarity in Korean restaurants. *International Journal of Hospitality Management, 29*(1), 2–13.

Kasimu, A., Zaiton, S. & Hassan, H. (2012). Hotels involvement in sustainable tourism practices in Klang Valley, Malaysia.

Keller, E. (2007). Unleashing the power of word of mouth: Creating brand advocacy to drive growth. *Journal of Advertising Research, 47*(4), 448.

Lee, J.-S., Hsu, L.-T., Han, H. & Kim, Y. (2010). Understanding how consumers view green hotels: How a hotel's green image can influence behavioural intentions. *Journal of Sustainable Tourism, 18*(7), 901–914.

Manaktola, K. & Jauhari, V. (2007). Exploring consumer attitude and behaviour towards green practices in the lodging industry in India. *International Journal of Contemporary Hospitality Management, 19*(5), 364–377.

Marriott International. (2007). Marriott helps "Clean up the world", from World Wide Web http://www.marriott.com/news/detail.mi?arrArticle=160342

Molina-Azorín, J.F., Claver-Cortés, E., Pereira-Moliner, J. & Tarí, J.J. (2009). Environmental practices and firm performance: An empirical analysis in the Spanish hotel industry. *Journal of Cleaner Production, 17*(5), 516–524.

Parasuraman, A., Zeithaml, V.A. & Berry, L.L. (1985). A conceptual model of service quality and its implications for future research. *Journal of marketing, 49*(4), 41–50.

Park, J. (2009). *The relationship between top managers 'environmental attitudes and environmental management in hotel companies*. Doctoral dissertation, Virginia Polytechnic Institute and State University. Retrieved from http://scholar.google.com.my/scholar?hl=en&as_sdt=–0,5&q=The+Relationship+between+Top+Managers%E2%80%99+environmental+Attitudes+and+Environmental+Management+in+Hotel+Companies

Rise, J., Sheeran, P. & Hukkelberg, S. (2010). The role of self-identity in the theory of planned behavior: A meta-analysis. *Journal of Applied Social Psychology, 40*(5), 1085–1105.

Saiful Bakhtiar, M.F., Mohd Zahari, M.S., Azhar, A.R. & Kamaruddin, M.S.Y. (2012). Hypermarket fresh foods' at-tributes towards customer satisfaction. In A. Zainal, S.M. Radzi, R. Hashim, C.Thamby Chik & R. Abu (Eds.), *Current Issues in Hospitality and Tourism Research and Innovations: Proceedings Conference of* the *International Hospitality and Tourism Postgraduate Conference (IHTPC 2013), Kuala Lumpur, Malaysia, 2–3 Sept 2013 (pp 141–145)*. London: Taylor & Francis.

Saiful Bakhtiar, M.F., Mohd Zahari, M.S., Azhar, A.R. & Kamaruddin, M.S.Y. (2014). A snapshot of environmental strategies in hotels: Locals' perspective in Malaysia. In N. Sumarjan, M.S. Mohd Zahari, S. Mohd Radzi, Z. Mohi, M.H. Mohd Hanafiah, M.F. Saiful Bakhtiar & A. Zainal (Eds.), *Hospitality and Tourism Synergizing Creativity and Innovation in Research: Proceedings Conference of the International Hospitality and Tourism Postgraduate Conference (IHTPC 2013), Shah Alam, Malaysia, 2–3 Sept 2013 (pp 437–441)*. London: Taylor & Francis.

Swanger, N.A., Benson, L.S. & Paxson, C. (2009). Ecotourism projects: Impact on environmental attitudes in introductory hospitality courses. *Journal of Hospitality & Tourism Education, 21*(2), 24–29.

Valls, J.F., Andrade, M.J. & Arribas, R. (2011). Consumer atti-tudes towards brands in times of great price sensitivity. *Innovative Marketing, 7*(2), 60–70.

Wastecare Corporation. (2013). Waste reduction and recycling tips for hotels, resorts and motels, from http://www.wastecare.com/Articles/Waste_Reduction_Re-cycling_Tips_Hotels.htm.

White, M.C. (2010). For hotels, eco-friendly ideas await a friendlier economy. *The New York Times*, from http://www.nytimes.com/2010/08017/business/7green.html.

Theory and Practice in Hospitality and Tourism Research – Radzi et al. (Eds)
© 2015 Taylor & Francis Group, London, ISBN 978-1-138-02706-0

Comparative research of Pro-Environmental Behavior (PEBs) in daily life and tourism circumstances

H.J. Kim & N.J. Kim
Hanyang University, Seoul, Republic of Korea

K.M. Yoo
Howon University, Jeollabuk-do, Republic of Korea

ABSTRACT: The objectives of the present research were to examine whether there is an inconsistency in pro-environmental behavior between daily life and tourism circumstance and compare the differences in such behaviors from the broad perspectives of the costs and benefits. As a national survey, 1,003 samples obtained through quota sampling, stratified by sex, age, administrative district and education level. The results are shown in two ways. First, there were significant differences in respondents' PEB between two settings. Respondents reported a far higher level of PEB in daily life setting than tourism's. Second, while the economic cost and benefit are key criteria for PEB in daily life, the level of inconvenience is the key criteria for selecting of PEB in tourism setting. Theoretical implications of inconsistencies in PEB are discussed, and offer guidelines about how to promote tourists' engagement in PEB in tourism settings as a managerial implications.

Keywords: Pro-Environmental Behavior, cost-benefit, rational choice theory, climate change

1 INTRODUCTION

According to the data of UNWTO (2007), tourism industry is estimated to contribute approximately 5 percent of the total carbon dioxide emission in the world and found to be responsible for the emission of the greenhouse gas to certain proportion. The worldwide global warming arising from the emission of the greenhouse gas is demanding the tourism industry to pursue changes into Pro-Environmental Behaviors (PEB) in order to alleviate the global warming.

However, while the researches on the PEB in daily life are very active, there are only limited researches in the area of tourism with even less researches on the relationship between the PEBS of these two different contexts (Barr, Shaw, Coles & Prillwitz, 2010; Hares, Dickinson & Wilkes, 2010). According to findings of prior studies, PEB is different according to the various settings. In particular, Given that the tourism settings are very different from the environment of the houses and that the results of several researches (Barr, et al., 2010; Miao & Wei, 2013) are reporting that it is much more difficult to behave environmentally under the tourism circumstances than daily life, it is important to compare the PEB under these two different contexts. When the focus is placed on the reasons for differences in these two settings, the research can discover the important variables that can promote PEB of tourists.

Numerous researches attempted to explain the PEB by setting variables such as the environmental attitude, value and belief as the independent variables. However, the relationship between the environmental attitude and the behavior is reported to be rather weak. Researches on the areas of tourism also failed to explain the reasons for the difficulties in performing the PEB in tourism because they applied same variables that are generally used to explain the PEB in other context.

Stern (2000) pointed out that the environmental behavior has the tendency to comply with some of the non-environmental motives, that is, to save money or the preference for personal comforts. However, majority of the researches failed to provide sufficient theoretical explanations for the reasons for the inconsistency of the environmental attitude and PEB, since they overlooked non-environmental motives.

This Study attempts to explain PEB from cost and benefit perspectives that the existing researches have neglected. More specifically, this Study empirically verify whether the pro-environmental behaviors in the daily life carry over to the tourism context, and compare the differences in such behaviors under the two circumstances from the broad perspectives of the costs and benefits.

2 PEB FROM THE PERSPECTIVE OF COST-BENEFITS

According to the rational choice theory, people assess the results of their behavior and act to attain the best results (Ajzen & Fishbein, 1980). When viewed from the perspective of cost-benefits, it can be forecasted that people will behave environmentally when the results of the behavior generates greater benefits than the cost.

Many researchers including tourism area have applied the planned behavior theory to explain PEB (Han, Hsu & Sheu, 2010), but themes of the PEB under the tourism circumstance require different approach. It cannot be approached simply from the perspective of economic cost-benefits as they are definitively different from the behaviors in daily life. Individuals under the tourism circumstance put the goals such as freedom and pleasure ahead of everything.

Then, how are the PEB of human beings decided? From the perspective of cost-benefits, the extent of the PEB will be increased with lower cost and the higher the benefits as a result of the behavior. The cost here is not limited to the economic cost, from broader meaning of the term, it encompasses time and efforts individuals put in. Such fundamentals are very faithful to the basic principles of the rational choice theory, because the rational choice signifies making the most satisfactory selection with the least cost. What is important here is that the satisfactory results under the daily life and the tourism circumstances may be extremely different. Miao and Wei (2013) conducted research on the differences in the PEB and the key motivations under the home and hotel environment. After having confirmed that the level of PEB under the hotel environment is lower than that under the home environment first, they verified the differences in the motivations under these two settings. As the result, while the normative motivation under the home environment was found to be the key motivation, the motivation for pleasure was the key motivation under the hotel environment. This means that tourists pursue values that are quite different from those in daily life.

Some of the researches explained the pro-environmental behavior from the cost-benefits perspective based on rational choice theory. Diekmann and Preisendörfer (2003) presented the 'low-cost hypothesis', asserting that the environmental concern imparts influence on the behavior most under the circumstances that incur low cost and induce no inconveniences to individuals. That is, the lower the cost pressure of the circumstance, the easier it is for the individual to convert to the behavior that concords with their attitudes. This signifies that if the cost of the PEB is high, the positive attitude towards the environment does not necessary result in pro-environmental behaviors.

Such principle can be inferred from the results of various prior researches (Hares, et al., 2010). Hares et al. (2010) conducted research on the awareness of the effect of tourism on climate change and the behavior of using airplane through the focus group interview, and the respondents of their research report that they will not be reducing the use of airplane while displaying high level of recycling activities. The authors interpreted that while the recycling activities accompanied by low cost and inconveniences are practiced well, the use of airplanes would not be reduced because of the high level of cost and inconveniences (Hares, et al., 2010). The finding of the study by Stern and Aronson (1984, p. 285) on energy consumption asserted that psychological variables such as attitude and individual norms impart greater influence on the energy conservation behaviors that are less costly and easy to perform. In these researches, typical energy conservation behavior that are easy to perform include 'lowering the room temperature', 'lowering the temperature of the boiler at home' and 'turning the lights off'. In addition, behavioral decision that include single key investments such as 'thermal insulation in the walls and the ceiling', and 'installation of new heating system' is determined mainly by the economic factors to be considered, while the attitude factors impart very little influence. Dunlap and Scarce (1991, p. 657) argued that the most popular behaviors are those that require the minimal efforts and personal costs.

On the basis of these discussion, it is reasonable to infer that the pro-environmental behavior under the tourism circumstance incurs greater cost (economic and inconvenience) with less benefits acquired than those under the daily life circumstance. In addition, from the perspective of cost and benefit, this study infers that the criteria for the selecting of the pro-environmental behavior under the daily life and tourism circumstances would be different. Since the best selection that people are pursuing under these two environments are fundamentally different.

Proposition 1: Pro-environmental behaviors are practiced to greater extent under the daily life circumstance than that under the tourism circumstance.

Proposition 2: The criteria of cost-benefits for selecting of the pro-environmental behavior are different between two circumstances.

1. The economic cost and benefit are key criteria for PEB in daily life.
2. In tourism setting, the extent of inconvenience is the key criteria for selecting of PEB.

3 METHODOLOGY

3.1 Sample and data collection

The population of this study is adults in South Korea over the age of 20 and the 1,003 national samples were collected. Quota sampling method was applied with the pro-panel composed on the basis of the administrative zones, gender, and age, and population ratio for each of the educational levels in accordance with the data from the population census by the National Statics Office. Specialized research institute was used for the collection of data. Data were collected over the period from June 18 to July 2, 2013 with 1,003 valid respondents collected.

3.2 Measurement items

Pro-environmental behavior items were composed by applying the items measured in previous studies (McKercher & Prideaux, 2011; McKercher, Prideaux, Cheung & Law, 2010; Miao & Wei, 2013). The measurements were expressed using slightly different descriptions by reflecting two different circumstances. And some items that may not coincide two circumstances were also included. For example, while the daily life circumstance included items such as 'replace with high efficiency lighting device', items such as 'use accommodation facility with environment-friendly certification' were included for the tourism circumstance. All the items were measured by 5 Likert scale with the range of 1 (not at all) ~ 5 points (all the time).

4 RESULTS

4.1 Characteristics of the respondents

It was confirmed that the sample were collected quite appropriately in accordance of the pro-panel of this study.

4.2 Results of principal component factor analysis

4.2.1 PEB in daily life
Four factors were extracted following a principal component factor analysis. The average values for each factor were in the order of the 'Life-type PEB (3.99)', 'Pro-environmental transportation mode (3.87)', 'Perseverance-type PEB (3.77)' and 'Planned PEB (3.52)'.

4.2.2 PEB in tourism circumstance
As the results of the factor analysis, three factors were deduced, which were in the order of 'Purchasing environment-friendly products' at 3.47,

followed by 'Energy conservation and reducing waste (3.38)' and 'Pro-environmental transportation modes (3.21)'.

4.3 Testing the research propositions

4.3.1 Testing the research proposition 1
In order to compare PEB in daily life and tourism settings, total average of the PEB in each environment were analyzed through the Paired t-test. As the result, there was significant difference between the PEB under the two circumstances. That is, the respondents in the daily life setting reported significantly higher levels of PEB than when they under the tourism setting. Then, Paired t-test was conducted for the 10 similar PEB in two different circumstances. As a result, significant differences were found in all items with the exception of the 'purchasing food in season'. As anticipated, PEB were practiced to greater extent under the daily life circumstance than under the tourism one.

4.3.2 Testing the research proposition 2
The second research proposition of this Study is that the cost-benefits criteria for selecting of the pro-environmental behavior are different between the daily life and tourism circumstances. Specifically, confirm whether the 'economic cost and benefits' for the daily life circumstance and the 'inconvenience level' for the tourism circumstance are the key criteria that determines the extent of the PEB.

To determine the cost of each pro-environmental behavior, several researches deemed that the cost of the behavior would be lower if the extent of the pro-environmental behaviors by reported respondents is greater. However, such method has the limitation of not being able to assess the cost and benefits specifically. In this Study, the economic cost and benefits, inconvenience level of the pro-environmental behaviors were measured with 10 Ph.D. degree holders in tourism who understand the detailed costs and benefits. These 2 costs and 1 benefit were measured by the 5 Likert scale with score ranging from 1 (very low) to 5 (very high). The results of the average costs and benefit for each item organized for pro-environmental behavior factors conducted above are as follows. The most PEBs in daily life circumstance accompany lower economic cost, higher economic benefits being returned, and low level of inconvenience in comparison to the tourism circumstance.

Then, graph was used to confirm which cost-benefits is the greatest criterion for the extent of the PEB under two circumstances. The average values of the economic cost, economic benefits, and inconvenience level were indicated on the Y-axis. And the life-type, Pro-environmental transporta-

Table 1. Demographic profile of the respondents
($n = 1,003$).

	Percentage (%)
Gender	
Male	48.9
Female	51.1
Age	
20-29	19.3
30-39	20.8
40-49	22.4
50-59	17.9
60 or over	20.4
Education background	
High school	58.0
College degree	39.6
Graduate degree	2.4

Table 2. Factor loading of PEB items in daily life.

Item	Factor loading
Factor 1: Life-type PEB	
Turning off the light	.741
Separate collection	.741
Reduce food waste	.597
Order suitably when eating out	.578
Remove water from food waste	.518
Select high energy efficiency product	.494
Factor 2: Planned PEB	
Installation water conservation type tap	.787
Replace with high efficiency lighting	.693
Cleaning air conditioner filter	.620
Purchasing food in season and vicinity	.522
Reduce the frequency of using laundry	.427
Factor 3: Perseverance-type PEB	
Maintain the room temperature (summer)	.800
Maintain the room temperature (winter)	.744
Reduce the hours of electrical use	.695
Factor 4: Pro-environmental transportation mode	
Use public transportation	.869
Walk or use bicycle	.750

Table 3. Factor loading of PEB items in tourism.

Item	Factor loading
Factor 1: Energy conservation and reducing waste	
Reduce the hours of electrical use	.801
Water saving in accommodation	.788
Water saving in destination	.781
Turning off the light in accommodation	.707
Reduce carbon emission in destination	.667
Select high energy efficiency product	.494
Reduce food waste	.659
Maintain the room temperature	.611
Reduce replace of towel	.582
Separate collection	.514
Factor 2: Pro-environmental transportation modes	
Use public transportation	.804
Walk or use bicycle	.758
Use low carbon transportation mode	.731
Use eco-friendly accommodation	.565
Car pool with travel companions	.540
Factor 3: Purchasing environment-friendly products	
Purchasing food in season and vicinity	.789
Purchasing product helping local resident	.693
Purchasing high energy efficiency product	.659
Reduce disposal product	.645

Table 4. Paired T-test of PEB between daily life and
tourism circumstance.

	Daily life	Tourism	t-value
Total average	3.79	3.35	30.457*
Turn off light	4.14	3.60	17.724*
Energy efficiency product	4.06	3.35	23.968*
Reduce using laundry	3.68	3.24	12.680*
Maintain room temperature	3.71	3.19	18.780*
Reduce electrical use	3.91	3.38	17.409*
Purchasing food in season	3.62	3.62	-.074
Use public transportation	3.84	3.28	17.086*
Walk or use bicycle	3.90	3.30	18.266*
Separate collection	4.27	3.58	23.310*
Reduce food waste	3.85	3.45	14.024*

*$p < .001$.

tion mode, perseverance-type and planned PEB indicated on the X-axis in the order of the higher performance level of average value (see Figure 1). As the results, pro-environmental behaviors are performed to greater extent with lower economic cost and higher economic benefits under daily life circumstances. On the other hand, the direction of inconvenience level does not coincide with the tendency of the level of the PEB.

The Figure 2 illustrates the cost-benefits of pro-environmental behaviors under tourism circumstance. Similar to the (see Figure 1), X-axis is marked with 'purchasing environment-friendly products', 'Energy conservation and reducing waste' and 'Pro-environmental transportation mode' in the order of the higher level of PEB from left to right. Y-axis indicates the average values

Table 5. Paired T-test of PEB between daily life and
tourism circumstance.

	Economic cost	Economic benefit	Inconvenience level
PEB in daily life			
Life type PEB	1	3.83	2.41
Planned PEB	2.6	3.3	2.9
Perseverance type PEB	1	3.67	3.67
Pro-environmental transportation mode	1	3.75	3.5
PEB in tourism circumstance			
Energy conservation and reducing waste	1.56	1.11	3.5
Pro-environmental transportation mode	2.9	2.7	4.1
Purchasing environment friendly products	3.63	2.63	2.63

352

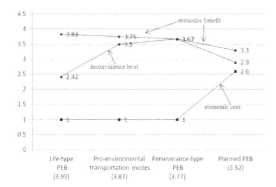

Figure 1.   The cost-benefit of PEB in daily life.

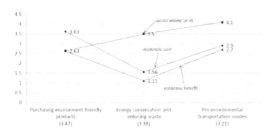

Figure 2.   The cost-benefit of PEB in Tourism.

for each of the items of cost-benefits. There was a trend of lower performance of pro-environmental behavior with greater 'inconvenience level' under tourism circumstance. The economic cost and benefits displayed tendency that is not in concordance with the extent of the pro-environmental behavior.

5   CONCLUSION

This study provides empirical evidence that people behave less environmentally in tourism circumstance than daily life. While majority of researches focused on variables such as the environmental attitude, value and knowledge, the present study attempt to explain the reason for lack of practicing pro-environmental behaviors under the tourism context by applying the perspective of cost-benefits. Results suggest that people value different cost and benefit criteria according to their situation.

This study offers the following implications to the marketers and managers in tourism industry. According to the result of the current study, all the pro-environmental behaviors under the tourism circumstance incur higher cost but offer lower benefits in comparison to those under the daily life circumstance. Therefore, modifying the context of the circumstances in which the pro-environmental

behaviors actually manifest rather than putting in efforts to modify the attitude, knowledge and value of tourists could be a more effective method.

As the results of this study, the most effective method must be to reduce the inconvenience experienced by behaving environmentally in the tourist destinations. This implies the need to apply environmental designs within the tourist destinations. Environmental designs are those that can reduce the stress and inconveniences of practicing environment-friendly behaviors and can be applied through the methods of adopting new facilities or devices. This had been proven in the preceding researches that reported reduction in the behavior of indiscriminant throwing out of waste by furnishing waste cans with more beautiful design and by placing them in more convenient locations (Geller, Brasted & Mann, 1979).

In addition, Consequence strategies could be an effective means of inducing PEB. It was found that people behave environmentally under daily life circumstances because of the low cost with high benefits. Therefore, including strategies such as discounts and coupons at the tourist destinations could be a means of increasing the benefits of PEB in order to promote such behaviors. Future studies in tourism area need to further expand the examination of PEB from the perspective of cost-benefits by transcending the domain of existing researches conducted from the perspective of ethics.

REFERENCES

Ajzen, I. & Fishbein, M. (1980). *Understanding attitudes and predicting social behaviour.* Englewood Cliffs, NJ: Prentice Hall.
Barr, S., Shaw, G., Coles, T. & Prillwitz, J. (2010). A holiday is a holiday: Practicing sustainability, home and away. *Journal of Transport Geography, 18*(3), 474–481.
Diekmann, A. & Preisendörfer, P. (2003). Green and Greenback The Behavioral Effects of Environmental Attitudes in Low-Cost and High-Cost Situations. *Rationality and Society, 15*(4), 441–472.
Dunlap, R.E. & Scarce, R. (1991). Poll trends: Environmental problems and protection. *Public Opinion Quarterly, 55*(4), 651.
Geller, E.S., Brasted, W.S. & Mann, M.F. (1979). Waste receptacle designs as interventions for litter control. *Journal of Environmental Systems, 9*(2), 145–160.
Han, H., Hsu, L.-T.J. & Sheu, C. (2010). Application of the Theory of Planned Behavior to green hotel choice: Testing the effect of environmental friendly activities. *Tourism management, 31*(3), 325–334.
Hares, A., Dickinson, J. & Wilkes, K. (2010). Climate change and the air travel decisions of UK tourists. *Journal of Transport Geography, 18*(3), 466–473.
McKercher, B. & Prideaux, B. (2011). Are tourism impacts low on personal environmental agendas? *Journal of Sustainable Tourism, 19*(3), 325–345.

McKercher, B., Prideaux, B., Cheung, C. & Law, R. (2010). Achieving voluntary reductions in the carbon footprint of tourism and climate change. *Journal of Sustainable Tourism, 18*(3), 297–317.

Miao, L. & Wei, W. (2013). Consumers' pro-environmental behavior and the underlying motivations: A comparison between household and hotel settings. *International Journal of Hospitality Management, 32,* 102–112.

Stern, P.C. (2000). New environmental theories: toward a coherent theory of environmentally significant behavior. *Journal of social issues, 56*(3), 407–424.

Stern, P.C. & Aronson, E. (1984). Energy use: The human dimension.

Theory and Practice in Hospitality and Tourism Research – Radzi et al. (Eds)
© *2015 Taylor & Francis Group, London, ISBN 978-1-138-02706-0*

The mediating effect of attitude between customer responsiveness and actual usage of "Touch 'n Go" card in foodservice outlets

S. Zurena & M.S.M. Zahari
Faculty of Hotel and Tourism Management, Universiti Teknologi MARA, Shah Alam, Malaysia

O. Ida-Rosmini
Faculty of Computer and Mathematical Sciences, Universiti Teknologi MARA, N. Sembilan, Malaysia

ABSTRACT: Restaurants affiliated with the foodservice industry plays an essential character in increasing revenue and in actual fact respond to customer attitudes. Empirically, this paper evaluates the extent to which customer's attitude mediates the relationship between customer responsiveness and actual usage of Malaysian "Touch' n Go" card in foodservice outlets. INDIRECT which employs bootstrap analysis was used to assess the mediating effect of attitude between customer responsiveness dimensions and actual usage. The results signify that 400 participated customer's revealed attitude mediates the relationship since the mean indirect effects of the bootstrap analysis for all customer responsiveness dimensions were positive and significant with a 95% bias corrected and accelerated confidence interval excluding zero. Obviously the amount of technology acceptance is reflected in the strength of attitude or intention toward using the technology. Given the absence of work in this area, this study is significant in terms of better design of marketing strategies, and in achieving an understanding of what customers need and expect when it comes to smart card usage in the restaurants.

Keywords: Customer responsiveness, actual usage, attitude, foodservice, "Touch' n Go"

1 INTRODUCTION

The use of electronic payments for the customers to conduct financial transactions, as well as the use of more narrowly defined "electronic money" has gained increasing attention in a variety of forums, along with government, business, corporate, and personal (Singh, 2009). Electronic money mechanisms were invented to replicate cash over the Internet (Anand & Madhavan, 2000; Dani & Krishna, 2001). Clearly, knowledge of the actual and potential demand is critical to assess the most likely scenario of the impact of electronic payment technology.

Bringing the evolution of technology in everyday life, there is a trend seen in many countries to move towards electronic payment (e-payment). Malaysia is not leaving behind on the world's modern technology, therefore, embarking on the e-payment via Touch' n Go cards.

It is interesting to note that today that the Touch 'n Go's growth from limited deployment for transport payment in the initial stages to payments for various alternative forms of business transactions for instance highway, transit, parking and retail, including restaurants (Euromonitor International, 2011).

Looking into restaurant industry, with stronger purchasing power and a higher standard of living today, Malaysia provides a significant pool of customers who hunt to modernize their payment lifestyles. The days of being weighed down by loose change are over. As Touch' n Go is a card for everyone, anytime and anywhere, it is the perfect alternative to cash with just a simple 'touch' the customer payment terminals and good to go. Despite the fact that all these initiatives are aimed in creating a cashless Malaysian society, there are many concerns as to the successful achievement of this target whereby it is acknowledged that there is still some opposition with regard to the perception and attitudes towards the use of Touch' n Go card in the restaurants. Consequently, it is interesting to evaluate the extent to which customer's attitude mediates the relationship between customer responsiveness and actual usage of Malaysian "Touch' n Go" card in restaurant outlets. The significance of this empirical study hinges upon the fact that it especially contributes knowledge in enhanced customer's model of behavior in Malaysian restaurant settings.

2 LITERATURE REVIEW

2.1 Smart cards

A smart card is a mechanism designed to store and, in most cases, the process data. They are very mobile (credit card size) and resilient (Lu, 2007), which makes them suitable for many applications involving identification, authorization, and payment. Since the creation of the cards in the 1970s, the technology has evolved, and many features have been further to the original concept (Shelfer & Procaccino, 2002).

Smart card technology has begun to penetrate the market, and attempts are being made to use them in many parts of business activity. Attoh-Okine and Shen (1995) reminded that Germans were using the smart cards for health care since 1992, and it has been adopted in France in support of postal, telephone, and telegraph services. In actual fact, smart cards (contactless or otherwise) are used in many sectors: health care, banking, government, human resources and, of course, transportation. This card is used to identify the store, biometrics, photographs, fingerprints, medical data, the results of DNA, religious affiliation, banking data, transport fares, and other individual data.

Transit agencies are keen in this type of technology, and a number of them are now using smart cards to substitute the traditional magnetic cards, or ticket, as a viable payment alternative (Blythe, 2004). It is seen as a safe method of user authentication and fare payment (Trépanier, Barj, Dufour & Poilpré, 2004). Eventually, it also makes the driver's job easier, because he no longer has to collect fares. Moreover, smart cards improve the quality of data, providing a more modern transit, and offer new prospects for innovative and flexible fare structuring (Dempsey, 2008). Although smart cards are mostly used in Europe and Asia, they have currently been implemented in Canada, especially in Québec where the entire major transit operator is equipped with smart card technology. The United States government has expressed the intention to establish world-class transit system in the country and the smart cards are expected to play a role in achieving this goal (United States Department of Transportation, 2010).

2.2 Touch 'n Go cards

Touch 'n Go cards is owned by CIMB Group Holdings Berhad, MTD Capital Berhad and PLUS Expressways Berhad. A smartcard that contains electronic cash is one of card-based e-money application and one of the alternative payments widely used in the transportation sector. Initially offering a contactless payment card for use on all the highways as the sole electronic toll collection operator, Touch 'n Go continually expanding its functionality (Euromonitor International, 2011). This card offered a wide range of applications including in the public transport, theme parks, parking lots, medical and health, movie, and restaurants as well. The cards may either be linked to one's credit or bank card with automatic reload facilities to a stipulated maximum level once the amount on the cards fall to a predetermined minimum level. The cards may also be reloaded at some of the toll booths or banks ATM terminals.

This card offers more convenient than currency in many settings; some consumers are likely to find cards physically easier to handle than coins and paper notes. The total number of card in circulation as at end 2006 was 6.5 million, and as at December 2007, the number had increased to 8.1 million (Bank Negara Malaysia, 2011).

2.3 Attitudes

Technology Acceptance Model or TAM is based on the Theory of Reasoned Action (TRA) which asserts that an attitude toward a specific behavior and subjective norm have an impact on behavioral intention, which in turn determines the behavior displayed (Ajzen & Fishbein, 1980; Fishbein & Ajzen, 1975). According to TAM, the amount of technology acceptance is reflected in the strength of attitude or intention toward using the technology (Davis, Bagozzi & Warshaw, 1989). An attitude can be characterized as a person's negative or positive evaluation of performing the target behavior. It has been suggested that attitude can be split into its cognitive and affective components (Fishbein, 1967). Intentions are assumed to capture the motivational factors that influence a behavior, and thus indicate how hard people are willing to try or to what extent they are planning to make an effort in order to perform the behavior (Ajzen & Fishbein, 1980).

In the context of technological adoption, attitude toward the act refers to the individual's assessment of the desirability of adopting and using the product. It is distinct from attitude toward the product and one's behavioral intention towards the product. However, in the consumer context, attitude has been found to have a direct and positive effect on intention to adopt technology (Bruner II & Kumar, 2005; Dabholkar & Bagozzi, 2002).

3 METHODOLOGY

Questionnaires were distributed to 8 participated restaurants which are listed in the Touch 'n Go directory for Food and Beverages outlets as to achieve an adequate response rate. Surveys amounting of

560 were distributed to the customers in the participated restaurants. Of the total number of respondents, resulting in an effective sample of 400 usable completed surveys and thus met the preliminary screening requirements, meanwhile representing for an overall response rate of 71.4 percent. Prior to the analysis, the questionnaires were screened for missing data, response patterns that indicated low discrimination ability and logic inconsistencies. A total of 160 questionnaires were removed from the data analysis since the questionnaires were incomplete and not suitable for use (Tabachnick & Fidell, 2007) leaving a sample of 400 respondents for the analysis. The data were analyzed using IBM SPSS Statistics version 21 for Windows together with INDIRECT macro to assess the mediating effect (Preacher & Hayes, 2008). The descriptive statistics performed in support of background information of the customers.

4 RESULTS AND DISCUSSION

The demographic characteristics of the respondents are displayed in Table 1. All information is presented in actual figures and percentages to facilitate interpretation.

The majority of the respondents were Malays (76%), followed by Indians (8.5%), Chinese (5.5%) and others (10%). More than half of the respondents were females (61%), while the remaining (39%) were males. As for age, the majority of the respondents were between 19 to 30 years old (66%). Merely 21.5 percent of the respondents had a first-degree qualification. The other 78.5 percent had Diploma, Certificate or other qualifications. The majority of the respondents work in other various occupations (39.5%). A total of 19 percent among the non-executive, executive (18.5%), professional (16.5%), homemaker (4%), senior management (1.5%), and retired (1%). A majority of the respondents earned between RM1001 to RM2000 per month (56%) and followed by RM2001 to RM3000 per month (19.8%).

In measuring the internal consistency of Customer Responsiveness dimensions (Perceived Usefulness, Perceived Ease of Use, Perceived Trust) and Attitude, Cronbach's alpha were used. Table 2 indicates that all dimensions in Customer Responsiveness had good internal consistency (0.835–0.880) while Attitude had excellent internal consistency of 0.913.

Table 3 summarized the bootstrap estimates based on 5,000 bootstrap samples. The mean indirect effects from the bootstrap analysis were positive for all customer responsiveness dimensions and significant (PU: $a \times b$=0.12, PEOU: $a \times b$=0.13, PT: $a \times b$ =0.26), with a 95% bias corrected and accelerated confidence interval excluding zero (PU: .04

Table 1. Distribution of respondents according to background information.

Background Information	Frequency	Percentage
Gender		
Male	156	39.0
Female	244	61.0
Age		
19 - 30	264	66.0
31 - 40	86	21.5
41 - 50	26	6.5
50 and above	24	6.0
Ethnic origin		
Malay	304	76.0
Chinese	22	5.5
Indian	34	8.5
Others	40	10.0
Highest level of education		
Secondary school	86	21.5
Diploma	118	29.5
Degree	152	38.0
Masters	22	5.5
PhD	4	1.0
Professional cert.	4	1.0
Others	14	3.5
Monthly income		
RM1001 - RM2000	224	56.0
RM2001 - RM3000	79	19.8
RM3001 - RM4000	34	8.5
RM4001 - RM5000	21	5.3
RM5001 - RM6000	22	5.5
RM6001 - RM7000	6	1.5
More than RM7000	14	3.5
Occupation		
Homemaker	16	4.0
Non-executive	76	19.0
Executive	74	18.5
Senior management	6	1.5
Professionals	66	16.5
Retired	4	1.0

Table 2. Reliability for *Customer Responsiveness* dimensions and *Attitude*.

Dimension	N	Cronbach's Alpha
Customer Responsiveness		
Perceived Usefulness	5	0.847
Perceived Ease of Use	5	0.835
Perceived Trust	5	0.880
Attitude	5	0.913

to 0.26, PEOU: .03 to 0.30, PT:.05, 0.47). In the indirect path, the respondents who indicated high agreement level of Perceived Usefulness, Perceived Ease of Use and Perceived Trust, were more likely to have better attitude (PU: a=0.21, PEOU: a=0.23, PT: a=0.45); while holding constant for *Customer Responsiveness*, the respondents who had better attitude, will also have a high level of actual usage (b=0.57). Based on the significant mean indirect

Table 3. Mediation on the effect of *Customer Responsiveness* on *Actual Usage* through *Attitude (5,000 bootstrap samples, N = 400)*.

Independent Variable (IV)	Perceived Usefulness (PU)	Perceived Ease of Use (PEOU)	Perceived Trust (PT)
Mediating Variable (M)	Attitude		
Dependent Variable (DV)	Actual Usage		
Effect of IV on M (a)	0.2134***	0.2305**	0.4471***
Effect of M on DV (b)	0.5720***		
Direct Effect (c')	0.2868	-0.2918	-0.1367
Indirect Effect (a*b)	0.1221	0.1319	0.2558
95% CI for a*b	0.0352, 0.2581	0.0274, 0.2952	0.051, 0.4664
Total Effect (c)	0.4088*	-0.1599	0.1191

$^*p < .05,^{**} p < .01,^{***} p < .001$.

effects, it can be concluded Attitude mediates the relationship between all Customer Responsiveness dimensions and Actual Usage.

5 CONCLUSION

Smartcards have been implemented to facilitate restaurant services in light of different needs. The results of the empirical findings contributed theoretically in which generalizes the theory of reasoned action and the technology acceptance model in undertaking research on customer perceptions of emerging payment technologies in the restaurant. Equally important, the findings have implications for management to implement advanced payment technology to ease restaurant services. Meanwhile, it is undoubtedly may also shed light on the importance of attitude in determining the customer's actual usage behavioral to use the payment technology and adjacent to that it promises substantial improvement to the restaurant industry as a whole. Hence, understanding customer attitudes towards the use of Touch 'n Go card, had better outcome in enhanced model of behavior, which is gradually more significant to researchers concentrated on the acceptance of individual behavior. Customers need to be continuously educated and motivated to change their payment habits by promoting the benefits of e-payments. This will eventually pave the way for the acceptance of e-payments on a national scale.

REFERENCES

Ajzen, I. & Fishbein, M. (1980). *Understanding attitudes and predicting social behaviour*. Englewood Cliffs, NJ: Prentice Hall.

Anand, R.S. & Madhavan, C.V. (2000). An online, transferable e-cash payment system *Progress in Cryptology—INDOCRYPT 2000* (pp. 93–103): Springer.

Attoh-Okine, N. & Shen, L.D. (1995). *Security issues of emerging smart cards fare collection application in mass transit*. Paper presented at the Vehicle Navigation and Information Systems Conference, 1995. Proceedings. In conjunction with the Pacific Rim TransTech Conference. 6th International VNIS.'A Ride into the Future'.

Bank Negara Malaysia. (2011). *Central Bank of Malaysia Financial Sector Blueprint 2011–2020*. Retrieved from www.bnm.gov.my.

Blythe, P.T. (2004). Improving public transport ticketing through smart cards. *Proceedings of the ICE-Municipal Engineer, 157*(1), 47–54.

Bruner II, G.C. & Kumar, A. (2005). Explaining consumer acceptance of handheld Internet devices. *Journal of Business Research, 58*(5), 553–558.

Dabholkar, P.A. & Bagozzi, R.P. (2002). An attitudinal model of technology-based self-service: moderating effects of consumer traits and situational factors. *Journal of the Academy of Marketing Science, 30*(3), 184–201.

Dani, A.R. & Krishna, P.R. (2001). An E-check framework for electronic payment systems in the web based environment *Electronic Commerce and Web Technologies* (pp. 91–100): Springer.

Davis, F.D., Bagozzi, R.P. & Warshaw, P.R. (1989). User acceptance of computer technology: a comparison of two theoretical models. *Management science, 35*(8), 982–1003.

Dempsey, P.S. (2008). Privacy issues with the use of smart cards. *TCRP Legal Research Digest*(25).

Euromonitor International. (2011). Consumer foodservice in Malaysia. Retrieved December 15, 2013 http://www.euromonitor.com/Consumer_Foodservie_in_Malaysia

Fishbein, M. (1967). *Attitude, theory and measurement*. New York, NY: John Wiley & Sons.

Fishbein, M. & Ajzen, I. (1975). *Belief, attitude, intention and behavior: An introduction to theory and research*. Addison-Wesley: Reading, MA.

Lu, H.K. (2007). Network smart card review and analysis. *Computer Networks, 51*(9), 2234–2248.

Preacher, K.J. & Hayes, A.F. (2008). Asymptotic and resampling strategies for assessing and comparing indirect effects in multiple mediator models. *Behavior research methods, 40*(3), 879–891.

Shelfer, K.M. & Procaccino, J.D. (2002). Smart card evolution. *Communications of the ACM, 45*(7), 83–88.

Singh, S. (2009). Emergence of payment systems in the age of electronic commerce: The state of art. *Global Journal Business, 2*(2), 17–36.

Tabachnick, B.G. & Fidell, L.S. (2007). *Using multivariate statistics* (5th ed.). Boston: Pearson/Allyn & Bacon.

Trépanier, M., Barj, S., Dufour, C. & Poilpré, R. (2004). *Examen des potentialités d'analyse des données d'un système de paiement par carte à puce en transport urbain (Examination of the Potential Use of Smart Card Fare Collection System in Urban Transportation)*. Paper presented at the Congrès annuel de 2004 de l'Association des transports du Canada, Québec.

United States Department of Transportation. (2010). Obama administration proposes major public transportation policy shift to highlight livability. In J.t. Press Release FTA 01–10, 2010 (Ed.).

Foodservice and food safety

Theory and Practice in Hospitality and Tourism Research – Radzi et al. (Eds)
© 2015 Taylor & Francis Group, London, ISBN 978-1-138-02706-0

Assessing plate waste in public hospital foodservice management

N.A. Zulkiply & C.T. Chik
Faculty of Hotel and Tourism Management, Universiti Teknologi MARA, Shah Alam, Malaysia

ABSTRACT: Food waste is one of the issues that become a global concern nowadays. In health institutions, the most significant food waste concern is on the plate issue, where the plate waste in hospitals is proven to contribute up to 50 percent of total waste stream. High level of plate waste will become an indicator to the level of malnutrition among patients and at the same time will affect financial problem to the management. Therefore, this study will be carried out by taking into consideration of qualitative and quantitative approach through visual estimation observation and semi-structured questionnaire. The future findings will be able to assess plate waste generated in public hospital foodservice in Selangor, and to find out the reasons for the matter as well.

Keywords: Waste, food waste, plate waste, food preference, public hospital

1 INTRODUCTION

Waste management, energy and water consumption were the most related issues in food and beverages industry (Lockwood, Alcott & Pantelidis, 2012). Waste can be categorized into four types which are municipal solid waste, industrial waste, agriculture waste and hazardous waste (United Nations Economic and Social Commission for Asia and the Pacific (UNESCAP), 2011). Municipal solid waste was the most related waste to food service establishment which commonly known as trash or garbage (Environmental Protection Agency (EPA), 2012). Food waste was one of the item included in trash from homes, schools, hospitals and businesses.

Food waste can be defined as any edible foods or leftover from some process, cooked or uncooked food (EPA, 2012). According to Gooch, Felfel, and Marenick (2010), 8 percent from food waste generated in Canada came from foodservice institution which include hotels, restaurants and institution food outlets such as hospitals. Hospitals were one of foodservice institution that generates several types of waste such as food waste and clinical waste.

Food waste was one of the global issues. This issue is not alien in a hospital setting as well. One of the important food waste generated in hospital were plate waste. Plate waste can be defined as proportion of edible portion of food that was discarded by people due to several reasons (Williams & Walton, 2011; Zakiah, Saimy & Maimunah, 2005). Williams and Walton (2011), stated in their study that 50 percent of the total waste generated

in a ward came from food waste. As studied by Zakiah et al. (2005), plate waste is a type of waste commonly reported in Malaysian hospitals and about 17 percent to 67 percent were contributed as plate waste. It was estimated that plate waste in hospital can add up to 30 percent from food costs (Zakiah et al., 2005).

Abdul Hamid, Ahmad, Ibrahim, and Nik Abdul Rahman (2012) stated in their study, about 7.34 million tons of municipal solid waste was generated and had been estimated to increase to 10.9 million tons in year 2020. From the huge amount, 60 percent are from municipal solid waste in a form of food waste. It contributed about 4.40 million tons and projected to increase to 6.54 million tons in 2020 (Abdul Hamid et al., 2012).

Therefore, this research was developed to assess the plate waste generated and investigate the causes that contribute to plate waste in foodservice hospital especially in the public hospitals.

2 LITERATURE REVIEW

2.1 *Public hospital in Malaysia*

Malaysia is a country that has mixed health care system which is public and private sector (Country Report, 2006). Private sector was mostly well-built hospital building in the town areas. Meanwhile, for public health care sector, the Ministry of Health (MOH) become the main health care provider through its broad network of primary care clinics and hospitals ranging from small district hospital without specialist to large hospitals at state capitals providing specialists and subspecialty services.

According to Angelopoulou, Kangis, and Babis (1998), patients in public hospital are more focused on medical and technical nature and do not concern about contextual or environment features of a hospital. Thus the patients in public hospital did not demand on features including food served by the institution. Their concerns are more on the medical services.

2.2 Hospital foodservice

However, Lau and Gregoire (1998) stated in their study that health care is a type of service industry and patients are becoming more discriminating about service quality. Since 30 years ago, with lots of demand from consumers, advances in technology and economic pressure, health care foodservice industry has changed (Assaf, Matawie & Blackman, 2008). In the past, the provision of food to patients was the responsibility of each individual hospital, which had its own kitchen facilities. Food was cooked, plated and served hot to patients. But this system required large number of staff to handle, and it was always a challenge to be working within tight schedules and at the same time staff must achieve high standard of quality (Assaf et al., 2008).

Besides, Engelund, Lassen, and Mikkelsen (2007), found that there are significant changes in food production systems in Danish hospitals since year 1995 to 2003. A change in employees' profile in kitchens also followed the trends. The educational background of employees resulted to an increase number of skilled employees. Plating system changed as well, with higher use of buffets and satellite kitchens and less use of central plating during the period. They also found out that increased used of cook-chill technology may possibly resulted in focusing more on the nutritional status of patients by offering more menus and buffet-style distribution.

2.3 Plate waste

Goonan, Mirosa, and Spence (2014) had stated that foodservice organization especially hospital were the largest producers of food waste. Williams and Walton (2011) referred to plate waste in hospitals as the served food that remains uneaten by patients. The implication of high level of plate waste will become an indicator to the level of malnutrition among patients and will affect financial problem to the management (Williams and Walton, 2011). In their study at 32 hospitals, the results for median plate waste in hospitals are 30 percent higher than other foodservice establishment. They also concluded that, hospitals that used bulk food delivery system have lower level of plate waste as compare to plated meal delivery.

There are several ways to measure plate waste. Comstock, Symington, Chmielinski, and McGuire (1979) in their research had listed out seven ways of data collection method to measure plate waste which are (i) aggregate nonselective plate waste; (ii) garbage analysis; (iii) food preference questionnaires; (iv) self-estimation; (v) individual plate waste; (vi) aggregate selective plate waste and (vii) visual estimation. According to Williams and Walton (2011), plate waste usually measured by weighing food or by visual estimation of the amount of food remaining on the plate. Therefore, this research will use visual estimation method where visual estimation scales were based on portion of the original serving which remained as waste (Comstock et al., 1979).

2.4 Food preferences

Some researchers found that presentation and varieties of food are the primary factor that influenced patient satisfaction towards meal services in hospital (Hartwell & Edwards, 2009; Hwang, Eves & Desombre, 2003). Patients satisfaction towards food served to them will lead to zero plate waste. Previous studies by DeLuco and Cremer (1990); Dubé, Trudeau, and Bélanger (1994); Gregoire (1994); Lau and Gregoire (1998) indicated that several items that are related to food preferences and quality such as meal taste, variety of foods, flavor and temperature of hot food and texture of meat and vegetables have the most influential overall foodservice satisfaction.

Monitoring client (patient) satisfaction with the items that related to food quality appears to be desirable since foodservice aspects is the most influential towards patients satisfaction (Wright, Connelly & Capra, 2006). Additionally, Dhingra, Sazawal, Menon, Dhingra, and Black (2007), Zakiah et al. (2005), Hartwell, Edwards, and Beavis (2007) stated that the reasons for high levels of plate waste can relate to the clinical condition of patients, food and menu issues such as poor food quality, inappropriate portion sizes, and limited menu choice), service issues including difficulty accessing food and complex ordering systems, and environmental factors such as inappropriate meal hour, interruptions, and unpleasant ward surroundings.

3 METHODOLOGY

As stated in research objectives, this study intent to assess the percentage of plate waste produced by using visual estimation method and to find the primary reason of wastage that could occur, therefore this study will apply a mixed methods study approach. Goerres and Prinzen (2012) stated that when the usage of qualitative approach and quantitative approach within one research project were employed the study is considered as mixed method approach. This study will be conducted in Selangor represented by two public hospitals, Hospital Klang and Hospital Selayang with the capacity of 893 and 960 numbers of beds respectively.

The total number of diet patients from two selected hospital has been taken. From the selected number of samples, researcher came out with several characteristics in order to select the respondents which are (i) Malaysian; (ii) adult patients age above 18 years old; (iii) normal diet patient; (iv) able to communicate and (v) admitted to the hospitals for at least two days. Then, systematic sampling technique will be adapted in collecting data process.

The instrumentation will consist of two parts which are visual estimation chart and semi structured questionnaire that are divided into three sections, which include (i) demographic; (ii) reason for plate waste and (iii) open-ended question for future explanation and suggestion.

4 SIGNIFICANCE OF STUDY

The findings of this study are expected to contribute to the extension of knowledge on the influences of food properties towards plate waste generation in public hospitals. By understanding the causes of plate waste by normal diet patients, it will help the organization to form strategies in improving their daily food system and quality of foods.

Moreover, in academic perspectives, the information in this study can be used as a baseline data for future research and create awareness on the need for contract revision for other researchers.

5 CONCLUSION

Plate waste is not good in any society. As mentioned before, plate waste will contribute to malnutrition among patients and it becomes an indicator of successful service for a health institution. Plate waste also will directly affect the amount of food cost in an organization. Therefore this study will provide a specific reason on how plate waste could happen, and it may help the organization to find the best way to reduce it.

ACKNOWLEDGEMENTS

Authors would like to thank Pn. Harizah Mohd Yaacob and Dr. Safina Mohamad from Hospital Selayang and Dr. Pukunan Renganathan from Hospital Tengku Ampuan Rahimah Klang for their support and assistance throughout the study.

REFERENCES

Abdul Hamid, A., Ahmad, A., Ibrahim, M.H. & Nik Abdul Rahman, N.N. (2012). Food Waste Management in Malaysia-Current situation and future management options. *Journal of Industrial Research & Technology, 2*(1), 36–39.

Angelopoulou, P., Kangis, P. & Babis, G. (1998). Private and public medicine: a comparison of quality perceptions. *International Journal of Health Care Quality Assurance, 11*(1), 14–20.

Assaf, A., Matawie, K. & Blackman, D. (2008). Operational performance of health care foodservice systems. *International Journal of Contemporary Hospitality Management, 20*(2), 215–227.

Comstock, E.M., Symington, L.E., Chmielinski, H.E. & McGuire, J.S. (1979). Plate waste in school feeding programs: individual and aggregate measures: DTIC Document.

Country Report. (2006). Malaysia. 4th ASEAN & Japan High Level Officials Meeting on Caring Societies: Support to Vulnerable People in Welfare and Medical Services, 28–31 August 2006.

DeLuco, D. & Cremer, M. (1990). Consumers' perceptions of hospital food and dietary services. *Journal of the American Dietetic Association, 90*(12), 1711.

Dhingra, P., Sazawal, S., Menon, V.P., Dhingra, U. & Black, R.E. (2007). Validation of visual estimation of portion size consumed as a method for estimating food intake by young Indian children. *Journal of health, population, and nutrition, 25*(1), 112.

Dubé, L., Trudeau, E. & Bélanger, M.-C. (1994). Determining the complexity of patient satisfaction with foodservices. *Journal of the American Dietetic Association, 94*(4), 394–401.

Engelund, E.H., Lassen, A. & Mikkelsen, B.E. (2007). The modernization of hospital food service–findings from a longitudinal study of technology trends in Danish hospitals. *Nutrition & Food Science, 37*(2), 90–99.

Environmental Protection Agency (EPA). (2012). Municiple Solid Waste, from www.epa.gov/epawaste/nonhaz/municipal/index.htm

Goerres, A. & Prinzen, K. (2012). Using mixed methods for the analysis of individuals: a review of necessary and sufficient conditions and an application to welfare state attitudes. *Quality & Quantity, 46*(2), 415–450.

Gooch, M., Felfel, A. & Marenick, N. (2010). Food waste in Canada: Opportunities to increase the competitiveness of Canada's agri-food sector, while simultenously improving the environment *Value Chain Management Centre*: George Morris Centre.

Goonan, S., Mirosa, M. & Spence, H. (2014). Getting a Taste for Food Waste: A Mixed Methods Ethnographic Study into Hospital Food Waste before Patient Consumption Conducted at Three New Zealand Foodservice Facilities. *Journal of the Academy of Nutrition and Dietetics, 114*(1), 63–71.

Gregoire, M.B. (1994). Quality of patient meal service in hospitals: delivery of meals by dietary employees vs delivery by nursing employees. *Journal of the American Dietetic Association, 94*(10), 1129–1134.

Hartwell, H. & Edwards, J. (2009). Descriptive menus and branding in hospital foodservice: a pilot study.

International Journal of Contemporary Hospitality Management, 21(7), 906–916.

Hartwell, H.J., Edwards, J.S. & Beavis, J. (2007). Plate versus bulk trolley food service in a hospital: comparison of patients' satisfaction. *Nutrition, 23*(3), 211–218.

Hwang, L.-J.J., Eves, A. & Desombre, T. (2003). Gap analysis of patient meal service perceptions. *International Journal of Health Care Quality Assurance, 16*(3), 143–153.

Lau, C. & Gregoire, M.B. (1998). Quality ratings of a hospital foodservice department by inpatients and postdischarge patients. *Journal of the American Dietetic Association, 98*(11), 1303–1307.

Lockwood, A., Alcott, P. & Pantelidis, I. (2012). *Food and Beverage Management* (5th ed.). New York: Routledge.

United Nations Economic and Social Commission for Asia and the Pacific (UNESCAP). (2011). Integrated Solid Waste Management, from http://www.unescap.org/

Williams, P. & Walton, K. (2011). Plate waste in hospitals and strategies for change. *e-SPEN, the European e-Journal of Clinical Nutrition and Metabolism, 6*(6), 235–241.

Wright, O.R., Connelly, L.B. & Capra, S. (2006). Consumer evaluation of hospital foodservice quality: an empirical investigation. *International Journal of Health Care Quality Assurance, 19*(2), 181–194.

Zakiah, L., Saimy, F. & Maimunah, A. (2005). Plate waste among hospital inpatients. *Malaysian Journal of Public Health Medicine, 5*(2), 19–24.

Theory and Practice in Hospitality and Tourism Research – Radzi et al. (Eds)
© 2015 Taylor & Francis Group, London, ISBN 978-1-138-02706-0

Complaint behavior on too long waiting or service delay: Analysis based on customer genders and occupations

N. Zainol, M.A.A. Bashir & A.R. Ahmad Rozali
Faculty of Hotel and Tourism Management, Universiti Teknologi MARA, Penang, Malaysia

M.S.M. Zahari
Faculty of Hotel and Tourism Management, Universiti Teknologi MARA, Shah Alam, Malaysia

ABSTRACT: Anecdotal evidence suggests that nearly two-thirds of service complaints in restaurants are time related or having waited too long for food to be served and time is in fact one of the major area concerns by restaurant customers. Many researchers noted too long waiting and service delay lead to the complaint and complaint behavior has also associated with customers' socio-demographics. This study further compares the complaints behavior between customer genders and occupation using Gerai as a contextual study setting. Through customers' self-reported experiences, the result revealed both males and females and regardless their occupation appears to be sensitive to service delays or too long waiting or intolerant to it. They are likely to react negatively or to be more aggressive in complaints not only dealing with service delays or too long waiting, but when they get a wrong food order, receive bad quality and too long waiting during the slow period. These findings have given significant implications for the current and potential Gerai operators in positioning themselves in the fast growing trend of food businesses.

Keywords: Gerai, customer, complaint behavior, gender, occupation

1   INTRODUCTION

In order to sustain in today's competitive environment, it is important for restaurant operators to satisfy their customer (Susskind, Kacmar & Borchgrevink, 2007). Restaurants provide services that are perceived in different ways by different customer (Ryu, Lee & Kim, 2012). The customer develops expectations of services that may not always match with what they receive; resultant in customer discontent and potential complaints seems to be inevitable in the restaurant industry (Ngai, Heung, Wong & Chan, 2007). Anecdotal evidence suggests that nearly two-thirds of service complaints in restaurants are time related or having waited too long to be served and time, in fact, is one of the major areas of concern for restaurant customers (Jones & Peppiatt, 1996). Customers in a restaurant who are miserable about their service received, such as long wait for seating or food service may complain about it (Fraser, Mohd Zahari & Othman, 2008). Moreover, customer complaint behavior with the unsatisfactory products or services may be varied (Ngai, et al., 2007).

Customer complaint behavior has attracted (Ngai, et al., 2007). Marketers agree that consumer complaints are constructive sources of information that help them identify sources of dissatisfaction (Ngai, et al., 2007). Consumers who were encouraged to complain reported greater increases in satisfaction and product evaluation compared to consumers who were not explicitly asked to complain. Ekiz and Au (2011) noted that attitudes towards complaining are characterized by the overall effect of "goodness" or "badness" of complaining to the sellers.

Many researchers noted that complaint behavior has a strong relationship with customers' socio-demographics (Chiang, 2007; Low, 2003; Yan, 2005). In the restaurant sector, Zainol Rozali, Usman, and Mohd Zahari (2014) revealed that besides dislike the service delay, the young and middle aged groups are more aggressive and easily get angry if they had to wait long for service beyond expectation compared more tolerable of the older group. Based on the result, this paper further compares the complaints behavior between customer genders and occupations using Gerai as a contextual study setting. The reasons of looking at this type of eating place is that besides other types of restaurants, Gerai to be one of the fastest growing types of eating place in the Malaysia in particular and remain popular as it offering a range of menu prices, particularly with cook to order foods (Ala carte), less expensive compare to full service and medium restaurants and attract frequent visit by majority of low, middle to high income customers. Most of this eating place in operation at dinner time, customer therefore are patronizing for the purpose of dining for their taste bud and filling the stomach.

2 LITERATURE REVIEW

2.1 *Customer complaint behavior*

The literatures on customer complaint behavior have indeed swelled over in the past decade (Ndubisi & Ling, 2007; Volkov, Harker & Harker, 2002). Crie (2003) defined consumer complaint behavior as a process that constitutes a subset of all possible responses to perceive dissatisfaction around a purchase episode, either during consumption or during possession of the goods or services. Thus, customer complaint behavior is known as complaint responses (Singh & Widing, 1991) and as the customer dissatisfaction response style (Singh, 1990). These responses or actions include switching patronage, telling friends and family and complaining to a customer agency. Moreover, action taken by customer is not only to complain to the seller, but also includes warning families and friends, stopping patronage, diverting to mass media, complaining to customer council and complaining by writing a letter to management (Heung & Lam, 2003; Ndubisi & Ling, 2007).

According to Singh and Widing (1991), three types of complaining behavior that can be found when dissatisfaction occurs: (1) Voice responses (seeking redress from the seller or no action); (2) Private responses (word-of-mouth communication); and (3) Third-party responses (implementing legal action). Broadbridge and Marshall (1995) indicated that there are three options in customer complaint behavior (1) To do nothing; (2) To take private action by switching brands or suppliers, boycotting the product or service, or warning family and friends; and (3) To take public action by seeking direct redress from the retailer or manufacturer, bringing legal action, complaining to the media or registering a complaint with a consumer association. In addition, Zeithaml, Bitner and Gremler (2000) identified four types of complainers in the context of service operations namely: (1) Passive: Those customers least likely to take any action, (2) Voicers: Customers who actively complain to the service provider but are less likely to spread negative word-of-mouth, switch patronage or go to third party for redressed of their complaints, (3) Irate: are most likely to complain to friends and relatives and switch over to another service provider and (4) Activists.

Many consumers do not actively complain but a serious problem or bad experience that a consumer could not tolerate could induce the action of complaint. Customer dissatisfied has led to a large-scale economic loss and there is a high probability of being compensated and the customer is likely to voice a complaint to the company or to take public action or private action (Ngai, et al., 2007). Public action refers to the direct complaint actions to the seller or a third party (e.g. consumer agency or government). Private action indicates that the complaint is privately done through negative word-of-mouth communications with family and friends or the decision not to repurchase the products or services again or boycott the products (Ndubisi & Ling, 2007). The public actions that could be taken by consumer included verbally complain to the retailer or manufacturer, write a comment card or complaint letters, write to newspapers or complain to consumer council (Heung & Lam, 2003; Ndubisi & Ling, 2006). For complaint customer, defection is often the last resort after complaint has failed (Kim, Kim, Im & Shin, 2003; Ndubisi & Ling, 2007).

3 METHODOLOGY

For empirical investigation, a quantitative research method using Gerai customer self-reported experience through questionnaire survey was carried out. The survey instrument used was developed and modified after review of literatures of how other researchers empirically measure the constructs. The questionnaire comprises two sections (demographic profile and complaint behavior). Respondents were asked to report their views based on the five-point Likert scale ranging from 1 = Strongly Disagree to 5 = Strongly Agree.

Owing to the large number of Gerai customers in the country only one contextual data collection setting which is Seberang Jaya, Pulau Pinang was chosen. Not to disturbing dining mood customer experiences was measured randomly at popular supermarkets in Seberang Jaya city rather than at the Gerai itself.

The surveys were conducted at three selected shopping complexes, Carrefour, Pacific and Sunway, Seberang Jaya City, Pulau Pinang. Respondents were randomly approached before entering the shopping malls. The respondents were informed that all information provided by them was strictly confidential and no individual would be identified. A total of 332 questionnaires were successfully distributed. The result showed that the instrument and items used were reliable with coefficient alpha value at 0.788.

4 RESULT AND ANALYSIS

4.1 Respondent profile

Male exceeded the female with 65.4 percent (n = 217) against 34.6 percent (n = 115). The majority of respondents were between 18 to 35 years old made up 45.8 percent of the total sample (n = 152) followed by 36 to 45 years old which represented 30.1 percent (n = 100) and 46 years and above represented around 24.1 percent (n = 80). 40.7 percent (n = 134) of respondents were among the students compared to 22.3 percent (n = 74) of government servants, 21.0 percent (n = 71) were self-employed and 16.0 percent (n = 53) the private sector employees.

4.2 Comparison on complaint behavior based customer gender

The independent sample t-test was used to identify significant differences in complaints behavior between customer genders. No statistically significant differences were found on any of 17 items used in the instrument. The result is shown in Table 1.

Looking at the table, both genders clearly reported themselves as somewhat agreeing that they do not care about the price (female, M = 3.54, p = 0.356 and males M = 3.44) and will start grumbling (females M = 4.17, p = .063 and male M = 3.96) if they have to wait long for their ordering food (female M = 3.53, p = 0.423 and male M = 3.63), especially when they dine with their family (female M = 3.73, p = 0.210 and males M = 3.59). Both genders also agreed that they will definitely get angry (females M = 3.63, p = 0.992 and male M = 3.64) if they have to wait too long for their meals, especially for a simple food (females M = 3.90, p = 0.477 and male M = 3.99) and do not get their order as expected time (females M = 3.44, p = 0.303 and male M = 3.55).

Both genders are having the view that they do not mind waiting but will definitely complaint if they get a wrong food order (females M = 3.63, p = .098 and male M = 3.82), receive bad quality of food (M = 3.99 for females, p = 0.687 and males M = 4.04) and to wait long during slow period (females M = 3.63, p = 0.297 and males M = 3.76) thus lose their mood of dining (female M = 4.11, p = 0.826 and male M = 4.13), do not feel like spending (females M = 3.56, p = 0.860 and male M = 3.58) and definitely force demand for compensation (female M = 3.32) and males (M = 3.28, p = 0.732) and might definitely walk out from the gerai (females M = 3.79, p = 0.605 and male M = 3.85).

On the other hand both genders expressed that they will to be more tolerant of delays when they

Table 1. Showing the items from the data collection where statistically significant differences between 'gender' were identified N (Male = 217, Female = 115).

Items	Gen	M	S.D	t-value	Sig 2-tail
I don't care about the price and I will straight away complaint if I have to wait long	M	3.44	.912	-.92	.356
	F	3.54	.901		
I will start grumbling if I have to wait long	M	3.96	.964	1.86	.063
	F	4.17	.954		
I will straight away complaint if I have to wait long for ordering food	M	3.63	1.02	-.80	.423
	F	3.53	.922		
I will complaint for the service delay, particularly when dine with my family	M	3.59	.953	-1.25	.210
	F	3.73	.911		
I will definitely complaint to the service staff if I have to wait long	M	3.64	1.01	.010	.992
	F	3.63	.985		
I will definitely get angry if I have to wait long for a simple food	M	3.99	1.03	-.712	.477
	F	3.90	1.13		
I straight away complaint if I don't get the order as expected time	M	3.55	.942	1.031	.303
	F	3.44	.881		
I do not mind waiting, but I will definitely complaint if I get a wrong food order	M	3.82	.934	-1.660	.098
	F	3.63	1.039		
I will definitely complain to the owner or manager after a long wait if I receive a bad quality of food	M	4.04	.932	.403	.687
	F	3.99	1.06		
I will definitely complain if I have to wait long during slow periods	M	3.76	1.01	-1.044	.297
	F	3.63	1.00		
Items	Gen	M	S.D	t-value	Sig 2-tail
I will lose my dining mood if I have to wait too long	M	4.11	.964	-.220	.826
	F	4.13	.960		
I do not feel like spending after I have been waiting	M	3.58	.974	.176	.860
	F	3.56	.938		
I would definitely demand for compensation if I have to wait long	M	3.28	.912	-.344	.732
	F	3.32	1.08		
I definitely walk out from the gerai if I have to wait too long	M	3.85	1.01	.518	.605
	F	3.79	.966		
I wouldn't complain if I see a lot of customers	M	3.32	1.00	-.303	.762
	F	3.36	900		
Despite too long waiting, I would not complain if I know why	M	3.47	1.02	.113	.910
	F	3.45	1.01		
Despite a long waiting, I will not complaint with an apology	M	365	.927	.095	.925
	F	3.63	.994		

see a lot of customers (female M = 3.36, p = 0.762 and male M = 3.32) and if they know why (female M = 3.45, p = 0.910 and male M = 3.47) and will not complain if they received an apology from the operators (females M = 3.63, p = 0.925 and male M = 3.65).

Table 2. Showing the items from the data collection where statistically significant differences between 'occupations 'were identified.

Items	Occ	n	(M)	S.D	Sig.	Sch
I will straight away complaint if I have to wait long for ordering food	STD SE GS PS	134 71 74 53	3.29 3.47 4.33 3.80	.987 .872 1.01 1.05	.003	GS > STD
I will complaint for the service delay particularly when dine with my family	STD SE GS PS	134 71 74 53	3.57 3.59 4.33 3.58	.844 1.00 1.08 .792	.043	GS > STD
I will definitely complaint if I have to wait long during slow periods	STD SE GS PS	134 71 74 53	3.54 3.93 5.00 4.07	1.00 .990 .951 .985	.000	GS > STD
I straight away complaint if I don't get the order as expected time	STD SE GS PS	134 71 74 53	3.40 3.46 4.17 3.59	.943 .930 .909 .774	.045	GS > STD
I do not mind waiting, but I will definitely complaint if I get a wrong food order	STD SE GS PS	134 71 74 53	3.51 3.59 4.83 3.69	1.00 .950 .914 .952	.001	GS > STD
I will definitely complain to the owner or manager after a long wait if I receive a bad quality of food	STD SE GS PS	134 71 74 53	3.80 4.04 4.33 4.32	.995 1.03 .829 .877	.004	GS > STD
I don't care about the price and I will straight away complaint if I have to wait long	STD SE GS PS	134 71 74 53	3.17 3.54 3.71 3.65	.983 .864 .725 .680	.019	GS > STD

Note: 1. Any significant differences are indicated by being bolded
2. Inter groups differences are based on Scheffé procedure.
3. STD (Student), SE (Self Employment), GS (Government Servant) and PS (Private Sector)

The overall findings indicate there is no convincing evidence to believe that the reaction towards service delay of male differ than female respondents. Both genders have a common reaction and attitudes toward the too long waiting and service delay.

4.3 *Comparison of complaint behavior based customer occupation*

One-way ANOVA and Scheffe post hoc multiple comparison test were applied to identify an item that has significant differences in complaints behavior between customer occupations. Only those items with statistically significant differences are reported in Table 2.

By means of a Scheffé multiple comparison test, most of the differences found between the government servants (GS) and students (STD) compared to self-employed (SE) and private sector employees (PS). The result showed that a service delay is a bad experience and the Government Servants (GS) are found to be more aggressive than the Students (STD). This is evident when Government Servants (GS) strongly reported themselves as somewhat agreeing that they will straight away complaint if they have to wait long for food ordering (M = 4.33, p = .003, compared M = 3.29, M = 3.47 and M = 3.80), especially when dine with their family (M = 4.33, p = .043 opposed to M = 3.57, M = 3.59 and M = 3.58). Most government servants expressed that will definitely complain if they had to wait too long for their meals, especially during slow periods (M = 5.00, p = .000) compared to student (M = 3.54), self-employed M = 3.93) and private sector employees (M = 4.07), if they do not get the order as expected time (M = 4.17, p = .045 opposed to M = 3.40, M = 3.46 and M = 3.59) and get a wrong food order (M = 4.83, p = .001 compared to M = 3.51, M = 3.59 and M = 3.69). Further, they also will definitely complain if they have to wait too long for a simple food (M = 4.67, p = .000 compared to M = 3.66, M = 4.13 and M = 4.26) and don't care about the price (M = 3.71, p = .019, compared M = 3.17, M = 3.54 and M = 3.65).

Despite these differences, the mean scores given by four occupational groups are in the same magnitudes which are mostly agreeing with the statement. These results suggest that although the government servant is more aggressive the other three occupational groups will also be impatient if they have to wait long beyond expected time when dining at the Gerai.

5 IMPLICATIONS AND CONCLUSION

It is evident from this study that both males and females appear to be sensitive to service delays or too long waiting. Both genders do not care about the price but will start grumbling, get angry, especially for a simple food and when they dine with their family if they have to wait too long for the ordering food. Males and females Gerai customers will definitely complain if they get a wrong food order receive bad quality and have to wait long during the slow period. Both genders also share similar feelings that they may lose their mood of dining, do not feel like spending, demand for compensation and walk out from the Gerai in a worst case scenario.

On the occupation, although intolerant with too long waiting or service delay and likely to react negatively by not revisiting, the government servants are found to be more aggressive than the students, private sector employee and the self-employed individuals with regard to complaints.

With these indicators Gerai operators therefore should not overlook these negative perceptions held by the customers, but rather improving their service by delivering more prompt and efficient to the customers and increase staff attitudes to an acceptable level. This approach would help the staff to better understand the importance service delivery in food business operation.

As a conclusion, it is hoped that the recommendation and information flow from this study could facilitate Gerai operators to better understand customer needs, therefore can well position themselves and be more competitive in the fast growing trend of restaurant businesses. Finally, managing customers waiting is becoming an important and critical as the world economy progressively turn into service orientation.

REFERENCES

Broadbridge, A. & Marshall, J. (1995). Consumer complaint behaviour: the case of electrical goods. *International Journal of Retail & Distribution Management, 23*(9), 8–18.

Chiang, C.-C. (2007). *The effect of experiencing outcome versus process service failures on Taiwan and US restaurant customers' service ratings: A case of college students.* (Unpublished Ph.D thesis), Alliant International University, San Diego.

Crie, D. (2003). Consumers' complaint behaviour. Taxonomy, typology and determinants: Towards a unified ontology. *The Journal of Database Marketing & Customer Strategy Management, 11*(1), 60–79.

Ekiz, E.H. & Au, N. (2011). Comparing Chinese and American attitudes towards complaining. *International Journal of Contemporary Hospitality Management, 23*(3), 327–343.

Fraser, R.A., Mohd Zahari, M.S. & Othman, Z. (2008). Customer reaction to service delays in Malaysian ethnic restaurants. *South Asian Journal of Tourism and Heritage, 1*(1), 20–31.

Heung, V.C. & Lam, T. (2003). Customer complaint behaviour towards hotel restaurant services. *International Journal of Contemporary Hospitality Management, 15*(5), 283–289.

Jones, P. & Peppiatt, E. (1996). Managing perceptions of waiting times in service queues. *International Journal of Service Industry Management, 7*(5), 47–61.

Kim, C., Kim, S., Im, S. & Shin, C. (2003). The effect of attitude and perception on consumer complaint intentions. *Journal of Consumer Marketing, 20*(4), 352–371.

Low, B.S. (2003). Gender differences: Biological bases of sex differences. *Cross-Cultural Research, 34*(1), 3–25.

Ndubisi, N.O. & Ling, T.Y. (2006). Complaint behaviour of Malaysian consumers. *Management Research News, 29*(1/2), 65–76.

Ndubisi, N.O. & Ling, T.Y.A. (2007). Evaluating gender differences in the complaint behavior of Malaysian consumers. *Asian Academy of Management Journal of Accounting and Finance, 12*(2), 1–13.

Ngai, E.W., Heung, V.C., Wong, Y. & Chan, F.K. (2007). Consumer complaint behaviour of Asians and non-Asians about hotel services: an empirical analysis. *European Journal of Marketing, 41*(11/12), 1375–1391.

Ryu, K., Lee, H.-R. & Kim, W.G. (2012). The influence of the quality of the physical environment, food, and service on restaurant image, customer perceived value, customer satisfaction, and behavioral intentions. *International Journal of Contemporary Hospitality Management, 24*(2), 200–223.

Singh, J. (1990). Identifying consumer dissatisfaction response styles: An agenda for future research. *European Journal of Marketing, 24*(6), 55–72.

Singh, J. & Widing, R.E. (1991). What occurs once consumers complain? A theoretical model for understanding satisfaction/dissatisfaction outcomes of complaint responses. *European Journal of Marketing, 25*(5), 30–46.

Susskind, A.M., Kacmar, K.M. & Borchgrevink, C.P. (2007). How organizational standards and coworker support improve restaurant service. *Cornell Hotel and Restaurant Administration Quarterly, 48*(4), 370–379.

Volkov, M., Harker, D. & Harker, M. (2002). Complaint behaviour: a study of the differences between complainants about advertising in Australia and the population at large. *Journal of Consumer Marketing, 19*(4), 319–332.

Yan, R.-N. (2005). *Waiting in service environments: Investigating the role of predicted value, wait disconfirmation and providers' actions in consumers' service evaluations.* (Unpublished Ph.D thesis), University of Orizona.

Zainol, N., Rozali, A.R.A., Usman, S. & Mohd Zahari, M.S. (2014). Service delay of cook to order food: How do Malaysia Gerai customers react? In S.N., M.S.M. Zahari, S.M. Radzi, Z. Mohi, M.H.M. Hanafiah, M.F. Saiful Bakhtiar & A. Zainal (Eds.), *Hospitality and Tourism: Synergizing Creativity and Innovation in Research* (pp. 437–441). UiTM Hotel, Shah Alam, Selangor, Malaysia: Taylor & Francis.

Zeithaml, V.A., Bitner, M.J. & Gremler, D.D. (2000). *Services marketing: Integrating customer focus across the firm* (2nd ed.). Boston: McGraw-Hill.

Theory and Practice in Hospitality and Tourism Research – Radzi et al. (Eds)
© 2015 Taylor & Francis Group, London, ISBN 978-1-138-02706-0

The impact of food quality and its attributes on customers' behavioral intention at Malay restaurants

M.A.A. Bashir, N. Zainol & A.M.F. Wahab
Faculty of Hotel and Tourism Management, Universiti Teknologi MARA, Penang, Malaysia

ABSTRACT: The foodservice business today is a dynamic industry. This can be seen from the assortment of foodservice establishments escalating in all sub-sectors of the industry, including restaurants. Food quality appears to be acknowledged as fundamental for the success of restaurants; however, most restaurant quality–related studies focused on atmospherics and service delivery, often neglecting the significance of food itself as the basis of a restaurant. The objectives of this study were to assess the effects of food quality on behavioral intentions and to further examine the contribution of individual food attributes on it. With regression analyses, the results demonstrated overall food quality significantly affects behavioral intentions; on top, food freshness was the utmost contributor to it. Hence, restaurateurs should identify that customer place a high value on freshness of food and this appears to be associated with crispness, juiciness, and aroma of the food.

Keywords: Food quality, food quality attributes, behavioral intentions, Malay restaurants

1 INTRODUCTION

Viewing at the phenomenon of increasing numbers of Malaysians dining out, entrepreneurs have taken this ideal opportunity to open businesses in foodservice industry, particularly restaurant. Subsequently, with increasing competition between restaurants, enticing new customers can no longer promise profits and success, but retaining existing customers is of more importance. Indeed, a competitive environment provides customers with more alternatives to choose from (Haghighi, Dorosti, Rahnama & Hoseinpour, 2012). As a result, the basis to sustainable competitive advantage lies in providing high-quality service that will in turn lead to repeat patronage. Relating to restaurant industry, total foodservice involves both tangible and intangible components. With regard on tangible components, this is the time when food quality plays an important role underneath restaurant business. It has been reported that in the restaurant industry, quality of food is the most critical factor for patrons' behavior (Jin, Lee & Huffman, 2011). Namkung and Jang (2007) as well asserted that quality of food is one of the best means to maximize success in the restaurant business. Despite the importance of food quality in restaurant businesses, according to Namkung and Jang (2007), a critical challenge facing restaurant industry these days is to offer quality food that is not only compelling for the customers but also can be better to business competitors. They further argued that although food quality has been noticed as important, most restaurant quality related studies have focused on atmospherics and service delivery, often overlooking the importance of food itself as the core competency of a restaurant. Therefore, the objectives of this study were to assess the effect of food quality on behavioral intentions and to further examine the contribution of individual food attributes on it.

1.1 Food quality and food quality attributes

Among the vital elements in consumer food perceptions and food choice decisions is the quality (Grunert, 2005). Generally, consumers favor products of high quality including the choice of food that they are consuming. Hence, it is necessary to know consumers' own perceptions of quality as consumers typically will be making purchasing decisions on these beliefs (Rijswijk & Frewer, 2008). Food quality has been commonly accepted as a fundamental element of the overall restaurant experience (Ryu, Lee & Kim, 2012; Ha & Jang, 2010; Namkung & Jang, 2007). According to Peri (2006), food quality is an essential condition to satisfy the needs and expectations of customers.

Despite the importance of food quality in restaurant business, there is no agreement on the individual attributes that represent food quality. Sulek and Hensley (2004) included all food attributes into only one variable, food quality, while Kivela, Inbakaran, and Reece (1999), who designed a model of dining satisfaction and return patronage, saw that food quality had numerous attributes: presentation, tastiness, menu item variety, and temperature. Yet, little has been done with the critical attributes of food quality in relation to customer satisfaction. A comprehensive review of the literature reveals that the general description of food quality among researchers focuses on (a) presentation, (b) variety, (c) healthy options, (d) taste, (e) freshness, and (f) temperature.

Presentation refers to how attractively food is presented and decorated as a tangible cue for customer perception of quality. Kivela et al. (1999) pointed out that the presentation of food is a key food attribute in modeling dining satisfaction and return patronage. In addition, Raajpoot (2002) described food presentation as one of the product and/or service factors in TANGSERV.

Variety entails the number or assortment of different menu items. Restaurants continuously develop new menus to attract diners, and several proactive restaurateurs have created a variety of food and beverage offerings. In previous studies, menu item variety was a crucial attribute of food quality in creating dining satisfaction (Raajpoot, 2002; Kivela et al., 1999).

Healthy options involve offering nutritious and healthy food. Johns and Tyas (1996) stated that healthy food could have a significant effect on customer perceived evaluation of the restaurant experience. Kivela et al. (1999) noted the importance of healthy foods in restaurants and suggested nutritious food as one of the core properties in dining satisfaction and return patronage. The more notable thing is that restaurant customers are increasingly interested in healthy menu items (Sulek & Hensley, 2004).

Taste is viewed as a key attribute in food in the dining experience (Kivela et al., 1999). Many customers have become knowledgeable about food; as a result the taste of food in restaurants has become ever more important (Cortese, 2003). It is thus not surprising that shabby restaurants with gourmet cooking are packed with customers. Thus, taste is usually believed to influence restaurant customer satisfaction and future behavior intentions (Kivela et al., 1999).

Freshness generally refers to the fresh state of food and appears to be correlated to crispness, juiciness, and aroma (Péneau, Hoehn, Roth, Escher & Nuessli, 2006). In previous studies, freshness of food has been cited as a key fundamental quality indication of food (Kivela et al., 1999; Johns & Tyas, 1996).

Temperature is also a sensory aspect of food quality (Kivela et al., 1999; Johns & Tyas, 1996).

According to Delwiche (2004), temperature influenced how the perceived flavor of food was evaluated, work together with other sensory properties such as taste, smell, and sight. Thus, temperature could be considered as one determinant enhancing enjoyment in the food experience.

1.2 Behavioral intentions

Behavioral intention refers to people's beliefs about what they intend to do in a certain situation (Ajzen & Fishbein, 1980). Fishbein and Ajzen (1975) first conceptualized behavioral intention as a substitute indicator of actual behavior. According to the theory of reasoned action from Fishbein and Ajzen, if behavior is volitional, the intention to perform an action correlates highly with the action itself.

Certain behaviors indicate that customers are connected with a company. Specific indicators of favorable post purchase behavioral intentions include saying positive things about the company to others (Boulding, Kalra, Staelin & Zeithaml, 1993), recommending the company or service to others (Reichheld & Sasser, 1990), and remaining loyal to the company (Rust & Zahorik, 1993; LaBarbera & Mazursky, 1983). Customer loyalty is obvious when customers express a preference for one company over others, continue to purchase from it, or increase business with it in the future (Zeithaml, Berry & Parasuraman, 1996). On the contrary, Hirschman (1970) identified conditions under which dissatisfied customers will complain or switch. Several aspects of adverse behaviors include different types of complaining behaviors (complaining to sellers, friends, or external agencies), considering switching to competitors, and decreasing the amount of business with a company (Zeithaml et al., 1996; Fornell & Wernerfelt, 1987).

1.3 Relationship between quality and behavioral intentions

While the connection between service quality and its outcome is neither straightforward nor simple, several researchers offer some evidence that delivery of high service quality positively influences behavioral intentions. For instance, Boulding et al. (1993) found strong links between service quality and behavioral intentions, including positive word of mouth and recommending the company to others. Parasuraman, Zeithaml, and Berry (1994) also reported a positive and significant relationship between customer perceptions of service quality and the willingness to recommend the company.

Although several studies showed that customers are willing to pay more for better service, other researchers stated that the nature and extent of the impact of service quality on customer behavioral intentions is still not clear (Zeithaml et al., 1996). In addition, service is not the only factor that a company sells to customers; additional conceptual and empirical research addressing these issues can develop our understanding of the behavioral consequences of quality features.

2 METHOD

2.1 Instrument development

A questionnaire was developed based on customer perceptions of food quality attributes and behavioral intentions in relation to the restaurant experience. The six selected food quality attributes were presentation, menu item variety, healthy options, taste, freshness, and temperature. The perceived quality of food attributes and behavioral intentions was measured on a 7-point scale ranging from 1 (extremely disagree) to 7 (extremely agree).

Before the questionnaire was finalized, two restaurant managers in full-service restaurants and two academic professionals in the hospitality industry reviewed the questionnaire to ensure content validity. Subsequently, a pilot study was conducted to ensure the reliability of each construct.

The reliability of measurements during pilot study was well above the recommended cutoff of 0.70, indicating internal consistency (Nunnally, 1978).

Care was taken to rephrase terms used in the questions to suit common usage in Malaysia. Feedbacks from respondents were considered and some changes were made before arriving at the final version of the instrument.

2.2 Data gathering procedure

The type of sampling was non-probability sampling. Data were collected based on convenience sampling since the respondents were selected mainly from Malay restaurants' customers in Shah Alam, Selangor. The selected Malay restaurants for the survey were chosen based on several characteristics, which are (1) offers Malay cuisines on the menu, (2) medium price range of RM10 – RM20 per person, and (3) offer full-service to the

customers. The researcher conducted the survey at five selected mid-scale Malay restaurants in Shah Alam area in the month of April 2013 during lunch and dinner time. Restaurant patrons were approached and requested to complete the survey and were informed that their individual responses were anonymous and confidential.

A total of 400 questionnaires were distributed. However, after performing data screening and cleaning procedures, only 330 usable questionnaires were obtained for further data analyses.

3 RESULTS

3.1 Sample profile

Among the respondents, 77.3% (n = 255) of them were in the range of 20 – 29 years old, 21.2% (n = 70) were 30 – 39 years old, while the remaining 1.5% (n = 5) was 40-49 years old. In regards to gender, there were a slightly high proportion of male customers with 56.4% (n = 186) as compared to female customers with 43.6% (n = 144). Based on respondents' ethnicity, a large proportion of 89.4% (n = 295) of the total respondents were Malay, followed by 9.1% (n = 30) was Chinese, and 1.5% (n = 5) was Indian. In terms of respondents' education level, a large proportion with 88.8% (n = 293) undergoes college or university education. This is followed by 7.9% (n = 26) from secondary school education, and the remaining 3.3% (n = 11) did not undergo any formal education. For respondents' reported monthly income, approximately 80.3% (n = 265) of respondents were having monthly income of RM3,000 and below, and the remaining respondents (19.7%, n = 65) were having monthly income of RM3,000 and above.

3.2 The effect of food quality on behavioral intentions

The relationship between food quality and behavioral intentions found to be significant with standardized coefficient of 0.81 (p < .001).

In addition, 65.4% of the variation for behavioral intention was explained by food quality, indicating the substantial effect of food quality on behavioral intentions.

Table 1. Reliability coefficient for each section of the questionnaire.

Sections	No. of Items	Cron Alpha
Section A: Food Quality	18	0.96
Section B: Behavioral Intentions	6	0.96

*Note: No. of respondents = 30.

Table 2. The effect of food quality on satisfaction.

Sections	B	SE B	β
Constant		–0.300	0.203
Food Quality	1.007	0.040	0.808***

*Note: R^2 = 0.654, ***p .001.

Table 3. The effects of food quality attributes on behavioral intentions.

Sections	B	SE B	β
Constant	0.537	0.275	
Food Presentation	0.296	0.043	0.337***
Food Variety	0.034	0.058	0.029
Healthy Options	0.030	0.048	0.033
Food Taste	0.153	0.063	0.126*
Food Freshness	0.325	0.060	0.356***
Food Temperature	0.024	0.050	0.022

*Note: $R^2 = 0.575$, *p .05, ***p .001.

3.3 The contribution of individual food attributes on behavioral intentions

Food quality attributes were able to explain 58% of variation in behavioral intentions. Among all of the attributes, three (food presentation, food taste, and food freshness) made statistically significant contributions and positive relationships with behavioral intentions.

In term of importance, food freshness ($\beta = 0.36$, $p < .001$) made the largest unique contribution to behavioral intentions. Alternatively, food variety, healthy options, and food temperature were discovered to be not significant predictors to behavioral intentions.

4 DISCUSSION AND CONCLUSIONS

Analyses for the relative importance of food attributes revealed that food freshness, food presentation, and food taste were the most significant contributors to customers' behavioral intentions. Restaurateurs should recognize that customer place a high value on fresh state of food and this appears to be related to crispness, juiciness, and aroma of the food. Additionally, it is noted that customer also put high expectation on the visual feature of the food, as it reflects tangible cue for customer perception of quality. Food taste also should not be overlooked because it is viewed as a key attribute in the dining experience. Moreover, customer are getting more knowledgeable about food, and as a result the taste of food in restaurants become ever more important.

Findings revealed that food temperature, food variety, as well healthy options are not focal attributes for customer in determining their behavioral intentions. The findings indicate that managers should pay more attention to popular menu offerings and allocate their resources to improve the quality of those items rather than developing various menu items. It also indicates that Malaysian, especially Malay people, is less concern on healthy options on the menu when visiting restaurants.

When considering the predictive power of food quality as a fundamental element of the restaurant experience, restaurateurs should not underestimate quality food a restaurant has to offer. In conclusion, from a managerial perspective, it might be useful to prioritize resources by focusing on the most important food quality attributes along with key attributes of service and atmospherics in order to sustain in the restaurant business.

5 LIMITATIONS AND FUTURE RESEARCH

This study does have some limitations. Initially, study findings may not be generalized. Data were collected from customers at five mid-scale Malay restaurants in Shah Alam. If the survey were expanded to include more states and countries, the magnitude and direction of the relationships among constructs may be different. Therefore, a more comprehensive sample, considering geographic dispersion, would ensure external validity. Similarly, the results may not directly apply to other segments of restaurant industry. Thus, future studies should target other segments of the restaurant industry.

Food quality was treated as a main independent variable in this study. However, many other factors may influence customer's restaurant experience, so it may be possible to provide deeper insight into the factors those restaurant owners and managers need to stress in their total offering. By considering these aspects, addition of more independent variables such as service and atmospherics may be desirable for future research to assess the relative influence of food quality compared to other factors for behavioral intentions. Moreover, future research could include a set of other variables that are related to one type of restaurant that would not be relevant for other types of restaurants. For instance, if the study were replicated for fast-food restaurants, location, accessibility to the restaurant, and convenience may also be critical in driving customers' behavioral intentions.

REFERENCES

Ajzen, I. & Fishbein, M. (1980). *Understanding attitudes and predicting social behavior.* Englewood Cliffs, NJ: Prentice Hall.
Boulding,W., Kalra, A., Staelin, R. & Zeithaml, V.A. (1993). A dynamic process model of service quality: From expectations to behavioral intentions. *Journal of Marketing Research, 30*(1), 7–27.

Fishbein, M. & Ajzen, I. (1975). *Belief, attitude, intention, and behavior: An introduction to theory and research.* Reading, MA: Addison-Wesley.

Fornell, C. & Wernerfelt, B. (1987). Defensive marketing strategy by customer complaint management: A theoretical analysis. *Journal of Marketing Research, 24*(4), 337–346.

Grunert, K.G. (2005). Food quality and safety: consumer perception and demand. *European Review of Agricultural Economics, 32*(3), 369–391.

Ha, J. & Jang, S.S. (2010). Effects of service quality and food quality: The moderating role of atmospherics in an ethnic restaurant segment. *International Journal of Hospitality Management, 29*(3), 520–529.

Haghighi, M., Dorosti, A., Rahnama, A. & Hoseinpour, A. (2012). Evaluation of factors affecting customer loyalty in the restaurant industry. African Journal of Business Management, 6(14), 5039–5046.

Hirschman, A.O. (1970). *Exit, voice, and loyalty: Responses to decline in firms, organizations, and states.* Cambridge, MA: Harvard University Press.

Jin, N.H., Lee, S.M. & Huffman, L. (2011). What Matter Experiential Value in Casual dining Restaurants?

Johns, N. & Tyas, P. (1996). Investigating the perceived components of the meal experience, using perceptual gap methodology. *Progress in Tourism and Hospitality Research, 2*(1), 15–26.

Kivela, J., Inbakaran, R. & Reece, J. (1999). Consumer research in the restaurant environment, Part 1: a conceptual model of dining satisfaction and return patronage. *International Journal of Contemporary Hospitality Management, 11*(5), 205–222.

LaBarbera, P.A. & Mazursky, D. (1983). A longitudinal assessment of consumer satisfaction/dissatisfaction: The dynamic aspect of the cognitive process. *Journal of Marketing Research, 20*(4), 393–404.

Namkung, Y. & Jang, S. (2007). Does Food Quality Really Matter in Restaurants? Its Impact On Customer Satisfaction and Behavioral Intentions. *Journal of Hospitality & Tourism Research, 31*(3), 387–409. doi: 10.1177/1096348007299924

Nunnally, J.C. (1978). *Psychometric theory.* New York: McGraw-Hill.

Parasuraman, A., Zeithaml, V.A. & Berry, L.L. (1994). Reassessment of expectations as a comparison standard in measuring service quality: implications for further research. *Journal of Marketing,* 111–124.

Peri, C. (2006). The universe of food quality. *Food Quality and Preference, 17*(1), 3–8.

Raajpoot, N.A. (2002). TANGSERV: A multiple item scale for measuring tangible quality in foodservice industry. *Journal of Foodservice Business Research, 5*(2), 109–127.

Reichheld, F.F. & Sasser, W.E., Jr. (1990). Zero defections: Quality comes to service. *Harvard Business Review, 68*(5), 105–111.

Rijswijk, W.V. & Frewer, L.J. (2008). Consumer perceptions of food quality and safety and their relation to traceability. *British Food Journal, 110*(10), 1034–1046.

Rust, R.T. & Zahorik, A.J. (1993). Customer satisfaction, customer retention, and market share. *Journal of Retailing, 69*(2), 193–215.

Ryu, K., Lee, H.R. & Kim, W.G. (2012). The influence of the quality of the physical environment, food, and service on restaurant image, customer perceived value, customer satisfaction, and behavioral intentions. International Journal of Contemporary Hospitality Management, 24(2), 200–223.

Sulek, J.M. & Hensley, R.L. (2004). The Relative Importance of Food, Atmosphere, and Fairness of Wait The Case of a Full-service Restaurant. *Cornell Hotel and Restaurant Administration Quarterly, 45*(3), 235–247.

Zeithaml, V.A., Berry, L.L. & Parasuraman, A. (1996). The behavioral consequences of service quality. *Journal of Marketing, 60*(2), 31–46.

Theory and Practice in Hospitality and Tourism Research – Radzi et al. (Eds)
© 2015 Taylor & Francis Group, London, ISBN 978-1-138-02706-0

Health is wealth: The significance of organic/slow food in the context of Indian people

A.K. Rai

School of Hotel Management, Manipal University, India

ABSTRACT: The word food, according to its ordinary connotation, is applied to any substance which, when taken into the body of an organism, can be used by that organism in the construction of new tissue. Food forms an important source of energy for human beings unfortunately in our culture the habit of taking junk food has become daily routine. Consequently, from children to adults, people face serious health problems. Obesity is one of those health issues which have become a challenge in the current generation. India is one the country in the world which has already been notorious for maximum number of patients suffering from diabetes thus nutritious diet is in need. It is said that the organic or slow food originated in Italy. Slow cooking is all about respecting the food and treatment of ingredients to retain the maximum natural flavors. It is related with organic cooking also as earlier the grains and vegetables were grown naturally without the usage of chemicals and pesticides. In today's world it is known as organic food. Organic food is grown without chemical fertilizers or pesticides, which can be leached from the soil and end up in water supplies. With the advancement of science and hunger for making profit, people started using pesticides and chemicals in order to enhance the productivity of the vegetable crops. On the contrary, there are people who still counting on slow food or organic food in this fast food era. Eating slow food means to have good and hygienic food. This practice should be maintained to reduce the effect of harmful chemicals and pesticides in the life of humans. The present paper discussed about organic food which could make people healthier and protects their lives by keeping them away from the diseases which are harmful to their health.

Keywords: Organic, slow food, fast food, biodiversity, homogeneity, fertilizers, globalization, metabolism

1   INTRODUCTION

Indian food market predominantly dominated by the traditional *dhabas* (known as Indian restaurants on the roadside), potential restaurants in the customer's colony and some restaurants in five star hotels until 1995. Having fast food for example burgers and pizza, was considered an option for eating outside homes. It was not at all synonymous with the American concept of fast food as a quick takeaway bite or a substitute for lunch. However the year 1996 witnessed a drastic change when it is considered to be the year of India's entry into the world food market. International giants such as McDonalds, Kentucky Fried Chicken (KFC), TGI Friday (TGIF), Dominos, and Pizza Hut all bombarded the Indian food market. Before these, UK-based joint called Wimpy's had established its chain in the country in 1990. By year 1996 it had about three to four joints established in Delhi. However, it did not pose much of a threat to because Wimpy's was looked at more of a hangout place rather than eating out with family in India.

It has been the American international giants such as McDonalds, and Pizza Hut, who have targeted their restaurants to the Indian families. Apart from the foreign and Indian fast food chains setting up shop, there is a range of specialty restaurants offering varied fare such as Chinese, Mexican, French, and Italian. These places are full of junk food and offering variety of fast food, offering range of items from burgers and pizzas but targeting Indian culinary senses and encouraging fast food culture in between the Indian families. Food is now supposed to remedy some of the health problems that food itself has produced and to serve as a quick fix for problems such as obesity, malnutrition, cardiovascular and oncological disorders, even sedentary lifestyle and repetitious work.

Prosperity has not produced easy access to and a larger choice of healthier and better food but a surfeit of non-nutritive, expensive, often seductive, cleverly marketed food that has created a scarcity of food that one can eat and enjoy without preplanning or thinking. As a result, a new stratum of experts has emerged, advising people on eating in general, and on eating out in particular. The adverse of this is a widespread decline of self-confidence common citizens, who feel that they, by themselves, cannot choose their food or eat it, either at home or outside, without professional help. These developments have induced subtle changes in the cultural status and meanings of Indian cuisines, taking place mostly outside the range of vision of nutritionists, ethnographers and columnists writing on food or restaurants. Few seem aware that the traditional concerns of ethnography of food—the cooked and the raw, the pure and the polluted, commensality and its absence, the sanctified and the profane—have merged now with a new, more fluid politics of food in countries like India. These politics are radically altering, perhaps for the first time, the relationship and the pecking order among cuisines that may have acknowledged each other's presence for centuries, but never as self-conscious, autonomous, well-bounded culinary traditions.

However, the fast food and junk food are a threat to our culture and culinary senses. Its increase fast food culture in urban area and decrease slow food cooking culture in families, thus, the Childhood obesity is increasingly being observed with the changing lifestyle of families with increased purchasing power, less work out and eating habits of fast food and junk food.

2 THE SLOW FOOD REVOLUTION

The famous multinational fast food joint, KFC has started a franchisee in Guwahati (India) recently. The launch was a major hit for the stall owners. It appears that there was a long queue for the KFC delicacies and many people felt they had scored bonus points for getting anything at all to take back home for their kids. That's the attraction of fast food in today's world. All of this reflects a drastic lifestyle change and of course, fast food is good for the taste buds but what lies beyond the taste and how that food metabolizes in our digestive systems is perhaps the pertinent question. We never saw so many obese kids and adults before. In fact, obesity used to be associated with the Americans, known to be big eaters. According to a research in America "This parallel increase in obesity rates and in the popularity of healthier foods with lower calorie and fat density has been noted in consumer research (Seiders & Petty, 2004) and in health sciences as "the American obesity paradox" (Heini & Weinsier, 1997) a survey done by Bharati, Deshmukh and Garg (2008) in Correlates of overweight and obesity among school going children of Wardha city in the Central India. According to Deshmukh et al. (2006), "All these factors are related with affluence and sedentary lifestyle. Overweight/obesity has classically been the disease of urban area in all age groups. Food in urban area has been replaced by high calorie snacks and junk food". Changed in eating habits have now suddenly become overweight issues, thus, Indians are in a frenzy to lose weight which has made gym business a success. The young generation can blame the McDonald culture but what about us adults? In an age where fast food has virtually taken over our culinary senses, it seems rather odd to be talking about slow food or slow cooking. But that is what precisely Carlo Petrini, an Italian, who started the Slow Food Movement in Bra, Piedmont, North Italy, was in Shillong (India) recently to share his philosophy. Petrini is one of the world's leading social movement leaders who has been promoting through the Slow Food Movement, his philosophy of what food sovereignty is all about.

The whole idea behind the Slow Food Movement is that it should be good (meaning tasty), clean (no chemicals) and fair (no exploitation) food. In 2004, he was named a "European Hero" by Time magazine. In January 2008, he was the only Italian to appear in the list of "50 People who could Save the World," drawn up by the prestigious British newspaper, The Guardian. Petrini has been mobilizing people worldwide about joining the Slow Food Movement because fast food chains promote wrong eating habits and is an imperialistic approach to food. He says fast food joints actually want all of us to eat the same food, cooked in the same way and to forget our indigenous cuisine which is known to be healthy, nutritious and which in turn promotes diversity and originality of food and benefit our farmers.

One of the things that globalization does is to create homogeneity in taste whether that be about fashion, about the brands we pursue and also the food we eat. The Slow Food Movement on the contrary fulfills need of the farming communities to produce endangered species of crops, which because of wrong policies are no longer produced as farmers think they are not viable. From Petrini's demonstration it was amazing to find how the global policy on food is resulting in huge wastage even as a large part of the world suffers from hunger and under-nutrition. The current food system is such that rich countries throw away more than half of the food produced. In Italy, about 4,000 tons of perfectly edible food is thrown away everyday. In the US, about 22,000 tons of food is discarded at food joints across the country and at homes. Obviously, this is not a sustainable food system and those who are worried about the future ought to be worried about this thoughtless consumption pattern which also promotes a culture of selfishness.

3   EFFECT ON EARTH

The slogan that countries need to grow more food has already affected mother Earth to the point of over-exploitation. Growing more food means killing the soil by using more and more fertilizers and pesticides. What happens in the future if the soil is unable to regenerate itself and produce food for the hungry millions? Studies show that 75 per cent of our water resources are used in intensive agriculture. The rest is polluted by chemicals. The present paradigm is bound to collapse unless we reverse the current process of exploitative agriculture. Petrini feels that we all need to take a hard look at the whole framework of agriculture. He exhorts everyone to give due respect to farmers and for governments and policy makers to stop talking down but to listen to the farmers voices for a change. After all, it is the farmer who actually knows how hard it is to produce crops under the present regime.

The underlying philosophy of the Slow Food Movement is respect for the biodiversity of people, of farming practices and crops. What has happened in the world today is that only the robust, genetically modified species of crops have a chance to survive even if they are not necessarily good for human consumption and have a whole range of adverse effects on the human system. But the loss of such small and weak species also means the loss of precious genetic resources. Today, nearly 70 percent of genetic resources of fruits and vegetables are already lost. This makes the earth poorer. Meantime we are putting more and more pressure on the earth by making it productive more food. The introduction of exotic species of crops and fish has resulted in the disappearance of indigenous varieties which used to feed the local population. Today, even the lakes are no longer owned by the community but by governments and multinationals and they are used to produce a single species which is supplied to the rich nations of the world.

4   BENEFITS OF ORGANIC FOOD

Organic foods are foods that are produced using methods of organic farming. That does not involve modern input such as pesticides and chemical fertilizers. Organic foods are also not processed using irradiation, industrial solvents, or chemical food additives. The organic farming movement arose in the 1940s in response to the industrialization of agriculture known as the Green Revolution. Organic food production is a heavily regulated industry, distinct from private gardening. Currently, the European Union, the United States, Canada, Japan, India and many other countries require producers to obtain special certification in order to market food as organic within their borders. In the context of these regulations, organic food is food produced in a way that complies with organic standards set by national governments and international organizations. Some of the benefits related with Organic foods are:-

(a) Free of Genetic Modification like genetic structure had not been engineered to withstand these chemical substances.
(b) Organic farming returns nourishment to the soil, which in turn creates better conditions for crops to thrive during droughts.
(c) Organic food is grown without chemical fertilizers or pesticides, which can be leached from the soil and end up in water supplies.
(d) Organic produce growers have paved the way with innovative research that has created ways to reduce our dependence on pesticides and chemical fertilizers
(e) Organic farmers have been collecting and preserving seeds for Increases Biodiversity as well as reintroducing rare or unusual varieties of fruits and vegetables
(f) Its Protects Family Farms and Rural Communities
(g) Fresher, Better Tasting, There's no argument that fresh food tastes better. Organic food often is fresher because is more perishable and has a shorter shelf life.
(h) Sustainable Seafood organic markets allow us to purchase seafood that is still abundant and fished or farmed in environmentally friendly ways.

These are some benefits of organic food which help us to free from so many diseases like obesity, and heavy weight.

5   FOOD SOVEREIGNTY

It is evident now that the younger generation no longer recognizes the herbs and vegetables that their parents grew up with. The tribes are known to be able to identify edible forest fruits and leaves, all of which had medicinal qualities. Today, with the forests being rapidly depleted, these indigenous species of medicinal plants and vegetables are no longer seen in the local markets. How many kids today would enjoy the dhekia sakh (fern), the lai-patta (mustard leaf) or the pui-sakh (a sort of spinach cooked with fish)? And there are many more leaves that are eaten in our daily diets that we have slowly forgotten about. Support for the inferential arguments can be found in the many studies showing that consumers generalize health claims inappropriately (Balasubramanian & Cole, 2002;

Garretson & Burton, 2000; Keller et al., 2003). The halo effects also apply to restaurant menus. According to Burton, Sheather, and Roberts (2003) found that adding a "heart-healthy" sign on a menu reduced the perceived risk of heart disease when objective nutritional information was absent, even though it was placed next to an objectively unhealthy menu item (lasagna).

First, as far as food is concerned, India can be claim the most diverse society in the world. Not merely that: Indians have borrowed heavily, unashamedly and openly from virtually every corner of the globe. The story of Indian food is often the story of the blatantly exogenous becoming prototypically authentic. Many food items that seem Indian came to India late in the day. Of course, there are always a few ultra-nationalists who claim that these preparations and ingredients are Indian and the foreigners, either dishonestly or out of ignorance, consider them to be originally theirs. Almost invariably such nationalists rely on some old texts where some of the analogues of cauliflowers and mushrooms are mentioned.

Attractive though such theories are, at least Indian vegetarian cuisine would have been devastatingly poorer if potato, tomato, French bean, sweat potato, tapioca, cashew nut, capsicum, maize, rajmah, papaya and, more recently, cheese and cocoa were not made a part of the cook's repertoire. Even more painful for the Indian food nationalist could be the fact that chilli, an inescapable part of Indian cuisine today, came to India from South America. So did pineapple, guava and *chiku*. Peach, pear, cinnamon, blackberry, lychee, cherry, and the ubiquitous tea came from China; cauliflower from Europe; onion from Central Asia. India is known for its spices, but some of its most important spices, including a few that are central not only to cuisine but to indigenous healing traditions, have come from outside. Among them are garlic, turmeric, fenugreek, ginger, cinnamon, and asafetida. Both India's diversity and uniqueness in the matter of food owe their vivacity to a certain cultural openness to the strange and the unknown.

We talk about political sovereignty but we often forget that we are losing our food sovereignty because we are eating what the market decides for us. People must look at the range of the fast food available in the supermarkets today. It makes us crazy even trying to choose our breakfast cereals. And how do we know a certain packaged food is good for us? It is what we read on the package that consists of calorie and nutritive contents. And how can we be sure whether something is good for our biology when we have not eaten it in the best part of our lives? Actually we have all become shopping robots. The moment we enter a supermarket we no longer use reason. In fact, it is our children who tell us what to buy and what not to. And we have become subservient, obliging parents. We have lost control over what we eat and drink. We have lost our food sovereignty. It is time to recover our lost heritage in food and go back to our indigenous food practices because they are healthy and tasty as well. Above all, they promote sustainable farming practices. Thus, on the basis of the above discussion, it can safely be said that in India (and other countries as well if they are taken into account) people are facing serious health diseases due to fast food. It is need of the hour to concentrate on the issues, problems which are dangerous to the growth of human beings and society in general. Therefore, people must be made aware of organic/slow food which makes their family healthy and wealthy.

REFERENCES

Balasubramanian, S.K. & Cole, C. (2002). Consumers' search and use of nutrition information: The challenge and promise of the nutrition labeling and education act. *Journal of marketing, 66*(3), 112–127.

Bharati, D., Deshmukh, P. & Garg, B. (2008). Correlates of overweight & obesity among school going children of Wardha city, Central India. *Indian Journal of Medical Research, 127*(6).

Burton, S., Sheather, S. & Roberts, J. (2003). Reality or perception? The effect of actual and perceived performance on satisfaction and behavioral intention. *Journal of Service Research, 5*(4), 292–302.

Deshmukh, P., Gupta, S., Bharambe, M., Dongre, A., Maliye, C., Kaur, S. & Garg, B. (2006). Nutritional status of adolescents in rural Wardha. *The Indian Journal of Pediatrics, 73*(2), 139–141.

Garretson, J.A. & Burton, S. (2000). Effects of nutrition facts panel values, nutrition claims, and health claims on consumer attitudes, perceptions of disease-related risks, and trust. *Journal of Public Policy & Marketing, 19*(2), 213–227.

Heini, A.F. & Weinsier, R.L. (1997). Divergent trends in obesity and fat intake patterns: the American paradox. *The American journal of medicine, 102*(3), 259–264.

Keller, H.H., Gibbs, A.J., Boudreau, L.D., Goy, R.E., Pattillo, M.S. & Brown, H.M. (2003). Prevention of weight loss in dementia with comprehensive nutritional treatment. *Journal of the American Geriatrics Society, 51*(7), 945–952.

Seiders, K. & Petty, R.D. (2004). Obesity and the role of food marketing: a policy analysis of issues and remedies. *Journal of Public Policy and Marketing, 23*(2), 153–169.

Theory and Practice in Hospitality and Tourism Research – Radzi et al. (Eds)
© *2015 Taylor & Francis Group, London, ISBN 978-1-138-02706-0*

Customer perceptions on Halal food quality towards their revisit intention: A case study on Chinese Muslim restaurants

N.H. Rejab, N.R.A.N. Ruhadi, A. Arsat, J. Jamil & H. Hassan
Faculty of Hotel and Tourism Management, Universiti Teknologi MARA, Shah Alam, Malaysia

ABSTRACT: Global populations of Muslims in the world have increased nowadays. Through that, the demand of foods and beverages with Halal logo were also increased. Apart from that, one of the famous parts of Halal logo and certification is food quality and it seems to be accepted as a fundamental component to satisfy restaurant customers. However, out of 12 Chinese Muslim restaurants in Shah Alam only 2 restaurants officially applied for Halal certification. These statements then raised some important questions among customers especially Muslim. Therefore, this research aimed to examine the relationship between customer perceptions of Halal food quality towards their revisit intention in Chinese Muslim restaurant. Thus, this study was carried out with the quantitative approach through questionnaires to the customer in Chinese Muslim restaurant. The future findings will access customer perceptions on Halal food quality such as, taste, ingredients used and appearance of the food towards their revisit intention in Chinese Muslim restaurant in Shah Alam.

Keywords: Halal, food quality, revisit intention, Chinese Muslim restaurant

1 INTRODUCTION

Islam is a way of life. It comes with comprehensive standard, protocols and guidelines to be followed by all Muslims. One of elements that received particular attention is regarding food and beverage. Thus, careful selection of food and beverages are emphasized in order to ensure the food and beverages are of standard dictated by Islamic law. One of the selection criterion is the food must be Halal. The word Halal comes from the Arabic word that means permitted by the legislation. The Halal concept came from Quran in Surah Al-Baqarah verse 168 mentioned:

> "O, mankind! Eat from the earth which is halal and tayyib and follow not the footsteps of the devil! He is an open enemy for you" (Surah al-Baqarah verse 168).

According to Syed Marzuki, Hall and Ballantine (2012) revealed that Halal has been recognized as a new benchmark for safety, hygiene and quality assurance even by non-Muslims. Jasimah (2000) stated Halal food is something good for the consumers to consume where they have gone through the selection process subjected to Halal and using Halal ingredients. Halal food quality not only covers ingredient but also cover taste and appearance of food.

The importance of having Halal certification cannot be taken for granted. This is a dominant value for those who want to serve in Halal market. One of the industries affected by Halal authenticity is restaurant industry. Mohamed and Backhouse (2014) explained that, in general, Muslim consumers usually seek for the Halal markings or logos and certifications for a sense of security when buying Halal food.

In Malaysia, restaurant owners are encouraged to apply the Halal certificate from Jabatan Kemajuan Islam Malaysia (JAKIM) in order to ensure and guarantee that all the ingredients, process, preparation, hygiene and cleanliness procedures comply with the Halal requirement and consistent with Hazard Analysis Critical Control Point (HACCP) and other quality assurance standards (Aziz & Nyen, 2013).

However, it was reported by the Director of JAKIM, that the department did not give accreditations to many restaurant in Malaysia because it was found that some of the restaurant use illegal ingredient item like alcohol in their food preparation (Halal Media, 2010). This is a worrying discovery when found that Malaysian frequently eat away from home at various level of food service establishments from low to high end across the country. One of the favorite spot to dine by some Muslims in Malaysia is Chinese Muslim restaurant. Therefore, this finding could affect the Chinese Muslim restaurant operation when the demand for their services increased.

Having said that, one interesting fact to divulge is that according to an interview with JAKIM, Saad (2014) explained that, out of 12 Chinese Muslim restaurants in Shah Alam, only 2 restaurants officially applied for Halal certification. This statement then lead to some important questions such as, does the purported Chinese Muslim restaurants really serves Halal food? Why does they purposely claiming serving Halal food? On what ground customers patronize the purported Chinese Muslim restaurant? What are their satisfactions towards Chinese Muslim restaurant? Or in other word, does Halal perception influence revisit intention in Chinese Muslim restaurant? Hence, those questions coupled with scant study done in area of Chinese Muslim restaurant warrant some empirical investigation in particular pertaining to customer perception of food quality that influence customer satisfaction to revisit intention at Chinese Muslim restaurant.

Due to the less implemented Halal logo and certification among ethnic restaurant especially Chinese Muslim restaurant, this study will be advantageous to them to be used as a guideline. Therefore, this research aims to examine the relationship between customer perceptions of Halal food quality towards their revisit intention in Chinese Muslim restaurant in Shah Alam.

2 LITERATURE REVIEW

2.1 *Chinese Muslim restaurant*

According to the interviewed session with several customers of the Chinese Muslim restaurant, they tend to eat at the restaurant because the food is quite healthy as the food prepare by using minimal oil and food serve hot. The differences between Chinese Muslim cuisines with other cuisines are the food is less sour and spicy compared to Thai cuisine and the food also less fat compared to Malay cuisine. The taste has become the attraction for their customers as the food cook in simple method and the children also like to eat Chinese Muslim food.

2.2 *Customer perception on Halal certification*

Based on Mohd Yusoff (2004), Halal certification could be obtained when the food has been verified as nutritious and prepared from permissible ingredients in a clean and hygienic manner. JAKIM is the department that responsible to encourage the food operators to apply Halal certification and ensure that their clients get the certificate (JAKIM, 2010). According to the study done by Siti Mashitoh, Norhayati Rafida and Alina (2013), the level of perception towards Halal certification among

customers in Malaysia is quite discouraging. Syed Marzuki et al. (2012) had stated that food quality is a part of the Halal certification attributes. The study also showed that the important of Halal certification signified highly valued characteristics to promote food and beverages.

2.3 *Food quality*

Grunert (2005) proposed that the food quality attributes will depend on food type and individual's food preference itself because due to the changes of the food attributes' formation over the time and also the changes of consumer's mind. Thus, in order to determine customer's perception on significant attributes of food quality, there is essential to link between customer's quality understandings with quality attributes. Moreover, Peri (2006) also proposed that an essential elements that restaurant must provide in order to fulfill the customer's needs and satisfaction is food quality.

Besides that, there are many researches were done on the food quality attributes such as food taste, appearance of the food and their ingredients used in order to prepare the food. Therefore, this research applied the food quality attributes in way by focusing on the importance of food taste, their ingredient used and food appearance toward their revisit intention to the restaurant.

2.4 *Revisit intention*

In order to revisit at the restaurant, usually customers have their own reason such as pressure from life of work, customers seek for quality, value and desirable environment that can keep them temporary free from stress and relax (Soriano, 2012).

Moreover, an excellent and memorable experience that received by customers from the restaurant will form a favorable intention such as recommending the restaurant to others, spread positive word of mouth or become a loyal customer will ultimately lead to revisit intention (Boulding Karla, Staelin & Zeithaml., 1993; Reichheld & Sasser, 1990).

Customer revisit intention depends on five attributes within a restaurant which include service quality, food quality, ambience quality, first and last impression, and comfort level of the restaurant (Kivela, Inbakaran & Reece, 1999). It is important for the restaurateur to identify factors that will form positive attitude among customers and influence their intention.

3 METHODOLOGY

As targeted in research objective, to examine the relationship between customer perceptions of Halal food quality towards their revisit intention in

Chinese Muslim restaurant the questionnaire development will be utilized. This study employed the use of a quantitative research design. The 7-point Likert scale questionnaires will be given to the target respondents' at Chinese Muslim restaurants in Shah Alam area. This is because Shah Alam has higher population and has many of races.

A cross sectional study will be use where data will be gathered just once at the same point of time. Then, convenience sampling technique will be adapted in collecting data process. Convenience sampling is most often used because it is the best way of getting some basic information quickly and efficiently (Hair Money, Samouel & Page, 2007).

Thus, in order to answer research objectives and research questions this study will use two main methods for data analysis which descriptive statistics and multiple regressions. The minimum number of sample sizes in this study will be between 384–400 total of respondents. In this research, the data that will be collected will be analyzed using the Statistical Package for the Social Science (SPSS) computer program.

4 SIGNIFICANCE OF STUDY

The findings of this study will contribute to the significance in two aspects. Hopefully, this study will be advantageous to foodservice operators or the owner of food establishments' especially Chinese Muslim restaurant as a guideline for them to examine what are the influences of customer perception on Halal food quality towards their satisfaction to revisit in Chinese Muslim restaurant. In the academic perspectives, the originality of this research could contribute to a new body of knowledge especially restaurant owners and extend the body of literature on the current issue and situation that recently happen in foodservice establishment.

5 CONCLUSION

Although this research is still ongoing, but in other studies found that Halal food quality is important in our society, especially for Muslim as it is compulsory to consume Halal food. Halal Industry Development and Corporation (2014) mentioned that Selangor become one of the states that were incorporated in the project of Halal Hub in Malaysia. Nowadays, demands for Chinese Muslim food are increasing. However, Halal food quality has become challenging for Chinese Muslim restaurant operator to retain the Halal certification. Thus, customers will give feedback and get

to satisfy their needs and preferences to revisit at the restaurant.

Hence, with scant study done in the area of Chinese Muslim restaurant warrant some empirical investigation in particular pertaining to customer perception of food quality that influence customer satisfaction towards their revisit intention at Chinese Muslim Restaurant.

ACKNOWLEDGEMENTS

This paper is sponsored by Research Acculturation Grant Scheme (RAGS 2013) Ministry of Education, Malaysia and Universiti Teknologi MARA (UiTM).

REFERENCES

Al-Quran
Aziz,Y.A. & Nyen,V.C., (2013). The Role of Halal Awareness, Halal Certification, and Marketing Components in Determining Halal Purchase Intention Among Non-Muslims in Malaysia: A Structural Equation Modeling Barber, N., and Scarcelli, J.M. (2009).Clean restrooms: how important are they to restaurant consumers?.*Journal of Foodservice*. 20(6):309e320.
Boulding, W., Karla, A., Staelin, R. & Zeithaml, V.A. (1993). A dynamic process model of service quality: From expectation to behavioral intentions. *Journal of Marketing Research, 30*(1), 7–27.
Grunert, K.G. (2005). Food quality and safety: Consumer perception and demand. *European Review of Agricultural Economics, 32*(3), 369–391.
Hair, J.F., Money, A.H., Samouel, P. & Page, M. (2007). Research Method for Business. England: John Wiley & Sons Ltd. Hashim, P. (2004). Food hygiene:Awareness for food business. Standard & Quality News, 11, 6–7.
Halal Industry Development and Corporation (2014). Malaysia as a Global Halal Hub. Retrieved onFebruary 26th 2014, fromhttp://www.hdcglobal.com/publisher/alias/?dt.driverAction=RENDER&pc.portletMode=view&pc.windowState=normal&pc.portletId=Newslatest.newsPortlet
Halal Media, 70% Ramadan buffets in Malaysian hotels suspected Haram, 24 Aug. 2010, fromhttp://halal-media.net/70-percent-ramadan-buffet-in-malaysian-hotels-suspected-haram/
JAKIM (2010). Halal Malaysia. JAKIM. Retrieved from:http://www.halal.gov.my/v2/.Retrieved date: January, 1st 2014
Jasimah, W., M., R., (2000). *Konsep Kesihatan Melalui Pemakanan: Pendekatan Islam dan Sains*. Utusan Publications.
Kivela, J., Inbakaran, R. & Reece, J. (1999). Consumer research in the restaurant environment, part 1: A conceptual model of dining satisfaction and return patronage. *International Journal of Contemporary Hospitality Management, 11*(5), 205–222.
Mohd Yusoff, H. (2004). Halal Certification Scheme. Standard & Quality News. 11:4–5.

Mohamad, N., & Backhouse, C., (2014). A Framework for the Development of Halal Food Products in Malaysia. Proceedings of the 2014 International Conference on Industrial Engineering and Operations Management Bali, Indonesia, January 7–9, 2014

Peri, C. (2006). The universe of food quality. *Food Quality and Preference, 17*(1/2), 3–8.

Reichheld, F.F. & Sasser, E.W. (1990). Zero defections: Quality comes to services. *Havard Business Review, 68*(5), 105–111.

Saad, F., (2014). JAKIM: Senarai restoren makanan cina muslim di Shah Alam. Interview date: February, 28th 2014.

Siti Mashitoh, A. Norhayati Rafida, A.R.& Alina, A.R.(2013).Perception Towards Halal Awareness and its Correlation with Halal Certification among Muslims. Middle-East Journal of Scientific Research 13(Approaches of Halal and Thoyyib for Society, Wellness and Health). Volume 1, (4). ISSN 1990-9233.

Soriano, D.R. (2002). Customers' expectations factors in restaurants: The situation in Spain. *International Journal of Quality & Reliability Management, 19*(8), 1055–1067.

Syed Marzuki, Hall & Ballantine (2012). Restaurant managers' perspectives on halal certification. *Journal of Islamic Marketing,* Vol. 3 Iss: 1 pp. 47–58.

Theory and Practice in Hospitality and Tourism Research – Radzi et al. (Eds)
© 2015 Taylor & Francis Group, London, ISBN 978-1-138-02706-0

Attributes to select casual dining restaurants: A case of customers in Klang Valley area, Malaysia

F. Ahmad, H. Ghazali & M. Othman
Faculty of Food Science and Technology, Universiti Putra Malaysia, Malaysia

ABSTRACT: The objective of this research was to rank the five attributes used by customers in selecting casual-typed restaurant when they dined out. Customers who had dined at casual dining restaurants within the past three months were conveniently selected. Twenty four attributes have been listed and respondents were asked to rank order the five attributes that influenced their decision in selecting a restaurant. There were a total of 384 usable questionnaires collected. Descriptive analysis was run using IBM SPSS Statistics version 21. Results found that the five attributes mostly influence customers' selections were: convenient location, the price of food, good taste of food, type of food and restaurant cleanliness. It is hoped the findings will help restaurateurs to fulfil consumers' preference when selecting a restaurant to dine out. Finally, restaurateurs may also use the information from this study to strategise their business and gain more profit in the future.

Keywords: Restaurant selection, important attributes, consumer preferences, dining out, Klang Valley

1   INTRODUCTION

1.1   *Background*

Trends of food consumption were changed due to the increasing of the disposable income. Overall review between the year of 2006 to 2011, the amount of per capita annual disposable income and expenditure are steady growth between 33.2 to 33.4 percent and the expenditure in most categories was double-digit growth rate which encourages people to dine out. Nowadays, Malaysians is found are eager to dine out and unsurprisingly, casual dining chains, cafés and informal restaurants had appealed their attention (Euromonitor International, 2013c). According to Habib, Dardak & Zakaria (2011), due to times shifting and influence of Western culture, the food consumption pattern among urban people were also changed even though Malaysians are still liking to have their home-cooked meals.

With that pattern, the casual-typed restaurants are gaining popularity especially among young affluent, professional worker and also students (Euromonitor International, 2013a). Casual dining restaurants are categorized in full-service type restaurants along with fine dining restaurants. However, the fine dining restaurants are less popular in Malaysia compared to casual dining restaurants (Euromonitor International, 2013b). Economic growth in Malaysia is highly giving an impact on the growth of casual dining restaurant. Despite the high

inflation rate in 2012 the value sales of full-service restaurants in Malaysia including the casual dining restaurant was increased 4 percent to RM10.6 billion (Euromonitor International, 2013b).

With the positive sales and the number of outlets opened this research is specifically focused on casual dining restaurants. In this sense, casual dining restaurant is the informal restaurants that are categorized as midscale restaurant, provide table service, offering a variety of menus and with modest prices (Barrows & Powers, 2009).

In line with the above notion, restaurant consumers may set a certain level of each attribute or characteristics before making a decision to which restaurant they wanted to dine in. Past researchers pointed several preferences used by customers to select a restaurant however those studies were undertaken in Hong Kong, India, United Kingdom and United States with little study looking at Malaysia restaurant customer's preferences. Thus, this study is therefore designed to access consumer's preferences attributes to select the casual dining restaurant.

2   LITERATURE REVIEW

2.1   *Eating out*

There are substantial numbers of researches looking at the factors that influence customer of eating out. Some factors are owing to busier lifestyle and dual-working families with children (Atkins & Bowler, 2001), less or no time to prepare the meal at home (Habib et al., 2011), higher disposable income, standard of living and leisure activities (Mehta & Maniam, 2002). Andaleeb & Conway, (2006) revealed that the increasing number of working women nowadays and time saving, eating healthy foods in a good environment also contributes to eating out habit (Ryu & Han, 2010). Warde & Martens (1998) contended eating out at the restaurant with good ambience and atmosphere not only creates different dining experience but developed social interaction among the customers.

2.2 Attributes that influence restaurant selection

Customers may have different preferences and expectation when choosing the type of restaurants to dine (Kim & Moon, 2009). Customers may prefer fast food restaurant because of the price, convenience and quick service (Akbay, Tiryaki & Gul, 2007; Farhana & Islam, 2011). In contrast, they may put an interest toward casual dining restaurant due to it food quality, service availability and atmosphere of the restaurant (Mattila, 2001; Namkung & Jang, 2008; Ryu & Han, 2010).

Ample evidences that there are attributes influence customers' choices. Clark & Wood (1998) found that quality of food, price, atmosphere and speed of service were highly influence customers in restaurant selection. Cullen (2005) revealed that quality of food is the most preferred attribute when customer decided to dine out for a social occasion and location is the most influential decision when they decided to eat out as substitution for cooking at home. Auty (1992) conducted a study on different occasion such as celebration, social occasion, convenience and business meal. Even there are different preferences among customers on each occasion, Auty's (1992) concluded that the types and quality of food are the most preferred choice for dining out in the restaurants. Quality of food is actually a group of several elements. Namkung & Jang (2007) categorised presentation, variety of menu, healthy options, taste, freshness and temperature into food quality elements. Joshi (2012) however divided food quality into sub- elements namely; menu variety, taste, presentation, healthy food options and familiar food.

3 METHODOLOGY

3.1 Sampling design

In this study purposive sampling technique was used to reach the casual dining restaurant's customers. This technique was chosen because of unavailability of sampling frame for casual dining restaurant's customers. The target population was the casual-type restaurant's customers who had dined at these restaurants within the past three months from the date the questionnaire is distributed. Klang Valley was chosen as the location of the study since the higher number of casual dining restaurants in this area (Euromonitor International, 2012). There were 384 respondents participated in this study and this was based on that voluntary basis.

3.2 Research instrument design

A questionnaire was used as an instrument to measure the variables of interest. Section A was designed to access information about eating out at casual dining restaurants such as frequency of dining, reasons of dining out, total spending per visit. In section B, respondents were asked to rank the importance attributes when selecting a restaurant. In this sense, respondents were asked to choose five out of twenty four listed attributes and rank them from one to five according to the level of importance. These attributes were adapted from Cullen (2005) and Kivela, Reece & Inbakaran (1999). Section C was created related to respondent's demographic profiles.

3.3 Data analysis

To access the top ranked attributes, descriptive analysis was run using IBM SPSS Statistics version 21. The attributes were all count using weighted calculation or total score method. Item that ranked at the first place were given the highest value (5) and the attributes that were in fifth place was given lowest value (1). The rating was then multiplied by the total responses received. The score of each attribute was summed up to get a total weighted score. Finally, the overall ranking was based on the weighted score of each attributes where the highest score received by each attributes was at the top ranking. The technique used to calculate the score was based on the research done by Mersha and Adlakha (1992).

4 RESULT

4.1 Demographic background

Table 1 showed the demographic profiles of respondents. Respondents were in average of 29 years old (mean 29.20) and more than half (67.7%) of them were female. Majority of the respondents were Malay ethnic and 46.9 percent having the undergraduate degree. More than half of the respondents (62.2%) were currently working in managerial field, professional and executives in the private sector. Half of the respondents (50.5%) were earned between RM1001 to RM3000 per month and only 1.6 percent received RM7001 and above per month. Regarding marital status, 57.6 percent were single while 40.9 percent of the respondents were married and 33.3 percent of married respondents were having children. 39.6 percent of total respondents were having their

Table 1. Demographic profiles of respondents.

Profile	Percentage (%)
Sex	
Male	32.3
Female	67.7
Ethnicity	
Malay	81.8
Chinese	12.8
Indian	3.1
Others	2.3
Highest Education Level	
Primary school	0.5
Secondary school	14.3
Certificate	5.5
Diploma	21.9
Degree	46.9
Postgraduate	10.9
Present occupation	
Managerial, Professional, Executives	48.7
Technical and supervisory	7.6
Clerical and related occupations	20.6
General workers	4.7
Self-employed	2.1
Unemployed	1.0
Student	14.1
Housewife	0.3
Others	1.0
Job sector	
Government	21.1
Private	62.2
Self-employed	2.6
Not related	14.1
Monthly income	
No income	9.1
RM 1000 and below	6.5
RM 1001 - RM 3000	50.5
RM 3001 - RM 5000	25.3
RM 5001 - RM 7000	5.5
RM 7001 and above	1.6
Others	1.6
Marital status	
Never married	57.6
Married	40.9
Divorced / widow	1.0
Presence of children (married)	
Household with no child	8.1
Household with child/children	33.3

meal at casual dining restaurants, two to three times a month with 36.5 percent were spent RM10 to RM20 per person for each visit.

4.2 *Ranking of influential attributes*

Table 2 showed the list of attributes considered when selecting a casual dining restaurant. While, Table 3 showed the ranking of five attributes chosen and total score calculated from respondent's ranked. Convenient location is the utmost important attributes that be a key decision attribute in making a choice to dine at casual-typed restaurant. Convenient location attributes leads the ranking

Table 2. List of attributes considered and total weighted score.

Ranking	Attributes	Total weighted score
1	Convenient location	808
2	Price of food	723
3	Good taste of food	693
4	Type of food	529
5	Restaurant cleanliness	433
6	Menu item variety	347
7	Speed of service	256
8	New meal experience	244
9	Ambience / atmosphere	243
10	Comfort level of restaurant	221
11	Freshness of food	198
12	Portion size of food	157
13	Clean staff	144
14	Good reputation of restaurant	129
15	Parking facilities	109
16	Nutritious food	93
17	Suitable for children also	88
18	Polite staff	79
19	Restaurant décor	59
20	Friendliness of staff	51
21	Wide restaurant	34
22	Staff that have menu knowledge	29
23	Handling of reservations	25
24	Prompt handling of complaint	17

Table 3. Top 5 ranking of attributes and total weighted score.

Ranking	Attributes	Total weighted score
1	Convenient location	808
2	Price of food	723
3	Good taste of food	693
4	Type of food	529
5	Restaurant cleanliness	433

with a total score 808 followed by the price of food (723), good taste of food (693), type of food (529) and restaurant cleanliness with a total score 433. This finding was similar with Auty (1992); Clark & Wood (1998); Cullen (2005) where location, cleanliness, price and type of food were in the top ranking of customers' choices before decided to dine at selected casual dining restaurants.

5 DISCUSSION

Respondents ranked convenient location as the most important attribute to be considered when selecting a casual dining restaurant. A report by Euromonitor International (2013a) stated young consumers like to spend their leisure time at shopping malls. Therefore, they are likely to have their meal near or in the shopping mall. Besides that, many restaurants are located in the malls. Hence, it may attract visitors to have their meal there.

In terms of price, consumers nowadays are more conscious on a price. A stronger inflation rate in Malaysia in 2012 cause consumers be more sensitive and conscious about spending the money on dining out activities (Euromonitor International, 2013b). The way customers valued the price is different among individuals. Perceived price depends on customer personal response to the product, interest in product, individual needs, expectations, personality and personal status (Ateljevic, 2000; Solomon, 1999). As most of the researcher agreed that the quality of food is the most important factors considered dining at certain restaurants, this research also found the same where good taste of food is much influence their restaurant selection. Cleanliness factors should not be overlooked since customers are also seeing this as one of the key decision attributes to select a restaurant. Cleanliness which include in atmosphere category is important because it also affects customer's emotion and contribute to the overall satisfaction toward the restaurant (Ladhari, Brun & Morales, 2008).

6 CONCLUSION

It is hoped this study findings gives a fruitful thought and view on current consumer's preference and behavior toward restaurant selection. Results showed that location as the main key attribute of selecting casual dining restaurants. Customers preferred to enjoy their dining out activities in a convenient location. With the persistent demand by customers and current trend of eating out, restaurateurs may strategize their location at the strategic places and offer a reasonable price, as it also be the attribute that customers consider after selecting the location.

Besides that restaurateurs must attract customers by providing a delightful taste of food and render a better service to customers apart of giving them a dandy experience in a comfortable environment while dining at the casual restaurant. It is such a symbiotic relationship between restaurateurs and customers whereby serving delightful foods with a good service and in better environment may fulfill customers' expectations and persuade them to come again. Indirectly, it will increase the sales and generate more profit of the restaurants. For academic perspective, this research may enrich the existing literature and strengthen the finding of the customer's preferences toward restaurants selection.

REFERENCES

Akbay, C., Tiryaki, G.Y. & Gul, A. (2007). Consumer Characteristics Influencing Fast Food Consumption in Turkey. *Journal of Food Control, 18*, 904–913.
Andaleeb, S.S. & Conway, C. (2006). Customer satisfaction in the restaurant industry: an examination of the transaction-specific model. *Journal of Services Marketing, 20*(1), 3–11.
Ateljevic, I. (2000). Tourist Motivation, Values and Perceptions. In A.G. Woodside, G. Crouch, J. Mazanee, M. Oppermann & M.Y. Sakai (Eds.), *Consumer Psychology of Tourism, Hospitality and Leisure* (pp. 139–209). Wallingford, UK: CABI Publishing. In Al-Sabbahy, H.Z., Ekinci, Y. & Riley, M. (2004). An Investigation of Perceived Value Dimensions: Implications for Hospitality Research. *Journal of Travel Research, 42*, 226–234.
Atkins, P. & Bowler, I. (2001). *Food in Society: Economy, Culture and Geography.* London: Arnold.
Auty, S. (1992). Consumer Choice and Segmentation in the Restaurant Industry. *The Service Industries Journal, 12*(3), 324–339.
Barrows, W. & Powers, T. (2009). *Introduction to Hospitality Industry.* New Jersey: Wiley.
Clark, M.A. & Wood, R.C. (1998). Consumer loyalty in the restaurant industry—a preliminary exploration of the issues. *International Journal of Contemporary Hospitality Management, 10*(4), 139–144.
Cullen, F. (2005). Factors Influencing Restaurant Selection in Dublin. *Journal of Foodservice Business Research, 7*(2), 53–85.
Euromonitor International. (2012). *Full-Service Restaurant in Malaysia.*
Euromonitor International. (2013a). *Consumer Lifestyles in Malaysia.*
Euromonitor International. (2013b). Full-Service Restaurant in Malaysia, (November).
Euromonitor International. (2013c). Income and Expenditure : Malaysia, (February).
Farhana, N. & Islam, S. (2011). Exploring Consumer Behavior in the Context of Fast Food Industry in Dhaka City Farhana & Islam. *World Journal of Social Sciences, 1*(1), 107–124.
Habib, F.Q., Dardak, R.A. & Zakaria, S. (2011). Consumers' Preference and Consumption Toward Fast Food: Evidences From Malaysia. *BMQR, 2*(1).

Joshi, N. (2012). A Study on Customer Preference and Satisfaction towardsRestaurant in Dehradun City. *Global Journal of Management and Business Research, 12*(21).

Kim, W. & Moon, Y. (2009). Customers' cognitive, emotional, and actionable response to the servicescape: A test of the moderating effect of the restaurant type. *International Journal of Hospitality Management, 28*(1), 144–156.

Kivela, J., Reece, J. & Inbakaran, R. (1999). Consumer research in the restaurant environment. Part 2: Research design and analytical methods. *International Journal of Contemporary Hospitality Management, 11*(6), 269–286.

Ladhari, R., Brun, I. & Morales, M. (2008). Determinants of dining satisfaction and post-dining behavioral intentions. *International Journal of Hospitality Management, 27*, 563–573.

Mattila, A.S. (2001). Emotional Bonding and Restaurant Loyalty. *Cornell Hotel and Restaurant Administration Quarterly, 42*(December), 73–79.

Mehta, S.S. & Maniam, B. (2002). Consumer and Marketing. *Marketing Determinants of Customers' Attitude Towards Selecting a Restaurant, 6*(1), 27–44.

Mersha, T. & Adlakha, V. (1992). Attributes of Service Quality: The Consumers' Perspective. *International Journal of Service Industry Management, 3*(3), 34–45.

Namkung, Y. & Jang, S. (2007). Does Food Quality Really Matter in Restaurants? Its Impact On Customer Satisfaction and Behavioral Intentions. *Journal of Hospitality & Tourism Research, 31*(3), 387–410.

Namkung, Y. & Jang, S.S. (2008). Are highly satisfied restaurant customers really different ? A quality perception perspective. *International Journal of Contemporary Hospitality Management, 20*(2), 142–155.

Ryu, K. & Han, H. (2010). Influence of the Quality of Food, Service, and Physical Environment on Customer Satisfaction and Behavioral Intention in Quick-Casual Restaurants: Moderating Role of Perceived Price. *Journal of Hospitality & Tourism Research, 34*(3), 310–329.

Solomon, M.R. (1999). *Consumer Behavior: Buying, Having, and Being* (4th ed.). Upper Saddle River, NJ: Prentice Hall. In Al-Sabbahy, H.Z., Ekinci, Y. & Riley, M. (2004). An Investigation of Perceived Value Dimensions: Implications for Hospitality Research. *Journal of Travel Research, 42*, 226–234.

Trochim, W.M.K. (2006). Nonprobability Sampling. *Research Methods Knowledge Base*. Retrieved May 10, 2014, from http://socialresearchmethods.net/kb/sampnon.php

Warde, A. & Martens, L. (1998). Eating out and the commercialisation of mental life. *British Food Journal, 100*(3), 147–153.

Theory and Practice in Hospitality and Tourism Research – Radzi et al. (Eds)
© 2015 Taylor & Francis Group, London, ISBN 978-1-138-02706-0

Customer behavioural intention: Influence of service delivery failures and perceived value in Malay restaurants

Z. Othman, M.S.M. Zahari & S.M. Radzi
Faculty of Hotel and Tourism Management, Universiti Teknologi MARA, Shah Alam, Malaysia

ABSTRACT: Service delivery failures can close down restaurants, if attention is not paid to the problem. In a restaurant, the customer experience process failure if the employee is distracted and outcome failure if the chosen menu item is not accessible. Knowing that failures will always happen, it is very important that the restaurant makes rations for recovery of these unfavorable occurrences. In spite of providing recovery in reaction of service delivery failures, customers' intention to re-patronize restaurants is also predisposed by the perceived value of the restaurant customers. If customers are satisfied with the service recovery, they may have intended to revisit or refer services to family and friends. This study seeks to investigate factors of service delivery failure, perceived values and customers' behavioural intention in Malay restaurants.

Keywords: Service delivery failure, perceived value, customer satisfaction, behavioural intention

1 INTRODUCTION

Similar with other industry, restaurant industry in Malaysia is filled in the past decade with diverse kinds of restaurants mushrooming all over the place spanning from fine dining, specialty, fast food, casual, theme, ethnic restaurants and even mixture of more than one category. Among all, ethnic restaurants is noted of having huge growth and development not only in the last 20 years but more prominent in the early 90s (Othman, 2007). The diversity of ethnic groups contributed to the positive growth of this restaurant with a superb blend of foods, service, ambience, design and colourful range of cuisines (Talib, 2009).

There are three major ethnic groups that constitute the integral construction of ethnic restaurants in Malaysia, namely Malay, Chinese and Indian. The Malays are the majority shareholder amount of sixty five percent of the population, with Islam as the official religion, Chinese make up for twenty percent who may be Buddhist or Christian and the rest is ten percent Indian which might be Hindu and Christian (Talib, 2009). The remaining of the group pact comprised of various indigenous and sizeable amounts of expatriate (Talib, 2009). Restaurants such as Malay, Chinese, Nyonya's, Kopitiam's, Indian and Mamak Restaurants are well accepted as most of these restaurants fuses cuisines from each ethnic group and introducing a unique Malaysian gastronomic heritage.

Nonetheless, in spite of the positive growth of ethnic restaurants in Malaysia, Chinese and Indian Hindu restaurants are more popular among their own ethnic groups. These most probably due to the Malays who represent a large Muslim population are very strict regarding 'halal' and 'haram' of food they consume. Thus, Malay customers are found to be less keen to dine at these types of restaurants. This is not to assume that all Chinese and Indian Hindu restaurants are selling 'non-halal' food. With this argument, Malay and Indian Muslim restaurants which are known as 'Mamak' restaurants control large 'halal' market segment by attracting most customers from various ethnic groups to patronize in their restaurant. This, in turn, flourishes the growth of both restaurants nationwide.

Of both the popularity and positive growth of these two competing restaurants, 'Mamak' found to be more popular and successful compared to Malay restaurants (Zahari & Othman, 2008; Othman, Zahari, Hashim & Ibrahim, 2009). Ramli and Ahmad (2003) contended that the success recipes of 'Mamak' restaurants in attracting new customers were by delivering faster service to their customers. Apart from that, Zahari and Othman (2008) revealed that 'Mamak' restaurants were better due to efficient and effective service delivery delivered to their customers although offering fewer varieties of foods. On the contrary, many Malay restaurants are found struggling to attract and retain customers, although selling numerous types of foods to their customers. Zahari and Othman

(2008) further noted that Malay restaurants were criticized of providing poor service delivery such as lack of service quality, poor service encounters, and poor employee interaction with customers. In turn, many Malay restaurants found to be unable to sustain their business operations for over a long period. Othman et al. (2009) argued that Malay restaurants are continuing to develop rapidly but are still having problems in terms of service delivery failures, particularly in waiting times for food to be delivered to the customers. For this study, Malay restaurant is characterize as restaurants that offer Malay meals at a medium price that customers perceive as "good value", with full service, buffets or limited service with customers ordering at the table and having their food brought to them.

Despite of repeated service delivery failures occurring in Malay restaurants, it is presumed that service recovery applied to apprehend the situation are not effectively executed or not even implemented at all. Thus, it is interesting to find out the underlying factors that still influenced customer to remain and loyal by patronizing the same Malay restaurants. Chen and Hu (2010) postulate that despite experiencing the service failure customers are still patronizing a particular restaurant due to specific values which is valuable to them. In other words, the role of perceived values which relate to experience or special values like price, location, food, ambience and few others might contribute to this causation (Chen & Hu, 2010). Although, this notion is made from a different restaurant setting, it is predicted that customers of the ethnic restaurants which in the context of this study Malay restaurant might have the same attitudes. It is therefore imperative to venture and further engrossing the issues and confirming the entire proposed dimension, especially in the medium class Malay restaurants in Malaysia. Thus, this study intends to examine the factors of service delivery failure, perceived values and behavioural intention in Malay restaurants in Malaysia and hypotheses that;

H₁: There is a significant relationship between service failure and customer behavioural Intention

H₂: Perceived value mediates the relationship between service delivery failure and customer satisfaction

2 LITERATURE

2.1 *The connection between service delivery failures, service recovery, perceived value and behaviour intention*

Undeniable that to be successful, restaurant industry must have the ability to deliver satisfying experiences to customers. Kong and Jogaratnam (2007)

highlighted that most successful restaurants compete on the basis of their ability to deliver superior and unexceptional service. Nevertheless, even in the luxurious restaurant with the best customer-oriented strategic plans, immaculate service delivery cannot be assured (Chiang, 2007). A restaurant, which involves a great amount of personal interaction between the staff and customers even, cannot avoid day-to-day errors, silly mistakes, unprecedented failures, as well as complaints in the process of service delivery. Service delivery failures, in fact, can make restaurants out of business by neglecting the cause of failure.

The initial work by Bitner, Booms and Tetreault (1990) introduce the examination of service failures through critical incident technique (CIT) classified failures into three broad groups: (1) employee responses to delivery system failures, (2) employee responses to customer needs or requests, and (3) unprompted and unsolicited employee actions. Even though, there exist many sub-categories to the aforementioned three groups that may be operational in nature, the three classes effectively indicate that failures usually link to customer evaluations of interaction with some aspect of the service organization. Service delivery failures occurred when service delivery performance does not meet the expectations of customers and can be classified as either pertaining to the outcome or the process (Smith, Bolton & Wagner, 1999). Smith (2007) highlighted that a process failure happened when the core service carried in a flawed or incomplete way, resulting in poor benefit to the customer such as status or esteem. Conversely, an outcome failure happens when a certain feature of the main service is not carried, instigating in the reduction of economic resources such as money and time to the customer.

Knowing failures will occur even in the finest restaurant, it is imperative therefore for the restaurant to make provisions for the recovery of these unfavorable instances and the provisions that a restaurant makes are known as service recovery. An organization's ability to recover from failure is an essential element of the whole service delivery system with significant implications on customer satisfaction (Church & Newman, 2000). Duffy (1998) stated that service recovery provides opportunities to decrease costs, the improve customer experience, and increase customer satisfaction. When customers are satisfied, they are more inclined to exhibit positive behavior toward the service provider (Kristen, 2008). It is no surprise that the satisfied customers are truly invaluable to an organization.

Despite providing recovery in response of service delivery failures, customers' intention to repatronize restaurants may also improve by having some values that may have some effect to the restaurant customers (Oh, Kim & Shin, 2004; Chen & Hu, 2010). Chen and Hu (2010) describes customer perceive values from two perspectives which is functional and symbolic dimensions. Functional value can be described as a thorough evaluation of value incorporating quality, the customary value for money and convenience characteristics. This type of value represents the customer's perception of quality in terms of goods and services received from the coffee outlet, the price paid for those goods and service, and the time saving to receive them. Symbolic value on the other hand can be thought as overall representations of experiential value perceived from the social, affection, the aesthetic, and reputation aspects. This value represents the customers' impression on others, perception of delight or pleasure, enjoyment of the visual appeal, and reputation of outlet, involved with the consumption experience. Customers who had a bad experience will still revisit and remain loyal to the restaurants that have positive perceived values. Positive perceived values will influence the customers to devour and repeat their purchase behavior without having treated with service recovery (Chen & Hu, 2010).

3 METHODOLOGY

In the early phase of the study, preliminary interviews session done in order to obtain the restaurant operators opinion and perception of the research issues especially in recognizing the existence of service delivery failures in Malay restaurants. Based on the feedback gathered, eighteen restaurant operators involved in the study stated that service failure takes place numerously on their foodservice establishment.

Self-reported questionnaire survey was subsequently developed as a mean to measure the restaurant customer experience in service failure and behavioural intention as well as the role of perceived value. Through convenience sampling technique, survey the experience of restaurant customers who had encountered service delivery failure in Malay restaurants around Klang Valley, Malaysia was undertaken. In other word, only Malay restaurant customers' with service delivery failure experience while dining are incorporated in this study. To have ample response, data were gathered during lunch hours as soon as the respondents finish their meals. This approach is chosen to guarantee the result gathered would be based on real experiences. To ascertain customers had gone through such failures, a couple of simple questions asked, for instance; *have you ever dined in Malay restaurants and have you ever experience service delivery failure.* If the response is yes, a set of questions was then preceded. The data collection process accomplished within a phase of three months (February-April 2012), and 481 respondents were successfully surveyed.

4 RESULT AND ANALYSIS

4.1 *Descriptive statistic*

Table 1 shows the overall mean score of each dimension. The majority of the respondents agreed that service delivery system failure is the most incidents occur in the Malay restaurant (M = 5.52, S.D, .121). With the frequent human interaction in the restaurant establishment especially during lunch and dinner, the failure that consists of "food out of stock", "unpleasant smell", "long waiting time", "unpleasant food presentation", "unbearable noise level", and "inconsistent service" rated highly by the respondents.

The next highest average rating given to implicit or explicit customer request (M = 5. 51, S.D, .116). This incident usually occurs because of the special request made by the customers, especially that involve changing the order and modification of a meal (less sugar/salt). Other than that, failure to provide baby chair and no menu card also categorize in this dimension. Nevertheless, respondents somewhat agree to unprompted and unsolicited employee's action (M = 5.17, S.D, .121). This rating contributed by unfriendly employees, inattentive employees, poor employee hygiene, and improper dress (not wearing the standard uniform).

There are several items that stood up in perceived values. This is an indication that a higher percentage of customers agree with the items pertaining to perceived value. From the table, it can be seen that Malay restaurant customers keep patronizing a Malay restaurant despite inconsistent service delivery due to several reasons such as food, price, environment, and convenient location with average mean of (M = 5.76, S.D, .152). There are two items that reflect on food as the basis of visitation to the restaurant namely due to Malay restaurant providing exquisite food, and due to the excellent taste of food. The customer also agrees that they patronize the Malay restaurant due to soothing environment, aesthetically appealing

Table 1. Mean and standard deviation for service delivery failure, perceived value and behavioural intention measurement.

Item	Measurement	M	S. D	N
1	Service failure	5.52	.121	481
2	Implicit or explicit customer request	5.51	.116	481
3	Unprompted and unsolicited employee's action	5.17	.121	481
4	Perceived value	5.76	.152	481
5	Behavioural intention	5.75	.124	481

Note: CI = Confidence interval; LL = lower limit, UL = upper limit Likert-scale. 1-strongly disagree to 7-strongly agree.

furnishing, and it is simply worth the time spent visiting a Malay restaurant. Apart from that, the customers also agree that convenient location motivates them to patronize the restaurant.

For behavioural intention, the average mean scores of the items are (M = 5.75, S.D, .124) with items such as customers will attempt to influence their friends and family to dine at a Malay restaurant with positive values. Satisfied customers said that they will convince friends and family to dine at Malay restaurant, dine out again at Malay restaurant, dine more often in Malay restaurant, recommend Malay restaurants to someone else, and lastly item indicates that the customers will consider Malay restaurant as their first choice for next visit.

4.2 Hypotheses testing

Structural Equation Modeling (SEM) was adopted in testing the paths which are hypothesized in the model. Table 2 shows the result of the two hypotheses.

4.2.1 H1: There is a significant relationship between service failure and customer behavioural Intention

The linkages between the service delivery failure and behavioural intention received strong support in a significant relationship (β:0.124, t:1.175, p < .001). This proposition is based on the belief when failure occurred, customers that experience that situation will eventually tend not to revisit the premises which the scenario took place. The result of this research showed a significant relationship between the service delivery failures and behavioural intention. This finding is consistent with previous studies that found that when the positive service delivery failure happened there will be

Table 2. Correlation between service delivery system failure and service recovery strategy.

No.	Structural Path	Standardized Estimate (β)	Critical Ratio (t-value)	Results	p
H_1	Service Failure → Behavioural Intention	0.124	1.175	Supported	***
H_2	Perceived value mediates the relationship between service delivery failure and customer satisfaction	0.276	4.006	Supported	***

Note: ***Significant at p < .001.

negative consequences on customers' behavioural intentions. This behavior may include negative word-of-mouth, and intention not to revisit (Tax, Brown & Chandrashekaren, 1996; Brown, 1997; Liu, Furrer & Sudharshan, 2001; Susskind, 2002). The results are similar to various studies (Zeithaml, et al., 1996; Alexandris, Dimitriadis & Markata, 2002) undertake in different settings thru out the industries.

4.2.2 H2: Perceived value mediates the relationship between service delivery failure and customer satisfaction

The result of (β:0.276, t:4.006, p < .001) showed a significant relationship between the service failures and the customer satisfaction as mediated by perceived value. Thus, essentially hypothesis H_6 is supported. This finding is consistent with previous studies that found there is a relationship between service delivery failures and customer satisfaction with perceived value as the mediating variables (Cronin & Taylor, 1994; Dabholkar, Sheperd & Thorpe, 2000; Akbar, Mat Som, Wadood & Alzaidiyee, 2010). This proposition suggested that the strength of the relationship between service delivery failures and customer satisfaction would be increased or decreased by the presence of perceived value. The test results reveal that perceived value mediates the relationship between service delivery failure and customer satisfaction.

5 DISCUSSION

Service delivery failures namely service failure, implicit or explicit customer requests and unprompted and unsolicited employee actions are found to be able to explain the variance in factor affecting the service delivery failures happened in Malay restaurant. It is found that factors such as unpleasant smell, unbearable noise level, crowded seating arrangement, insufficient seating arrangement, failure to provide individual menu card, and failure to comply to customer's specific food request are among the items receiving critical attention from the customers as being the service failures typically encountered in Malay restaurants.

The other factor assumed contributing to positive customer satisfaction and behavioural intention is the roles of perceived value. This study introduced perceived value as the new potential variable to explain the service recovery-satisfaction mismatch in Malay restaurant setting. Interestingly, it demonstrated the presence of perceived value able to shed some light on the otherwise daunting and confusing service recovery-satisfaction phenomenon. Perceived value is also found to mediate the relationship between service delivery failures and customer satisfaction. The result obtained posits that customers who experience service delivery failures will still be satisfied if the customers perceived some elements in the restaurant to be worth the visit. These perceived values may include tasty foods, exquisite foods, cheaper price, convenient location, and soothing environment. With this finding, the restaurant operators should not therefore ignore or being ignorant of perceived value, but highly conversant with it as those elements could be used in retaining the customers despite having a slight service failure and slacking in service recovery.

6 CONCLUSION

As this research is an initial investigation, future research should replicate and extend the study to other states in Malaysia or research settings with wider and comprehensive sample across other restaurants. This will definitely extend the knowledge in the restaurant industry, especially the ethnic restaurants segmentation.

REFERENCES

Akbar, S., Som, A.P.M., Wadood, F. & Alzaidiyeen, N.J. (2010). Revitalization of Service Quality to Gain Customer Satisfaction and Loyalty. *International journal of business & management*, 5(6).

Alexandris, K., Dimitriadis, N. & Markata, D. (2002). Can perceptions of service quality predict behavioural intentions? An exploratory study in the hotel sector in Greece. *Managing Service Quality*, 12 (4), 224–231.

Bitner, M.J., Booms, B.H. & Tetreault, M.S. (1990). The service encounter: diagnosing favorable and unfavorable incidents. *Journal of Marketing*, 54 (1), 71–84.

Brown, S.W. (1997). Service recovery through information technology: Complaint handling will differentiate firms in the future. *Marketing Management*, 6(3), 25–27.

Chen, P.T. & Hu, H.H. (2010). How determinant attributes of service quality influence customer-perceived value: an empirical investigation of the Australian coffee outlet industry. *International Journal of Contemporary Hospitality Management*, 22(4), 535–551.

Chiang, C.C. (2007). The effect of experiencing outcome versus process service failures on Taiwan and U.S. restaurant customer's (Doctoral dissertation, Alliant International University).

Church, I. & Newman, A.J. (2000). Using simulations in the optimization of fast food service delivery. *British Food Journal*, 102(5/6), 398–405.

Cronin Jr, J.J. & Taylor, S.A. (1994). SERVPERF versus SERVQUAL: Reconciling performance-based and perceptions-minus-expectations measurement of service quality. *Journal of marketing*, 58(1).

Duffy, D.L. (1998). Customer loyalty strategies. *Journal of Consumer Marketing*, 15(5), 435–448.

Dabholkar, P.A., Shepherd, C.D. & Thorpe, D.I. (2000). A comprehensive framework for service quality: an investigation of critical conceptual and measurement issues through a longitudinal study. *Journal of retailing*, 76(2), 139–173.

Kong, M. & Jogaratnam, G. (2007). The influence of culture on perceptions of service employee behavior. *Managing Service Quality*, 17 (3), 275–297.

Kristen, A.R.K. (2008). *The effects of service recovery satisfaction on customer loyalty and future behavioral intentions: An exploratory study in the luxury hotel industry.* (Doctoral dissertation). Retrieved from ProQuest Dissertations and Theses (UMI No. 3317342).

Liu, B.S.C., Furrer, O. & Sudharshan, D. (2001). The relationships between culture and behavioral intentions toward services. *Journal of Service Research*, 4(2), 118–129.

Oh, H., Kim, B.Y. & Shin, J.H. (2004). Hospitality and tourism marketing: recent developments in research and future directions. *International Journal of Hospitality Management*, 23(5), 425–447.

Othman, Z. (2007). *Service Delivery System and Customer Patronization: A Comparison Of Ethnic Restaurants In Shah Alam* (Master dissertation, Universiti Teknologi Mara).

Othman, Z., Zahari, M., Hashim, R. & Ibrahim, S. (2009). Do Thai foods outshine Malaysians foods locally and internationally? *Journal of Tourism, Hospitality & Culinary Arts*, 23–34.

Ramli, A.S. & Ahmad, R. (2003). Factors influencing customers patronizing Mamak restaurants. Paper presented at the Proceeding of the 2003 Tourism Educators of Malaysia Conference.

Smith, A.K., Bolton, R.N. & Wagner, J. (1999). A model of customer satisfaction with service encounters involving failure and recovery. *Journal of Marketing Research*, 356–372.

Smith, J.S. (2007). *An examination of the relationship between service recovery system structure, service operating environment, and recovery performance.* (Doctoral dissertation) Retrieved from Proquest Dissertation and Theses database (UMI No. 3272492).

Susskind, A.M. (2002). I told you so! Restaurant customers' word-of-mouth communication patterns. *The Cornell Hotel and Restaurant Administration Quarterly*, 43(2), 75–85.

Talib, S.A. (2009). *Modeling satisfaction and behavioural intention of fine dining restaurant consumers' with different purchasing orientation* (Unpublished doctoral dissertation). Universiti Teknologi Mara, Shah Alam, Malaysia.

Tax, S.S., Brown, S.W. & Chandrashekaran, M. (1998). Customer evaluations of service complaint experiences: implications for relationship marketing. *The Journal of Marketing*, 62 (2), 60–76.

Zahari, M.S. M & Othman, Z. (2008). Customer reaction to service delays in Malaysian ethnic restaurants. *South Asian Journal of Tourism and Heritage*, 1 (1), 20–31.

Zeithaml, V.A., Berry, L.L. & Parasuraman, A. (1996). The behavioural consequences of service quality. *Journal of marketing*, 60(2).33- 43.

Theory and Practice in Hospitality and Tourism Research – Radzi et al. (Eds)
© *2015 Taylor & Francis Group, London, ISBN 978-1-138-02706-0*

Customer awareness and attitude towards restaurant grading system in Shah Alam

N.H. Mohd-Rejab, N.N. Ahmad-Jazuli, J. Jamil, A. Arsat & H. Hassan
Faculty of Hotel and Tourism Management, Universiti Teknologi MARA, Shah Alam, Malaysia

ABSTRACT: Restaurant grading system may assist as a warning for food handlers to avoid from being downgraded and further lead to possible legal action. The restaurant grading have been implemented to maintain and keep the food premises clean and safe place to eat. This system acts as the indicator representing the level of food premises cleanliness. The purpose of restaurant grading system is to improve the standard of cleanliness of food premises in order to restrain incidence of food poisoning. This paper will discuss the benefits of food grading to the restaurants operators in forms of quality, safety, freshness, cleanliness and integrity of the operators.

Keywords: Attitude, restaurant grading system, perception, food premises, awareness

1 INTRODUCTION

Grading system is the indicator that most country used to show the cleanliness and safety of food in restaurants. The system has been implemented in many countries such as USA and Singapore. Moreover, local health department is responsible of inspecting restaurants to ensure the compliance of the established hygiene standards (Musa et al., 2010). In Malaysia, the local authorities are responsible to give the grade to the restaurants. Despite their acceptance of these programs as standard public health practices, their effectiveness in preventing foodborne disease remains unclear. Local authorities' roles are divided into three categories. Firstly, to plan and develop infrastructure at urban centers which is including shop units for food services. Secondly, they are responsible to produce licenses to food operators through formal application in accordance to shop units available and handling courses for food operators comply with regulations. Thirdly, to ensure the quality of food services rendered to the public by setting rules for food operators to avoid from food contamination. Restaurant grading system will assist the food handlers in terms of cleanliness and hygiene to avoid from being downgraded and possible legal action. Every food handlers including school canteens, government office canteens, food stalls and licensed fast food restaurants will be graded according to the system.

In Shah Alam, Majlis Bandaraya Shah Alam (MBSA) had introduced the restaurant grading system in June 2008 (Mstar Online, 2008). The implementation of restaurant grading system is to maintain and keep the food premises clean as public nowadays are concerned about cleanliness. In order to cope with the public demands, restaurant operators need to provide clean premises and ensure that the restaurant comply with existing regulations. Local authorities implement grading system in order to create awareness among customers in promoting customers in selecting cleaner premises.

Therefore, displaying the grades will assist customers in selecting restaurant, as well as serving as an encouragement for restaurant operators to keep their premise clean consistently. In line with this notion, this proposed study is meant to look into the attitude and awareness of customers towards restaurant grading system. In addition, this research also strives to identify the impacts of the implementation of the grading system towards customers' perception.

2 LITERATURE REVIEW

2.1 *Customer attitude*

As defines by Perner (2010), consumer attitude simply as a composite of a consumer's belief, feeling and behavioral intentions towards some object within the context of marketing. Consumers have the right to hold the negative or positive feeling towards a product or services. In general, the most important reason in selecting the food is safety, followed by its nutrition, health and taste. The environmental corncerns are less important, whereby the personal benefits of safe food are more important (Liu, Pieniak,& Verbeke,2013).

The attitude and behavior are also important and necessary when it comes to preparing and handling food hygiene. Customer can select eating-place with better grading but mostly they do not care about it (Walpuck, 2013). Customer simply eats at stall that does not showed grade card or some does not apply the grading inspection (Gregory & Kim, 2004). Not only the customer attitude, food handler attitude also played the important rules in keeping food safety.

2.2 Customer awareness on hygiene factors in food premises

Recently, Sheth, Gupta and Ambegaonkar (2011) stated that an unhygienic handling of food can causes a critical risk for food safety. Such as poor hygiene shown it leads to detection of pathogens like Salmonella Enteritidis on hand towel samples, Staphylococcus aureus, and Escherichia coli in the working equipment (Sheth, et al, 2011). The poor personal hygiene also could lead to the foodborne illness.

The combination of theoretical and practical training of food safety had shown to lead to a lower level of hand contamination (Soares, Garcia-Diez, Esteves, Oliveira & Saraiva, 2013). Practicing the steps of hand washing and the workers attitude among food handlers is important to keep the lower impact of foodborne illness. More over the workers are given proper training and knowledge about food safety and personal hygiene (McIntyre, Vallaster, Wilcott, Henderson & Kosatsky, 2013). Apart from that, several studies report that increased food safety knowledge does not always assure that good hygiene practices are implemented (Buccheri et al., 2010; Park, Kwak & Chang, 2010; Soares et al., 2012).

As state by Leach, Woodell-May, and Higgins (2001) from previous study, customers highlighted factors that are most important when consuming food is the works hygiene issues, cleanliness of dining area and equipment, control of food temperature and food is protected from flies. Aksoydan (2006) revealed that hygiene factors that influence people to dine out are workers, food preparation area, dining area and conditions of equipment and food. Food handlers play important role in food businesses. They can cause foodborne disease if they neglected proper food handling practices (Abdul Mutalib, Abdul Rashid, Mustafa, Nordin, Hamat & Osman, 2012). A must to develop food establishment, there is a demand in providing hygienic food premises market have to supply to the consumer.

2.3 Customers awareness towards restaurant grading system

The definition of awareness can be said as the process in which a person select, arrange and interpret stimuli, these stimuli are filtered and adjusted to become one's own view of the world. This is same as stated by Walters and Bergiel (1989) which by perception is an entire process by an individual becoming aware of its environment and interpret it according to their experience. As for Hanna and Wozniak (2013), explained, "Awareness is the process of selecting, organizing, and interpreting sensations into a meaningful whole". While Van der Walt, Strydom, Marx, and Jooste (1996) added that a process of awareness occurs when sensory receive stimuli via brain and then categories them according their past experience, beliefs, like and feeling. This means that even though two persons exposed to the same thing, in the same environment, they will never experience the same. Our brain takes in and processes only a small number of all these stimuli (Solomon, Bamossy, Askegaard & Hogg, 2006).

3 METHODOLOGY

This study will employ the use of quantitative research design with a semi-structured questionnaire to utilize the information gathered. The descriptive research design and correlation approach will be used. A cross sectional study will be used where data will be gathered just once at the different restaurants. The population of this research would be the restaurant customer's within area in Seksyen 7 (Commercial) and the sample would be collected among respondents dining at selected restaurants in Seksyen 7. This planning stage the data collection setting will be focused on the restaurants operate in Seksyen 7 (Commercial), Shah Alam as it would give a solid groundwork for the proposed research.

4 SIGNIFICANCE OF STUDY

This research will reveal the awareness and attitude of customers towards restaurant grading system and the benefit to the restaurant operators as it symbolized the quality, safety, freshness, cleanliness and integrity of the operators. The positive outcomes are expected from this research as it will definitely help other restaurant operators in scoring a good restaurant grade. The findings of the research is expected to give limelight to government authorities, especially MBSA and Ministry of Health on enhancing their restaurant grading system, because this research are implement from consumer's point of view. From academic view, this research will be contributed to the body of knowledge by adding the literature on the respective topic, which might serve as a reference for future research.

5 CONCLUSION

In conclusion, the implementation of the restaurant grading system is important to maintain the cleanliness of food. The displayed grades will assist customers in selecting restaurant, as well as serving as an encouragement for restaurant operators to keep their premise clean consistently. However, the customers still need to be educating with the importance of food safety and the grading system as a threshold to select any restaurant by any potential customer locally or internationally.

ACKNOWLEDGEMENTS

This paper is sponsored by Research Acculturation Grant Scheme (RAGS 2013) Ministry of Education, Malaysia and Universiti Teknologi MARA.

REFERENCES

Abdul-Mutalib, N.A., Abdul-Rashid, M.F., Mustafa, S., Amin-Nordin, S., Hamat, R.A. & Osman, M. (2012). Knowledge, attitude and practices regarding food hygiene and sanitation of food handlers in Kuala Pilah, Malaysia.*Food Control, 27*(2), 289–293.

Aksoydan, E. (2007). Hygiene factors influencing customers' choice of dining-out units: findings from a study of university academic staff. *Journal of Food Safety, 27*(3), 300–316.

Buccheri, C., Mammina, C., Giammanco, S., Giammanco, M., Guardia, M.L. & Casuccio, A. (2010). Knowledge, attitudes and self-reported practices of food service staff in nursing homes and long-term care facilities. *Food control,21*(10), 1367–1373.

Gregory, S. & Kim, J. (2004). Restaurant choice: the role of information. 7 (1), 243?247. *Journal of Foodservice Business Research, 7*(1), 243–247.

Hanna, N. & Wozniak, R. (2013).Consumer Perception. In *Consumer Behavior: An Applied Approach* (4th ed., pp. 73–108). Iowa, US: Kendall Hunt Publishing Company.

Leach, M., Woodell-May, J.E. & Higgins, J. (2007). *U.S. Patent No. 7,179,391* Washington, DC: U.S. Patent and Trademark Office.

Liu, R., Pieniak, Z. & Verbeke, W. (2013). Consumers' attitudes and behaviour towards safe food in China: A review. *Food Control, 33*(1), 93–104.

McIntyre, L., Vallaster, L., Wilcott, L., Henderson, S.B. & Kosatsky, T. (2013). Evaluation of food safety knowledge, attitudes and self-reported hand washing practices in FOODSAFE trained and untrained food handlers in British Columbia, Canada. *Food Control, 30*(1), 150–156.

Mstar Online (2008). *Sistem demerit buat pengendali makanan di Selangor mulai bulan depan.* Retrieved from:http://mstar.com.my/berita/cerita.asp?file=/2008/5/27/TERKINI/Mutakhir/

Musa, M., Jusoff, K., Khalid, K., Patah, M.O.R.A., Anuar, J. & Zahari, H. (2010). Food Borne Illness Risk Factors Assessment in UiTM Shah Alam, Malaysia. *World Applied Sciences Journal, 8*(7), 864–870.

Park, S.H., Kwak, T.K. & Chang, H.J. (2010). Evaluation of the food safety training for food handlers in restaurant operations. *Nutrition research and practice, 4*(1), 58–68.

Perner, L. (2010). *Consumer behavior: the psychology of marketing.* Retrieved from, http://www.consumerpsychologist.com/

Sheth, M., Gupta, A. & Ambegaonkar, T. (2011). Handlers' hygiene practices in small restaurants of Vadodara. *Nutrition & Food Science,* 41(6), 386e392.

Soares, K., Garcia-Diez, J., Esteves, A., Oliveira, I. & Saraiva, C. (2013). Evaluation of food safety training on hygienic conditions in food establishments. *Food Control,* 34, 613e618.

Solomon, M.R. (2010). *Consumer behaviour: a European perspective.* Pearson education.

Walpuck, D. (2013, June 4). *The ABCs of New York's Restaurant Grades | Food Safety News.* Retrieved from http://www.foodsafetynews.com/2013/06/the-abcs-of-new-yorks-restaurant-grades/

Walters, C. G. & Bergiel, B. J. (1989). *Consumer behavior: A decision-making approach* Cincinnati: South-Western Pub. Co.

Van der Walt, A., Strydom, J. W., Marx, S. & Jooste, C. J. (1996). *Marketing Management* (3rd ed.). Cape Town, South Africa: Juta.

Theory and Practice in Hospitality and Tourism Research – Radzi et al. (Eds)
© 2015 Taylor & Francis Group, London, ISBN 978-1-138-02706-0

Food temperature knowledge and practice among food handlers at school canteens

A. Ena-Arzairina, R.S. Raja-Saidah & M.D. Hayati
Faculty of Business Management, Universiti Teknologi MARA, Shah Alam, Malaysia

A.A. Saidatul-Afzan
Faculty of Hotel and Tourism Management, Universiti Teknologi MARA, Shah Alam, Malaysia

ABSTRACT: Reducing the risk and occurrence of foodborne illness is a priority for the foodservice industries. Food handlers play an important role in preventing the occurrence of foodborne illness and meeting the objective of serving harmless food. This paper is to identify food temperature knowledge amongst food handlers in school canteens. This study had used pretest-posttest questionnaire approach. The respondents are amongst food handlers at school canteens. To choose the sample size, purposive sampling is use for collecting data. The data had been collected by using a structured questionnaire. Result shows that the pretest and posttest method are effective in which the training given had more input for food handlers on the knowledge and thus evade from food borne illness.

Keywords: Food temperature knowledge; school canteens; hygiene practice; training

1 INTRODUCTION

Nowadays, foodborne diseases are extensive devel-oped in developing countries and more so especially at school canteens. Sharifa Ezat, Netty & Sangaran (2013) reported that food poisoning cases is on the rise as the evident by the incident rate of 62.47 cases per 100,000 population in 2008 and 36.17 in 2009 in Malaysia.

However, there was in fact an increase in number of episodes of foodborne outbreak reported by vari-ous states in Malaysia commonly outbreaks occurring in schools (Maizun Mohd Zain & Nyi Nyi Naing, 2002; WHO, 2008; Sharif & Al-Malki, 2010). These numbers established that school chil-dren have been the foremost victim in many food poisoning cases mainly in Malaysia.

Schools are key settings that will provide chil-dren with any health promotion programs that planned to make the children start healthy dietary and physical activity behaviors (Cullen & Watson, 2007). Stu-dents are captive customer which who is usually in-competent to purchase food from exter-nal sources within six hours of learning (Sanlier & Konaklioglu, 2012). Thus, it is vital for schools to realize their shared responsibility to provide safe food and nutri-tious to their students. Hertzman and Barrash (2011) explained that there is a still insufficient study done on foods served in schools, especially by education scholars.

Furthermore, Centre Disease Control and Preven-tion (CDC, 2012) identified five risk factors of food handling that promote to food-borne illnesses which include cross contamination between raw and fresh ready to-eat foods, lack of hygiene and sanitation by food handlers, tempera-ture abuse during storage and improper cooking procedure.

Basically, temperature is the most important factor in delaying or retaining the shelf life of perishables food (Rojers, 2012). Chaug and Wei (2013) stated that the fresh foods (perishables item) demand strict temperature to avoid less quality because of product characteristics itself have short shelf life when fluctuating temperature. Therefore, temperature is the most important factors and has an impact on the storage life of food products (Myo & Yoon, 2014).

The purpose of this study was to identify food temperature knowledge amongst food handlers. Previous studies have assessed the important of knowledge and practices among food handlers.

2 MATERIAL AND METHODS

2.1 *Target respondents*

The respondents for this study were among the food handlers in a school canteen. This is a case study based on the school that went through an experience of food poisoning case in 2013. The total sample sizes were eight food handlers that handle the school canteen. Purposive sampling technique is use for collecting data.

2.2 Data collection method

The study is using experimental design, which is pretest, and posttest experimental group design. It also known as before-and-after design. The use of this method is to find out the outcome of before and after the food, handlers go for training on food temperature. The pretest measured baseline knowledge, and the posttest-measured knowledge gained by the respondents after training using the food temperature topic.

The questionnaires adapted and developed in simple words and easy to answer. The questionnaires are using objectives questions, which is more on food temperature. The questionnaire format would be the closed-ended questions and it contains 10-items that designed to obtain information on food safety. Direct questions along with demographic characteristics of the respondents (age, gender and education level) were included in the front page of the questionnaire.

First, the respondents need to answer the questionnaires on food temperature. Then, after a week, authors gave them a short training about food temperature. Five days later, authors gave the same question as before respondents have their training to know either respondent improve their food temperature knowledge from the short training program.

2.3 Data analysis

Table 1. Demographic characteristics of the respondents.

Characteristics		a(n)%
Gender	Male	(2) 25%
	Female	(6) 75%
Age	20	(2) 25%
	21-30	–
	31-40	(3) 37.5%
	41-50	(3) 37.5%
	>50	–
Nationality	Malaysian	(5) 62.5%
	Indonesian	(3) 37.5%
	Bangladesh	–
	Vietnam	–
Race	Malay	(5) 62.5%
	Indian	–
	Chinese	–
	Jawa	(3) 37.5%

(*Continued*)

Table 1. (Continued).

Characteristics		a(n)%
Gender	Male	(2) 25%
	Female	(6) 75%
Age	20	(2) 25%
	21-30	–
	31-40	(3) 37.5%
	41-50	(3) 37.5%
	>50	–
Nationality	Malaysian	(5) 62.5%
	Indonesian	(3) 37.5%
	Bangladesh	–
	Vietnam	–
Race	Malay	(5) 62.5%
	Indian	–
	Chinese	–
	Jawa	(3) 37.5%
Level of education	No formal education	–
	Primary school	(2) 25%
	Secondary school	(4) 50%
	Diploma	(2) 25%
Working experience	1 year	(6) 75%
	1-5 years	(2) 25%
	6-10 years	–
	10 years	–
Attending 'Kursus Pengendali Makanan'	Yes	(6) 75%
	No	(2) 25%
Got Typhoid injection	Yes	(6) 75%
	No	(2) 25%
Know use of thermometer	Yes	(6) 75%
	No	(2) 25%

a(n)%—(number of respondents) percentage of respondents.

3 RESULTS

3.1 Demographic information

According to Table 1, there were 75% female and 25% male respondents. The majority of the respondents were the age group of 31 to 40 and 41 to 50 years old (37.5%). Approximately, 37.5% of the food handlers are Indonesian and majority of them are Malaysian with 62.5%. Result also showed that 50% education level are among secondary level with 75% working experience less than 1 year as food handlers.

Finally, 75% of the food handlers had their formal training on food handling certified by the Ministry of Health, Malaysia with 75% had their typhoid injection and 75% have knowledge on how to use a thermometer.

Table 2. Before and after knowledge questions and answers for food handlers.

Statements	Before Correct answers (n)	After Correct answers (n)
- The correct temperature for storing frozen food	1	8
- Ideal temperature for bacteria to multiply	3	3
- How many times can heat up the leftovers	8	8
- True facts about bacteria	3	3
- Ideal temperature for refrigerator	0	3
- The following is the most likely to contain bacteria	8	8
- Very important to provide safe food for	8	8
- The main factor causing bacteria to flourish	0	0
- The bacteria that cause food poisoning will only multiply in foods with a pH below except	0	8
- Cooked food should be discarded after	0	8

Table 3. Cross tabulation between education level and the use of thermometer.

		Use of thermometer No	Use of thermometer Yes	Total
Education level	Primary school	1	1	2
	Secondary school	1	3	4
	Diploma	0	2	2
Total		2	6	8

Table 4. Cross tabulation between working experience and the use of thermometer.

		Use of thermometer No	Use of thermometer Yes	Total
Working Experience	Below 1 year	2	4	6
	1-5 years	0	2	2
Total		2	6	8

Table 5. Cross tabulation between nationality and the use of thermometer.

		Use of thermometer No	Use of thermometer Yes	Total
Nationality	Malaysian	0	5	5
	Indonesian	2	1	3
Total		2	6	8

3.2 Knowledge transfer

Table 2 shows level of ability on answering the question before and after the training. There are 10 same questions which are answered by the same respondents before and after the training. All the food handlers have a hard time to answer question B8 which is before and after training, no one give the right answer for question B8. It might be the hardest question of all. However, all respondents answering questions B10, B1, B3, B7 and B9 true which is before the training not every respondent answer correctly especially for question B9 and B10 which is before the training, every respondent cannot answer correctly.

Hence, the food handlers did not answer right to question B5 before the training but after training, there are two people who can answer it right. However, question B2 and B4 remain, either before or after training. There is only three people answer the question correctly. They know how to answer the question might be because of their education level and also from the training at Kursus Pengendali Makanan.

The easiest questions are B3, B6 and B7, which all the respondents can answer correctly before and after the training. As can be seen, a respondent who answers many questions correctly got is a diploma holder. However, he did not attend to the 'Kursus Pengendali Makanan' and his working experience is under 1 year. Other than that, the other respondent who got high mark is studying until secondary school. It can be concluded that the respondent who study higher education can adapt more new knowledge than others.

Overall performance, basically the food handler has improved and the training is effective towards food handler knowledge. It can be seen by the in-creasing of answering true question after the train-ing.

3.3 Cross tabulation

Cross tabulation is one of the analytical tools that are very popular used in market research. The interrelationship can be detected by comparing the per-centages across the different categories obtained from a cross tabulation (Lua, Halilah & Cosmas, 2007). The used of cross tabulation is because to know in depth about the respondent and knowledge.

Table 3 shows the education level and the knowledge of either know about how to use a thermometer or not. There is just one food handler who only knows how to use a thermometer. Then, three respondents that have an education level until secondary school who knows how to use a thermometer and there are two respondents which both of them study until diploma level knows how to use a thermometer.

Total respondent who knows how to use thermometer are 6 people. It shows that the food handlers have basic knowledge on temperature control although they did not have proper understanding on using thermometer.

Table 4 show the cross tabulation between working experience and knowledge of using a thermometer. As can be seen, a respondent who has working experience below than 1 year do not know how to use thermometer. There are only 4 people who know how to use it. However, respondents who have been working experience within 1-5 years both know how to use thermometer. It might be because of the experience and knowledge gain from training program. Therefore, food handlers can learn the usage of thermometer and understanding the temperature, which are critical factors that can cause bacteria easily to multiply.

Table 5 indicates the cross tabulation between nationality and the knowledge of using thermometer. The table shows that only one Indonesian respondent know how to use the thermometer. Rather than Malaysian respondent which all of them know how to use thermometer. From the result shows that Malaysian respondents are more knowledgeable on food temperature.

3.4 *Discussion*

The study have illustrated that the food handlers initially already have the basic knowledge on food safety. Consequently, food handlers were lacking of knowledge on understanding the importance of food temperature in food handling. The food temperature is vital for every food handlers to understand and applying it to avoid from any cross contamination and food temperature abuse occurred. Furthermore, table 2 proves the effectiveness of ongoing training. From the training, it explains that food handlers' knowledge increased. Therefore, ongoing training have increased the knowledge of food handler pertaining to the food temperature knowledge of every food handlers especially in using thermometer for monitoring purposes.

3.5 *Conclusions*

As a conclusion, to ensure food handlers comply with safe food handling, more research is required to ensure that the most effective types of training and interventions can occur. There is also a need for improvements in the reporting of intervention studies in this area, which allows for the identification of bias in the results, and provides a guide to how representative the sample may be to the wider population. Thus, attention need to focus not only to the food handlers but also to the stakeholders and parents involvement relating with school canteen activities to ensure the students gets safe and healthy food.

REFERENCES

Bond, T.G. & Fox, C.M. (2007). Applying the Rasch Model: Fundamental Measurement in the Human Science. New Jersey: Lawrence Erlbaum Associates.
CDC (Centers for Disease Control, Prevention), 2012. Foodborne Outbreak Online Database (FOOD). Retrieved Disember 20, 2013 from http://wwwn.cdc.gov/foodborneoutbreaks

Chaug, I.H. & Wei,T.C. (2013). Optimizing fleet size and delivery scheduling for multi-temperature food distribution. Applied Mathematical Modeling Journal, 1–15.
Cullen, K.W. & Watson, K. (2007). Measuring school foodservice workers perceptions of organizational culture. The Journal of Child Nutrition and Management.
Hertzman, J. & Barrash, D. (2011). An assessment of food safety knowledge and practices of catering employees. British Food Journal, 109 (7), 562–576.
Kim, D.I., Barton, K., Hill, M. & Choi, S. (2010). Sample size impact on screening methods in the lllRasch Model. McGraw Hill.
Lua, P.L., Halilah, H., Cosmas, G. & Nurul, H.M.N. (2007). The impact of demographic characteristics on Health-Related Quality of Life Profile of Malaysian epilepsy Population. Applied Research in Quality of Life, 2(4):247–271.
Maizun Mohd Zain & Nyi Nyi Naing. (2002). Soci demographic Characteristics of Food Handlers and Their Knowledge, Attitude and Practice towardslFood Sanitation: A Preliminary Report. Southeast Asian J Trop Med Public Health, 33(2), 410–417.
Myo, M.A. & Yoon, S.C. (2014). Temperature management for the quality assurance of a perishablelllfood supply chain. Food Control, 40(2), 198–207.
Rojers, L.K. (2012). Cold storage, modern materials handling. Retrieved December 19, 2013 from http://mmh.com/images/site/MMH1201_BestPrac_ColdStorage.pdf.
Sanlier, N. & Konaklioglu, E. (2012). Food safety knowledge, attitude and food handling practices of students. British Food Journal, 114(4), 469- 480.
Sharif, L. & Al-Malki, T. (2010). Knowledge, att tude and practice of taif university students on food poisoning. Food Control, 21, 55–60.
Sharifa Ezat W.P., Netty D., Sangaran G. (2013). Paper review of factors, surveillance and burden of food borne disease outbreak in malaysia. Malaysian Journal of Public Health Medicine 2013, Vol. 13 (2):98–105.
WHO. World Health Organisation (2008). Foodborne disease outbreak:guidelines for investigation and control. Retrieved December 25, 2013 from http://www.who.int/publications/2008.

Theory and Practice in Hospitality and Tourism Research – Radzi et al. (Eds)
© *2015 Taylor & Francis Group, London, ISBN 978-1-138-02706-0*

Attracting and sustaining customers through antecedent satisfaction of dining experience

S. Ismail
Faculty of Education, Universiti Teknologi Malaysia, Johor Bahru, Malaysia

M.S. Mohd-Shah
Faculty of Hotel and Tourism Management, Universiti Teknologi MARA, Shah Alam, Malaysia

ABSTRACT: Restaurant industry is a competitive business in Malaysia. Attracting new customers is no longer guarantee profits. However, study on how to attract potential customers and sustain first visit and regular customers is still limited. Therefore this study was conducted to identify how the antecedents of satisfaction in dining experience can attract new customers and sustain the repeat customers. Data of this qualitative case study were gathered mainly from the interviews and supported by close observation, participant observation and documentary evidence to provide strong and well-grounded findings of the 6 Malay upscale family restaurants in Johor Bahru, Malaysia. Findings of this study showed that antecedents of dining experience has two major dimensions, which are previous experience of the repeat customers and first time customer knowledge of the restaurant. Menu variety, delicious food, attractive restaurant appearance and promotions were among satisfaction determinants for previous experience of the repeat customers while convenient access and restaurant promotion programs are determinant satisfaction for first time customer knowledge of the restaurant. This study suggested for the restaurateurs to pay attention on antecedents of dining experience to attract the potential customer and to ensure the first visit customers repeat their visit while the regular customers sustaining their visit.

Keywords: Customer satisfaction, antecedent experience, repeat customer, first visit, expectation

1 INTRODUCTION

Tourism industry has become second source of income for Malaysia after manufacturing industry, which contributed continuous positive growth and sustained it position as the 9th most popular country in the world for tourist destination since year 2009. In the year of 2010, the profit was reported RM56.5 billion that involved 24.5 million tourists (The United World of Tourism, UNWTO, 2012). The number of profit increased to RM58.3 billion with the involvement of more than 24.7 million tourists in the year 2011. The first six months of year 2012, number of tourist visited Malaysia was 11.6 million, which increased by 2.4 percent compare to only 11.4 million for the same period of year 2011 (Tourism Malaysia, 2012). The number is expected to reach 36 mil-lion from the year 2014 until 2020.

The increment of 4 percent tourist visiting Malaysia in 2012 has generated RM26.8 billion to the national GDP compare to RM25.7 billion in year 2011. Under the Tourism Transformation Plan 2020, the Globe Shopper Index reported that the number of tourist visiting Malaysia is expected 36 million with the expectation profit RM168 billion. The expectation can be achieved due to many reasons. Besides well known as a safe country to be visited due to stability in political, plenty of preserved historical places, finding of Fordes Online research that been conducted towards 5,339 respondents from 97 countries crossed the world showed that Malaysia was found as the 10th best country for hospitality, the 8th best country (out of 18 Asia Pacific countries) for modern service provided and the 4th world best country for shop-ping and Kuala Lumpur was found as the 2nd best capital city in Asia Pacific (Economy Investigation Unit, 2012).

Food service is one of the important elements in tourism industry (Tourism Malaysia, 2013). In the third quarter of 2010, the accommodation and restaurants sector contributed 5.1 percent to the Malaysian GDP. The percentage increased from 3.8 percent compared to the similar period in 2009 (Department of Statistics Malaysia Quarter Report, 2010). Sales within the Malaysian food and beverage retail industry were forecast to reach of US$15.69 billion in 2011 and are predicted to grow to US$21.17 billion by 2015 (Exporter Guide, 2012).

Since restaurant industry has played an important role in Malaysian economic growth, the industry requires much attention and more awareness to increase or maintain the economic growth particularly. Thus, this study was conducted with the aim to identify how the antecedents of satisfaction in dining experience can attract new customers and sustain the repeat customers. Three main research objectives developed, which were to:

a) identify dimensions of dining experience antecedents.
b) investigate how dining experience antecedents attracting first time customers?
c) identify how dining experience antecedents sustaining repeat customers?

2 LITERATURE REVIEW

2.1 The concept of satisfaction

Basically, there are two general conceptualisations of satisfaction existing in the literature: transaction-specific satisfaction and cumulative satisfaction. A transaction-specific concept of satisfaction provides valuable insights into a particular and short-run product or service encounter. Hence, satisfaction as a cumulative concept describes the customer's total consumption experience with a product or service (Anderson and Fornell, 1993) over more than one experience. Therefore, the satisfaction goes beyond an expected utility to encompass post-purchase consumption utility. To sum up, the satisfaction concept is not a transient perception of how happy a customer is with a product or service at any given point in time, but it is the customer's overall evaluation of their purchase and consumption experiences (Anderson and Sullivan, 1993).

Customer satisfaction based on this concept is a fundamental indicator of current and long run performance because it directly affects customer loyalty and subsequent business profitability. Therefore, consumer and marketing researchers should give more attention to the management of customer cumulative satisfaction (Cronin and Taylor, 1992). Furthermore, those researchers had not looked at the 'management' of satisfaction.

2.2 The significance of customer satisfaction in providing an understanding of the background to the study

The phenomenon of customer satisfaction, based on customer experience, was found to be of high interest not only to researchers but also to marketers (Cardozo, 1965). This has been reflected in the constant growth of social science literature on customer experience over the last four decades

(Kivela et al., 2000). Customer satisfaction has been recognised as essential factors leading to the success of most service industries including restaurants. It can determine the restaurateur's profit (Gustafsson and Johnson, 2004) by providing high quality service (Stevens et al., 1995).

2.3 The need for sustainable customer satisfaction management

Cumulative satisfaction is a fundamental indicator of current and long run performance because it directly affects customer loyalty and subsequent business profitability. These are some of the reasons why consumer and marketing researchers have given more attention to cumulative satisfaction and have such a great interest in the management of customer satisfaction (Cronin and Taylor, 1992).

3 METHODOLOGY

This qualitative case study was conducted at six up-scale Malay family restaurants in Johor Bahru, Malaysia. Data were gathered mainly from in-depth interviews with 108 different demographic restaurant customers and 28 restaurant staff from various positions. The data was also supported by close observation, participant observation and documentary evidence to provide strong and well-grounded findings and establish validity of the data. Theoretical, snowball, convenience and purposive sampling techniques were used to determine the case, sample size and sample within the case. Data was analysed using coding process that assisted by NUDIST computer software.

4 RESULT

Finding of this study found that antecedent of dining experience has two major dimensions, which are previous experience of the repeat customers and first time customer knowledge of the restaurant with it sub dimensions of convenient access and restaurant promotion programs.

4.1 Dimensions of dining experience antecedents

First time customers get to know about the restaurants and are influenced to visit them because of convenient access (strategic locations of the restaurant to passers-by), word-of-mouth recommendations from other customers and paid-for promotion programmes such as the TV programme and radio programme and non-paid promotion programme such as distributing food samples at shopping complexes.

Repeat customers, on the other hand, are largely influenced by their direct experiences of dining at the restaurant. Most of those repeat customers will, presumably, have had good experiences on previous visits, in terms of having had the choice of a varied menu, delicious food and because of the attractive appearance of the restaurant. Some of them, however, are also influenced by restaurant promotion programmes.

4.2 How dining experience antecedents attracting first time customer?

Convenient access and restaurant promotion programmes were the factors influencing first time customers' knowledge of the restaurants.

4.2.1 Convenient access

This study has indicated that convenient accessibility to the restaurant is one of the important factors that influence first time customers to visit the restaurants. The restaurants were easy to access either because of they were located at a strategic location such as private and government offices that were visible from the roadside, or close to many landmarks, or they were situated in a residential area. The transportation links such as roads to access the restaurants was an issue. The findings also show that the demographics, traffic patterns and first-hand knowledge of the site and the surrounding area are all parts of the site selection equation, which, lead the customers to come back to the restaurants.

4.2.2 Restaurant promotion programmes

Findings of this study showed that customers who were satisfied with their dining experience tended to convey their experience with others through positive word-of-mouth communication. As a result, those who received the information (potential customers) started to develop hopes and expectations of receiving the same dining experience.

However, during the dining experience the restaurants managed to deliver products and services that fulfilled their hopes and expectations. As a result, the customers met other satisfaction factors such as attractive restaurant appearance, air-conditioning, having a presentable buffet table, a unique eating space and delicious food alongside a variety of menus. These satisfaction factors then have led the customers to repeat their visit.

4.3 How dining experience antecedents sustaining the repeat customers

Customers made repeat visits to the restaurants due to their satisfaction with aspects of their previous meal experience based on their evaluation of

what they received from the restaurants in relation to their purpose for visiting the restaurant. The restaurants managed to provide (confirm) the customers' needs in terms of an appropriate variety of food options. For instance the restaurants sell almost all types of food including western foods and eastern foods that not only the customers' common favorite food but also traditional foods that are rarely found in restaurants nowadays; freshness of food ingredients and delicious taste of the food.

This study found that the actual reason for customer meal experience satisfaction was because of the food quality (fresh ingredients and taste of the food) that confirmed their expectation. Delicious and authentic foods, due to fresh ingredients, were also found as indicators of food taste quality.

The quality of food had influenced customers to revisit the restaurant even though some of them had to drive long distances to get to the restaurant and had to pay more for similar types of food, which was being offered by other restaurants

Besides, the large number of customers at the restaurant succeeded in drawing a potential customer's attention. The number of customers in a restaurant was seen to provide an indication of the quality of the food to potential customers.

The study also found that the restaurants' interior decoration through beautiful décor with illuminated and colourful menu photos and different lighting effects that made the restaurants look stylish were another attraction in the meal experience.

The data thus suggested that the factor influencing a customer's satisfaction with a previous dining experience was whether what the customers' experienced in the restaurant was better than their expectation and/or whether the products offered by the restaurant confirmed their expectations, which resulted in them developing loyalty towards the restaurant and as a result revisiting the restaurants.

5 DISCUSSION

This study found that the antecedent experience stage is a stage where repeat customers developed their dining experience at the restaurants and how first time customers gain knowledge of the restaurant and the first time customers got their knowledge about the restaurants. Findings are consistent with the study of Ismail and Mohd Shah (2014) who found that the antecedent of dining experience is one of the stages in dining experience process (antecedent, reservation, arrival, seating, meal, payment, departure). Therefore, to ensure customers' loyalty, restaurateurs need to ensure all stages in the dining process satisfy customers who dine at the restaurant.

First time customers got their knowledge about the restaurants through commercial promotion programs (such as television, radio, food samples and pamphlet distributions). Demographics, traffic patterns and first-hand knowledge of the restaurant and the surroundings were achieved through convenient accessibility to the restaurant such as being located at a strategic location (for example close to private and government offices that were visible from the roadside, or close to many land-marks, or situated in a residential area for the transportation links such as roads to access the restaurants).

For repeat customers, they developed their loyalty towards the restaurant, and as a result revisited the restaurants because of their experience in the restaurant were better than their expectation and/or the products offered by the restaurants confirmed their expectations. For instance, satisfaction with the dining experience was achieved when the product performance of the restaurants confirmed the customers' expectation through providing them with the opportunity to choose from a varied menu (traditional menus, international menus and customers' common favorite menus), confirmed customers' needs through providing fresh food ingredients that gave the authentic food flavor; and an attractive and stylish restaurant interior.

In addition, during the dining experience, satisfaction was achieved when the restaurants managed to deliver products and services that fulfilled customers' hopes and expectations through an attractive restaurant appearance, air-conditioning, having a presentable buffet table, a unique eating space and delicious food alongside a variety of menus. This finding is consistent with Pliner and Hobden (1994) who viewed individuals as being more likely to try new foods if they first see others eating and enjoying them. Bashir et al., (2013) also found that food taste was greatest contribution in satisfying customers in Malay restaurants. For repeat customers, these satisfaction factors then influenced them to develop loyalty to come back to the restaurants. A study by Nazri and Zainal (2013) also found that Malay restaurants used quality of ingredients and pay more attention on the dishes to fulfil customer expectation.

Those customers (first visit and repeat customers) then shared their satisfaction with the pre-meal experience with others by conveying positive word-of-mouth communication. Those (potential customers) who received the information started to develop hopes and expectations of receiving the same dining experience. The factors influencing customer satisfaction with a previous dining experience influenced the customers' level of satisfaction with the dining experience based on whether what the customers' experienced in the restaurant was better than their expectation and/or whether the products offered by the restaurant confirmed their expectation.

In the antecedent dining experience, those customer satisfaction factors were consistently available over time, and later these developed first visit customers' expectations that they would get a similar experience in future visits. After consuming the dining experience, many customers showed positive behavioural changes through repeat visits to the restaurants, repurchase of restaurants' products and/or services, and they conveyed positive word-of-mouth recommendations about the restaurants to others.

The customers chose the restaurants as their first preference, or had it in mind as their first choice for the place of their next dining out activity, and they revisited the restaurants with the intention to repurchase a similar memorable (customers' unexpected experience that emerged prior to visiting the restaurant) dining experience. The customers also conveyed positive word-of-mouth recommendations about their memorable experience to others. Those who had been influenced by the dining experience of others would also pass through similar stages of the dining experience process. If they were satisfied with the dining experience offered by the restaurants, they then might show similar loyalty to the restaurants by conveying positive word-of-mouth recommendations, choosing the restaurant as their first choice, repeat the purchase and repeat the visit. The satisfied customer was usually influenced to repeat the dining experience satisfaction. The previous dining experience is an antecedent for their future visits to the similar restaurants or to repurchase the restaurant's products and or service, or for price tolerance, or to choose the restaurant as the first choice for dining out, or to convey positive word-of-mouth recommendations to others. Finding of this study is in line with the study of Bradya and Robertson (1998), who aimed to test the relationship between service quality and satisfaction in or-der to determine whether service quality should universally be considered as an antecedent of satisfaction. The results of these authors' study confirmed that the effect of service quality on behavioral intentions is mediated by consumer previous dining experience and the customers' level of dining experience satisfaction.

Yuksel et al., (2006) focused on the evaluation of consumers' continuous attitude at the post-purchasing point. These authors suggested that the customers' over-all attitude towards a restaurant can be developed before or after purchasing and satisfaction mediates the relationships between its antecedents and consequences. The antecedents of dining experience can be in a form of self-congruence (intention to return) and service quality (physical quality and staff behavior).

6 CONCLUSION

In conclusion, successful restaurants use customers' needs and expectations as a starting point, developing proposals around their customers' needs and expectations, also meeting other corporate imperatives. Managing satisfaction therefore has to do with managing services and/or products, but also with managing expectations and perceptions of the customer. Understanding customers in this way is something that restaurants can no longer ignore. If restaurant want their services to be used and interventions to succeed, they need to meet the public on their terms and manage needs and expectations more clearly along the way to see the results in satisfaction. Measuring satisfaction seems to be just one element in this overall satisfaction management approach. Through this study, the place and the role of antecedent customer satisfaction dining experience become clearer and Malaysian Malay restaurants will have some practical guidelines on the way to Customer Satisfaction Management through attracting potential customers, and sustaining first visit and repeat customers.

REFERENCES

Anderson, G. and Fornell, C., 1993. A Customer Satisfaction Research Prospectus. In: Oliver, R.L., and Rust, R.T. (eds.), *Service Quality: New Directions in Theory and Practice,* Newbury Park, CA, Sage, pp. 241–268.

Anderson, E.W., and Sullivan, M.W., 1993. The Antecedents and Consequences of Customer Satisfaction for Firms, *Marketing Science*, 12, Spring, pp. 125–143.

Bashir, M.A.A., Ariffin, H.F., Baba, N., and Mantihal., S., (2013), The impact of Food Quality and its at-tributes on satisfaction at Malay Restaurants, Proceeding of "the InternationÒl Hospitality and Tourism Postgraduate Conference: Synergizing creativity and innovation in Research, 2–3 September 2013, pp. 175–180.

Cardozo, R., 1965. An Experimental Study of Consumer Effort, Expectations and Satisfaction, *Journal of Marketing Research*, 2, August, pp. 244–249.

Cronin, J., and Taylor, S.A., 1992. Measuring Service Quality: A Re-examination and Extension, *Journal of Marketing*, 58 (1), pp. 125–131.

Department of Statistics Malaysia Quarter Report, 2010 Economy Investigation Unit, 2012

Exporter Guide, 2012

Gustafsson, A., and Johnson, M.D., 2004. Determining Attribute Importance in a Service Satisfaction Mod-el, *Journal of Service Research*, 7, pp. 125–140.

Ismail, S., and Mohd Shah, M.S., 2014, Dining Experience Process in Restaurant, Proceeding of 2nd World Conference on Islamic Thought & Civilization: Rise and Fall of Civilization: Contemporary States of Muslim Affairs, 18–19 August 2014.

Kivela J., Inbakaran R., and Reece J., 2000. Consumer Research in the Restaurant Environment, Part III Analysis, Findings and Conclusion, *International Journal of Contemporary Hospitality Management,* 12 (1), pp. 13–30.

Nazri, A.R.M., and Zainal, A., 2013, The relationship between perceived service performance and customer satisfaction in Malay Upscale restaurants in Kuala Lumpur, Proceeding of the International Hospitality and Tourism Postgraduate Conference: Synergizing creativity and innovation in Research, 2–3 September 2013, pp. 181–186.

Stevens P., 1995. Dineserve: A Tool for Measuring Service Quality in Restaurants, *Cornell Hotel and Restaurant Administration Quarterly*, 36 (2), pp560–580.

The United World of Tourism, UNWTO, 2012 Tourism Malaysia, 2012

Tourism Malaysia, 2013

Yuksel, E., Dawes, L.P., and Massey, R.G., 2006. An Extended Model of the Antecedents and Consequences of Consumer Satisfaction, *European Jour-nal of Marketing,* 42 (1/2), pp. 35–68.

Theory and Practice in Hospitality and Tourism Research – Radzi et al. (Eds)
© *2015 Taylor & Francis Group, London, ISBN 978-1-138-02706-0*

Evaluation of hygiene practices on microbiological quality of tuna sandwiches

N.A. Mahyuddin, H. Nadia & W.I. Wan-Zunairah
Faculty of Food Science and Technology, Universiti Putra Malaysia, Malaysia

ABSTRACT: The safety of ready-to-eat food is a very important issue. Improper handling of ready-to-eat food may result in foodborne outbreaks. The main aim of the present work was to study the influence of hygiene practices of food handler's on microbiological quality of tuna sandwiches produced in a catering unit. The tuna sandwiches were prepared with two hygiene practices and stored at different temperatures. Both sandwiches were analyzed for different microbiological indicators such as total plate count, *E.coli* and coliform. With good hand hygiene, the presence of presumptive *E.coli* was only observed when the samples were kept at 30°C at day 6. Otherwise, the *E.coli* count remained unchanged from day 0 of storage. Presumptive *E.coli* was present in both temperatures when the samples were prepared with poor hand hygiene. Good hand hygiene practices obviously affected the microbiological quality of the samples during a period of 6 days of storage at 4°C. Results of the study clearly indicated the importance of practicing good hand hygiene and temperature control in the preparation of tuna sandwiches to minimize the risk of bacterial contamination.

Keywords: Ready-to-eat foods, tuna sandwiches, hygiene practice, storage temperature, total plate count, *E. coli,* coliform

1 INTRODUCTION

Ready-to-eat (RTE) food is defined as any food for consumption without further treatment or processing, or minimally processed foods. RTE foods are also known as foods which are intended to be consumed as they are. This definition covers both open—and pre-wrapped ready-to-eat products and is intended to apply whether the RTE food may be consumed hot or cold, purchased in the store to be eaten elsewhere. They are not normally includes the take-away and fast food. These foods are usually stored in refrigeration or at room temperature (Schaub, 2010).

The demand for the RTE food is now expanding very rapidly worldwide. According to a global online ACNielsen consumer survey (2006), the demand for the food was reported to be higher in Asian region, compared to the Western countries, especially in Thailand, China, Taiwan and Malaysia. The result of the survey also revealed that the biggest reason for buying ready-to-eat food is convenience. This is unsurprising as more women in Asia are working and therefore they do not have enough time for cooking. Another important reason is that some consumers believe that the RTE food is cheaper than buying all the ingredients and preparing the food from scratch.

The RTE foods are prepared from different types of raw materials depending on target group. It may be prepared by using ingredients obtained from canned food, fermented food, freshly cooked food or fresh, minimally processed food. The nature of the ingredient may consist of those prepared from grains and cereals such as breads and rice; animal and dairy products such as meats, surimi, cheese, egg and yogurt; seafood such as canned tuna, slices of salmon, shrimps; and fruits and vegetables such as freshly cut-fruits, lettuce, cucumber and tomatoes.

As the RTE food is gaining its popularity and most of them are characterized as having high nutritive contents, high water activity and near neutral pH, the microbiological quality has becoming a major concern to both service provider and consumer. Due to this, much effort has been done to identify the source of contaminants, as well as to understand those conditions that would appear to extend the shelf-life the food and minimize the growth of microbes associated with food-borne illness (Gibson and Roberts, 1986; Sheridan and McDowel, 1998; McCann et al., 2003). Therefore, a lot of studies have been carried out to evaluate the effects of environmental conditions on the growth of foodborne pathogen. Beuchat (1996) reported that food containing fish product must be stored at 50°C or below in order to deter the growth of bacteria. Proper cooking will kill certain bacteria. The growth of *E.coli* and the risk of developing a food-borne illness from properly cooked foods are low. However, in the case of RTE foods, since the food will not be cooked again, there is no way to kill bacteria that get on the food after its initial cooking process. Therefore, it is essential that RTE protein-containing foods be properly refrigerated to prevent bacteria growth (NIAID, 2011).

Considering the growing importance in food safety in RTE food and the current global issues, the present study aim to evaluate the microbiological quality of tuna sandwich stored at different temperatures. This study would generate information on the importance of practicing good hand hygiene in the preparation of RTE as well as keeping the food at the appropriate temperature in ensuring acceptable microbiologically safe food. Information generated from this study could become the basis for further investigation on the issues as to the best of our knowledge, there is no information available on the effects of hand hygiene on bacterial contamination on sandwich in Malaysia.

2 MATERIALS AND METHOD

2.1 *Experimental design*

All the ingredients required were purchased from a local supermarket one day before preparation of the sandwiches. Lettuce, cucumber and tomatoes were kept refrigerated prior to use. The vegetables were thoroughly washed in running tap water before cut into small pieces. Both cucumber and tomatoes were cut at about 2 mm thick while lettuce was shredded with a clean knife and surface. The tuna and salmon spreads were used and the cans were opened just before the preparation.The weight of each piece sandwich was standardized to 25 g in room with a temperature of 24°C. The preparation of the sandwiches for this was conducted at the Food Service Complex, Faculty of Food Science and Technology, Universiti Putra Malaysia. Then the samples were stored at 4 and 30°C for 6 days.

dilutionwas performed for further microbiological analyses which were TPC, fecal coliform count, and presumptive E. coli detection and count.

2.2 *Microbiological analysis*

2.2.1 *Total plate count*

Total plate counts were determined using ISO 4833:2003 (E) procedures. 25 g of sandwich were added to 225 ml of 1% peptone water in a stomacher bag and shaken for 60 seconds. The homogenized (1 ml) was sampled from the bag and it was serially diluted to obtain 10^{-1}, 10^{-2} and 10^{-3} dilution.The plates then were incubated at 30°Cin inverted condition. After 72 ± 3 h of incubation, numbers of colonies per plate were counted using a colony counter.

2.2.2 *Detection and enumeration of coliform*

Detection and enumeration of coliforms were performed following ISO 7521:2005 (E) procedures which employing Most Probable Number (MPN) technique using a three tubessystem.According to this procedure, a sufficient number of dilutions were prepared to ensure all the tubes for final dilutions will yield a negative result.The number of tubes which shows positive result were counted. Concentrations of coliforms were estimated using MPN table as in Table B.5 in ISO 7218: 2007 (E) (Appendix 1).

2.2.3 *Detection and enumeration of E.coli*

Detection and enumeration of *E.coli* were performed following ISO 7251:2005 (E) employing MPN technique using a three tubes system.The presence of *E.coli* was indicated by the formation of red colour in the solution (Figure 3.3). The number of tubes with red colour formation were counted as '+ve result' and the concentration of *E.coli* were estimated using the MPN technique according to ISO 7218: 2007 (E).

2.3 *Statistical analysis*

Data on total plate counts were subjected to analysis of variance (ANOVA) using Statistical Analysis System Ver. 9.2. (SAS Institute, Cary, North Carolina, USA). Mean differences between treatment were compared using Duncan's Multiple Range Test at P<0.05. For total coliforms and presumptive *E. coli*counts, MPN values were generated based on MPN table using a three tubes system as in ISO 7218: 2007 (E).

3 RESULTS AND DISCUSSION

3.1 *Microbiological quality for tuna sandwiches*

3.1.1 *Total plate count*

By overall TPC of samples prepared with poor hand hygiene were higher (1.22E + 07cfu/g) than those prepared with good hand hygiene (2.19E + 06cfu/g) (Figure 1).Results showed that TPC in tuna sandwich were markedly affected by levels of hygiene practiced but it was dependent on storage temperature. At 4°C storage temperature, samples prepared with poor hand hygiene had a

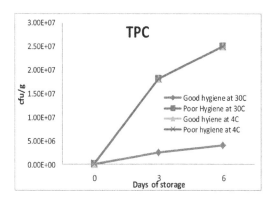

Figure 1. Changes in Total Plate Count (TPC) on tuna sandwich prepared under two hygiene prac-tices and stored at 4°C and 30°C.

Table 1. Most Probable Number (MPN) of total coliform of tuna sandwich prepared using two levels of hand hygiene during 6 days of storage at two temperatures.

Storage temperature	Days of storage	Good hand hygiene (MPN/g)	Poor hand hygiene (MPN/g)
4°C	0	0.3×10^1	1.1×10^1 $(0.7 \times 10^1$ $-4.6 \times 10^1)$
	3	0.9×10^1 $(0.7 \times 10^1$ $-4.4 \times 10^1)$	3.8×10^1 $(2.9 \times 10^1-$ $1.4 \times 10^2)$
	6	0.3×10^1 $(2.9 \times 10^1$ $-1.1 \times 10^2)$	2.3×10^1 $(1.8 \times 10^1$ $-1.1 \times 10^2)$
30°C	0	0.3×10^1	7.5×10^1 $(5.8 \times 10^1$ $-2.7 \times 10^2)$
	3	1.1×10^3 $(0.9 \times 10^3$ $-5.1 \times 10^3)$	2.4×10^2 $(2.0 \times 10^2 -$ $1.2 \times 10^3)$
	6	2.4×10^2 $(2.0 \times 10^2$ $-1.2 \times 10^3)$	$> 1.1 \times 10^3$

higher TPC value compared to TPC in samples prepared hygienically (P < 0.05). The overall mean of TPC of samples prepared with good hygiene practices was 1.39E + 04 cfu/g and the corresponding TPC of samples prepared with poor hygiene was 2.34E + 04 cfu/g. However, the TPCs were not significantly affected by storage period. In the study, TPC has been seen to decrease when the samples were kept at 4°C and not at 30°C, although the condition of culture media were not at optimum condition, as it it became acidic. This is well explained by Zymslowka (1999) who pointed out that some bacteria such as *Bacillus cereus* and *B. subtilis* may have a higher survival rate as these two bacteria strains were resistant to unfavorable temperature and environment because they have the ability to form spores. Similar findings were also reported by Terzieva *et al.* (1991).

3.1.2 Coliform count

The amount of coliform on all sandwiches increased markedly from day 0 to day 3 when stored at 30°C. The MPN of coliform was maximum at day 6 of storage for sandwich prepared with poor hand hygiene. However, amount of coliform decreased from day 3 to day 6 when the sandwich prepared with good hand hygiene was stored at 30°C. The progressive decrease in coliform number as the duration of storage extended is expected as coliform were reported to be easily inactivated or die-off naturally (USEPA, 2006). The presence of coliform in the sandwich may come from the environment as they are common inhabitants of the soil and surface water (USEPA, 2006). In this study, coliform may be present in vegetable used (lettuce, cucumber and tomato) as the vegetables may collect the coliform from the soil where they were planted and from the water source that has

been used to water the crops. The vegetables could also have been contaminated with coliform during harvesting, transportation, and marketing or by consumer itself prior of usage.

3.2.3 Presumptive E.coli

Presumptive *E.coli* on sandwich stored at 30°C shows different result from those of sandwich kept at 4°C for both hygiene levels (Table 2). At day 0 and day 3 of storage, *E.coli* was not detected but its presence was seen at day 6 for sandwich prepared with good hand hygiene. This result indicates that temperature had played important role in determining the growth of *E. coli*. The high dependency of bacterial proliferation on temperature is a general phenomenon and such results have been reported in many studies (Gibson and Roberts, 1986; McCann et al., 2003; Lopez-Velasco et al., 2010). When the samples were stored at 30°C, the presumptive *E.coli* in sandwich prepared with poor hand hygiene was detected at day 0 and similar results were continuously detected until day 6 of incubation. Such response was not observed for samples stored at 4°C. Lack of positive growth of *E.coli* at low temperature is expected and has been frequently reported in similar studies. Velasco et al. (2009) found that there was a slight decreased in *E.coli* 0157:H7 numbers observed after 15 days at 4°C, while storage at 10 oC resulted in a small increase in *E.coli* 0157:H7 count. Results obtained in this study also agreed with those reported by Raouf et al. (2006) which indicated that the survival

Table 2. Presence of *E.coli* tuna sandwich prepared using two levels of hand hygiene during 6 days of storage at two temperatures.

Storage temperature	Days of storage	Good hygiene (MPN/g)	Poor hygiene (MPN/g)
4°C	0	Absent	Present
	3	Absent	Present
	6	Absent	Present
30°C	0	Absent	Present
	3	Absent	Present
	6	Present	Present

and growth of *E.coli* 0157:H7 on lettuce and parsley decreased in number when the vegetables were kept at 4 - 5°C but the population has increased when the products were stored at 12°C and 21°C. Generally, the base temperature for *E.coli* is 10°C (Roberts, 1995). However, there would be some variation in the base temperature as some fraction of *E.coli* cells may elongate during incubation even at 2 - 4°C, and the elongated cells subsequently may be divided at 4°C when temperatures fluctuates occasionally (Jones et al., 2004). Jones et al.(2004)also reported that the elongated cells were divided when the incubation temperature fluctuated at 6 hour intervals, but not when temperatures were constant.

4 CONCLUSION

The results showed that tuna sandwich stored at high temperature with poor hand hygiene practice in the study exhibited unsatisfactory and unacceptable microbiological quality that posed health risks for consumers. Results of the study clearly indicate the importance of practicing good hand hygiene in the preparation of RTE food to minimize the risk of bacterial contamination. Besides, the RTE food prepared must be kept at low temperature, as the food will not undergo anymore treatment prior to consumption. In addition, the storage temperature is considered vital to the quality of tuna sandwich. The knowledge of good hand hygiene practice is very important for handler to avoid any risk of foodborne outbreak.

REFERENCES

ACNielsen Consumer Survey. (2006). Consumer and Ready–to–Eat Meals: A global ACNielsen Report December 2006. United States, ACNielsen.

Beuchat L.R., (1996). Listeria monocytogenes: Incidence on Vegetables. *Food Control*, 7, 223–228.

International Standard, 2003.Microbiology of food and animal feeding stuff—Horizontal method for the enumeration of microorganisms—Colony-count technique at 30 oC. Reference number ISO 4833: 2003. Geneva, Switzerland.

International Standard, 2005.Microbiology of food and animal feeding stuff—Horizontal method for the detection and enumeration of presumptive Escherichia coli—Most probable number technique. Reference number ISO 7251:2005. Geneva, Switzerland.

Jones, T., Gill, C.O., McMullen, L.M., 2004. The behavior of log phase Escherichia coli at temperature that fluctuates about the minimum for growth. Letters in Applied Microbiology 39, 296–300

López-Velasco, G. Davis, M. Boyer, R.R. Williams, R.C. Ponder, M.A. 2010 Alterations of the phylloepiphytic bacterial community associated with interactions of Escherichia coli O157:H7 during storage of packaged spinach at refrigeration temperatures. *Food Microbiology*, 27, 476–486

Terzieva, S.I. and McFeters, A. 1991 Survival and injury of Escherichia coli, Campylobacter jejuni and Yersinia enterocolitica in stream water. Canadian *Journal of Microbiology* 37, 785–790.

McCann, J., Stabb, E.J., Millikan, D.S. and Ruby E.G. 2003.Population dynamics of Vibrio fischeri during infection of Euprymnascolopes.*Applied Environmental Microbiology*, 69(10), 5928–5934.

National Institute of Allergy and Infectious Diseases.US Department of Health and Human Services, National Institute of Heath.Retrieved from http://www.niaid.nih.gov/Pages/default.aspx.

Raouf, A.U.M., Beauchat L.R., Ammar, M.S., 2006. Survival and growth of Escherichia coli O157:H7 on salad vegetables. *Applied and Environmental Microbiology*.59, 1999–2006.

Roberts, T.A. 1995. Microbial growth and survival: Development in predictive modelling. *International Biodeterioration and Biodegradation*, 96.297–309.

Roberts, T.A. and Gibson, A.M. 1986. Chemical Method for controlling Clostridium botulinum in processed meats. *Food Technology Champaign*, 40(4). 163–171

Schaub, K. 2010. What is Ready to Eat Food. Retrievedfromhttp://www.livestrong.com/article/100736-readytoeat/.

Sheridan, J.J and McDowel, D.A. (1998).Factors affecting the emergence of pathogens on foods.*Meat Sciences*, 49.151–167.

USEPA, (2006). Total Coliform Rule: A Handbook for Small Noncommunity Water Systems serving less than 3,300 persons.One of the Simple Tools for Effective Performance (STEP) Guide Series. EPA United States Environmental Protection Agencies. Retrieved from http://www.epa.gov/ ogwdw/disinfection/tcr/pdfs/ stepguide_tcr_smallsys-3300.pdf

Velasco, G.L., Davis, M., Boyer, R.R.,Williams R.C.,Ponder, M.A., 2010. Alteration of the phylloepiphytic bacterial community associated with interactions of Escherichia coli 0157:H7 during storage of packaged spinach at refrigeration temperature. *Food microbiology*, 27, 476–486.

Zymslowska, D.L., 1999. Survival of Bacteria Strains in Fish Feed Stored at Different Temperature. *Polish Journal of Environmental Studies* Vol. 8, 447–451.

Theory and Practice in Hospitality and Tourism Research – Radzi et al. (Eds)
© *2015 Taylor & Francis Group, London, ISBN 978-1-138-02706-0*

Consumers' understanding and perception on the use of food premise grading system

H. Hassan, J. Jamil, A. Arsat, A.A. Saidatul-Afzan & F.S. Chang
Faculty of Hotel and Tourism Management, Universiti Teknologi MARA, Shah Alam, Malaysia

1 INTRODUCTION

Grading of food premise was done annually when the premise starts in operation and was issued with a grading certificate, which will be required to be displayed in a prominent position on the premise. There are several objectives of grading system for food premise which includes facilitating user to make informed choices of food premises in public areas, to increase the level of cleanliness, food quality and safety, to instill values and practices of self-cleanliness among food handler and to enable food premises to consistently maintain high cleanliness standards.

In addition, the grading system, which was first introduced in the 1980's involving public participation, especially customer who will choose the premise they like based on its cleanliness. According to Auckland Council (2011), the food grading system is a program designed to improve food safety. Therefore, all food premises that provide food to the public must apply and display the license to show their grade category. In addition, the council also stated that consumer can judge how well the food premise is complying with the rules and regulations that govern food safety. Nonetheless, the grade of the food premises will only make sense of consumer perception of the various level of food grading (Majlis Perbandaran Seberang Perai (MPSP), 2009).

Poor personal hygiene and sanitation performance are serious hazards in any food businesses and also could harm customers. Personal hygiene means conditions or practices of maintaining good health habits including bathing, washing hair, wearing clean clothing and frequent hand washing while sanitary means healthful of hygienic where it involves reducing the number of disease-causing microorganisms on the equipment surfaces and utensils to an acceptable public health level.

As reported, there were 11,226 cases from January till September 15 in 2007, a 100 percent increase compared to the same period in 2006. Despite a perceived higher awareness about food hygiene, the country continues to be plagued with food poisoning cases in 2010, which about 50 percent involved were school canteens. Food poisoning in Malaysia increased every year. For instance in 1995, the numbers of cases were about 1438 as compared to 8640 cases in 1999. Moreover, Health Minister, Datuk Seri Liow Tiong Lai said the 311 cases was an increase of 29 cases from the previous year which in 2009 and mostly due to contamination of food during preparation, transportation as well as serving (Bernama, 2011). In 2000, World Health Organization (WHO) claimed that, many developing countries do not have a comprehensive food safety program, meanwhile for food-borne disease are often not reported. From the statement, it was revealed that food premises grading system is important and one of the ways to guide consumer to select a better restaurant to avoid food poisoning. Hence, this study investigated the consumer understanding on food premises grading system.

2 LITERATURE REVIEWS

In Malaysia, the food premise grading systems were certificated by Ministry of Health. This means the food premises are audited on a routine basis to check if the premises meet the current rules and regulations that were set by Ministry of Health. According to Majlis Perbandaran Seberang Perai (MPSP) (2009), on auditing, the inspection focuses on several important aspects including the physical condition of the premises at the time of the last grading inspection, the conduct of the operator and staff, cleaning and sanitizing of the premises, training of staff and food safety procedures in place. The level of cleanliness of the food premises are classified by grade marks which are 'A' for 80 percent to 100 percent, 'B' for 65 percent to 79 percent, 'C' for 50 percent to 64 percent while 'D' for 0 percent to 50 percent. Under Section 11 in Food Act 1983 the food premises that obtain grade D need for immediate closure not exceeding 14 days.

2.1 *Food premise grading system: Malaysian perspective*

Generally in Malaysia, through the government enforcement on food grading system, it can be seen

Table1. Qualification in food premises grading system (MBSA, 2009).

Gold A	Awarded for excellence. This is given to premises that achieved a very high level of compliance with the regulations. In addition, they consistently demonstrate best practice to ensure that food is safe.
Grade A	Grade A is given to premises that achieved a high level of compliance with the regulations.
Grade B	Grade B is given to premises that achieved a good level of compliance with the regulations.
Grade D	Grade D is given to premises that have a poor level of compliance with the regulations, and/or have repeated faults from the previous inspection.
Grade E	Grade E is given to premises that have an unsatisfactory level of compliance.
Grade Pending	Grade pending is given to a new premise or a premise that has recently transferred ownership.
CI	This premise holds a certificate of inspection. It does not require registration and is not included in the food grading scheme. These premises still need to meet the standards set by the Food Hygiene Regulations 1974 and they are inspected annually.

clearly that either the case of food poisoning by the state in Malaysia had increase or decrease. If the poisoning cases were increase, MOH need to improve the food premises grading system. Afterward, in Malaysia, the food grading systems are given by the council city (MPSP, 2009) which is under the supervisory of Ministry of Health. That is the reason for food grading system designed in order to improve food safety in food premises. Next, according to the food grading system in MPSP, 2009 the grades include 'A', 'B', 'C', as well as 'D'. However, for this study, it will be focusing in Shah Alam, Selangor area which is under the supervision of the Majlis Perbandaran Shah Alam (MBSA). Under foodservice area, Food Premises Grading Program is implemented in order to achieve several objectives, which are to facilitate users to make inform choices of food premises in public areas, to increase the level of cleanliness, food presentation, safety and quality of food prepared, instill values and practice of self-cleanliness among traders and handlers of food and lastly is to enable food premises to consistently maintain high cleanliness standards (MPSP, 2009).

2.2 Consumer perception

A consumer perception is one of the factors that drive on food premises choices (Worsfold, 2006). He further stated that, nearly all consumers claimed that the standard of food hygiene was crucial for them when deciding where to dine out. Most of the food premises neglect the importance of training and understanding of hygiene and sanitation among food operators. According to Simon, Leslie, Run, Jin, Reporter, Aguirre and Fielding (2005), the idea that gives consumers education will change their perception towards food

premises grading systems which related to food illness issues.

Through food premises grading system, there are the logos or label 'A', 'B', and 'C' that give consumer information about the level of cleanliness at particular food premises. From these grades, as expectation the food premises that received 'A' would have higher consumer demand compared with 'B' and 'C'. While the government enforcement plays a significant role that contributed to the food grading system, food premises operators shall make sure that food handlers are supervised and instructed and/or trained in food safety activities in order to get "high score" in grading system, thus, consumers are becoming more educated and fussy in going places to dine in.

3 METHODOLOGY

The population of Shah Alam is approximately about 646, 890 peoples and have 56 sections (MBSA Website, 2012). Among all, Sections 2, 6, 7, 8, 13, 14, 18, 19, 23, 25 and new townships in Shah Alam (Setia Alam and Kota Kemuning) were chosen as the contextual setting. The respondents for the study embraced 150 customers that dine in at food premises that received grade 'A', 'B' or 'C' in selected areas. Stratified sampling technique was used in capturing sufficient data to reduce sampling error and ensures the greatest representation, saves money due to this method often requires smaller sample. This sampling method took 40 percent of customers from 'A' grade food premises, 30 percent of customer from 'B' grade food premises and 30 percent of customer from food premises that received 'C' grades. Therefore, this study focused on the food premises in Shah Alam areas due to lack of studies carried out in the context in Malaysia especially in those particular areas. During the data collection process, respondents were informed that their participation is voluntary basis and all the information given will be kept private and confidential. The data was analyzed using Statistical Package for Social Sciences (SPSS 18).

4 RESULTS AND ANALYSIS

4.1 Respondents' profile

In this study, the total numbers of respondents are 150. There are 57 male (38%) and 93 females (62%). Six main categories for age, which are 18–20 years old (n = 13, 9%), 21–25 years old are about 42 respondents (28%), for 26–30 years old are about 36 (24%), for 31–35 years old are about 27 respondents (18%), 36–40 years old are about

Table 2. Basic knowledge of food safety towards food premises grading system.

	Mean	Std. Deviation
Sufficient information food safety	2.29	.824
Important food grading system	2.65	1.227
Perception A	3.44	1.052
Perception B	3.36	1.172
Perception C	3.28	1.142
Grade your food	2.05	.961
Logo	2.69	1.466

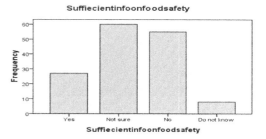

Figure 1. Sufficient information on food safety.

20 respondents (13%). For the age group above 41 years old, the there are 12 people (8%). For single, there are 65 respondents (43%) and 85 of them are married (57%). For education level, two of the respondents have SRP/PMR (1%), while 27 respondents have SPM. Meanwhile, diploma holders were represented by 48 number of respondent (32%). For degree holder, there are 62 respondents (41%). Lastly for other educational level, there are 11 respondents (7%). Among all, 24 of the respondents are students (16%), government servants (68 respondents, 45%), private sector (29 respondents, 19%), self-employed (21 respondents, 14%) and non-employed (8 respondents, 5%). For income categories, the respondents that has below RM 999 is 28 (19%), RM 1000–1500 (n = 13, 9%), RM 1600–2000 (n = 17, 11%) RM 2100–2500 (n = 33, 22%) and RM 2600 (n = 59, 39%).

4.2 Basic knowledge of food safety towards food premises grading system

4.3 Sufficient information on food safety

Most of the respondent (n = 55) claimed that they do not have sufficient information, 27 of them have sufficient information and the others are not sure (n = 60) while eight of them do not know either they have sufficient information or not. The mean of this section is 2.29 and the standard deviation (e^2) is 0.824.

4.4 The most important of food premises grading system

For question asked on "what the most important about food grading system", most of the respondent (n = 56) answered that food grade show that the premises is good in all aspects. While the others (n = 39) said that it can reduce the number of food poisoning cases. Other opinions, some (n = 31) said it will enhance customer satisfaction and lastly (n = 24)

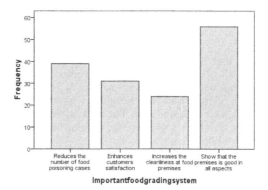

Figure 2. Important of food premises grading system.

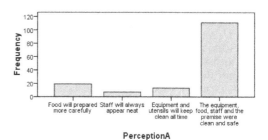

Figure 3. Perception about food premises that have Grade A.

said that it can increase the cleanliness of the food premises. For this question the mean is (M = 2.65) while for standard deviation is (e^2 = 1.227).

4.5 Perception about food premises that have grade 'A'

Next, for question about "your perception on Grade 'A', most of the respondent (n = 111) said that if the food premises received 'A' grade, meaning that the equipment, food, staff and the premises were clean and safe. Other than that, 19 respondents

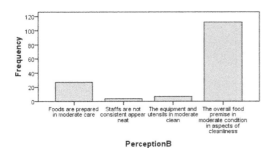

PerceptionB

Figure 4. Perception about food premises that have Grade B.

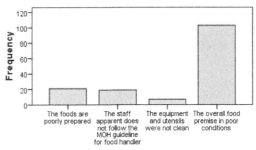

PerceptionC

Figure 5. Perception about Food Premises that have Grade 'C'.

said that the food was prepared more carefully and 13 of them stated that the equipment and utensils were kept clean all the time. Lastly, only seven of them said that the staff or food handler was always appeared neat. The mean score for this question is (M = 3.44) while for standard deviation is (e^2 = 1.052).

4.6 Perception about food premises that have Grade 'B'

Next, for question about "your perception on Grade 'B', most of the respondent (n = 112) said that if the food premises have 'B' grade, meaning that the overall food premises in moderate condition in aspects of cleanliness. Other than that, (v = 27) said that the food were prepared in moderate care and (v = 7) stated that the equipment and utensils in moderate cleanliness. Lastly, only a few of the respondents (v = 4) said that the staff or food handler does not appear neat consistently. The mean score for this question is (M = 3.36) while for standard deviation is (e^2 = 1. 172).

4.7 Perception about food premises that received 'C' Grade

For question about "your perception on Grade 'C', most of the respondent (v = 103) said that if the food premises have 'C' grade, meaning that the overall food premises is in poor conditions. Other than that, (v = 21) said that the food are poorly prepared and (v = 19) stated that the staff apparent does not follow the MOH guideline for food handler. Lastly, a few respondents (v = 7) said that the equipment and utensils were not clean. The mean score for this question is (M = 3.28) while for standard deviation is (e^2 = 1.142).

Only 30 male said that they see the label 'A', 'B', 'C' or 'D' at the food premises. Meanwhile for the

female, only 49 respondents said that they have seen the label. The total respondents said yes are 79 respondents are equal to 52.7 percent. Other than that, 15 male and 18 female respondents said that they are not sure which equal to 22 percent are about 10 respondents for male and 17 respondent for female which made up 18 percent from the overall percentage. Lastly, for the respondent whom answers those do not know are about 7.3 percent, which include 2 male and 9 female respondents.

4.8 Food premise grading system: Consumer perception

In this study, the consumer perception on the food premise grading system which is they are dependent on the food premise grades 'A', 'B' or 'C' as a guide in selecting the food premise is slightly different of the distribution agreement between agree and disagree. The agreed respondent was highest which is 34.7 percent (n = 52), strongly agreed 11.3 percent (n = 17), followed with total of disagreed 28.7 percent (n = 43) and neutral 25.3 percent (n = 38).

Majority of the respondents agreed with the statement of grades 'A', 'B' or 'C' of food premises show the level of cleanliness and safety. The total of agreed respondents was 87.3 percent (n = 131). In this study, cross tabulation analysis revealed that the relationship between demographic which is level of age between the consumer's understanding and concern about food premise grading system is shown that there are about half of the total respondents which are 72 respondents agreed on the statement. They are from the age of 21 to 25 (24 respondents), from age of 26 to 30 (20 respondents) and the rest from other level of age. While, about 27 respondents were disagreed on that statement and 51 respondents were neutral.

5 CONCLUSION

The food premise grading system is important in order for the local authorities to provide information in a transparent way to the consumer. Knowledge also plays significant roles that can influence customer purchase behavior. If there are greater number of study in the field of food premises grading in Malaysia, it would bring and give positive impact on society especially to create the awareness and indirectly reduces number of disease that were related to food consumption. There are limitations of this study due to the location as this research was only covered in Shah Alam area which cannot be generalized to the whole Malaysia population. Therefore, it is suggested to carry out the similar nature of study in different contextual setting.

ACKNOWLEDGEMENTS

This paper is sponsored by Research Acculturation Grant Scheme (RAGS 2013) by Universiti Teknologi MARA and Ministry of Education, Malaysia.

REFERENCES

Auckland Council. (2011). from http://www.aucklandcity.govt.nz/council/services/foodsearch/grading.asp

Bernama. (2011). Half of Food Poisoning Cases Involve School Canteens. http://www.bernama.com.my/bernama/v6/newsgeneral.php?id=574886

Majlis Perbandaran Seberang Perai (MPSP). (2009). from http://www3.mpsp.gov.my/index.php?option=com_content&view=article&id=141:program-pengredan-premis-makanan-26112009-&catid=11:2009&Itemid=58&lang=en

Simon, P.A., Leslie, P., Run, G., Jin, G.Z., Reporter, R., Aguirre, A. & Fielding, J.E. (2005). Impact of restaurant hygiene grade cards on foodborne-disease hospitalizations in Los Angeles County. *Journal of Environmental Health, 67*(7).

Worsfold, D. (2006). Consumer information on hygiene inspections of food premises. *Journal of Foodservice, 17*(1), 23–31.

Theory and Practice in Hospitality and Tourism Research – Radzi et al. (Eds)
© 2015 Taylor & Francis Group, London, ISBN 978-1-138-02706-0

Analysing passengers' expectation towards KTMB railways catering

S.M.A. Azid, M. Ramli, R.P.S. Raja-Abdullah, L. Benedict-Jipiu & N. Baba
Faculty of Hotel and Tourism Management, Universiti Teknologi MARA, Shah Alam, Malaysia

ABSTRACT: In Peninsular Malaysia, the railway sector especially train as a mode of transportation is relevant with our current economic situation. The rail sector in Malaysia can be divided into 3 main areas which are the Freight Services, Keretapi Tanah Melayu Berhad (KTMB) Intercity Passenger Services and KTM (Keretapi Tanah Melayu) Services. Other than business of rail transportation, maintaining and managing the system to serve, KTMB has also involved in their own foodservice catering operation that includes preparing and transporting food for off-site catering. In its catering situation, the challenging aspects of food delivery add to the necessity of proper food handling. Operating as a catering business, responsibility for ensuring safe food with high quality products is important for the consumption of the passengers on board. This research aims to analyze passenger's expectation towards the KTMB food on board railways catering.

Keywords: KTMB, catering, expectation, passengers'

1 INTRODUCTION

The KTMB services are even more challenging where it operates 24 Intercity Passenger Trains daily where 16 express trains and 8 local trains were involved. In addition, there are 206 Commuter Train Services operated daily between Monday and Friday, 213 services on Saturday and 177 services operated on Sunday and public holidays, 39 stations and halt along between the Rawang-Seremban and Sentul-Port Klang sectors. KTMB also operates Freight Trains supporting key maritime industry players in the international container movement with the launching of containerized Land Bridge Services between Klang Valley and Bangkok.

Law (2011) stated that the food catering service for KTMB is more challenging than the other catering services because the service providers have to secure the expected service level with the material availability which were involves with the long distance travelling. Hence, all foods need to be kept proper by avoiding the occurrence of food contamination on board. Passengers' satisfaction is said to be very important and should be given more priority. Based on an interview with Mr. Roslan Bin Rasali the Manager of Food and Beverage in KTMB, the company have a catering license, however, starting from year 2010 until to-date, they have only one single catering operation in the coach. The KTMB have yet to apply HACCP certification thus they will definitely improve in developing a new catering center onboard train in the future.

Besides that, the food production is being outsourced by other food contractors and suppliers to provide meals for the passengers. He added that some passengers who board the train were complaining on the types of food sold which does not meet their expectations. Therefore, in order to overcome this problem, as a Manager of Food and Beverage he is responsible for choosing and finding the newest types of menu that will be sold on board train.

He stated that he had to go through several stages of processes before could come out with a valid menu. Rasali and Mustapha (2013, November 11) reported that each contractor that undersigned a contract with KTMB does not last more than a year due to unprofitable food business.

This study helps to relate the customers' expectations from boarding the train with the independent variables such as i) types of food served which determined the variety of menu offered, price, food quality and food packaging and; ii) atmospheric elements which identified the suitability of color, lighting, design, layout and temperature criteria. It also helps to attract passengers not only among the local people, but also tourists from abroad.

2 LITERATURE REVIEWS

2.1 *Food*

People generally choose rice because it is a form of carbohydrate and also can fill up one's stomach (Babu, Subhasree, Bhakyaraj & Vidhyalakshmi,

2009). Rasali and Mustapha (2013, November 11) described few reasons as to why rice is suitable to be served on board, in Malaysia, it is because eating rice is part of our culture and many recipes have been developed based on rice. Therefore, rice has become the daily food for people and it is suitable to be eaten on board. In their major study, Kulkarni, Desai, Ranveer and Sahoo (2012) identified that noodles are easy to cook and there are different types of noodle available and people can choose to eat them instead of rice, and also it has a longer shelf life.

2.2 *Atmospheric elements*

Grayson and Mcneill (2009) mentioned that atmospherics can be said to be one of the most important aspects to be considered when managing business objectives and customer's expectations.

Besides that, atmospherics have been defined as the effort to design buying environments to produce specific emotional effects in the buyer that enhance his purchase probability (Ariffin, Bibon & Raja Abdullah, 2011). Therefore, there are five atmospheric elements such as color, lighting, design, layout and temperature.

Marcovic, Raspor and Segaric (2010) highlighted that customers will seek new dining experiences that will satisfy their ever-changing expectations. They also added that it is important to know, understand and meet customer's expectations. Besides that, the atmosphere is most important when we visit a place. Therefore, the element of an atmosphere should be chosen carefully, especially in terms of color specification.

The theme for dining area of the on-board train is very suitable because in an on-board train, there are too many passengers with different culture and religions, so, it is suitable to create a themed dining area on board such as family theme. Passengers will feel like having a family activity, eating together and at the same time strengthening their relationship.

Zijlstra and Mobach (2011) described that a functional layout can help to improve the efficiency of operations. They added that following this line of thought, customers may perhaps benefit from a central position of a good display in the layout given the interdependencies with the wait.

An adequate supply of fresh air with suitable temperature and humidity is necessary for passengers' comfort and productivity in the dining room (McSwane, Rue & Linton, 2005). Finally, they stated that modern heating, ventilation and air conditioning (HVAC) systems will filter, warm, humidify and circulate the air in the dining area and maintain a desirable and a comfortable temperature to the passenger.

3 METHODOLOGY

This study seeks to analyze the relationship between types of food served and the atmospheric elements towards passengers' expectation. A quantitative approach is considered to be the most appropriate method for the data collection process, with a structured questionnaire as the research instrument. A correlational study will be used in order to determine the extent of a relationship between all independent and dependent variables using data collection/statistical data. Researchers will also self-administer in distributing and collecting the questionnaire. This study will be conducted on KTMB on-board, which involve both journeys between Intercity and Electric Train Service. It is assumed that the data collection will be collected in two weeks time. All passengers that are on board train at that particular time will participate. Employees at KTMB will also be assured that any information given during the interview and in the questionnaire will be kept confidential and no respondents will be identified.

The instrumentation will consist of four sections that include food, atmospheric elements, passenger expectation and personal detail.

4 SIGNIFICANCE OF STUDY

The outcome of this study does not only focus on the aspects of Foodservice provided by KTMB, but it is also important for enhancing the economy in Malaysia as it can boost up the tourism industry. From an academic perspective, the researchers want to determine the factors affecting customer expectation towards on-board railway catering in KTM Berhad. Moreover, in practical perspective, the researchers want to find out the weaknesses that catering railway are facing so that KTMB management could be improved more in terms of upgrading and modernizing of the system for passenger satisfaction.

5 CONCLUSION

Even though this study is yet to be completed, theoretically researchers found that on board train is preferred because of its good accessibility to the terminal. Railway transportation systems has provided the most efficient ways for moving a large number of people, especially in densely populated urban centers. Therefore, customers' satisfaction is one of the key to measure the quality of a service.

REFERENCES

Ariffin, H.F., Bibon, M.F. & Raja Abdullah, R.P.S. (2011). Restaurant's atmospheric elements: What the customer wants. *Journal of Asian behavioural studies, 1(2), 33–44, 1*(1), 33–44.

Babu, P.D., Subhasree, R., Bhakyaraj, R. & Vidhyalakshmi, R. (2009). Brown rice-beyond the color reviving a lost health food: A review. *American-Eurasian Journal of Agronomy, 2*(2), 67–72.

Grayson, R.A. & McNeill, L.S. (2009). Using atmospheric elements in service retailing: understanding the bar environment. *Journal of Services Marketing, 23*(7), 517–527.

Kulkarni, S., Desai, A., Ranveer, R. & Sahoo, A. (2012). Development of nutrient rich noodles by supplementation with malted ragi flour. *International Food Research Journal, 19*(1), 309–313.

Law, K.M. (2011). Airline catering service operation, schedule nervousness and collective efficacy on performance: Hong Kong evidence. *The Service Industries Journal, 31*(6), 959–973.

Markovic, S., Raspor, S. & Segaric, K. (2010). Does restaurant performance meet customers' expectations? An assessment of restaurant service quality using a modified Dineserv approach. *Tourism & Hospitality Management, 16*(2), 181–195.

McSwane, D.Z., Rue, N.R. & Linton, R. (2005). *Essentials of food safety and sanitation* (4th ed.). United States of America: Pearson Prentice Hall.

Rasali, R. & Mustapha, Z. (2013, November 11). [Personal interview].

Zijlstra, E. & Mobach, M.P. (2011). The influence of facility layout on operations explored. *Journal of Facilities Management, 9*(2), 127–144.

Theory and Practice in Hospitality and Tourism Research – Radzi et al. (Eds)
© *2015 Taylor & Francis Group, London, ISBN 978-1-138-02706-0*

The relationship between Perceived Organizational Supports (POS) and intention to leave a job among employees of casual dining restaurants in Klang Valley area

H. Ghazali, N. Mohamad-Nashuki & M. Othman
Faculty of Food Science and Technology, Universiti Putra Malaysia, Malaysia

ABSTRACT: In restaurant industry, employee turnover is considered as part of its nature and is inevitable. However, due to the high costs the organization had to suffer, upper managements always look for the root of the problem and seek answers to minimize the damage. In this research, the relationship between Perceived Organizational Support (POS) and intention to leave among the employees of casual dining restaurants in Klang Valley area were studied. In addition, the difference between the leaving intention between managerial and non-managerial staff were also examined. The result obtained showed that POS and intention to leave were negatively correlated, while non-managerial staff had higher tendency in leaving the restaurant they were currently employed. As conclusion, it is suggested that the upper management continue to show support and rewards their employees' efforts so that the employees do not consider any job alternatives and leave.

Keywords: perceived organizational support, intention to leave, restaurant industry, managerial employees, non-managerial employees

1 INTRODUCTION

Casual dining restaurant is one of the popular eating spot among Malaysians. It is a moderate-up-scale dining place that placed details on themes in menu, service, and décor (Rivera, DiPietro, Murphy & Muller, 2008). Euromonitor International (2012) reported that Malaysian casual dining restaurants are divided according to their price range; (i) higher price range where the average guest cheque is RM40-50, (ii) lower price range where the average guest cheque is RM10-20. Unfortunately, behind its popularity and growth, restaurant industry suffers from high rate of turnover. It is reported that in a survey done between June 2010 and July 2011 by Malaysian Employers Federation (MEF), Hotel/Restaurant industry was placed third as the industry with highest annual average turnover rate at 32.4 percent, only behind IT/Communication (75.7%) and Associations/Societies (33%) (Goh, 2012). For example, Ryan, Ghazali, and Mohsin (2011) reported that turnover rate among non-managerial staff in Malaysian fastfood industry was 100 percent.

Turnover is defined as the movement of people in and out of employment within an organization (Abdullah et al., 2010). The most well-known effect of employee turnover is cost (Ingram & Jones, 1998; Kahumuza & Schlechter, 2008; Moncarz, Zhao & Kay, 2009; Tuzun & Kalemci, 2012; Wildes, 2005). Costs of turnover could involve direct replacement expenses like advertising, head-hunter fees, and employee development and indirect opportunity costs like lost sales, lower productivity, and customer defections (Hay, 2002). Since it is considered as a threat to the company's success (Mansfield, 2007), efforts have been taken by the managements to reduce turnover among their employees.

Intention to leave is the employee's intention of leaving the organization they are currently employed (Cho, Johanson & Guchait, 2009). Although it does not define an actual turnover (Mobley, Horner & Hollingsworth, 1978), it is said to be the better predictor of employee's turnover action in the future compared to job satisfaction (Ghiselli, La Lopa & Bai, 2001; Tuzun & Kalemci, 2012). Due to its troubling trait that continues to cause problem to hospitality establishment (Ryan, et al., 2011), a lot of studies were done to investigate the cause of turnover. Unfortunately, it is still hard to predict who would have the tendency to leave the organization they are working for. In Malaysia specifically, study on turnover intention is rather limited (Ponnu & Chuah, 2010) as well as academic studies in human resources in hospitality industry despite the industry's growth (Ahmad & Zainol, 2011). One of the highly studied antecedents of intention to leave was perceived organizational support (POS). Therefore, the objectives of this study are two-fold:

i) To determine the relationship between POS and intention to leave
ii) To find out if there is a difference between managerial and non-managerial staff's intention to leave.

2 LITERATURE REVIEW

2.1 *Perceived organizational support and intention to leave*

The study of perceived organizational support (POS) began when Eisenberger, Huntington, Hutchison, and Sowa (1986) extended the study of employee's Organizational Commitment (OC) by Mowday, Steers, and Porter (1979) into social exchange approach. They focused on employee's beliefs of the commitment showed by the organization to them compared to the employee's individual commitment to the organization as centred in the study of OC. Thus, they suggested that to determine the organization's readiness to reward increased work effort and to meet needs for praise and approval, employees develop global beliefs concerning the extent to which the organization values their contributions and cares about their well-beings. Higher POS resulted in greater effort in helping the organization to achieve their goals (Arshadi, 2011) while reducing the tendency to look for and accept alternatives jobs (Allen, Shore & Griffeth, 2003).

In the relation between POS and behavioral intentions, it was found that there were actual withdrawal behaviors like tardiness, absenteeism, and voluntary turnover (Arshadi, 2011; Rhoades & Eisenberger, 2002). Allen et al. (2003) stated that the increment of POS decreases the intention to leave because POS creates the feeling of obligation towards the organization. Other studies also found negative relationship between the two variables (Jawahar & Hemmasi, 2006; Kahumuza & Schlechter, 2008; Wayne, Shore & Liden, 1997). Kahumuza and Schlechter (2008) found a strong relationship between POS and intention to leave ($r = -0.521$, $p < .01$) thus stressing on the importance of POS to oppose intention to leave. Studies in hospitality industry showed no difference in the relationship pattern as both Cho et al. (2009) and Blomme, Van Rheede, and Tromp (2010) found that the employees of hospitality industry tend to have lower leaving intention if they have higher POS. Thus, the following hypothesis is proposed:

H1: POS will have negative relationship with intention to leave a job.

2.2 *Managerial and non-managerial employees and intention to leave*

Ghiselli et al. (2001) reported that turnover rate among managerial staff in restaurants were between 33 to 100 percent. In comparison, it was reported that due to the boredom associated with highly routinized labor, poor pay and working condition, and employers and employees' agreement for short-term employment due to lower pay and personal convenience, turnover rate for non-managerial fast food restaurants were exceptionally high in the US, at 300% (Ryan, et al., 2011). In regards to turnover intention and job position, Khatri, Fern, and Budhwar (2001) found that managerial employees have better leaving intention than non-managerial employees although they did mention that previous studies found the contrast (Price & Mueller, 1981; Tai & Robinson, 1998). As the number of research comparing between managerial and non-managerial employees' intention to leave are limited, the hypothesis of this study will be proposed based on Khatri and colleagues' study. Thus, the following hypothesis is proposed:

H2: Managerial employees have higher intention to leave compared to non-managerial employees.

3 METHODOLOGY

Data for this study were collected among the managerial and non-managerial staff of casual dining restaurants in Klang Valley area. Questionnaire was used as the instrument for this research. The questionnaires were handed to the managers or supervisors in charge for them to distribute among their staff and themselves. Collection of the questionnaires yielded a result of 717 usable sets of data which were later used in analyses. Convenience sampling was used as sampling method for this research. All respondents were advised that their participations were voluntary and that their answers will be kept anonymous.

To measure POS, 16-item construct of Survey of Perceived Organizational Support (SPOS) developed by Eisenberger et al. (1986) were used, which is the shorter version of the original 36-item. Seven point Likert scale was used to indicate the employee's extent of agreement to each statement (1 = strongly disagree, 7 = strongly agree). An example of item is *"The company cares about my general satisfaction at work"*. The reliability of this construct measured by Cronbach's alpha was 0.875.

Intention to leave was measured by using five item construct designed by Wayne et al. (1997). "I am seriously thinking of quitting my job" was one of the items that the respondents needed to indicate their level of agreements using seven point Likert scale. Cronbach's alpha for this construct was 0.806.

Respondents were also asked to specify their demographic profile variables. They needed to state their gender, age, marital status, level of education, work position, working hour per week, working duration in the industry, and working duration in the restaurant.

Analyses of the data were done according to the objectives of this research. Firstly, to determine the relationship between POS and intention to leave, Pearson's correlation test was done. Next, t-test was done to see whether there were differences in managerial and non-managerial employees' intention of leaving. On top of those analyses, descriptive test was also done to get the descriptive results of the demographic variables.

4 RESULTS AND DISCUSSION

4.1 Descriptive results

In the descriptive results, it was found that the majority of the respondents were male employees which made up of 62.5 percent of the total respondents, leaving only 37.5 percent of female respondents. The average age of working employees were 24.1 years old. Most of the respondents were single (81.3%), while the highest group of staff was secondary school students or SPM leavers where they made up more than half of the total respondents

Table 1. Means, standard deviations, and correlations of variables *(N = 717)*.

	Mean	SD		Intention to leave	POS
Intention to leave	4.12	1.44	Pearson Correlation	1	
			Sig. (2-tailed)		
POS	4.70	0.94	Pearson Correlation	−.464**	1
			Sig. (2-tailed)	.000	

** Correlation is significant at the .01 level (2-tailed).

(54.4%) during the collection of the data. Non-managerial staff outnumbered the managerial staff with 75.9 percent and 23.4 percent respectively. In average, the staff worked for 54.3 hours a week, had 17.4 months of experience in the restaurant they were currently employed, and 35.3 months working experience in the restaurant industry.

4.2 Correlations between variables

Table 1 below showed the means, standard deviations, and correlations between POS and intention to leave. The result showed that there was significant relationship between the two variables with moderate strength (r = −.464, p = .01). It indicated that individual with higher POS would have lower intention of leaving the restaurant they currently employed. This finding is in line with previous studies between POS and intention to leave where it was found that the two variables showed significant negative relationship (Allen, et al., 2003; Blomme, et al., 2010; Cho, et al., 2009; Jawahar & Hemmasi, 2006; Kahumuza & Schlechter, 2008). Therefore, hypothesis 1 is supported. When employees perceived that the organizational they worked for support their values as well as the well-beings and would reward the efforts they put into their works, it became easier for them to work alongside the company to reach the latter's goals. With the appreciation showed towards their efforts, the employees would have lesser tendency to leave their jobs and find alternatives.

4.3 T-test results

An independent sample t-test was conducted to compare the intention of leaving between managerial and non-managerial staff in order to achieve the second objective. In the test, it was found that there was a significant difference in the intention among managerial staff (M = 3.89, SD = 1.35) and non-managerial staff (M = 4.21, SD = 1.46), t(710) = −2.52, p = .012 as shown in Table 2 below. The results suggest that the employees work positions in the restaurant affects their tendency of having leaving intention. Specifically, the result of this study showed that non-managerial staff has higher intention to leave compared to the managerial

Table 2. Means, standard deviations, and independent samples test of variables.

	Managerial		Non-managerial		Levene's Test		t-test		
Intention to leave	Mean	SD	Mean	SD	F	Sig	t	df	Sig (2-tailed)
	3.888	1.348	4.206	1.462	1.219	.270	−2.516	710	.012*

* Correlation is significant at the .05 level (2-tailed).

staff. Therefore, hypothesis 2 is rejected. The higher turnover intention among non-managerial staff could be linked to the high percentage of secondary school students or SPM leavers. A lot of them might find working in the restaurant as a temporary job before continuing or pursuing their studies, thus increasing the tendency of non-managerial staff to have greater turnover intention.

5 CONCLUSION

The result of this study showed that with the increment of employee's POS, the staff of casual dining restaurants in Klang Valley would have lesser intention to leave their restaurants. Therefore, it is important for the upper management to show more support and rewards their employees' hard works in order to make sure that the employees would stick around. This is essential because if an employee believes that the management does not appreciate them, the chance is that they will look for other jobs as alternatives and may result in more turnover rate for the organization. On the other hand, in comparison between managerial and non-managerial staff, the latter were indicated to have higher leaving intention compared to the former. As non-managerial staff are the main workforce for a restaurant, retaining them is considered crucial, especially the skilled and experienced staff.

REFERENCES

Abdullah, R., Alias, M.A.M., Zahari, H., Abdul Karim, N., Abdullah, S.N., Salleh, H. & Musa, M.F. (2010). The Study of Factors Contributing to Chef Turnover in Hotels in Klang Valley, Malaysia. *Asian Social Science, 6*(1), 80–85.

Ahmad, R. & Zainol, N.A. (2011). *What it takes to be a manager: The case of Malaysian five-star resort hotels.* Paper presented at the Proceedings of the 2nd international conference on business and economic research.

Allen, D.G., Shore, L.M. & Griffeth, R.W. (2003). The role of perceived organizational support and supportive human resource practices in the turnover process. *Journal of management, 29*(1), 99–118.

Arshadi, N. (2011). The relationships of perceived organizational support (POS) with organizational commitment, in-role performance, and turnover intention: Mediating role of felt obligation. *Procedia-Social and Behavioral Sciences, 30*, 1103–1108.

Blomme, R.J., Van Rheede, A. & Tromp, D.M. (2010). Work-family conflict as a cause for turnover intentions in the hospitality industry. *Tourism and Hospitality Research, 10*(4), 269–285.

Cho, S., Johanson, M.M. & Guchait, P. (2009). Employees intent to leave: A comparison of determinants of intent to leave versus intent to stay. *International Journal of Hospitality Management, 28*(3), 374–381.

Eisenberger, R., Eisenberger, R., Huntington, R., Hutchison, S. & Sowa, D. (1986). Perceived organizational support. *Journal of applied psychology, 71*(3), 500–507.

Euromonitor International. (2012). Full-service restaurants in Malaysia *Euromonitor Passport GMID database.*

Ghiselli, R., La Lopa, J. & Bai, B. (2001). Job satisfaction, life satisfaction, and turnover intent among food-service managers. *The Cornell Hotel and Restaurant Administration Quarterly, 42*(2), 28–37.

Goh, L. (2012, February 19). Why job-hoppers hop, *The Sunday Star.*

Hay, M. (2002). Strategies for survival in the war of talent. *Career Development International, 7*(1), 52–55.

Ingram, H. & Jones, S. (1998). Teamwork and the management of food service operations. *Team Performance Management, 4*(2), 67–73.

Jawahar, I. & Hemmasi, P. (2006). Perceived organizational support for women's advancement and turnover intentions: The mediating role of job and employer satisfaction. *Women in Management review, 21*(8), 643–661.

Kahumuza, J. & Schlechter, A.F. (2008). Examining the direct and some mediated relationships between perceived support and intention to quit. *Management Dynamics: Journal of the Southern African Institute for Management Scientists, 17*(3), 2–19.

Khatri, N., Fern, C.T. & Budhwar, P. (2001). Explaining employee turnover in an Asian context. *Human Resource Management Journal, 11*(1), 54–74.

Mansfield, S.L. (2007). *The relationship of CEO's and top leadership teams' hope with their organizational followers' job satisfaction, work engagement, and retention intent.* Regent University.

Mobley, W.H., Horner, S.O. & Hollingsworth, A.T. (1978). An evaluation of precursors of hospital employee turnover. *Journal of applied psychology, 63*(4), 408–414.

Moncarz, E., Zhao, J. & Kay, C. (2009). An exploratory study of US lodging properties' organizational practices on employee turnover and retention. *International Journal of Contemporary Hospitality Management, 21*(4), 437–458.

Mowday, R.T., Steers, R.M. & Porter, L.W. (1979). The measurement of organizational commitment. *Journal of Vocational Behavior, 14*(2), 224–247.

Ponnu, C. & Chuah, C. (2010). Organizational commitment, organizational justice and employee turnover in Malaysia. *African Journal of Business Management, 4*(13), 2676–2692.

Price, J.L. & Mueller, C.W. (1981). A causal model of turnover for nurses. *Academy of Management Journal, 24*(3), 543–565.

Rhoades, L. & Eisenberger, R. (2002). Perceived organizational support: a review of the literature. *Journal of applied psychology, 87*(4), 698–714.

Rivera, M., DiPietro, R.B., Murphy, K.S. & Muller, C.C. (2008). Multi-unit managers: Training needs and competencies for casual dining restaurants. *International Journal of Contemporary Hospitality Management, 20*(6), 616–630.

Ryan, C., Ghazali, H. & Mohsin, A. (2011). Determinants of intention to leave a non-managerial job in the fast-food industry of West Malaysia. *International Journal of Contemporary Hospitality Management, 23*(3), 344–360.

Tai, T.W.C. & Robinson, C.D. (1998). Reducing staff turnover: a case study of dialysis facilities. *Health Care Management Review, 23*(4), 21–42.

Tuzun, I.K. & Kalemci, R.A. (2012). Organizational and supervisory support in relation to employee turnover intentions. *Journal of Managerial Psychology, 27*(5), 518–534.

Wayne, S.J., Shore, L.M. & Liden, R.C. (1997). Perceived organizational support and leader-member exchange: A social exchange perspective. *Academy of Management Journal, 40*(1), 82–111.

Wildes, V.J. (2005). Stigma in food service work: how it affects restaurant servers' intention to stay in the business or recommend a job to another. *Tourism and Hospitality Research, 5*(3), 213–233.

Theory and Practice in Hospitality and Tourism Research – Radzi et al. (Eds)
© *2015 Taylor & Francis Group, London, ISBN 978-1-138-02706-0*

Physicochemical changes and oxidative stability of nanostructured ginger rhizome marinated spent hen during chill storage

A. Norhidayah
Faculty of Hotel and Tourism Management, Universiti Teknologi MARA, Shah Alam, Malaysia

A. Noriham
Faculty of Applied Sciences, Universiti Teknologi MARA, Shah Alam, Malaysia

M. Rusop
NANO—ScienceTech Centre, Institute of Science, Universiti Teknologi MARA, Shah Alam, Malaysia

ABSTRACT: The effect of nanostructured ginger rhizome marination on meat quality was determined. The samples were divided into five groups: control (no antioxidant); 3 percent w/w of micron size ginger (MG); 3 percent w/w Submicron Ginger (SM), 3 percent w/w Nanostructured Ginger (NG) rhizome powder and BHA:BHT (100ppm:100pm). Peroxide value was used as lipid oxidation's indicator while changes in pH and colors indicate the physicochemical deterioration due to lipid oxidation. Significant lower PV value ($p < .05$) were exhibited in treated samples compared to control. There were significant increase in pH and decreased in L, a* and b* value for all tested samples with a control sample posses the highest pH and lowest L, a* and b* value throughout the storage period. NG significantly reduced the lipid oxidation, maintained the pH and caused slight changes in the L, a* and b* value but those activities slightly lower than BHA: BHT. Thus nanotechnology enhanced the preservative property and delayed the oxidation process in meat.

Keywords: Ginger, marination, meat quality, nanotechnology, oxidation

1   INTRODUCTION

Lipid oxidation is the major problems that take place during processing and storage of meat. It is important parameters that influence the quality and acceptance of meat and chicken. Lipid oxidation is commonly influence by the phospholipid composition, polyunsaturated fatty acid amounts, presence of metal ions, heam pigments and others. Poultry meat has become a mass consumer product throughout the world. In Malaysia alone, the poultry annual per capita consumption were reported to be the highest for the last 20 years as compared to other types of meat. It is due to the government subsidized and control price which has made poultry affordable and become the most homogeneous meat among Malaysian (Jang, Ko, Kang & Lee, 2007). It is an indicator for drastic demand as compared to other types of meat. Spent hen are rich in protein and lipids which made them prone to oxidation (Suradkar, Bumla, Maria, Sofi2 & Wani, 2013,). Furthermore, it contain significantly higher ($p < .05$) fat content compared to broiler which increased its potential to oxidized (Chueachuaychoo,

Wattanachant & Benjakul, 2011). Fat content increase with age thus synthetic antioxidants such as butylated hydroxyanisole (BHA) or butylated hydroxytoluene (BHT) have been applied in meat products to decrease the lipid oxidation that promote the discoloration and other physicochemical deterioration. Synergistic effect of BHA and BHT was also confirmed on antioxidation. However, their application in foods had been restricted due to potential health effect and toxicity. Concerning on the safety implication due to synthetic antioxidant, demand for natural antioxidant riscd recently (Simirgiotis & Schmeda-Hirschmann, 2010). Ginger (*Zingiber officinale*) Rosc is a globally used herb that contains active constituents 6-gingerol and its derivatives, which have a high antioxidant activity (Stoilova, Krastanov, Stoyanova, Denev & Gargova, 2007) thus make it a good source of natural antioxidant. Moreover, ginger were also consumed as food additive to enhance the sensorial acceptance (Naveena, Mendiratta & Anjaneyulu, 2004) during storage. Smoked buffalo treated with ginger extract showed better in appearance, color and juiciness compared to control. Lower shear force values were observed in ginger extract treated yak meat and goose meat (Gao et al., 2011). However, natural antioxidants are less effective compared to synthetic antioxidants (Fasseas, Mountzouris, Tarantilis, Polissiou & Zervas, 2007). It is due to the fact that plant's active compound which consists of flavonoids and lignans grouped are poor water soluble compounds thus limit the absorption rate (Yen et al., 2008).

Recently, the nanonization of herbal medicines become an emerging approach and has a continual growth technology globally. Nanoparticles is a colloidal systems varying in particle size from 10 nm to 1000 nm (Cao et al., 2013). It is being reported that nanonization increased the *Rhizoma Chuanxiong*'s extractiaon yield and enhanced its bioactivity. It is due to the cellular tissues breakage that allowed the active constituent to disperse well compared to its coarse particle. Furthermore, Ma et al. (2009) also found that the physicochemical and medicinal characterization of *Liuwei Dihuang* was maximized after ground to nanoscale size. Thus it showed that size reduction into nanoscale enhanced and improved the bioactivity of herbs. When the plant medicine undergoes nanonization the cell membrane and cell walls were crushed into pieces, thus the active constituents could directly contact with the outer solvent. On the other hand, the contact area of the plant medicine in the form of nanoparticle with body fluid were also increased thus promotes faster absorption and higher bioavailability (Liu, Chen, Shih & Kuo, 2008).

Numerous works being conducted in other type of herbs, but to our knowledge, there is very limited research conducted on nanostructured ginger rhizome powder (Ma et al., 2009; Su, Fu, Quan & Wang, 2006; Yen, Wua, Lin, Cham & Lin, 2008). Additionnaly, the effect of nanoparticle herbs in preserving the physicochemical property of meat during chill storage is scarce in the literature. Although not yet reported in the literature, the nanotechnology distinctly enhances ginger rhizome bioactivity to a certain extend and could therefore be incorporated in spending hen as a source of natural antioxidants to prolong quality and stability. In the present study, the potential of nanostructured ginger rhizome powder as preservative and marinating ingredient was evaluated and compared to synthetic antioxidant (BHA/BHT).

2 METHODOLOGY

2.1 *Preparation of material*

Fresh ginger (*Zingiber officinale* Rosc.) rhizome variety Bentong was obtained from Bentong,

Pahang, Malaysia. Voucher specimens (SK 2049/12) of these plants were deposited at the Herbarium Institute of Bioscience, University Putra Malaysia. The sample weight before and after drying were noted. The dried ginger rhizome obtained was ground using food processor (Panasonic, 176, China) for 5 min, sieved and namely as micron ginger (19.54 μm), the submicron ginger was prepared by milling using a hammer mill at 2890 rpm and sieved using 250μm sieve with particle size was relatively 4.12 μm. While nanostructured ginger rhizome was prepared until the particle size was relatively around 160.5 nm in size (Norhidayah, Noriham & Rusop, 2013).

Spent hen chicken thighs (aged 69-75 weeks of lying cycle) were obtained from local market in Shah Alam area. The purchased chicken thighs were personally carried to the experimental lab in the ice box. Chicken thighs were received free from visible blood and in the ranged from 100 to 130 g of weight and 6.0 to 6.2 in pH. Any remaining visible fat was physically removed and rinsed to remove any contaminant.

The marination technique was based on Yusop et al. (2012) which consists of 3 percent marinate solution (3 gram in 100 ml of distilled water). The mixtures were homogenized using an ultrasonic processor (Hielscher, UP400S) for 15 minutes and kept in an airtight bottle before use. All together, 50 chicken thighs were allocated to five groups (10 chicken thighs per group). The first group was marinated with 10 mL of distilled water (control), while the other three groups were marinated with 10 mL of micron, submicron and nanostructured size solution per 100 g chicken thigh respectively. The combination of BHA: BHT (100ppm:100ppm) were used as comparison. The entire treated spent hen chicken thigh underwent vacuum packaging using polyethylene bag prior to chill storage at 4°C for 12 days.

2.2 *Instrumental color*

The colour of the raw marinated chicken thigh (boneless; 2 cm X 2 cm) which consisted of L* (muscle lightness), a* (muscle redness) and b* (muscle yellowness) values were evaluated using a Chromatometer (Konica Minolta, CR-400) that was previously calibrated with a white tile provided with the instrument.

2.3 *pH determination*

The muscle pH was determined using an MP120 pH meter fitted with a combined glass electrode InLab427 (Mettler-Toledo, GmbH, 8603), previously calibrated at pH 4.0 and 7.0.

2.4 Oxidative stability

2.4.1 Fat extraction

Lipid extraction was conducted based on Kinsella et al. (1977) with some modification. The obtained fat was directly used for analysis or stored in the airtight amber bottle and flushed in N2 gas and stored at 4°C until used. (Kinsella, Shimp, Mai & Weihrauch, 1977)

2.4.2 Peroxide Value (PV)

The Peroxide Value (PV) was conducted according to the method by (AOAC, 2002). The peroxide value was calculated as follows:

Peroxide value (PV) =
$Vs - Vb \times T \times 10^3 \times 2$ Weight of sample (g)

Where: Vs = Volume in ml titration for sample
Vb = Volume in ml titration for blank
    T = Molarity of sodium thiosulphate
    The PV expressed as miliequivalents of peroxide oxygen per kilogram of sample (mEq/kg).

2.5 Statistical analysis

All experiments were carried out in triplicate and presented as mean and standard deviation. One-way analysis of variance (ANOVA) was used to analyze the data using SPSS statistical software (version 21 SPSS Inc., USA). The means were compared with Duncan's multiple comparison test (DMCT) and $p < .05$ was considered as statistical significance.

3 RESULTS AND DISCUSSION

3.1 pH changes

Meat pH has a profound influence on meat quality since changes in pH relatively caused protein denaturation and affect the quality attributes such as color and water holding capacity. The comparative pH value of tested samples is presented in Table 1.

Initially the muscle pH value were in the ranged of 6.09 to 6.46 which indicate that all pH values were within the expected range for normal spent hen. The ultimate pH of raw spent as reported by Chueachuaychoo et al. (2011) was around 6.10. Generally, at the early stage of storage there were significantly lower pH values (in range of 6.09 to 6.42) for treated samples compared to control sample (6.46 ± 0.007). It showed that ginger rhizome powder and synthetic antioxidant BHA: BHT decreased the initial pH of marinated chicken thigh. Kim et al. (2009) reported lower pH in the tested sample that contains dietary garlic compared to control sample. However, the pH values of

Table 1. Effect of nanostructured ginger rhizome and synthetic antioxidant marination on pH value.

Day/Sample	0	3	6	9	12
C	6.46aE	6.64aD	6.76bC	6.92aB	7.17aA
MG	6.37cD	6.49bC	6.78aB	6.85bA	6.85bA
SM	6.42bD	6.42cD	6.46cC	6.56cB	6.73cA
NG	6.09eE	6.16eD	6.21eC	6.25eB	6.33eA
BHA/BHT	6.11dE	6.30dD	6.37dC	6.51dB	6.67dA

Value are expressed as mean ± standard deviation (n = 3). Values marked by the different superscript letters within a column (a-e) and (A-E) within a row denote statistically differences (p .05).

all the samples gradually increased during storage at 4°C for 12 days of chill storage. The pH value showed significantly ($p < .05$) higher values in C sample and the lowest observed in NG marinated chicken thigh throughout the study period. The sample that depicted higher pH value may attribute to higher degree of oxidation (Lingaiah & Reddy, 2001). Chicken thigh treated with nanostructured ginger rhizome powder (NG) had the best effect with highest pH value of 6.33 while that of control reaching pH of 7.12 after 12 days of chilled storage. The pH increment could be due to microbial activity in meat. Nanostructured ginger rhizome treatment exhibited lowest pH among all treated samples during chilled storage.

3.2 Color changes determination

Effect of treatment and days of storage on color stability of marinated chicken thigh is shown in Table 2. In general, the L* value was in the range of 51.68 to 57.44 which was the expected lightness value for spent hen meat. This indicates the freshness for all tested sample at 0 day of chilled storage. Perhaps, the L* value obtained were comparable to the one reported previously, where L value for spent hen *Pectoralis major* muscle was around 52.51 (Chueachuaychoo, et al., 2011). However, obviously higher degree of lightness showed in NG sample compared to other samples ($p < .05$). Nanostructured ginger rhizome used in this study was very fine (nanoscale) and light cream in color. Therefore, it was expected that chicken thigh treated with nanostructured ginger rhizome ha d different color from the other samples.

The L* value slightly decreased throughout the storage period but no differences were detected between 3 to 9 days of chilled storage, but drastically reduced after storage for 12 days with C sample exhibiting lowest L value with the rate of color degradation were in between 16.57 percent to 21.66 percent for treated samples and 25.33 percent in C

Table 2. The lightness value (L*) of marinated spent hen chicken thigh.

Day/Sample	0	3	6	9	12
C	53.7aC	49.5bB	48.0bcAB	45.6cA	40.1dB
MG	51.68aD	49.8bB	47.2cB	45.1dA	41.98eB
SM	53.70aC	51.2bB	48.9cAB	47.2cA	44.9dA
NG	57.4aA	53.1bA	50.4cA	46.7cA	45.1dA
BHA/BHT	54.9aB	51.5bAB	47.2cB	45.4cdA	44.1dA

Value are expressed as mean (n = 3). Values marked by the different superscript letters within a sample (a-d) and (A-D) between a sample denote statistically differences (p .05).

Table 3. The redness (a*) of marinated spent hen chicken thigh.

Day/Sample	0	3	6	9	12
C	17.8aA	13.8bC	10.8cC	9.2cdB	7.4dB
MG	16.5aA	14.3bBC	11.6cBC	8.6dB	7.8dB
SM	17.9aA	15.5bAB	13.1cAB	10.0dB	6.3eB
NG	18.0aA	15.3bAB	12.8cAB	11.9cdA	10.3dA
BHA/BHT	17.5aA	16.3aA	13.7bA	13.1bA	10.5cA

Value are expressed as mean (n = 3). Values marked by the different superscript letters within a sample (a-d) and (A-C) between a sample denote statistically differences (p .05).

sample. Hence it is suggested that the ginger rhizome powder and BHA:BHT can be used to preserve the color of chicken thigh. The decrease in color values during storage was also probably because of the oxidation of pigments (Ahmed, 2004).

Table 3 showed the redness value (a*) as affected by marination process. In general, the a* value for tested chicken thigh were in the range of 16.45 to 18.03 with no difference among samples during the 0 day of chilled storage (p > .05). It showed that all tested sample were in the same redness level. Basically, addition of ginger rhizome powder did not give any difference in the redness value during the 0 day of storage. However, the redness value for all tested sample gradually decreased as a function of chill storage. The meat redness was drastically reduced for the first 9 days of storage in all treated sample except for spent hen chicken thigh marinated with synthetic antioxidant (BHA/ BHT). BHA:BHT reasonably stabilized the a* value up to day 3 and maintained up to 9 days of chill storage and only deteriorated after 12 days of storage period. Among all, BHT/BHA treated sample showed the highest a* value up to nine days of storage period but no significant different (p > .05) from the one marinated with nanostructured ginger rhizome at the end of storage period. Thus it shows the comparable effectiveness between nanostructured ginger rhizome powder and synthetic antioxidant.

While yellowness value (b*) illustrated in Table 4. Primarily the yellowness value (b*) for the entire

Table 4. The yellowness (b*) value of marinated spent hen chicken thigh.

Day/Sample	0	3	6	9	12
C	11.7aA	6.5cC	7.3bA	5.8dB	4.6dBC
MG	14.0aA	11.1bAB	7.6cA	6.6cB	4.2dC
SM	14.1aA	10.1bB	8.8cA	8.5cA	6.6dA
NG	13.0aA	10.3bB	8.7cA	8.1cA	5.3dAB
BHA/BHT	12.7aA	11.8abA	9.0cA	8.7cA	5.9dAB

Value are expressed as mean (n = 3). Values marked by the different superscript letters within a sample (a-d) and (A-D) between a sample denote statistically differences (p .05).

tested samples was comparable to each other (in the range of 11.70 to 14.10) but drastically reduced after day three of chilled storage.

The dramatic decreased can be seen in the control sample when the b* value reduced to almost 45 percent after three days of storage and reduced up to 60.34 percent at the end of storage period. No significant different can be seen in the MG with control sample after 12 days. But surprisingly, the b* value was significantly higher in the other tested sample. The yellowness of SM and NG found to be equivalent to the BHA/BHT sample. Thus it showed that size reduction technology has the potential to enhance the freshness of marinated chicken samples.

Table 5. The peroxide value (PV) of marinated spent hen chicken thigh during chilled storage.

Day/Sample	0	3	6	9	12
C	7.38aD	7.68aD	10.16 aC	13.33aB	16.77 aA
MG	1.45cD	4.27bC	6.89bB	7.23bB	14.41bA
SM	1.69cE	2.71cD	3.55cC	5.05dB	11.54cA
NG	2.25bD	2.67cC	2.84dC	6.11cB	8.89dA
BHA/BHT	1.68cC	1.86dC	2.34eC	3.68eB	6.75eA

Value are expressed as mean (n = 3). Values marked by the different superscript letters between a sample (a-e) and (A-E) within a sample denote statistically differences (p .05).

3.3 Peroxide Value (PV)

Peroxide Value (PV) is the primary products of lipid oxidation which is generated by oxygen attacking on the fatty acids double bond. Since spent hen relatively high in fatty acid double bond, therefore, it is reasonable to determine the concentration of peroxide in treated samples to clarify the effect of various particle size of ginger rhizome powder in delaying the oxidation process. The lower value indicated the effectiveness of test material to inhibit the oxidation process which leads to better quality of meat product. Table 5 shows the peroxide value (PV) of marinated spent hen chicken thigh during 12 days of chilled storage.

The PV was found to be storage time dependent whereby the PV increased throughout the chilled storage period with C sample showed significantly higher PV compared to the other samples. Initially, significantly lower PV (p < .05) can be detected in treated spent hen chicken thighs compared to C sample which showed the effectiveness of ginger rhizome in protecting the spent hen chicken thigh from lipid oxidation. However, the value gradually increased with prolongs storage time (p < .05) with highest PV showed in all samples at 12 days of chilled storage. Samples treated with BHA/ BHT were found to be more stable towards oxidative rancidity as shown by the increased in PV only after nine days of chilled storage and the lowest among others throughout the storage period. Meanwhile, NG sample showed relatively less effective to delay lipid oxidation when the value is 24.07 percent higher compared with BHA/BHT sample at 12 days of chilled storage but significantly better compared to other treated samples.

4 CONCLUSION

This study revealed that the nanosize ginger rhizome powder were able to maintain pH and color changes and also significantly delayed lipid oxidation in marinated chicken thighs during chilled storage. Hence this study revealed the potential of nanostructured ginger rhizome powder to be used as food preservative in chicken.

ACKNOWLEDGEMENT

The authors are grateful to the Ministry of Higher Education Malaysia for the scholarship and Ministry of Science Technology and Innovation (MOSTI) Malaysia for grant (06-01-01-SF0390).

REFERENCES

Ahmed, J. (2004). Rheological behaviour and colour changes of ginger paste during storage. *International Journal of Food Science and Technology 39*,325–330.

AOAC. (2002). *Determination of peroxide content. 965.33. In Official methods of analysis* (17 ed.). Gaithersburg, Maryland: Association of Official Analytical Chemists.

Cao, Y., Gu, W., Zhang, J., Chu, Y., Ye, X., Hu, Y. & Chen, J. (2013). Effects of chitosan, aqueous extract of ginger, onion and garlic on quality and shelf life of stewed-pork during refrigerated storage. *Food Chemistry, 141*,1655–1660.

Chueachuaychoo, A., Wattanachant, S. & Benjakul, S. (2011). Quality characteristics of raw and cooked spent hen Pectoralis major muscles during chilled storage: Effect of salt and phosphate. *International Food Research Journal, 18*,601–613.

Fasseas, M.K., Mountzouris, K.C., Tarantilis, P.A., Polissiou, M. & G., Z. (2007). Antioxidant activity in meat treated with oregano and sage essential oils. *Food Chemistry 106*, 1188–1194.

Gao, D. & Zhang, Y. (2010). Comparative antibacterial activities of extracts of dried ginger and processed ginger. *Pharmacognosy Journal, 2*,41–44.

Jang, I.S., Ko, Y.H., Kang, S.Y. & Lee, C.Y. (2007). Effect of a commercial essential oil on growth performance, digestive enzyme activity and intestinal microflora population in broiler chickens. *Animal Feed Science Technology, 134*,304–315.

Kim, J., Jung, D.H., Rhee, H., Choi, S.-H., Sung, M.J. & Choi, W.S. (2008). Improvement of bioavailability of water insoluble drugs: Estimation of intrinsic bioavailability. *Korean J. Chem. Eng., 25(1)*,171–175.

Kinsella, J.E., Shimp, J.L., Mai, J. & Weihrauch, J. (1977). Fatty acid content and composition of freshwater finfish. *Journal of the American Oil Chemists' Society* *54*,424–429.

Lingaiah & Reddy, P. (2001). Quality of chicken meat patties containing skin and giblets. *Journal of Food Science Technology—Mysore, 38*(4),400–401.

Liu, J.R., Chen, G.F., Shih, H.N. & Kuo, P.C. (2008). Enhanced antioxidant bioactivity of Salvia miltiorrhiza (Danshen) products prepared using nanotechnology. *Phytomedicine, 15*(1–2),23–30

Ma, P.Y., Fu, Z.Y., Su, Y.L., Zhang, J.Y., Wang, W.M., Wang, H., Zhang, Q.J. (2009). Modification of physicochemical and medicinal characterization of Liuwei Dihuang particles by ultrafine grinding. *Powder Technology, 191*,194–199.

Naveena, B.M., Mendiratta, S.K. & Anjaneyulu, A.S.R. (2004). Tenderization of buffalo meat using plant proteases from Cucumis trigonus Roxb (Kachri) and Zingiber officinale roscoe (Ginger rhizome). *Meat Science, 68*,363–369.

Norhidayah, A., Noriham, A. & Rusop, M. (2013). Antioxidant Activities of Nanostructured Ginger (Zingiber officinale Roscoe) Rhizome as Affected by Different Milling Time. *Advanced Science Letters, 19*,3572–3575.

Simirgiotis, M.J. & Schmeda-Hirschmann, G. (2010). Determination of phenolic composition and antioxidant activity in fruits, rhizomes and leaves of the white strawberry (Fragaria chiloensis spp. chiloensis form chiloensis) using HPLC-DAD-ESI-MS and free radical quenching techniques. *Journal of Food Composition and Analysis, 23*,545–553.

Stoilova, I., Krastanov, A., Stoyanova, A., Denev, P. & Gargova, S. (2007). Antioxidant activity of a ginger extract (Zingiber officinale). *Food Chemistry, 102*(3), 764–770.

Su, Y.I., Fu, Z.Y., Quan, C.J. & Wang, W.M. (2006). Fabrication of nano Rhizama Chuanxiong particles and determination of tetramethylpyrazine. *Transactions of Nonferrous Metals Society of China, 16*(Supplement 1),s393-s397.

Suradkar, U.S., Bumla, N.A., Maria, A., Sofi2, A.H. & Wani, S.A. (2013,). Comparative Quality of Chicken Nuggets Prepared from Broiler, Spent Hen and Combination Meats. *International Journal of Food Nutrition and Safety, 3*,119–126.

Yen, F.L., Wua, T.H., Lin, L.T., Cham, T.M. & Lin, C.C. (2008). Nanoparticles formulation of Cuscuta chinensis prevents acetaminophen-induced hepatotoxicity in rats. *Food and Chemical Toxicology 46*,1771–1777.

Yusop, S.M., O'Sullivan, M.G., Preuß, M., Weber, H., Kerry, J.F. & Kerry, J.P. (2012). Assessment of nanoparticle paprika oleoresin on marinating performance and sensory acceptance of poultry meat. *LWT—Food Science and Technology 46*,349–355.

Perceptions and current practices in bakery industries from halal perspective

S. Bachok & C.T. Chik
Faculty of Hotel and Tourism Management, Universiti Teknologi MARA, Shah Alam, Malaysia

M.A. Ghani & M.K. Ayob
Faculty of Science and Technology, Universiti Kebangsaan Malaysia, Malaysia

ABSTRACT: Malaysia is being recognized as an Islamic country although there are multiracial population such as Malay, Indian, and Chinese. In Malaysia and other countries where most people are Muslims, consumers are very concerned in food related issues, halal food production and food services. From Muslim perspectives, products with halal logo are much more meaningful and crucial compared to those with International Organization for Standard (ISO) recognition for Halal food. In line with the emergent of Muslim population all over the world, the awareness for halal food are also increasing. This study was conducted to identify the intention, level of awareness and practices in implementation of halal in bakery and confectionery industry in Malaysia. It is also beneficial for Muslim and public to know the status of food taken especially bakery product in their daily lives. The objective of the study was to know the consumer awareness and trust on product. In obtaining the information of the topic of interest, quantitative approach through questionnaire survey with bakery operators had be conducted. The findings showed that Halal logo is an influencing factor in determining the consumer confident before making informed choice of their purchase. The results also suggested a few guidelines for manufacturer to improve their production in attracting Muslim consumer to purchase bakery products. In producing halal bakery product, all of the raw ingredients must be halal and must not cross contact with any equipment that has been used for non-halal food. For the industry, this study can be used as a guideline to produce halal product and food according to syariah. It also acts as reference for any future study in the related field mainly in Malaysia.

Keywords: Non-halal ingredient, halal practices, bakery. confectionery

1 INTRODUCTION

Halal and haram issue in food has been discussed quite extensively because food is consumed in our body. The term *halal* is from (halla), Halal food is food that is prepared, stored, manufactured, slaughtered and served in a manner required by the *Syariah Law*. The term *haram* is an arabic word which means prohibited or unlawful. Eating and drinking haram food and drinks is forbidden for every Muslim. The ingredients and the preparation of foods must be permitted based upon Islamic teachings: the equipment used in preparing and processing the food must be clean from the Islamic point of view, in addition, the food is stored and served in a manner required by Syariah. Bakery product is a new emergent business in Malaysia. Baked goods are enjoyed whether as a substitute for our staple food rice or also as an item for leisure in the afternoon with coffee. Even bakery home business has mushroomed and this adds to the bakery consumption among local consumer.

Various incidents of breach of regulations has been seen among the food operators, who openly seems not to care and do not respect the rules that have been prescribed in Islam. Issues of food mixed with haram substances such as involving swine and its derivatives has been heard and reported in the news. In line with the issue, consumers often received anonymous news regarding certain food products that suspected contained haram substances that lead to the dilemma among the Muslim consumers. This is questionable or suspicious food and more information is needed to categorize them as Halal or Haram. They are often referred to as *Mashbooh*, which means doubtful or questionable. Majlis Ugama Islam Singapura (MUIS) (1991) stated that Halal foods are good because of all the sources of the food are lawfully good. In other words, the concept of halal must be parallel with the concept of 'Toyibbah' (Talib, Ali, Anuar & Jamaludin, 2008). A verse in Quran which means: "*O mankind, eat only lawful and good (halal toyyiba) from what is on earth, and do not follow the steps of evil (Satan): for evil (Satan) is an enemy to you*" (Surah Al Baqarah, 2:168). Thus it is understandable that each of halal food must also be good in terms of materials used, clean processes and can benefit human body (Abdul Rahman, 1999).

Naturally bakery products were added and contained chemical. The chemical is an additive added either directly or indirectly during processing, storage or packaging. It helps to improve life expectancy of bread storage, nutritional value or flavor. They are often referred to as Mashbooh, which means doubtful or questionable (Majlis Ugama Islam Singapura, 1991). For example emulsifiers were widely used in the improvement of bread to control the size of gas bubbles. It allows dough to trap more gas and therefore the batter will expand larger and more powder to make soft. Emulsifiers also reduce the rate of bread to develop off odor. Some types of emulsifiers are decreasing agents and preservative such as L-cysteine hydrochloride (E920). Cysteine is the amino acids that occur naturally in making bread dough especially to bun burgers and French bread. It may be obtained from animal hair and human hair. Enzyme is protein that accelerates metabolic reactions, and it is derived from plants, animals, fungi and bacteria. For example, Chymosine an enzyme that were used in milk to produce cheese, obtained either from calf stomach which is rennet or synthetically through genetic engineering. Enzymes are also used in baking, known as "processing aid," however it need not be specified on the product label. Not surprisingly, most people do not know that their bread contains added enzymes. Glycerine and gelatin and some more similar material are also not stated whether the source is of animals or plants origins. From various ingredients used in making bread, the origins of the bread dough or its content must be known. First is the source of fat or shortening used which the materials can derive from fats or oils of plant or animal. Fat derived from animals usually invite suspicion, whether it is lawful or unlawful.

In commercial bakery industry in Malaysia most substances used as shortening derived from palm oil shortening. However, there is bread made with animal fat content because the material can provide a fluffier and softer feel. Fat derived from animal fats usually beef fat (tallow), lard (lard) or milk fat (cream). Lard clearly passed as a non-halal product. While fat cow, although this animal is lawful for Muslims, it remains doubtful whether it is slaughtered according to Islamic law or not.

Having highlighted the issue of Halal, the understanding are not only limited to food and

Table 1. Reliability statistics.

Attributes	Cronbach's alpha	N of items
Bakery information	0.844	5
Individual owner information	0.752	5
Process	0.710	7
Inspection	0.720	7

drink as we commonly have known. But it also includes every act or acts of man. In the food industry, processing standards, food production and handling must be taken with careful attention. A good and proper handling cover from receiving raw materials, processing, storage and distribution of finished products but a more specific code of practice should be implemented in certain sectors. Halal certification requires a few criteria that have been set by JAKIM. Halal must follow all lawful standards that have been outlined in Islamic law. Food, equipment used and the manufacturing of food premises must be carried out using utensils and equipment that are free from the Islamic filth and with no pollution or cross contamination. It also should be in accordance with guidelines of Hazard Analysis Critical Control Point (HACCP), Good Hygiene Practice (GHP) and Good Manufacturing Practice (GMP) prescribed by Health Ministry of Malaysia. Thus the bakery operators are part of the important elements in maintaining and producing halal bakery products in Malaysia.

2 MATERIAL AND METHODS

Questionnaire related to the knowledge of the bakery and confectionery shop operators in Halal issue was distributed to know the extent of their knowledge and practices in food processing. The questionnaire consisted of 4 parts; bakery information, individual information, processes and inspection exercise.

Sixty-five questionnaires were distributed but only 50 usable questionnaires came back. The data were analysed using SPSS version 17.0 software package.

Referring to Table 1 above, all of the values lie between 0.70 which is near to 1.0. Hence, in this study, the sets of questions in every section of questionnaire are reliable as they correlated with each other.

3 RESULTS AND DISCUSSION

A total of 50 bakery and confectionery operators were involved in this study. There are 26 female respondents and 24 of them were male. From

Table 2. Bakery shop owner.

Owner	No	Percentage
Malay	14	28.0
Cina	35	70.0
India	1	2.0
Total	50	100.0

50 bakery and confectionery shop operators, 68 percent (34) of them worked in private business while 12 of the owners manage branches (24%) and about 8 percent managing the bakery and confectionery shop in the commercial sector.

Most of the bakery and confectionery industry are owned by Chinese with a total of 70 percent (35) respondents, followed by the Malays 28 percent (14) and the rest are Indian owners (2%). Bakery shops in Malaysia are still dominated by the Chinese and this can cause doubt among the users of bakery products in the country.

In this country the concept of halal in Islam, the management and the practice of halal bakery is only considered as satisfactory. This can be seen when the number of operators in the country mostly pioneered by non-Muslim. A survey showed that in main city for each state more than 50 percent of the bakery shop predominantly owned by non-Muslim especially Chinese.

However, there are still concerns among the non-Malay businessmen regarding halal bakery products produced by their companies to be able to penetrate the global market. This shows the level of awareness on the implementation of halal bakery operators is increasing from time to time. They believe halal products will bring confidence in buyers because consumers are now wise to choose the content of the products they buy. Results from this study indicate that all bakery and confectionery operators stated that their products are also suitable for Muslim and non-Muslim customers.

Results of the surveys and questionnaires conducted also found that most shops selling bakery ingredients and has succeeded in producing good quality product in terms of appearance, taste and texture as they have extensive experience tailoring course. Most of the operators believe the quality of the product is satisfactory (74%) followed by 12 percent industrialists strongly agree with this statement. However, only 2 percent of the operators do not agree with the quality of the product.

Based on observations on the responses given, it was found that the level of knowledge and preparation of bakery and confectionery operators on the product is clear and they are able to provide quality and affordable affluent to attract customers. They believe that customers are satisfied with the quality of the products they sell. Forty two percent of bakery and confectionery outlet operators believe that most customers who came are very satisfied of the services and products offered, while 34 percent of them strongly agreed with the statement. However, there are 24 percent operators stayed neutral on this statement. Based on the survey, 46 percent of respondents agreed that all employees are satisfied with their work environment, while 18 percent of employers strongly agree with this statement.

There are 26 percent and 38 percent of the respondents strongly agree and agree respectively that they practice self-hygiene and are neat according to the standards. However, 36 percent of employers are neutral on this issue.

For production area, 18 percent strongly agree, while 68 percent agree that their product do not contain non halal ingredient while 18 percent are neutral. This can be evaluated when the majority of the respondents had less than satisfactory response to prove they understand the concept of eating halal food in Islam. In addition, Chinese and Indian respondents were not clear on the concept of halal, haram and syubhah. For entrepreneurs who are Muslims, they are quite confused on the materials used and do not know the concept of syubhah.

Sixty percent of employers agree that they have a proper system for each processing for bakery products while 18 percent strongly agree that their employers have a systematic system in each job processing. However, 22 percent of employers reacted neutral on this question.

Based on the result, about 72 percent of the bakery and confectionery operators agree that they prepared many kinds of products for customers to choose. Fourteen percent of the operators strongly agreed with the statement, while 14 percent operators gave neutral answer to this question.

From the descriptive statistic of Halal practices, it shows all employees in the bakery and confectionery premises clearly understand Halal system standards ($M = 4{:}00 \pm 0.73$) because the operators of the premises provide training to all employees ($M = 3.96 \pm 0.81$). Additionally, employees understand the importance of food safety and hygiene ($M = 4.16 \pm 0.74$) through the monitoring conducted by JAKIM, JAIN and MAIN ($M = 4{:}12 \pm 0.69$). Monitoring and inspection covers the implementation of Halal, Halal certification ($M = 4.24, \pm 0.43$), processes, raw materials, storage and delivery of products ($M = 4.08 \pm 0.63$). All premises gave their cooperation during the monitoring process conducted ($M = 4{:}58 \pm 0.50$). Halal label is an important factor for Muslim in choosing product and as source of information of the contained ingredients.

4 CONCLUSION

The study sought has been conducted as part of the understanding on the perception and practices of Halal in Bakery and Confectionery industry. Based on the analysis of questionnaires and conclusions that have been made, some action and joint attention should be taken to overcome the ambiguity and lack of sensitivity of the respondents in some areas that have been identified. Some ambiguities identified in the survey are respondents' preferences and perceptions in Halal food regarding the issue of basic materials and additives as well as the issue of scientific ambiguity of the concept of Halal food as a whole.

Individuals and employers should strive to improve their knowledge in nutrition-related issues in terms of halal concept. This can be done by following the seminars, conventions, exhibitions, lectures and classes organized by parties either formal or informal. The operators should consult and discuss the problems they face on processes and ingredients with authoritative parties. Association and establishments involved among such as JAKIM, Consumer Association, non-governmental organization, communities within the society should prepare educational facilities and guidance for society. The lack of knowledge of operators and staff on scientific issues (additives in food) should be thoroughly and promptly resolved by the parties involved in providing education to the people especially JAKIM. This monitoring should be carried at all times to ensure every bakery operators and users are always aware of halal, all relevant information can be channeled, their problems are identified, monitored and enforcement of the law can be implemented on the target audience successfully.

Most of the weakness of association and authoritative bodies in this country is in terms of monitoring and this is evident from the issues raised and hotly debated and discussed about halal food nowadays. Monitoring should be conducted from a small community to the largest community. Complete organizational structure and the frequency of monitoring are necessary to ensure that the role and function of the body and the associations involved are on track. Hopefully this study has given all the benefits to the Malaysian especially Muslim to be aware on the type of products they consumed. In addition, it had been as a reference to Malaysia's Halal certification policy and procedure. Furthermore, this study perhaps can be a reference for any future study in the related field mainly in Malaysia. This study will benefit all Malaysian especially Muslim to be more particular on the type of product they consume. Additionally, it could be use as a reference to Malaysia's Halal certification policy and procedure. Furthermore, this study perhaps could also be a reference for any future study in the related field mainly in Malaysia.

REFERENCES

Abdul Rahman, M. (1999). *Halal Food Practice: Malaysia Experience*. Unpublished paper presented by Director of Research Division, Department of Islamic Development Malaysia (DIDM). Kuala Lumpur.

Majlis Ugama Islam Singapura. (1991). *Muslim diet: Halal food and haram drink*.

Talib, A., Ali, M., Anuar, K. & Jamaludin, K.R. (2008). *Quality assurance in halal food manufacturing in Malaysia: A preliminary study.* Paper presented at the International Conference on Mechanical & Manufacturing Engineering, Johor Bahru, Malaysia.

Theory and Practice in Hospitality and Tourism Research – Radzi et al. (Eds)
© *2015 Taylor & Francis Group, London, ISBN 978-1-138-02706-0*

Nutrient content of menu for basic recruit training in military foodservice

A. Nurhazwani, A. Nurul-Aziah & M.R. Aikal-Liyani
Faculty of Entrepreneurship and Business, Universiti Malaysia Kelantan, Kelantan, Malaysia

ABSTRACT: The study was to evaluate the nutrient content of menu provided to recruits in the Basic Recruit Training Centre, Port Dickson, Malaysia. From a 4-week cycle menu, the menu during the 3rd week was randomly chosen to obtain an average nutrient content. Nutrient content of menu was evaluated for 7 days from three main meals (breakfast, lunch, dinner) and snacks (morning and tea breaks) and was compared with the Recommended Nutrient Intakes (RNIs) of Malaysia and the military dietary recommendation for Malaysian Armed Forces, *Perintah Majlis Angkatan Tentera* (PMAT) to determine the nutritional adequacy of the menu for male recruits. The energy from the menu was within the range of 2772 to 2843 kcal, and met the energy recommendation of RNIs for general male aged 19 to 29 years but it did not meet the energy recommendation for recruits in training based on military dietary recommendation, PMAT. Protein from the menu has exceeded the RNIs and PMAT recommendation (17.4% to 18.3%). Fat content of the menu exceeded the recommended range of 20% to 30%. Carbohydrate was not sufficiently provided by the catering within the recommended carbohydrate range of 55% to 70%. Nutrient values for vitamin A, vitamin C, iron, thiamine, riboflavin, niacin, sodium and zinc met the recommendation of RNIs (100% of RNIs). However, the levels of calcium and dietary fiber in the menu were suboptimal. Therefore, corrective action should be taken to ensure the nutrient content of menu provided is sufficient for the physical activity level of recruits during training especially from carbohydrate food sources.

Keywords: Nutrient content, menu, contract catering, basic recruit training, military foodservice

1   INTRODUCTION

Basic recruit training is the prerequisite before joining Malaysia military for recruit's age between 18 to 24 years old (Ministry of Defense Malaysia, MINDEF, 2011). Recruits are in the transition phase from a civilian to a soldier and are not yet a soldier. They are trained in the basic recruit training centre, Port Dickson where the basic health needs for soldiers are regulated by the military institution (Marshall & Meiselman, 2006). This training center is a semi closed food system, in which the contract catering serves as the primary foodservice for the recruits and other ranking officers. It is compulsory for the recruits to eat within the catering facilities while the other ranks can select to eat elsewhere (Zainon, Menu Planner, personal communications, December 13, 2010). Thus, energy and nutrient intakes as well as food choices of recruits are greatly influenced by the food system (Smith, Davis-street, Rice & Nillen, 2001). Hence, this presents a challenge to the contract catering, especially the menu planners, who must compete by items offered that appeal to the soldiers for sufficient energy intake during training. They need to know if the menu provides an adequate nutrient for basic recruit training.

Starting from 1999 onwards, some military cookhouses in Malaysia were subcontracted to a contract catering in order to improve the facilities, food, and services, including the military cookhouses in the basic recruit training center. The types of dishes provided by the contract catering are more or less the same with the cookhouses, with some new dishes incorporated. However, while the cookhouses used fully fresh food ingredients to cook the menu, the contract catering envisioned the ready-made paste, packed in flexible plastic retort pouches with some freeze-dried components, as main cooking ingredients, added with fresh food materials (Kor Perkhidmatan Diraja, KPD, 2010). Menu was planned based on the regulation in the Order of the Military Council, *Perintah Majlis Angkatan Tentera* (PMAT) Malaysia (PMAT, 2003). The contract catering offers a 4-week cycle menu for the soldiers. They have the standard food portions when serving the food to the soldiers according to the military dietary recommendation for Malaysian Armed Forces, *Perintah Majlis Angkatan Tentera* (PMAT, 2003).

Reports in the literature suggested that the soldiers lacked of energy and nutrient intake, especially during basic recruit training (Moore, Friedl, Kramer, Martinez-Lopez, Hoyt, Tulley, DeLany, Askew & Vogel, 1992). It is questioned whether the energy intake was adequate for energy requirements at each Phase 1, Phase 2 and Phase 3 during basic recruit training. Lack of energy and nutrient intake cause body mass loss and lowered the soldier's performance, which may lead to attrition during basic recruit training (Lee,McCreary & Villeneuve, 2011; Niebuhr, Scott, Li, Bedno, Han & Powers, 2009). The earliest study on Malaysian recruits was done by Isa (1991) where he found that recruits ate less and had low energy intake during the military cookhouse operation in basic recruit training centre. Another problem with the menu provided is, the training intensity varies in each training phase, thus energy requirements are different for each training phase. Phase 2 basic recruit training needs higher level of energy requirements because the recruits consistently go for field training (Isa, 1991). Hence, it is questioned whether the food provided is sufficient for each training phase, especially at Phase 2 training. The recruits may spend more time on physical training at Phase 2 BT yet still consume the same amount of food provided by the menu at Phase 1 basic recruit training (Jesse, Training Cell Warrant Officer I, personal communication, July 30, 2011). There are limited studies or scholarly research that has explored the nutrient content of menu provided by the contract catering in the military foodservice. The objectives of this study were to evaluate the energy, macronutrients and micronutrients content of menu and to compare the nutrient values of the menu to the dietary recommendation, the Recommended Nutrient Intakes (RNIs) of Malaysia and the military dietary recommendation, (PMAT) for nutritional adequacy of male recruits participating in basic recruit training.

3 METHODS

3.1 *Study population*

Basic recruit training centre, Port Dickson is the only training centre for recruits in Malaysia. The catering in this training centre participated in the study, had one main cookhouse for cooking purposes (B1) and three satellite kitchens with dining facilities (B2, B3 and B4) for food receiving, reserved for male recruits. For one satellite kitchen, the feeding strength was around 200 to 300 diners per meal. All of the dining facilities provide same menu for all recruits. Dining facilities for male recruits were selected due to high proportion of male recruits and to avoid gender bias during

the study. This study had approval from the ethics committee of the Medical Research Ethics Committee of the Putra University of Malaysia and permission to conduct this study was granted from the Malaysian Armed Forces.

3.2 *Evaluation of menu*

Executive chef of the contract catering was requested to provide standard recipes and standard food portions of foods and beverages served in the basic training centre. Energy and nutrients for standard food portion of each food and drink during breakfast, snacks, lunch and dinner were obtained with Nutritionist Pro™ (First Data Bank, USA, 2003) with reference to the Malaysian Food Composition Tables (Tee, Mohd Ismail, Mohd Nasir & Khatijah, 1997). To obtain average nutrient information on a 4-week cycle menu, the menu from the 3rd week of a 4-week cycle menu was randomly chosen to analyse the energy and nutrients and compared with the RNIs of Malaysia (2005) and PMAT (2003) to determine the nutritional adequacy of the menu for the recruits.

3.3 *Statistical methods*

Nutrient content of menu was analysed using IBM SPSS Version 21 for the mean, standard deviation, and percentages. The average nutrient data of energy, protein, carbohydrate, fat and selected micronutrients were compared to the Recommended Nutrient Intakes (RNIs) of Malaysia for male aged 19 to 29 years (National Coordinating Committee on Food and Nutrition, 2005) and the military dietary recommendation for recruits in training (PMAT, 2003). No current RNI exists for sodium, however, an intake of 2400 mg/day or less is recommended by the Institute of Medicine (Institute of Medicine, IOM, 2005).

4 RESULTS

Macronutrients and micronutrients analysis of menu is summarized in Table 3.1. The nutrients content of menu was evaluated on 7 days from three main meals (breakfast, lunch, dinner) and snacks (morning and tea breaks). The menu was analysed for the energy and nutrients based on the detailed recipe information provided by the executive chef of the catering. The menu usually had two options whereby soldiers may choose the types of entrees and vegetables whereas the starches, desserts and sides provided had only one option. The energy from the menu provided was within the range of 2772 to 2843 kcal, and met the energy recommendation of RNIs but it did not meet

Table 1. Average daily energy and nutrients content of menu compared to nutrient recommendation of RNIs and PMAT.

Nutrient	RNI 19–29 years male	PMAT	Mean±SD		% RNI		% PMAT	
			Menu Option 1	Menu Option 2	Menu Option 1	Menu Option 2	Menu Option 1	Menu Option 2
Energy (kcal)	2440	3200	2843	2772	116.52	113.61	88.84	86.63
Protein (g)	62	60	123.48	126.59	199.17	204.17	205.80	210.98
Kcal (%)	10–15	–	17.4	18.3	–	–	–	–
Fat (g)	–	71	103.64	97.34	–	–	145.97	137.10
Kcal (%)	20–30	–	32.8	31.6	–	–	–	–
Carbohydrate (g)	–	–	353.46	346.87	–	–	–	–
Kcal (%)	55–70	–	49.8	50.1	–	–	–	–
Vitamin A (μg)	600	–	1175.31	1209.01	195.88	201.51	–	–
Vitamin C (mg)	70	–	152.83	141.20	218.33	201.72	–	–
Calcium (mg)	800	–	587.54	575.13	73.44	71.89	–	–
Iron (mg)	14	–	23.40	24.67	167.16	176.21	–	–
Thiamin (mg)	1.2	–	1.48	1.51	122.91	125.58	–	–
Riboflavin (mg)	1.3	–	2.39	2.23	183.76	171.38	–	–
Niacin (mg) NE	16	–	19.40	20.25	121.25	126.57	–	–
Sodium (mg)*	2300	–	4289.68	4074.80	178.73	169.78	–	–
Zinc (mg)	6.7	–	7.80	8.19	116.40	122.26	–	–
Dietary fibre (g)	38	–	13.39	13.37	35.23	35.18	–	–

*Institute of Medicine (2005)

the energy recommendation for soldiers in training based on military dietary recommendation, PMAT for male recruits. Protein from the menu has exceeded the RNIs and PMAT recommendation (17.4 to 18.3%). Fat content of the menu exceeded the recommended range of 20 to 30%. Carbohydrate was not sufficiently provided by the catering within the recommended carbohydrate range of 55 to 70%. Nutrient values for vitamin A, vitamin C, iron, thiamine, riboflavin, niacin, sodium and zinc did meet the recommendation of RNIs (100% of RNIs). However, the levels of calcium and dietary fiber in the menu were suboptimal.

5 DISCUSSION

The research question discussed here was whether the menu provided by the catering meet the recommended level of nutrient intake for the male army recruits during basic recruit training. The menu offered in the basic training centre is a 4-week cycle menu, a series of daily menu on a basis of four weeks, after which the cycle is repeated. 'Cycle menu are used to offer variety with some degree of predictability for ordering, budgeting and production scheduling' (Spears and Gregoire, 2010). In many military institutional training, the soldiers were dependent on the catering to provide large proportion of the food consumed and they were at nutritional risk than the general population

(Williams, 2009) due to the effect of the military exercises. Hence, the recommended nutrient intakes (RNIs) for Malaysia may not be applicable to these recruits but the RNIs at least can assess the level of nutritional adequacy among the recruits in the training centre in order to avoid diet insufficiency rather than go for optimal diet during basic recruit training. The Malaysian Armed Forces have military dietary recommendation when providing food for the soldiers, PMAT (2003) with energy provided must at the minimum of 2700 kcal/day with intake of 60g from protein and 60g from fat. However, the dietary recommendation was meant for general soldier involved with military support activities. Soldiers and recruits in training centre as well as in the field operation are subjected to normal combat rations with energy provided must be at minimum level of 3200 kcal/day with energy contribution of 60 g from protein and 71g of fat (PMAT, 2003).

The current menu provided by the catering easily provided daily energy requirements of energy for general Malaysian male and general soldier population based on the RNIs and PMAT recommendation. However, the energy provided by the menu from the catering did not meet the minimum of intake values for recruits in training based on PMAT recommendation for soldiers in military training (3200 kcal/day). The energy needs for Phase 1 basic recruit training (3257 kcal), Phase 2 basic recruit training (3588 kcal) and Phase 3 basic recruit training (3492 kcal) exceeds the minimum recommendation of energy intake for soldiers in training by PMAT. Thus, the catering should have provided additional foods to meet the energy demands during basic recruit training.

Comparison of energy from the menu by the contract catering was compared with energy from the military cookhouses (Isa 1991) in the basic training centre. Energy value provided by the military cookhouse was 3070 kcal and this means that the catering served less energy from the food than the previous military cookhouse, with energy value 2772 to 2843 kcal. Energy contribution of macronutrients from the menu by the catering was 17.4 to 18.3% from protein, 31.6 to 32.8% from fat and 49.8 to 50.1% from carbohydrate meaning that the protein in the menu of the catering was within recommendation of 10 to 15% of the RNIs of Malaysia. Fat content in the menu was shown to exceed the recommended range of fat intake of between 20 to 30% while carbohydrate did not achieved the 50 to 70% of recommended carbohydrate intake of the RNIs. The catering should supply more carbohydrate foods and reduced the fat content in the menu for soldiers during basic recruit training. The menu were nutritionally adequate for these nutrients; vitamin A, vitamin C, iron, thiamine, riboflavin, niacin, sodium and zinc except for calcium and dietary fibre. Menu should be monitored for high fat, high sodium, low calcium and low dietary fiber that may impact the health of soldiers. Thus, corrective action is necessary to improve the diet for recruits.

As a conclusion, menu provided by the catering did not meet the military dietary recommendation (PMAT) for male recruits of minimum 3200 kcal/day and calcium and dietary fiber provided in the menu were not adequate for the male recruits. Menu that did not meet the nutrition recommendations may place the recruits at increased risk of health consequences such as fatigue due to low intake of carbohydrate food and high body fat percentage due to high fat intake. This study is important as it will provide the menu planners of the catering with current benchmark of nutrient content of the menu, which are required to the formulation of managerial policies for menu planning in the Malaysian Armed Forces. Future studies should compare the nutritional adequacy of the menu with female recruits as this population is at nutritional risk during basic recruit training.

REFERENCES

Institute of Medicine (US). Panel on Dietary Reference Intakes for Electrolytes & Water. (2005). *DRI, dietary reference intakes for water, potassium, sodium, chloride, and sulfate*. Natl Academy Pr.

Isa. (1991). A study of energy requirement of the Malaysian Armed Forces, Dissertation Master of Science, Universiti Kebangsaan Malaysia

Jesse, Training Cell Warrant Officer I, personal communication, July 30, 2011).

Kor Perkhidmatan Diraja, KPD. (2010), Taklimat penswastaan perkhidmatan katering Tentera Darat—Dasar dan konsep pelaksanaan penswastaan perkhidmatan catering di kompleks sajian TD [PowerPoint slides]. KPD Cawangan sajian

Lee, J.E., McCreary, D.R. & Villeneuve, M. (2011). Prospective Multifactorial Analysis of Canadian Forces Basic Training Attrition. *Military Medicine, 176*(7), 777–784.

Marshall, D.W. & Meiselman, H.L. (2006). Limited choice: An exploratory study into issue items and soldier subjective well-being. *Journal of Macromarketing, 26*(1), 59–76.

Ministry of Defence, MINDEF. (2011), Cawangan Sumber Manusia (Sel Promosi), Pengambilan Perajurit muda Lelaki dan Wanita Tentera Darat Malaysia, Kementerian Pertahanan, Kuala Lumpur.

Moore, R.J., Friedl, K.E., Kramer, T.R., Martinez-Lopez, L.E. & Hoyt, R.W. (1992). *Changes in soldier nutritional status and immune function during the Ranger training course* (No. T13–92). ARMY RESEARCH INST OF ENVIRONMENTAL MEDICINE NATICK MA.

National Coordinating Committee on Food and Nutrition, NCCFN.(2005). Recommended Nutrient Intakes for Malaysia (2005). Ministry of Health Malaysia, Kuala Lumpur.

Niebuhr, D.W., Scott, C.T., Li, Y., Bedno, S.A., Han, W. & Powers, T.E. (2009). Preaccession fitness and body composition as predictors of attrition in US Army recruits. *Military medicine, 174*(7), 695–701.

Perintah Majlis Angkatan Tentera,PMAT. (2003). Bilangan 7 Tahun 2003, keluaran 427, Kandungan: Sukatan Rangsum Angkatan Tentera (Pembatalan PMAT Bilangan 7 Tahun 1990). Pusat Latihan Kor Perkhidmatan (PULMAT) Cawangan Bekalan PMAT 7/2003,ms 1–35

Smith, S.M., Davis-Street, J.E., Rice, B.L., Nillen, J.L., Gillman, P.L. & Block, G. (2001). Nutritional status assessment in semiclosed environments: ground-based and space flight studies in humans. *The Journal of nutrition,131*(7), 2053–2061.

Spears, M.C. & Gregoire, M.B. (2010). *Foodservice Organizations: A Managerial And Systems Approach*: Pearson Prentice Hall.

Tee, E.S., Ismail, M.N., Nasir, M.A. & Khatijah, I. (1997). *Nutrient composition of Malaysian foods.* Kuala Lumpur: Institute Medical for Research.

Williams, P.G. (2009). Foodservice perspective in institutions. *Meals in science and practice: Interdisciplinary research and business applications*, 50–65.

Zainon, Menu Planner, personal communications, December 13, 2010).

Theory and Practice in Hospitality and Tourism Research – Radzi et al. (Eds)
© *2015 Taylor & Francis Group, London, ISBN 978-1-138-02706-0*

Assessment of food handlers' knowledge, attitude and practices on food hygiene in Serdang, Selangor

N.A. Mahyuddin & Z. Zainon
Food Safety Research Centre (FOSREC), Universiti Putra Malaysia, Malaysia

U.F. Ungku-Zainal-Abidin
Faculty of Food Science and Technology, Universiti Putra Malaysia, Malaysia

ABSTRACT: This paper reported a research on food hygiene knowledge, attitude and practices of food handlers from restaurants in Serdang, Selangor. A total of 200 questionnaires were distributed among the food handlers at food premises selected using convenience sampling technique. The findings from this study showed that food hygiene knowledge related to hand washing procedure (72.50%), temperature control of food, and microbial (66.0%) were lacking among the food handlers. Attitude towards learning more about food hygiene and sanitation has the lowest rating ($M = 4.55$), which suggest food handler may not perceive this as important to them. Respondents reported the lowest practices for hand washing before using glove ($M = 3.92$), which demonstrated that food handlers in Serdang might be lack of knowledge regarding the correct use of glove. This study provided data about the current knowledge, attitudes and practices of food handlers from restaurant operations. The results showed that further continuous effort should be invested in food hygiene education and enforcement for food handlers from commercial foodservice operations such as restaurant.

Keywords: Food hygiene, knowledge, attitude, practices, food handlers

1 INTRODUCTION

Food safety is very important in today's food industries as people have become increasingly concern about the risks and quality related to the foods. This issue is not only a threat to the low income countries but also as a challenge to the medium and high income nations. Statistics of food borne illness cases shows that the most common bacteria which cause food borne diseases are Salmonella (378 cases in 2003), Escherichia coli (74 cases in 2003), Campylobacter (268 cases in 2002) (Sun & Ockerman, 2005). Incidents of food poisoning outbreaks continue to rise from malpractice of food hygiene regardless an increasing number of food handlers receiving food safety training (Clayton, Griffith, Price & Peter, 2002)

The improper food handling practices among food handlers may be implicated in 97% of all food borne illness associated with catering outlets (Howes, McEwan, Griffiths & Harris, 1996). Research showed lack of knowledge pertaining to time and temperature control, prevention of cross contamination and sources and modes of transmission of food poisoning bacteria has contributed to poor practices among the food handlers (Albrecht, 1995).

Good sanitation practices in restaurants are important not only to reduce cross contamination of food but also to increase the morale of workers (Al-Khatib & Mitwalli, 2009). All food handlers should be given some training to enable them to perform their job without any risk of causing food poisoning. According to Ribel, Mathias, Campbell, and Wiens (1994), training had an impact on examination scores in the short-term and their study showed restaurants with staffs having food handler education significantly perform well in food hygiene practices than those staffs that have no such education.

Many food premises involved in food poisoning cases due to food handlers poor handling practices. They have the responsibility of ensuring the production of safe foods, and their knowledge, attitudes and practices play a major role in the occurrence of food poisoning cases. The objectives of this study were three folds: 1) to determine the level of knowledge on food hygiene practices among the food handlers in the Serdang, 2) to evaluate food hygiene attitudes and practices among food handlers, and 3) to self-reported food hygiene practices among food handlers.

1.1 Food safety in the foodservice industry

One of the most important aspects in foodservice operations but usually receives the smallest amount of visibility and attention is food safety (Manask, 2002). Food service is almost in the last part of food preparation, it guides the final linkage of food safety for public concern. Every food should be assured that it is safe to be consumed and especially from farm to fork which means the raw materials that we get must be from good sources and free from any contaminations. Foodborne illness episodes have been commonly associated with foodservice and the greatest number of food poisoning outbreak also arises from this segment of industry (Cavalli & Salay, 2004; Clayton & Griffith, 2004).

Every foodservice operation has the potential to cause foodborne illness through errors in purchasing, receiving, storing, preparing and serving food (National Restaurant Association, 1992). A wide range of common contributing factors to the foodborne illness in foodservice has been documented (Coleman & Griffith, 1998; Ryan et al., 1996; Weingold et al., 1994). Worldwide epidemiological research identified inadequate heat treatment, inappropriate storage of foods, infected food handlers and cross-contamination as the major contributing factors (Griffith, 2000). Malpractices of food hygiene are consistently suggested as the underlying cause of foodborne illness in foodservice industry.

A study on the CDC report of foodborne outbreaks between 1988 and 1992 found improper holding temperature of food and poor personal hygiene of employees reported in 59% and 36% of outbreaks, respectively (Bean, Goulding, Daniels & Angulo, 1997). In a more recent study, employees' poor safe food handling practices associated with bare-hand contact and handling of food by infected person were identified as contributing factors in foodservice operations implicated with foodborne illness outbreaks (Hedberg et al., 2006).

1.2 Personal hygiene for food handlers

Food handler is anyone who is employed in the production, preparation, processing, packaging, storage, transport, distribution and sale of food. The area of personal hygiene that of utmost importance includes hands and skin, cuts, boils, septic spots, hair, ear, nose and mouth, smoking, wearing jewelry, perfume, protective clothing, general healthcare and reporting of illness and most importantly is the hygiene education itself. Food handlers can spread food borne illnesses as their hands and other body parts may harbor microorganisms and their actions as well, may compromise the chain of safety 'from farm to fork"(Akonor & Akonor, 2013). Therefore, food handlers have critical roles

and responsibilities in preventing foodborne illness outbreaks (Howellset et al., 2008).

1.3 Factors affecting food hygiene practices

Numerous studies have investigated factors that influence food handlers' safe food handling practices with the overarching goal to enhance current interventions strategies and help address current challenges in managing food safety. Factors affecting food handlers' practices are multidimensional and include their food safety knowledge. A number of studies have investigated the role of knowledge and attitudes on food handlers' safe food handling practices in the foodservice industry (Abdul-Mutalib et al., 2012; Bas, Ersun & Kivanc, 2006; Ko, 2012; Martin, Hogg & Otero, 2012). Knowledge about and attitudes toward food safety are important affecting food handlers' practices. Attitudes an important factor besides knowledge and enforcement, ensure a downward trend of food borne illnesses (Howes et al., 1996). Several researchers have applied behavioral theories to understand underlying factors influencing food safety practices (Ball, Wilcock & Aung, 2010a; Brannon, York, Roberts, Shanklin & Howells, 2009).

2 METHODOLOGY

2.1 Population and study sample

The study population for this research consisted of food handlers, who were employees working in restaurants or food premises in Serdang, Selangor. Sample of the study was selected from the registered or listed restaurants given by the Department of Health in Majlis Perbandaran Subang Jaya (MPSJ). The sample size was 200 food handlers in Serdang area and group administered questionnaires method was employed to recruit respondents for data collection. Study sample was located in 17 medium-size restaurants including several fast food restaurants in Serdang area. Convenience sampling technique was used for selecting respondents because of limited cooperation obtained from the study population.

2.2 Research instrument

A self-administered survey questionnaire was used for data collection. The research instrument was constructed by adapting items from the Food and Drug Administration (FDA) outlines, Malaysia Food Hygiene Regulations (2009) and also the Malaysian Ministry of Health Guidelines in Food Hygiene (Ministry of Health, 2009). There were 41 questions in the questionnaire divided into 4 sections: 1) Respondents demographic information, 2) food handler knowledge about food hygiene and sanitation, 3) food handler attitudes towards food hygiene and sanitation, and 4) food handler self-reported food hygiene and sanitation practices.

Knowledge items were measured using nominal scale with yes or no response. Respondents' knowledge was calculated by the percentage of correct response for all the knowledge items. A 5-point Likert scale was used for measuring food hygiene attitudes (1 = Strongly disagree, 5 = Strongly agree). Similarly, food handler self-reported food hygiene practices were measured using 5-point scale (1 = Never, 5 = Always).

2.3 Pilot testing

According to Bourque and Fielder (1995), all questions should be pre-tested or pilot test to check whether the ideas in designing the questions are reliable and easy to understand by the respondents. A pilot test of the survey questionnaire was conducted by involving food handlers (n = 30) from medium-size restaurants or food premises in Serdang area.

2.4 Data collection procedures

An approval letter from Universiti Putra Malaysia was granted to conduct this study. The questionnaires were distributed from September 2013 until end of October 2013.The approval letter was basically to prove to the owners or managers of the restaurants that the study conducted was legitimate. After researcher received the permissions from the restaurant owners, the questionnaires were distributed among the food handlers and it require only 10 to 15 minutes for the respondents to complete the questionnaires. Respondents were also given three days to complete the questionnaires if they have no time to finish it before the researcher return to collect the questionnaire.

2.5 Data analysis

The data for this research was analyzed using Statistical Packages for Social Sciences (SPSS) version 20. Descriptive statistics including mean, standard deviation, frequency, and percentage were used to summarize the data.

3 RESULTS AND DISCUSSION

3.1 Respondents profile

Slightly more than half of the respondents were female food handlers (56.5%). The majority of

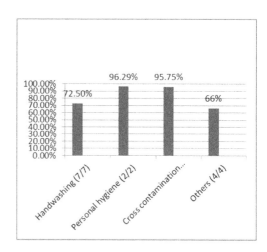

Figure 1. Percentage of respondents who obtained 100% score for different category of food hygiene knowledge questions.

respondent age between 21 to 30 years old (54.0%) followed by those age less than 20 years old (26.5%). Possibly, some of the respondents were the nearby university students who work part time job during their studies. Most of the respondents were single (71.0%) and the highest educational level for most of them were either SPM/STPM (40.0%) or Certificate (Diploma) (40.0%). Respondents with less than five years work experience in the foodservice industry constituted 75.5% of total respondents followed by those who have between six and ten years experience (17.5%).

3.2 Food hygiene knowledge

Figure 1 presents the level of knowledge among the food handlers in Serdang for different aspects of food hygiene. Respondents were assessed on their knowledge about hand washing (2 items), personal hygiene (7 items), cross-contamination (2 items) and other aspect such as storage temperature of foods and chemical materials and microorganism (4 items). Overall, respondents' knowledge was highest on the aspect of personal hygiene. A total of 96.29% respondents answered correctly all questions related to personal hygiene followed by cross-contamination prevention (95.75%) and hand-washing (72.5%). Specifically, questions on personal hygiene include keep finger nail short and smoking, eating and drinking are not allowed at food preparation area, coughing or sneezing is not allowed at food preparation area, and food handlers must wear clean overalls, gloves and hats while preparing the foods.

Questions for cross contamination was related to use of different chopping board for raw and

Table 1. Mean score rating for respondents' food hygiene attitudes (n = 200).

Attitudes	Mean (± SD)[a]
1. Learning more about food hygiene and sanitation is important to me.	4.55 (± 0.78)
2. Refrigerator must be kept clean and the inside surfaces should be washed regularly to avoid from any contamination of foods.	4.56 (± 0.76)
3. Older foods must always be used first and any food that has been stored longer than the recommended time should be destroyed.	4.58 (± 0.68)
4. Safety food handling is an important part of my job responsibilities	4.61 (± 0.67)
5. Using caps, masks, gloves and adequate clothing can reduce the risk of food contamination.	4.62 (± 0.61)
6. Raw foods should be kept separately from cooked foods.	4.66 (± 0.73)

[a]Rating scale: 1 = Strongly disagree; 5 = Strongly agree.

Table 2. Mean scorerating for respondents' self-reported food hygiene practices (n = 200).

Practices	Mean[a]
Hand-washing	
1. Wash hands after touching unwrapped cooked foods	4.15
2. Wash hands before touching unwrapped cooked foods	4.17
3. Wash hands before touching unwrapped raw foods	4.20
4. Wash hands after touching unwrapped raw foods	4.26
Glove uses	
1. Wash hands before using gloves	3.92
2. Wash hands after using gloves	4.11
Cross-contamination prevention	
1. Use protective clothing when touch or distribute unwrapped foods	4.10
2. Wear cap while preparing foods	4.16
3. Aware the need to store food away before dusting/sweeping food preparation area	4.23

[a]Rating scale: 1 = Never; 5 = Always

cooked food and wearing jewelry while working in the kitchen. Finally, hand-washing knowledge was assessed on the need to scrub hands for five seconds, and wash hands every time after touching their hair, face or body, after handling waste food, after using the toilet or after handling raw meat, poultry, vegetables and eggs.

It can be concluded from the results that the respondents have low level of knowledge about hand washing procedure. Respondents also reported low level of knowledge related to other aspect of food hygiene (e.g., microorganism related to foodborne illness and temperature control) possibly due to lack of understanding on the terminology used in the questionnaire as most of them have low educational level.

3.3 Food hygiene attitudes

Table 1 shows the mean score for respondents' attitudes towards food hygiene. Attitudes toward ensuring "raw foods should be kept separately from cooked foods" has the highest mean score ($M = 4.66$) followed by "using caps, masks, gloves and adequate clothing can reduce the risk of food contamination" ($M = 4.62$) and "safety food handling is an important part of my job responsibilities"($M = 4.61$). The lowest mean scores for attitudes was towards "learning more about food hygiene and sanitation is important to me"($M = 4.55$). The findings indicated that even respondents have positive attitudes toward safe food handling as part of their job responsibilities, they may not perceive learning more about food hygiene and sanitation is important to them.

3.4 Food hygiene practices

Respondents' self-reported food hygiene practices are shown in Table 2. In terms of hand washing, respondents reported the lowest practices was before and after touching cooked food compared to raw foods. This finding indicated poor safe food handling practices among respondents because unwrapped cooked food must be handled after hand washing. The result also showed poor practices related to correct use of glove. Respondents reported the lowest practices for hand washing before using glove ($M = 3.92$), which demonstrated that food handlers in Serdang might be lacking of knowledge regarding the correct use of glove.

The lowest practices for cross contamination prevention among Serdang food handlers was the "use protective clothing when touch or distribute unwrapped foods" ($M = 4.10$).This states that the food handlers seldom use complete protective clothing while preparing foods, which includes wearing clean overalls, protective shoes, apron, cap and gloves while preparing the foods.

4 CONCLUSION

In general, the findings of this study suggest food hygiene knowledge, attitude and practices amongfood handlers from restaurants in Serdang are relatively good except for a few areas of concern. Knowledge on food hygiene among food handlers was quite low particularly in the area of hand washing procedure and temperature control of food. Although food handlers generally showed positive attitudes towards food hygiene, their perception on the importance of learning more about food hygiene and sanitation is considered low. Finally, this study showed food handlers in Serdang have poor practices of food hygiene related to correct use of glove in food preparation. Particularly, poor practices were reported in hand washing before using glove. Authorities of local areas could emphasize these areas of concern to improve food hygiene knowledge, attitudes and practices among food handlers in Serdang.

REFERENCES

Abdul-Mutalib N-A., Abdul Rashid, M-F., Mustafa, S., Amin-Nordin, S., Hamat, R.A., Osman, M. 2012. Knowledge, attitude and practices regarding food hygiene and sanitation of food handlers in Kuala Pilah, Malaysia.*Food Control, 27*, 289–293.

Akonor, P.T., & Akonor, M.A. 2013. Food safety knowledge: The cases of domestic food handlers in Accra. *European Journal of Nutrition and Food Safety, 3*(3), 99–111.

Al-Khatib, I.A. & Mitwalli, S.M. 2009. Food sanitation practices in restaurants of Ramallah and Al-Bireh district of Palestine. *Eastern Mediterranean Health Journal, 15*(4).

Ball, B., Wilcock, A. & Aung, M. 2010. Background factors affecting the implementation of food safety system.*Food Protection Trends, 30*(2), 78–86.

Bas, M., Ersun, A.S. & Kıvanç, G. 2006.The evaluation of food hygiene knowledge, attitudes, and practices of food handlers in food businesses in Turkey.*Food Control, 17,* 317–322.

Bean, N.H., Goulding, J.S., Daniels, M.T. & Angulo, F.J. 1997.Surveillance for foodborne disease outbreaks—United States, 1988–1992.*Journal of Food Protection, 60*, 1265–1286.

Brannon, L.A., York, V.K., Roberts, K.R., Shanklin, C.W. & Howells, A.D.2009. Appreciation of food safety practices based on level of experience.*Journal of Foodservice Business Research, 12*(2), 134–154.

Cavalli, S.B. & Salay, E. 2004. Food quality and safety control activities in commercial foodservice in the cities of Campinas and Porto Alegre, Brazil. *Foodservice Research International,* 14(4), 223–241.

Clayton, D.A. & Griffith, C.J. 2004. Observation of food safety practices in catering using notational analysis. *British Food Journal,* 106(2/3), 211–227.

Clayton, D.A., Griffith, C.J., Price, P. & Peters, A.C. 2002. Food handlers'beliefs and self—reported practices. *International Journal of Environmental Health Research,12*,25–39.

Coleman, P. & Griffith, C.J. 1998. Risk assessment: A diagnostic self-assessment tool for caterers. *International Journal of Hospitality Management,* 17(3), 289–301.

Griffith, C.J. 2000. Food safety in catering establishment. In J.M. Farber & E.C.D. Todd (Eds.), *Safe handling of foods.*New York: Marcel Dekker. (pp. 235–256)

Hedberg, C.W., Smith, S.J., Kirkland, E., Radke, V., Jones, T.F., Selman, C.A. & the EHS-Net Working Group.2006. Systematic environmental evaluations to identify food safety difference between outbreak and nonoutbreak restaurants. *Journal of Food Protection,* 69, 2697–2702.

Howells, A., Roberts, K., Shanklin, C., Pilling, V., Brannon, L. & Barrett, B. 2008. Restaurant employees' perceptions of barriers to three food safety practices. *Journal of the American Dietetic Association, 108,* 1345–1349.

Howes, M., McEwen, S., Griffith, M. & Harris, L. 1996. Food handler certification by home study: Measuring changes in knowledge and behavior. *Dairy, Food and Environmental Sanitation,* 16(11), 737–744.

Ko, W.H. 2012. The relationship among food safety knowledge, attitudes and self-reported HACCP practices in restaurant employees.*Food Control, 29,* 192–197.

Manask, A.M. 2002. The complete guide to food service in cultural institutions. New York: John Wiley & Sons.

Ribel, P.D., Mathias, R.G., Weins, M., Cocksedge, W., Hazlewood, A.,Kirschner, B. & Pelton. 1994. Routine restaurant inspection and education of food handlers: Recommendation based on critical appraisal of literatureand survey of Canadian jurisdictions on restaurant inspectors and food handlers. *The Canadian Journal of Public Health.* July-August, 67–70.

Sun, Y.M. & Ockerman, H.W. 2005. A review of the needs and current applications of hazard analysis and critical control point (HACCP) system in foodservice areas.*Food Control,* 16(4), 325–332.

Gastronomy

Theory and Practice in Hospitality and Tourism Research – Radzi et al. (Eds)
© 2015 Taylor & Francis Group, London, ISBN 978-1-138-02706-0

Social influence and construction of postpartum food intake among Malay women

M.S.Y. Kamaruddin, A.A. Azdel, N.A. Ahmad & M.F.S. Bakhtiar
Faculty of Hotel and Tourism Management, Universiti Teknologi MARA, Shah Alam, Malaysia

S.A. Wahab
Faculty of Communication and Media Studies, Universiti Teknologi MARA, Melaka, Malaysia

ABSTRACT: Food without doubt has been a central axis in understanding religion, society as well as culture. It is one of the most important elements in one society in determining what is allowed and prohibited. Globally, the function of food during the postpartum period is inherited from mother to daughter for generations. For Malay women in yesteryear, the social circle who advices the food intake during this respective period was determined by Malay shamans known as *bidan*. Thus the food must be followed without questions However, the expansion of experience and extension of knowledge results in the postpartum food is no longer a rigid intake. Qualitative approach were employed through semi structured interview, the finding shows factors that bidan is no longer become an ultimate source of food knowledge during this period and Malay culture in this abstinence period is no longer stagnant but moving to more scientific well proven knowledge. However, the movement is still slow because some women still believe in a construction and accumulation of anecdotal evidence.

Keywords: Influence, social, postpartum, Malay, foodways

1 INTRODUCTION

The Chinese called the person as *pei yue* who sometimes a women's own mother, mother in law, a paid professional to cook for mother and child during abstinence period (Koon, Peng & Karim, 2005). Unlike Chinese, the aboriginal strongly hold to shaman words (Rahimi, Fatimah, Rahimah, Sarah & Marlia, 2003; Ramle & Beri, 1993). Balwi (2003) and Ali and Howden-Chapman (2007) stated the Malays especially in rural area still believe in shamans. The main roles of bidan are to handle childbirth and responsible on women's health during postpartum. The aforementioned peoples who determine the actions of the women to go through these forty days. The action permitted and action prohibited includes sexual prohibition, contact with decease family, and housework until dietary guidance (Katz, 1996).

Generally, one of the utmost factors that affect health is food (Balwi & Koharuddin, 2003) which Stefanello, Nakano and Gomes (2008) contended women are instructed about care and the family spreads beliefs, habits, attitudes and behaviors during confinement. In this period, women are encouraged to eat hot food to keep the body warm (Ali & Howden-Chapman, 2007). The food can be divided into few categories or elements which are hot, cold, poisonous (Ali & Howden-Chapman, 2007; Katz, 1996; Stefanello, et al., 2008) and windy (Balwi & Koharuddin, 2003). Each element of food is constructed by cultural belief and the consumption might have positive or a negative effect to the mother. The same goes to foodways (preparation to consumption) which are tailored by particular customs which most of it will incline to bad luck to women. In Malay community, the practices were strongly influenced by Hinduism when the Hindus were brought into Malaysia in the early centuries until about 1400's (Ali & Howden-Chapman, 2007). They posited this traditional food and foodways knowledge practiced heretofore is the intersection between Hindu belief and Islamic religion.

Plethora of past literature has looked into food during confinement period after childbirth, which most of it was focusing more on the nutritional aspect. Only few studies (such as Hishamshah et al., 2011; Kamaruddin, Mohd Zahari, Muhammad, Amir & Azdel, 2014; Matthey, Panasetis & Barnett, 2002) emphasize on the social aspect. Current study stated the word "traditional practices" in the postpartum period evolve with time and technology (Kamaruddin, et al., 2014). Therefore, this paper will look into the social circle influencing Malay Women in food intake during the postpartum period. By understanding this, it will show the fluidity of culture to be molded along with the advancement of technology and social.

2 LITERATURE REVIEWS

2.1 Malay food and foodways during postpartum period

Freimer and Echenberg (1983) articulated food practices are subject to pressure for change by environmental and acculturative. Manderson and Matthew (1981) in Freimer and Echenberg's (1983) study showed postpartum period in hospital disallowed to follow certain traditional food practices thus cause minor ailments. Hafez and Yakout (2010) revealed that women do not practice postpartum food guidelines because of incomplete knowledge of it. Looking at the consumption of the food, Chmielowska and Shih (2007) highlighted it is important to feed the postpartum women with small amount of food. Because in this confinement period when women are considered physically and symbolically ill (Stefanello, et al., 2008), they adopt an array of food practices. Malay was strongly influenced by Hinduism, thus their set of rules is accepted as local culture until Islam arrived (Ali & Howden-Chapman, 2007). It is then accepted by the Malay as local culture until the Islam belief arrived (Ali & Howden-Chapman, 2007). They reported 46 types of *pantang* that are practiced which one of the most five common *pantang* is hitting or killing animal. This *pantang* not only exclusive to women but also to men-husband—though he slaughtering and fishing for food preparation. Failing to follow the food practices would negatively influence the health of newborn (Barennes et al., 2009). Chmielowska and Shih (2007) listed it is considered taboo for women to do household duties in the kitchen.

This dietary is believed to be associated with medical methods that was within Malay medical books known as *kitab-kitab* which the most popular is called *kitab Mujarrabat Melayu* (Balwi & Koharuddin, 2003). He further explained these medical books explained any physical and spiritual-or supernatural-disease that includes incantation and magic. In understanding the how food affects the health of human, these books explained four elements of human which is soil, water, fire and wind. The balance of these four elements will have an effect on human health which, if one of the elements dominates any of the three, human will become unhealthy (Balwi & Koharuddin, 2003). When many scholars stated a vulnerable state of women after delivering a baby and fall under the state of unhealthy (Ali & Howden-Chapman, 2007; Cheung, 1997; Huang & Mathers, 2001; 1996; Matthey, et al., 2002; Stefanello, et al., 2008; Wong, 1994), Balwi and Koharuddin (2003) expounded food is the most significant factors to affect the balance of the human elements. Moreover, foodways (preparation up to consumption) is believed to affect the human's health (Ali & Howden-Chapman, 2007).

In Malay food context, medicinal food is derived from both flora and fauna. The derivation of food from flora is obtained from roots, flowers and leaves either cooked or raw which are called *jamu*. Food component derived from animal includes exotic ingredients such as *hempedu gajah, hempedu sawa, lemak landak, and tulang sotong* (Balwi & Koharuddin, 2003) to treat inner and outer disease. From both derivations of foods, those can be categorized as cooling, windy, hot, sharp, poison or itchy in food confinement period (Shahar, Earland & Rahman, 2000). Traditionally, this is conducted with the shamanistic methods. In addition, the effectiveness of food in treating the unhealthy condition (in this case, postpartum mother) must be followed with certain pantang larang.

Rashiqah (July 23, 2009) explained that Malay community still believe and follow *pantang larang* even though it comes to the third millennium. Rahimi et al. (2003) reported that there is no significant difference in the belief of *pantang larang* between women in the urban and rural area. Recent study articulated food belief is still held firmly by both urban and rural women nevertheless the only difference is the amount of food practices being applied during confinement period (Yusof, Zulkifli, Shaidan & Kamaruddin, 2013).

3 METHODOLOGY

The qualitative approach is well suited in this context of study as suggested by Esterberg (2002) whereby belief and practice are subjective to an individual who therefore, this approach is appropriate to understand and scrutinize the social process in context. This also helps to increase the depth of understanding on how Malay women belief and practice their food and foodways during the forty crucial days of confinement period. As this study is looking on the belief and practice of food and foodways during postpartum period, using qualitative approach in this study was justified. Mother with at least one child is chosen in this study. These samples are believed to experience the food and food practices during postpartum period and would provide relevant and valuable information for this study. Semi structured interviewing method suggested by Esterberg (2002) was employed for data collection process and the interviewing process. The interviewing method offers the researcher to access women ideas, thoughts and memories in their own words instead of researcher's word (Reinharz & Chase, 2003). Interview pretest question were conducted to see the formality of the language, the sense-or non-sense of the question and familiarity of the question need to be tested as noted by Esterberg (2002) to avoid pitfall during data collection process. The questions were revised and reworded as followed to form an open ended question as women are more spontaneous to it (Reinharz & Chase, 2003). The actual data collection took approximately thirty to forty minutes.

4 FINDINGS AND DISCUSSION

4.1 Social influences on food intakes during the postpartum period

It is not surprising to hear that all informants give an almost similar answer on who is determining the food consumption during the postpartum period.

> "My mother and mother-in-law [determining food intakes] during my first baby"

Yet, women are becoming more independent to select food that is suitable with their body needs when they undergone the next postpartum period of the following child as Informant 1 explained:

> "For my second baby, still my mother and mother-in-law determine my food intake but I also decide it for myself based on the experience I gathered on my first child postpartum period and knowledge I obtained through reading"

She added further that she personalized her dietary intake during the postpartum period due to the factor of allergy to certain food:

> "After I deliver my second baby, I eat food that would be based on my body tolerance because I am allergic to certain food such as nuts."

Informant 2 added relatives played their part in easing the postpartum period:

> "My mother and mother-in-law [decide the food] during my first baby. Also, my cousins who had undergone this painstaking food preparation and consumption during postpartum period"

Unlike Informant 1, the young generation mother who has the eagerness to personalize her

own food intakes, Informant 3, the older generation mother really depending on her mother from the first until the fourth child when she articulated in short:

> "Definitely my mother...."

The husband does not really play their role during the postpartum period when Informant 3 later added on her deceased husband:

> "my late husband doesn't really care because he said that there is no such thing as 'pantang larang' in food...everything is good"

Informant 2 did not explain directly about how hospital plays a role in constructing her first food intake during postpartum but it is understood in a same notion as previous informant:

> "During my time in hospital, I just ate whatever that is served by nurses and food that were brought by my mother and mother-in-law since I have no idea about food that is beneficial for my health"

Informant 1 demystified it:

> "I have to admit that I obtained the information relating to postpartum guidelines firstly from nurses who took care of me"

Informant 3 pondered and recalled her memory which offers an insight the limitation of reading material that lead to why old generation mother accept every food intake and practice from her mother's experience during her first confinement period:

> "...following my mom and grandma's experience should offer me an advantage"

Informant 2 took an easy understanding thus creating her own opinion by looking on the condition of her mother and mother in law. Nevertheless, she did state the differences of dietary practices from both mother and mother in law but decided to follow anything that would benefit for her health:

> "Most of the traditional belief that has been told by my mother and mother-in-law are good. ...there are so many traditional beliefs which in my experience based on my mother and mother-in-law...different set of traditional beliefs but I cannot say that one is better than the other"

Informant 2 provides a ground for a slight disagreement when she said:

"I do not really believe and follow articles or magazine relating to postpartum care and food because it is simply an ideology of various authors. Instead of following what was written in the article, why I don't follow my mother's guideline that has living proof which is me [smiling]"

5 CONCLUSION

While literature emphasizing on the cultural knowledge inheritance, especially from the shamanistic point of view, the study showed the shaman's role are less influenced the belief and practice of postpartum food among Malay women. Instead, they are most likely to be influenced by their closest social circle. From the lens of the role of food during the postpartum period, they believed anything that has been passed from generation to generation about the food after childbirth is more reliable. Nevertheless, due to the evolution of social status whereby women seek for status equality (Counihan, 1999), this study confirms women who are driven to seek a higher education ergo the tendency of questioning the traditional belief is somehow rather appeared. Chmielowska and Shih (2007) contended traditional dietary and dietary taboo are still exercised and obeyed by modern people, but this study showed traditional belief is somehow dissolved and improved in line with the advancement of social and technological. Nonetheless, the outcome of this study proves differently from what Huang and Mathers (2001) reported. The dependability on the closest social circle becomes the key factor to most of their postpartum foodways. Hishamshah et al. (2011) highlighted the practice of abstinence food is due to self-belief and only few stated because of social pressure, this study showed social pressure do not exist through this duration of postpartum, but more likely to guide what is best for the benefit of the mother and also the baby. Through this respective lens, the changes or modifications of such customs in modern society are inevitable, which the culture changes simultaneously with physical characteristic of the population (Chmielowska & Shih, 2007). One thing for sure, there is no longer a set of rules in practicing postpartum foodways due to customization of the foodways from different aspect such as food allergy, acculturation of practices and involvement of scientific knowledge as opposed to anecdotal evidence.

ACKNOWLEDGEMENTS

This paper is sponsored by Research Intensive Faculty (RIF) Grant by Universiti Teknologi MARA Malaysia. Grant Number 600-RMI/DANA 5/3RIF (106/2012).

REFERENCES

Ali, A. & Howden-Chapman, P. (2007). Maternity services and the role of the traditional birth attendant, bidan kampung, in rural Malaysia. *Journal of Public Health Management and Practice, 13*(3), 278–286.

Balwi, M. & Koharuddin, M. (2003). Sains perubatan naturalistik Melayu: satu kajian awal. *Jurnal Kemanusiaan, 17*(2), 1–8.

Barennes, H., Simmala, C., Odermatt, P., Thaybouavone, T., Vallee, J., Martinez-Ussel, B., Newton, P. & Strobel, M. (2009). Postpartum traditions and nutrition practices among urban Lao women and their infants in Vientiane, Lao PDR. *European journal of clinical nutrition, 63*(3), 323–331.

Cheung, N.F. (1997). Chinese zuo yuezi (sitting in for the first month of the postnatal period) in Scotland. *Midwifery, 13*(2), 55–65.

Chmielowska, E. & Shih, F.-s. (2007). *Folk customs in modern society: 'Tradition of Zuoyuezi' in Taiwan. An phpysical anthropology perspective.* Paper presented at the Conference EATS IV Stockholm, Sweden.

Counihan, C.M. (1999). *The anthropology of food and body: Gender, meaning and power.* New York: Routledge.

Esterberg, K.G. (2002). *Qualitative methods in social research.* Boston: McGraw-Hill.

Freimer, N., Echenberg, D. & Kretchmer, N. (1983). Cultural variation-nutritional and clinical implications: In cross cultural medicine. *Western Journal of Medicine, 139*(6), 928–933.

Hafez, S.K. & Yakout, S.M. (2010). Early postpartum dietary practices among a group of Saudi women. *Journal of American Science, 6*(11), 990–998.

Hishamshah, M., bin Ramzan, A., Sirri, M., Rashid, A., Mustaffa, W. & Haroon, R. (2011). Belief and practices of traditional post partum care among a rural community in Penang, Malaysia. *Internet Journal of Third World Medicine, 9*(2).

Huang, Y.C. & Mathers, N. (2001). Postnatal depression-biological or cultural? A comparative study of postnatal women in the UK and Taiwan. *Journal of Advanced Nursing, 33*(3), 279–287.

Kamaruddin, M., Mohd Zahari, M.S., Muhammad, R., Amir, A.F. & Azdel, A.A. (2014). *Food beliefs and practices among Malay women in postpartum period.* Paper presented at the Hospitality and Tourism: Synergizing Creativity and Innovation in Research.

Katz, E. (1996). Recovering after childbirth in the Mixtec highlands (Mexico). *L'approche Ethnopharmacologique*, 99–111.

Koon, P.B., Peng, W.Y. & Karim, N.A. (2005). Postpartum dietary intakes and food taboos among Chinese women attending maternal and child health clinics and maternity hospital, Kuala Lumpur. *Malaysian Journal of Nutrition, 11*(1), 1–21.

Manderson, L. & Mathews, M. (1981). Vietnamese behavioral and dietary precautions during pregnancy. *Ecology of Food and Nutrition, 11*(1), 1–8.

Matthey, S., Panasetis, P. & Barnett, B. (2002). Adherence to cultural practices following childbirth in migrant Chinese women and relation to postpartum mood. *Health Care for Women International, 23*(6–7), 567–575.

Rahimi, H., Fatimah, A.M., Rahimah, I., Sarah, Y. & Marlia, M.S. (2003). Adakah pantang larang pemakanan di kalangan Orang Asli mempengaruhi tahap kesihatan ibu dan anak: pengalaman daerah Kuala Lipis. *Malaysian Journal of Public Health Medicine, 3*(1), 73–77.

Ramle, A. & Beri, S. (1993). *Komuniti Orang Asli di Terengganu*. Terengganu: Percetakan Yayasan Islam Terengganu.

Rashiqah, I.A.R. (July 23, 2009). Pantang larang Melayu: Sesetengahnya ada bukti saintifik tetapi tidak disedari masyarakat., *Berita Harian*.

Reinharz, S. & Chase, S.E. (2003). Interviewing women In J. Holstein & J.F. Gubrium (Eds.), *Inside interviewing: New lenses, new concerns* (pp. 73–90). California: Sage.

Shahar, S., Earland, J. & Rahman, S.A. (2000). Food intakes and habits of rural elderly Malays. *Asia Pacific Journal of clinical nutrition, 9*(2), 122–129.

Stefanello, J., Nakano, A.M.S. & Gomes, F.A. (2008). Beliefs and taboos related to the care after delivery: their meaning for a women group. *Acta Paulista de Enfermagem, 21*(2), 275–281.

Wong, L. (1994). Apart from the Sesame Oil Chicken: The Chinese women 'doing the month. *Dou-Shin Ltd, Taipei*.

Yusof, S., Zulkifli, S., Shaidan, S. & Kamaruddin, M. (2013). *The differences of food consumption and practices during confinement between urban and rural Malay women*. Paper presented at the Hospitality and Tourism: Synergizing Creativity and Innovation in Research, London.

Theory and Practice in Hospitality and Tourism Research – Radzi et al. (Eds)
© *2015 Taylor & Francis Group, London, ISBN 978-1-138-02706-0*

The influence of Malay food characteristics on customer purchase intention

J. Nur-Syuhada
Universiti Pendidikan Sultan Idris, Perak Darul Ridzuan, Malaysia

M.S. Fadzilah, M.A. Khairunnisa & M. Rosmaliza
Faculty of Hotel and Tourism Management, Universiti Teknologi MARA, Shah Alam, Malaysia

ABSTRACT: Malay as one of the major ethnic in Malaysia owns numerous of unique and delicious traditional food. The main characteristics of Malay cuisine are strong, spicy and aromatic, combining the rich taste of many herbs and spices. This authentic Malay cuisine not only prepared at home but also available in variety types of restaurant businesses, ranging from hawker stall to upscale restaurant and also being patronized not only by Malays but multicultural customer. However, the emergence of new types of cuisines and menu choices have partly caused the changes in consumers' desires for new taste and eating habits. The growing numbers of other ethnic restaurants such as Thai, Korean, Japanese, Arab foods have altered customer's food consumption. This phenomenon has slightly given impact towards Malay food stalls and restaurants. Therefore, the purpose of this study is to identify whether the characteristics of Malay foods itself such as appearance, taste, smell and texture, as well as cooking method influence customer purchase intention. A quantitative approach was used and the results revealed that Malay food characteristics had a significant impact on the customer purchase intention. Thus, it is important for the Malay restaurateur and food handler to improve the quality food served to capture more customers.

Keywords: Malay, food characteristics, customer, purchase intention

1 INTRODUCTION

Each ethnic delicacy in Malaysia has its own interesting and unique stories along with the exclusive characteristics. Back in the glory, people are proud and feel appreciate of their traditional food that's full of uniqueness and techniques that differ from any other ethnic. The traditional foods for each ethnic have marked the differences between the cultures exist in this country. Malay as one of the major ethnic owns plenty of unique and delicious foods. The main characteristics in its traditional Malay cuisine can be described as strong, spicy and aromatic that combining the rich tastes of many herbs, roots and spices. Coconut milk, for example, is another favorite ingredient in most of Malay dishes in giving the heavy, rich and creamy texture. In terms of cooking methods, Hassan (2011) mentioned that Malay cuisines are quite similar to life of Malay villages, slow and laid back as most authentic Malay delicacies cooked on low heat for a long time as compared to Chinese food. In relation to this, the authentic Malays foods not only prepared at home, but also available in varying types of restaurant businesses, ranging from a hawker stall to upscale restaurants and also being patronized not only the Malays but also multicultural customers (MBSA, 2006). However, the various multinational restaurants expanding into new geographical areas along with the emergence of new types of cuisines and new menu choices have partly caused the consumers desires for new alternatives and eating habits (Kaynak, Kucukemiroglu and Aksoy, 2006). According to Appadurai (1996), the growing numbers of other ethnic restaurant in Malaysia such as Thai, Korean, Japanese, Arab and *mamak* foods have altered the customers' food consumption decision thus making restaurants businesses becoming increasingly competitive. This phenomenon is slightly giving impact to Malay restaurants. Yoshino (2010) mentioned that Malay cuisine is basically home-cooking and that Malays have not come up with any particular ways of presenting it to consumers as "ethnic cuisine" in a commercial setting. In line with that, many of the Malay restaurants now do a little bit of everything especially to the foods itself to keep up with changing trends and to satisfy customers' wide ranging of taste. The purpose of this study is therefore to identify whether the Malay foods characteristics such as appearance, taste, smell and texture as well as cooking methods influence the customer purchase intention.

2 LITERATURE REVIEW

2.1 Malay food

Malay cuisine is a combination of social habits, history, tradition, climate or a summary of all those unique things combined based on great products from all states (Nummedal and Hall, 2006). Historically, Malay cuisines were strongly influenced by various ethnic groups including Siamese, Javanese, Sumatran, Minangkabau and others (Travellers worldwide.com, 2011). The rich historical heritage has evidently resulted in its exotic cuisine. Varieties of ingredients used are often described as spicy and flavorful as it used a combination of spices, herbs and roots. Strong, tangy and flavorful ingredients such as *serai* (lemon grass), *pandan* (screwpine), *daun limau purut* (kaffir lime leaves), *kemangi* (a type of basil), *kesum* (polygonum or laksa leaf), *buah pala* (nutmeg), *kunyit* (turmeric) ketumbar (*coriander*) *buah pelaga* (cardamom), *bunga acengkih* (cloves), *biji sawi* (mustard seeds), *halba* (fenugreek) and *buah pala,* (nutmeg) are often used in Malay dishes. *Santan* (coconut milk) is a basis of many Malay dishes to create rich and creamy texture. In terms of cooking methods, it usually takes longer time to cook as most authentic Malay delicacies cooked on low heat such as *mereneh* (simmer) and *mengukus* (steaming) (Mohamed, Mohamad & Hussain, (2010). Depending on the main basic ingredient, Malay dishes can be more or less, distinguished into several styles of cooking such as *masak lemak* (coconut), *masak pedas* (hot chillies), *masak assam* (tamarind), *masak merah* (tomato sauce), *masak hitam* (dark soy sauce), *masak assam pedas* (chillies and tamarind). These basic styles of cooking can be applied to a variety of food from meat, poultry, fish, seafood and vegetables. Normally, these dishes are served with white rice for lunch and dinner. Some of other famous and unique Malay foods are *rendang, gulai, laksa, satay, otak-otak, lontong, tempoyak, nasi kerabu,* and many others.

2.2 Food characteristics

The flavor is an important characteristic of a food. Actually, the flavor is a blending of taste and aroma. Sometimes the words flavor and taste are used synonymously. According to Salleh (2010), taste is only one part of the flavor. Taste involves the sensations produced through stimulation of the taste buds on the tongue. It is generally accepted that there are only five primary taste sensations: sweet, sour, bitter, salty, and umami (Nielsen, 2005). But the perceived flavor of a food involves to a considerable extent there is the sense of smell along with the taste sensations while texture is the term used to describe the characteristics of a finished food product. This is related to the order in which the ingredients are added, the way of mixing and the method of cooking affect the resulting the final products (Kpratishnair, 2007). To Drewnowski (1997), the taste, smell, and texture of foods help determine food preferences and eating habits. In reality, there are multiple links between taste perceptions, taste preferences, food preferences, food choices and the amount of food consumed. In terms of cooking technique, there are some basic methods can be used. Commonly, the basic cooking method can be divided into two general groups; the dry heat methods and moist heat methods. The dry heat methods usually conducted without moisture while the moist heat method is conducted by water or water–based liquids. Different cooking methods are suited to different kinds of foods (Gisslen, 2003). For preparing Asian foods for example, types of cooking methods are basically used include stir-frying, deep-frying, steaming, stewing and blanching (Ramly, Ahmad & Ahmadin, 2003).

2.3 Purchase intention

Intention is the cognitive representation of a person's readiness to perform a given behavior, and it is considered to be the immediate antecedent of behavior (Fishbein and Azjen, 1975). Intention represents one of the most important parts of human behavior where several cognitive and behavioral factors can vary sharply between individuals. It is a difficult task because the customer' intention or decision depends on many factors that cannot be directly controlled (Nielsen, 2005). It is also an individual motivation to engage in a specific action and is the best predictor of the actual action (Guido et al., 2009). In line with that, intention to purchase refers to the general predisposition of consumers towards the act of purchase (Ajzen (1991). It is the implied promise to one's self to buy the product again which depicts the impression of customer retention. Thus, intention to purchase could help the food manufacturer and marketer to predict the purchase behavior of customer and understand the market. Study by Sirohi (1998) shown that purchase intention is something that most food manufacturers and restaurateur focus on since it help them to identify the behavior and perception of their customer about the product so that they can provide and improve their products to address customer satisfaction and retention.

3 METHODOLOGY

The primary objective of this study is to identify factors that influence costumers' purchase intention of Malay food. A quantitative approach was considered as the suitable method to validate the theoretical framework and the hypotheses of this study. The questionnaire was distributed to the targeted respondents at the local and private colleges and universities, shopping complexes, business offices and restaurants within Klang Valley, Malaysia. A time frame of one month was set and a total of 300 questionnaires was successfully collected and 285 of them were found valuable and beneficial in this study. The descriptive analysis and multiple regressions were used to describe the characteristics of the sample and to explore the relationship between one continuous dependents variable and a number of independent variables.

4 ANALYSIS AND RESULT

4.1 Food characteristics

The food characteristics factor of the purchase intention on Malay food consisted of a sub-variables, including appearance, food taste, smell and texture and cooking method.

4.1.1 Appearance

Table 1 shows that most of the respondents slightly agreed that the Malay local food usually served cold and dry, makes them change to another type of food for example, the Thai's food. The respondents also slightly agree that the shape and color of Malay local food are attractive and can influence them to consume it. They also fairly agree that food presentation of Malay food makes it attractive to consume. However, the respondents somewhat agreed that the local food presentation usually unpleasant compared to the other type of food.

4.1.2 Taste, smell and texture

There are five items used by the researcher to evaluate whether the taste, smell and texture influence the respondent's purchase intention of Malay food. By referring to the Table 2, it is clearly shown that the taste, smell and texture are the important element to influence the respondent's purchase intention. Most of the respondents slightly agreed that the local food taste better when served hot and the aroma is very tempting and it influences them to consume. They also agreed that the local food served is usually saltier and spicier than other food. The respondents somewhat agreed that the texture of the Malay local food makes a huge impact on them in choosing what to eat; for example; easy to chew, mushy, mouthful and so on.

Table 1. Mean score for the appearance factor of Malay food.

Items	Mean	S.D
I found that Malay food often served cold and dry, makes me change to another type of food (for example; Thai's food).	2.90	.886
The shape and color of Malay food are attractive and can influence me to consume it.	3.37	.959
The Malay food presentation makes it desirable to consume.	3.54	.940
The Malay food presentation usually unpleasant.	3.96	.436

Scale: 1 = Strongly Disagree, 2 = Disagree, 3 = Neutral, 4 = Agree and 5 = Strongly Agree

Table 2. Mean score for the taste, smell and texture factor of Malay local food.

Items	Mean	S.D
The local food served usually salty and spicy.	3.61	.554
The local food more salty or spicier than other food and I like it.	3.76	.563
The local food taste better when served hot.	4.12	.576
The aroma of the local food is very tempting and it can influence me to eat.	3.99	.529
The texture of the local food makes a huge impact on me in choosing what to eat. (For example; easy to chew, mushy, mouthful and etc.).	3.70	.534

Scale: 1 = Strongly Disagree, 2 = Disagree, 3 = Neutral, 4 = Agree and 5 = Strongly Agree.

Table 3. Mean score for the method of cooking factor of Malay food.

Items	Mean	S.D
The cooking method used in local food preparation is often made the food become unhealthy; but it tastes better.	3.67	.780
The method of cooking used in local food preparation usually adding excessive amounts of fat or salt in the food (Deep-fried, pan-fried, braised, and etc.)	3.74	.771
I believe that the cooking method used in local food preparation basically is the best; in capturing the flavor and nutrients of the food.	2.58	.967

Scale: 1 = Strongly Disagree, 2 = Disagree, 3 = Neutral, 4 = Agree and 5 = Strongly Agree.

4.1.3 Method of cooking

Table 3 displays the items used to measure respondents' opinion towards the cooking method used in the Malay food. Most of the respondents quite agreed that the cooking method used in Malay local food preparation is often made the food look unhealthy but it tastes better. They also relatively agreed that the method of cooking used in Malay food preparation usually adding excessive amounts of fat or salt such as deep-fried, pan-fried, braised, and others. However, the respondents slightly

Table 4. Mean score for purchase intention.

Items	Mean	S.D
I will switch to other than Malay food.	2.28	.588
I will dine out again at local (Malay) restaurants in the future	3.56	.978
I will recommend my friends and family to eat local (Malay) food.	3.62	0.754

Scale: 1 = Strongly Disagree, 2 = Disagree, 3 = Neutral, 4 = Agree and 5 = Strongly Agree.

Table 5. The result of the multiple regression analysis.

Independent Variable (IV)	Dependent Variable (DV) Customer purchase intention
	Beta Coefficients and Significance Levels
Appearance	.41*
Taste, Smell and Texture	.33***
Method of cooking	.36 ***
R²	.61.1
	.0.34*
F-Change	52.25***

Note: * P < .05, ** P< .001, *** p < .000.

disagreed that the cooking method used in Malay local food preparation is the best in capturing the flavor and nutrients of the food.

4.2 Purchase intention of Malay food

This section discusses behavioral intention toward Malay food that can be used to foretell the actual purchase behavior on Malay food. Table 4 indicates that the majority of the respondents slightly disagreed that they will switch other than Malay local food. They also will dine out again at Malay restaurants in the future and will recommend to their friends and family to eat Malay local food.

4.3 The relationship of malay food characteristics and purchase intention

To test the relationship between the Malay local food characteristics and purchase intention, multiple regressions was used.

The Table 5 signifies that Malay food characteristics were able to clarify 61.1 percent the variance in intention to purchase. There was a positive and a moderate correlation between Malay food characteristic and customer purchase intention ($\beta = .34$, $p = 0. 05$). The beta value under standardized coefficient was also evaluated by comparing the contribution of each of the independent variables to the dependent. The appearance ($\beta = .41$, $p = 0.05$) have a moderate influence on the consumer purchase intention of Malay local food. Followed by beta

value for method of cooking ($\beta = .36$, $p = 0.000$) and the least influence consumer intention to purchase Malay food was the taste, smell and texture with standard beta coefficient ($\beta = .33$, $p = 0.000$). This result indicates that all of the three variables significantly influence customer purchase intention on Malay food.

5 IMPLICATION, RECOMMENDATION AND SUGGESTION

From the overall summary, it is proven that the food characteristic pertaining to appearance, taste, smell and texture, and method of cooking have given significant affect on customers purchase intention toward Malay food. It is important therefore for the Malay restaurateurs or food handlers to improve the quality of the food served. The ingredients used to prepare the food must be fresh. The presentation of the Malay local food must be pleasant in order to attract more consumers and to make the food to be the most preferable product to purchase. In conclusion, the findings of the current study can provide intuitive information that supports the food marketer and restaurateur to capture more customers to choose Malay food and helps them to create better strategies and planning to increase the sales and revenue.

REFERENCES

Ajzen, I. (1991). The theory of planned behavior. *Organizational Behavior and Human Decision Processes, 50*, 179–211.
Appadurai, A. (1996). *Modernity at Large: Cultural Dimensions of Modernity*. Minneapolis: University of Minnesota Press.
Drewnowski, A. (1997). Taste Preferences and Food Intake. *Annual Review of Nutrition, 17, 237*-253.
Fishbein, M., and Ajzen, I. (1975). *Belief, Attitude, Intention, and Behavior: An Introduction to Theory and Research.* Reading, MA: Addison-Wesley.
Fox, R. (2003). Food and Eating: an Anthropological Perspective. *Social Issues Research Centre.* Retrieved October 28, 2011 http://www.sirc.org
Gifford, S.R. & Clydesdale, F.M. (1986). The psychophysical relationship between color and sodium chloride concentrations in model systems. *Journal of Food Protection, 49, 977*–82.
Gisslen, W. (2003). *Professional cooking.* 5th ed. New York: John Wiley & Sons Inc.
Guido G.,Prete M.I., Peluso A.M, Maloumby-Baka R.C.,Bufa C., (2009). The role of ethics and product personality in the intention to purchase organic food products: a structural equation modeling approach, *International Review of Economics*, 57 (1), 79–102.
Hassan, S.H. (2011). Consumption of functional food model for Malay Muslims in Malaysia. *Journal of Islamic Marketing, 2*(2), 104–124.

Imram, N. (1999). The role of visual cues in consumer perception and acceptance of a food product. *Nutrition & Food Science, 99(5)*. 224–228.

Kaynak, E., Kucukemiroglu, O. and Aksoy, S. (2006). Consumer Preference for Fast Food Outlet in a Developing Country. *Journal of Euromarketing, 5(4), 99–113.*

Kpratishnair. (2007). *Sensory characteristics of food.* Retrieved October 20, 2011 from http://www.ifood.tv/blog/sensory_characteristics_of_food

Majlis Bandaraya Shah Alam.(2006). *Report of restaurants in Shah Alam City.* Malaysia, Selangor: Majlis Bandaraya Shah Alam.

Mohamed, A., Mohamad, S. & Hussain, H. (2010). Food gifts in Malay Weddings: Custom and Interpretation. *Journal of Social Studies, Development and Environmental, 5* (1) 103–115.

Nielson. (2005). Asians the World's Greatest Fast Food Fans [Press release]. Retrieved from http://blog.apastyle.org

Nummedal, M., and Hall, M. (2006). Local food and tourism. *Tourism Review Journal, 9, 365–378.*

Ramly, M.A.S., Ahmad R. & Ahmadin S.N. (2003). *Factors influencing customers patronizing mamak restaurants—a survey.* (Technical report). Institute of Research, Development and Commercialization: Universiti Teknologi Mara.

Roscoe, J.T. (1975). *Fundamental Research Statistics for Behavioral Science (*2nd Ed.). New York: Holt, Rinehart and Winston.

Salleh, M.M. (2010). Consumer's perception and purchase intentions towards organic food products: exploring attitude among academician. *Canadian Social Science, 6(6),* 119–129.

Sekaran, U. (2006). *Research Method for Business: A Skill-Building Approach,* 4th Ed., New York: John Wiley and Sons, Ltd., Publication.

Sirohi , N. (1998). A model of Consumer Perception and Store Loyalty Intention For supermarket Retailer. *Journal Of Retailing,* 74 (2), 223–245.

Travellers Worldwide (2011). *About food in Malaysia.* Retrieved from http://www.travellersworldwide.com

Verbeke, W. & Lopez, G.P. (2005). Ethnic food attitudes and behaviour among Belgians and Hispanics living in Belgium. *British Food Journal, 107(11),* 823–840.

Yoshino, K. (2010). Malaysian Cuisine: A Case of Neglected Culinary Globalization. *In Globalization, Food and Social Identities in the Asia Pacific Region,* ed. James Farrer. Sophia University Institute of Comparative Culture.

Theory and Practice in Hospitality and Tourism Research – Radzi et al. (Eds)
© 2015 Taylor & Francis Group, London, ISBN 978-1-138-02706-0

A study of food tourism: The potential for activating business in Malaysia

S.M. Hashemi
School of Housing, Building and Planning, Universiti Sains Malaysia, Pulau Penang, Malaysia

N. Hosseiniyan
School of Business, Universiti Sains Malaysia, Pulau Penang, Malaysia

ABSTRACT: Food is a rudimentary and physical necessity of human beings and a basic theme for business and marketing. Food is also, particularly, related to tourist activities and plays its primary function in entertaining and fulfilling the tourists' basic, natural, and inevitable requirement to provide energy as well as to promote business. Thus, food and tourism are increasingly being combined together in order to stimulate business. The purpose of this article is to review the contribution of food in attracting the tourists and visitors. It also intends to review the contribution of food in tourism with particular reference to the importance of food tourism for business point of view with the Malaysian perspectives. The issue was discussed based on documentary analysis of the existing academic materials with reference to food tourism, tourism business and factors in order to investigate the truth. The content analysis revealed that there was a great potential of food tourism in Malaysia due to its fabulous, delicious, and a variety of food items to activate business in Malaysia. Some practicable implications of the findings are furnished for food policy makers, industrialists, academicians, research scholars, and the people affiliated with food and business tourism and hospitality industry in order to utilize the existing potential of food and business tourism in Malaysian.

Keywords: Food tourism, attractions, tourists, Malaysia

1 INTRODUCTION

Food is known as one of the significant attractions amongst tourists to enrich their tourist activities. It also plays its vital role in tourist decision making and measuring their level of satisfaction. It is an effective instrument to develop and promote tourism business and its contribution to destination development. Hotels, shopping malls, and food stalls offer a variety of foods to entertain the tourists. The taste and flavor of the food, as well as the decorative styles presenting this food, attract and capture the tourists' attention instinctively. Consequently, food assists to promote tourism industry to a great extent. Food is not only a source of energy for tourists but also provides motivation for tourists to be in a more enjoyable and memorable holiday atmosphere beyond their scheduled.

The current paper aims to review the food factor contributing to attract the tourists from different corners of the world as a little is done regarding the current issue of food in tourism with the Malaysian perspectives. Furthermore, its purpose is to review the contribution of food in tourism with particu-

lar reference to the importance of food tourism to sustain business with the Malaysian contexts. A lot of potential of food tourism exits in Malaysia, particularly in the rural areas, as an extensive study needs to be carried out to fill the existing gap (Liu, 2006). The findings of a study conducted by Ling, Karim, Othman, Adzahan and Ramachandran (2010), revealed that that Malaysia has the potential of being a food tourism destination as the country is viewed as a melting pot of cultural food variety at reasonable prices as compared to other tourist destinations. In summary, Malaysia is a good food tourism destination; however, it is the need to ascertain and highlight the significance of food tourism to promote business.

Therefore, this study will provide directions and guidelines to enhance the value, quality, and safety of food tourism as it has bright prospects for tourism and business is still underdeveloped. However, the potential in food tourism requires quality standards, and presenting and decorating styles of food for tourists. The term of food tourism used in the current research is associated with hospitality business affiliated food services.

2 LITERATURE REVIEWS

Food tourism can be defined as visitation to food festivals, food producers, restaurants and specific location for which food tasting and experiencing are associated (Hall & Mitchell, 2005). Food tourism may also be regarded as an example of gastronomy, culinary, gourmet or cuisine tourism. Based on the literature, tourism and food is recognized as very importance roles, and also close relationship with each other as each tourist needs to eat food during travelling, visiting and staying at visiting destination (Henderson, 2009). Besides, delicious food plays its role as a motivator that fulfills the physiological need of the human body (McKercher, Okumus & Okumus, 2008).

In conjunction with the food tourism, Malaysia is celebrating its fourth Visit Malaysia Year (VMY2014) in the year 2014 with the theme 'Celebrating 1Malaysia Truly Asia'. It is a grand and marvelous activity with hundreds of fabulous foods to entertain 36 million prospectus tourists from all over the world. Delicious and fabulous food would capture the attention of tourists as Malaysia is known as a haven for delicious food for tourists (Ministry of Tourism and Culture Malaysia, 2014). The Malaysian foods are so tasty that no visitors leave Malaysia without tasting its gorgeous food and drinks (Yen, 2010). Food also provides enthusiastic entertainment and pleasant pleasure for the eaters (Wei & Nakatsu, 2012). Dinning habits of the tourists according to their culture is amazing not only for the local people, but also for other visitors and tourists hailing from different countries (Yüksel & Yüksel, 2002). Therefore, it would help to the practitioners associated with food production, food business, and tourism planning and marketing as well, to develop and promote food preparation, preservation and presentation system for the tourists in order to satisfy their aesthetic appeal (Horng & Tsai, 2010).

3 MALAYSIAN FOODS

The previous studies (for example Leong et al., 2009; Lim, Norzan & Mohd, 2009) revealed that tourists are much interested in Malaysian foods because of cultural, social, business, and educational interests. For instance, a study by Leong et al. (2009) in Malaysia revealed that 82 percent of the respondents (i.e., tourist) are likely to enjoy Malaysian food. The finding by Leong et al. (2009) also suggested that the tourist are satisfied with the food availability, food quality and meal experience in Malaysia. Malaysian foods and drinks are popular among tourists and visitors due to their unique taste, flavor, and essence. Popular Malaysian foods

such as *asam laksa, asam pedas, bak kut teh, beef rending, char kuey teow, fish head curry, ikan bakar, mee goring, mee rebus, nasi lemak, pasembur, roti canai* are recited by the tourists. Different type of seafood dishes such as *clam, crab, fish, lobster, octopus, oyster, shrimp,* and *squid* are also very popular amongst tourists (Kwoczek, Szefer, Hac & Grembecka, 2006), are normally found in seafood restaurants in Malaysia and the resort menu is a mixture of foreign and local dishes (Venugopal, 2005).

In addition, fast food is also being a significant attraction and popular among the tourists in Malaysia, as they enjoy the diverse flavors at fast food outlets during their tourist activities. World fame international fast food chains such as McDonald, KFC, and Pizza Hut are not only popular among tourists but also amid locals. The tourists prefer hygienic, nutritional, accessible and culturally accepted food (Pendergast, 2006). Different fruits such as pineapple, watermelon, papaya, and *rambutan* are often served as dessert at Malaysian restaurants to the tourists after lunch and dinner (Ibrahim & Rashid, 2010).

Malaysian drinks and beverages such as *teh tarik* (pulled tea), *white and black coffee, ginger tea, milo, horlicks, jus epal* (juice apple), *air belimbing* (starfruit juice), *air tembikai* (watermelon juice), *limau air* (lime juice), *air kelapa* (coconut), and *air soya* (soy milk), are very popular among locals as well as tourists (Abdullah & Asngari, 2011). The majority of the tourists like soft and sweet drinks as well as hot and cold tea to quench their thirst that easily available at any restaurant (Boo et al., 2010).

Western-food and Thai food are also available in Malaysia. Overall, Malaysia is an ideal destination for tourists due to its potential to be 'food capital' around the world and culinary and tourism destination among Asian countries (Karim, Chua & Salleh, 2009), thus, food tourism can open new horizons for enterprises to boost up business in Malaysia (Ashley & Haysom, 2006).

4 FOOD PRESENTATION

In addition to the peculiar tastes of food and drinks, the way of presenting food to the customers and the tourists attracts a lot. Generally, the visual presentation of food is made by the chefs working in the restaurants. They prepare the food in different stages. The typical style of cutting, tying, sewing, chopping, and slicing of meat, fruit, and vegetables are commonly used in the Malaysian restaurants. The food and fruit, sometimes, are decorated with iced cakes, drizzled with sauces, topped with ornamental sculptural consumable, sprinkled with powders, edible seeds, and other topping.

Food plating, also known as the overall style and arrangement of food in plate to be served to the customers, captures the tourists' attention (Styler, 2006). Some popular styles of plating include a classic arrangement, and stacked arrangement. The main item is put in front of the plate, whereas the vegetables or starches are placed in the back side of the plate in a classic arrangement of platting. In conjunction with the presentation style, catering is another factor that attracts the attentions of the tourists, visitors, and customers.

With reference to the food, attractions and presentation, different stakeholders such as the tourism industry, public and private sector organizations, and tour operators are considered very important for the selection of foods, destinations and accommodations of tourists (Telfer & Wall, 1996). Food stall keepers decorate their stalls in an attractive way to capture the attention of the tourists. The food hawkers found in the streets and at beaches sell different types of edible items. They also contribute in developing food tourism and the economy of the country (Boyne, Hall & Williams, 2003).

5 PROMOTION OF FOOD IN TOURISM

Malaysia is a heaven for tourists for delicious and fabulous food. Food can play a key role in attracting the locals as well as international tourists. Food is a basic element like transportations, accommodations, destination attractions that fascinate the tourists. Therefore, hygienic foods, drinks, and beverages should be promoted in conjunction with the promotion of other facilities for tourists (Boyne, et al., 2003). A systematic, organized, and solid framework and guidelines are required to be regulated and implemented in order to provide food facilities to the tourists. Proper marketing strategies can play a vital role to organize the system.

Restaurants are known as the places to facilitate and provide food items to the customers, visitors, and tourists. Therefore, standardized and quality oriented restaurants situated on or near the tourist destinations that fulfill the tourists' food requirements should be encouraged. Different promotional strategies are useful to promote food and business tourism. The internet has become an effective advertising and promotion as it has developed rapidly through the world. Various websites can be developed to upload different pieces of information along with the colored photos of food dishes and informative audios and videos of food related materials. The tourists can utilize this facility in purchasing foods, drinks, goods and services online (Boyne & Hall, 2004).

According to Richards (2012), social media such as Facebook, twitter, and blogs are effective instruments to promote food. A considerable number of social media users create upload and exchange their personal experiences and tour related activities with their friends, colleagues, and acquaintances. In this way, the people get information regarding foods, drinks, and beverages and tourist destinations as well. You Tube is also significant to get information regarding foods and drinks. Useful audios and videos can be uploaded for tourist purposes.

Print media such as newspapers, magazines, periodicals, booklets, multi-colored brochures, flyers, pamphlets, handouts are also effective sources of food publicity amongst locals and tourists (Jones & Jenkins, 2002). Electronic media such as television and radio play very effective role to develop foods, drinks, and beverages amongst tourists (Dore & Crouch, 2003). In a nutshell, an integrated and result oriented approach should be used to promote and publicize the food items amongst the locals and foreign tourists. Market philosophy approach, in this respect, is a suitable strategy to promote food items and regional food related sectors. However, consumer behavior should also be studies deeply keeping in mind what the consumers seek (Hunter, 2002). Subsequently, the high sale of food, drinks, and beverages would stimulate tourism induced economic activity in the processing sector and local production to contribute to the economy of the country (Boyne, et al., 2003).

Malaysia, being a multi-racial country, represents many races. Malaysian people prepare and present several traditional foods at different festivals such as Hari Raya, Chinese New Year, Deepavali, Hari Gawai celebrated by Muslims, Chinese, Indians, and Kadazan and Iban people, respectively. Dances and songs associate with food are also performed on these occasions. In conjunction with the festivals and celebrations, different festivals such as KL Coffee and Tea Festival 2007, Malaysia International Gourmet Festival (MIGF) 2009, Epicure Malaysia 2008, Malaysia International Halal Food Showcase (MIHAS) 2010 and so on played a vital role in promoting Malaysian food and tourism business in Malaysia (Ministry of Tourism and Culture Malaysia, 2013a).

In connection with the Malaysian food tourism, Ministry of Tourism, Malaysia organizes festivals such as Fabulous Food 1Malaysia (FF1M), Asian Food Heritage, Street Food Festival and Restaurant Food Festival every year throughout Malaysia to promote Malaysian versatile food. Malaysian culture week programs are not only held in Malaysia but also in other country like the UK, where, the first culture week was held in 2006 and similar programs were subsequently held in 2007, 2008 and 2012 to promote Malaysian culture, heritage, traditional music, and crafts particularly delicious and mouthwatering Malaysian food that, definitely, mesmerizes the tourists and visitors (Ministry of Tourism and Culture Malaysia, 2013b). Ministry of Tourism, Malaysia also distributes Best Fabulous Food 1 Malaysia every year (Ministry of Tourism and Culture Malaysia, 2013a). Furthermore, the ministry is also launching corporate stamps depicting various Malaysian tourist landmarks as well as its fabulous food such as *dim sum, nasi lemak, roti canai,* and *satay* to promote tourist activities in Malaysia that will also generate revenue. Therefore, such features present the essence of Malaysia vigorously, which are popular amongst local and foreign tourists as well (Ministry of Tourism and Culture Malaysia, 2012). New Voyage Magazine recognizes Malaysia as the most charming and fascinating destination in 2013 for tourists and visitors. One of the most popular visiting sites, Penang Island, was recognized due to its marvelous architecture, rich history and delicious food and also a UNESCO world heritage site (Ministry of Tourism and Culture Malaysia, 2013a).

Another healthy activity; The Malaysian Rice Plates Project launched by Ministry of Tourism, captured the attention of 60 Malaysian artists who showed their worth by creating individual artwork on rice plates inspired by the Malaysian currency. Such type of contests and activities encourages the food stallers to enhance and promote their food through food tourism (Ministry of Tourism and Culture Malaysia, 2012).

Moreover, food marketing encourages entrepreneurship that stimulates the food production activities to sell the local products to the locals and tourists. For example hawker food industry in Malaysia, currently, is flourishing rapidly. Therefore, more improvements can be made to elevate the strength and reduce the weakness, particularly in promotion strategies in line with the current trends of food and tourism business even though the Ministry of Malaysia is already aggressively promoting the Malaysia's unique local cuisine via the Ministry's Fabulous Food 1 Malaysia (Ministry of Tourism and Culture Malaysia, 2013a).

6 CONCLUSION AND IMPLICATIONS

Food is a basic need for human beings provide energy. It fulfills the basic requirements to obtain nutrients for the human body to keep it healthy, fit, and physically strong. Food supply regulates the various processes in the human body. With regard to the role of food in tourism, it is a significant element to attract and entertain the tourists in the

Malaysian contexts. Millions of visitors and tourists visit Malaysia every year. They enjoy Malaysian foods and drinks during their stay in Malaysia. They love Malaysian food uniqueness. Therefore, a variety of fabulous and delicious foods in Malaysia are one of the factors to promote tourist activities and tourism business as well.

In view of the above discussion, it is the need of the hour to improve the standards and quality of food items to be provided to the tourists during tourist activities so that the tourists may achieve the desired excitement, relaxation, escapism, status, and lifestyle. Therefore, the government should show its effectiveness and efficiency in regulating food quality and safety and providing food safety regulations/ guidelines, training, licensing, and relocation in order to facilitate the tourists in a better way.

Innovative food trials should be planned based on regions, communities, and distinct specialties of state in order to attract tourists for local delicacies and promote tourist activities in Malaysia. The food preservation process should also be upgraded in this respect. The Malaysian culinary, with the mixture of Malay, Chinese, and Indians, is a significant aspect of tourism because it has a potential to enhance tourism and business in Malaysia (Ling, et al., 2010). Therefore, it should be promoted to a great extent along with the tourist destinations in order to generate revenue and to boost up the Malaysian economy. Hawker food industry should also be improved that can contribute to boost up tourism and the Malaysian economy.

Finally, similar studies should be carried out in order to investigate the other aspects of tourism such as activities, engaging tourists in visiting the museum, going shopping, attending music and/or film festivals and participating in common outdoor recreation in conjunction with food tourism that will assist to promote tourism, business and hospitality industry in Malaysia. Other studies such as food tourism in urban and rural areas and consumer behavior should also be done to improve the food tourism and tourism business in Malaysia.

REFERENCES

Abdullah, L. & Asngari, H. (2011). Factor analysis evidence in describing consumer preferences for a soft drink product in Malaysia. J. *Applied Sci, 11,* 139–144.

Ashley, C. & Haysom, G. (2006). From philanthropy to a different way of doing business: Strategies and challenges in integrating pro-poor approaches into tourism business. *Development Southern Africa, 23*(2), 265–280.

Boo, N., Chia, G., Wong, L., Chew, R., Chong, W. & Loo, R. (2010). The prevalence of obesity among clinical students in a Malaysian medical school. *Singapore medical journal, 51*(2), 126.

Boyne, S. & Hall, D. (2004). Place promotion through food and tourism: Rural branding and the role of websites. *Place Branding, 1*(1), 80–92.

Boyne, S., Hall, D. & Williams, F. (2003). Policy, support and promotion for food-related tourism initiatives: A marketing approach to regional development. *Journal of Travel & Tourism Marketing, 14*(3–4), 131–154.

Dore, L. & Crouch, G.I. (2003). Promoting destinations: An exploratory study of publicity programmes used by national tourism organisations. *Journal of Vacation Marketing, 9*(2), 137–151.

Hall, C.M. & Mitchell, R. (2005). Gastronomic tourism: Comparing food and wine tourism experiences. *Niche tourism: Contemporary issues, trends and cases*, 73–88.

Henderson, J.C. (2009). Food tourism reviewed. *British Food Journal, 111*(4), 317–326.

Horng, J.-S. & Tsai, C.-T. (2010). Government websites for promoting East Asian culinary tourism: A cross-national analysis. *Tourism management, 31*(1), 74–85.

Hunter, C. (2002). Sustainable tourism and the touristic ecological footprint. *Environment, Development and Sustainability, 4*(1), 7–20.

Ibrahim, Y. & Rashid, A. (2010). Homestay program and rural community development in Malaysia. *Journal of Ritsumeikan Social Sciences and Humanities, 1*(2), 7–24.

Jones, A. & Jenkins, I. (2002). A taste of Wales-Blas Ar Gymru: Institutional malaise in promoting welsh food tourism products. *Tourism and gastronomy*, 115–132.

Karim, M.S.A., Chua, B. & Salleh, H. (2009). Malaysia as a culinary tourism destination: International tourists' perspective. *Journal of Tourism, Hospitality & Culinary Arts, 1*(33), 63–78.

Kwoczek, M., Szefer, P., Hac, E. & Grembecka, M. (2006). Essential and toxic elements in seafood available in Poland from different geographical regions. *Journal of agricultural and food chemistry, 54*(8), 3015–3024.

Leong, Q., Abdul Karim, S., Selamat, J., Mohd Adzahan, N., Karim, R. & Jamaluddin, R. (2009). Perceptions and acceptance of 'belacan'in Malaysian dishes. *International Food Research Journal, 16*(4), 539–546.

Lim, V.L., Norzan, N.N. & Mohd, Z. (2009). *Tourists' Perceptions towards Malaysian Foods*. Paper presented at the Proceedings of 2nd National Symposium on Tourism Research, Universiti Sains Malaysia, Penang, Malaysia 18 July 2009 (pp. 191–202). Theories and Applications.

Ling, L.Q., Karim, M.S.A., Othman, M., Adzahan, N.M. & Ramachandran, S. (2010). Relationships Between Malaysian Food Image, Tourist Satisfaction and Behavioural Intention. *World Applied Sciences Journal, 10*, 164–171.

Liu, A. (2006). Tourism in rural areas: Kedah, Malaysia. *Tourism management, 27*(5), 878–889.

McKercher, B., Okumus, F. & Okumus, B. (2008). Food tourism as a viable market segment: it's all how you cook the numbers! *Journal of Travel & Tourism Marketing, 25*(2), 137–148.

Ministry of Tourism and Culture Malaysia. (2012). *Government of Malaysia*. Retrieved from http:// www.motac.gov.my/en/download/viewb/category/60-isu-okt-dis-2012.html

Ministry of Tourism and Culture Malaysia. (2013a). *Government of Malaysia*. Retrieved from http://www.motac.gov.my/en/download/view/category/65-isu-jan-mac-2013.html

Ministry of Tourism and Culture Malaysia. (2013b). *Government of Malaysia*. Retrieved from http://www.motac.gov.my/en/download/view/category/69-isu-jul-sept-2013.html

Ministry of Tourism and Culture Malaysia. (2014). *History of visit Malaysia year*. Retrieved from http://www.vmy2014.com/aboutvmy2014/history-of-visit-malaysia-year

Pendergast, D. (2006). Tourist gut reaction: Food safety and hygiene issues. In J. Wilks, D. Pendergast & L. P (Eds.), (pp. 143–157). London: Elsevier Ltd.

Richards, G. (2012). *An overview of food and tourism trends and policies*. Paper presented at the OECD Studies on Tourism Food and the Tourism Experience The OECD-Korea Workshop: The OECD-Korea Workshop.

Styler, C. (2006). *Working the plate: The art of food presentation*. New Jersey: John Wiley & Sons.

Telfer, D.J. & Wall, G. (1996). Linkages between tourism and food production. *Annals of Tourism Research, 23*(3), 635–653.

Venugopal, V. (2005). *Seafood Processing: Adding value through quick freezing, retortable packaging and cook-chilling*: CRC Press.

Wei, J. & Nakatsu, R. (2012). Leisure food: derive social and cultural entertainment through physical interaction with food *Entertainment Computing-ICEC 2012* (pp. 256–269): Springer.

Yen, N.Y. (2010). Sharing the best food Malaysia has to offer, from http://puteraadika.blogspot.com/2010/01/sharing-best-food-malaysia-has-to-offer.html.

Yüksel, A. & Yüksel, F. (2002). Market segmentation based on tourists' dining preferences. *Journal of Hospitality & Tourism Research, 26*(4), 315–331.

Theory and Practice in Hospitality and Tourism Research – Radzi et al. (Eds)
© *2015 Taylor & Francis Group, London, ISBN 978-1-138-02706-0*

Understanding and usage of Malay food terminologies among young Malay culinarians

M.A. Khairunnisa, M.S.M. Zahari, M. Rosmaliza & M.S.M. Shariff
Faculty of Hotel and Tourism Mangement, Universiti Teknologi MARA, Shah Alam, Malaysia

ABSTRACT: Nearly every culture and language has contributed to the culinary language. Including Malay, there are abundant of unique food names terminologies that can be found in Malay delicacies that typically named after the appearance of the food, the way food is prepared, places, people and certain events or incidents. Although Malay culinary is rich in terminologies, much of the terminology is believed no longer being learned not only among the Malay youngsters but also among the young Malay professional chefs and culinarians. Much of the Malay food names terminologies are believed of getting ignored and gradually disappeared and in fact some of the words do no longer exist. This study is empirically investigating the level of understanding of Malay food names terminologies and its impact to the usage among the young Malay culinarians in their daily cooking activities. A quantitative research approach was used and the information requires were obtained from the young Malay culinarians who previously had undergone formal culinary education and working directly in the food sectors within the Kelang Valley, Malaysia. The results from descriptive and multiple regression analyses revealed that the majority of the respondents have a relatively low understanding of Malay food names terminologies and only use a few of it in their daily cooking activities.

Keywords: Malay food, food terminologies, and culinar

1 INTRODUCTION

Language is like anthropomorphic organisms with lives and existence of it is dependent on the communicative activities and behaviors of their speakers (Lehmann, 2006). Language is not a static process but acts as a vehicle through which culture is expressed, conveyed from one generation to others (Grenoble & Whaley, 1998). The preservation of a native language acts as a stronger connection to their community because the societies who embrace their language will encourage their children and the youngster to use the language more often in the society (Mulia, 2003). Lazear (1995) mentioned that language also develops within all fields of profession which act as a mechanism to pass a specific knowledge and information through language terminologies. Michaud (2008) expressed that, language through its terminology denotes and represent specific meanings of something. These terminologies exist in just about every kind of area and cooking with no exception. Despite the universal terminologies established within particular disciplines, every ethnic culture or country are also having their own food languages or terminologies which signifying certain meaning to either in the preparations, methods of cooking, equipment, eat-

ing decorum and others (AlTamimi, 2011). American, for instance, although embraced, accepted and use certain French culinary terms, they also created, possessed and practices their own food terminologies using American English language (Sundari, 2008). These include the essential cooking methods like *stew, simmer, poach, deep-fry*, and food names such as *Caesar salad, corn dog, jacket potato*, and many others. Spanish also practices their own culinary terms like *blando, cocer al horno, rociando, escalfado* denoting to methods of cooking, whereas *cazuel, plancha, comal* are referring to types of utensils and equipments.

Within the Malay community, there are numerous and valuable food terminologies created and used since the olden days (Sharif, Supardi, Ishak & Ahmad, 2008). Although Malay culinary is rich in terminologies, much of them are believed no longer being used. According to Sariyan (2010), the biggest single threat to the Malay food language is the modern culture and society that marginalizing the tradition of the past. This phenomenon occurs not only among the ordinary Malay youngsters but also goes to young Malay professional chefs and culinarians. Many of Malay food terminologies are fading out of use and being replaced by others foreign language such as English and French that they learn from their culinary studies. Owing to the lack of usage, much of the Malay food terminology are believed of getting ignored and gradually disappeared, in fact some of the words are no longer exist. With that, this study is empirically investigating the level of understanding of Malay food names terminologies and its impact to the usage among the young Malay culinarians in their daily cooking activities.

2 LITERATURE REVIEW

2.1 Food terminology

Nagaral (2009) noted that food language is expressed through its numerous terminologies. Like many other crafts, cooking has developed highly specialized terminologies for describing its various operations (Cook, 2010). The food terminologies are the standardized way to communicate specifically with defined cooking techniques and methods that refer to the preparation of foods. Lowinsky (1992) mentioned that food terminologies also can be learned through food tradition which be passed down from generation to generation. In relation with this notion, language of every culture or country possessed their own unique and different food terminologies that may resemble the same meaning with other food terminologies from other countries (Cusack, 2000). For instance, the spice *turmeric* in Britain and America is known as *buried* in South Africa. *The chickpea* is also known as *Chana* in Hindi and *garbanzo* bean in Spanish. For that reason, ethnic food language represents the collected terminologies and words of many generations of people who have learned how to produce, prepare and pass on their cooking skills, recipes, methods in food preparation.

With regard to the study, Michaud (2008) measuring knowledge and understanding toward cooking found that young adults generally have a low level of food terminology knowledge. He noted that people who do not understand the cooking terms such as the use of proper equipment's, appropriate ingredients make inappropriate substitutions for ingredients that may end up with substandard food product even if the recipe and preparation instructions were accurately prepared. Levy and Auld (2004) argued understanding of cooking terminologies can be so intimidating that people may become uneasy and discouraged them to cook. Therefore, defining cooking terminologies in the recipe instructions, providing suggestions for acceptable substitutions, or describing in detail the proper ways to measure ingredients can help to increase food terminology knowledge and to reduce errors resulting from insufficient cooking knowledge.

2.2 Food names terminologies

In terms of food names terminologies, each of it has its own history and origins. Most of the food names terminologies formation was developed either from the names of places, noble people or characters of a special occasion to common history associated with the production of the food itself. Malays possessed its own terminologies of food that embrace from its preparation, method of cooking, and numerous unique food names (Omar, 2004). Muhammad, Mohd Zahari, Othman, Jamaluddin, and Rashdi (2009) claimed that a Malay food terminology has been shaped by cultural transmission over many generations. Muhammad et al. (2009) also stated that, in the Malay community, there are abundant of unique food names that can be found in food delicacies that typically named after the appearance of the food, the way food is prepared, places, people and by certain events or incidences (Yoshino, 2010). Some of the famous and unique Malay food names include *nasi tumpang, rendang tok, tahi itik, laksa, cek mek molek, beriani gam, cakar ayam, nasi dagang* and many others.

2.3 Usage of terminologies

The ways people use terminologies convey a great deal of information about themselves, their audience, and the situations they are in and indicates their social status, age, sex, and motives. Further, constantly use and practice the linguistic and terminologies is important to preventing them from being forgotten especially the crucial ones created by the older generations (Theophano, 2003). Sundari (2008) looks the usage of terminologies formation, terminologies structure, and type of food names terminologies in the restaurants. He suggests the restaurant owners to better understand and frequently practice of English language structure if they want to use English food menu names in their restaurant. Agbo (2009) affirmed that the understanding of the food terminologies is important because the more food terminologies and related words they understand, the better there will be able to communicate and use it in their daily conversation.

3 METHODOLOGY

As this study examines the understanding and usage of Malay food terminologies among the young Malay culinarians, a quantitative research approach through a questionnaire survey is opted. The information requires were obtained from young Malay culinarians who previously had undergone formal culinary education. The age range of Malay culinarians from 22 to 30 years old of who are working directly in the food sectors like in independent restaurants, hotels restaurants, cafes and cafeterias within the Kelang Valley area were chosen as a sample. Two weeks were spent on the survey and, as a result of the positive feedback a total of 200 usable questionnaires were obtained. Statistical Packages for Social Science (SPSS) was utilized for analyzing the collected data.

4 ANALYSIS AND RESULTS

4.1 The level of understanding of Malay food names terminologies

The result of the descriptive statistic in Table 1 shows that the mean score is ranging from 2.42 to 3.57 which indicate the young Malay culinarians is having fairly good understanding of Malay food names terminologies. As such the majority of the young Malay culinarians possessed a fairly understanding on Malay food names terminologies in general.

They also do not have a good understanding on how Malay food acquired its names. This can be seen from a mean score which indicate that they

Table 1. Mean score for the level of understanding of Malay food terminologies.

Items	Mean	S. D.
My understanding of Malay food names terminologies in general.	2.85	.901
My understanding of Malay food names created from the special event or incident.	2.44	.939
My understanding of Malay food names derived from people's name or title.	2.66	1.025
My understanding of Malay food names acquired from the appearance of the dish or how food is made.	2.86	.994
My understanding of food named Nasi dagang	3.57	.836
My understanding of food named Beriani Gam.	3.42	.904
My understanding of food named rendang tok.	3.54	.890
My understanding of Malay food called Cek Mek Molek.	3.50	1.017

Scale: 1 = Very poor, 2 = Poor, 3 = Fair, 4 = Good, 5 = Very good.

Table 2. Mean score for the level of usage of Malay food terminologies.

Items	Mean	S. D.
Level of usage of Malay food names terminologies in general.	2.64	.920
Level of usage of Malay food names terminologies during my culinary studies or exposure.	2.53	.987
Level of usage of Malay food terminologies in recipes and menu writing.	2.59	.925
The usage of Malay food names terminologies among my working colleagues.	2.71	.938
The use of Malay food names terminologies in my daily cooking activities.	2.84	.925

Scale: 1 = Never, 2 = Few, 3 = Sometime, 4 = Often, 5 = Very often.

admitted of having a fair understanding of Malay food names created from the special event or incident, fair understanding that Malay food names derived from people's name or title or Malay food Malay food names acquired from the appearance of the dish or how food is made. However, based on the given examples of food names terminologies, the mean scores indicate that the young Malay culinarians admitted of having a relatively good understanding on the terms Nasi Dagang, Beriani Gam, Badak Berendam, Cek Mek Molek. This result is anticipated as the young Malay culinarians are already familiar and aware of those foods in the restaurants but this does not indicate that they know about the origin of the foods or how the foods are actually acquiring its name.

4.2 The level of usage of Malay food names terminologies

Most likely due to low understanding on the Malay food names terminologies, result of this section analysis obviously revealed a low usage on Malay food names terminologies among the young Malay culinarians in general. For that reason, it is not surprising that the level of usage of Malay food names terminologies in their daily cooking activities is relatively low. The young Malay culinarians also admitted that they only sometimes use Malay food names terminologies during their culinary studies or culinary exposure, in recipes and menu writing and among my working colleagues.

4.3 Correlations between the level of understanding of Malay food names terminologies and the usage

To test the correlation between the understanding of Malay food names terminologies and the usage,

Table 3. Pearson correlations result.

Independent Variable	Dependent variable (Usage)
Understanding of Malay Food names terminologies	.278**

**Correlation is significant at the .01 level (2-tailed).

Table 4. Multiple regression results.

Independent Variable (IV)	Dependent Variable (DV) Usage of Malay Food Names Terminologies
	Beta Coefficients and Significance Levels
Food Names	.28 ***
R²	.19
F-Change	52.25***

Note: * P < .05, ** P < .001, *** p < .000

Pearson Moment Correlation was used. The result revealed that the understanding of Malay food names terminologies significantly influences the usage ($p < .000$). Therefore, this signifies that there is a relationship between two variables. However, the beta value ($\beta = 0.278$) indicated that the understanding of Malay food names terminologies causes relatively low in it usage. This accords with Sundari (2008) that although peoples are familiar with the food names, but it does not mean they know about the origins or how the foods are named.

4.4 Relationship between Malay food terminologies and the usage

The relationship between Malay food names terminologies and the usage of it among the young Malay culinarians was further confirmed with standard multiple regression. The beta value under standardized coefficient was evaluated by comparing the contribution of each of the independent variables to the dependent. The result signifies that Malay food names terminologies was able to clarify 19 percent (F-change $= 52.25$, $p < .001$) of the variance in the usage of it. By looking at the beta value, food name terminologies ($\beta = 0.28$, $p = .000$) had a slight impact on the usage of Malay food terminologies. This result indicates that the understanding of food names terminologies significant influences on the usage of it. In other words, poor understanding of food names terminologies leads to poor usage on it.

5 IMPLICATION, CONCLUSION, RECOMMENDATION

Finding of this study clearly witnessed that the young Malay culinarians are having a poor understanding of Malays food terminologies in general that lead to minimize usage and practices of it in their daily cooking activities. Owing to this phenomena and not exaggerate that some the Malay food terminologies is probably getting faded or dying out. Therefore, preserving and enriching the Malay food terminologies is crucially important for all parties. The relevant authorities, either in the public or private sector including culinary institutions should help to increase the awareness and disseminating the information on the importance of preserving the Malay food heritage through the media, food events and many others. Such actions would at least encourage younger generations, especially the culinarians, to better understand and appreciate Malay foods, language and heritage. Besides that this also can be done through increasing the use of traditional Malay food terminologies in Malay recipe books as a source of references for future generations.

As a conclusion, the uses of Malay food terminologies should be encouraged and should not only be used for everyday communication or vernacular language, but as professional or technical terminologies like French, English and Japanese which already have a strong position in technical field including the culinary field.

ACKNOWLEDGMENT

This research was funded by the Ministry of Higher Education, Malaysia through Universiti Teknologi MARA under RIF grant: 600-RMI/DANA 5/3/RIF (599/2012).

REFERENCES

Agbo, M. (2009). The Syntax and Semantics of Verbs of Cooking in İgbò. *Journal of Theoretical Linguistics, 6*(2), 70–82.

AlTamimi, S. (2011). Food for Thought: The Universal Language of Food, from http://sailemagazine.com

Cook, G. (2010). *Sweet talking: Food, language, and democracy*. Open University United Kingdom: Cambridge University Press.

Cusack, I. (2000). African cuisines: Recipes for nationbuilding? *Journal of African Cultural Studies, 13*(2), 207–225.

Grenoble, L.A. & Whaley, L.J. (1998). *Endangered languages: Language loss and community response*. Cambridge: Cambridge University Press.

Lazear, E.P. (1995). Culture and language. *Journal of Political Economy*, 1–50.

Lehmann, C. (2006). On the value of a language. *European Review Journal*(14), 151–166.

Levy, J. & Auld, G. (2004). Cooking classes outperform cooking demonstrations for college sophomores. *Journal of nutrition education and behavior, 36*(4), 197–203.

Lowinsky, N.R. (1992). *Stories from the motherline: Reclaiming the mother-daughter bond, finding our feminine souls*: Jeremy P. Tarcher, Inc.

Michaud, P. (2008). *Development and evaluation of instruments to measure the effectiveness of a culinary and nutrition education program*. (Unpublished master dissertation), Clemson University.

Muhammad, R., Mohd Zahari, M.S., Othman, Z., Jamaluddin, M.R. & Rashdi, M.O. (2009). *Modernization and ethnic festival food*. Paper presented at the International Conference of Business and Economic, Kuching, Sarawak.

Mulia, A. (2003). *Language development and revitalization in a South East Asian community: An insider's perspective* Paper presented at the Conference on Language Development, Language Revitalization and Multilingual Education, Bangkok, Thailand.

Nagaral, V. (2009). A linguistic analysis of the cookery language. *Journal of International Referred Research, 2*, 21–59.

Omar, A.H. (2004). *The encyclopedia of Malaysia: Languages and literature* (Vol. 9). Singapore: Didier Millet.

Sariyan, A. (2010). *Realizing Malay language as a world language.* Paper presented at the Arif Budiman Conference, Ministry of Higher Education Malaysia-Beijing Foreign Studies University, London.

Sharif, M., Supardi, A., Ishak, N. & Ahmad, R. (2008). *Malaysian Food as Tourist Attraction.* Paper presented at the 1st Malaysian Gastronomic–Tourism Conference.

Sundari, W. (2008). *The process of formation of English food menu names at Simpang Lima restaurant.* (Unpublished master dissertation), Diponegoro Semarang University.

Theophano, J. (2003). *Eat my words: Reading women's lives through the cookbooks they wrote.* New York: Palgrave Macmillan.

Yoshino, K. (2010). *Malaysian Cuisine: A Case of Neglected Culinary Globalization.* Paper presented at the Proceeding In Globalization, Food and Social Identities in the Asia Pacific Region Conference.

Theory and Practice in Hospitality and Tourism Research – Radzi et al. (Eds)
© *2015 Taylor & Francis Group, London, ISBN 978-1-138-02706-0*

Determinants of food heritage in Malaysia context

A.M. Ramli & M.S.M. Zahari
Faculty of Hotel and Tourism Management, Universiti Teknologi MARA, Shah Alam, Malaysia

ABSTRACT: Identity formation is a central issue and plays a central role in determining the image of a nation. Identity formations are even critical in multiracial and multicultural nations and many countries in fact are still struggling in developing their identity and social integration. Malaysia without doubt is experiencing the urgency of having its own food identity when sharing food cultural background is becoming central issues between the neighboring countries. The adaptations of food from various ethnic groups by the dominant or majority ethnic group in multi-racial and cultural nations is believed will lead to common acceptable cuisines and longitudinally forming the national food identity. The endorsement, certification or gazetting of the heritage and traditional food is not only mechanisms to preserve the treasures but importance element toward country identity formation. This research proposal is therefore aiming to describe on the role of food heritage determinants through the process of endorsement and certification toward nation food identity formation.

Keywords: Food, heritage, determinants, nation, identity, formation

1 INTRODUCTION

Not to exaggerate that identity formation is a central issue and plays a central role in determining the image of a nation. Identity formations are even critical in multiracial and multicultural nations and many countries in fact are still struggling in developing their identity and social integration (Alba et al., 2000). Social theories seek to explain the identity formation, how it develops, what factors facilitate and inhibit it, and what results from it (Matsunaga, et al. (2010). Identity formation extends not only to the individual but also to groups, organizations, societies, politics and countries with no exception of food. Similar to other matters, food identity is another central issue for a country (Spurrier, 2010).

In line with this, Malaysia is experiencing the urgency of having its own food identity when sharing food cultural background is becoming central issues between the neighboring countries. Countries like Singapore and Indonesia which are sharing a common historical roots and cultural heritage with Malaysia are disputing over some of the traditional food when each country pursues to validate those traditional foods as their identity. Owing to this issue and according to Chong (2009) each country is becoming more aggressive to defence and protect such heritage as theirs to safeguard the country as heritage is the core of a people's and country identity. Some clear examples can be seen on *Rendang, Nasi Lemak, Laksa, Chili Crab, Bak Kut Teh, Hainanese Chicken rice* and many others (Chong, 2012; Jb, 2012; Star, 2009; Wo, 2009).

In promoting gastronomic tourism for the international tourists, Singapore for instance insistently proclaimed some of the common Chinese, Indian, Malay and Peranakan food as their local heritage. This can clearly be seen on their tourism promotional packages under the Singapore Tourism Board (STB) classified dishes like *Ayam Buah Keluak, Bak Kut Teh, Cendol, Char Kway Teow, Chili Crab, Laksa, Nasi Lemak, Rendang, Teh Tarik and others* as Singapore 'iconic' dishes (Chaney & Ryan, 2012; Star, 2012a).

Not to be overstated that issues pertaining to food ownership or even worse on the Indonesia counterpart. A long debated history with the neighboring countries not only occurs in the territorial boundaries but also in cultural ownership like traditional dances, music instruments, arts and culinary treasures. In this context, the examples of proclaiming cultures are the textile art of '*batik*', '*wayang kulit*', the ceremonial dagger 'the *Keris*', folk song '*Rasa Sayang*', music instrument '*Gamelan, angklung*' and traditional dances '*Pendet*' and culinary treasures like '*rendang*' and a few others (Chong, 2012; Hussain, 2011). These issues to that extent sparks a nationalistic outrage, protest, street demonstration and staged rallies which branding Malaysia as "a nation of thieves" by the irresponsible group. These resurgent disputes have given significant impact on the relationship between two countries which led to a discussion among the highest authorities from both countries (Chong, 2009).

Those incidences have open Malaysian eyes particularly the government on the importance to at least having our own cultures like oral tradition, languages, festive events, rites and beliefs, music and songs, the performing arts, traditional medicine, literature, traditional sports and games as well as traditional cuisine identity (Lim, 2012), although sharing the fundamental basis of it cannot be avoidable. Former Minister of Tourism, Dato Seri Ng Yen Yen for instance stressed the importance of having strong nation's culture and food identity that leads to the contribution of destination branding and image for our country.

Scholars recognized the foundation that take place in the processes of constructing the food identity within ethnic groups in the multicultural countries in particular is reflecting all the way through a sharing common accepted cuisine (Appadurai, 1988; Spurrier, 2010; Wilk, 2006). Cross-culturing of food among the ethnic groups through acculturation and assimilation processes gradually creates or form food identity in the multicultural countries (Fox & Ward, 2008; Helland, 2008). It is interesting to note that the adaptations of food from various ethnic groups by the dominant or majority ethnic group in multi-racial and cultural nations will also lead to common acceptable cuisines and longitudinally forming what could be called national food identity (Fox & Ward, 2008). Besides this, Manaf (2008) noted that gazetting and certification the heritage and traditional food in the long run strengthening the nation food identity formation. Wo (2009) on the same note stressed that Gazetting is not only mechanisms to preserve food heritage but importance element toward country identity formation therefore government should stimulate public awareness about it.

This statement concurred with the Commissioner of Heritage, Professor Emeritus Datuk Zuraina Majid which stated that;

"Instead of keeping national heritage under lock and key, the governments want to promote the list" (Wo, 2009).

In fact, the seriousness of the issues can clearly be seen through the former Minister of Information Communication and Culture statement that;

"The ministry views the declaration of national heritage as a catalyst to fuel our spirit of patriotism and nationalism. It will also boost the spirit of conservation and preservation" (Star, 2012c)

In line with what have been mentioned, Malaysian government presently is taking serious action

in preserving it gastronomic treasures through various initiatives. Besides many initiation processes such the promoting the transferring of traditional food knowledge (TFK), encouraging acculturation, assimilation and adaptation of culture within the ethnic group (Ishak et al., 2012; Md Nor et al., 2012), gazetting or certification of Malaysian food heritage by Department of National Heritage become one of the significant national agenda. Almost all states also taking steps in preserving their food heritage (Bernama, 2010a). Despites these, some question arises. What are main purpose and the criteria's used in selecting national food heritage? What are the main attributes used in determining food heritage? Is the criterion such history elements, food characteristics, value of uniqueness, practice and integration are considered? Which agencies responsible for the endorsement and most importantly what are the impacts of the endorsement or certification on preserving foods and towards food identity formation? These questions in fact need to be revealed. In fact, there still limited empirical studies on the relationship between the endorsement and certification of food heritage and food identity formation (Hergesell, 2006) and none of those available studies looking in Malaysia perspectives. This highlighted issue therefore warrants for an investigation.

2 LITERATURE REVIEW

2.1 *Food heritage in Malaysia context*

In the Malaysian context, Wahid, Mohamed, and Sirat (2009) associate food heritage with classical and traditional foods which are continuously practicing by all generations without major altering of the original flavours. Food heritage can be reflected from the environment history, belief, ideology and food technology of society in an era or period of time (Utusan, 2010). The Heritage Commissioner of National Heritage Department, Prof. Datuk Zuraina Majid states that food heritages are based on two categories. The first category refers synonymous or common foods which are part of our lives whereas the second consists of foods that are almost extinct in other words it were once part of our culture but are slowly dying out (Wahid et al., 2009; Wo, 2009).

Auxiliary with the above statement, the first category refers to those food that represent the nation, places or ethnic groups or food that are unique or signature dishes such *nasi dagang, masak lemak, assam laksa* and many others. This is also includes the popular dishes that are goodwill or accepted to the Malaysian society for example, *satay, nasi lemak, kuih bakul* and *cendol* which are preparation skills are not limited to one race only. For the second category, food like pulut *kukus periuk kera* (glutinous rice cooked in monkey pot plants) and *ikan panggang tanah liat* (grilled fish wrapped in clay) are some examples that are slowly fading away owing to peculiarity in the method of cooking, recipes and technique of preparing those dishes (Wo, 2009) and some did not exist anymore especially the indigenous cuisines (Abdul latib, 2009).

As the unique of food to a particular culture are closely related to the ingredients, preparation methods, dishes, or eating decorum there is a great concern and attention given to the preservation of traditional cuisine. This cultural heritage is difficult to preserve and measures than a physical object as it associated with values, beliefs, behaviors and rules of the society. The need of continuity and preserving food heritage is being considered as conditions of comparative advantage in maintaining a local food culture in the face of homogenizing pressures from the outside (Shariff et al., 2008) and continuation of preserving creating valued products especially the traditional cuisines (UNESCO, 2008).

2.2 *Determinants of food heritage*

Food heritage determinants can be associated with the historical elements, food characteristics, value of uniqueness, practice and integration element as mentioned by Ramli et al. (2013). Each of the determinants is being explained by researchers. Guerrero et al. (2009) stated dimension of traditional foods consist of elements of habit, natural, origin and locality. Habit and natural associated traditional food with something anchored in the past to the present, transmitted from one generation to another or food that has been consumed from the past or existed for a long time that has always been part of the consumers' life. Hjalager and Corigliano (2000) identified historic resources as focal points of food festivals and special events that attract tourists and local residents. They further asserted that the culture of the food and eating festivals can promote local culinary traditions, lifestyles, and gastronomic heritage. Lin, Pearson & Cai (2011) revealed that the origin of food is the most important information in aiding international tourists to recognize the authenticity of a nation. Staple, flavouring and preparation are the food characteristics closely related to the heritage. Belasco (2008) stated that staple foods or basic food that have unique value and very meaningful to communities ranging from meat and potatoes, stew and *fufu* (porridge) and many others could be classified as heritage. Flavouring, which has a distinctive way of seasoning dishes, unique flavour and combinations serve as important group "markers" closely associated with heritage. For instance, culinary

identity in parts of China may expressed through the combination of soy sauce, garlic, and sesame oil, while a mix of garlic, tomato, and olive oil may signal "southern Italian. Specific cuisines favour or distinct manner of preparing food such as such as stir frying in China; stewing for Mexico and other food characteristics is also associated with heritage. Guerrero et al. (2009) highlighted that taste as one of factors in recognizing the authenticity of a food product or cuisine. Karim et al. (2011) posited that staple food, cooking methods and taste should be preserved and sustain as it representing identity of the community or ethnic and considered as country's food heritage and it the way representation of country food identity.

Guerrero et al. (2009) associate value of uniqueness as innovation with food that are new or unusual ingredients; new combinations of products; different processing systems or elaboration procedure, including packaging' coming from different origin or cultures' being presented and/or supplied in new ways; and always having temporary validity has significant role in heritage.

Cross-cultural processes through the acculturation, assimilation, adaptation plays an important role in the practices and integration in the multiple ethnic groups and these also closely associated with heritage (Kwik, 2008). The cross cultural process consists of food knowledge which referred to the cultural tradition of sharing food, recipes and cooking skills and techniques and passing down the collective wisdom through generations (Cleveland, Laroche, Pons & Kastoun, 2009).

3 METHODOLOGY

This proposed study intends to describing the role of determinants of food heritage through the process of endorsement and certification toward food identity formation from the relevant authority and individual or public perspective The mix approaches which combine the qualitative and quantitative will be used to gather information. Specifically, the qualitative approach as the main paradigm will be dealing with the interview using semistructured questions with the relevant authorities that closely involved with the endorsement of Malaysia food heritage while the quantitative through questionnaire survey with the public will be used to support the qualitative findings. For qualitative, the authority of government agencies and food experts as appointed panels by the Jabatan Warisan Negara that responsible for endorsement and certification of nation's food heritage will be the key informants for this study. For quantitative, a self-completed questionnaire will be used and the target population will be among the public which consists of major ethnic groups like Malay, Chinese and Indian. The reason of choosing the three major ethnic groups is because some of their major foods are becoming common and acceptable among the Malaysian hence being classified as Malaysian food and qualified to be endorsed as food heritage. A preliminary screening process will be carried out to select suitable respondents set by the researcher such as age (30 years and above), having knowledge on ethnic foods and experience of preparing them. Those fulfilled the criteria's will be proceeding with the questionnaire, this selected criteria is being recognized as purposive sampling. The contextual location for the data collection will be in Kuala Lumpur, Shah Alam, Petaling Jaya and Klang area.

4 CONCLUSION

Undertaking study on the role of determinants of food heritage through the process of endorsement and certification toward food identity formation will definitely extend the existing literatures and creating to a new body of knowledge particularly in Malaysia context. In other words, besides proliferate of food studies the insights obtain lay the fundamental basis, develop and strengthen the understanding on the role of food toward cultural heritage and identity of the nation which permit other researcher to further explore on this issue. A benchmark of this study in addition will create a foundation for the future direction of other studies of a similar nature.

In the practical perspective, as previously mentioned the endorsement and certification of determinants criteria of food heritage are still not fully developed particularly in Malaysia. Determining and understanding the determinants of food heritages without doubt will directly benefit many parties like the government authority such as the Ministry of Tourism and Culture, states government agency and non-government organization (NGO). Determinants that will be obtained from this study besides creating the foundation and practicable guidelines but aids in developing standard mechanisms or standardize tools for certification and endorsement of food heritage that can used by all the parties. It is also will help the relevant authorities' assists in creating food cultural preservation and documentation for future references and generation.

In the other context, recognizing the historical elements, food characteristics, value of uniqueness, practices and integrations elements as food heritage attributes will helps the nation in preserving and sustaining the ethnic foods potpourri for future generation, in the long run contributes toward nation food identity formation or nation food image not only among the Malaysian but also in the international arena. Preserving and creating nation food identity on the other hand catalyzing and fueling the patriotism and nationalism among the younger generation not only food but others elements as well without tear down each ethnic tradition identity.

REFERENCES

Abdullatib, S. (2009). 100 Makanan Tradisi, *Kosmo Online*.Retrieved from http://kosmo.com.my

Alba, R., Portes, A., Kasinitz, P., Fonari, N., Anderson, E. & Glazer, N. (2000). Beyond the Melting Pot 35 Years Later: On the Relevance of a Sociological classic for the Immigration Metropolis of Today. *The International Migration Review, 34*(1), 243.

Belasco, W. (2008). *Food: The key concepts.* Oxford: Berg Publishers.

Bernama. (2010a). Memastikan Kesinambungan Warisan, *Bernama*.From http://www.kpkk.gov.my

Bernama. (2010b). Yen Yen bent on temptin tourists with food from http://updated.internalinsider

Chabrol, D. & Muchnik, J. (2011). Consumer skills contribute to maintaining and diffusing heritage food products., retrieved from Anthropology of food http://aof.revues.org/6847

Chaney, S. & Ryan, C. (2012). Analyzing the evolution of Singapore's World Gourmet Summit: An example of gastronomic tourism. *International Journal of Hospitality Management, 31*, 309–318.

Choi Tuck Wo. (2009). Dish branding, *thestar*. Retrieved from http://thestar.com.my/lifestyle/story.

Chong, J.W. (2009). The Indonesia-Malaysia Dispute Over Shared Cultural Icons And Heritage. *ILSP Law Journal*, 177–186.

Chong, J.W. (2012). " Mine, Yours or Ours?": The Indonesia-Malaysia Disputes over Shared Cultural Heritage. *SOJOURN: Journal of Social Issues in Southeast Asia, 27*(1), 1–53.

Cleveland, M., Laroche, M., Pons, F. & Kastoun, R. (2009). Acculturation and consumption: Textures of cultural adaptation. *International Journal of Intercultural Relations, 33*(3), 196–212.

Crocetti, E., Rubini, M. & Meeus, W. (2008). Capturing the dynamics of identity formation in various ethnic groups: Development and validation of a three-dimensional model. *Journal of Adolescence, 31*(2), 207–222.

Elis, S. (2009). Our rich 'food' heritage, *Bernama*. Retrieved from http: //blis2.bernama.com

Fox, N. & Ward, K.J. (2008). You are what you eat? Vegetarianism, health and identity. *Social Science & Medicine, 66*(12), 2585–2595.

Guerrero, L., Guardia, M.D., Xicola, J., Verbeke, W., Vanhonacker, F., Zakowska-Biemans, S., Hersleth, M. (2009). Consumer-driven defination of traditional food products and innovation in traditional foods. A qualitative cross-cultural study. *Appetite, 52*, 345–354.

Helland, S.H. (2008). *Chinese Malaysian flavours: an anthropological study of food and identity formation in Penang.* University of Oslo.

Heritage. (2012). Senarai Warisan Kebangsaan-Warisan Makanan 2012 dan 2009 Jabatan Warisan Negara.

Hjalager, A.-M. & Corigliano, M.A. (2000). Food for tourists—determinants of an image. *International Journal of Tourism Research, 2*(4), 281–293.

Hussain, Z. (2011). Jakarta Cooks Up 'Rendang Diplomacy', *The Jakarta Globe*. Retrieved from http://www.thejakartaglobe.com

Ishak, N., Zahari, M.S.M., Sharif, M.S.M. & Muhammad, R. (2012). *Acculturation, Foodways and Malaysian Food Identity*. Paper presented at the Current Issues in Hospitality and Tourism and Innovations, Kuala Lumpur.

Jb, S.B. (2012). No need to patent 'rendang' dish: William Wongso, *The Jakarta Post*. Retrieved from http://www.thejakartapost.com

Kwik, J.C. (2008). *Traditional Food Knowledge: Renewing Culture and Restoring Health*. Master University of Waterloo, Published Heritage Branch. Library and Archives Canada databas.

Lim, Y. (2012). KL central cultural makeover, *The Star online*. Retrieved from http://thestar.com.my

Lin, Y.-C., Pearson, T.E. & Cai, L.A. (2011). Food as a form of destination identity: A tourism destination brand perspective. *Tourism and Hospitality Research, 11*(1), 30–48.

Lyons, L.T. (2012). *The Food Heritage of Virginia An Untapped Asset of Community & Economic Development*. University of Virginia. Retrieved from http://www.virginia.edu

Manaf, Z.A. (2008). Establishing the national digital cultural heritage repository in Malaysia. [Research paper]. *Library Review, 57*(7), 537–548.

Matsunaga, M., Hecht, M.L., Elek, E. & Ndiaye, K. (2010). Ethnic Identity Development and Acculturation: A Longitudinal Analysis of Mexican-Heritage Youth in the Southwest United States. *Journal of Cross-Cultural Psychology, 41*(3), 410–427.

McCoy, L. (2012). *Food Heritage Planning Proposals: Planning For Food Heritage Celebrations In Central Virginia*. Retrieved from http://www.virginia.edu

Md Nor, N., Sharif, M.S.M., Zahari, M.S.M., Salleh, H.M., Isha, N. & Muhammad, R. (2012). The Transmission Modes of Malay Traditional Food Knowledge within Generations. *Procedia—Social and Behavioral Sciences, 50*(0), 79–88.

Ramli, A., Zahari, M.M., Ishak, N. & Sharif, M.M. (2013). Food heritage and nation food identity formation. *Hospitality and Tourism: Synergizing Creativity and Innovation in Research*, 407.

Rozin, P. (2006). The integration of biological, social, cultural and psychological influences on food choice. In R. Shepherd & M. Raats (Eds.), *The psychology of food choice* (pp. 19–39). UK: CAB International.

Shariff, N.M., Mokhtar, K. & Zakaria, Z. (2008). Issues in the Preservation of Traditional Cuisines: A Case Study in Northern Malaysia. [Article]. *International Journal of the Humanities, 6*(6), 101–106.

Spurrier, C.T. (2010). *Cassava, Coconut and Curry: Food and National Identity in Post-Colonial Fiji*. Master of Arts in Anthropology, University of South Carolina, ProQuest.

UNESCO. (2008). *World Heritage Information Kit*. France: UNESCO World Heritage Center Retrieved from http://whc.unesco.org.

Utusan. (2010). Hayati warisan menerusi makanan, minuman, Newspaper, *Utusan Malaysia Online*.

Vanhonacker, F., Verbeke, W., Guerrero, L., Claret, A., Contel, M., Scalvedi, L., Hersleth, M. (2010). How European consumers define the concept of traditional food: evidence from a survey in six countries. [Article]. *Agribusiness, 26*(4), 453–476.

Wahid, N.A., Mohamed, B. & Sirat, M. (2009). *Heritage food tourism: bahulu attracts?* Paper presented at the Proceedings of 2nd National Symposium on Tourism Research, Universiti Sains Malaysia, Penang, Malaysia 18 July 2009.

Wilk, R. (2006). *Home cooking in the global village: Caribbean food from buccaneers to ecotourists*. Oxford: Berg Publishers.

Theory and Practice in Hospitality and Tourism Research – Radzi et al. (Eds)
© *2015 Taylor & Francis Group, London, ISBN 978-1-138-02706-0*

Does food besides tourism core products contribute to Sabah destination image?

M.Z.N. Adilah, M.S.M. Zahari & A. Emaria
Faculty of Hotel and Tourism Management, Universiti Teknologi MARA, Shah Alam, Malaysia

ABSTRACT: The tourism core products either natural or artificial resources are found to have strong influenced on the image of a tourism destination. However, there is still unclear understanding of how and to what extent the local food of particular tourism destination besides it tourism core products contribute to a tourism destination image. This paper conceptually discusses the contribution of local food in addition to tourism core products toward destination image from a general perspective to the specific issue related to state of Sabah, Malaysia.

Keywords: Food, tourism core products, Sabah, destination image

1 INTRODUCTION

Besides psychological factor like perceptions, beliefs, learning, attitudes and social factor like leisure, family, disposal income, and status that inspire or influence tourists choose a destination; the impression or image of a destination itself cannot be ignored. The ability of a country to develop and maintain their destination image not only become a fundamental of attracting but determines the success or sustainability of a destination among the international tourists (Laws, Scott & Parfitt, 2002). A destination with a good and positive image will have a better success rate of being chosen as a tourist destination compared to a destination with negative image (Beerli & Martin, 2004). In this context, although there are many interpretations, a general term of the destination image is referred to individual perception or impression regarding particular place and emotional describes and portray of the destination (Ibrahim & Gill, 2005) with the other definitions are highlighted in the literature section.

There are two factors that influence how an image is developed for a respective tourist destination as delineated in the Tourism Marketing frameworks (Lin, Morais, Kerstetter & Hou, 2007). The first factor relates to person emotionality and rationality. This is occurring when tourists give or perceive a major importance to the image of tourism destination and ultimately influence their final choice or behavioral intention. Bigne, Sanchez and Sanchez (2001) contended that tourism destination assessment can assist tourism management to identify strengths and weaknesses in turn able to predict tourists' behavior and intentions. The clear example can be looked at the Eiffel Tower in France whereby the tower itself has long been the image of this country.

The second factors are the outcome emerging of two major aspects or image dimensions which are perceptual and cognitive and affective. In perceptual and cognitive image, tourists evaluate the destination resources, its attractions and judging them based on the value perceived and the importance of it to them. The higher value of the destination to them the higher chance for them to visit the destination (Beerli & Martin, 2004). In this context besides others, Kenya as a country is able to attract a substantial number of international tourists and continuously develop its image owing to its vast tourism resources, especially the unique wildlife safari plus pristine tropical beaches, culture attraction and pleasant climate and most of the international tourists felt those resources are important and benefits they (Akama & Kieti, 2007).

Affective in other words indicates the emotions and feeling experienced by tourists at the destinations. The tourist emotions and feeling are normally influenced by their personal touch or emotional components that a destination could offer and attract them (Beerli & Martin, 2004). In this perspective, Spain is one of the countries that develop the destination image from the personal touch and emotional components. High values of caring, good hospitality, human kindness, safety, freedom and friendliness of its peoples become the strengths of Spain's as a tourism destination (Pot, 2005).

In short, Page and Connel (2006) stated that resources either natural like wildlife, beaches, weather and culture or artificial resources such as entertainment, events and transportation, just to name a few are closed or strongly influenced the image of a tourism destination They also postulate that mixture of tourism products, experiences and other intangible items are associated with the image of a tourist destination. In fact, the characteristics of tourism core products and services also linked to the destination image.

Based on the above notion, Jalis, Zahari, Zulkifly and Othman (2009) deduce besides tourism core products, the local attributes such as history, heritage including food could contribute or increase the image of a destination. Auxiliary with this argument, it is assumed that food also plays in important role in strengthening the image of a particular destination. Goeldner and Ritchie (2006) contended that food is one of the important elements that cannot be ignored by tourists and it influenced their travelling mood toward a destination. Some travelers travel to other countries to try and consume food and beverages that the country offers (Boniface, 2003). In other word, the relationship between food of the particular destination and it destination image should not be ignored and overlook.

Hong Kong is one of the country that attract a lot of Taiwanese, Japanese, Singaporean and other country travelers not only because of shopping, sky scraping, amusement parks, marina and culture, but foods without exception moderate their destination image (Kivelä & Crotts, 2006). Besides engaging other travel and leisure activities tourists took a chance to savor the Honky food products during their stayed. Similar scenario goes Singapore as it becomes the most popular hawkers' food and cuisines among the international tourists than upscale restaurants (Henderson, 2000). With that, this paper conceptually discussing the contribution of tourism core products and food to destination image from a general perspective to the specific contextual setting that is Sabah, Malaysia.

2 LITERATURE

2.1 Tourism core products

In tourism, the core products usually come in a variety of forms. Medlik and Middleton (1973) postulate that, tourism core products as a bundle of physical products such as plant, services, and activities at the touristy location that create the whole tourism experiences. One of the elements in tourism core products is physical plants that are a site, natural resources and facilities such as a waterfall, wildlife, or resort. Besides that, physical plants can be a fixed property likes hotel or cruise ship or the physical environments that consist of water quality, marines, natural resources, cultural resources, and the condition of the facilities, weather, equipment's, and buildings.

Komppula (2001) posited tourism core products are the collection of tangible and intangible features where all of the features related to activities that tourists join or participate during their visit at the particular tourism destination. In this sense, tangible features refer to physical plants like marinas, natural resources, cultural resources while intangible features relate to hospitality, services, peoples as well as the communication among tourists and the local peoples.

Scholars' conceptualized destination at particular country is a geographic location that combined the tangible and intangible features (Murphy, Pritchard & Smith, 2000). In line with this, tourism providers are offering tangible and intangible features for tourists since both features complement each other and influence the travel experience (Albayrak, Aksoy & Caber, 2013).

As attractions are part of the tourism core product, Dolnicar and Huybers (2010) stated that type of attraction in the destination consists of natural (e.g. national park, animal park, general natural beauty and scenery), cultural/historical (e.g. museum, architecture, wineries) and marines (e.g. snorkeling). The image of attraction like natural is synonymous with relaxation activities. Tourists perceive natures as part of the wilderness and mountain therefore the image in the destination are determined by nature, enjoyment, relaxation and wilderness.

Chai (2011) emphasized that cultural attraction as new niche products that can be offered in tourism development henceforth represent the place and a sense of the destination. In addition, culture provides a way for the country to boast its status or reputation at the same time makes the country look real. Ashworth and Larkham (2003) asserted culture can be divided into two concepts that are "location and culture" and "boundaries and frontier" where these two concepts can give an effect to the possibilities and volume of cultural tourism. There are eight examples of culture elements: (a) archeological sites and museums, (b) architecture, (c) art, sculpture, crafts, galleries, festivals, events, (d) music and dance, (e) drama, (f) language and literature study, tours, events, (g) religious festivals, pilgrimages, and (i) complete (folk or primitive) cultures and sub-cultures.

2.2 Relationship between food and tourism

Scholars start to argue that food is one of the important elements that cannot be ignored by tourists and it influenced their travelling mood toward a destination. On the connection of foods and tourism, Long (2004) used the term culinary tourism to express the idea of experiencing food and beverage at particular country. According to Wolf (2002) foods at the particular country encouraged the pursuit of travel in the quest for the enjoyment of prepared food, drinks and other related food activities that resulting in a great memorable gourmet experience. These experiences have the power to modify eating and drinking habits and tastes as well as imbue the tourists' experiences of the people of the new locations and countries being visited (Johns & Kivelä, 2001).

Kivelä and Crotts (2006) contended seasoned tourists attached a great value to the local food experience of the countries that being visited. Some travel organizations, in fact regularly offer gourmet or culinary holidays with Italy and France as top destinations as well some countries in Asia. Holiday with cooking feature regularly offers in destinations like Tuscany, Provence in Europe, Melbourne and the Sydney Napa Valley in Australia and the Sonoma Valley in California. These places now become premier food tourism destinations (Kivelä & Crotts, 2006).

Kivelä and Chu (Kivelä & Chu, 2001) posited that tourists not only dine out in search of new tastes in the choice of food and beverages, but at the same time are on the lookout for new local food experiences when visiting a country. Intrepid Travel Agency reported that tourism and holiday operators from Australia, the United States, much of Europe and Asian countries like China, India, Thailand, Malaysia, Vietnam and Japan now offer foods tour packages which combine shopping with side-trips to sample the local foods that are available. Scarpato (2002) noted that local food can add value to the traditional tourism experience; especially for those who yearn for more and are constantly searching for new products and experiences. He further contended that local foods have created demand for short and/or weekend holidays that comprise this lifestyle pursuit.

2.3 Destination image

Despite the wide use in the empirical context, destination image does not have a solid conceptual structure and as such, its definition is still rather loose. This assertion is found in Gallarza, Gil and García (2002) who lament that there are almost as many definitions of image as scholars devoted to its conceptualization. For instance, tourism images are defined as an individual's overall perception or total set of impressions of a place or as the mental portrayal of a destination.

The concept of destination image can be better understood by looking at a proposition by Gallarza et al. (2002) who developed a comprehensive theoretical framework, defining image in terms of its four features: 1) complexity (it is not unequivocal), underlining an analytical dimension 2) multiplicity (in elements and processes), providing an action dimension 3) relativistic (subjective and generally comparative) translating destination image as a strategic tool and 4) dynamic (varying with time and space), allowing for tactical decisions based on destination image.

Interestingly, Dann (1996) suggests that destination image is made up of two distinct but hierarchically interrelated components: 1) cognitive and 2) evaluative. The cognitive component is viewed as the sum of beliefs and attitudes of an object, leading to some internally accepted picture of its attributes (external forces, pull attributes). On the other hand, the affective component of image is related to motives in the sense that it is how a person feels about the object under consideration (internal forces, push attributes). Emanating from a review of the existing literature, Beerli and Martin (Beerli & Martin, 2004) classified all attributes influencing image assessments into nine dimensions: 1) natural resources, 2) tourist leisure and recreation, 3) natural and marine environment, 4) general infrastructure, 5) culture, history and art, 6) social environment, 7) tourist infrastructure, 8) political and economic factors and 9) atmosphere of the place.

3 TOURISM CORE PRODUCTS, FOOD AND SABAH DESTINATION IMAGE

Sabah is one of the states in Malaysia which aggressively promoting its tourism sector. The diversity of landscapes comprising of marine, natural resources and culture attraction is making Sabah as a one of the important tourism spots in the region (Khaled, Sebastian & Elmar, 2013). Not to exaggerate that having a vast of interesting tourism products has increased Sabah as a destination among the local and international tourists. This is supported by positive growth of tourist receipts both from domestic and international from 2,300,428 in 2008 to 2,875,761 in 2012 and the figure is expected to continually flourish (Sabah Tourism Report, 2012).

Som, Marzuki, Yousefi and Abu Khalifeh (2012) contended a mixture of various resources and attraction like natural landscapes, uniqueness of diverse cultures, beautiful beaches, untouched flora and fauna not only provide Sabah as a relaxation, peaceful and recreational place but increase its tourism destination. Page and Connel (2006) explicitly noted that in order to attract more tourists, tourism destination should combine tourism products with the activities, experiences and other intangible aspects. In line with this, an excellent range of tourism products together with the natural resources activities, marine and cultural with the promotion and advertisements directly establishing Sabah's image as one of the world-class tourist destination (Sabah Tourism Report, 2012).

As tourists cannot leave their taste bud while travelling, Goeldner and Ritchie (2006) therefore argue food and accommodation cannot be ignored by tourists. Food, in fact, is considered as one of the most integral elements to tourists when they are on traveling. Food has become a 'focal point' for travel decision-making and the hallmarks attraction of a number of destinations around the world (Hall & Sharples, 2003). Numerous tourism regions, in fact, see foods as an essential part of the local heritage and that can attract tourists. In this sense, tourists or travelers in addition to traveling activities, will also encounter or consume the various types of food in the country or destination that they are visiting. In other word, during the period of their vacation, they will definitely experience the varieties of food and beverages, as well as learn about the food culture and heritage of that particular country. Lertputtarak (2012) strongly argue that food at the destination is the fundamental and essential for tourists' destination selections thus indubitable that food besides other main tourism products could create the destination image for that particular country.

With the above nation and in the case of Sabah, the international tourists who visited this state are assumed to have tasted or consumed the local food or delicacies in addition to experiencing the marine, natural resources and cultural events as the core tourism products. In line with this, some questions related to local food and the international tourists could be probed? Do Sabah food besides the marine, natural resources and cultural contribute to Sabah destination image and what are the levels of consumption of Sabah local food among the international tourists? In other words, what are moderating effects of local food through international tourists' level of consumption toward Sabah as destination image?

To date, there is still unclear understanding of how and to what extent the Sabah local food contributes it as a tourism destination as compared to other attributes. There has been in fact a very limited analysis looking at the relationship between foods and tourism destination (Scarpato, 2002) and none of those available studies looking at Malaysia perspectives. This highlighted issue is warrants empirical investigation and the finding yet to be revealed with the investigation is still in the process.

4 CONCLUSION

There is still unclear understanding of how and to what extent the local food of particular tourism destination besides it tourism core products contributes to tourism destination image as still limited analysis looking at it. It is hope understanding on the stipulated issue will create awareness and in the context of this study Sabah state government, tourism authority, hotel and food operators on the importance of local food to the state tourism development besides it tourism core products. In addition, besides strengthening its core tourism products, gastronomic products which represent all ethnics could be promoted and in the long run this human necessity product is gradually being recognized internationally and at the same time creating its image and continuously boost up the state economy.

REFERENCES

Akama, J.S. & Kieti, D. (2007). Tourism and socio-economic development in developing countries: A case study of Mombasa Resort in Kenya. *Journal of Sustainable Tourism, 15*(6), 735–748.

Albayrak, T., Aksoy, S. & Caber, M. (2013). The effect of environmental concern and scepticism on green purchase behaviour. *Marketing Intelligence & Planning, 31*(1), 27–39.

Ashworth, G.J. & Larkham, P.J. (2003). The convergence process in heritage tourism. *Annals of Tourism Research, 30*(4), 795–812.

Beerli, A. & Martin, J.D. (2004). Factors influencing destination image. *Annals of Tourism Research, 31*(3), 657–681.

Beerli, A. & Martín, J.D. (2004). Tourists' characteristics and the perceived image of tourist destinations: A quantitative analysis a case study of Lanzarote, Spain. *Tourism management, 25*(5), 623–636.

Bigné, J.E., Sanchez, M.I. & Sanchez, J. (2001). Tourism image, evaluation variables and after purchase behaviour: inter-relationship. *Tourism management, 22*(6), 607–616.

Boniface, P. (2003). *Tasting tourism: Travelling for food and drink*. Hampshire, England: Ashgate Publishing, Ltd.

Chai, L.T. (2011). Culture heritage tourism engineering at Penang: Complete the puzzle of "The Pearl Of Orient". *Systems Engineering Procedia, 1*, 358–364.

Dann, G. (1996). Tourist images of a destination: An alternative analysis. In D. Fesenmaier, J.T. O'Leary & M. Usysal (Eds.), *Recent advances in tourism marketing research* (pp. 45–55). New York: The Haworth Press.

Dolnicar, S. & Huybers, T. (2010). Different tourists: Different perceptions of different cities consequences for destination image measurement and strategic destination marketing. In J.A. Mazanec & K. Wöber (Eds.), *Analyzing international city tourism* (pp. 127–146). Vienna/New York: Springer.

Gallarza, M.G., Saura, I.G. & García, H.C. (2002). Destination image: Towards a conceptual framework. *Annals of Tourism Research, 29*(1), 56–78.

Goeldner, R. & Ritchie, J.R.B. (2006). *Tourism: principles, practices, philosophies* (10th ed.). Hoboken: Wiley & Sons Inc.

Hall, C.M. & Sharples, L. (2003). The consumption of experiences or the experience of consumption? An introduction to the tourism of taste. *Food tourism around the world: Development, management and markets*, 1–24.

Henderson, E. (2000). Rebuilding local food systems from the grassroots up. In F. Magdoff, F. Bellamy, J. & F.H. Buttel (Eds.), *Hungry for profit: The agribusiness threat to farmers, food, and the environment* (pp. 175–188). New York: Monthly Review Press.

Ibrahim, E.E. & Gill, J. (2005). A positioning strategy for a tourist destination, based on analysis of customers' perceptions and satisfactions. *Marketing Intelligence & Planning, 23*(2), 172–188.

Jalis, M.H., Zahari, M.S.M., Izzat, M. & Othman, Z. (2009). Western tourists perception of Malaysian gastronomic products. *Asian Social Science, 5*(1), 25–36.

Johns, N. & Kivelä, J. (2001). Perceptions of the first time restaurant customer. *Food Service Technology, 1*(1), 5–11.

Khaled, Y., Sebastian, E. & Elmar, G.P. (2013). *A decompilation framework for static analysis of binaries.* Paper presented at the 8th IEEE International Conference on Malicious and Unwanted Software, Fajardo, Puerto Rico, USA.

Kivelä, J. & Crotts, J.C. (2006). Tourism and gastronomy: Gastronomy's influence on how tourists experience a destination. *Journal of Hospitality & Tourism Research, 30*(3), 354–377.

Kivelä, J.J. & Chu, C.Y.H. (2001). Delivering quality service: Diagnosing favorable and unfavorable service encounters in restaurants. *Journal of Hospitality & Tourism Research, 25*(3), 251–271.

Komppula, R. (2001). *New-product development in tourism companies-case studies on nature-based activity operators.* Paper presented at the 10th Nordic Tourism Research Symposium, Vaasa, Finland.

Laws, E., Scott, N. & Parfitt, N. (2002). Synergies in destination image management: a case study and conceptualisation. *International Journal of Tourism Research, 4*(1), 39–55.

Lertputtarak, S. (2012). The Relationship between Destination Image, Food Image, and Revisiting Pattaya, Thailand. *International Journal of Business & Management, 7*(4).

Lin, C.-H., Morais, D.B., Kerstetter, D.L. & Hou, J.-S. (2007). Examining the role of cognitive and affective image in predicting choice across natural, developed, and theme-park destinations. *Journal of travel research, 46*(2), 183–194.

Long, L.M. (2004). *Culinary tourism.* Lexington: University Press of Kentucky.

Medlik, S. & Middleton, V.T. (1973). Product formulation in tourism. *Tourism and marketing, 13*, 173–201.

Murphy, P., Pritchard, M.P. & Smith, B. (2000). The destination product and its impact on traveller perceptions. *Tourism management, 21*(1), 43–52.

Page, S. & Connell, J. (2006). *Tourism: A modern synthesis* (2nd ed.). London: Cengage Learning EMEA.

Pot, C. (2005). *An evaluation of Spain's marketing campaign 'smile you are in Spain', through an image analysis in the Dutch market.* Ma European Tourism Management Bournemouth University, Spain.

Sabah Tourism Report. (2012). Attraction of Sabah.

Scarpato, R. (2002). Gastronomy as a tourist product: The perspective of gastronomy studies. *Tourism and gastronomy*, 51–70.

Som, A.P.M., Marzuki, A., Yousefi, M. & Abu Khalifeh, A.N. (2012). Factors Influencing Visitors' Revisit Behavioral Intentions: A Case Study of Sabah, Malaysia. *International Journal of Marketing Studies, 4*(4), 39–40.

Wolf, E. (2002). Culinary tourism: A tasty economic proposition. *Retrieved in June, 25,* 2007.

Theory and Practice in Hospitality and Tourism Research – Radzi et al. (Eds)
© *2015 Taylor & Francis Group, London, ISBN 978-1-138-02706-0*

Food and culture in tourism: Where does Penang Island stand?

D. Mohamad & S.I. Omar
Sustainable Tourism Research Cluster, Universiti Sains Malaysia, Penang, Malaysia

B. Mohamed
School of Housing, Building and Planning, Universiti Sains Malaysia, Penang, Malaysia

ABSTRACT: Tourism industry is complex and contested. As a sector that provides benefits in the form of economic development, tremendously and rapidly, the tourism sector has been long studied from various aspects, especially factors related to travel motivations. This paper centers on the relationship between gender and travel motivations to Penang Island, in relation to food and cultural aspects. As a field, both food and cultural tourism sectors are gaining in importance and in popularity. The findings reported by this paper were based on 801 international responses collected during the four-month data collection (August to November 2012). Results indicate that gender plays an unimportant role in impacting the travel motivations in addition insignificance of income in relation to preferences of food services propensity. Nevertheless, interesting results were observed for the travel motivations' variables, in terms of tourism experience and the image portrayed.

Keywords: International tourist, Penang Island, travel motivation, gender, food, cultural aspects

1   INTRODUCTION

Publicly known, tourism sector serves the purpose of travel, entertainment, education and business; where all these can be further encapsulated into expectation and perception. As the tourism sector grows and expands, tourism sectors that are highly depending on limited tourism resource are putting notable effort in brandishing their tourism products. While this suggests that much have been achieved, this also indirectly translated into much more needed to be done. It is therefore, the tourism operators are in urgent pursuit of providing tourism products that offer distinctiveness.

In line with this, the tourism industry envisages the potential holds by gastronomy in meeting the tourists' expectation and demand. Gastronomy or else known as a food tourism has been receiving attention from tourism stakeholders where, according to USA Today (2014), food tourism was first coined by the growing demand for local food and wine tourism that was first made famous back in 2001. Meanwhile, Kirshenblatt-Gimblett (2003) illustrates the relationship between food and tourism as follows:

Even when food is not the main focus of travel, one must eat, regardless of whether or not a memorable experience is the goal. Indeed, the tourism and hospitality industries design experiences, including culinary ones, within the constraints of the tourist's time, space and means. "culinary tourism is a space of contact and encounter, negotiation and transaction, whether at home or abroad" (para. 3)

Shenoy's (2005) finding demonstrates the positive relationship between gender and food tourism, of which, further enhances the credibility of food tourism in developing the tourism industry. Given its ability in promoting the local cuisine as one of the enjoyable tourism activities, food tourism has received the international recognition as it diversifies the tourism experience, dynamically and creatively (World Tourism Organization, 2012).

On the other hand, culture aspect has been employed as one of tourism products as early as the end of the 20th century (Organisation for Economic Co-Operation and Development, 2009), due to the demand and supply chain. Compare to food tourism, culture and tourism have been enjoying a long tradition as culture is a product that begins with a snapshot of people, their lifestyle and beliefs (United Nations Educational, Scientific and Cultural Organization (UNESCO) 2004). Simply stated, other than communicating with people, culture can be experienced through observation. Although one's culture may be have similar to one another, the art of experiencing and enjoying culture lies in the form of one's ability in interpreting the connectivity between their cultural resources and the host community's cultural resources (Failte Ireland, undated).

Rather than viewed separately, both food and cultural aspects should be studied in parallel as stated by Public Broadcasting Service (PBS) (2005, para. 1) who quoted gastronome Jean Anthelme Brillat-Savarin: "Tell me what you eat, and I'll tell you who you are". Other than well-known for its two bridges (13.5km and 24.0km, respectively) and Komtar building, the Penang Island has a rich resource both on food and culture. Against this background, the main objective of this paper is to find the role played by gender in influencing travel motivations as well as the connection between food, culture and tourism; if any.

2 RESEARCH METHOD

This paper is motivated by the Malaysia Government and the Penang State Government incentives in promoting the Visit Penang campaign. Prior to the actual data collection, a pilot study was conducted in order to test the data collection instrument reliability and validity. Results from 20 responses shows that the instrument is ready to be used, with minor correction to its sections paraphrasing as well as word replacement and reduction to avoid ambiguity and confusion. Data collection was done over a four-month timeframe starting from August to November 2012, where the self-administered survey targets the international tourists with the minimum age of 18.

Out of 930 questionnaires handed out, 801 questionnaires were retrieved back, which was amounting to 86 percent response rate. Out of 801 responses (Asia = 438, non-Asia = 363), it is learned that 719 respondents have received tertiary education and 55.3 percent was contributed by male respondents (amounting to 443 questionnaires), with 419 respondents aged between 18 and 54. This paper, however, did not impose the importance of male respondents over female respondents (333 respondents aged between 18 and 54).

3 FINDINGS

This paper employs a number of analysis techniques (cross-tabulation, correlation, regression and correspondence analysis) in order to study the relationship between gender and travel motivations to the Penang Island (food and culture). In relation to the food construct, the variables studied inclusive tour package inclusive food services, the Penang Island perceived local cuisine image, chances to experience the local cuisine and preferences of items bought. In the context of the culture construct, attention was given to the Penang Island perceived cultural image, historical uniqueness,

Table 1. Income and tour package (food service) relationship.

		Package include food & beverages	Income group
Package in-clude food & beverages	Pearson Correlation	1	-.081
	Sig. (2-tailed)		.514
	N	161	67
Income group	Pearson Correlation	-.081	1
	Sig. (2-tailed)	.514	
	N	67	351

cultural distinctiveness and chances to attend cultural performances. Previously mentioned, results from three analyses (correlation, regression and correspondence analysis) indicate that gender plays an unimportant role in impacting the travel motivations.

This paper runs a three-phase analysis where the first analysis is to find the existence of a relationship between variables (correlation analysis). Analysis between variables for each construct present significant relationships where the lowest value recorded by the Penang Island perceived cultural image and historical uniqueness (.072) and the highest value recorded by chances to attend cultural performances and the historical uniqueness (0.397), with both relationships (culture construct) were significant at .01 level. In the context of food construct, positive relationship is only recorded by chances to experience the local cuisine and the Penang Island perceived local cuisine image (0.246, significant at .01 level). This paper further learned that the tourists' tour package inclusive food services propensity is unrelated to income (see Table 1). Of interest, significant relationships were also detected in the following relationships; [1] tour package inclusive food services and the Penang Island perceived cultural image (.170) and [2] the Penang Island perceived local cuisine and cultural images (.090), with both relationship significant at .05 level.

The second analysis involves regression analysis where this serves the purpose of studying the impact of a variable on the remaining variables in addition to finding a sturdier connection between variables (due to the low significance level presented by the correlation analysis). Differing from the first phase, observation shows that only the food construct presents the significant relationships. From results, it is learned that the 'Penang Island perceived local cuisine image' variable plays a major influence on both 'tour package inclusive food services' and 'preferences of items bought' variables, where the relationship is significant at .018. Interestingly, an in-depth examination reveals that this relationship can be further improved by paying less attention to

Table 2. The Penang Island images and tour package (food services) relationship.

		Penang Island perceived culture image	Penang Island perceived local cuisine image	Tour package inclusive food services
Penang Island perceived culture image	Pearson	1	.090[*]	.170[*]
	Sig.		.011	.031
	N	800	800	161
Penang Island perceived local cuisine image	Pearson	.090[*]	1	.093
	Sig.	.011		.240
	N	800	800	161
Tour package Inclusive food services	Pearson	.170[*]	.093	1
	Sig.	.031	.240	
	N	161	161	161

'preferences of items bought' variable (a decrease contributes to an increase of 2.457 units for the relationship). Within this paper's interest, this could be translated into offering food-based tourism products that have a long life cycle as well as easy to be transported.

In correspondence analysis, this paper searches for the dependency level between each construct's variables and between constructs. Each construct reveals significant relationship except for the Penang Island perceived local cuisine image and tour package inclusive food services (0.237) and the Penang Island perceived cultural image and historical uniqueness (0.286). Despite the similar significant value (.000), only one relationship presents a total dependency (the Penang Island perceived cultural image and cultural distinctiveness), while the remaining dependency levels amounted below than 25 percent (except for the Penang Island perceived local cuisine image and chances to experience local cuisine = 70 percent). Additionally, results have brought attention to relationships present by the Penang Island perceived images and tour package inclusive food services variables. Despite the lower significant level (.031) compared to the Penang Island perceived local cuisine images (.011), the Penang Island perceived cultural image is observed to be more dependent on tour package inclusive food services variable (29 percent) (as supported by the correlation analysis, significant at .05) (see Table 2).

4 DISCUSSIONS AND CONCLUSIONS

Based on the above mentioned findings, it is learned that little impact on travelling to/visiting the Penang Island propensity has been contributed by the food construct. Although 'Penang food tourism' keyword shows as a highly research content via means of the Internet (77.6 million results) and according to Tourism Malaysia (2012), the

Penangites (and the other visitors) are artful and a pampered lot when it comes to food; limited number of respondents (56) placed priority and preference towards tour package that include food services. This finding translated into the inability of food construct serving as a travel motivation to the Penang Island. This is further strengthened by the fact that the well-famed Penang Island dishes (Ab Karim, K'ng, Othman, Ghazali & Abd. Halim 2012) could only offer a low degree impact as cultural landscape. In other words, the connection between the Penang Island and food tourism is yet to be recognized by the international tourists. In light of this situation, the Malaysian Exhibition Services Sdn Bhd (2014) is upholding the connection by organizing a conference on food, hotel and tourism that will be held in November 2014; in conjunction with promoting Penang as a place that offer the state-of-the-art business platform.

Within this paper knowledge, results in relation to 'preferences of items bought' variable is strongly described by the Tourism Malaysia's (2012) documentation on the Penang Island's famous dishes preferred. Here, it is clearly shown that dishes that are most enjoyed by visitors (for example Penang Laksa, Char Kuey Teow, Hokkien Mee and Nasi Kandar) exclusively come in the form of dine-in and/or takeaway, with limited life cycle and is unsuitable for air and/or long travel. Logically, the food products that are suitable for air travel (which also can serve as souvenirs purpose) should have a lower reaction to chemical and climate change (Natural Resources Defense Council, 2007). The food may come in the form of crackers, granola bar nibbles, cookies and muffins. Therefore, this finding shall play a major role in directing the food-based business players in improving and upgrading their products' quality. Simply stated, the products, more importantly, should meet the following requirement, but not limited to: long life cycle, ability to be repacked, easily transported convenient size and considered as a whole meal on its own. Other than bringing the famous dishes into the international market by means of powdered base/stock, attention can be allocated towards the Malaysia's dried dessert for example bahulu, sagun and maruku.

On the other hand, despite the positive relationships documented by the culture construct, variables exhibited a notably low dependency level (below 25 percent) except for the Penang Island perceived cultural image and cultural distinctiveness relationship (100 percent dependency level). The findings are resulted from the existence of multicultural society where parts of this society are living within the historical sites area and still conserving (directly and/or indirectly) the culture settings as documented by Lim (2011). More specifically, 'chances to attend cultural perform-ances' variable contributes to the culture con-struct's outstandingly low dependency level where both recorded at 7.7 percent (cultural distinctive-ness) and 2.1 percent (the Penang Island perceived local culture image), respectively. The cultural tourism of the Penang Island can be made well-known by taking advantage over the information and communication technology rapid develop-ment and the individual high dependency level on computers. One of the approaches undertaken to relate the Penang Island to cultural tourism is by designing a user-centered tourism website, as sug-gested by Goh (2008). Apart from taking persona attributes into consideration (see Goh, 2008), this paper suggests uploading the cultural-related vid-eos on social medium (for example, Facebook, Twitter, Instagram, Youtube and Dailymotion). More importantly, the uploaded videos are to be combined with in-trend videos. This action serves two purposes where first is to secure viewership and second is to promote peer-to-peer sharing. Additionally, this paper also proposes the tourism players to employ the spam technique in distribut-ing and spreading the cultural-related videos.

ACKNOWLEDGEMENT

The authors would like to extend their apprecia-tion to the following institutions that made this study and paper possible.

- Penang Global Tourism for granting the research grant called Penang International Travelers Sur-vey 2012 (Grant No. U527); and
- Universiti Sains Malaysia for granting the Research University Grant called Tourism Capacity and Impact Studies (Grant No. 1001/PTS/8660011).

REFERENCES

Ab Karim, M.S., K'ng, Y.W., Othman, M., Ghazali, H. & Abd. Halim, N. (2012). *Sustaining Penang street food culture and the reasons for its property*. Proceed-ings of the UMT 11th International Annual Sym-posium on Sustainability Science and Management, 9–11 July 2012, Terengganu, Malaysia, pp. 920–926. Retrieved from http://fullpaperumtas2012.umt.edu.my/files/2012/07/SSE24-ORAL-PP920-926.pdf

Failte Ireland. (undated). *Cultural tourism making it work for you: a new strategy for cultural tourism in Ireland*. Retrieved from http://www.aoifeonline.com/uplds/cultural-tourism.pdf

Goh, C.H. (2008). Developing a Penang cultural tourism website prototype: a user-centered approach. *Wacana Seni Journal of Arts Discourse, 7*, 91–109.

Kirshenblatt-Gimblett, B. (2003). Foreword. In L.M. Long (Ed.), *Culinary tourism*. Kentucky: The Uni-versity Press of Kentucky. Retrieved from http://books.google.com.my/books?id=SrQo2-qXO9oC&printsec=frontcover&dq=isbn:0813143780&hl=en&sa=X&ei=oHVCU4TsAovArAfCiYHIDw&redir_esc=y#v=onepage&q&f=false

Lim, T.C. (2011). Culture heritage tourism engineering at Penang: complete the puzzle of "The Pearl of Orient'. *System Engineering Procedia, 1*, 358–364.

Malaysian Exhibition Services Sdn. Bhd. (2014). *Food & Hotel Penang 2014*. Retrieved from http://www.foodandhotelpenang.com/images/FHP2014.pdf

Natural Resources Defense Council (2007). *Food miles: how far your food travels has serious consequences for your health and the climate*. Retrieved from http://food-hub.org/files/resources/Food%20Miles.pdf

Organisation for Economic Co-Operation and Devel-opment (OECD) (2009). *The impact of culture on tourism*. Retrieved from http://www.tava.gov.lv/sites/tava.gov.lv/files/dokumenti/petijumi/OECD_Tour-ism_Culture.pdf

Public Broadcasting Service. (PBS) (2005). *Food and culture*. Retrieved from http://www.pbs.org/opb/meaningoffood/food_and_culture/

Shenoy, S.S. (2005). *Food tourism and the culinary tour-ist* (Doctoral dissertation, Clemson University). Retrieved from http://www.clemson.edu/centers-insti-tutes/tourism/documents/Shenoy2005.pdf

Tourism Malaysia. (2012). *Penang best 12 foods: must eat*. Retrieved from http://www.tourismmalaysia.or.jp/gourmet/area-etc-penang.pdf

United Nations Educational, Scientific and Cultural Organization. (UNESCO). (2004). *IMPACT: The effects of tourism on culture and the environment in Asia and the Pacific: tourism and heritage site man-agement in Luang Prabang, Lao PDR*. Bnagkok, Thailand: UNESCO. Retrieved from http://www2.unescobkk.org/elib/publications/IMPACT_Luang-Prabang/impact.pdf

USA Today. (2014). *What is culinary tourism?* Retrieved from http://traveltips.usatoday.com/culinary-tourism-1910.html.

World Tourism Organization. (UNWTO). (2012). *Global report on food tourism*. Madrid: UNWTO.

Theory and Practice in Hospitality and Tourism Research – Radzi et al. (Eds)
© 2015 Taylor & Francis Group, London, ISBN 978-1-138-02706-0

Sociocultural factors, female students body internalization, food choices and eating patterns

M.N.A. Akbarruddin, M.S.M. Zahari & Z. Othman
Faculty of Hotel and Tourism Management, Universiti Teknologi MARA, Shah Alam, Malaysia

ABSTRACT: Body internalization is the process by which an individual accepts or internalizes the image of an ideal body figure captured from various sources namely socio-cultural factors like family, peers and media thus applies it to the individual self. Body internalization along with some other factors such awareness of thin ideal and perceived pressures to be thin are significantly related to body image dissatisfaction. In addition, body image dissatisfaction can also be strong predictors of eating disorder. This paper reviewing and discusses on the socioculture factors that associate with female students body internalization, the effect of it on food choices and the eating pattern and relate the specific issue related to Malaysia context which is under investigtion

Keywords: Sociocultural, body internalization, food choices, eating pattern

1 INTRODUCTION

Body image is a construct with various elements which mirror how a person thinks, sees, acts and feels about his outward body appearance (Humenikova & Gates, 2008). Body image is a multidimensional effect includes a person's sensations, awareness, thoughts, feelings, judgments and behavior of her body. Similarly, body image satisfaction and dissatisfaction could also be categorized as affective elements of body image since it point out of the way that people thought and feel about themselves (Davidson & McCabe, 2006). In addition, body image could be either a trait in which it is stable and applicable in any number of circumstances or a state where it can fluctuate due to the context or a mood with most of the researches are focusing on the trait aspect and its contributors (Colautti, McCabe, Skouteris, Wyett, Fuller-Tyszkiewicz & Blackburn, 2011). Instead of having a negative impact toward psychosocial and health of a person's body image dissatisfaction can also be a strong predictor of eating disorder (Stice, 2002). Considerable evidences found that body dissatisfaction as a predictor of the development of eating pathology (Shroff & Thompson, 2006).

Besides body dissatisfaction, research also shown that the sociocultural model helps to explain disordered eating in which family, peers, and the media provides pressure for an individual to reach a certain ideal body image (Stice, 1994). Thompson, van den Berg, Roehrig, Guarda, and Heindberg (2004) states that when women internalize these pressures negative consequences can occur. It appears that in Western countries which have high socioeconomic levels, the body image dissatisfaction phenomenon among girls and young women seem to become a typical matter (McCabe, Ricciardelli & Karantzas, 2010; Halliwell, 2013). Because of that, Lai (2000) posit that body dissatisfaction is an occurrence which only applies to Westerners. Furthermore, there have been cross-cultural studies which indicate higher level of eating pathology and body image dissatisfaction occurrence in a Western context compare to non-Western societies (Mahmud & Crittenden, 2007). The study also found that there is significant culturally related variation in what constitutes as perfect body appearance. Some non-Western societies ordinarily regard heavier body size as the perfect figure while Western respondents favor thin and slimmer figure (Brewis & McGarvey, 2000).

However, these presumptions have gradationally been disputed among scholars. Some have dismissed the notion that body dissatisfaction and eating pathology as exclusively Western phenomenon (Gordon, 2001). Lee and Lee (2000) contended, as non-Westerners are facing fast and rapid growth in term of socioeconomic, the pressures exerted by socio-cultural factors such as exposure to mainstream media and peers that advanced the notion of thin ideal body would eventually led to rise in levels of body dissatisfaction.

2 LITERATURE

2.1 Sociocultural factors

In sociocultural theory, human behavior and learning may be explained by both society and culture (Berger, 2004). Depending on the specific society in which one lives in, there is a culture within that helps individuals mold their behavior and learn what is right and what is wrong.

Sociocultural theory, as it relates to disorder eating and body image dissatisfaction denotes that Western society posits unrealistic expectations for attractiveness that most individuals do not possess. Thus, these individuals endure body dissatisfaction which may eventually lead to behaviors to alter the physical. Some of these methods may include chronic dieting and binge eating, which typically lead to disordered eating (Halliwell& Harvey, 2006).

To better understand the function of socio cultural influences that might lead to disordered eating, a model was developed by Stice (1994). This model explains the development of eating disorders among young women. In addition, this model has been found to support adolescent male disordered eating and body dissatisfaction (Smolak, Levine & Thompson, 2001). According Stice's model, there are diverse factors that contribute to the pressure related to appearance, namely the media, family and peers. The media has been suggested to be one of the most frequent sociocultural factors contribute to disordered eating and body image dissatisfaction among individuals, especially adolescents followed by family and peer influences (Dittmar, 2009). These factors may contribute to the problem of disorder eating through the internalization of body dissatisfaction and cultural ideals (Stice, 1994). Finally, these factors have been individually studied in relation to body image dissatisfaction and disordered eating. However, there are still researches combining these three socio cultural factors (McCabe & Ricciardelli, 2001).

2.2 Peer influence

Since most individuals live in a connected society, more often than not they will follow certain norms that are accustomed to that particular social group. Peer influence is a crucial part in shaping and developing identity and personality of a person and because of that the development of ideal body image may be attributed to social pressure and other personal life events. Peer influence on social, personality and behavior development start to receive attention among researchers (Stice & Whitenton, 2002; Schutz & Paxton, 2007).

Stice and Whitenton (2002) examined whether an increase in body dissatisfaction overtime was predicted by early menstrual occurrence, rise in body mass, perception of pressure to be thin and deficient social support namely peer influences. The study found that adolescents' perception of pressure to be thin, the ideal body image internalization and lacks in societal support, weight-related teasing and depression like symptoms would able to foretell elevated body dissatisfaction. Another predictive factor that increased body dissatisfaction was the pressure to be thin.

Griffiths and McCabe (2000) examined the effect of peer influence on body image dissatisfaction and disordered eating among young female adolescents. The relationships with others (peers) or the impact of significant others may predict body dissatisfaction. It appears that girls use reference points with their friends in regards to body dissatisfaction and might use cultural standards in terms of their ideal body image and might endure body dissatisfaction. In addition, individuals with increased body dissatisfaction tended to have stricter food restriction and/ or bulimic eating.

2.3 Media influence

The media also plays an important role in the development of disordered eating and body image dissatisfaction. Levine & Harrison (2004) revealed there is positive relationship between the exposure to media which depicted an idealized image of a perfect body figure and the emergence of body image dissatisfaction. Durkin & Paxton (2002) posited that media images are processed in people's mind coupled with pre-existing individual characteristic could moderate and/or mediate the effect of media exposure in the development of negative body image. In investigating the media influenced on adolescent girls and boys, Lawrie, Sullivan, Davies & Hill (2006) revealed that media strongly influenced young adolescent girls to be slimmer. However, in gaining muscle both adolescents were having the similar view on media influence.

2.4 Parent influence

It is widely accepted that family has a major role in shaping and influencing the way a person behave and the personality that they carried within. The fact that the family is the first group of people that a person encounter in his life could explain a lot of variance in why people eat, what they eat, speak the language that they speak and behave the way they do. Parents influences have also proven to be particularly relevant in the process of body dissatisfaction formation (Stice, 1994; McCabe & Ricciardelli, 2003) and their impact on the development of body image concerns has been verified in many cross-sectional studies and a few prospective and experimental investigations (McCabe& Ricciardelli, 2001; Agras, Bryson, Hammer & Kraemer, 2007; Rodgers, Paxton & Chabrol, 2009). The most important factor in the contribution of disordered eating was also a family and mothers appeared to be more influential than the fathers in communication about dieting (Wertheim, 2002).

2.5 Body internalization

In sociology, internalization refers to the process of acceptance of a set of norms and values established by people or groups which are influential to the individual through the process of socialization. It involves the integration of attitudes, values, standards and the opinions of others into one's own identity or sense of self. Dittmar & Howard (2004) state that body internalization refers to the process by which an individual accepts or internalizes the image of an ideal body figure captured from various sources namely socio-cultural factors and applies it to herself. Body internalization along with some other factors such awareness of thin ideal and perceived pressures to be thin are significantly related to body image dissatisfaction (Thompson & Stice, 2001).

Body internalization was found strongly related to body image dissatisfaction. The study found that when women were made aware of the subtle body image messages in advertising, they were likely to be dissatisfied with their bodies (Cusumano & Thompson, 1997; 2001). A meta-analysis name three constructs that have been linked to the development of body image dissatisfaction: awareness of the thin ideal, internalization of the thin ideal, and perceived pressures to achieve this ideal (Thompson & Stice, 2001). Out of these, internalization and perceived pressure to be thin were found to be the most strongly related to body image dissatisfaction and this was true for all ages and cultural groups.

Internalization of the thin beauty ideal also has been found to be a mediating factor between exposure to media images and the development of body image dissatisfaction (Stice, Schupak-Neuberg, Shaw, and Stein, 1994). Comparing of the three factors, exposure to media, awareness of media and internalization of media, the later one (internalization) was the most closely linked to body image dissatisfaction (Cusumano and Thompson, 1997).

2.6 Food choice and eating pattern

Food choice is a complex decision that is affected by the interaction between food, people, and the environment. Furst, Conners, Bisogni, Sobal & Falk (1996) developed a conceptual model to describe the food choice process which involved life course, influences (ideals, personal factors, resources, social

framework and food context), personal system (value negotiations, sensory perceptions, quality, managing relationships, monetary considerations, convenience, health and nutrition), and strategies.

According to Shepherd (1989), food choice factors can be classified into three main groups. The first includes product-related factors, which comprise the intrinsic (physical and chemical properties of foods, sensory aspects, nutritional content, etc.) and extrinsic properties (packaging, convenience, price, brand, labels, etc.) of a product. The second is consumer-related factors which consist of demographic, psychographic, psychological and physiological factors. The third group is environmental context which includes economic and social factors.

Eating patterns on the hand can be regarded as the occasions of eating and the context of eating occasions (de Castro, 2009). In other word, eating pattern can be seen as when, how much and what foods are eaten by certain individuals. It was discovered that eating patterns could be connected with energy intake and body weight (Keim, Van Loan, Horn, Barbieri & Mayclin, 1997).

3 RELEVANT ISSUES IN MALAYSIA CONTEXT

Malaysia is still regarded as experiencing nutritional and lifestyle shift, inadvertently unable to escape from experiencing rapidly increasing rate of obesity in conjunction with a micro—nutrient deficit (Ismail, Chee, Nawawi, Yusoff, Lim & James, 2002; Khor, 2005). Researcher postulated that these factors in combination with the ubiquity of Western ideals of attractiveness have resulted in more negative body image among Malaysian women (Swami, 2006). This notion has been supported through studies which have informed the existence of body dissatisfaction in a huge percentage of pre-pubescent girls, adolescents and adults (Shariff & Yasin, 2005; Leong, Poh & Ng, 2004; Dev, Permal & Fauzee, 2008 and Fatimah, Idris, Romzi & Fauziah, 1995).

Besides the aforementioned age categories, Erol, Toprak, and Yazici (2006) pointed that eating disorder behavior is also prevalent among female university students compared to adolescent girls. It is also noticeable that female students carry higher risk of experiencing high level of depression, anxiety and stress due to biopsychosocial factors such as social roles and body image concerns (Zaid & Chan, 2007). Such concern is also proven to influence adolescents' food choice and eating patterns (Neumark-Sztainer, Story & Perry, 1999). Although there are studies on body dissatisfaction conducted in Malaysia (Shariff & Yasin, 2005; McDowell & Bond, 2006; Mellor, McCabe, Ricciardelli, Yeow,

Daliza & Hapidzal, 2009) to what extent sociocultural factors influence body dissatisfaction and the effect of it on food choices and the eating pattern among Malaysian female university students are not known since little investigation undertaken to date (Gan, Mohd Nasir, Zalilah & Hazizi, 2011). With that, the issue related to it is still under investigation and yet to reveal the answers.

4 CONCLUSION

It is hoped that reviewing and investigating the relationship between sociocultural factors, body internalization and the effect of it on food choices and the eating pattern would explain further variation in the body dissatisfaction and strengthening the current knowledge related to food choice and eating pattern particularly among female students. In addition, information generated could be useful to related practitioners in refining the existing diagnostic and treatment.

REFERENCES

Agras, W.S., Bryson, S., Hammer, L.D. & Kraemer, H.C. (2007). Childhood risk factors for thin body preoccupation and social pressure to be thin. *Journal of the American Academy of Child & Adolescent Psychiatry*, 46(2), 171–178.

Berger, K.S. (2004). *The developing person through the life span (6 ed.)*. New York: Worth Publishers.

Brewis, A.A. & McGarvey, S.T. (2000). Body image, body size, and Samoan ecological and individual modernization. *Ecology of Food and Nutrition*, 39(2), 105–120.

Colautti, L.A., Fuller-Tuszkiewicz, M., Skouteris, H., McCabe, M., Blackburn, S. & Wyett, E. (2011). Accounting for fluctuation in body dissatisfaction. *Body Image*, 8(4), 315–321.

Cusumano, D.L. & Thompson, J.K. (1997). Body image and body shape ideals in magazines: Exposure, awareness, and internalization. *Sex Roles*. 37. 701–721.

Davidson, T.E. & McCabe, M.P. (2006). Adolescent body image and psychosocial functioning. Journal of Social Psychology, 146, 15–30.

Dev, R.D. O, Permal, V. & Fauzee, M.S.O. (2009). Rural–urban differences in body image perception, body mass index, and dieting behavior among Malay adolescent Malaysian schoolgirls. *European Journal of Scientific Research*, 34, 69–82.

Dittmar, H. (2009). How do "body perfect" ideals in the media have a negative impact on body image and behaviors? Factors and processes related to self and identity. *Journal of Social and Clinical Psychology*, 28(1), 1–8.

Dittmar, H. & Howard, S. (2004). Thin-ideal internalization and social comparison tendency as moderators of media models' impact on women's body-focused anxiety. *Journal of Social and Clinical Psychology*, 23(6), 768–791.

De Castro, J.M. (2009). When, how much and what foods are eaten related to total daily food intake. *British Journal of Nutrition, 102(8)*, 1228–1237.

Erol, A., Toprak, G. & Yazici, F. (2006). Psychological and physical correlates of disordered eating in male and female Turkish college students. *Psychiatry and Clinical neurosciences*, 60(5), 551–557.

Fatimah, A., Idris, M.N.M., Romzi, M.A. & Faizah, H. (1995). Perception of body weight status among office workers in two government departments in Kuala Lumpur. *Malaysian Journal of Nutrition*, 1, 11–19.

Furst, T., Connors, M., Bisogni, C.A., Sobal, J. & Falk, L.W. (1996). Food choice: a conceptual model of the process. *Appetite*, 26(3), 247–266.

Gan, W.Y., Mohd, N.M., Zalilah, M.S. & Hazizi, A.S. (2011). Differences in eating behaviours, dietary intake and body weight status between male and female Malaysian university students. *Mal J Nutr*, 17(2), 213–228.

Gordon, R.A. (2001). Eating disorders East and West: A culture-bound syndrome unbound. In M. Nasser, M.A.Kazman & R.A. Gordon (Eds.), *Eating disorders and cultures in transition* (1–16). New York: Brunner-Routledge.

Halliwell, E. (2013). The impact of thin idealized media images on body satisfaction: Does body appreciation protect women from negative effects? *Body image*, 10(4), 509–514.

Halliwell, E. & Harvey, M. (2006). Examination of a sociocultural model of disordered eating among male and female adolescents. *British Journal of Health Psychology*, 11(2), 235–248.

Humenikova, L. & Gates, G.E. (2008). Body image perceptions in Western and post-communist countries: A cross-cultural pilot study of children and parents. *Maternal & Child Nutrition*, 4, 220–231.

Ismail, M.N., Chee, S.S., Nawawi, H., Yusoff, K., Lim, T.O. & James, W.P.T. (2002). Obesity in Malaysia. *Obesity Reviews*, 3, 203–208.

Keim, N.L., Van Loan, M.D., Horn, W.F., Barbieri, T.F. & Mayclin, P.L. (1997). Weight loss is greater with consumption of large morning meals and fat-free mass is preserved with large evening meals in women on a controlled weight reduction regimen. *Journal of Nutrition, 127*, 75–82.

Khor, G.L. (2005). Micronutrient status and intervention programs in Malaysia. *Food and Nutrition Bulletin*, 26, S281–285.

Lee, S. & Lee, A.M. (2000). Disordered eating in three communities of China: a comparative study of female high school students in Hong Kong, Shenzhen, and rural Hunan. *International Journal of Eating Disorders*, 27(3), 317–327.

Leong, S.H., Poh, B.K. & Ng, L.O. (2004). Study on body image perception, weight loss behaviour and practices among adolescents girls. *Malaysian Journal of Nutrition*, 10, 72.

Levine, M.P. & Harrison, K. (2004). Media's Role in the Perpetuation and Prevention of Negative Body Image and Disordered Eating.

Mahmud, N. & Crittenden, N. (2007). A comparative study of body image of Australian and Pakistani young females. *British Journal of Psychology*, 98(2), 187–197.

McCabe, M.P., Ricciardelli, L.A. & Karantzas, G. (2010). Impact of a healthy body image program among adolescent boys on body image, negative affect, and body change strategies. *Body image, 7*(2), 117–123.

McCabe, M. & Ricciardelli, L. (2001). Parent, peer and media influences on body image and strategies to both increase and decrease body size among adolescent boys and girls. *Adolescence, 36*(142), 225–240.

Mellor, D., McCabe, M., Ricciardelli, L., Yeow, J., Daliza, N. & Hapidzal, N.F.M. (2009). Sociocultural influences on body dissatisfaction and body change behaviors among Malaysian adolescents. *Body Image, 6*, 121–128.

Neumark-Sztainer, D., Story, M., Perry, C. & Casey, M.A. (1999). Factors influencing food choices of adolescents: findings from focus-group discussions with adolescents. *Journal of the American Dietetic Association, 99*(8), 929–937.

Rodgers, R.F., Paxton, S.J. & Chabrol, H. (2009). Effects of parental comments on body dissatisfaction and eating disturbance in young adults. A sociocultural model. *Body Image, 6*(3), 171–177.

Schutz, H.K. & Paxton, S.J. (2007). Friendship quality, body dissatisfaction, dieting and disordered eating in adolescent girls. *British Journal of Clinical Psychology, 46*(1), 67–83.

Sekaran, U. & Bougie, R. (2010). *Research Methods for Business: A Skill Building Approach.* John Wiley & Sons.

Shariff, Z.M. & Yasin, Z.M. (2005). Correlates of children's Eating Attitude Test scores among primary school children. *Perceptual and Motor Skills, 100*, 463–472.

Shepherd, R. (1989). Factors influencing food preferences and choice. *Handbook of the psychophysiology of human eating*, 3–24.

Shroff, H. & Thompson, J.K. (2006). The tripartite influence model of body image and eating disturbance: A replication with adolescent girls. *Body Image,3*(1), 17–23.

Smolak, L., Levine, M.P. & Thompson, J.K. (2001). The use of the Sociocultural Attitudes Towards Appearance Questionnaire with middle school boys and girls. *International Journal of Eating Disorders, 29*(2), 216–223.

Stice, E. (1994). Review of the evidence for a sociocultural model of bulimia nervosa and an exploration of the mechanisms of action. *Clinical Psychology Review, 14*, 633–661.

Stice, E. (2002). Risk and maintenance factors for eating pathology: A meta-analytic review. Psychological Bulletin, 128, 825–848.

Stice, E. & Whitenton, K. (2002). Risk factors for body dissatisfaction in adolescent girls: a longitudinal investigation. *Developmental psychology, 38*(5), 669.

Stice, E., Schupak-Neuberg, E., Shaw, H.E. & Stein, R.I. (1994). Relation of media exposure to eating disorder symptomatology: an examination of mediating mechanisms. *Journal of abnormal psychology, 103*(4), 836.

Thompson, J.K., van den Berg, P., Roehrig, M., Guarda, A.S. & Heinberg, L.J. (2004). The Sociocultural Attitudes Towards Appearance Scale-3 (SATAQ-3): Development and validation. *International Journal of Eating Disorders, 35*, 293–304.

Wertheim, E.H. (2002). Parent influences in the transmission of eating and weight related values and behaviors. *Eating Disorders, 10*(4), 321–334.

Theory and Practice in Hospitality and Tourism Research – Radzi et al. (Eds)
© *2015 Taylor & Francis Group, London, ISBN 978-1-138-02706-0*

The effect of displacement on indigenous tribes' socio-culture and food practices

N.M. Shahril & M.N. Syuhirdy
Faculty of Hotel and Tourism Management, Universiti Teknologi MARA, Penang, Malaysia

M.S.M. Zahari & A.H. Hamizad
Faculty of Hotel and Tourism Management, Universiti Teknologi MARA, Shah Alam, Malaysia

ABSTRACT: Displacement inevitably causes many changes to the indigenous tribes. In the new places or new settlement, tribes without doubt encountering many facets of their life, including the economy, education, lifestyles, belief, religion and many others. However, the available studies on Malaysia Orang Asli as indigenous tribes were mostly focusing on the socio-economy, such as household income and the compensation value paid after the acquisition of the land by the authority with little looking at socio-culture and traditional food practices. This paper reports the empirical findings of displacement effects on Orang Asli socio-culture and traditional food practices. Displacement benefits the indigenous tribes and positively improved their socio-culture in term economy, knowledge and understanding through education. Displacement also has given the impact on Orang Asli food practices. These positive indications have given significant implication not only to the indigenous groups itself, but also to the responsible authorities.

Keywords: Displacement, indigenous, socio-culture, food practices

1   INTRODUCTION

Orang Asli are the peninsular Malaysia indigenous tribes' which accounted approximately nineteen culturally and linguistically distinct groups representing 0.6 percent of the total National population. Among all, Semai, Temiar, Jakun (Orang Hulu), and Temuan are the largest groups of Orang Asli scattered throughout the peninsular (AITPN, 2008). For long period or to be exact before the Independence Day, Orang Asli has been living in the remote forest area isolated from civilization and development process with nomadic behavior. However, to ensure that they are not left behind from the waves of modernization and benefits from modern development process the displacement scheme was introduced by the government since late sixties (Nicholas, 2000).

According to AITPN (2008), displacements of Orang Asli were occurred since 1948 with three major reasons. The first reason was to curb the spreading of communist ideology. Orang Asli were displaced in a new designated place by the British. Pengkalan Hulu in Perak and Gua Musang in Kelantan are the examples of the earliest displacement area established during that pre independent period. The second reasons were to control the deforestation through a nomadic behavior through swidden farming (Nicholas, 2000). In this sense, the swidden farming is an unplanned process whereby the Orang Asli opens and closes forest area for agricultural purposes that destroyed forest area, flora and fauna. Third reasons relate to the planned infrastructure development projects such as electric hydro and dam construction (Kenyir Dam in Terengganu), highway (North and South highway PLUS) and land acquisition projects (FELDA and FELCRA) (Gomes, 2004).

Akpanudoedehe (2010) argued that displacement inevitably causes social cultural changes to the indigenous peoples or tribes. In the new places or new settlement, tribes are without doubt encountering and adjusting many facets of their life, including the economy, education, lifestyles, belief, religion and many others (Woube, 2005). Jamal (1996) noted that indigenous tribes have to adapt and to blend in with the locality or new peoples surrounding them. With this, it is argued that Orang Asli through displacements program are believed to have altered their socio-culture elements like domestic economy, household income, education, religious belief, cultural including food practices. Nevertheless, the available studies on Orang Asli were mostly focusing on the socio-economy such as household income and the compensation value paid after the acquisition of the land by the authority (Akpanudoedehe, 2010; Gomes, 2004). Not to exaggerate that there is no available study specifically looking at the impact of displacement on Orang Asli traditional food practice. In other words, to what extent the displacement program besides the economy, lifestyle and others altered or influence the Orang Asli traditional food practices are yet not discovered. This is also related to the type of the food prepared, method of cooking, utensils, ingredients and eating decorum. This study therefore aims to reveal such issues and hypothesizes that;

H1: There is a significant relationship between the effect of indigenous tribe social culture and food practices.

2 LITERATURE REVIEW

2.1 *Displacement and food practices*

Indigenous peoples' food practices contain treasures of knowledge from long-evolved cultures and patterns of living in local ecosystems ein (Kuhnlein, Erasmus & Spigelski, 2009). The dimensions of nature and culture that define food practices of an indigenous culture contribute to the whole health picture of the individual and the community not only on the physical health, but also the emotional, mental and spiritual aspects of health, healing and protection from disease (Erasmus, 2009; Spigelski, 2009). Indigenous people are those who retain knowledge of the land and food resources rooted in historical continuity within their region of residence. The local food practices are defined as "traditional food practices" which invariably include some foods that may be used by many outside of the indigenous culture (Kuhnlein, et al., 2009). Erasmus (2009) and Spigelski (2009) state that in the context of indigenous peoples "traditional foods" are referring to food that can be accessed locally, without having to purchase them, and within traditional knowledge and the natural environment from farming or wild harvesting.

Martinez (1998) noted food like living things that are sensitive or evolve with changes, alteration and modernization, urbanization are found to have a major impact on it and strongly link to social change (Cwiertka, 2000). With that, some behavioral shifts are occurring in the preparation and consumption of food, including the traditional ones and this largely associated with modernization. Modernization in fact not only effects on daily food practices but also in traditional events, celebrations as well as ceremonies (Gillette, 1997). Modernization is also a structural change that involves technology development and adoption representing materials (Inkeles & Smith, 1974) The production, processing, distribution of food were extensively shaped by modernization (Sobal, 1999). Jussaume (2001) posited that the modernization of food production, distribution as well as consumption is encouraged by business and policy makers. Food changes are also influenced by the new trends in the consumption which Miele (1999) described as the emergence of a post-modern circuit of food and the rise of a new culture of consumption among the consumers. The convenience concept of food which comprises of three components like time, physical energy and mental energy significantly contributes to the alteration or changes (Buckley, Cowan & McCarthy, 2007).

Berry (2000) argued that as the world move, a slight change occurs on food ingredients, methods of preparation, cooking and eating decorum. Similar goes into the cooking equipment's and methods of cooking. If in the medieval days, people used equipment's made of clay, metals and ceramic in the preparation of festival customary food however were gradually modified along with human civilization. A range of stoves, ovens with the gas burners and electric coils, equipment's of stainless steel are used to roast, bake, poach, simmer and fry are among the example (Bakalian, 1993). In sum, the changes of food practices intricately related to the complexities of social and economic circumstances through the force of globalization. Kuhnlein et al. (2009) states that with increasingly more of the produced, processed food and marketed on a global scale the less practice of the local food traditions among the urban society and the indigenous are apparent.

3 METHODOLOGY

A descriptive research design using a quantitative approach through a cross sectional study was applied to a self-reported and self-administered questionnaire. As this study will establish a fundamental basis for other similar future studies, two tribes of Orang Asli (Kensui and Kintaq) in Hulu Perak resettlement area were chosen as a sample population. Based on the data from the Department of Orang Asli Affairs (JHEOA), there are around 200 families with the population of 600 Orang Asli of both tribes resettled in two displacement areas. As the majority of the Orang Asli did not understand English well the original questionnaire (English version) was first translated into Bahasa Malaysia (target language) by a senior English lecturer at one of the Malaysian public universities who is also proficient in Bahasa Malaysia.

Before the actual survey, permission from the Department of Orang Asli Affairs (JHEOA), Hulu

Perak and the tribe leaders were first obtained. With five research assistants, the survey was conducted on 4 weekends (8 days). Given the fact that the researcher and research assistants had direct access to the respondents after a series of community programs and with the help of the tribe leaders and officer from JHEOA, the survey was successfully conducted. In light of the positive feedbacks and the absence of any obvious problem with either the instrument or the process, good responses with a total of 150 questionnaires were collected.

4 ANALYSIS AND RESULT

Based on frequency test, 80 percent (n = 120) of respondents were males compared to 20 percent (n = 30) females. 20 percent (n = 30) were above 60 years old while 53.3 percent (n = 80) was in the range of 40–59 years old (n = 20) and 26.6 percent were between 30–39 years old. 60 percent (n = 90) were farmers or self employed working at the vegetables farm, fruit orchard or selling forestry products and 40 percent (n = 60) were working in mining and refinery factory. 54.7 percent (n = 82) of the was displaced since 1990 and 45.3 percent (n = 68) since 1992.

4.1 Displacement effects on Orang Asli socio-culture

Looking at the displacement effect on Orang Asli socio-culture, descriptive statistic revealed that the majority of respondents agreed that displacement have changed their lifestyle, educational level and their standard of living. Displacement well improved their economy by providing many economic activities, creating job opportunities and generate more incomes. On top of that, displacement provides better religious practices and does not ruin their inherited custom and traditional ritual practices. It is interesting to note displacement give them a good environment for living, healthy lifestyle with the better planned residential layout. A proximity which refers to the distance between resettlement areas to other places also contributed to changes in social-culture. In this sense, they believe that closed distance between the displacement area and town promote better economic activities and offers vast job opportunities. Closed distances of housing in the neighborhood between each other's encourage the social and religious activities.

4.2 Displacement effects on Orang Asli food practices

The magnitude of mean score between 3.33 and 4.02 indicates that the majority of respondents

Table 1. Mean score for displacement effects on Orang Asli socio-culture.

	Items	Mean	S.D
SC1	Displacement changed my lifestyle than before	3.96	.732
SC2	Displacement has changed my educational level than before	4.17	.823
SC3	Displacement improved my standard of living.	3.98	.737
SC4	Displacement provides many economic activities (shop, selling agricultural product)	3.53	.564
SC5	Many job opportunities available in my new displacement area.	3.89	.677
SC6	Displacement enables me to earn more income.	3.73	1.246
SC7	Religion worshiping is better practicing in the new displacement area.	3.92	1.223
SC8	Many religions have been practicing since the displacement	4.24	.702
SC9	Displacement does not ruin inherit custom and traditional ritual practicing	4.33	.650
SC10	Displacement not much ruined the inherit festive	4.34	.703
SC11	Displacement provides good environment and health.	4.12	.694
SC12	Displacement provides a comfortable life with well planned residential layout.	4.27	.644
SC13	The closed distance between the displacement area and town promote better economic activity.	3.65	1.165
SC14	The closest distance between the displacement area and the town offers vast job opportunities.	3.87	1.154
SC15	Closed neighbourhoods encourage social and religious activity.	3.72	1.221

Scale: 1 = strongly disagree, 2 = disagree, 3 = slightly agree, 4 = agree, 5 = strongly agree

agreed with most items. As such the improvement of lifestyles has slightly altering Kintaq and Kensui tribes' knowledge and food practices. They believed that new lifestyles and educational levels improved their food knowledge and cooking practices. Majority of Orang Asli believed that displacements not only transformed their religious beliefs, but teaches them on the new understanding of permissible food knowledge and ways of preparing food especially among the Muslim. The improvement of the knowledge of food also introduced them to a new variety of food ingredients, understanding of the usage of ingredients and their usage of modern utensils.

4.3 The relationship between socio-culture and food practices

A single-step multiple regression was conducted to test the socio-culture variables as predictor against the criterion variable relates to food practices (see Table 3).

Looking at the Table 3, the socio-culture factors manage to explain only 12 percent ($R^2 = 0.12$,

Table 2. Mean score for displacement effects on Asli food practices.

	Items	Mean	S.D
FP1	New lifestyles improve my cooking practices	3.43	1.083
FP2	Displacement exposed me to new food knowledge and practices.	3.45	1.293
FP3	Educational level improved my food knowledge	3.37	1.251
FP4	Religion understanding teaches me on permissible food knowledge.	3.55	1.245
FP5	Religion understanding teaches me on the permissible way to prepare the food.	3.45	1.196
FP6	Improved income also improved my cooking practices.	3.56	1.212
FP7	New lifestyles introduced me to a variety of ingredients.	3.57	1.200
FP8	Displacement exposed me to a new array of ingredients available.	3.55	1.282
FP9	Improved educational levels increased my understanding on usage of ingredients.	3.39	1.067
FP10	Improved income level altered my choices on using ingredients.	3.33	1.245
FP11	Displacement changed a variety of ingredient usage in preparing custom and traditional ritual food.	3.98	.798
FP12	New lifestyle exposed me on the use of modern utensils in cooking.	3.87	.753
FP13	Displacement introduced me to modern utensils in preparing food.	3.98	.690
FP14	Displacement teaches me the variety of cooking methods using modern utensils	4.02	.807

Scale: 1 = strongly disagree, 2 = disagree, 3 = slightly agree, 4 = agree, 5 = strongly agree

Table 3. Results of multiple regressions of the socio-culture dimensions with food practices.

Predictor	Criterion	Model 1 Std. B
Step1:Model Variable		
	Socio-culture - Food practices	.34***
R^2		.12
Adj. R^2		.11
R^2 Change		.12
F-Change		19.499***

*Note. *$p < .05$, **$p < .01$, ***$p < .001$*

F-change = 19.499, $p < .001$) of the variance in the Orang Asli food practices. The socio-culture was found significantly and positively influenced the food practices. The value of $\beta = 0.34$, $p < .000$ demonstrated that elements of displacement have given impact on the food practices among the Orang Asli. In actual facts this holds through from the researcher observation a slight change occur in Orang Asli socio culture and food practices owing to the advancement of lifestyle, education and economy. In sum, the hypothesis is supported.

5 IMPLICATION AND CONCLUSION

This study finding clearly revealed that displacement undertaken by the government has given significant benefits to the Orang Asli. Displacement has positively improved their socio-culture, economy, knowledge and understanding through education. In addition, displacement promotes healthy and giving more organized lifestyle and proper religious practices of this indigenous group. Not harsh to say that displacement bringing the Orang Asli into the mainstream or not left them behind the development. These positive indications have given significant implication not only to the Orang Asli itself but also to the responsible authorities.

In view of government authorities, the Jabatan Hal Ehwal Orang Asli (JHEOA) who are responsible for the development of the Orang Asli in Malaysia should continuously improving the economy, education and lifestyle of this indigenous group. This can be done by continuously upgrading basic infrastructure such as water, electricity, school and creating more job opportunities for this group. This agency should play an important role by ensuring all children of Orang Asli are not let behind on education therefore the future generation will not left behind and getting equal opportunities with other ethnics group in Malaysia.

In addition to above changes, slight changes are occurring particularly on their traditional food practices. Despite practicing of their traditional food new knowledge gained through displacement has implicates and improved their usage of modern utensils, ingredients and methods of cooking. In other words, the displacement through economy, lifestyle and education not only increased the Orang Asli food knowledge, but at the same time teach them the healthy food practices. This is not to say that their traditional food practices is not healthy, but improving their style of cooking using modern technique enhancing their food potpourri. With this it clearly indicates that the Orang Asli is not totally rejected other food, but willing to accept it and therefore the authority, other government and non-government bodies should continually educate this indigenous group with proper and acceptable food knowledge without neglecting their traditional food practices.

REFERENCES

AITPN. (2008). *The Department of Orang Asli affairs, Malaysia: An agency for assimilation*. Retrieved from http://www.aitpn.org

Akpanudoedehe, J. (2010). *Socio-economic and Cultural Impacts of Resettlement on Bakassi People of Cross River State Nigeria*. Paper presented at the Unpublished Seminar Paper Presented at the Department of Sociology University of Calabar, Calabar–Nigeria.

Bakalian, A.P. (1993). *Armenian-Americans: From being to feeling Armenian*. New Brunswick, New Jersey: Transaction Publisher.

Berry, J.W. (2000). Cross-cultural psychology: A symbiosis of cultural and comparative approaches. *Asian journal of social psychology, 3*(3), 197–205.

Buckley, M., Cowan, C. & McCarthy, M. (2007). The convenience food market in Great Britain: Convenience food lifestyle (CFL) segments. *Appetite, 49*(3), 600–617.

Cwiertka, K.J. (2000). Encounters and Transitions in Foodways: Japan and the West. *Food Culture*, 8–11.

Erasmus, B. (2009). *Indigenous Peoples' food systems: The many dimensions of culture, diversity and environment for nutrition and health*. Quebec, Canada: FAO Publishing.

Gillette, B. (1997). *Contemporary Chinese Muslims (Hui) remember ethnic conflict: Stories of the late 19th century "Hui Uprising" from Xian*. Paper presented at the Association for Asian Studies meeting, Chicago, United State of America.

Gomes, A.G. (2004). The Orang Asli of Malaysia. *International Institute for Asian Studies Newsletter, 35*, 10.

Inkeles, A. & Smith, D.H. (1974). *Becoming modern: Individual change in six developing countries*. Havard, MA: Harvard University Press.

Jamal, A. (1996). Acculturation: The symbolism of ethnic eating among contemporary British consumers. *British Food Journal, 98*(10), 12–26.

Jussaume, J.R.A. (2001). Factors associated with modern urban Chinese food consumption patterns. *Journal of Contemporary China, 10*(27), 219–232.

Kuhnlein, H.V., Erasmus, B. & Spigelski, D. (2009). *Indigenous peoples' food systems: the many dimensions of culture, diversity and environment for nutrition and health*: Food and Agriculture Organization of the United Nations (FAO).

Martinez, R.O. (1998). Globalization and the social sciences. *The Social Science Journal, 35*(4), 601–613.

Miele, M. (1999). Short circuits: New trends in the consumption of food and the changing status of meat. *International Planning Studies, 4*(3), 373–387.

Nicholas, C. (2000). *The Orang Asli and the contest for resources: indigenous politics, development and identity in Peninsular Malaysia*: International Work Group for Indigenous Affairs Copenhagen.

Sobal, J. (1999). Food system globalization, eating transformations, and nutrition transitions. In R. Grew (Ed.), *Food in global history* (pp. 171–193). CO: Westview Press.

Spigelski, D. (2009). *Indigenous Peoples' food systems: The many dimensions of culture, diversity and environment for nutrition and health*. Quebec, Canada: FAO Publishing.

Woube, M. (2005). *Effects of resettlement schemes on the biophysical and human environments: The case of the Gambela Region, Ethiopia*: Universal-Publishers.

Theory and Practice in Hospitality and Tourism Research – Radzi et al. (Eds)
© *2015 Taylor & Francis Group, London, ISBN 978-1-138-02706-0*

Traditional Hari Raya food: An insight from three Malay women generations

M.S.M. Sharif, M.S.M. Zahari, R. Muhammad & N.M. Nor
Faculty of Hotel and Tourism Management, Universiti Teknologi MARA, Shah Alam, Malaysia

ABSTRACT: Festival food refers to the food prepares and served particularly during culture and religious celebrations. A specific Malay traditional food is seen to be a major contributor in making the Malay Muslim religious celebration like *Hari Raya Aidil Fitri* and *Hari Raya Aidil Adha* be more lively and happening besides other religious activities. It is becoming a long tradition that woman regardless generations are the important individuals besides man who are dealing with traditional Hari Raya food preparation. Traditional Hari Raya food is in fact to have brought significant meaning to the Malays women generations in celebrating the religious festive celebration. Through the interview, this paper qualitatively describes the insightful meaning of traditional Hari Raya food from three Malay women generations that were obtained from a study on understanding the process of Malay traditional Hari Raya food knowledge transfer within the three generations Malay women or mothers. Two specific themes on the meaning of traditional Malay Hari Raya food were identified from the response during the interview session. Women regardless generations strongly attached to traditional Hari Raya food which were long being practiced and commonly accepted by the Malay communities for generations.

Keywords: Malay ethnic, Malay generation, Malay traditional food, women

1 INTRODUCTION

Ethnic festival foods have long been recognized as the integrative force that enhances solidarity and alliances between individuals and kin groups (Gunaratnam, 2001). It becomes a bonding mechanism not only for the family but also among the communities. These foods were created by the older generations using the natural resources or any consumable things available in their area with different types of cooking methods, equipments and eating decorum (Chapman, Ristovski-Slijepcevic & Beagan, 2010). These food treasures in turn have contributed to the uniqueness of ethnic food festival for the later generations. Festival's food provides communities with opportunities to build the relationship and togetherness, promote the area of utilizing for particular foods (Dalessio, 2007), ingredients, cooking styles and local environments.

In Malaysia, Jalis, Zahari, Zulkifly, and Othman (2009) posited that most the traditional food for each of the major and minority ethnic groups are derived or synonym with either religious festival or cultural events and activities such as Hari Raya, Chinese New Year, Thaipusam, Deepavali and Gawai, Kecamatan, wedding reception and many others. In this sense, festival food refers to the traditional food prepared and served particularly during either culture, and religious festivals or celebrations. In the context of Malay ethnic, festival food refers to the traditional food prepared and served for the Hari Raya celebration. Muhammad et al. (2009) asserted a specific Malay traditional food are seen to be a major contributor in making the Malay Muslim religious celebration like *Hari Aidil Fitri* and *Hari Raya Aidil Adha* more lively and happening besides other religious activities. In fact, Hari Raya will look dull and monotonous without the present of traditional Hari Raya food (Kamaruddin, Zahari, Radzi & Ahmat, 2010). Not to exaggerate that it is becoming a long tradition for woman regardless generations, they were the important individuals besides man who are dealing or responsible with the traditional Hari Raya food preparation. They are in fact as a central figure that leads the family for traditional Hari Raya food preparation (Muhammad, Zahari, Kamaruddin & Ahmat, 2013). Based on this statement, the traditional Hari Raya food is argued to have brought significant meaning to the Malays women generations. This paper is therefore qualitatively describes the meaning of traditional Hari Raya food from three different Malay women generations.

2 LITERATURE REVIEW

2.1 *Malay traditional foods*

Traditional food referred to foods that have been prepared and consumed for many generations which includes all indigenous food plants found in that region or in a specific locality (European Union, 2007). Trichopoulou, Soukara & Vasilopoulou (2007) pronounced traditional food as an expression of one culture, history and lifestyle reflects the cultural heritage and have left their prints to contemporary dietary patterns to subsequent generations. Traditional foods are those foods originating locally in an area with respect to the country, region, district or sub district (Ohiokpehai, 2003) with some ingredients and traditional food preparation represent an intrinsic part of the identity of regional foods and by association with the people who consume them (Fajans, 2006). Jordana (2000) stated that traditional food is a food that is differentiated through particular qualitative aspects and has a specific cultural identity, while Kwik (2008) contended that traditional food may be interpreted as describing a process that does not change.

The Malay food on the other hand has its own history of existence and the origin of its results in its unique characteristics to these days. There was unclear written evidence as to when the tradition of Malay food culture exists (M.S.M. Sharif, Zahari, Nor & Muhammad, 2013). However, from the popular beliefs and based on the earliest record that Malay food culture started to emerge in the 15th century during the Malacca Sultanate and Malacca itself act as an important trade center in the Malay Archipelago (Hooker, 2003). Besides Sultanate of Malacca, the active participation in the spice trade which involved ingredients, cooking methods, recipes and eating decorum have created an important legacy in the Malay culinary or food culture traditions. In addition, the ethnic composition of Malay, Chinese descendants especially the Baba Nyonya and Eurasian catalyst the development of culinary culture tradition. Anom (1995) argues that the formation potpourri of Malay food culture tradition is also resulted from the acculturation processes among the Javanese, Bugis and Minangkabau.

In the north of Malaysia, Malay food culture has been mostly influenced by the Indian flamboyance during the sixth century under the auspices of the Srivijaya kingdom. There was a large Hindu temple built by Indian traders and missionaries in the northern districts (Su-Lyn, 2003). Again, the active participation in the spice traders contributes a value added to Malay culinary culture traditions (M.S.M. Sharif, Nor & Zahari, 2013). In addition, the Malay food culture was also heavily convinced by the Pattani culture of Thailander whereby in the eighteenth century the northern states such Kedah, Perlis, Pulau Pinang and Perak including the two east coast states namely Kelantan and Terengganu were under the influence of Siamese monarchy which immensely powerful at that time (Hooker, 2003). From these incidences the vast potpourris of Malays food culture were apparent.

2.2 *Malay traditional Hari Raya foods*

Hari Raya (Aidil Fitri and Aidil Adha) mean festivals or celebration days throughout the world for Muslims. Of the two, Aidil Fitri is the most important festival in the Islamic calendar which marks the end of Ramadan (which is the fasting month) and the arrival of Shawwal and it is a grand celebration in Malaysia (Munan, 1990). The practice, besides others, during this celebration is the appearance of traditional food, as part of the festival day is spent visiting relatives and friends and food is the integral elements in connecting relatives and friends (Anon, 1995). Md. Nor et al. (2012) contended it is a tradition for the Malay community to travel back to their hometown or Kampong to get together with families and relative to celebrate the Hari Raya. *'Balik Kampong'* is become the trends and culture of the Malay communities and the eve of Eid is the final preparation of the whole family members and ask forgiveness among their both parents and other family members in the morning of Hari Raya before performing Eid prayers (Muhammad, Zahari, Othman, Jamaluddin & Rashdi, 2009).

As rice is the main staple food for the Malays, therefore some of the festive food is made out of it such as *ketupat, ketupat palas, lontong,* and *lemang* (Marlia Musa, 2011). Typically, the Malay festive food would be different from what they always consumed during the usual days. Throughout this celebration, beef and chicken dishes are the main dish used to accompany ketupat, nasi impit and lemang (Anon, 1995). Besides these other typical Malay delicacies like kueh mueh or biscuits plus the modern ones are apparent. That is why Malay communities are normally looking forward for the traditional dishes during Hari Raya owing to some of them were only prepared and served during this celebration (Md. Nor et al., 2012; M.S. Sharif, Supardi, Ishak & Ahmad, 2008).

3 METHODOLOGY

In establishing the meaning of Malay traditional Hari Raya food among the three Malay women generations, a qualitative approach was employed.

The rationale of conducting a qualitative study is because of it's subjective in nature and reflecting on the perception in understanding the social and human activities (Merriam, 2009). The informants for this study were among three generations of Malay women, namely grandmother (age between 71 to 85 years old), mother (age between 50 to 70 years old) and daughter (age between 20 to 40 years old) within a family. A Malay residential area in Selangor, Negeri Sembilan, Melaka and Johor are the contextual setting for the study. To guide the informants to share their knowledge, experience and opinion on the meaning of traditional Hari Raya food, a set of open ended interview questions were developed.

Before the actual interviews, eight groups of informants (grandmother, mother and daughter) were first identified through relatives and head of the Malay community. All respective groups were contacted via telephone asking for permission, date, times and venues for the interviews were to be conducted. Subsequently, the interviews were successfully conducted within two weeks. All the information was analyzed using thematic analysis.

4 RESULT OF THE INTEPRATATION

4.1 *Informants' profile*

As previously mentioned, eight groups of Malay women informants involved in this study. They consisted of grandmother, mother and daughter from selected families. For the interpretation, they are recognized as Group informants. In short, the group of informants is simplified and presented in Table 1.

It is worth mentioning, in the subsequent section the response given by the first generation or grandmother is indicated by Group Informant A while information addressed by second generation or mother is specified by Group Informant B and information given the third generation or daughter is indicated by Group Informant C. Below were some of the responses given by the informants.

4.2 *The meaning of traditional Hari Raya food*

Besides other important questions the one question used in this paper is asking the meaning of traditional Malay Hari Raya food. As this question asked the individual women from different generations, different views are also expected. However, the majority of three generation women as key informers were having similar or identical views.

Two specific themes were identified from the response received during the interview session with regard to meaning of Malay traditional Hari Raya

Table 1. Informants' profiles.

Inform-ant	Age	Educa-tion	Occupa-tion	Chil-dren	State
Group Informant 1					
G.Mother	67	Primary School	House-wife	6	Johor
Mother	50	SPM	House-wife	5	Johor
Daughter	31	Diploma	Banker	3	Johor
Group Informant 2					
G.Mother	69	MCE	Retiree	5	Johor
Mother	52	SPM	Clerk	3	Johor
Daughter	27	SPMV	Supervi-sor	2	Johor
Group Informant 3					
G.Mother	69	Primary School	House-wife	7	Melaka
Mother	45	SPM	Clerk	5	Melaka
Daughter	24	Diploma	Asst.Adm in.	1	Melaka
Group informant 4					
G.Mother	68	Primary School	House-wife	1	Melaka
Mother	49	Stand-ard 6	House-wife	6	Melaka
Daughter	25	STPM	Supervi-sor	2	Melaka
Group Informant 5					
G.Mother	71	Primary School	House-wife	5	N.Sembil an
Mother	50	SPM	Clerk	4	N. Sembilan n
Daughter	29	Degree	Bank Officer	3	N. Sem-bilan
Group Informant 6					
G.Mother	66	Primary School	House-wife	5	N. Sem-bilan
Mother	45	Diplo-ma	Teacher	5	N. Sem-bilan
Daughter	28	Degree	Teacher	3	N. Sem-bilan
Group Informant 7					
G.Mother	72	Primary School	House-wife	6	Selangor
Mother	51	SPM	Asst.Adm in	4	Selangor
Daughter	27	Degree	Teacher	2	Selangor
Group Informant 8					
G.Mother	77	Primary School	House-wife	4	Selangor
Mother	56	Master	Lecturer	5	Selangor
Daughter	34	Degree	Teacher	3	Selangor

food 1). Food that attached with a religious festival and 2). Common acceptable food for Malays or family.

4.2.1 *Food that attached with religious celebration*

The majority of the three Malay generation women refers the traditional Malay Hari Raya as foods or dishes that are closely related to the Malay communities during the religious celebration. The foods are so long attached and well connected with this ethnic and specially prepared for the celebration. The traditional Hari Raya foods were carried by the Malay generations or old folks for ages or have been practicing since generation. Most of the informers named rendang, *ketupat, nasi impit, lemang, masak lodeh, peanut sauce* are the examples of the traditional Malay Hari Raya food that commonly prepared and served for this religious celebration. Some of the verbatim answers are as follows.

"Malay traditional Hari Raya food based from my understanding refers to traditional foods or dishes that are closely related to the Malay Muslim communities during the religious celebration like Aidil Fitri and Aidil Adha. The foods were specially prepared for the celebration and I could say celebration slightly dull with the absence of those foods. To name a few…. chicken and beef rendang, ketupat, kuah lodeh and peanut sauce are the obvious ones"

(Informant 2A)

"Based on my observation, Malay traditional Hari Raya foods are the dishes that the Malays adored to prepare and served during religious celebration and it is a too attached to the Malay communities in Malaysia, Singapore, Brunei, Indonesia and the Southern Thailand. In other word, these were the hereditary food that created and practices, especially by the Malay Muslim in this region. In sum, I would say foods like rendang, lemang, ketupat, masak lodeh, kuah kacang are synonymous with Malay Muslim festive celebration"

(Informant 1B)

"Looking at my mom and my mother in-law cooking practices… I understand that Malay traditional Hari Raya foods were usually attached with religious celebration, which were carried by the Malay generations and old folks for ages… ketupat, rendang, peanut sauce and kuah lodeh were usually preferred as the Malay traditional Hari Raya dishes for Malay especially in Selangor, Negeri Sembilan, Melaka and Johor while beside those foods pulut palas, kerutuk, nasi dagang and serunding are so intact with the Malay in Kelantan, Terengganu, Pahang or east coast of Malaysia"

(Informant 3C)

4.2.2 *Common acceptable food*

Besides food that attached with the religious celebration, some informants' relate the Malay traditional Hari Raya foods as common dishes or foods that are prepared or cooked by the Malays families for festive celebrations, although the styles and procedures for the preparation might slightly differ from one family to another. The informants also

believed that traditional Hari Raya food comes from great ancestors and passing down through generation line and continually being practiced. Some of the comments received;

"To me… when talking about Malay traditional Hari Raya food… it is a common dish or foods that are prepared or cooked by the Malay families for festive celebrations, although the styles and procedures of the preparation process may be slightly differ from one family to another. This can clearly be seen on rendang. Negeri Sembilan family may prefer the spicier type of it as opposed to Selangor family. For whatever they do…… rendang is still rendang and commonly accepted for the whole nation. Can you see any house in any states in Malaysia celebrating Hari Raya without rendang, ketupat or lemang and kuah kacang. In fact, these are few examples of Malay traditional Hari Raya foods that are common and become as family tradition…"

(Informant 3A)

"Traditional foods are long time food which was passed down by our great—great grandparents. Even though the dishes were common, but the taste and texture of the dishes were different from one family to another because each family has its own styles and procedures of doing it…"

(Informant 8B)

"When we say traditional, it comes from great grandparents down to our mothers and to us who practice the same thing. That is what I understood… Because that is our family secret and tradition which being passed and practiced for generations…"

(Informant 7C)

5 DISCUSSION & CONCLUSION

Although using only one question, the finding revealed that women regardless generations strongly attached to traditional Hari Raya food by sharing the same view of it. This clearly indicates that the Malay generations' women have played their significant role in preserving traditional food including the Hari Raya dishes by continually transferring the knowledge and practicing it. The involvement of grandmother, mothers, daughters and other family member as can be seen during Hari Raya food preparation not only strengthening the understanding, acceptance but developing a sense of belonging to Malay traditional food among the young generation. As a conclusion, the investigation of traditional Malay food in relation to modernization should continuously be undertaken.

ACKNOWLEDGEMENT

This research was funded by the Ministry of Higher Education, Malaysia through Universiti Teknologi MARA under RAGS grant: 600-RMI/RAGS 5/3 (179/2012).

REFERENCES

Anon, E. (1995). *Malaysia Festive Cuisine*. Kuala Lumpur: Berita Publishing Sdn. Bhd.

Chapman, G.E., Ristovski-Slijepcevic, S. & Beagan, B.L. (2010). Meanings of food, eating and health in Punjabi families living in Vancouver, Canada. *Health Education Journal, 101*(3), 246–250.

Dalessio, W.R. (2007). *Are We What We Eat? Food Preparation, Food Consumption, and the Process of Identity Formation in Contemporary Ethnic American Literatures*. (Doctor of Philosophy), University of Connecticut.

European Union. (2007). *European Research on Traditional Foods*. Belgium: Publications.Europa.Eu.

Fajans, J. (2006). Regional food and the tourist imigination in Brazil. *Appetite, 47*(3), 389.

Gunaratnam, Y. (2001). Eating into multiculturalism: hospice staff and service users talk food, 'race', ethnicity, culture and identity. *Critical Social Policy, 21*(3), 287–310.

Hooker, V.M. (2003). *A Short History of Malaysia: Linking East and West*. Singapore: Allen & Unwin.

Jalis, M.H., Zahari, M.S., Zulkifly, M.I. & Othman, Z. (2009). Malaysian gastronomic tourism products: Assessing the level of their acceptance among the western tourists. *South Asian Journal of Tourism and Heritage, 2*, 31–44.

Jordana, J. (2000). Traditional foods: challenges facing the European food industry. *Food Research International, 33*(3–4), 147–152.

Kamaruddin, M.S.Y., Zahari, M.S.M., Radzi, S.M. & Ahmat, N.H.C. (2010). *Modernization and Malay Matrimonial Foodways in the Rural Area*. Paper presented at the TEAM Conference, Taylor's University.

Kwik, J.C. (2008). Traditional food knowledge: A case study of an Immigrant Canadian "foodscape". *Environments, 36*(1), 59–74.

Marlia Musa, N.O., Fazleen Abdul Fatah. (2011). Determinants of Consumers Purchasing Behavior for Rice in Malaysia. *American International Journal of Contemporary Research, 1*(3), 8.

Md. Nor, N., Md. Sharif, M.S., Mohd. Zahari, M.S., Mohd. Salleh, H., Ishak, N. & Muhammad, R. (2012). The Transmission Modes of Malay Traditional Food Knowledge within Generations. *Procedia—Social and Behavioral Sciences, 50*(0), 79–88.

Merriam, S.B. (2009). *Qualitative Research: A Guide to Design and Implementation*. San Francisco: Jossey-Bass.

Muhammad, R., Zahari, M.S.M., Kamaruddin, M.S.Y. & Ahmat, N.h.C. (2013). The Alteration of Malaysian Festival Foods and Its Foodways. *Procedia—Social and Behavioral Sciences, 101*(0), 230–238.

Muhammad, R., Zahari, M.S.M., Othman, Z., Jamaluddin, M.R. & Rashdi, M.O. (2009). *Modernization and Ethnic Festival Food* Paper presented at the International Conference of Business and Economic, Kuching, Sarawak.

Munan, H. (1990). *Cultures of the World Malaysia*. Singapore: Times Books International.

Ohiokpehai, O. (2003). Promoting the Nutritional Goodness of Traditional Food Products. *Pakistan Journal of Nutrition, 2*(4), 267–270.

Sharif, M.M., Zahari, M.M., Nor, N.M. & Salleh, H.M. (2013). The adaptations of Malay food knowledge among Malay generations in Kuala Lumpur, Malaysia. *Hospitality and Tourism: Synergizing Creativity and Innovation in Research*, 395.

Sharif, M.S., Supardi, A., Ishak, N. & Ahmad, R. (2008). *Malaysian Food as Tourist Attraction* Paper presented at the The 1st Malaysian Gastronomic—Tourism Conference.

Sharif, M.S.M., Nor, N.M. & Zahari, M.S.M. (2013). The Effects of Transmission of Malay Daily Food Knowledge on the Generation Practices. *Procedia—Social and Behavioral Sciences, 85*(0), 227–235.

Sharif, M.S.M., Zahari, M.S.M., Nor, N.M. & Muhammad, R. (2013). Factors that Restrict Young Generation to Practice Malay Traditional Festive Foods. *Procedia—Social and Behavioral Sciences, 101*(0), 239–247.

Su-Lyn, T. (2003). *Malaysia & Singapore: World Food*: Lonely Planet.

Trichopoulou, A., Soukara, S. & Vasilopoulou, E. (2007). Traditional Foods: A Science and Society Perspective. *Trends in Food Science & Technology, 18*(8), 420–427.

Theory and Practice in Hospitality and Tourism Research – Radzi et al. (Eds)
© *2015 Taylor & Francis Group, London, ISBN 978-1-138-02706-0*

Social interaction among Malaysian ethnics: Its impact on food preparation and consumption

N. Ishak & M.S.M. Zahari
Faculty of Hotel and Tourism Management, Universiti Teknologi MARA, Shah Alam, Malaysia

ABSTRACT: A social interaction with other people is very important in a society. Through social interactions with others, people begin to develop relationship, understanding and establish an identity. Social interaction among various ethnic groups either directly or indirectly through social events when food is present is believed to influence each ethnic to prepare and consume each other food. Furthermore, when interacting with each other through event ethnics not only learn the appropriate social behaviors, respecting the culture but as well understand their food. This study evaluates the extent to which social interaction among the Malay, Chinese and Indian through food event is giving impact to the preparation and consumption of each ethnic food. Using questionnaire survey among the individuals who experienced in preparing and consuming the three Malaysia major ethnic foods, namely Malay, Chinese and Indian some useful insight obtained from this study. Social interaction through event gives the exposure and the ability of ethnic groups to at least prepare an acceptable each other foods.

Keywords: Social interaction, Malaysian ethnics, food, preparation, consumption

1 INTRODUCTION

Food holds the key to any culture and people make many assumptions about other people's diets based on who they claim to be and people also make many assumptions about who people are, based on what they eat (Counihan & Van Esterik, 1999). Mintz (1996) noted that preparation and consumption is always conditioned by means that the food we eat and the manner in which we preserve, prepare, and serve them and all devote with meaning at some levels. Consumption or eating in all cultures is an expression of beliefs, systems and social distinctions that exist within groups and society. The food that we eat closely links to cultural codes, and it is precisely enables it to become a good indicator of identity (Crouch & O'Neill, 2000). Kittler and Sucher (2004) claim consumption and preparation of food gives valuable insights on community that performs the acts. Food through its preparation and consumption work like a sign system as of language that can transcend time and space.

Scholars believed that besides others, cross-culturing in food preparation and consumption among ethnic groups through food knowledge, food media and social interaction or events are unavoidable in multi-racial countries (Cleveland, Laroche, Pons & Kastoun, 2009; Kim, Laroche & Tomiuk, 2001; Shalom & Horenczyk, 2004). In fact, the precursor that takes place in the process of constructing the nation food identity within ethnic groups in the multicultural/ethnic countries is reflecting all the way through sharing a common accepted cuisine.

As the central focus of this study, social interaction among various ethnic groups either directly or indirectly through social events when food is present is believed to influence each ethnic to prepare and consume each other food. Furthermore, when interacting with each other through event ethnics not only learn the appropriate social behaviors, respecting the culture but as well understand their food. With this notion, this study aims to evaluate the extent to which social interaction among the Malay, Chinese and Indian through food event is giving impact to the preparation and consumption of each ethnic food. The objectives are to a) perception of social interaction toward food preparation and consumption, b) comparison between ethnics' perception of social interaction toward food preparation and consumption. In addition, it is hypothesized that:

H1: There is a significant relationship between social interaction and preparation, consumption of each other ethnic food.

2 LITERATURE REVIEW

2.1 Social interaction

A social interaction with other people is very important in a society. Through social interactions with others, people begin to establish an identity, develop relationship and understanding. Social interactions refer to particular forms of externalities, in which the actions of a reference group to affect an individual's preference. The reference group depends on the context and is typically an individual's family, neighbors, friends or peers. In social science, a social interaction is any relationship between two or more individuals.

Social interaction issues have received significant attention among the academic scholars. Sobel (2002) re-evaluated the macroeconomic models with the argument that social interaction builds different capital, human or social capital while Horst (2010) examined the basis of how social interaction comes into existence over time. Byman, Järvelä, and Häkkinen (2005) revealed that information is useful for developing a new pedagogical model for web-based learning and enhancing the quality of virtual interaction. On the hand, the patterns of young children's social interaction that occurred in the computer area were described as parallel play, verbal conflicts, sociable interaction, knowledge construction through positive and negative processes, and non-verbal communication (Lim, 2012). Manski (2000) and Putnam (2000) urged economists to understand more about how social interaction affects outcomes and to gather more usable data on how individuals interact within their communities.

On food, Meiselman (2001) suggests that there are at least four major concurrent contexts that can alter the perception of food and beverages during consumption. 1) its function as a meal component 2) social interaction during consumption, 3) the environment in which food is selected and 4) consumed and food choice freedom. The increasing popularity of foods from different ethnics, groups and countries besides other elements probably also due the increasing of social interaction through the festival, gathering, partying and even eating in restaurants lead to exposing more of new food options through consumption (Meulenberg & Viaene, 1998).

2.2 Food preparation and consumption

Many scholars refer to the actions or behaviors which involved the preparation, presentation and consumption of food that are characterized by individuals and society as foodways (Cusack, 2003; Freeman, 2002; Gutierrez, 1999).

Gutierrez (1999) posited the elements of foodways such as food preparation and consumption are important elements in any celebrations, occasions and events and it symbolize values and important meanings in specific religions, cultures and ethnicity (Murphy, 2000). Gutierrez (1999) also noted that most of the elements of food preparation and consumption are passed down from generation to generation or informally transmitted orally and through demonstration. The New England Clambake is one of the examples of food and consumption practices which have long been a tradition of family and community in New England. The practices were passed down from the Natives Americans to the colonists and further to their descendants. As the process of preparing and cooking of this baking product takes at least two or three days to complete, it therefore gives the opportunity to the community to share stories and connect with one another.

Spiro (1955) posited the preparation, presentation and consumption of ethnic food are important in any event because they are an expression of ethnic identity. Ethnic festival foods are a whole interrelated system of food conceptualization and evaluation, procurement, distribution, preservation, preparation, consumption and nutrition shared by all members of a particular nation or society (Thomas, 2004). The ethnic festival food preparation and consumption are the most resilient of cultural forms and have become a powerful medium of social exchange (Kroll, 1994).

Nevertheless, the preparation, presentation and consumption of food or cuisines are varied in some communities or ethnic groups and depend on the occasions. Some ethnic groups, such as Hindus, Buddhist and Jews have structured their preparation and consumption by combining certain occasions with their religion (Vu, 2009). The foods served at festivals, for example, eloquently express family reactions to the transgression of boundaries. Therefore, through the festival, food preparation and consumption not only creates the social interaction among the family members but among the community (Vu, 2009).

3 METHODOLOGY

Survey questionnaire is chosen for this research and the population was among individuals who experienced in preparing and consuming the three Malaysia major ethnic foods namely; Malay, Chinese and Indian. Restaurant customers are believed the best individuals to provide reliable information. The huge populations of Malay, Chinese and Indians restaurant customers, however, limit researchers to collect the desired information throughout the country. Thus, Kelang Valley (Lembah Kelang) comprises of Kuala Lumpur, its suburbs and adjoining cities and towns in the state of Selangor, such as Shah Alam, Petaling Jaya, and Klang is the data collection setting. This study employed, self-administered questionnaire and questions are designed to measure levels of agreement relevant to issue investigated with seven points Likert scale.

Surveyed with experienced respondents was undertaken at seven selected popular restaurants among the Malay, Chinese and Indian in Kuala Lumpur, Shah Alam, Petaling Jaya and Kelang in five weekends. Respondents randomly approached after finishing dined with a screening question whether they had experienced in preparing and consuming all three major ethnic foods. Those answered "yes" were preceded with the question-naire. With a positive response, 392 questionnaires were successfully collected.

4 ANALYSIS AND RESULT

Based on frequency, Malays constituted around 60.2 percent (n = 236) followed by Chinese by 21.7 per cent (n = 85), Indian with 16.6 per cent and other ethnic groups with 1.5 percent (n = 6). This proportion did provide a reasonable representa-tion of ethnic ratio in Malaysia which consists of 60:30:10 (Malay, Chinese and Indians).

4.1 *Perception of social interaction toward food preparation and consumption*

This section analyzes perception of social interac-tion toward food preparation and consumption. As shown in Table 1, the magnitude of mean scores above 6.00 to below 7.00 indicates that the majority of respondents in the sample agree with the state-ment related to social interaction as the point that they understand and learnt each other ethnic food.

The respondents agrees that parties/social gath-ering creates their understanding of each other eth-nic cultures (M = 6.45, item CP1), food (M = 6.32, item CP2) and ingredients (M = 6.38, item CP3). Their knowledge and understanding on how other ethnic foods being prepared also enhanced and this can be seen through mean the score (M = 6.30, item CP4) given this item. Similarly, the respond-ents agree parties/social event/gatherings/festivals attending contribute their understanding of how some other major ethnic being served (M = 6.36, item CP5) and consumed (M = 6.17, item CP6) and further influence them to consume the foods (M = 6.41, item CP7). Parties/social events/gather-ings/festivals also built their interest to consume (M = 6.44, item CP8) and to prepare (M = 6.17, item CP9) some of the popular other ethnic foods. Parties/social events/gatherings/festivals also give

Table 1. Results of mean and standard deviation of social interaction through event (n = 392).

	Items	M	S.D
CP1	Parties/ social events/ gatherings/ festivals contribute to my understanding of other two ethnic cultures	6.45	1.18
CP2	Parties/ social events/ gatherings/ festivals contribute to my understanding of other two major ethnic foods	6.32	1.06
CP3	Parties/ social events/ gatherings/ festivals create my understanding of ingredients on some of the other two popular major ethnic foods	6.38	1.13
CP4	Parties/ social events/ gatherings/ festivals contribute to my understanding of how some other two popular major ethnic foods being prepared	6.30	1.09
CP5	Parties/ social events/ gatherings/ festivals attending contribute to my understanding of how some other two popular major ethnic being served	6.36	1.09
CP6	Parties/ social events/ gatherings/ festivals contribute to my understanding of how some other two popular major ethnic beings consumed	6.17	1.07
CP7	Parties/ social events/ gatherings/ festivals influence me to consume some of the other two popular major ethnic foods	6.41	0.14
CP8	Parties/ social events/ gatherings/ festivals built my interest to consume some of the other two popular major ethnic foods	6.44	1.16
CP9	Parties/ social events/ gatherings/ festivals built my interest to prepare some of the other two popular major ethnic	6.17	1.09
CP10	Parties/ social events/ gatherings/ festivals give me the exposure and the ability to at least prepare an acceptable some of the oth-er two major ethnic foods	6.20	1.99
CP11	Parties/ social events/ gatherings/ festivals catalyst the sharing of the other two major ethnic foods	6.36	1.25

Scale: 1 = Strongly Disagree, 2 = Disagree, 3 = Slightly Disagree, 4 = Neutral, 5 = Slightly Agree, 6 = Agree, 7 = Strongly Agree.

them the exposure and the ability to at least pre-pare an acceptable some the other ethnic foods (M = 6.20, item CP10) thus catalyst the sharing of the other ethnic foods (M = 6.36, item CP11).

4.2 *Comparison between ethnics perception of social interaction toward food preparation and consumption*

Analysis of Variance (ANOVA) was undertaken looking at the differences between perception of three ethnic groups (Malay, Chinese and Indian) on the social interaction toward food preparation and consumption. In addition, post hoc multiple comparisons using the Scheffe test was also applied to see the underlying differences within the ethnics group.

Ethnics respondent strongly agrees that social events, gatherings and festivals contribute to their understanding of each ethnic food culture (Malay, M = 6.55, Indian, M = 6.52 and Chinese, M = 6.52,

Table 2. Showing the items of 'Social Interaction' where statistically significant differences between ethnic groups were identified.

Items	Eth	N	M	SIS.D	Sig	Sch
	Mly	166	6.55	.577	.868	
CP1	Ind	101	6.52	.610		
	Chn	125	6.52	.604		
	Mly	166	5.54	1.00	.064	
CP3	Ind	101	5.27	1.04		
	Chn	125	5.13	1.10		
CP7	Mly	166	5.26	1.16	.152	
	Ind	101	5.49	1.08		
	Chn	125	5.20	1.17		
CP9	Mly	166	6.08	1.00	.610	
	Ind	101	6.20	.872		
	Chn	125	6.13	.942		
	Mly	166	5.82	1.00	.074	
CP10	Ind	101	5.48	1.10		
	Chn	125	5.33	1.05		
	Mly	166	6.14	1.01	.371	
CP11	Ind	101	6.29	.876		
	Chn	125	6.28	.964		

1. Scale: 1 = strongly Disagree, 2 = Disagree, 3 = Slightly Disagree, 4 = Neutral, 5 = Slightly Agree, 6 = Agree, 7 = Strongly Agree.
2. Inter groups differences are based on Scheffé procedure.

Table 3. Results of Multiple Regression of social interaction against preparation and consumption.

Predictor	Model 1/Std. β
Step 1: Model Variables (Social Interaction)	.347***
R^2	.120
Adj. R^2	.118
R^2 Change	.120
F – Change	53.432

Note: $*p < .05$, $**p < .01$, $***p < .001$

$p < .000$ demonstrated that elements of social interaction through events have given a slight impact on the preparation and consumption of each other ethnic group. In actual facts this situation holds true whereby from the researcher observation and the random interviewed with selected respondents revealed that the involvement or social interaction among the ethnic groups through social events have given a slight impact on the preparation and consumption of each other ethnic food. In sum, the hypothesis is slightly supported.

$p = .868$, item CP1). Ethnics respondent agree that social events, gatherings and festivals contribute to their understanding of the ingredients of some popular each ethnic food (Malay, $M = 5.54$, $M = 5.27$ and Chinese $M = 5.13$, $p = .064$, item CP3). Ethnic respondent sharing the same view that social gatherings/festivals influence them to consume some of the other two popular major ethnic foods (Malay, $M = 5.26$, Indian, $M = 5.49$ and Chinese, $M = 5.20$, $p = 0.152$, item CP7). With that feeling they also believed that social events, gatherings/festivals built their interest to prepare some of the popular major ethnics food (Malay, $M = 5.82$, Indian, $M = 5.48$ and Chinese, $M= 5.33$, $p = .610$, item CP9), give them the exposure and ability to at least prepare an acceptable some of the major ethnic foods (Malay, $M = 5.82$, Indian, $M = 5.48$ and $M = 5.33$, $p = .074$, item CP10) thus believed gatherings and festivals catalyst the sharing of the each ethnic foods (Malay, $M = 6.14$, Indian, $M = 6.29$ and Chinese, $M = 6.28$, $p = 0.371$, item CP11).

4.3 Relationship between social interaction and food preparation and consumption

Results of a single standard multiple regressions is exhibited in the Table 3.

Looking at the table, the social interaction manages to explain only 12 percent ($R^2 = 0.120$ F-Change = 53.432, $p < .001$) of the variance in food preparation and consumption. Social interaction was found significantly influenced the preparation and consumption. The value of $\beta = 0.347$,

5 IMPLICATION AND CONCLUSION

This study has witnessed how food in the multicultural/ethnic country like Malaysia evolves through cross culturing processes. Besides others, social interaction through events when food is present slightly provides a significant role for the ethnic groups in understanding and learning on each other food and cuisines. In other words, social gathering/events creates understanding of each other ethnic cultures, ingredients, other ethnic foods being prepared, served and consumed. Most importantly, social interaction through events gives the exposure and the ability to at least prepare some of each other ethnic foods. These optimistic indications therefore carry varying consequences and implications for those individuals who are closely associated with food preparation and the relevant authorities. As conclusion, do not look food or cuisine only from culinary perspective or the art and science but incorporating the elements of gastronomy offers a path towards an understanding the culture aspect of it. In other word, food undoubtedly has great relevance to the society and culture and plays a part in the wider economy and nation construction.

ACKNOWLEDGMENT

This research was funded by the Ministry of Higher Education, Malaysia through Universiti Teknologi MARA under RIF grant: 600-RMI/DANA 5/3/ RIF (515/2012)

REFERENCES

Byman, A., Järvelä, S. & Häkkinen, P. (2005). What is reciprocal understanding in virtual interaction? *Instructional Science, 33*(2), 121–136.

Cleveland, M., Laroche, M., Pons, F. & Kastoun, R. (2009). Acculturation and consumption: textures of cultural adaptation. *International Journal of Intercultural Relations, 33*(3), 196–212.

Counihan, C.M. & Van Esterik, P. (1999). *The anthropology of food and body: Gender, meaning and power.* New York: Routledge.

Crouch, M. & O'Neill, G. (2000). Sustaining identities? Prolegomena for inquiry into contemporary foodways. *Social Science Information, 39*(1), 181–192.

Cusack, I. (2003). Pots, pens and 'eating out the body': Cuisine and the gendering of African nations. *Nations and Nationalism, 9*(2), 277–296.

Freeman, E. (2002). *The wedding complex: Forms of belonging in modern American culture.* Durham: Duke University Press.

Gutierrez, C.P. (1999). *Cajun foodways.* Jackson: Univ. Press of Mississippi.

Horst, U. (2010). Dynamic systems of social interactions. *Journal of economic behavior & organization, 73*(2), 158–170.

Kim, C., Laroche, M. & Tomiuk, M.A. (2001). A measure of acculturation for Italian Canadians: Scale development and construct validation. *International Journal of Intercultural Relations, 25*(6), 607–637.

Kittler, P.G. & Sucher, K.P. (2004). *Food and culture* (4th ed.). Canada: Thomson Wadsworth.

Kroll, D. (1994). Prepared ethnic foods: trends and developments. *Business Communications, July* (5), 9–14.

Lim, E.M. (2012). Patterns of kindergarten children's social interaction with peers in the computer area. *International Journal of Computer-Supported Collaborative Learning, 7*(3), 399–421.

Manski, C.F. (2000). Economic analysis of social interactions. *Journal of Economic Perspectives, 14*(3), 115–136.

Meiselman, H.L. (2001). Criteria of food quality in different contexts. *Food Service Technology, 1*(2), 67–77.

Meulenberg, M. & Viaene, J. (1998). Changing food marketing systems in western countries. In W.M.F. Jongen, M.T.G. Meu-lenberg & W. Pers (Eds.), *Innovation in food production systems.*

Mintz, S.W. (1996). *Tasting food, tasting freedom.* Boston: Beacon Press.

Murphy, C. (2000). Piety and Honor: The Meaning of Muslim Feast in Old Delhi. In S. Khare & M. Rao (Eds.), *Food, society and culture: Aspect in South Asian food systems.* North Carolina: Carolina Academic Press.

Putnam, R.D. (2000). *Bowling alone: The collapse and revival of American community.* New York: Simon and Schuster.

Shalom, U.B. & Horenczyk, G. (2004). Cultural identity and adaptation in an assimilative setting: Immigrant soldiers from the former Soviet Union in Israel. *International Journal of Intercultural Relations, 28*(6), 461–479.

Sobel, J. (2002). Can we trust social capital? *Journal of economic literature, 40*(1), 139–154.

Spiro, M.E. (1955). The acculturation of American ethnic groups. *American Anthropologist, 57*(6), 1240–1252.

Thomas, M. (2004). Transitions in taste in Vietnam and the diaspora. *The Australian journal of anthropology, 15*(1), 54–67.

Vu, V. (2009). *The Changing Foodways of Vietnamese Americans in Orange County, California.* (Unpublished Master Dissertation), California State University, California, Fullerton.

Relevant areas in hospitality and tourism

Theory and Practice in Hospitality and Tourism Research – Radzi et al. (Eds)
© *2015 Taylor & Francis Group, London, ISBN 978-1-138-02706-0*

Assessing medical tourists' destination choice behavior: A conceptual perspective

A.A. Suki, L. Putit, J.M. Yusof & N.R.M. Khan
Universiti Teknologi MARA, Shah Alam, Malaysia

ABSTRACT: Over the recent years, medical tourism has become one of the niche sectors that significantly contributed towards a country's economic development. The purpose of this paper is two-fold: firstly, it aims to assess medical tourists' attitude, subjective norm and perceived behavioral control in affecting their intention towards adopting destination choice behavior. Secondly, it further attempts to determine potential significant linkages among the tourists' perceived destination image, perceived value and perceived behavioral control. Data collection procedures would involve initial focus group interview followed by survey questionnaire design and its subsequent distribution to 600 potential target respondents. The unit of analyses would be tapped on medical tourists visiting a progressively developing nation for the purpose of seeking medical services. Data would be analyzed for its reliability and validity prior subsequent testing of relationships using structural equation modelling statistical approach. Expected contributions of this study include an extended knowledge building on a new conceptual research framework of destination choice behavioral adoption within the medical tourism sector. It extends the application of theory of planned behavior by exploring its relationship with destination image and perceived value towards the provision of medical services. Subsequently, it is expected that there exists a potentially increased level of awareness amongst medical tourists' towards these services, and that more foreign medical tourists would identify Malaysia as a medical destination choice of theirs.

Keywords: Destination choice behavior, theory of planned behavior, perceived behavioral control, perceived value, destination image

1 INTRODUCTION

Tourism has become one of the major industries that could contribute to a country's development. Medical tourism is one of these promising industries because it has significant economic potential, involves trade in services and represents the union of at least two divisions: medicine and tourism (Bookman & Bookman, 2007).

Several past researches have been addressed in medical tourism and hospitality sectors. Jotikasthira (2010), for example, analyzes the factors that determine the motivation and behavior of potential medical tourists in choosing a destination, with particular emphasis on the role played by destination image. The study reveals that the image of a destination with regard to hygiene and its image with regard to safety are also important in choosing a medical-tourism destination.

The setting of the research is Malaysia since the country has gained worldwide attention and positive destination image (Mohamad, Abdullah & Mokhlis, 2012). According to Medical Tourism Association 2010 Survey, Malaysia is amongst the top ten popular medical tourism destination with Prince Court Medical Center, Kuala Lumpur has been qualified as number one 2013 world's best hospital for medical tourism by Medical Travel Quality Alliance (MTQUA).

On predicting the probability of behavior, Theory of Planned Behavior (TPB) as proposed by (Ajzen, 1991) has been applied extensively over the past years. It suggests three influencing factors that determine individuals' behavioral intention, and that include attitude, subjective norm and perceived behavioral control. In this study, these socio-psychological factors can be used to predict the probability of a consumer's choice behavioral intention.

Drawing from literature, this paper develops a conceptual model assessing medical tourists' intention to adopt Malaysia as destination choice in medical tourism. Specifically, the proposed model aims to: (1) assess the robustness of TPB with attitude, subjective norms and perceived behavioral control in affecting medical tourists' intention towards adopting a Medical Tourism destination choice behavior, and (2) to determine linkages between, perceived destination image, perceived value and perceived behavior control.

2 LITERATURE REVIEW

2.1 Medical tourism

Tourism has become one of the major industries that could contribute to country's development. Melanie (2009) refers tourism as the activities of visiting or travelling to a destination that exceeds fifty miles from the residence for no more than a year due to recreational purposes or fulfilling other non-formal agendas. Middleton and Clarke (2012) suggest the main tourism offering are destination attractions; destination facilities and services; accessibility of the destination; images, brands and perceptions; and price to the visitor.

Sen Gupta (2004) view medical tourism as a provision of cost-effective medical care to patients in collaboration with the tourism industry. Connell (2006) further defined medical tourism as constituting a form of popular mass culture, to which individuals travel long distances to obtain medical, dental, or surgical services while being holidaymakers in a more conventional sense. Recently, medical tourism is viewed as the process of patients travelling abroad for medical care and procedures as a result of unavailability or unaffordability of certain medical procedures in their own respective countries (Voigt et al., 2010). The plausible reasons for tourist searching out medical care in another country include cost, access, expertise, quality and service (Hume & Demicco, 2007).

2.2 Intention

Ajzen (1991) stated that intentions refer to the motivational factors that influence a person's behavior. Thus, it indicates how willing a person is to attempt a behavior, and how much effort he is likely to exert toward that behavior. Individual intentions are an indication of the readiness to undertake a given behavior, and are assumed to precede actual behavior (Ajzen, 2002).

Diverse models of consumer behavior are building upon the theory of reasoned action which emphasizes intention as an immediate antecedent to actual behavior (Fishbein & Ajzen, 1975). Alegre and Cladera (2009) have pointed out that several studies of consumer intentions, for instance, to make repeat purchases or visit a destination have focused on the factors that determine this intention. These suggested determinants are tourist motivations, a satisfactory prior experience, perceived quality and a previous repeat visit to a destination.

2.3 Theory of planned behavior

The TPB was developed by Ajzen (1985), and it provides a well-documented and rigorously tested framework for investigating goal-oriented behaviors. In brief, this theory states that attitude towards the act; subjective norm and perceived behavioral control are indirectly linked to behavior via intention.

According to the TPB, attitude refers to evaluations of both favorable and unfavorable behavior. Subjective norm describes the perceived pressure from others to commit such behavior. Perceived behavioral control refers to how easy or difficult it would be to carry out the act. Although the TPB has been used extensively in social psychological research to explain a variety of human behaviors, this model has received only limited attention in the destination behavior in medical tourism sector.

Martin, Ramamonjiarivelo and Martin (2011) on the other hand, has used TPB model in developing a scale of medical tourism, and the result shows that all the three variables of TPB, i.e. attitude, subjective norms and control, has shown positive significant impacts on intention to participate in medical tourism. Hence, the proposed study intends to minimize the gap by drawing on TPB in the adoption of destination choice behavior in medical tourism.

3 RESEARCH MODEL AND HYPOTHESIS

3.1 Attitude

Ajzen (1991) stated that attitude can be considered as a sum of beliefs. Person will have an attitude toward that phenomenon based on the overall evaluation of his or her beliefs (Ajzen, 1991). The author also states that people can simultaneously hold two different attitudes categorized as implicit and explicit attitudes. The explicit attitudes are retrieved in favor of an implicit evaluative response by means of motivation and capacity. In medical tourism, attitude is developed based on the motivational belief to travel to seek for medical treatment. In this study, attitude towards behavior generally reflect a consumer's overall affective evaluation of the outcomes of adopting destination choice behavior in medical tourism.

Putit (2008) stated that attitude has a positive and significant relationship with intention to transact online for travel services.

Similar results was also found by Reza Jalilvand, Samiei, Dini, and Yaghoubi Manzari (2012) found significant relationship of electronic word of mouth, destination image, tourist attitude toward destination and travel intention using TPB approach. In this proposed study, it is also posited that attitude significantly influences medical tourists' intention towards adopting destination choice behavior. The hypothesis developed as following:

H1: There is a positive significant relationship between attitude and intention to adopt destination choice behavior.

3.2 Subjective norm

Subjective norm (SN) or social norms deals with a consumer's motivation to perform a behavior, and is constructed to incorporate the expectations of what the important people in their life (e.g., family, friends, and significant others) think about performing that particular behavior (Ajzen, 1991).

Taylor and Todd (1995a) decompose subjective norms into two aspects, that is, societal and social norms. Societal norms refer to adherence to the larger, societal fashions involving social community or mass media. Jalilvand et al. (2012) emphasized on the socio-demographics characteristics influence using electronic word of mouth in assessing the travel intention. The role of social community such as the doctor, nurses and the club plays vital role in develop the intention in medical tourism.

On the other hand, social norms reflects adherence to opinions from family, friends, and peers (Taylor & Todd, 1995). In medical tourism, social norm plays an important role towards developing behavioral intention of medical tourists. The medical tourists normally tend to travel along with family or friends in seeking for medical treatment. It is thus posited here subjective norm significantly affects intention to adopt destination choice behavior; hence the following hypothesis:

H2: There is a positive significant relationship between subjective norm and intention to adopt destination choice behavior.

3.3 Perceived destination image towards perceived behavioral control

Perceived Destination Image refers to a sum total of the images of individual element or attributes that make up the tourism experience (Milman & Pizam, 1995). The destination image could be conceptualized as a formation of multi-dimensional impression from tourist's perception of the functional and psychological attributes of a destination. They also develop a conceptual model that explain destination image holistically by integrating three levels of image such as country image, island image and resort image.

If travelers are highly familiar with a destination, they may not collect any additional information from external sources (Snepenger & Snepenger, 1993). Previous study by Huang (2009) attempted to explain travelers' behavioral intentions, by using a model which was developed based on existing

human behavior theories of reasoned action and planned behavior respectively. The study revealed that destination image and subjective norm positively impact behavioral intentions. However, the literature does not cover attitude and perceived behavior control as predictor.

Meanwhile, Reza Jalilvand et al. (2012) also found that the relationship of destination image that positively affects tourists' attitude. If the destination country of medical tourism has a positive image, it will lead to positive attitude towards developing behavioral intention in medical tourism. Since Malaysia has gain a positive destination image (Mohamad, et al., 2012), it is believed that perceived destination image will become one of the most important predictors in medical tourism. As such, this study proposes the role of destination image in towards perceived behavioral control respectively, hence the following hypotheses:

H3: There is a positive significant relationship between perceived destination image and perceived behavioral control.

3.4 Perceived value towards perceived behavior control

Kotler and Keller (2006, p. 133) defined perceived value as "the difference between the prospective customer's evaluation of all the benefits and all the costs of an offering and the perceived alternatives. Perceived value is a context-specific construct that may drive customers' attitudes and behaviors. In particular, little empirical research has been carried out to examine the perceived value of medical tourism services.

The major driving force of medical tourism industry is the provision of cost effective private medical care combined with the attraction of visiting exotic sites in the destination countries (Sen Gupta, 2004).

Han and Hwang (2013) indicated that perceived benefits were generally associated with perceived value and behavioral intentions, that value had a significant mediating impact, and that national culture had a significant moderating role in the proposed relationships. Alternatively, the perceived medical quality, service quality and enjoyment were critical components that significantly influenced the perception of value. Regarding sacrifice, the effects of perceived risk on perceived value were significant. Therefore, in this context, this proposed study aims to investigate the relationship between perceived values towards perceived behavioral control. Hypothesis developed as following:

Hypothesis 4: There is a positive significant relationship between perceived value and perceived behavioral control.

3.5 Perceived behavior control

According to Ajzen (1991), it is expected that those who perceive more behavioral control have more intention of performing that behavior. Perceived behavioral control (Ajzen, 1985, 1991; Ajzen & Madden, 1986) reflects the person's belief about his or her own ability to perform the intended behavior.

Ko (2011) found that medical tourism system has significant influence on a patient's perceived behavioral control. This model explains several main elements of medical tourism as types of medical tourists, medical tourists generating regions, medical tourist destination regions, types of medical service providers, medical agencies, medical tourism products, areas of medical tourism and types of relevant human resources.

In medical tourism, many non-volitional factors may obstruct a successful outcome and this includes language and cultural barriers, differences in financial currencies. In these cases, the possibility of failure due to factors perceived to be beyond the direct control of a person attempting to perform certain behavior will impact the likelihood that she or he will actually attempt to perform that activity (Martin, et al., 2011). Therefore, this study will further investigate on the direct relationship between perceived behavioral control and intention to adopt destination choice behavior. Hypothesis developed as per below:

H5: There is a positive significant relationship between perceived behavioral control and intention to adopt destination choice behavior.

Drawing from the above literature, this paper develops a conceptual model assessing medical tourists' intention to adopt medical destination choice behavior. Specifically, the proposed model aims to: (1) assess the robustness of TPB with attitude, subjective norms and perceived behavioral control in affecting medical tourists' intention towards adopting a medical tourism destination choice behavior, and (2) to determine linkages between perceived destination images, perceived value and perceived behavior control. Following the discussion, the conceptual framework generated as shown in Figure 1.

4 METHODOLOGY

Both qualitative and quantitative approach will be used to develop the framework of research. The first phase will use the unstructured methods to confirm existing hypothesized constructs and/or to identify potentially additional constructs that may be relevant to the group being studied. The

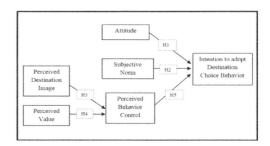

Figure 1. Conceptual framework.
Sources: Adapted from Ajzen (1985), Huang (2009), and Han and Hwang (2013).

method to be utilized in this phase is in the form of focus groups; interactive interviews will also be carried out to gather detailed information regarding respondents' purpose of medical tourism visit. The purpose in conducting qualitative is to confirm the framework develops and adding any input from the first phase before proceeds to the second phase.

The second phase involves quantitative approach comprising of structured survey questionnaires tapping on tourists' intention to adopt Malaysia as a medical tourism destination choice. The cross sectional technique will be used in collecting the data. A series of a 7-point Likert scale will subsequently be used to measure these constructs. A total of 600 tourists will be identified as target respondents for this study; judgmental sampling will be used to assure representatives, and the unit of analysis will be tapped on medical tourists visiting Malaysia for seeking medical procedures in private hospitals that registered under Malaysia Health care Tourism Council and Joint Commission International accredited.

Data analysis will be utilized by using Social Package Statistical Sciences (SPSS) version 20.0 and structural equation modelling (SEM) that provides statistical procedures that include cluster analysis and factor analysis, evaluation of the measurement model and structural model development respectively.

5 CONCLUSION

In essence, the research extends the application of the theory of planned behavior by exploring its relationship with destination image, perceived value in adoption of destination choice. Understanding individual's behavior is a result of multiple traits beyond the disciplines of socio-psychology that permeates within his/her internal driving forces.

Other than that, the study assess on the robustness of the theory of planned behavior, attitude, subjective norm and perceived behavior control. Specifically, it expects to contribute towards new knowledge building on the destination choice behavioral adoption model. It extends the application of the theory of planned behavior by exploring its relationship with destination image and perceived value towards the provision of medical services.

Subsequently, it is expected that there exists a potentially increased level of awareness amongst medical tourists' towards these services, and that more foreign medical tourists would identify Malaysia as a medical destination choice of theirs. It also could potentially contribute towards the development of marketing strategies' and subsequently set a benchmark for how any nations, including Malaysia as a forefront service provider in this sector, and thus generating higher economic status

ACKNOWLEDGMENTS

This research was supported by the Fundamental Research Grant Scheme of the Ministry of education of Malaysia.

REFERENCES

Ajzen, I. (1985). From intentions to actions: A theory of planned behavior. In J.K. Beckman (Ed.), *Action control: From cognition to behavior*. Berlin Heidelberg: Springer.

Ajzen, I. (1991). The theory of planned behavior. *Organizational Behavior and Human Decision Processes, 50*(2), 179–211.

Ajzen, I. & Madden, T.J. (1986). Prediction of goal-directed behavior: Attitudes, intentions, and perceived behavioral control. *Journal of experimental social psychology, 22*(5), 453–474.

Alegre, J. & Cladera, M. (2009). Analysing the effect of satisfaction and previous visits on tourist intentions to return. *European Journal of Marketing, 43*(5/6), 670–685.

Bookman, M.Z. & Bookman, K.R. (2007). *Medical tourism in developing countries*. New York, NY: Palgrave Macmillan Ltd.

Connell, J. (2006). Medical tourism: Sea, sun, sand and... surgery. *Tourism management, 27*(6), 1093–1100.

Fishbein, M. & Ajzen, I. (1975). *Belief, attitude, intention and behavior: An introduction to theory and research*. Boston.: Addison-Wesley.

Han, H. & Hwang, J. (2013). Multi-dimensions of the perceived benefits in a medical hotel and their roles in international travelers' decision-making process. *International Journal of Hospitality Management, 35*(December), 100–108.

Huang, Y.-C. (2009). *Examining the antecedents of behavioral intentions in a tourism context*. Doctoral of Philosophy, Texas A&M University. Retrieved from http://scholar.google.com.my/scholar?hl=en&as_sdt=0,5&q=Examining+the+Antecedents+of+Behavioral+Intentions+in+a+Tourism+Context

Hume, L.F. & Demicco, F.J. (2007). Bringing hotels to healthcare: A RX for success. *Journal of Quality Assurance in Hospitality & Tourism, 8*(1), 75–84.

Jotikasthira, N. (2010). *Salient factors influencing medical tourism destination choice*. DBA thesis, Southern Cross University, Lismore, NSW. Retrieved from http://scholar.google.com.my/scholar?hl=en&as_sdt=0,5&q=%27Salient+factors+influencing+medical+tourism+destination+choice%27

Ko, T. (2011). Medical tourism system model. *International Journal of Tourism Sciences, 11*(1), 18–52.

Kotler, P. & Keller, K. (2006). *Marketing Management* (12th ed.). Singapore: Pearson Education.

Martin, D.S., Ramamonjiarivelo, Z. & Martin, W.S. (2011). MEDTOUR: A scale for measuring medical tourism intentions. *Tourism Review, 66*(1/2), 45–56.

Middleton, V.T. & Clarke, J.R. (2012). *Marketing in travel and tourism* (3rd ed.). Butterworth Heinemann, Oxford: Routledge.

Milman, A. & Pizam, A. (1995). The role of awareness and familiarity with a destination: The central Florida case. *Journal of travel research, 33*(3), 21–27.

Mohamad, M., Abdullah, A.R. & Mokhlis, S. (2012). Tourists' evaluations of destination image and future behavioural intention: The case of Malaysia. *Journal of Management & Sustainability, 2*(1), 181–189.

Putit, L. (2008). Consumers' E-Transaction Behaviour Adoption: An Exploratory Study. *Journal of Human Capital Development, 1*(1), 41–46.

Reza Jalilvand, M., Samiei, N., Dini, B. & Yaghoubi Manzari, P. (2012). Examining the structural relationships of electronic word of mouth, destination image, tourist attitude toward destination and travel intention: An integrated approach. *Journal of Destination Marketing & Management, 1*(1), 134–143.

Sen Gupta, A. (2004). Medical tourism and public health. *People's Democracy, 28*(19).

Snepenger, D. & Snepenger, M. (1993). *Information search by pleasure travelers*. New York: Van Nostrand Reinhold.

Taylor, S. & Todd, P. (1995). Decomposition and crossover effects in the theory of planned behavior: A study of consumer adoption intentions. *International Journal of Research in Marketing, 12*(2), 137–155.

Voigt, C., Laing, J., Wray, M., Brown, G., Howat, G., Weiler, B. & Trembath, R. (2010). *Health tourism in Australia: Supply, demand and opportunities*: CRC for Sustainable Tourism.

Theory and Practice in Hospitality and Tourism Research – Radzi et al. (Eds)
© 2015 Taylor & Francis Group, London, ISBN 978-1-138-02706-0

Establishing Shariah-compliance hotel characteristics from a Muslim needs perspective

M.S. Nor-Zafir
Universiti Teknologi Malaysia, Johor, Malaysia

ABSTRACT: The purpose of this paper is to develop the characteristics of Shariah-compliance hotel (hereafter SCH) from the perspective of Muslim needs. SCH is a newly innovative services introduced by hotels to satisfy the Muslim travelers. Nonetheless, lack of study found in the current literature on SCH characteristics that has been developed base on Muslim needs and the implementation of SCH. This study focuses on developing a model of SCH and issues or challenges in implementing it. This is a preliminary study and uses in-depth, open-ended interviews through snowball sampling. Interviews were conducted with six experts in the hospitality industry and data were recorded, transcribed and categorized in order understand the issues of implementing SCH. The difficulties to obtain Halal certificate and the absence of SCH standard affected the practice of SCH in Malaysia. Along with this, two Malaysian Standards (MS1900:2005 & MS1500:2004) have been identified as a tool to implement SCH.

Keywords: Halal, Islamic hotel, Muslim needs, Shariah compliant hotel, qualitative

1 INTRODUCTION

The Halal revolution has created a high awareness and demand among Muslims to consume Halal products or Islamic services for example the Shariah-compliance hotel (hereafter SCH). SCH is a new innovative and creative services offered by hotels around the world (Nor Zafir, Abu Bakar & Hashim, 2014). Recently, the existence of Islamic hospitality or Shariah hotel is mushrooming beyond the Gulf countries until Asian countries (Tarrant, 2010). This development can be traced not limited to Malaysia and Indonesia only which the Muslims are their majority population, but also to the countries that the Muslims are the minority residents, such as Thailand, Philippine and China (Anonymous, 2011; Lee, 2010).

However, there is a lack of comprehensive studies found regarding to Halal or Shariah compliance practices in Malaysian hospitality industry (Zailani, Omar & Kopong, 2011). As such, this study will contributes significantly on the literature of Shariah-compliance hotel concept in Malaysia. The purposes of this paper are to develop the SCH characteristics based on the daily Muslim requirements and to provide better understanding of SCH concept by exploring the issues and challenges in implementing it. This study adopted snowball sampling technique and in-depth interviews with six experts of hotel industry including government officials, hoteliers and hotel associations that

represent hotels industry in Malaysia. The first section will describe the Muslim tourists' needs, SCH development, SCH concept and its characteristics.

2 LITERATURE REVIEW

2.1 Determining the needs of Muslim travelers

The needs of Muslim during travelling could be developed based on the Pillars of Islam and the Article of Iman. This is because a Muslim must perform and belief on the Pillars of Islam and the Article of Iman otherwise it is a sin. As Islam is a way of life, a Muslim is required to perform the religious duties everywhere (Begg, 1997). For example, a Muslim traveler needs to perform his or her religious routine such as prayer (Salaah) five times a day and consume Halal food during travelling.

On the supply side perspective, hotels could provide Islamic services that comfort the Muslim travelers to carry out their religious duties. For example, a hotel could put up the sign on the direction of Mecca in its rooms, offer prayer mat upon checking-in, provide Halal foods and information about the prayer time (Hashim, Murphy & Mohammad, 2006). In the view of some experts, the analysis of religious needs is vital for hotels to provide hotel services and create a winning marketing strategy (Muhamad & Mizerski, 2010; Weidenfeld, 2006). Thus, it is worthwhile for hotel managers to understand how Muslims should perform their religious routines in order to satisfy their needs during travelling. This knowledge will provide a better understanding to the hotel managers on a new service that could give greatest impact on potential Muslim travelers' choice. Table 1 presents the suggested hotel services based on Muslim travelers' needs.

Table 1 shows the guidelines of Shariah Law compliance hotel services. The first column is the daily religious duties that a Muslim has to perform and rules for them to obey base on Al-Quran and Sunnah or example, solat 5 times a day, fasting during Ramadhan month and segregation between man and woman. The second column presents hotel services that hotel could provide to Muslim guests facilitating their religious routine such as qiblah direction and prayer mat in guest room, availability of Halal food in hotel premise and solat time. The last column shows the supplier of the stated Islamic services.

2.2 *The evolution of SCH concept*

As Islam is a way of life, Islamic hospitality is embedded in Islamic religion, culture and experience. It is extensively practiced and observed in Middle East where Islam is originated. It offers traditional principles and custom for Muslim travelers (Kaaki, 2008).

The concept of Shariah hotel has evolved from fulfilling religious needs to lifestyle option which emphasizing on family and health (Anonymous, 2009). In the Middle East, it is believed that many hotels offer basic Islamic services that address the Muslim's religious needs since more than 30 years ago. This trend seems to experience a shift from providing basic services to full services, also extend the target market from only Muslims to non-Muslims by striking on the selling proposition of family and health oriented concept (Henderson, 2010). However there is lack of empirical researches or

Table 1. The hotel services, Muslim needs and related hotel function.

Routine	Hotel services	Hotel functions
Solat Requirements:		
Ablution or Wudu'	Shower hose or bidet in the toilet.	Room
Clean prayer place	Prayer mat on loan or prayer room.	Housekeeping
Covering body part or Awrah	Telekung or kainsarung or prayer cap on loan.	Housekeeping
Solat time	Solat time schedule for easy reference in the room. Azan on the radio or speaker in the room.	Front Office/Housekeeping
Facing Qiblah	Provide Qiblah direction in the room.	Room
Congregational prayer	Provide imam for the 5 obligatory prayers at prayer room/surau.	Management
Siyam or The Fast		
Sahoor (Take food before dawn)	Available food at room service for Sahoor.	Food and Beverage
Break fast	Availability of dates in the restaurant.	Food and Beverage
Taraweeh prayer	Host the congregational prayer of Taraweeh.	Management/Marketing
Studying Quran	Availability of Quran in the room.	Housekeeping
Adab	No adult entertainment channel.	Room
Az-Zakah		
Payment of Zakah	Invite Religious Department to open a collection counter of Zakah.	Management/Marketing
Perform Hajj		
Enhancing knowledge of Hajj	Offer or arrange classes of Hajj to the guests upon request. Brochure of Hajj is available.	Management Marketing
General needs of a Muslim traveler:		
Food	Provide Halal food.	Food and Beverage
Facilities	Separate female & male gym, swimming pool, game room, function room and spa.	Recreation
Room entertainment	Provide appropriate channels for the guest, no adult entertainment.	Room
Recreation	Arrange with travel agent, the Islamic tourism packages.	Marketing

Source: Original

articles written that acknowledge the underdeveloped concept of SCH (Henderson, 2009; Rosenberg & Choufany, 2009; Wilkinson, 2007).

The literature provides several proposals on guidelines and constructs of SCH. However, the agreement on a common set of construct or guideline does not appear to exist (Abdul Hamid, 2010). Most of the authors provide guidelines that narrowly focused to Shariah-compliant services offered by hotels thus fail to see the actual needs of Muslim guests and ignore the management aspect of SCH. Perhaps, this is due to the scholars' background where none of them are Muslim and most of the writers stay at non-Muslim countries. In addition, the scope of previous researches only limited to Middle East area that consists of minority Islamic population (Henderson, 2010). Since, the SCH guidelines are conceptual, an empirical research on how hotels practice Shariah-compliance concept is required to provide a strong guideline for future reference by hotel internationally.

Therefore, it is important to understand what has been practiced by Malaysia hotels in terms of Islamic services to develop a better picture on this matter and give insights on the practice of SCH at the Muslim populated country. This paper report initial findings of SCH practices and contributes to the Malaysia hospitality industry by providing information on the implementation of SCH by Malaysia hotel.

2.3 The growth of SCH

Several factors have contributed to the growth of Shariah compliant hotels. These include: 1) The increase efforts to develop tourism industry among Organization of Islamic Conference (OIC) countries and enhance travelling within its members; 2) The attractiveness of high spending and lucrative market of the Middle East (ME) travelers that sparks the interest of many hotels to provide Islamic services to fulfill their needs; 3) Stringent security measures and difficulties to get visa approval in the West shifted the ME travelers to the East and increase demand on Islamic services in tourism industry; 4) Halal revolution has created high awareness by Muslim on their religious needs and it creates an opportunity to serve the Muslim tourist untapped needs; 5) The growth of Islamic banking and finance increase investors' interest to invest in Halal products and services such as SCH; and 6) The fast growing Muslim market is estimated at USD$500 billion annually and Halal market worth USD$2.1 trillion in the world and this increase the popularity and visibility of the SCH (Abdul Hamid, 2010; Henderson, 2009; Henderson, 2010; Kaaki, 2008; Zailani, et al., 2011).

However, according to many researchers, the main factor that impact the SCH development is the absence of SCH standard and requirements (Abdul Hamid, 2010; Birchall, 2009; Henderson, 2010; Rosenberg & Choufany, 2009; Siddiqui, 2011). Henderson (2010), Rosenberg and Choufany (2009) and Birchall (2009) mentioned that without an established standard of SCH, the practice of SCH will be different from one hotel to another (Anonymous, 2009).

3 RESEARCH METHODOLOGY

This preliminary study adopted in-depth, open-ended interviews through snowball sampling to understand the practice and generate a set of SCH attributes in Malaysia. Snowball sampling was applied as the population is small and specialized. The interview process began with two experts in Malaysia's hospitality industry who later recommended other experts to provide more information regarding the subject matter. In total six experts from diverse background were interviewed. The interviews were recorded, transcribed and categorized within a week, based on interview questions. The thematic analysis was employed to identify relevant themes after examination of the data. The respondents were asked about their opinion on issues in implementing SCH. Several issues emerged and discussed in the next section.

4 DISCUSSION

4.1 Government regulation

From the interviews conducted, several issues were highlighted by the respondents including the government regulation, the application to get Halal certificate, and the Shariah-compliant standard hotel. The encouragement for hotel to obtain Halal certificate for its restaurant or food and beverage outlet is seen as a good one at the right time. Recently, Malaysian government enforce the regulation that only hotel with Halal certificate for restaurant can cater the government functions. The effect of this is tremendous as many hotels were forced to or would like to obtain Halal certificate for having their restaurant certified Halal enable them to host government events.

4.2 Halal certificate

Delayed process to obtain Halal certificate by JAKIM, a body that is responsible to grant Halal certification in Malaysia, lower the motivation of hotels to obtain Halal certification. Moreover, the requirements for restaurant to be Halal certified are not friendly to hospitality industry, and these resulting difficulties for hotels to apply. Majority of the respondents have the same opinion that JAKIM needs to simplify the process and requirement of obtaining Halal certificate. Respondent A and B highlighted that currently, JAKIM is addressing this issues and continuous discussion among stakeholders of hospitality industry is going on to improve the Halal application system. This move will boost Malaysia image as one of top tourist country among Muslim tourists for example from Middle East.

4.3 Shariah compliant hotel standard

The absence of Shariah-compliant standard is an important category emerged in this study. As respondent B highlighted, "Many hotels claimed that they are SCH without understand the term Shariah itself. They promote SCH as their marketing strategy but not as the genuine concept of the hotels. In my opinion, the SCH concept should be developed based on the Islamic perspective of doing business and not as a tag line in the promotion campaign". Another respondent explained that, "in our hotel, we adopted two Malaysia Standards such as MS1500:2009 (Malaysian Standard of Halal Food-Production, Preparation, Handling and Storage by JAKIM) and MS1900:2005 (Malaysian Standard for Quality Management System-Requirements from Islamic Perspectives) to implement SCH". In contrast, respondent E mentioned that, "the Shariah-compliant concept is only applicable in the restaurant". It is clear that hotels implement varied concept of SCH in Malaysia. Thus, it is timely for a government to consider developing a standard guideline for SCH.

4.4 Muslim needs perspective SCH model

The respondents were asked on the suitability of SCH model established for this study and majority of them agreed it should be based on the Muslim needs. All of the respondents highlighted that the application of MS1500:2009 which is Malaysian Standard of Halal Food-Production, Preparation, Handling and Storage (Halal certification by JAKIM) could be used as a tool to apply SCH into practice. In addition, one respondent suggested that the MS1900:2005 which is Malaysian Standard for Quality Management System-Requirements from Islamic Perspectives could be adopted in implementing SCH in the hotel operation. Thus the development of SCH model in Malaysia should be based on the Muslim's needs and include several standards like MS1500:2009 and MS1900:2005.

5 CONCLUSION

The universal standard of SCH is urgently needed to encourage more hotels in Malaysia to seriously consider applying it in their operations. The benefits of SCH are many, one of it is SCH could be a competitive advantage for the hotel as the demand on Islamic goods and services are increasing. In addition, the influx of Middle East tourists to Malaysia provides immense opportunities for SCH hotels to be able to attract these tourists to stay in their hotels and help to increase their profits. As mentioned earlier, the Middle East tourists are a lucrative market because they spend more than other tourists.

Nevertheless, the decision to adopt SCH must be made as according to the hotel's target market so as not to affect their financial gain. The composition of hotel's target market should determine whether the hotel should apply basic SCH or full service SCH. Therefore, the application of SCH needs a careful consideration for hotel operated in a multiracial country like Malaysia. Future research may need to address this issue from the perspective of profit gained and innovation services addressing different religious needs in applying SCH in Malaysia.

As a conclusion, this paper identified several issues in relation to the practice of SCH including the government regulation, difficulties to obtain Halal certification and the increasing need of SCH standard. This paper ended by proposing the characteristics of SCH for practice in Malaysia. The proposed characteristics of SCH are based on the Muslim travelers' needs and the hotel services for SCH (Table 2).

Table 2. The proposed characteristics of SCH.

Muslim Travellers' Needs (Customer focus)	Hotel Services (Shariah-compliant)
1. Obligatory Religious Activities • Solat Requirement • The Fast or Siyam • Az-Zakah • Perform Hajj	1. Special services for religious activities: • Shower hose, Prayer mat, mosque location, prayer dress, solat time, Qiblah, Imam. • Sahoor and Breakfast food, Taraweeh, Quran. • Payment of Zakah • Knowledge of Hajj
2. General needs of a Muslim traveler: • Food • Facilities • Room entertainment • Recreation	2. Services for general needs: • Halal food • Separate male and female. • No adult entertainment • Islamic tourism packages
3. General rules for Islamic values in hotel management: • Management • Finance • Operation • Interior design and decoration	3. Hotel Management practice: • MS1900:2005 & MS1500:2004 • No Riba • Dress code, Separate floor or for male and female • Toilet face away from Kaaba, no art deco on human part

Source: Original

REFERENCES

Abdul Hamid, I. (2010). *Islamic compliance in hotel and restaurant business.* Paper presented at the Asia-Euro Conference: Transformation and modernisation in tourism, hospitality and gastronomy.

Anonymous. (2009). Demand for Shariah-compliant hotels rising. *Emirates, 24*(7).

Anonymous. (2011). Crescent rating: Halal friendly hotel rating system, from http://dinarstandard.com/marketing/muslim-travel-guides

Begg, M. (1997). Islam and travelling Retrieved July 13, 2011, from www.islamiccentre.org

Birchall, L. (2009). Laying down sharia law, from http://www.hoteliermiddleeast.com/5750-laying-down-sharia-law/1/print/

Hashim, N.H., Murphy, J. & Mohammad, N. (2006). Tourism and Islam: Understanding and embracing opportunities. *From The Experts.*

Henderson, J. (2009). Islamic tourism reviewed. *Tourism Recreation Research, 34*(2), 207–211.

Henderson, J.C. (2010). Sharia-compliant hotels. *Tourism and Hospitality Research, 10*(3), 246–254.

Kaaki, L. (2008). A halal environment: Demand for Islamic hotels on the rise, *Arab News.*

Lee, G. (2010). Manila hotel 1st to get halal certificate. *Manila Bulletin.*

Muhamad, N. & Mizerski, D. (2010). The constructs mediating religions' influence on buyers and consumers. *Journal of Islamic Marketing, 1*(2), 124–135.

Nor Zafir, M.S., Abu Bakar, A.H. & Hashim, N.H. (2014). The practice of Shariah-compliant hotel in Malaysia. *International Journal of Trade, Economics and Finance, 5*(1), 26–30.

Rosenberg, P. & Choufany, H. (2009). Spiritual Lodging– the Sharia-Compliant Hotel Concept. *HVS Global Hospitality Services–Dubai.*

Siddiqui, R. (2011). Opinion: Shariah compliant hotels. *Halal Focus.*

Tarrant, C. (2010). *Middle East hotel gusts express high interest in sharia compliant hotels.* United Kingdom:: BDRC.

Weidenfeld, A. (2006). Religious needs in the hospitality industry. *Tourism and Hospitality Research, 6*(2), 143–159.

Wilkinson, G. (2007). Dry hotels-old hat or new fad? *Arabian Business. com.*

Zailani, S., Omar, A. & Kopong, S. (2011). An exploratory study on the factors influencing the non-compliance to Halal among hoteliers in Malaysia. *International Business Management, 5*(1), 1–12.

Theory and Practice in Hospitality and Tourism Research – Radzi et al. (Eds)
© *2015 Taylor & Francis Group, London, ISBN 978-1-138-02706-0*

Education and training as a challenge facing Small to Medium Tourism Enterprises (SMTEs) of the Eastern Cape (EC) province of South Africa (SA): An empirical study

D. Vallabh & O. Mhlanga
Walter Sisulu University, East London, South Africa

ABSTRACT: Small businesses contribute prominently to the economy through creating more employment opportunities and generating higher production through entrepreneurship education and training. The study is focusing on Eastern Cape Province (EC), economically the poorest province in South Africa. The study identifies the barriers to education and training in small to medium tourism enterprises. Further, the study investigates the level of education and training in the sector. Quantitative research was deemed appropriate for the study, whereby systematic random sampling was employed to select a sample of 332 respondent organizations. The largest proportion of managers (37.4%) had obtained a national diploma as the highest education qualification and only 10.2 percent held a postgraduate qualification. Less than 30 percent of managers had training or experience in Human Resource Management, Strategic Management and Project Management. It is recommended that information technology, computer training and internet access be a priority in the school curricula.

Keywords: Economy, education and training, information and technology, small medium tourism enterprises

1 INTRODUCTION AND LITERATURE REVIEW

The South African (SA) tourism and hospitality industry is one of the country's leading economic growth sectors, not only because of its contribution to the gross domestic product, but also because of its creation of sustainable jobs, particularly, in the small business sector of the country (Vallabh, 2014). The majority of these new jobs are expected with the small, micro and medium enterprises (SMMEs), across the broader SA travel and tourism economy (Rogerson, 2005; Thomas, 2004). Hence, greater consideration should be given to SMTEs to help them be innovative, adopt systems that will allow them to explore their potential. It is therefore argued that in order to achieve this, SMTEs need to maintain and understand the importance of education and training initiatives.

There is consensus among policy makers, economists and business experts that small and medium enterprises (SMEs) are drivers of economic growth. A healthy SME sector contributes prominently to the economy through creating more employment opportunities, generating higher production volumes, increasing exports and introducing innovation and entrepreneurship skills (Tassiopoulos, 2011). According to Nguyen and Waring (2013), SMEs are the first step towards development in economies towards industrialization. Thurik and Wennekers (2004), however, estimated that the SME contribution to employment generation was 39% in South Africa.

The application of education and training by SMTEs can help them overcome the management issues that often confront these small businesses. Skelton (2013) has demonstrated that integrating education and training into the business process and operating in the tourism and hospitality industry can offer SMTEs many advantages in operational, tactical and strategic management. Despite this acknowledged importance and SME contribution to economic growth, SMEs across the globe and in South Africa in particular, are still faced with numerous challenges that inhibit entrepreneurial growth. Consequently, a number of key constraints limit the effectiveness of the tourism and hospitality industry from playing a more meaningful role in the national economy. The major constraints include:

- the inadequate funding of resources by government;
- limited involvement by the private sector;
- the limited integration of local communities and previously neglected groups into tourism;

- the inadequacy or absence of tourism training, education and awareness;
- inadequate protection of the environment through environmental management;
- generally poor levels of service standards within the industry;
- lack of infrastructure in rural areas;
- lack of appropriate institutional structures; and
- prevalence of violence, crime and inadequate security as they affect tourists and tourism (Republic of South Africa, 1996)

Given the constraints facing the tourism and hospitality sector nationally, the focus of this study is the EC, economically, the poorest province in SA. The EC is largely a tourism destination and is the third most popular destination for domestic tourists in SA (Tourism Enterprise Partnership (TEP), 2010). It is characterized by high levels of unemployment, poverty and crime, which impact negatively on the economy and are problematic to the sustainability of SMTEs (Vallabh & Radder, 2010). Tourism and hospitality enterprises are therefore important to the economic development of the EC because of the positive contribution it can make to job creation, poverty alleviation and growth.

2 PROBLEM STATEMENT

A study conducted by Baumeister (2002) estimates that 91 percent of the formal business entities in South Africa are SMEs of which they contribute between 52 to 57 percent to GDP and account for approximately 61 percent of employment. However, the failure rate for SMEs in South Africa is considered to be higher than the average failure rate for other businesses. Brink and Berndt (2009) estimated that 30 percent of SMEs fail during the first year of operation, although claims are made that it is closer to 90 percent. Cameron and Miller (2008) noted that SA has one of the lowest SMEs survival rates in the world. Cameron and Miller (2008) identify the lack of education and training in SMEs as one of the main reasons for their failure. Chiliya and Roberts-Lombard (2012) found that in developing countries education is a major drawback for the success of small firms.

Given the myriad of challenges facing small businesses, a study to determine education and training as a challenge facing small, medium tourism enterprises (SMTEs) of the Eastern Cape (EC) province of South Africa (SA) was conducted. The results of the study could have significant implications towards job creation and generate sustainable and equitable growth in the Eastern Cape Province of South Africa. SMEs are a panacea for improving the standards of living in a society and the stability of a country.

2.1 Research objectives

The primary objective if this study is to provide a better understanding to education and training as a challenge facing small, medium tourism enterprises (SMTEs) of the Eastern Cape (EC) province of South Africa (SA). In order to achieve this primary objective, the following secondary objectives have been set:

- To identify the barriers to education and training in medium tourism enterprises (SMTEs) of the Eastern Cape (EC) province
- To investigate the level of education and training on small, medium tourism enterprises (SMTEs) performances of the Eastern Cape (EC).

3 RESEARCH METHODOLOGY

Quantitative research was deemed appropriate for this study. Systematic random sampling was employed to select a sample of 332 respondent organizations from the 2012 database of the Eastern Cape Parks and Tourism Board consisting of formally registered small to medium tourism enterprises. A total of 310 usable questionnaires were finally obtained in September 2012. Descriptive statistics was used in the study.

This study focuses on two major challenges, namely, education; and training, skills and managerial competence.

4 RESULTS AND DISCUSSION

The results of the current study provided some interesting information regarding the most senior managers' educational qualifications as reflected in Table 1. The largest proportion of managers (37.4%) had obtained a National Diploma as their highest education qualification and 10.2 percent held a postgraduate qualification.

The largest proportion of managers were trained and experienced in General Management. Less than 30 percent had training or experience in Human Resources Management, Strategic Management and Project Management (Table 2).

Herrington, Kew and Kew (2008) noted that "South Africa's inadequately educated workforce is cited as the most problematic factor for doing business in the country". Education is important to the growth and development of SMTE owners as it plays a major role in helping them cope with the problems which may beset them and can assist to correct shortcomings in business training. The ability to communicate with customers and deal with people is essential in any entrepreneurial business. SMTEs are not utilizing information technology in

Table 1. Educational qualifications.

Level of education	f	%
Grade 10 or lower	7	2.4
Grade 12 or matriculation	65	22.1
National Diploma	110	37.4
Degree (3 to 4 year qualification)	82	27.9
Postgraduate	30	10.2
Total	294	100.0

Table 2. Experience and formal training managers (N = 310).

Subject areas	Experience		Formal training	
	f	%	f	%
General Management	194	62.6	130	42.0
Technical / Practical skills	143	46.1	73	23.5
Marketing Management	113	36.4	80	25.8
Tourism Business Studies	112	36.1	100	32.2
Financial Management	105	33.8	91	29.3
Human Resource Management	88	28.3	47	15.1
Strategic Management	87	28.0	50	16.1
Project Management	74	23.8	52	16.8

their organizations to its full potential. Managers view the internet as a mechanism for promoting their hotel, rather than or training and development purposes (Ateljevic, 2007). SMTEs therefore need to educate themselves on technology and other critical areas. De Klerk, Kroon, Lesomo and Van der Walt (2005) highlighted the need for higher learning institutions to equip students explicitly for entrepreneurship and to tailor these competencies to specific local situations and needs.

Chiliya and Roberts-Lombert (2012) found the level of education of the owner impacting on the financial performance of the business. According to McCartan-Quinn and Carson (2003) management education and marketing education for SMTE owners is inherently problematic. Similarly, Tassiopoulos (2011) holds that SMTE managers have few formal qualifications and have limited previous experience in tourism. These problems need to be addressed by encouraging additional training in tourism businesses in order to improve the state of small businesses.

The most frequently cited reason for business failure is perceived to be a direct consequence of SMME management competence (Ateljevic, 2007). McCartan-Quinn and Carson (2003) defined managerial competence in the specific context of the small business as "a body of knowledge, or of skill/ability, personal qualities/characteristics, set of awareness, attitudes or outlooks, or motivations/drives, that may, in their various ways, positively and constructively contribute to effective business/

managerial thought or action". Nguyen and Waring (2013) emphasized that the role, competence, characteristics, attitudes and values of the business owner all have a significant impact on the success of the organization.

Another challenge highlighted is business management training, which is considered to be inadequate among SMTEs (SEDA, 2011). It is further noted that many tourism operators lack the relevant practical skills and experience of the industry. This includes relevant business, technical and management skills. Further, owners of the business rarely have formal training and education, which becomes the source of managerial problems. Training courses for tourism businesses in South Africa as a whole have been weak, often generic and have failed to provide the broader skills that are required by tourism owners (SEDA, 2011). Training and skills development is therefore a critical challenge that needs to be addressed by the sector. Similar findings have been found in other studies (De Klerk, et al., 2005; Mbedzi, 2011; Rogerson, 2005; Tassiopoulos, 2010). This lack of training and experience poses a core challenge in terms of human resources for the tourism sector when seeking skilled employees to meet future needs of the industry.

Based on the preceding points, SMTEs need to realize that education and training can be used for not only for operational purposes but for tactical and strategic management too (Nguyen & Waring, 2013). This can help to empower SMTEs to achieve economies of scale. Economies of scope can also be acquired through increased knowledge of internal organizational capabilities. Therefore, education and training in the hospitality industry is considered essential if SMTEs are to increase their organizational efficiency. Therefore, it should become a priority for SMTE managers to incorporate training into their overall strategy.

5 RECOMMENDATIONS FOR ENTREPRENEURIAL DEVELOPMENT

Recommendations for the growth of SMTEs included the following:

- Create a strategic vision for building SMTEs which includes a better working relationship between government, intervention agencies and target communities.
- Government need to review the restrictive employment laws which are seen as an obstacle to business growth.
- Trainers and consultants must be well-trained and experienced in specific areas of expertise that they offer, for example, marketing or accounting.

- Improvements in the quality of education at schools in the EC needs to be addressed with some urgency
- Information technology, computer training and internet access must be a priority in the school curricula. Adult education programmes must embrace such training.
- A more enabling environment that encourages individuals to see entrepreneurship as a financially viable employment option should be fostered.
- Higher Education Institutions can provide access to expertise, technology and resources that could assist small business owners.

6 LIMITATIONS OF THE STUDY

The study only focused on SMTEs of the EC, thus making it a regional-specific study. SMTEs from other provinces of SA, which were outside the targeted geographical area, were excluded due to time and financial constraints.

REFERENCES

Ateljevic, J. (2007). Small tourism firms and management practices in New Zealand: the Centre Stage Macro Region. *Tourism management, 28*(1), 307–316.

Baumeister, H. (2002). *Customer relationship management for SMEs.* Paper presented at the Proceedings of the 2nd Annual Conference eBusiness and eWork e2002, Prague, Czech Republic.

Brink, A. & Berndt, A. (2009). *Relationship Marketing and Customer Relationship Management*: Juta and Company Ltd.

Cameron, L. & Miller, P. (2008). Enhancing HRM practice in SMEs using the concept of relationship marketing.

Chiliya, N. & Roberts-Lombard, M. (2012). Impact of Level of Education and Experience on Profitability of Small Grocery Shops in South Africa. *International Journal of Business Management & Economic Research, 3*(1), 462–470.

De Klerk, S., Kroon, J., Lesomo, T. & Van der Walt, J.L. (2005). *Profile of micro businesses in a residential area in South Africa.* Paper presented at the ICSB, Washington.

Herrington, M., Kew, J. & Kew, P. (2008). Global Entrepreneurship Monitor (GEM) South African Report. Cape Town.

Mbedzi, K.P. (2011). *The role of government agencies in promoting SMME's in Limpopo: A critical assessment.* Unpublished Masters Dissertation, Stellenbosch: University of Stellenbosch.

Mc Cartan-Quinn, D. & Carson, D. (2003). Issues which impact upon marketing in the small firm. *Small Business Economics, 21*(2), 201–213.

Nguyen, T.H. & Waring, T.S. (2013). The adoption of customer relationship management (CRM) technology in SMEs: An empirical study. *Journal of Small Business and Enterprise Development, 20*(4), 824–848.

Rogerson, C.M. (2005). Tourism SMMEs in South Africa: A case for separate policy development. *Trade and Industry Policy Strategies: Working Paper, 2*, 1–28.

SEDA. (2011). Small enterprises in the tourism sector: Opportunities and challenges. Pretoria.

Skelton, R. (2013). Africa changes its tune to genuine digital downloads. Africa in Fact: The Journal of Good Governance. Retrieved 12 June 2013 from www.gga.org

Tassiopoulos, D. (2010). *An investigation into the co-producers of preferred strategic behaviour in small, micro and medium tourism enterprises in South Africa.* Unpublished thesis, Stellenbosch: University of Stellenbosch.

Tassiopoulos, D. (2011). *New Tourism Ventures* (2nd ed.). Cape Town: Juta.

Thomas, R. (2004). *Small firms in tourism international perspectives.* London: Elsevier.

Thurik, R. & Wennekers, S. (2004). Entrepreneurship, small business and economic growth. *Journal of Small Business and Enterprise Development, 11*(1), 140–149.

Tourism Enterprise Partnership (TEP). (2010). *Towards 2010 and beyond.* Johannesburg: Acumen Publishing Solutions.

Vallabh, D. (2014). *Customer relationship management in small to medium tourism enterprises (SMTES) in the Eastern Cape Province, South Africa.* Unpublished Thesis, Nelson Mandela Metropolitan University (NMMU).

Vallabh, D. & Radder, L. (2010). *Using CRM as a vehicle to grow small medium tourism enterprises (SMTEs) in a de-veloping economy: Eastern Cape, South Africa. Towards a conceptual framework.* Paper presented at the 55th International Council for Small Business (ICSB), Cincinnati, Ohio.

Theory and Practice in Hospitality and Tourism Research – Radzi et al. (Eds)
© *2015 Taylor & Francis Group, London, ISBN 978-1-138-02706-0*

Service recovery and satisfaction: The moderating role of religiosity

M.H.A. Rashid
Universiti Teknologi MARA, Puncak Alam, Malaysia

F.S. Ahmad
Universiti Teknologi Malaysia, Kuala Lumpur, Malaysia

ABSTRACT: The aim of this paper is to discuss the role of religiosity on the relationship between service recovery and recovery satisfaction. This conceptual paper is based on reflecting the relevant scholarly discussions in various conferences and available published literatures. This paper identifies that religiosity plays a significant role on the relationship between service recovery and recovery satisfaction. This is due to the notion that highly religious individual tend to be more forgiving in the event of service failure. However, this argument which is theoretical in nature needs to be statistically validated and hence proposed by this research. Additionally, extant studies demonstrated that service recovery is critical in enhancing customer satisfaction. Therefore, the three dimensions of justice theory namely distributive justice, procedural justice and interactional justice should be considered if companies plan to embark on service recovery efforts. The discussion offered in the paper is expected to be valuable for service provider seeking ways to win back upset customers in the event of service failure. The discussion established that fair compensation, reasonable policies/procedures, and effective communication process during service recovery are the key components in promoting satisfaction.

Keywords: Service failure, service recovery, recovery satisfaction, justice theory, religiosity

1   INTRODUCTION

Every company in the world strive their best to win customers' heart. However, it is nearly impossible to deliver a perfect service with zero-defect. Even a giant corporations such as Starbucks, Toyota, Sony and General Electric experienced service failure in delivering their services (Lusch & Vargo, 2006). Service failure may contribute to customer defection, negative word of mouth or the customers will straight away complain to the service provider (Kim, Kim & Kim, 2009). Considering the negative influence of service failure, service providers shall strategize effective ways to overcome the problems. Service recovery is the only way to rectify the situation. The level of compensation given should be equal to customer's loss. According to Riscinto-Kozub (2008), service recovery is vital and it is one of the key components in developing long term relationship, fortifying customer loyalty and promotes positive behavioral intentions.

Justice theory has been gaining popularity in studies related to service recovery (Nikbin, Ismail, Marimuthu & Jalalkamali, 2010). Justice theory was developed based on social exchange theory and equity theory (Ok, 2004). There are three dimensions of justice theory include distributive justice (compensation), procedural justice (policies and procedures) and interactional justice (interpersonal communication). Extant literatures claimed that the application of justice theory in service recovery have been investigated in industries such as tourism (Bernardo, Llach, Marimon & Alonso-Almeida, 2013); restaurant (Ok, 2004); airlines (Nikbin, Armesh, Heydari & Jalalkamali, 2011) and a few other industries. However, its application is still limited in Asian's service recovery context.

Of late, there is an emerging trend of religious awareness in current market (Swimberghe, Sharma & Flurry, 2009). While most service recovery studies stresses on customer's emotions and other related outcomes, fewer attention has been given to the role of customer's characteristics (Tsarenko & Tojib, 2012). Therefore, religiosity is believed to play a significant role in studies related to service failure. This is due to the notion that highly religious people tend to be more forgiving compared to less religious people in the event of transgression (Tsarenko & Tojib, 2012).

2 LITERATURE REVIEW

2.1 Service failure

Service failure is inescapable and it can jeopardize company's reputation. According to Patterson, Cowley, and Prasongsukarn (2006), service failure is defined as a problem that happens in exchange where a customer perceives a loss due to a failure on the part of service provider. However, Komunda and Osarenkhoe (2012) argued that service failure is the failure of the company's core service. It may include failure to withdraw money from the Automated Teller Machine, or the failure of product/ service provided by the company.

Service failure may happen due to a number of reasons such as new staff, newly-introduced technology, or new customers (Michel, 2001). In the event of service failure, the action taken by the company is crucial to either fortify the existing relationship or turn the situation into a major problem (Dong, Evans & Zou, 2008). Thus, it is vital for the company to ensure that they take immediate action in resolving customer's problem.

Service failure can result in negative word of mouth, problematic relationship between the customer and company, and negative future behaviors (Ha & Jang, 2009). Unresolved customer's problem will only bring negative impact to the company's reputation. Customers will become more dissatisfied and they will share the unhappy experiences with others (Ha & Jang, 2009). Therefore, good service recovery efforts are critical to avoid such problems.

2.2 Service recovery

Service recovery refers to the actions taken by the organization in responding to a service failure (Gronroos, 1988). Service failure and recovery is the 'moment of truth' in testing the strength of relationship between the company and customers (Smith, Bolton & Wagner, 1999). Customer will evaluate the recovery efforts taken by the company following the service failure. This is critical especially if it involves long term or loyal customers.

Service recovery is important to return upset customers to a state of satisfaction. This can be done if the employees act quickly, being friendly, express empathy and demonstrate generous manner in resolving customer's problem (Michel, 2001). A good recovery effort will enhance customers' opinions towards the company, promote positive word of mouth, improve customer satisfaction and develop long term relationship (Michel & Meuter, 2008). Service failure and recovery should not be viewed as an obstacle, rather it should be seen as an opportunity to improve weaknesses and learning from mistakes. Komunda and Osarenkhoe (2012) stated that service recovery may enable the tracking of common complaints and a database could be developed to better manage it. As a result, the company will become aware of the problems and it can be avoided from occurring again.

2.3 Recovery satisfaction

Generally, customer satisfaction is defined as a customer's judgment towards a particular product or service. It is a judgment that a product or service provides a pleasurable level of consumption-related fulfillment (Oliver, 1997). Customer satisfaction is a critical issue in the area of marketing and consumer behavior (Ghalandari, Babaeinia & Jogh, 2012). It has become a key component in measuring business performance and guiding principle in the development in new product or service (Feng & Yanru, 2013). Satisfied customers will enhance company's reputation by sharing positive experiences with others.

Service failure is inevitable and therefore, companies will face challenging time to ensure satisfied customers will remain loyal to them. Poor service recovery will lead to double-deviation and may threaten the relationship that has been developed for years. An excellent recovery effort will improve customer's overall satisfaction, will promote brand loyalty and positive word of mouth (Choi & La, 2013). In specific, Kim et al. (2009) described recovery satisfaction as a positive emotion perceived by the customers as a result of service recovery efforts taken by the company.

2.4 Justice theory in service recovery

Justice theory states that a customer evaluates a service recovery attempt as fair or unfair (DeWitt, Nguyen & Marshall, 2008). A number of researches in western countries have considered the application of justice theory in service recovery. According to Patterson et al. (2006), justice theory was derived from the social exchange and equity theory. It can be categorized into three dimensions namely distributive justice, procedural justice, and interactional justice. Justice theory has been used in a number of service recovery studies including airline industry (Chang & Chang, 2010); restaurant (Ok, 2004); retail industry (Lin, 2012); and hotel industry (Prasongsukarn & Patterson, 2012).

Previous service recovery studies demonstrated that the three dimensions of justice theory influences recovery satisfaction. According to Ok (2004), the three dimensions of justice have positive effects on recovery satisfaction in restaurant setting. Effective service recovery will not only improve satisfaction, however it can lead to trust and re-patronage intentions (Wen & Chi, 2013). Ha and Jang (2009) claimed that an effective service recovery will transform upset customers to be satisfied which can promote long term relationship. Therefore, the following proposition is derived from the aforementioned discussion:

P1:Service recovery will influence recovery satisfaction.

2.4.1 Distributive justice

Wen and Chi (2013) described distributive justice as the outcome that the customer expects to receive during service recovery and it should be equal to the customer's loss. In specific, Weun, Beatty, and Jones (2004) defined distributive justice as the tangible end result given to the initially frustrated customer. Typical end results include a discount, cash refund, replacement, and amendment (Wen & Chi, 2013). Prasongsukarn and Patterson (2012) claimed that distributive justice was found to affect recovery satisfaction in multi industry settings such as retail, hospitality and auto repair. These findings proved that monetary rewards are important to satisfy upset customers (Ha & Jang, 2009). Therefore, based on the preceding discussion, the following proposition is developed:

P1a: Distributive justice will influence recovery satisfaction.

2.4.2 Procedural justice

Procedural justice concerns with the procedures, policies, processes and rules involved in service recovery (Smith, et al., 1999). However, del Río-Lanza, Vázquez-Casielles, and Díaz-Martín (2009) argued that procedural justice deals with aspects such as accessibility, speed, process control, delay and flexibility in dealing with service failure. Based on both definitions, we defined procedural justice as the policies and procedures that will help to solve customer's problem in timely manner. Previous research demonstrated that procedural justice can influence recovery satisfaction. Studies in airline industry by Nikbin smail, Marimuthu, and Salarzehi (2012) and Chang and Chang (2010) indicated that procedural justice influences recovery satisfaction. Therefore, the following proposition is derived based on the previous discussion:

P1b: Procedural justice will influence recovery satisfaction.

2.4.3 Interactional justice

Interactional justice refers to the customers' perception regarding the way they are treated during the service recovery process which includes respect, caring, honesty and willingness to help (Wen & Chi, 2013). Sparks and McColl-Kennedy (2001) argued that interactional justice concerns with the human interactions during service recovery. Based on both definitions, we summarized interactional justice as the treatment and communication process involved during service recovery process. Extant studies claimed that interactional justice can influence customers' judgment towards company's recovery effort. A study in banking and home construction industries shows that interactional justice influence customer satisfaction (Maxham III & Netemeyer, 2002). In addition, a study in airline industry by Chang and Chang (2010) also found that interactional justice affects recovery satisfaction. Based on the preceding discussion, the following proposition is developed:

P1c: Interactional justice will influence recovery satisfaction.

2.5 The role of religiosity

In recent years, there is an emerging concern pertaining to religiosity in global market (Swimberghe, et al., 2009). A number of issues pertaining to religiosity are still impending given the fact that this area is still maturing (Tsarenko & Tojib, 2012). According to Worthington et al. (2003), religiosity (also called as religious commitment) is defined as the degree to which an individual obey to their religious belief and practice it in daily life.

Religious people is claimed to be more honest, fair and nice compared to people without religious orientation (Morgan, 1983). In service failure context, Tsarenko and Tojib (2012) argued that highly religious individual tend to be more forgiving compared to less religious individual. Therefore, it is expected that being religious will influence the manner an individual behave when service failure occurs. As a result, it may affect their level of satisfaction and future re-patronage intentions. This is supported by Swimberghe et al. (2009) claiming that further research is required to examine the buying behavior of high and low religious people. This is due to the notion that dissatisfied customers may perform one of the following behaviors: stop buying from the company; share negative word of mouth with others; and complain to the business owner or third party. The emerging trend of religious awareness in the global market evidenced that it is critical to further explore this area, specifically in service failure perspective. Therefore, the following proposition is developed:

P2: Religiosity will moderate the relationship between service recovery and recovery satisfaction.

P2a: Religiosity will moderate the relationship between distributive justice and recovery satisfaction.

P2b: Religiosity will moderate the relationship between procedural justice and recovery satisfaction.

P2c: Religiosity will moderate the relationship between interactional justice and recovery satisfaction.

3 CONCLUSION

The central issue discussed in this paper is the integration of religiosity in service recovery studies. While the area of religiosity has been examined in other marketing studies, less attention has been directed to its role in service recovery context. The developing trend of religious awareness demonstrated that this study is critical to be conducted. Theoretically, this study will contribute to the body of knowledge in service recovery related area. Practically, this study will help service provider to be more alert in treating frustrated customers when service failure occurs. Low or highly religious people may have different perception towards company's efforts in rectifying the problem. As mentioned earlier, highly religious people tend to be more forgiving when they experienced service failure compared to the less religious people (Tsarenko & Tojib, 2012).

This paper also discusses the role of service recovery towards recovery satisfaction. Attaining customer satisfaction following service failure is a challenging task for the organization. A successful recovery may promote loyalty; however poor recovery will lead to double-deviation and bad reputation. The concept of justice theory in service recovery was deliberated in this paper. Distributive justice (tangible compensation), procedural justice (policies and procedures), and interactional justice (communication process) are the three dimensions of justice theory that are believed to influence recovery satisfaction.

4 RECOMMENDATIONS FOR FUTURE RESEARCH

This paper introduces the moderating role of religiosity on the relationship between service recovery and recovery satisfaction. Future research may explore other potential variable such as personality type to be examined. While most service recovery studies were conducted in western countries, less attention has been given in Asian region. Therefore future researcher is suggested to conduct such studies in their country which may yield different findings. Future research is recommended to conduct service recovery studies in other industries

that have never been examined. Typical areas that have been investigated include restaurant, online service, retailing, hotel and banking sector.

ACKNOWLEDGEMENT

The work was supported by the Ministry of Higher Education (MOHE) via Exploratory Research Grant Scheme of Universiti Teknologi Malaysia (UTM). Research name: Integrating Intercultural Competence for Superior Service Satisfaction: A Structural Equation Modeling in Dynamic Economy of Malaysia and Turkey, grant no. PY//2012/01358- Q.K130000.2563.04H80.

REFERENCES

Bernardo, M., Llach, J., Marimon, F. & Alonso-Almeida, M.M. (2013). The balance of the impact of quality and recovery on satisfaction: the case of e-travel. *Total Quality Management & Business Excellence, 24*(11–12), 1390–1404.

Chang, Y.-W. & Chang, Y.-H. (2010). Does service recovery affect satisfaction and customer loyalty? An empirical study of airline services. *Journal of Air Transport Management, 16*(6), 340–342.

Choi, B. & La, S. (2013). The impact of corporate social responsibility (CSR) and customer trust on the restoration of loyalty after service failure and recovery. *Journal of Services Marketing, 27*(3), 223–233.

del Río-Lanza, A.B., Vázquez-Casielles, R. & Díaz-Martín, A.M. (2009). Satisfaction with service recovery: Perceived justice and emotional responses. *Journal of Business Research, 62*(8), 775–781.

DeWitt, T., Nguyen, D.T. & Marshall, R. (2008). Exploring Customer Loyalty Following Service Recovery The Mediating Effects of Trust and Emotions. *Journal of Service Research, 10*(3), 269–281.

Dong, B., Evans, K.R. & Zou, S. (2008). The effects of customer participation in co-created service recovery. *Journal of the Academy of Marketing Science, 36*(1), 123–137.

Feng, J. & Yanru, H. (2013). Study on the relationships among customer satisfaction, brand loyalty and repurchase intention. *Journal of Theoretical & Applied Information Technology, 49*(1).

Ghalandari, K., Babaeinia, L. & Jogh, M.G.G. (2012). Investigation of the effect of perceived justice on post-recovery overall satisfaction, post-recovery revisit intention and post-recovery word-of-mouth intention from airline industry in Iran: The role of corporate image. *World Applied Sciences Journal, 18*(7), 957–970.

Gronroos, C. (1988). Service quality: The six criteria of good perceived service quality. *Review of business, 9*(3).

Ha, J. & Jang, S.S. (2009). Perceived justice in service recovery and behavioral intentions: The role of relationship quality. *International Journal of Hospitality Management, 28*(3), 319–327.

Kim, T.T., Kim, W.G. & Kim, H.-B. (2009). The effects of perceived justice on recovery satisfaction, trust, word-of-mouth, and revisit intention in upscale hotels. *Tourism management, 30*(1), 51–62.

Komunda, M. & Osarenkhoe, A. (2012). Remedy or cure for service failure? Effects of service recovery on customer satisfaction and loyalty. *Business Process Management Journal, 18*(1), 82–103.

Lin, W.-B. (2012). The determinants of consumers' switching intentions after service failure. *Total Quality Management & Business Excellence, 23*(7–8), 837–854.

Lusch, R.F. & Vargo, S.L. (2006). Service-dominant logic: reactions, reflections and refinements. *Marketing theory, 6*(3), 281–288.

Maxham III, J.G. & Netemeyer, R.G. (2002). Modeling customer perceptions of complaint handling over time: the effects of perceived justice on satisfaction and intent. *Journal of retailing, 78*(4), 239–252.

Michel, S. (2001). Analyzing service failures and recoveries: a process approach. *International Journal of Service Industry Management, 12*(1), 20–33.

Michel, S. & Meuter, M.L. (2008). The service recovery paradox: true but overrated? *International Journal of Service Industry Management, 19*(4), 441–457.

Morgan, S.P. (1983). A research note on religion and morality: Are religious people nice people? *Social Forces*, 683–692.

Nikbin, D., Armesh, H., Heydari, A. & Jalalkamali, M. (2011). The effects of perceived justice in service recovery on firm reputation and repurchase intention in airline industry. *African Journal of Business Management, 5*(23), 9814–9822.

Nikbin, D., Ismail, I., Marimuthu, M. & Jalalkamali, M. (2010). Perceived justice in service recovery and recovery satisfaction: The moderating role of corporate image. *International Journal of Marketing Studies, 2*(2), 47.

Nikbin, D., Ismail, I., Marimuthu, M. & Salarzehi, H. (2012). The Relationship of Service Failure Attributions, Service Recovery Justice and Recovery Satisfaction in the Context of Airlines. *Scandinavian Journal of Hospitality and Tourism, 12*(3), 232–254.

Ok, C. (2004). *The effectiveness of service recovery and its role in building long-term relationships with customers in a restaurant setting.* Kansas State University.

Oliver, R.L. (1997). Satisfaction: A behavioral perspective on the customer. *New York.*

Patterson, P.G., Cowley, E. & Prasongsukarn, K. (2006). Service failure recovery: the moderating impact of individual-level cultural value orientation on perceptions of justice. *International journal of research in marketing, 23*(3), 263–277.

Prasongsukarn, K. & Patterson, P.G. (2012). An extended service recovery model: the moderating impact of temporal sequence of events. *Journal of Services Marketing, 26*(7), 510–520.

Riscinto-Kozub, K.A. (2008). *The effects of service recovery satisfaction on customer loyalty and future behavioral intentions: An exploratory study in the luxury hotel industry*: ProQuest.

Smith, A.K., Bolton, R.N. & Wagner, J. (1999). A model of customer satisfaction with service encounters involving failure and recovery. *Journal of marketing research, 36*(3), 356–372.

Sparks, B.A. & McColl-Kennedy, J.R. (2001). Justice strategy options for increased customer satisfaction in a services recovery setting. *Journal of Business Research, 54*(3), 209–218.

Swimberghe, K., Sharma, D. & Flurry, L. (2009). An exploratory investigation of the consumer religious commitment and its influence on store loyalty and consumer complaint intentions. *Journal of Consumer Marketing, 26*(5), 340–347.

Tsarenko, Y. & Tojib, D. (2012). The role of personality characteristics and service failure severity in consumer forgiveness and service outcomes. *Journal of Marketing Management, 28*(9–10), 1217–1239.

Wen, B. & Chi, C.G.-q. (2013). Examine the cognitive and affective antecedents to service recovery satisfaction: A field study of delayed airline passengers. *International Journal of Contemporary Hospitality Management, 25*(3), 306–327.

Weun, S., Beatty, S.E. & Jones, M.A. (2004). The impact of service failure severity on service recovery evaluations andpost-recovery relationships. *Journal of Services Marketing, 18*(2), 133–146.

Worthington, J.E.L., Wade, N.G., Hight, T.L., Ripley, J.S., McCullough, M.E., Berry, J.W., Schmitt, M.M., Berry, J.T., Bursley, K.H. & O'Connor, L. (2003). The Religious Commitment Inventory-10: Development, refinement, and validation of a brief scale for research and counseling. *Journal of Counseling Psychology, 50*(1), 84.

Theory and Practice in Hospitality and Tourism Research – Radzi et al. (Eds)
© 2015 Taylor & Francis Group, London, ISBN 978-1-138-02706-0

Postgraduate students, reading comprehension, writing skills and thesis completion

A.H. Hamizad & M.S.M. Zahari
Faculty of Hotel and Tourism Management, Universiti Teknologi MARA, Shah Alam, Malaysia

N.M. Shahril & M.N. Shuhirdy
Faculty of Hotel and Tourism Management, Universiti Teknologi MARA, Penang, Malaysia

ABSTRACT: Owing to the complexities, academic scholars extensively regard the thesis as the last safeguard for postgraduate students in accomplishing their studies. Many postgraduate students however are reported of failing to submit their thesis work on time owing to various factors. In fact, accomplishing thesis is one of the daunting challenges facing by the postgraduate students in many universities. Lack of information seeking skills, writing skills and other factors could impede and delaying postgraduate thesis completion. This paper empirically investigates the relationship between reading comprehension, writing skills and thesis completion among the hospitality post graduates students.

Keywords: Postgraduate, student, reading comprehension, writing skills, thesis

1 INTRODUCTION

In almost all universities besides other core and non-core management subjects, thesis is one of the core requirements for the postgraduate students to accomplish before graduating. Thesis is a substantial piece of work, written with a view to proving or disproving something with the purpose of adding to, or creating new knowledge of a specific study field. Thesis requires student to demonstrate a mastery of the subject area being researched as well as a comprehensive understanding of the research methodology being used (Clewes, 1996). It encompasses both intellectual and skills development and for the vast majority of postgraduate students and the thesis by far is the most challenging piece of academic work (Bruning, Schraw, Norby & Ronning, 2004). Carrying out a thesis project can become equivalent to a full-time job with no obvious immediate benefits and can take several years to complete.

A thesis project should grasp more consideration and attention than an examination or assignment and setting postgraduate's research on the back burner can deter their advancement in the field (Clewes, 1996). Maintaining a steady progress will help students to avoid the unfortunate circumstance of having an incomplete research project after they have finished their coursework. Thesis on the other hand is a kind of academic project and this academic project marks the transition

from student to a researcher or intellectual (Knight & Sutton, 2004). Therefore, there is unrelenting pressure on the universities to provide adequate research training both in the field of expertise by which postgraduate students will need to demonstrate a significant and original contribution to knowledge (Clewes, 1996).

Owing to its complexities, academic scholars extensively regard the thesis as the last safeguard for students in accomplishing their studies (Abel, 2002; Paltridge, 1997). Many postgraduate students are reported of failing to submit their thesis work on time. In Canada, for instance, the completion rates are varying with 40 percent in arts, 60 percent in the humanities and life sciences (Feigenbaum, 1994) whereas in the UK the completion rates are between 51 percent to 64 percent in the humanities and sciences (Wright & Cochrane, 2000). In fact, accomplishing thesis or research project is one of the most daunting challenges facing by the postgraduate students in many universities.

The attrition and low completion rate among the postgraduate students without exception is also a major problem facing by the Faculty of Hotel and Tourism Management, University Technology MARA, Shah Alam, Malaysia. Report from this Faculty revealed that low graduation rates are owing to the failure of students in completing their final thesis or research project although passing through all other coursework. Out of 70 students enrolled in the program approximately only 30–40 percent graduated on time. This phenomenon creates significant questions what are the underlying reasons that cause the causations. According to Knight and Sutton (2004), even with all the resources supports, many postgraduate students are still struggling to understand, integrate, apply the literature, theory, writing and other factors particularly when thesis is to be written inEnglish. With this notion, empirical evidences needs to be obtained thus this paper empirically investigates the relationship between reading comprehension, writing skills and thesis completion among the hospitality post graduates students by hypothesizing that;

H1: Lack of reading comprehension and writing skill significantly delaying postgraduate thesis project completion.

2 LITERATURE

2.1 *Post-graduate thesis*

Generally, a thesis is referring to a study on an exacting topic in which innovative research has been done, presented by students or a proposition stated for consideration, particularly one to be discussed and proved or maintained against opposing views (Clewes, 1996). A thesis or research project conducted is actually a mission of knowledge and understanding through experimentation, investigation and attentive search with the aimed at finding and interpretation or analysis of new knowledge at resolving debatable existing knowledge (Knight & Sutton, 2004). Thesis is normally associated with higher-level degree of academic pursue involving postgraduates, master or doctoral students. Clewes (1996) further asserted that postgraduate thesis is a piece of academic work using systematic procedures which demand a substantial amount of work to complete.

Barras (1991), Ary, Jacobs, Razavieh, and Sorensen (2010) and Shields and Tajalli (2006) deduced that whatever research projects conducted it requires specific components or structures. Scholars agreed that the most common and acceptable components of the thesis consist of the background of the study or introduction, review of literatures, methodology, analysis and findings and conclusion (Ary, et al., 2010; Nyawaranda, 2005; Shields & Tajalli, 2006). Besides all those components, a good or poor progression of the thesis project is depending on the research processes or attributes (Nyawaranda, 2005). In other word, a research project will not be accomplished without supporting attributes such as information, writing skill and other factors and these attributes could become obstacles for students

in completing their research work (Lessing & Schulze, 2002).

2.2 *Reading comprehension and writing skills*

In the thesis, writing skills and reading comprehension seemed to be one of the many attributes that need to be taken into consideration by most of the postgraduate students. Kaur (2009) states the ability of speaking and writing well in English is crucial for the postgraduate students to complete their thesis. A number of studies reported that there are high proportions of postgraduate student who struggle to complete their studies within the specific time given (Zainal Abiddin & Ismail, 2009). Burns (2000) states many factors could contribute to the pressures of undertaking and coping with the requirements of postgraduate work such as lack of writing skills using appropriate language, reading, and comprehending academic texts in a critical manner. Lack of knowledge in research skills includes linkages in sentence formation or redundancy facts in writing construction may also affect postgraduate thesis achievements (Burns, 2000).

3 METHODOLOGY

A quantitative study through cross-sectional approach using an individual unit of analysis was applied in this study. The information requires were obtained through self-reported and self-administered questionnaire survey and the samples were among graduated postgraduate students from the Faculty of Hotel and Tourism Management, University Technology MARA, Malaysia. As the intention also to tap the experiences, the final year students who were in the process or in the verge of submitting their thesis were also included.

The closed-ended questions using a summated rating scale or interval scale were opted in soliciting respondent feelings and opinion. The instrument consisted of three sections (A, B and C). Questions in each section were mostly adapted from the previous similar studies with slight modification made to suit the objectives.

Prior to the actual survey, cellular telephone numbers and electronic mail addresses of the graduates were obtained from the Faculty of Hotel and Tourism Management office. Sixty graduated students were then contacted to acquire their willingness to participate in the study. Subsequently, the survey questionnaire was e-mailed and one week was given to them to respond. As some of the graduates were among the researchers friend, an overwhelming response with 50 completed questionnaire received.

For final year post graduate students the data collection process was undertaken around two weeks. Data gathering was personally administered by the researchers. In light of the positive feedback, good responses with a total of 51 completed questionnaires were collected. A total of 101 responses were successfully obtained from both methods of data collection.

The reliability tests were undertaken for Section B and C. Items used in stipulated dimensions were reliable with a coefficient alpha value of 0.82 for Section B and 0.91 for Section C.

4 RESULT AND ANALYSIS

4.1 Respondent profile

The majority of the respondents were part-time students with 59.4 percent (n = 60) compared to 40.6 percent (n = 41) of full time students. This is anticipated as the majority of students who enrolled in postgraduate study in many universities are among the working individuals as opposed to full time students. Almost equal proportion of female students with 50.5 percent (n = 51) compared to 49.5 percent (n = 50) of male students. Among those respondents, 87.1 percent (n = 88) in the range of 20–30 years old while the remaining 12.9 percent (n = 13) was in the range of 31–35 years old.

4.2 Reading comprehension and writing skills

A descriptive statistic was undertaken looking respondent toward reading comprehension and writing skills. The range of mean scores from 1.0 to 2.4 of all items in this section analysis posited that students were having difficulties in reading comprehension and writing skills. Difficulties are expressed in developing an adequate reading and comprehension skills (M = 2.26, C1), traced in understanding particular concepts or text materials (M = 2.38, C2), connecting components of a larger strategic reading process (M = 2.39, C3). Similar difficulty was experienced in recognizing the suitable phases from text material to be used (M = 2.40, C4), developing narrative summaries from text material (M = 2.37, C5) and understanding academic or research language (M = 1.82, C6).

Compare to reading comprehension, writing elements is the most difficult part experiencing by students in accomplishing their thesis work. Most difficulty were experiencing in determining the appropriate writing styles (M = 1.30, C7), to come up and formalize the ideas in writing (M = 1.16, C8), connecting between statements and paragraph (M = 1.01, C9), arguing findings in an

Table 1. Reported mean scores for reading comprehension and writing skills (n = 101).

	Items	M	SD
C1	Developing adequate reading and comprehension skills	2.26	.607
C2	Understanding particular concepts or text materials	2.38	.570
C3	Connecting components of a larger strategic reading process	2.39	.496
C4	Recognize the suitable phases of texts material to be used	2.40	.609
C5	To develop narrative summaries from text material	2.37	.683
C6	Understanding academic / research language	1.82	.805
C7	Determine the appropriate writing styles	1.30	.609
C8	To come up and formalize the ideas in writing	1.16	.745
C9	Connecting between statements and paragraphs	1.01	.742
C10	To argue the findings in an analytical manner that generates new knowledge and insight	1.01	.794
C11	Grammatical accuracy and appropriateness	1.00	.800
C12	Using proper and appropriate vocabularies	1.32	.761
C13	Constructing the references with the right format	1.31	.761
C14	Using the suitable verb phrase and the nouns	1.36	.639
C15	Expressing thoughts clearly in English	1.39	.799
C16	To consider a possible structure for writing my thesis	1.41	.763
C17	Writing interpretation of the results derived from the analysis	1.28	.631
C18	Formulating and justifying empirical research questions for my study	1.18	.774
C19	The overall writing phase/process of a thesis	1.17	.744

Scale: 1 = Most Difficult, 2 = Difficult, 3 = Slightly Difficult, 4 = Easy, 5 = Easiest).

analytical manner that generates new knowledge and insight (M = 1.01, C10), grammatical accuracy and appropriateness (M = 1.00, C11), using proper and appropriate vocabularies (M = 1.32, C12), constructing the references with the right format (M = 1.31, C13) and using the suitable verb phrase and the nouns (M = 1.36, C14).

Most difficulty were also responding in expressing thoughts clearly in English (M = 1.39, C15), consider a possible structure for writing their thesis (M = 1.41, C16), writing, interpretation of the results derived from analysis (M = 1.28, C17), formulating and justifying empirical research questions for their study (M = 1.18, C18) and the overall writing phase/process of a thesis (M = 1.17, C19).

4.3 Experiences in thesis completion

The last descriptive analysis was looking at the overall experience of post graduate students in accomplishing their thesis project. The mean score ranges between 3.84 and 4.72, indicate that the majority of students agree and strongly agree with the statements probed in the questionnaire survey. They agree that lack of writing skills (M = 3.93,

Table 2. Reported mean scores for overall experience on thesis completion *(n = 101)*.

	Items	M	SD
E1	Lack of writing skills delayed my thesis accomplishment	3.93	.604
E2	Less able to formalize the ideas from reading into writing delaying my thesis completion	3.91	.602
E3	Less able to recognize the suitable phases of study material affects my accomplishment	3.84	.543
E4	Reading comprehension and writing is the most daunting challenges in my thesis accomplishment	4.52	.502
E5	Less able in expressing thoughts clearly in English causing delay in thesis completion	4.55	.505
E6	Less ability in interpreting results derived from analysis cause delayed in my thesis accomplishment	4.54	.573
E7	I realized that reading comprehension and writing is a very tough process in thesis completion	4.72	.667

Scale: 1 = Strongly Disagree, 2 = Disagree, 3 = Neutral, 4 = Agree, 5 = Strongly Agree).

E1), less able to formalizing the ideas from reading into writing (M = 3.91, E2) and less able to recognize the suitable phases from study material affects M = 3.84, E3) delaying their thesis completion. With that feeling, it is not surprising that the majority of the postgraduate students strongly agrees that reading comprehension and writing is the most daunting challenges in their thesis accomplishment (M = 4. 52, E4), agree that less able in expressing thoughts clearly in English (M = 4.52, E5) and less ability in interpreting results derived from analysis cause (M = 4.54, E6) delayed in their thesis accomplishment therefore realizing that reading comprehension and writing is a very tough process in thesis completion (M = 4.72, E7).

4.4 The relationship between reading comprehension, writing skills and thesis project completion

The section evaluates the relationship between reading comprehension, writing skills and the thesis completion. In particular, this analysis is to see how strong writing skills affected post graduate students thesis completion. A single-step multiple regression was conducted and the result is tabulated in Table 3.

The result shows that reading comprehension and writing skills were able to explain 67 percent (R^2 = 0.51, F-change = 12.494, p<.001**) of the variance in the thesis completion. This shows that reading comprehension and writing skills significantly contribute to the prediction of the thesis completion. The value of β = 0.67, *p*<.001** indicates writing skills significantly and positively affect the thesis project completion, therefore the hypothesis H1 vigorously supported.

Table 3. Results of multiple regressions of reading comprehension and writing skills on the thesis completion.

Predictors	Model 1 Std. β
Model 1	
Reading compression and Writing Skills	.67***
R^2	.51
Adj. R^2	.40
R^2 Change	.51
F-Change	12.494

Note: *p < .05, **p < .01, ***p < .001.

5 IMPLICATIONS AND CONCLUSION

No doubt thesis is one of the most the important components in postgraduate studies and in fact acts as the last safeguard for students accomplishing their studies. By far, it is the most challenging piece of academic work as it involved many contributory factors which determine their rate of completion or graduation. With this challenge, maintaining steady progress in all factors or aspects therefore will help the individual student to accomplish their mission and will avoid any unfortunate circumstances while neglecting one of those contributory factors might lead to the deferment of thesis completion and extending their graduation date.

A very clear picture emerged in this paper that postgraduate students are having problem relating to reading comprehension and facing difficulties in developing an adequate and acceptable academic writing skills. This depressing indication has given some implications not only for the students but also to supervisor and faculty as a whole.

For the students, they should not think that the postgraduate studies are similar to those undergraduates which are not really required substantial amount of efforts in many aspects. Their undergraduate mentality in all facets should be transformed or in other word should shift their line of thought and be more competent, critical and analytical. The ability to anticipate, decide and solving matters should also be practiced.

Postgraduate students in this study contact the post-graduate students of the Faculty of Hotel and Tourism Management therefore must avoid any elements of procrastination but provide the milestone or schedule, targeted date and seriously follow the schedule. Same goes to the supervisor. Mutual understanding between student and supervisor should be developed. Regular meeting with supervisor plus presenting a good piece of work should always be practiced by the individual student. In other words, presenting a bit and pieces of their work to the supervisor is more meaningful rather than showing the bulk of it at the end of the period. For the supervisor, their roles of supporting, encouraging, monitoring and commenting on their students work should frequently be done.

On academic writing, postgraduate student must take a serious concern in improving their writing skills especially during post graduate studies. Studies proven that practicing are the most effective ways in acquiring and polishing the writing skill. The individual student without any other options should keep themselves writing if they want to improve their English writing skill. Students' ability on this matter could not be achieved without a proper writing guidance from the program lecturers through class assignments, essay tests and examinations.

As a conclusion, the above highlighted issues could not be resolved without a direct involvement of the faculty as a whole. Having clear insight on this issue faculty therefore should take a proactive action by providing a necessary support to the students. The writing skills program should frequently be organized. Supervising program particularly for a new supervisor need continually be done and most importantly colloquium in every stage of student thesis should be organized by post graduate department.

ACKNOWLEDGMENT

This research was funded by the Ministry of Higher Education, Malaysia through Universiti Teknologi MARA under RIF grant: 600-RMI/DANA 5/3/RIF (599/2012).

REFERENCES

Abel, C.F. (2002). Academic success and the international student: Research and recommendations. *New directions for higher education, 2002*(117), 13–20.

Ary, D., Jacobs, L.C., Razavieh, A. & Sorensen, C. (2010). *Introduction to research in education* (8th ed.). Belmont, CA.

Barrass, R. (1991). *A guide to better writing for scientists, engineers and students.* London, UK: Chapman & Hall.

Bruning, R.H., Schraw, G.J., Norby, M.M. & Ronning, R.R. (2004). *Cognitive psychology and instruction* (4th ed.). Upper Saddle River, NJ: Pearson.

Burns, R. (2000). Realizing the university in an age of supercomplexity: Buckingham: Society for Research into Higher Education and Open University Press.

Clewes, D. (1996). Multiple perspectives on the undergraduate project experience. *Innovations-the learning and teaching Journal of Nottingham Trent University*, 27–35.

Feigenbaum, A.V. (1994). Quality Education and America Competitiveness. *Quality Progress, 27*(9), 83–84.

Kaur, S. (2009). *A qualitative study of postgraduate students' learning experiences in Malaysia.* Universiti Sains Malaysia, Penang, Malaysia.

Knight, C. & Sutton, R. (2004). Neo-Piagetian theory and research. *Enhancing pedagogical practice for educators of adults, 2*(1), 47–60.

Lessing, A. & Schulze, S. (2002). Postgraduate supervision and academic support: students' perceptions. *South African Journal of Higher Education, 16*(2), p. 139–149.

Nyawaranda, V. (2005). Supervising Research Projects/Dissertations. *A Paper delivered at the ZOU, Bindura: Mashonaland Central Regional Centre, 12.*

Paltridge, B. (1997). *Genre, frames and writing in research settings* (Vol. 45): John Benjamins Publishing.

Shields, P.M. & Tajalli, H. (2006). Intermediate theory: The missing link in successful student scholarship. *Journal of Public Affairs Education, 12*(3), 313–334.

Wright, T. & Cochrane, R. (2000). Factors influencing successful submission of PhD theses. *Studies in higher education, 25*(2), 181–195.

Zainal Abiddin, N. & Ismail, A. (2009). Identification of Resource Needs in Postgraduate Studies. *Research Journal of Social Sciences, 4*, 33–44.

Theory and Practice in Hospitality and Tourism Research – Radzi et al. (Eds)
© *2015 Taylor & Francis Group, London, ISBN 978-1-138-02706-0*

The influence of Arabic language communication practices by tourism operators on destination loyalty: Survey on Arab Muslim visitors

S. Wahab
Faculty of Business Management, Universiti Teknologi MARA, Puncak Alam, Malaysia

M. Yusoff
Academy of Language Studies, Universiti Teknologi MARA, Shah Alam, Malaysia

ABSTRACT: Destination loyalty has been mentioned as one of the most important subject in tourism studies. Many factors might influence tourist's loyalty to the visited destination. Foreign language practices among tourism operators might be one of them. In order to test the proposed hypotheses, a structured questionnaire was administered to tourists of Middle East tourist around Kuala Lumpur City, Malaysia. From 200 set of questionnaires distributed randomly to Arab tourist, 100 were valid to analyzed. The results seem to show that Arabic language communication practices by tourism operators, effect tourist enjoyment and destination loyalty. These conclusions may well prove crucial for the future use of the variety foreign language in the promotion of tourism destinations. The results of the study lend support to the importance of the Arabic language as an information source in the promotion of tourism destinations, which contributes in particular to attracting new visitors especially from the Middle East. The study focuses on empirically testing the advantages that, from a merely conceptual perspective, are becoming evident in the use of the variety communication language in the tourist sector.

Keywords: Tourism, communication skills, Arabic language, destination loyalty, Middle East

1   INTRODUCTION

Tourism may be as the sum of the process, activities, and outcomes arising from the relationships and the interaction among tourist, tourism suppliers, host governments who are involved in attracting, transporting, hosting and managing of tourists and other visitors (Weaver, 2010). Given the current and potential future importance of the verbal communication as an information source in the tourism sector, it appears necessary to analyze the consequences of tourist satisfaction with the information obtained through the verbal communication. If the positive effect of the former on destination satisfaction were to be demonstrated, it would provide overwhelming empirical support for the inclusion of that information source in the promotional mix of tourism destinations. Moreover, the present study focuses on the moderation effect of destination experience as well as communication experience on this relationship. These variables may well alter the propose defect due to the typology of the product under analysis (experience product) and the necessary experience in the use of the communication medium. Consequently, the study will focus in the main on empirically testing the advantages that,

from a merely conceptual perspective, are becoming evident in the use of Internet in the tourist sector. Destination loyalty has been mentioned as one of the most important subject in many academic studies. Many factors can influence tourists to become loyal to the destination they visited. Language is a factor why tourist travel and become loyal to the destination (Yoon & Uysal, 2005). In visiting destination, tourist satisfaction is the logical areas to be studied in order to understand more about customer experience and to destination's repeat visitation (Allen, Rodriguez & Fraiz, 2007). Hence, language used and tourist satisfaction are significantly important towards ensuring the tourists' loyalty towards particular destination.

According to Opperman (2000), research on destination loyalty has not been thoroughly investigated. In previous academic studies, there are many factors can affect the tourist to revisit the destination. That makes the competition in the tourism industry become stiffer. Many countries especially in Asian such as Thailand and Indonesia have various beautiful places that can attract to visit their country. Therefore, Malaysia need to develop their marketing destinations' strategy ensures the loyalty of tourists to visit the same destination again. In order to ensure the success of the markets destinations' strategy, it has to be guided by a thorough analysis of tourist motivation be its interaction with tourist satisfaction and loyalty.

A review of tourism literature reveals an abundance of studies on language and satisfaction, but destination loyalty has not been thoroughly investigated. The lacking of conceptual clarification, distinctions and logical linkages among the constructs has therefore created a gap in the said study. Thus, the aim of this paper is to explore the influence Arab Language communication practices, may have on destination satisfaction. More specifically, the study will analyze the relationship between the satisfaction with the search for holiday related information offered by all tourism entities on the verbal communication and the satisfaction with the destination selected.

2 LITERATURE REVIEW

2.1 *Definition of destination loyalty*

Recommendations to other people and repeat purchases are most usually refer to tourist loyalty. According to Flavian et al. (2001), the concept and degree of loyalty is one of the critical indicators used to measure the success of marketing strategy. Travel destinations can be considered as products or services and tourist have a loyalty to revisit and give suggestion to the other potential tourist such as friends or family. Jacoby and Chestnut (1978) mention loyalty can been measured in the behavior approach, the attitudinal approach and the composite approach.

Information about the tourist loyalty is important for our country since tourism is a main source for state income. The information such as a positive experience or service, products and other resources provided by tourism destinations could produce repeat visits as well as positive word of mouth effects to potential tourist such as friends and family (Yoon & Uysal, 2003). Destination loyalty is reflected by tourist intention to revisit the destination and in their recommendation to other.

According to Hsu (2007), loyal customers prefer word-of-mouth advertising. They prefer word-of-mouth advertising because is free. Repeated purchases or recommendations to other people are most usually referred to as consumers' loyalty in the marketing literature. Based on the concept and degree, loyalty is one of the critical indicators used to measure the success of marketing strategy (Yoon & Uysal, 2003).

2.2 *Attitude on language and tourist experience*

While consumers interact with a product or service towards which they have developed an attitude, they are subject to two sets of forces (Yi 1990). On the one hand, new experiences produce forces towards change and the existing attitude creates forces towards stability (e.g. resistance to change). An attitude may change with product experience. As a result, an attitude may be affected by the previous attitude (Castaneda, Frias & Rodriguez, 2007). Furthermore, prior attitude could influence cognitive evaluations of consumption experience (consumer satisfaction). This is consistent with the findings by Lord et al. (1979) that an attitude guides the judgment of relevant evidence or information processing and with the results obtained by Oliver (1980). Following Oliver, the nature of the mediatorial process is predicted by adaptation level theory (Helson, 1964). Customer attitudes explain by their past experience of the service. Hence, the tourist's previous attitude to the destination can affect his or her satisfaction with the destination after the holiday (Castaneda et al., 2007).

The knowledge not only of languages but also of the culture of others is an absolute must (Leslie & Russel, 2006; Monod, 1992). This advocacy of the need for foreign language skills and correlating support by tourism employers brings into focus the place of foreign language study in tourism courses. Recognition of this leads to a study to investigate this area (Leslie, Russell & Forbes, 2002), the findings of which catalyzed research, as noted above, into the views and needs of tourism employers. To further the study, research was then undertaken to investigate the position of foreign language studies in tourism courses across Europe (Leslie & Russel, 2006). It is found that foreign language communication were influence customer experience and value creator for the visits. They suggest that tourism student should be train with foreign language skill in order to success in to enhance employment capacity and more competent in cross cultural service. Hence, it is necessary for future research to investigate the value of foreign language skills in the promotion and delivery of tourism services which leads on to establishing support for and the benefits attributable to foreign language capability in business generally and tourism specifically.

2.3 *Conceptual framework*

A tourism destination is essentially an experience good. The tourist's information search will always take place internally such as when previous experiences and knowledge are used as the basis for planning a repeat visit (Fodness & Murray, 997; Vogt & Fesenmaier, 1998; Chen & Gursoy, 2001). When the internal search provides sufficient information for making a trip decision, external search is obviously unnecessary (Beatty & Smith, 1987). Therefore, the importance of internal information sources (consumption experience motivated by a previous visit to the destination) will be far greater than the importance of external information sources (Leslie & Russel, 2006; Peterson, et al., 1997). This may imply that in tourists with previous experience of the destination the effect that language communication may have on destination satisfaction will be small as the latter will be determined mainly by the previous knowledge of the destination. However, in individuals with no previous destination experience, external sources of information will be the ones providing the information the tourist needs to plan the holiday period.

In general, the aim of the above hypothesis is to test the effect of Internet satisfaction on destination loyalty.

3 METHODOLOGY

The researcher will use non-experimental in this study by utilizing correlational research as the research design. Salkind (2007) states correlation research examines the relationship between variables. On other words, Salkind (2007) mention correlation and prediction examine association but no causal relationship, where change in one factor is directly influences a change in another at one time.

The population of the study involves international tourists from Middle East who visited Kuala Lumpur as holiday destination. They require to response on the question regarding their experience when communication with tourism operators along their visit to selected destination. 200 set of questionnaires distributed at tourist spot such as KL International Airport, Bukit Bintang Shopping Complex and KL Convention Centre. The population of the respondents at Kuala Lumpur was too large and cannot identify by specific number of tourists visited Kuala Lumpur every day. According to Hair et al. (2006) they said 25 respondents to each independent variable, 200 is more than enough number of questionnaires distributed.

4 ANALYSIS AND FINDINGS

4.1 *Respondents' background*

From 100 respondent, one (4%) of them are from Bahrain, Iraq and Lebanon, six (24%) are from Iran, three (12%) from Turkey and United Arab

Table 1. Percentage for origin of respondents' country.

Country	Frequency	Percentage (%)
Bahrain	1	4
Iran	6	24
Iraq	1	4
Jordan	2	8
Lebanon	1	4
Turkey	3	12
Ukraine	4	16
United Arab Emirates	3	12
Qatar	4	16
TOTAL	25	100

Table 2. Percentage for age of respondents.

Range	Frequency	Percent (%)
Less than 20	7	28
20 – 30 years old	4	16
31 – 40 years old	6	24
41 – 50 years old	3	12
51 – 60 years old	4	16
More than 60 years old	1	4

Table 3. Correlations between Arabic language communication practices and destination loyalty *(n=200)*.

Correlations

		Mean Push Factors	Mean Destination Loyalty
Mean Push Factors	Pearson Correlation	1	.152*
	Sig. (2-tailed)		.032
	N	200	200
Mean destination loyalty	Pearson Correlation	.152*	1
	Sig. (2-tailed)	.032	
	N	200	200

*. Correlation is significant at the .05 level (2-tailed).

Emirates and four (16%) are from Qatar and Ukraine.

Most of them are below 30; 11 out of hundreds (44%). Eight are (34%) between 41 to 60 years, are the second higher group.

From the analysis it is found that p value is .032 showing a significant relationship between language usage and destination loyalty. Therefore research hypothesis is accepted.

5 CONCLUSION AND RECOMMENDATION

It was found that there is a significant relationship between language use and destination loyalty. This finding supports the previous research by Yoon and Uysal, (2005) and Allen et al. (2007). Based on the findings of this study, there are several suggestions and recommendations made to promote Malaysia as a destination for international tourists.

Government must do an intensive campaign to increase foreign language skills among tourist operators in order to attract foreign tourist to visit Malaysia.

The future research should use different ethnic of respondents, areas and location of visit in order to understand their culture and intensity during destination selection. The future research can focus on a bigger area of population so that the result is more generalized.

ACKNOWLEDGEMENT

Researcher would like to appreciate Research Management Unit of Universiti Teknologi MARA (UiTM) for sponsoring this research work managed by Research Management Institue (RMI) through Research Intensive Faculty (RIF) [File No: 600-RMI/DANA 5/3/RIF (892/2012)].

REFERENCES

Beatty, S. and Smith, S. (1987), "External search effort: an investigation across several product categories", Journal of Consumer Research, Vol. 14, pp. 83–95.

Castan˜eda J.A Frı´as D.M., Rodrı´guez M.A. (2001) The influence of the Internet on destination satisfaction, Internet Research Vol. 17 No. 4, 2007 pp. 402–420

Chen, J. and Gursoy, D. (2001), "An investigation of tourists' destination loyalty and preferences", International Journal of Contemporary Hospitality Management, Vol. 13 No. 2, p. 79.

Flavian, C., Martinez, E., and Polo, Y. (2001). Loyalty to grocery stores in the Spanish market of the 1990s. Journal of Retailing and Consumer Services, 8, 85–93.

Fodness, D. and Murray, B. (1997), "Tourist information search", Annals of Tourism Research, Vol. 24 No. 3, pp. 503–23.

Hair, J.F. Jr. Black, W.C., Babin, B.J. Anderson, R.E. and Tatham, R.L. 2006. Multivariate data analysis. 6th ed. New Jersey: Prentice Hall.

Helson, H. (1964), Adaptation-Level Theory, Harper & Row, New York, NY.

Hsu, S—H. (2007). Developing on index for online customer satisfaction: Adaption of American Customer Satisfaction index. Experts system with Applications 34, 3033–3042.

Jacoby, J., and Chesnut, R.W. (1978). Brand loyalty measurement and management. New York: Wiley.

Leslie, D and Russel, H 2006) The importance of foreign language skills in the tourism sector: A comparative study of student perceptions in the UK and continental Europe, Tourism Management 27 (2006) 1397–1407.

Leslie, D., Russell, H. & Forbes, A. (2002). Foreign language skills and tourism management courses in the UK. Journal of Industry and Higher Education, 16(6), 403–414

Lord, C., Ross, L. and Lepper, M. (1979), "Biased Beatty, S. and Smith, S. (1987), "External search effort: an investigation across several product categories", Journal of Consumer Research, Vol. 14, pp. 83–95.

Lord, C., Ross, L. and Lepper, M. (1979), "Biased assimilation and attitude polarization: the effects of prior theories on subsequently considered evidence", Journal of Personality and Social Psychology, Vol. 37, pp. 2098–109

Monod, D. (1992) cited in Embleton, D. & Hagen, S. (1992). Languages in International Business: A practical guide. London: Hodder & Stroughton.

Oliver, R.L. (1980), "A cognitive model of the antecedents and consequences of satisfaction decisions", Journal of Marketing Research, Vol. 17 No. 4, pp. 460–9.

Oppermann, M. (2000). Tourism destination loyalty. Journal of Travel Research, 39, 78–84.

Peterson, R.A., Balasubramanian, S. and Bronenberg, B.J. (1997), "Exploring the implications of the Internet for consumer marketing", Journal of the Academy of Marketing Science, Vol. 25 No. 4, pp. 329–46.

Salkind, N.J., (2007). Exploring Research. United Kindom: Pearson Education.

Vogt, C. and Fesenmaier, D.R. (1998), "Expanding the functional information search model", Annals of Tourism Research, Vol. 25 No. 3, pp. 551–78.

Weaver, David B., (2010). Tourism Management 4th Edtion. Australia: John Willy and Son Ltd.

Yi, Y. (1990), "A critical review of consumer satisfaction", in Zeithaml, V.A. (Ed.), Review of Marketing, American Marketing Association, Chicago, IL, pp. 68–123.

Yoon, Y. and Uysal, M. (2003). An Examination of the effects of motivation and satisfaction on destination loyalty: a structural model. Tourism Management 26, 45–56.

Yoon, Y., Uysal, M. (2005). An Examination of the effects and satisfaction on destination loyalty: a structural model. Tourism Management 26, 45–56.

Theory and Practice in Hospitality and Tourism Research – Radzi et al. (Eds)
© 2015 Taylor & Francis Group, London, ISBN 978-1-138-02706-0

Employability mismatch in hiring disabled workers in hospitality industry

M.M. Shaed, C.T. Chik, N. Sumarjan & S.A. Jamal
Faculty of Hotel and Tourism Management, Universiti Teknologi MARA, Shah Alam, Malaysia

ABSTRACT: Employment nowadays is very important to everyone. People need a job to survive. As our living cost increase, more money is needed to sustain daily expenses. Disabled people are not excluded, they too need income to support their life. However to some employer, disability is seen as limited ability and thus they are put aside in staff selection. Although Malaysia human resource policy encourage disabled people to enter main stream employment, their numbers are still limited. This paper reviewed the employability mismatch between the employer expectation and the employee's ability and the hopes of the disabled community in getting employed, especially in the hospitality industry.

Keywords: Disability employment, employability mismatch, hospitality

1 INTRODUCTION

Disability is described via medical model as destruction of mental or physical that can limit the activities in person's life (Thanem, 2008). It is also defined as a limitation or insufficient skill to do activity in a normal way. Usually, trauma, accident, genetics, pathology as destructions through disease can cause disability. Magdalene, Ang, Ramayah, and Vun (2013) stated that based on data by UNESCAP (2003), in total, there are over 650 million people who are disabled in this world, while 400 million from them came from the Pacific and Asian region.

For Malaysia, in year 2013, Social Welfare Department reported that there are a total of 477,549 of disabled people who registered in Malaysia. The previous year reported that 445,006 disabled people in Malaysia which clearly shows that the numbers of disabled people are increasing each year due to the increment of total population. According to the data, majority of the disabled people had learning disability 174,454 people, followed by physical disability 157,144 people, deaf 56,901 people, blind 44,332 people, mental disability 18,122 people, speech disability 3,530 and also others 23,066 people. When the number is divided according to ethnics, Malay people have the highest number of disabled people which are 295,132 in number while Chinese at about 95,307, Indian 48,694 and others 38,416 people. The "others" category are also included ethnics from Sabah and Sarawak. However, it is still remain unknown the exact number of disabled people in Malaysia because the registration for them are done voluntary.

The reason for this research is because many disabled people in Malaysia are being ignored and

have the problems of unemployment or underemployment especially in hospitality industry. Underemployment for people with disability in Malaysia means that they receive salary that does not reflect with their education level. Therefore, it is important to know the employability mismatch between the disabled workers and the management itself. In addition, Khoo, Tiun and Lee (2013) stated that there is empirical evidence that showed since 1990s, underemployment and unemployment have happen in advanced countries such as the United Kingdom. Furthermore, this problem is not only affected Malaysia but also to the broader economy in terms of calculating economic losses and the opportunity costs of inactivity. World Bank calculated that in 1993, Canada had lost about US\$55.8 billion (7.7%) of their GDP due to the disability.

The disabled person should not be treated differently, socially ignored or marginalised in any workforce. They should have equal rights to be employed and consequently get fair opportunities in training and development. Thus, in this study, researcher wants to explore the relation or mismatch of attributes, attitudes and legislations towards the employability and hiring disabled workers in hospitality industry. It is to make known that they can contribute to the economic development and nation's growth.

2 LITERATURE REVIEWS

2.1 *Attributes*

2.1.1 *Gender*
Sometimes people treat others based on their gender. Especially for women, sometimes they are

treated lower than men. Haq (2003) stated that usually women with disabilities are considered as weak, inactive, inefficient, and not suitable in economically productive roles. Magdalene, Ang, Ramayah & Vun (2013) also stated that according to Roberts (1996), in some ethnic, women are similar to disabled people because they usually will get misconceptions, prejudice and also discrimination. Additionally, it is also hard for women to get promotion opportunities and they usually receive lower income.

In terms of employment, Haq (2003) stated that other than their disability, women also faced double discrimination, because of their gender. In Malaysia, usually the income for single women with disabilities is $470 less than women without disabilities, which are $1,000 per month for disability women worker and at the same time $300 less per month than men who has disabilities. However for married women who have disabilities, their income is $100 less than disabled married men who have disabilities. As compared to men with disabilities, mostly in terms of employment, women with disabilities work in managerial, professional occupations and also in higher ranking in service with higher rates in wages.

Other than that, Haq (2003) also stated that until recently, there are no statistics of gender specific disabled worker available in Malaysia. According to Department of Welfare Services, there are 73, 353 people who has different types of disabilities, registered in their department. They estimated that about 210,000 or 1% of the total 21 million Malaysia populations have disabilities. In 1997, it is recorded about 538 people that have disabilities were employed in public sector. While in 1998, according to the Labour Department, 3,309 people employed in private sector which conclude that within this time, only 5.24% of people with disabilities are employed in Malaysia. Other than that, it is predicted that the rate of employment for men with disabilities is higher than women with disabilities.

It is also believed that compared to women, more men with disability are hired in workforce rather than women with disability whether they are part-time worker or under employment. Haq (2003) stated that according to Bowe (1984) about 42% of men with disabilities are in the workforce meanwhile only 24% of women with disabilities are in the workforce.

This is clearly unfair when more women with disabilities are unemployed, or they only gain part-time job with few benefits or low wage works. Despite of being disabled, employment are also very important to women with disabilities because of three reasons. First, employment is important for their living cost and their own saving. Second, employment is important for their dignity and as their contribution to the society and lastly for the integration into the mainstream non-disabled community.

2.1.2 *Physical appearance*

According to Poria, Reichel and Brandt (2011), for anyone who had disabilities, physical appearance also had been identified as one of the important and meaningful explanation towards attitudes and communication in hospitality and tourism industry. This is supported by Gröschl (2006) which already stated that self-presentation skills, aesthetic, appearance and physical looks of customers towards employees are very important in this industry. It is different towards those employees in hard or technical industries. Gröschl (2006) also concluded that sometimes it is one of barrier in selection criterion as people with disabilities who being limited in their aesthetic and self-presentation skills to work in hospitality industry.

2.1.3 *Misconceptions*

Attitudes cannot be predictable in every person. Some people have different attitudes reflected by their condition, emotion and others. Some people that have disability will be ignored and will become victims in crime such as harassment by others. Commissioner to the Australian Network on Disability (2011) stated that one of the important obstacles to employment is misconceptions about disability and negative attitudes by employers.

Based on journal from Davis and Olson (2012), disability persons always face the problem of being ignored. They did not get the chance to prove their skills in specific job. Employers usually assume that this kind of person cannot perform the job thus the disabled person did not manage to show their skills. Other than that, according to Haq (2003) sometimes the attitudes of the employer can cause barrier to employment.

According to Accessible Ontario Customer Service (2012), when employers hiring disabled workers, they need to discuss with the disabled workers about changes that were required. Employers need to think about the changes that suits to the disabled workers and also to other workers. In this case, they cannot assume that disabled workers must not do certain work. Instead, employer needs to carefully observe the disabled workers ability to do their job. Other than that, as an employer, they cannot force their workers to tell that they have disability but at the same time, workers need to inform their limited capability if it can affect their job.

Other than that, Accessible Ontario Customer Service (2012) also gives some examples on how employers can make changes when they hire disabled workers. Some of the changes can be made is easy access to a building for wheelchair user. Employers also have the authority to give some of the disabled worker's task to other workers. Furthermore, the working hours for disabled workers also can be change to suit the disabled workers condition. Lastly, they also should give the disabled person time off if they have appointments with doctors or any healthcare workers.

McLaughlin, Bell & Stringer (2004) stated that some female employer can have more positive attitudes towards disabled person rather than male employer. Additionally, they also reported that female employer has less unfavourable judgement towards disabled workers rather than male employers.

Attitudes were also affected by myths. For example, according to Accessible Ontario Customer Service, some employer does not want to hire people with disabilities due to the myths that disabled workers will have high turnover rate and they also take sick days more often than workers without disabilities.

In other words, Magdalene, Ang, Ramayah & Vun (2013) mentioned that the intention to hire disabled workers will be higher if the employers' attitude towards hiring disabled worker is more favourable. Plus, it is believed that between men and women, women have more favourable attitudes in terms of hiring disabled people. Schur, Kruse, Blasi & Blanck (2009) stated for those who had previous contact with the disabled person, they will have more positive acceptance and react positively towards the disabled people.

2.2 Legislations

2.2.1 Anti-discrimination laws

Currently, there are no anti-discrimination laws in Malaysia to protect people with disabilities. Magdalene, Ang, Ramayah, and Vun (2013) stated that a study carried out by British disability employment website, Ready, Willing and Able (RWA) which shows that most employers still did not hire disabled workers even though the anti-discrimination laws had been introduced in England ten years ago. There is no evidence about positive employment after the introduction of the disability act. Khoo, Tiun and Lee (2013) stated that these laws included all groups of people with disabilities because they needed to be treat equally like other humans.

In addition, according to The British Institute of Human Rights (2012), there is also Human Rights Act which important to disabled people. It is not just about the law but in encourages the way public services are delivered to disabled peo-

ple. Any public services such as educational bodies, hospitals, or even residential homes need to make sure that they did not breach the human rights of people with disabilities.

3 METHODOLOGY

This particular study reviewed literature from year 2000 until 2014. Journals such as Asia Pacific Disability Rehabilitation Journal, The British Institute of Human Rights, Deloitte Access Economics, North Dakota Center for Disability, The Journal of International Social Research, International Journal of Contemporary Hospitality Management, Industrial Relations and also Equal Opportunities International Journals had been referred. The information were exhausted using Ebcohost and Science Direct search engines using keyword such as 'disability', 'worker with disability', 'hospitality employment', 'hospitality mismatch' and 'disability employment'. The Statistics on disability worker were taken from Malaysia Welfare Department.

Unofficial personal interview with the Manager of Bengkel Daya Klang, Mr. Mohmad Azib Bin Hasan had also been taken to gauge some information on the numbers of disabled employment inside and outside of Welfare Department. Some general information had also been downloaded from several international and local agencies engaging in disability people opportunities and activities.

4 CONCLUSIONS

For the conclusions, researcher realizes that there are many disabled people especially in Selangor with 71,236 people (Malaysia Welfare Department, 2014) specifically as compared to other states in Malaysia. Therefore future research is viable to be done in Selangor to gauge the misconception and probable discriminations in employment for the disabled worker. Those misconceptions and discriminations by others had become one of the barrier why disabled people are not hired especially in hospitality industry.

Having known the reasons for mismatch between the disabled workers and potential employers could help in better opportunities and working environment for the disabled workers in the future especially in the hospitality industry.

ACKNOWLEDGEMENTS

This paper is sponsored by grant from Ministry of Education and Universiti Teknologi MARA.

REFERENCES

Accesible Ontario Customer Service. (2012). Tips on Serving Customers with Disabilities, *Ontario Education Services Corporation.*

Bowe F. (1984). Disabled women in America: A statistical report drawn from census data. *Washington, D.C: President's Committee on Employment of the Handicapped.*

British institute of human right. (2012). Your Human Rights: A Guide for Disabled People. *The British Institute of Human Rights.*

Commissioned by the Australian Network on Disability. (2011). The economic benefits of increasing employment for people with disability. *Deloitte Access Economics.*

Davis, L. & Olson, D. (2012). Communicating Effectively With People Who Have a Disability, *North Dakota Center for Disability.*

Hassan, M.A. (personal communication, May 15, 2014)

Haq, F.S (2003). Career and Employment Opportunities for Women With Disabilities in Malaysia. *Asia Pacific Disability Rehabilitation Journal.* 14(1).

Gröschl, S. (2006). An Exploration of HR Policies And Practices Affecting the Integration of Persons with Disabilities in the Hotel Industry in Major Canadian Tourism Destinations.

Khoo, S.L., Tiun, L.T. & Lee, L.W. (2013). Unseen Challenges, Unheard Voices, Unspoken Desires: Experiences of Employment by Malaysians with Physical Disabilities. *Kajian Malaysia,* 31(1), 37–55.

Magdalene, C.H., Ang, Ramayah T. & Vun, T.K. (2013). Hiring Disabled People in Malaysia: An Application of the Theory of Planned Behavior, *The Journal of International Social Research,* 6(27).

Malaysia welfare Department (2014). Retrieved From http://www.jkm.gov.my/index.php?pagename=utama&lang=bm

Mclaughlin, M.E., Bell, M.P., Stringer D.Y.(2004). Stigma And Acceptance Of Persons With Disabilities. Understudied Aspects Of Workforce Diversity. *University Of Texas At Arlington Group & Organization Management,* 29 (3), 302–333

Poria, Y., Reichel, A. & Brandt, Y. (2011). Dimensions of hotel experience of people with disabilities: an exploratory study. *International Journal of Contemporary Hospitality Management, 23*(5), 571–591.

Sekaran, U. & Bougie, R. (2010). Research Methods for Business: A Skill Building Approach. *UK: John Wiley & Sons.*

Schur L., Kruse D., Blasi J. & Blanck P. (2009). Is Disability Disabling in All Workplaces? Workplace Disparities and Corporate Culture. *Industrial Relations,* 48 (3).

Thanem, T. (2008). Embodying disability in diversity management research. *Equal Opportunities International,* 27(7), 581–595.

Theory and Practice in Hospitality and Tourism Research – Radzi et al. (Eds)
© 2015 Taylor & Francis Group, London, ISBN 978-1-138-02706-0

Finding value in a heritage landscape: The visitors' perception study

K.B. Shuib & H. Hashim
Universiti Teknologi MARA, Malaysia

N.K. Bahrain
Universiti Malaysia Sarawak, Malaysia

ABSTRACT: There is no doubt that heritage landscape like Bukit Melawati in Kuala Selangor is rich in history and possesses unique character that attracts many local visitors and foreign tourists. However, several literature has highlighted that there is a general public neglect of these valuable heritage due to ignorance which has led to some historical and heritage sites being abandoned or damaged. This paper presents the findings from a study on visitors' perception on Bukit Melawati as a significant heritage tourism destination in the state of Selangor. Some 215 visitors responded to the questionnaire survey carried out at the site. The findings from the study highlighted the tourism potentials and also the perceptions of the visitors about the various monuments, relics and historical remains available at Bukit Melawati. Several suggestions for ensuring a more sustainable tourism of the site are presented at the end of the paper.

Keywords: Visitors' perception, historic and heritage landscapes, sustainable tourism

1 INTRODUCTION

The Kuala Selangor District Council's website (n.d.) described the history of Bukit Melawati in Kuala Selangor. The Sultan Ibrahim of Selangor had built a fortress on the hill in order to protect the state from the Dutch invaders at the end of the 17th century. The Dutch forces at that time had already landed further south and conquered Malacca. A heritage landscape such as Bukit Melawati is significant in the historical context of a particular area and has its own background and story that makes it what it is today. However, it is sad to know that such historical places have not been getting the deserved attention and interest from the people who visited them. The visitors are neither aware nor concerned about the existence of Bukit Melawati as an important heritage site that needed to be valued and appreciated. Generally, the public tend to ignore these places due to the lack of interest and knowledge. This paper is aimed at understanding visitors' perception for the heritage landscape at Bukit Melawati and to come up with some proposed solutions to increase the potential of tourism for such places. Visitors include residents, local and foreign tourists. Bukit Melawati steeped in history is significant and quite popular especially among local visitors who comprised 86 percent of the respondents in this study. However, being a local hotspot destination does not mean that the visitors are aware of the importance in protecting such valuable heritage.

2 BACKGROUND OF BUKIT MELAWATI

Bukit Melawati or Melawati Hill is located in Kuala Selangor, a small town on the west coast of Peninsular Malaysia. This coastal town is about 65 kilometers northwest of Kuala Lumpur and 45 kilometers northwest of Shah Alam, the state capital of Selangor.

Administratively, Bukit Melawati functions as the Kuala Selangor district capital. It is also the major service center providing goods and services for the surrounding residents, visitors and tourists. Historically, Kuala Selangor was the stronghold of the Selangor Sultanate in the late 18th Century.

Melawati Hill is important in the history of the state because it has the advantage of overlooking the Straits of Malacca, one of the major shipping routes in East Asia. Some of the historical attractions of Bukit Melawati are the 200-year old fort and the lighthouse at the top of the hill. Several canons could also be found on the hilltop. Other heritage monuments and remains on the hill are the royal mausoleum, a legendary 100-step structure, an old rest house, an execution block (used for beheading offenders) and a poisonous well. A museum was built at Bukit Melawati to showcase the various historical and cultural background of Selangor and prominent heritage events of the state. Besides the historical relics, visitors to Bukit Melawati were also entertained by the existence of surprisingly tame monkeys—the *Silver Leaf* and

Figure 1. The location of Bukit Melawati.

Figure 2. The lighthouse at the top of the hill.

Figure 3. View of the Straits of Malacca.

Long Tail Macaques – which roam freely on the heritage hill. Many take the opportunity to feed these monkeys. Visitors to Bukit Melawati are not allowed to drive their cars on weekends and holidays. Kuala Selangor District Council operates trams to bring visitors up and down the hill and to the nearby nature reserve called *Taman Alam*.

Figure 4. The tram ferrying visitors up and down the hill.

Figure 5. The crowd in Bukit Melawati on weekend.

3 BACKGROUND LITERATURE

Orbasly (2000) highlighted that tourism is "a unique economic opportunity" and causes significant lifestyle change among the local communities. Besides economic and cultural dynamism, tourism in heritage areas also involves appreciation, preservation and conservation of the various elements or features.

Heritage, as derived from the word inheritance is defined in Oxford Dictionary as 'that which has been or may be inherited'. According to Millar (1989), heritage is about a special sense of belonging and of continuity that is different for each person. Akagawa and Sirisrisak (2008: 177) quoted the definition from the UNESCO World Heritage Center as "our legacy from the past, what we live with today, and what we pass on to the future generations..." Similarly, Chapman (2004) stated that heritage is everything that we inherit from the earlier generations. McKercher and du Cros (2002) quoted the definition of heritage from the International Council of Monuments and Sites (ICOMOS) (1999) as a broad concept that includes tangible assets, such as natural and cultural environments, encompassing landscapes, historic places, sites, and built environments, as well as intangible assets, such as collections, past and continuing cultural practices, knowledge and living experiences. Petzer and Ziesemer (eds, 2008: 196-197) stated that heritage consists of "those things that are inherited or inheritable. It includes those that we inherit from other people as well as those that come from the past in general. Heritage includes not only tangible (physical) objects, but also intangible ideas, responses and skills". Therefore, heritage has been regarded as accumulated experience, an educational encounter and a contact with previous generations.

The GARLAND Guidelines (2007) quoted by Petzer and Ziesemer (eds, 2008: 196) defined landscape as "a concept, a real or imaginary environment, place, image or view in which the land, and natural and semi-natural elements, are prominent, dominant ..."landscapes may, and often do, includes human and man-made components as well. They are the product of the appearance, uses and perceptions of places that are part of the outdoor environment". Heritage landscapes can be cultural i.e. places or buildings made by people or natural such as mountain ranges or reefs. They can be in the form of buildings that are easily recognizable (Chapman, 2004, Petzer & Ziesemer eds, 2008) and these places are constructions of history, owing their distinctiveness to the past (Herbert, 1995).

The concept of place meaning is related to people's experiences and attachment to the place (Tuan, 1974 and Tuan, 1975). Several researchers has shown that the perceptions, values, attitudes and emotions that form the meanings attributed to places were dependent upon how people interacted with these places (Tuan, 1974, Kaltenborn & Bjerke, 2002, Brown 2005, Raymond & Brown, 2006). The value as perceived by people is derived from a place or an object's various meanings and those meanings are closely tied to a person's value about the object or the place. The Australia ICOMOS (1999) stated that meaning "denote what a place signifies, indicates, evokes or expresses" and the meanings which are associated with tangible and intangible properties of the landscapes, define the place (Kaltenborn and Bjerke, 2002, Memmott and Long, 2002).

4 RESULTS AND DISCUSSION

This study was carried out to seek visitors' perception towards Bukit Melawati through a questionnaire survey. A total of 215 respondents answered

Table 1. Respondents' profile (n = 215).

Variables	Attributes	Percent
Gender	Male	34%
	Female	66%
Age	15-20	24%
	21-25	64%
	26-30	6%
	31-35	0
	36-40	2%
	41-45	1%
	46-50	2%
	51 and above	0%
Race	Malay	86%
	Chinese	4%
	Indian	3%
	Others	7%
Marital Status	Single	88%
	Married	12%
Place of Residence	Kuala Selangor	11%
	Other districts in Selangor	25%
	Other States in Malaysia	61%
	Other countries	2%
Occupation	Government	3%
	Private	8%
	Self-employed	4%
	Student	82%
	Others	2%

the questionnaire. The sample was selected using purposive sampling, which is a non-probability sampling method. The survey was carried out on several days including working days and weekends. The respondents' profile is summarized in Table 1.

It can seen from the data in Table 1 that heritage places such as Bukit Melawati attracts the younger generation who find the environment pleasing for recreation and leisure. The high percentage of students who visited Bukit Melawati reflects the popularity of Bukit Melawati among teachers who would organize trips for their students to learn about the historical occurrence that had taken place such as the fight against the Dutch invaders.

Among the 215 respondents, 65 percent were first time visitors to the heritage hill. For those who had been to Bukit Melawati more than once, 63 percent had come less than five times.

Respondents were asked about their perception of Bukit Melawati using a four point Likert-scale, ranges from "strongly disagree" to "strongly agree". Table 2 summarizes the findings. Interestingly, the respondents showed great interest towards the place. The reason behind this is because the majority of respondents had a good perception towards Bukit Melawati as a heritage landscape. This was evident in the results and findings shown which asked respondents regarding their perception on Bukit Melawati.

However, some points may need to be highlighted from the results whose score are lower than 3. These include the need to publicize the heritage

Table 2. Respondent's perception (*n* = 215).

Variables	Strongly disagree	←···→		Strongly agree	Mis-sing	Mean
	1	2	3	4		
Interested about the history of Bukit Melawati	0	8	131	73	3	3.31
Attracted to come to Bukit Melawati	4	21	122	62	6	3.15
Concerned about the history of Bukit Melawati	2	6	158	40	9	3.14
Aware about Bukit Melawati as a heritage site	6	26	117	60	6	3.10
Informative and attractive museum	5	23	131	41	15	3.04
Informative and attractive leaflet	8	25	118	47	17	3.03
Informed about the existence of Bukit Melawati	11	24	121	52	7	3.03
Good presentation of history	8	37	101	56	13	3.01
Aware about Bukit Melawati's important people	5	32	112	41	25	2.99
Attractive and informative interpretive signage	10	37	115	38	15	2.91
Attractive entrance	12	38	109	41	15	2.90
Likes the guidebook provided	14	38	108	32	23	2.82
Bukit Melawati being publicized in the media	12	56	91	38	18	2.79
Helpful and informative staff/guide	15	50	90	31	29	2.74

landscape in the media such as television, the need to provide guidebooks and the need for a more attractive main entrance. The improvements to be made on these factors will therefore increase the people's knowledge and awareness of Bukit Melawati which will in the end also increase its tourism potential.

5 RECOMMENDATIONS

According to Orbasli (2000), heritage management is the management of visitors in the interest of the historic fabric and the enhancement of visitor appreciation and experience. Visitor management is not only a matter of traffic or pedestrian flow management but involves imaginative solutions to enhance visitors experience, maintain a favorable reputation for the destination, and ensure a high quality environment for residents to live and work, and visitors to enjoy. The management of tourism in historic towns includes considerations for information and interpretation, commerce, cultural experience, traffic/transport and pedestrian. Some of the recommendations from this study include the following:

4.1 Promotion and publicity in media

As mentioned by Ziegler and Kidney (1980) about the role of newspapers in publicizing and supporting community-based historic preservation projects, it becomes clear that a place like Bukit Melawati also needs to be publicized in the media such as newspapers and television. It would be a good idea if local newspapers could write articles about Bukit Melawati and highlight its history especially the story on Selangor's Sultan Ibrahim built a fortress on the hill of Bukit Melawati in an effort to resist the invading Dutch forces. This is a story worth being highlighted in the media in order to stimulate interests from people and visitors alike.

4.2 Improved facilities

In the findings of the research some 7 percent of respondents mentioned about improving facilities at Bukit Melawati as one of the ways to stimulate more visitors' interest towards the place. This is a small figure if we only look at the number. However, the researcher would like to point out that in reality, or at the site, there were many facilities that have been damaged or broken due to vandalism and other factors. These facilities include benches, rock stairs and lamp posts. Not only these facilities could not be used by visitors, they also affected the view at Bukit Melawati. This problem could degrade the place's aesthetic values if no action is taken to repair or replace the damaged facilities.

4.3 Enhanced signage

Fifty eight percent of respondents agreed that the interpretative signages at Bukit Melawati are informative and attractive. This is significant if we only look at the number. However at the site, there were many signage's which are worn out and damaged to the point that the texts could hardly be read by visitors let alone understand its meanings. Shackley (2000) highlighted the function of interpretative signage at heritage landscape sites to aid the visitors in discovering its history. Hence, without interpretative signage with clear texts on them, it would be difficult for visitors to read and discover the history of the place.

4.4 *Designation as a cultural heritage site*

Under the National Heritage Act of 2005 (Malaysian Government, 2006), a place or an area having significant value to the community or group or the state can be nominated as a heritage landscape site. This place which has historical, cultural and natural values could be proposed as a cultural landscape site. The feedback that this research has received regarding the heritage landscape site is enough to suggest that this place must be declared a national treasure.

5 CONCLUSION

Indeed, the people's perception and awareness are growing regarding the importance to protect Bukit Melawati as a heritage site. Visitors are concerned about heritage and wanted the place to be preserved as effectively as possible. The potentials of Bukit Melawati can be seen in terms of commercial and tourism opportunities; whereby informal traders operate their businesses and many foreign tourists were eagerly taking photos to capture the unique experience at the site. Although this is a positive trend towards promoting tourism in historic places, care must also be taken to ensure that the site's originality and uniqueness remain unchanged in order to avoid people from harming it in terms of vandalism, littering and so on. The people also need to be educated in terms of understanding the importance of working together in order to protect the site from threats.

ACKNOWLEDGMENT

The authors wish to acknowledge the generous support given by Universiti Teknologi MARA (UiTM) from its RMI's Principal Investigator Support Initative (PSI) grant in making this paper possible.

REFERENCES

Akagawa, N. & Sirisrisak, T. 2008. Cultural landscapes in Asia and the Pacific: Implications of the world heritage convention. *International Journal of Heritage Studies* 14(2): 176–191.

Australia ICOMOS 1999. *The Burra Charter*. Sydney: Australia ICOMOS Inc. viewed 22 July 2007, <http://www.icomos.org/australia/>.

Brown, G. 2005. Mapping spatial attributes in survey research for natural resource management: methods and applications, *Society & Natural Resources*. 18(1): 1–23.

Chapman, H. 2004. *Heritage and places*. Port Melbourne: Heinemann Library.

Herbert, D.T. 1995. *Heritage, tourism and society*. London: Mansell Publishing Limited.

Kaltenborn, B.P. & Bjerke, T. 2002. Associations between landscape preferences and place attachment: a study in Roros, Southern Norway. *Landscape Research* 27(4): 381–396.

Kuala Selangor District Council. n.d. *Malawati Hill*. viewed 23 November 2013. http://www.mdks.gov.my/web/guest/bukit-melawati.

Kuala Selangor District Council n.d. *Kuala Selangor Map*. viewed 23 November 2013. http://www.mdks.gov.my/en/web/guest/peta-kualaselangor1.

McKercher, B. & du Cros, H. 2002. *Cultural tourism*, New York: The Haworth Hospitality Press.

Memmott, P. & Long, S. 2002. Place theory and place maintenance in Indigenous Australia. *Urban Policy and Research*. 20(1): 39–56.

Millar, S. 1989. Heritage Management For Heritage Tourism. *Tourism Management* 10(1): 9–14.

Mowforth, M. & Munt, I. 2003. *Tourism and sustainability: development and new tourism in the third world*. 2nd edition. London: Routledge.

Malaysian Government. 2006 *National Heritage Act of 2005 (Act 645)*. Kuala Lumpur: International Law Book Services.

Orbasli, A. 2000. Tourists *in historic towns: urban conservation and heritage management*. London: E & FN Spon.

Petzer, M. & Ziesemer, J. 2008. *Heritage at risk—ICOMOS world report 2006/2007 on monuments and sites in danger*. Altenburg: E. Reinhold-Verlag.

Raymond, C. & Brown, G. 2006. A method for assessing protected area allocations using a typology of landscape values. *Journal of Environmental Planning and Management*. 49(6): 797–812.

Shackley, M. 2000. Visitor management: case studies from world heritage sites. Oxford: Butterworth-Heinemann.

Tuan, Yi-Fu. 1974. *Topophilia: a study of environmental perception, attitudes and values*. New Jersey: Prentice Hall.

Tuan, Yi-Fu. 1975.'Place: an experiential perspective, *Geographical Review* 65(2): 151–165.

Ziegler Jr, A.P. & Kidney, W.C. 1980. *Historic preservation in small towns*. Nashville: American Association for State and Local History.

Theory and Practice in Hospitality and Tourism Research – Radzi et al. (Eds)
© *2015 Taylor & Francis Group, London, ISBN 978-1-138-02706-0*

Motivation towards satisfaction among international sport events volunteers

M.R. Norhidayah, M. Hairunnisa & M.A. Norafifah
Faculty of Business Management, Universiti Teknologi MARA, Shah Alam, Malaysia

N. Othman
Faculty of Hotel and Tourism Management, Universiti Teknologi MARA, Shah Alam, Malaysia

ABSTRACT: A success in organizing international race event potentially open the door to other events and opportunities as well branding Malaysia as an international event organizer. Volunteers are the vital sources for sport events where in many cases their involvement can ensure the survival of the event. Therefore, majority of the sports organization relies on volunteers to ensure the smoothness of an event without understanding the motivation and satisfaction that will lead to re-volunteering for future events. Hence, the purpose of this study is to investigate the volunteer motivation towards satisfaction among international event volunteer by using Volunteer Motivation Scale for International Sporting Event (VMS-ISE) consist of expression of value, patriotism, career orientation, love of sport, interpersonal contact, personal growth and extrinsic reward. Meanwhile, Volunteer Satisfaction Index (VSI) will measure the level of satisfaction among international sport event volunteers. Therefore, understanding the motivation factors that lead to the satisfaction among volunteers will develop a successful volunteer retention strategy for international sport event organizers.

Keywords: Volunteerism, sport event, Volunteer Motivation Scale for International Sport Event (VMS-ISE), and volunteer satisfaction index (VSI)

1 INTRODUCTION

1.1 *Volunteers of race event*

Recently, the volunteering activities became a worldwide trend and the numbers of volunteers have been increasing rapidly especially in the area of sports. Pi (2001) noted that the concept of volunteerism was introduced in the twentieth century in the field of sports. Previous studies stated, the success of sport event was depending on the numbers of volunteer in ensuring the sustainability of the Games, especially in the mega sports events such as the Olympic Games, Paralympic Games (Khoo & Engelhorn, 2007) and Formula One Grand Prix race. According to Twynam, Farrell and Johnston (2002), the genre of event plays an important role to influence the motivation of special event volunteers to commit and attach to the events.

Formula One race event is considered as a high profile international sport event (Henderson, Foo, Lim & Yip, 2010) that able to attract the media representative as well as spectators' worldwide (Horne & Manzenreiter, 2006). As the most watched race events, Formula One event is predicted to generate revenue of the countries (American University

Washington DC, 2011) by giving a strong boost to the tourism sector. According to Dolles and Soderman (2008), there are several advantages to the host countries such as projection of images as a tourism destination as well introducing their uniqueness of culture and society worldwide. Special event volunteers have not been well explored even though there are a lot of researches that have been done on volunteer motivation (Farrell, Johnston & Twynam, 1998). The limited literature on volunteer in sports (Strigas & Jackson, 2003) shows the significance of this study. Furthermore, Bang and Ross (2009) noted that the Volunteer Motivation Scale for International Sporting Event (VMS-ISE) should repeatedly be tested using different samples of special sporting event in order to develop and validate the scale among a broader population of volunteers.

The VMS-ISE are made up of seven factors, namely; i) Expression of Value, ii) Patriotism, iii) Career Orientation, iv) Love of Sport, v) Interpersonal Contact, vi) Personal Growth, and vii) Extrinsic Reward (Bang & Chelladurai, 2003; Bang & Ross, 2009). Understanding the motivation and satisfaction of the volunteers will make it possible to remain and influence them to re-volunteering in future events. Due to that, the researcher use Volunteer Satisfaction Index (VSI) developed by Galindo-Kuhn and Guzley (2002) as cited by Pauline (2011) (Volunteer Satisfaction Factors) to measure the satisfaction among volunteers in international sport event.

2 LITERATURE REVIEW

2.1 *Volunteerism*

Motivation plays as an essential factor for volunteer to contribute their time to undergo training in order to remain as a volunteer (Kim & Chelladurai, 2008). Volunteerisms have been defined as unpaid help provided in an organized manner to parties to whom the worker has no obligation (Wilson & Musick, 1997). A sports volunteer is defined as the Individuals that offer their help to others in a sport, without receiving neither remunerations nor expenses (Sport England, 2004). Thus, it is important to understand and match the volunteer preferences that will motivate them (Clary & Snyder, 1999) to enhance their involvement (Munro, 2003). Volunteers are considered as a valuable source of human capital (Pauline, 2011), where certain organizations rely on volunteers to run the daily or special tasks (Cuskelly, 2004) such as health, press, public relations, accreditation, technology and telecommunications, transport, access control and catering (Khoo & Engelhorn, 2007).

2.2 *Volunteer motivation scale for international sporting event (VMS-ISE)*

By using the VMS-ISE model, this study explores the motivation towards satisfaction among international sport event volunteers. All of the dimensions were used as a main construct and the items for each construct based on past research on volunteer motivation in international sport events.

2.2.1 *Expression of value*

This factor focuses on a concern for others where people became a volunteer in order to express or act on important values such as helping and being an element of successful events (Bang & Ross, 2009). This element is similar to what Farrell et al. (1998) mentioned where people are willing to do valuable things for the community, the event and intend to contribute something to the society (Bang & Chelladurai, 2003; Bang & Ross, 2009). This statement is in line with Strigas and Jackson (2003), where purposive involved motives are related to the desire of volunteers to benefit their action to sports organization, and contribute to the sport events and the community.

2.2.2 *Patriotism*

Patriotism is considered as one of the important motivating factors among international sport volunteers (Bang & Chelladurai, 2003). Previous literature stated that motivational factors for large scale event was different from the volunteers in other contexts (Bang & Ross, 2009) where local volunteers at international sporting events are being motivated by their patriotism factors to support their country The event itself became a main reason for volunteering in the event rather than the simple reason of helping others.

2.2.3 *Career orientation*

Career orientation expresses the gaining experience (Bang & Chelladurai, 2003; Clary et al., 1998) and career contacts where people who volunteer at the events (Ancans, 1992) aim to acquire knowledge and gain career-related experience through volunteering. This element is similar with element of understanding postulated by Clary et al. (1998) in Value Function Inventory (VFI). The understanding functions involve the opportunity that would be gained by volunteers in creating new learning experience, exercise knowledge, skills and abilities (Clary, et al., 1998).

2.2.4 *Love of sport*

According to Bang and Ross (2009), Love of Sport became a strong motivational factor and reason why people are willing to travel in order to volunteer at international sporting events (Bang & Ross, 2009) such as Formula One Grand Prix, Sepang.

2.2.5 *Interpersonal contact*

Interpersonal contact focuses on social development where people meet and communicate with each other for networking (Bang & Chelladurai, 2003; Bang & Ross, 2009). Solidary stage under VFI was related to the social context, group identification and networking (Farrell, et al., 1998; Williams, Dossa & Tompkins, 1995). Hence, volunteering activities benefitted the individual volunteer in terms of networking as well as gaining experience and field skills that lead to better employability.

2.2.6 *Personal growth*

Personal Growth can be explained through gaining new perspectives, feeling important and needed, high self-esteem and confidence level (Bang & Chelladurai, 2003; Bang & Ross, 2009). The types of volunteer opportunities undertaken by respondents were often seen as opportunities for personal growth. In several cases, personal growth intended as a part of future life plans, lifelong learning and personal improvement (Clary, Snyder & Ridge, 1992).

2.2.7 Extrinsic rewards

It was originally thought that extrinsic rewards would enhance volunteer motivation to engage in volunteerism (Green & Chalip, 1998; Weinberg & Gould, 2003). Extrinsic rewards focus on tangible items such as merchandise, food vouchers, monetary, and memorabilia (Bang & Ross, 2009). It is in contrast to intangible rewards that more focus on internal factors such as the feel of satisfaction through volunteering activities (Strigas & Jackson, 2003).

2.3 Volunteer satisfaction and intention to re-volunteering

Volunteer satisfactions depend on either volunteer's experience match their expectations (Davis, Hall & Meyer, 2003). The positive experience gained during the event would increase the probability of re-volunteering activities (Twynam, et al., 2002) (Downward & Ralston, 2006) and lead to the positive commitment to the organization (Bang & Ross, 2009). Due to that, understanding the determinant of satisfaction, specifically in volunteer job duties will influence future intention to re-volunteer (Pauline, 2011). Conversely, the negative outcomes that being experienced by the volunteer became a major reason to quit volunteering (Doherty, 2009) and affect the number of people who are willing to volunteer in future events. This issue was critical because majority of the sports organizations were heavily reliant on volunteers and it is quite impossible for the sports and recreational events to run smoothly without the help of volunteers. As a result, organizations that fail to retain their current volunteers will need to take additional efforts such as recruiting and re-training the new volunteers that is costly and time consuming (Pauline, 2011).

2.4 Volunteer satisfaction factors

This study explores the satisfaction among volunteers using the Volunteer Satisfaction Index (VSI) designed by Galindo-Kuhn and Guzley (2002) as cited by Pauline (2011). The VSI is made up of five dimensions consist of i) Communication Quality, ii) Organizational Support, iii) Participation Efficacy, iv) Work Assignment, and v) Group Integration.

2.4.1 Participation efficacy
Participation efficacy became a strong dimension that leads to the satisfaction among volunteers (Galindo-Kuhn & Guzley, 2002). Volunteer feel satisfied through their participation that will give a positive impact and bring good changes to client (Wong, Chui & Kwok, 2011) other than themselves (Galindo-Kuhn & Guzley, 2002; Pauline, 2011).

2.4.2 Work assignment
Previous literature stated that role assigned to the volunteer emerged as second factors that lead to the satisfaction among volunteers especially on job-fit for skills and convenience (Galindo-Kuhn & Guzley, 2002). It is important to ensure that volunteers are given the task that match their skills and expertise).

2.4.3 Communication quality
Communication quality refers to the basic nature of communication that received by the volunteers from the organization that they are willing to volunteer (Wong, et al., 2011). Communication with other volunteers, recognition from organizations (Farrell, et al., 1998), and the clarity of information regarding job description provided by the organization during the recruitment were found as factors that lead to volunteers' satisfaction (Galindo-Kuhn & Guzley, 2002).

2.4.4 Organizational support
Satisfaction can be gained through organizational support that includes educational and emotional resources provided by the organization (Wong, et al., 2011). Educational support refers to the training attended by the volunteer that bring more satisfaction (Ozminkowski, Supiano & Campbell, 1990), and improve the quality of experience (Pauline, 2011). Meanwhile, emotional support refers to the relational environment that exists between organizational members and volunteers (Galindo-Kuhn & Guzley, 2002; Pauline, 2011).

2.4.5 Group integration
Group integration refers to a development of social relationship among volunteers, other volunteers and paid staff (Pauline, 2011; Wong, et al., 2011). Hence, this relationship will provide a social aspect of the volunteer experience that leads to volunteer job satisfaction (Galindo-Kuhn & Guzley, 2002).

3 CONCEPTUAL FRAMEWORK

3.1 Proposed conceptual framework

Figure 1 illustrates the proposed framework in examining the relationship between VMS-ISE towards satisfaction among international sport event volunteers.

For the purpose of this study, VMS-ISE is served as independent variable (IV) which comprises of seven major constructs: Expression of Value, Patriotism, Career Orientation, Love of Sport, Interpersonal Contact, Personal Growth and Extrinsic Reward. Meanwhile, satisfaction

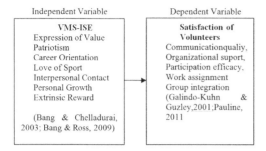

Figure 1. Proposed conceptual framework.

is the dependent variable (DV) that predicts the remaining of volunteers in future events by using VSI model designed by Galindo-Kuhn and Guzley (2002).

3.2 *Hypothesis*

Based on the framework above, several hypotheses are proposed:

H1: There is a significant relationship between volunteer motivation scale for international sporting event (VMS-ISE) and satisfaction among international sport event volunteers.

H1a: There is a significant relationship between expression of value and satisfaction among international sport event volunteers.

H1b: There is a significant relationship between patriotism and satisfaction among international sport event volunteers.

H1c: There is a significant relationship between career orientation and satisfaction among international sport event volunteers.

H1d: There is a significant relationship between love of sport and satisfaction among international sport event volunteers.

H1e: There is a significant relationship between interpersonal contact and satisfaction among international sport event volunteers.

H1f: There is a significant relationship between personal growth and satisfaction among international sport event volunteers.

H1g: There is a significant relationship between extrinsic reward and satisfaction among international sport event volunteers.

4 CONCLUSION AND IMPLICATION

Volunteering is an important economic resource for organizations that replace the value of money into energy, time and expertise that people provide in volunteer work. Volunteers play an important role to ensure the success of the event and sometime their involvement will ensure the survival of the games.

When volunteers experience a positive outcome from their contribution of previous events, a possibility to stay for the future events is high. Thus, it is important to understand volunteers' motivational factors and satisfaction in order to develop a successful volunteer retention strategy for international sport event organizers. It is hoped that event industry can use the VMS-ISE model as a guideline for recruiting and retaining the volunteer. The findings of this study will also contribute to expanding the body of knowledge about motivation among volunteers of international sporting events in Malaysia. Consequently, this study will benefit several stakeholders in event industry such as academician and event practitioner specifically international sport event organizer.

ACKNOWLEDGEMENT

The authors gratefully acknowledge contribution by the Dean of Faculty Business Management, Universiti Teknologi MARA, Malaysia and their colleagues for the support and guidance which have significantly contributed to the quality of this study.

REFERENCES

American University Washington DC. (2011). Formula 1 Racing: The economy and the environment TED case study Retrieved October, 2011, from http://www.american.edu/

Ancans, I.S. (1992). Why People Volunteer: A report to the voluntary action directorate multiculuralism and citizenship Canada, Ottwa: Ottawa: Volunteer Center Ottawa-Carleton.

Bang, H. & Chelladurai, P. (2003). *Motivation and satisfaction in volunteering for 2002 World Cup in Korea.* Paper presented at the conference of the North American Society for Sport Management. Ithaca, New York.

Bang, H. & Ross, S.D. (2009). Volunteer motivation and satisfaction. *Journal of Venue and Event Management, 1*(1), 61–77.

Clary, E.G. & Snyder, M. (1999). The Motivations to Volunteer Theoretical and Practical Considerations. *Current directions in psychological science, 8*(5), 156–159.

Clary, E.G., Snyder, M. & Ridge, R. (1992). Volunteers' motivations: A functional strategy for the recruitment, placement, and retention of volunteers. *Nonprofit Management and Leadership, 2*(4), 333–350.

Clary, E.G., Snyder, M., Ridge, R.D., Copeland, J., Stukas, A.A., Haugen, J. & Miene, P. (1998). Understanding and assessing the motivations of volunteers: A functional approach. *Journal of personality and social psychology, 74*(6), 1516–1530.

Cuskelly, G. (2004). Volunteer retention in community sport organisations. *European Sport Management Quarterly, 4*(2), 59–76.

Davis, M.H., Hall, J.A. & Meyer, M. (2003). The first year: Influences on the satisfaction, involvement, and persistence of new community volunteers. *Personality and Social Psychology Bulletin, 29*(2), 248–260.

Doherty, A. (2009). The volunteer legacy of a major sport event. *Journal of Policy Research in Tourism, Leisure and Events, 1*(3), 185–207.

Dolles, H. & Söderman, S. (2008). Mega-sporting events in Asia: Impacts on society, business and management: An introduction. *Asian business & management, 7*(2), 147–162.

Downward, P.M. & Ralston, R. (2006). The sports development potential of sports event volunteering: Insights from the XVII Manchester Commonwealth Games. *European Sport Management Quarterly, 6*(4), 333–351.

Farrell, J.M., Johnston, M.E. & Twynam, G.D. (1998). Volunteer Motivation, Satisfaction, and Management at an Elite Sporting Competition. *Journal of Sport Management, 12*(4), 288–300.

Galindo-Kuhn, R. & Guzley, R.M. (2002). The volunteer satisfaction index: Construct definition, measurement, development, and validation. *Journal of Social Service Research, 28*(1), 45–68.

Green, C. & Chalip, L. (1998). Sport volunteers: Research agenda and application. *Sport Marketing Quarterly, 7*(2), 14–23.

Henderson, J.C., Foo, K., Lim, H. & Yip, S. (2010). Sports events and tourism: The Singapore formula one grand prix. *International Journal of Event and Festival Management, 1*(1), 60–73.

Horne, J. & Manzenreiter, W. (2006). An introduction to the sociology of sports mega-events. *The Sociological Review, 54*(s2), 1–24.

Khoo, S. & Engelhorn, R. (2007). Volunteer motivations for the Malaysian Paralympiad. *Tourism and Hospitality Planning & Development, 4*(3), 159–167.

Kim, M. & Chelladurai, P. (2008). Volunteer Preferences for Training Influences of Individual Difference Factors. *International Journal of Sport Management, 9*(3), 233–249.

Munro, J.A. (2003). *Motivations and enduring involvement of leisure program volunteers: A study of the Waterloo Region Track 3 Ski School.* Paper presented at the Leisure Research Symposium: A Leisure Odyssey, May, Waterloo, Canada.

Ozminkowski, R.J., Supiano, K.P. & Campbell, R. (1990). Volunteers in nursing home enrichment: A survey to evaluate training and satisfaction. *Activities, Adaptation & Aging, 15*(3), 13–44.

Pauline, G. (2011). Volunteer satisfaction and intent to remain: an analysis of contributing factors among professional golf event volunteers. *International Journal of Event Management Research, 6*(1), 10–32.

Pi, L.-L. (2001). *Factors affecting volunteerism for international sports events in Taiwan, Republic of China.* (Unpublished PhD thesis), United States Sports Academy Daphne, AL.

Sport England. (2004). Sport England: Volunteer policy Retrieved May 29, 2007, from http://www.sportengland.org/se-volunteer-policy.pdf

Strigas, A. & Jackson, E.N. (2003). Motivating volunteers to serve and succeed: Design and results of a pilot study that explores demographics and motivational factors in sport volunteerism. *International Sports Journal, 7*(1), 111–123.

Twynam, G.D., Farrell, J.M. & Johnston, M.E. (2002). Leisure and volunteer motivation at a special sporting event. *Leisure/Loisir, 27*(3–4), 363–377.

Weinberg, R.S. & Gould, D. (2003). *Foundations of sport and exercise psychology*: Human Kinetics.

Williams, P.W., Dossa, K.B. & Tompkins, L. (1995). Volunteerism and special event management: A case study of Whistler's Men's World Cup of Skiing. *Festival Management and Event Tourism, 3*(2), 83–95.

Wilson, J. & Musick, M. (1997). Who cares? Toward an integrated theory of volunteer work. *American Sociological Review, 62*, 694–713.

Wong, L.P., Chui, W.H. & Kwok, Y.Y. (2011). The volunteer satisfaction index: A validation study in the Chinese cultural context. *Social indicators research, 104*(1), 19–32.

Theory and Practice in Hospitality and Tourism Research – Radzi et al. (Eds)
© *2015 Taylor & Francis Group, London, ISBN 978-1-138-02706-0*

Analyzing the ranking of security challenges of tourism development: A case study of Mazandaran Province, Iran

Z. Sharifinia
Department of Geography and Tourism Planning, Sari Branch, Islamic Azad University, Sari, Iran

ABSTRACT: The purpose of this study is to rank the security challenges and strategies of tourism in Mazandaran, Iran using a Delphi method. Mazandaran has a lot of tourist attractions. Every year it has millions of tourists. Unfortunately, it faces some security challenges. These challenges may cause unsustainability for the tourism activities if not managed appropriately. Thus, this study is proposed to answer the following questions: What are the challenges of tourism development in terms of security in Mazandaran? What strategies that can be used to overcome these challenges? The population includes the members of the permanent center of province trip facilities and tourism experts. Twenty seven people were recruited as the sample using the Snowball sampling technique. The results indicated that the most important security challenge was the 'high costs of residence which left the tourists stranded outside and so decreased their security level'. The most efficient strategies to improve security were 'promoting cooperation between the police and the travel agencies', 'increasing the safety of the transportation system', and 'familiarizing the tour guides with security challenges'.

Keywords: Delphi method, Mazandaran, security, tourism, tourism development

1   INTRODUCTION

'Security' means avoiding any stress and anxiety which leads to disturbing the peace of mind. Security establishment includes providing desirable situation and normal conditions in the society by police forces in the framework of the law and by the objective of protection of ethnic-social values and protecting individual and social benefits (Hezarjaribi 2011).

From the view point of tourism, security includes all affairs that are conducted to preventing disturbance in tourism activities. It also means as establishing conditions that help tourists to visit attractions with the sense of safety (Khoshfar et al. 2013). Sustainable growth of tourism depends on good performance of tourism cycle and several elements that interact together to form tourism system. Each element is important and is associated with other components. Security is one of the most important elements. Security and tourism are related to each other in both internal and external relations (Lotfifar & Yaghfuri 2013).

Any country has its own difficulties and challenges based on the cultural, social, geographical, economic and political feature (Rabani et al. 2012).

War, terrorism, insecurity, border guarding, severe regulations about visa, ideological conflicts among countries were challenges about tourism in the past and these challenges will remain in the future (Pishgahifard & Jahanian 2011). Thus contributory factors to the security disturbance in tourism sector must be identified and classified (Seydaii & Hedayatimoghadam, 2010). Karamidehkordi et al. (2013) stated that identifying the security challenges is one of the primary factors which tourism planners must consider. Many researchers have warned about the possibility of serious impact of natural phenomena or human factor on tourism industry. Natural phenomena such as earthquake and flood, and also human risks like political riot, revolt, terrorism, insurrection, crime and war have impacted visitors' behavior, especially when they are shown in the mass media (Asli et al. 2009).

George (2003) evaluated whether tourists, who visited Cape Town, felt safe while staying in the city. The findings revealed challenges of using the city's public transport after darkness, nationality of tourists and their previous experience of crime affected the respondent's perceptions of security.

Baker and Stockton (2014) have examined the effects of crime on health of tourists in two cities of the USA (Honolulu and Las Vegas) that engaged in mass tourism. In Honolulu, an inverse relationship between the number of visitors and violent crimes was found while in Las Vegas, a significant correlation between the number of visitors and crime was established. The findings also improved the relationships between increasing law enforcement employees and crime reduction.

Reza and Jawaid (2013) have investigated the impact of terrorism activities on tourism in Pakistan using the annual time series data from the period of 1980 to 2010. The results indicated a significant negative impact of terrorism on tourism both in the long and short run. The result of rolling window estimation method showed that terrorism has negative coefficients throughout the sample period.

Anuar et al. (2012) worked on safety aspect of the urban tourism using Safe City Programme in Putrajaya since year 2004. The results showed that respondents believed crime prevention steps in Safe City Programme are important approaches to ensure the tourist's safety.

Milman et al. (1999) studied on central Florida's tourists regarding their overall sense of perceived safety and the impact of various safety devices on their overall feelings of safety while on vacation. Previous exposure to crime had an impact on the perception of safety. The findings approved certain physical and behavioral devices such as deadbolt locks, closed-circuit TV cameras, door view ports, caller screening by telephone operators, locked side entrances into the hotel, and routine visits to the hotel by police provided tourists with a greater sense of safety and security.

Despite of the great virgin and gifted tourism attractions like unique climate conditions, attractive and beautiful four seasons natural landscapes, thick forests, numerous watering places and thermal springs, extensive and beautiful Caspian sea shore, sociable and cultured people, neighboring with countries around Caspian sea, and relation with middle east, Mazandaran has some challenges in the area of tourism security. Therefore, the present study attempts to find the answers to the following questions: What are the challenges of tourism development in terms of security in Mazandaran? What strategies can be opted to overcome these challenges?

The aim of this study is to identify challenges of tourism security in Mazandaran and to identify strategies to overcome the security challenges using the Delphi method. The findings of this study will be of interest to researchers, policy makers, and planners of the tourism industry in Iran.

2 MATERIAL AND METHODS

The present study is descriptive-analytic and the data is collected through library and field research. The Delphi method was applied in four steps to rank the security challenges of tourism development in

Mazandaran. The population was the members of Permanent Center of Province Trip Facilities (governor generals, mayors, governors and their representatives) and tourism experts. Using the Snowball sampling technique, twenty seven people were selected as sample. A questionnaire was used in the four steps. The first questionnaire included 2 open questions: 1) in your opinion, what are the challenges of tourism development in terms of security in Mazandaran? 2) What strategies can be used to overcome these challenges? In the first step, twenty three out of twenty seven members of the panel answered the questions (a return rate of 85%). In the second step, the panel members combined the similar questions into one. Finally, 10 questions were left to assess the challenges and 6 questions to assess the strategies that could be used to increase the security of tourism development. Then the questionnaire was returned to the panel members in the form of Likert Scale (a little = 1, little = 2, average = 3, much = 4, very much = 5) in order to rank the questions. Questions with Standard Deviation less than 1 (SD < 1) were considered as consensus questions by the panel members. Twenty three questionnaires were designed to reach consensus. In the third step the panel members were asked whether they agree or disagree with the rankings of the questions in the questionnaires of the second step. Those who agreed returned the questionnaires without any explanation. But those who disagreed explained their reasons and then returned the questionnaires. In total fourteen questionnaires were used with a response rate of approximately 62 percent.

The fourth step was the final summing-up step regarding the adjusted questions of the questionnaire in the third step. To analyzes the data, descriptive statistics and SPSS were employed. The validity of the questionnaires in each step was proved by some professors of tourism planning. Regarding the reliability of the questionnaires with Delphi method, Dalkey (1969) says when the number of a Delphi group is more than thirteen experts; the reliability coefficient is also more than 80 percent (a ≥ %80).

3 CASE STUDY

Mazandaran is in the north of Iran located on the southern coast of the Caspian Sea and surrounded by Golestan, Semnan, Tehran, Alborz, Qazvin, and Gilan provinces. Sari is the largest city and the capital of the province. It is one of the most densely populated provinces in Iran and has diverse natural resources, especially large reservoirs of oil and natural gas. The province's four largest cities are Sari, Babol, Amol and Qaemshahr. Founded as a province in 1937, Mazandaran was declared the second modern province after the neighboring Gilan. The diverse nature of the province features plains, prairies, forests and rainforests stretching from the sandy beaches of the Caspian Sea to the rugged and snowcapped Alborz mountain range, including Mount Damavand, one of the highest peaks and volcano in Asia, which at the narrowest point (Nowshahr County) is 5 miles. Mazandaran is a major producer of fish, and aquaculture provides an important economic addition to the traditional dominance of agriculture. Another important contributor to the economy is the tourism industry, as people from all over Iran enjoy visiting the area.

4 RESULT

The results are presented in the form of summaries and tables.

4.1 The first step of Delphi method

4.1.1 Opinions of panel members about challenges of tourism security in Mazandaran

In the first step, the panel members proposed 21 challenges as possible challenges of tourism development in terms of security. Similar challenges were combined and 10 remained. Table 1 shows these challenges. According to Table 1, the most frequent challenges (frequency distribution of 23) are: 'lack of security insurance of tourists by the investors of tourism facilities',' inadequate cooperation of the local communities with the police', 'illegal tourism facilities', and 'high costs of residence which left the tourists stranded outside and so decreased their security level'.

4.1.2 Opinions of panel members about strategies to increase tourism security in Mazandaran

In this step, panel members proposed seventeen strategies to increase the tourism security. Some of them were combined, and six strategies remained. Table 2 shows these strategies. According to this table, the most frequent strategies (frequency distribution of eleven) are: 'promoting cooperation between the police and the travel agencies', and 'familiarizing the tour guides with security challenges'.

4.2 The second step of Delphi method

4.2.1 Agreement level of panel members with the ranking of questions

In the second step, panel members recognized that from ten challenges that they had introduced in the first step, 'High costs of residence which left the tourists stranded outside and so decreased their

Table 1. Challenges of tourism security in Mazandaran, arranged according to importance level in the first step of Delphi method.

Row	Security challenges of tourism	Number
1	High costs of residence which left the tourists stranded outside and so decreased their security level	23
2	Illegal tourism facilities	23
3	Lack of security insurance of tourists by the investors of tourism facilities	23
4	Inadequate cooperation of the local communities with the police	23
5	Unfamiliarity of tour leaders with security issues	22
6	Lack of a comprehensive and strategic program, and the tourism aims not defined in the form of an executive program	21
7	Qualitative and quantitative weakness of land transportation facilities decreasing the financial and life security	21
8	Lack of security for investors to invest on tourism development	20
9	Lack of full support of tourist sites on the part of the police because of their abundance and distribution	18
10	Relative weakness to provide security for historical texture tourists in not crowded and low traffic hours	12

Table 2. Strategies to increase tourism security in Mazandaran in the first step of Delphi method.

Row	Strategies	Number
1	Performing different conferences and seminars to develop tourism security	5
2	Familiarizing the tour guides with security challenges	9
3	Establishing security in tourist sites especially for eco—tourism tours	8
4	Seriously punishing those who guide tourists illegally	6
5	Promoting cooperation between the police and the travel agencies	9
6	Increasing the safety of the transportation system	7

security level' was the most important in tourism security of Mazandaran. According to Table 3, in five questions there was a strong agreement among panel members (SD<1). However no agreement was found on the following question: 'relative weakness to provide security for historical texture tourists in not crowded and low traffic hours' (average of 2.01).

Table 3. Challenges of tourism security in Mazandaran, arranged according to importance level in the second step of Delphi method.

Row	Security challenges of tourism	Average	SD	Coefficient of variance
1	High costs of residence which left the tourists stranded outside and so decreased their security level	4.28	0.89	0.21
2	Illegal tourism facilities	3.96	0.67	0.16
3	Lack of security insurance of tourists by the investors of tourism facilities	3.94	0.64	0.20
4	Inadequate cooperation of the local communities with the police	3.80	0.59	0.19
5	Unfamiliarity of tour leaders with security issues	3.73	0.81	0.13
6	Lack of a comprehensive and strategic program, and the tourism aims not defined in the form of an executive program	3.64	1.18	0.35
7	Qualitative and quantitative weakness of land transportation facilities decreasing the financial and life security	3.68	1.15	0.31
8	Lack of security for investor to invest in tourism development	3.55	1.09	0.24
9	Lack of full support of tourist sites on the part of the police because of their abundance and distribution	3.47	1.27	0.38
10	Relative weakness to provide security for historical texture	2.01	1.16	0.56

Table 4. Strategies to increase tourism security in Mazandaran in the second step of Delphi method.

Row	Strategies	Number	SD
1	Promoting cooperation between the police and the travel agencies	4.37	0.51
2	Increasing the safety of the transportation system	3.87	0.45
3	Establishing security in tourist sites especially for eco—tourism tours	3.75	1.03
4	Seriously punishing those who guide tourists illegally	3.50	1.06
5	Familiarizing the tour guides with security challenges	3	0.75
6	Performing different conferences and seminars to develop tourism security	3	1.19

effective to increase tourism security of Mazandaran. Also, according to Table 3, in three strategies, there was a strong agreement among panel members (SD < 1).

4.3 The third step of Delphi method: Agreement level of panel members with the determined standards

In the third step, the questions from the first and second questionnaires were given to the panel members in the form of a questionnaire in order to be summed up. Table 5 summarized their agreement level with the ranking.

According to Table 5, agreement percentage with nine questions is more than 50% are the main security challenges of tourism in Mazandaran. Panel members do not consider 'relative weakness to provide security for historical texture tourists in not crowded and low traffic hours' as one of the main challenges of tourism security in the province (with agreement percentage of less than 50%).

In this step, the Standard Deviation of the members' answers regarding the strategies used to increase tourism security decreased from 83 percent in the second step to 69 percent in the third step. Thus, it can be said that panel members reached an agreement (consensus).

4.4 The fourth step of Delphi method: Agreement level with determined standards (final summing-up or consensus)

The fourth questionnaire included the adjusted questions of the third step was given to the panel members to make the final summing-up

4.2.2 Ranking opinions of panel members about strategies to increase tourism security

In this step, panel members recognized that from among the six strategies that they had introduced in the first step, 'promoting cooperation between the local communities and the police' is the most

Table 5. The agreement level of panel members with the challenges of tourism security in Mazandaranin the third step of Delphi method.

Row	Security obstacles of tourism	Agreement percentage	Disagreement percentage
1	High costs of residence which left the tourists stranded outside and so decreased their security level	89	11
2	Illegal tourism facilities	85	15
3	Lack of security insurance of tourists by the investors of tourism facilities	84	16
4	Inadequate cooperation of the local communities with the police	77	23
5	Unfamiliarity of tour leaders with security issues	74	26
6	Lack of a comprehensive and strategic program, and the tourism aims not defined in the form of an executive program	73	27
7	Qualitative and quantitative weakness of land transportation facilities decreasing the financial and life security	71	29
8	Lack of security for investor to invest in tourism development	66	34
9	Lack of full support of tourist sites on the part of the police because of their abundance and distribution	64	36
10	Relative weakness to provide security for historical texture tourists in not crowded and low traffic hours	41	59

(consensus). In this step, because both the first to ninth ranks obtained from the results of the questionnaires and Standard Deviation of members' answers regarding strategies to increase tourism security were similar to the third step, the tables were not presented. Therefore, the general agreement (consensus) about the most important challenges and strategies to increase tourism security is available.

5 CONCLUSION

The present study was an attempt to answers the following questions:

– *What are the challenges of tourism development in terms of security in Mazandaran?*

Delphi method was used to answer the above question. Panel members reached an agreement about the 10 challenges as the most important challenges of tourism security in Mazandaran. The question of 'high costs of residence which left the tourists stranded outside and so decreased their security level' was the most important security challenge of tourism development with the highest amount of average (4.28) and 89 percent of agreement of panel members.

– *What strategies can be used to overcome these challenges?*

From the Delphi method, panel members reached an agreement on the six strategies as the most important strategies of tourism security in Mazandaran. The strategies of 'promoting cooperation between the police and the travel agencies', 'increasing the safety of the transportation system', and 'familiarizing the tour guides with security challenges with a standard deviation less than 1 were identified as the most important strategies to increase tourism security in Mazandaran.

Mazandaran, located in the north of Iran, has a lot of historical, cultural, and natural tourism attractions. It has millions of intra and international tourists every year. Mazandaran is recognized as the biggest tourism center in Iran. But unfortunately it faces some security challenges. These challenges may cause unsustainability in the tourism process and activities if not managed correctly.

The implications hold for the Iranian Organization of Cultural Heritage, the governors, mayors, travel agencies, tour leaders working in Iran. Using the findings of this present study, all tourism planners and managers can promote the sustainable development of tourism in the province and even at the national level. They can achieve this by considering the most important security challenges and the most effective strategies found by the study to increase tourism security in Mazandaran.

REFERENCES

Anuar, A. et al. 2012. The effectiveness of safe city programme as safety basic in tourism industry: case study in Putrajaya. *Procedia—Social and Behavioral Sciences* 42: 477–485.

Asli, D.A. & Boylu, Y. 2009. Cultural comparison of tourists'safety perception in relation to trip satisfaction. *International Journal of Tourism Research* 12(2): 179–192.

Baker, D. & Stockton, S. 2014. Tourism and crime in America: apreliminary assessment of the relationship between the number of tourists and crime, two major american tourists cities. *International Journal of Safety and Security in Tourism* (5): 1–25.

Dalkey, N (ed.) 1969. *The Delphi method: an experimental study of group opinion*. Santa Monica CA: the rand corporation.

George, R. 2003. Tourist's perception of safty and security while visiting Cape Town. *Tourism Management* 24(5): 575–585.

Hezarjaribi, J. 2011. Social security feel from tourism development viewpoint. *Geography and Environmental Planning Journal* 22(42): 121–143.

Karamidehkordi, M. et al. 2013. Recognizing and ranking thechallenges of rural tourism development in security area using Delphi method, case study: ChaharmahalBakhtiyari Province. *Journal ofStrategic Researches of Security and Social Order* 4(1): 74–59.

Khoshfar, GH. et al. 2013. Considering the social and individual security feel from tourists viewpoint and the factors effecting on it, case study: Naharkhoran and Olangdareh in Gorgan. *Journal of Planning and Tourism Development* 2(6): 181–202.

Lotfifar, M. & Yaghfuri, H. 2013. Role of security in tourism development, case study: Chabahar. *1st National Conference of Role of Security in Tourism Development in Chabahar*: 1–8.

Milman, A. et al. 1999. The impact of security devices on tourist's perceived safety: the central Florida example. *Journal of Hospitality & Tourism Research* 23(4): 371–386.

Pishgahifard, Z. & Jahanian, M. 2011. Tourism security and legality in the Persian Gulf:viewpoint of I.R. Iran. *Journal of the Persian Gulf* 2(4): 51–61.

Rabani, R. et al. 2012. Role of police in security making and tourism attraction, case study: Esfahan. *Journal of SocialSecurity Studies* 3(26): 39–60.

Reza, S.A. & Jawaid, S.T. 2013. Terrorism and tourism: a conjunction and ramification in Pakistan. *Economic Modelling* 33: 65–70.

Seydaii, E. & Hedayatimoghadam, Z.2010. Role of security in tourism development. *Journal of Social Science* 4(8): 97–110.

Theory and Practice in Hospitality and Tourism Research – Radzi et al. (Eds)
© 2015 Taylor & Francis Group, London, ISBN 978-1-138-02706-0

Tourist's gender and income differences on perceptions and intentions of online information

R. Radzliyana & P.H. Khor
Faculty of Sports Science and Recreation, Universiti Teknologi MARA, Perlis, Malaysia

A.A. Azlan
Faculty of Mathematic and Statistic, Universiti Teknologi MARA, Perlis, Malaysia

K.C. Lim
College of Law, Government and International Studies, Universiti Utara Malaysia, Kedah, Malaysia

ABSTRACT: This study examined the perceptions of online information within leisure and among tourists particularly from gendered and income perspectives. The sport website acceptance model was utilized to develop items consisted in the survey questionnaire. The participants of the study were 386 local and international tourists involved in Kuala Lumpur Standard Chartered Marathon 2013 (228 males and 158 females), and they were selected by using the simple random sampling technique. Perceptions of online information and intention to use websites comprised of four components namely accessibility, flexibility, interactivity and reliability. The *t*-test results indicated that there was significant difference between mean scores of perceptions of online information between genders particularly for accessibility item. ANOVA results also indicated that there was significant difference between mean scores of perceptions of online information for different income groups. Post-hoc Tukey HSD analysis revealed significant difference of mean scores on perceptions of online information between different income groups particularly for items under accessibility component. Findings of this study suggested that future sport tourism organizations may include other attributes apart from gender and income groups when developing and providing online information since it can be easily reached by the related stakeholders as well as the society.

Keywords: Gender, income, leisure, tourist, online information

1 INTRODUCTION

1.1 *Background of the study*

The Internet was recognized as an extremely valuable marketing and communication tool due to its specific characteristics compared to other form of existed communication channels (Jere & Davis, 2011). Research reported that these specific characteristics of online marketing include accessibility, flexibility, interactivity and reliability (Radzliyana, Khor & Lim, 2012). Initially, Rafaeli and Sudweeks (1997) stated that the Internet was an interactive channel of communication while Berthon, Pitt, and Watson (1996) claimed that the Internet service was reachable despite boundaries. Therefore, the Internet enables businesses to minimize costs but maximize audience needs and segmentation.

Previous researchers had reported that the Internet was gaining prominence among sports events tourists due to its specific characteristics which include convenient, more efficient and a less time consuming service (Heung, 2003; Shilbury, Quick & Westerbeek, 2003). Further, Cai, Feng, and Breiter (2003) revealed that some tourists and potential tourists often used the Internet involved in the purchase decision, making decision on the tourism destination as well as seeking for additional information in significant to their travels. According to Jang (2004), seeking information from the Internet had offered numerous benefits to both travelers and the tourism marketers in terms of cost-effective and it allows real-time communication for both parties.

From time to time, the Internet becomes a major source of information throughout the globe (Peterson & Merino, 2003), thus it is important for respective tourism organizations to provide as much information as might be required by the potential tourists on the website. For instance, as referred to the tourism destination, an event is an experiential product and it relies on graphic and verbal representations and description, therefore, optimal information provision is regarded as crucial for success (Cai et al., 2003).

Previous research identified sport and tourism as the most chosen leisure activities (Ritchie & Adair, 2004) which regarded by many as the world's biggest social phenomenon (Kurtzman & Zauhar, 2003). With regards to sport tourism as world's biggest social phenomenon, this study intended to investigate differences on perceptions of online information and intention to use the websites from gender and income perspectives. Previous studies however, revealed that the relationship between gender and the technology adoption is inconclusive. Nonetheless, gender is often considered a critical factor because male and female sports tourists tend to consider technology use for achieving different ends (Chang & Melbourne, 2008). For instance, in adoption studies pertaining to mobile technologies, it was found that women often have less access to cell phones in developing countries than their male counterparts (Zainudeen, Iqbal & Samarajiva, 2010). Kwon and Chon (2009) furthermore found that gender is a significant determinant in mobile TV adoption. Other studies have shown some dissimilarity between the genders vis-à-vis online platform preferences and motives. In terms of income, previous findings indicated that earlier adopters to technologies tend to have more years of formal education and higher social status (income) than later adopters (Rogers, 2003).

Findings of the study offered supplementary knowledge on the theoretical perspectives of leisure and tourists behavior towards perceptions of online information in terms of gender and income differences. Besides, the obtained information could assist future marketing and communication managers in providing effective and efficient online information in order to cater numerous needs and wants of potential tourists or consumers.

2 LITERATURE REVIEW

The revolutionary development of online information has dramatically changed society and people's everyday lives, including the way travelers or tourists search for information and plan trips. Previous researches indicate that the Internet which provides online information has become one of the most important information sources for travel information acquisition (Lake, 2001). Many studies have indicated that the major purpose of information search is to support decision-making (i.e. reduce risk and uncertainty) and product choice in which the information search behavior strengthens the decision-making and choice behavior (Bettman,

1979; Bloch, Sherrell & Ridgway, 1986; Moorthy, Ratchford & Talukdar, 1997). Previous researchers stated that, for tourists, information acquisition is necessary for choosing a destination and for onsite decision such as selecting accommodations, transportation, activities and tours (Fodness & Murray, 1998; Gursoy & Chen, 2000; Snepenger, Meged, Snelling & Worrall, 1990).

Normally, consumers acquire a satisfactory amount of information about services before they are able and willing to use them (Moorman, Diehl, Brinberg & Kidwell, 2004). Typically, consumers want to learn about specific characteristics of online information services such as possibilities of new services, service characteristics and pricing issues (Kleijnen, de Ryter & Wetzels, 2007). Other researcher stated that intention to use the website is defined as the consumer's intent to engage in an online exchange relationship with the website (Zwass, 1998). Earlier research findings indicated that in many aspects, tourist information processing is different from that of other consumers. The differences are mainly due to structural reasons (Schertler, Schmid, Tjoa & Werthner, 1995). Tourists generally obtain online information from the respective tourism websites and they tend to rely on that form of communication channel due to some specific features or characteristics of the websites which include accessibility, flexibility, interactivity and reliability (Mircheska & Hristovska, 2010; Radzliyana et al., 2012). Online information accessibility occurred when it provides permanent exposure and global market reach. For instance, accessibility is extremely important especially dealing with international trade where business is conducted across different time zones. In addition, accessibility happened when online information not only provides virtually unlimited access for numerous consumers but also delivers unlimited amount of information as there is practically no restrictions in terms of the "space of advertisement" (Mircheska & Hristovska, 2010).

In terms of tourist's behavior, they have to leave their daily environment, having to move to geographically distant places to consume the tourism product. In fact, Werthner and Klein (1999) added that the tourism product normally cannot be tested and controlled in advance. Although it is commonly believed that in modern times the difference between the travel patterns of men and women are much less pronounced than before, gender differences related to travel and tourism still remain substantive (Collins & Tisdell, 2002). In terms of information processes, Krugman (1966) reported that women engaged in greater elaboration of advertisements than did men, regardless of whether the advertisements focused on contents considered of more interest to men or to women. Meanwhile, other researchers found greater stimulus elaboration among women than men when subjects were given adequate time to process information (Rosenthal & DePaulo, 1979). Despite several arguments that gender differences are not significant (O'Keefe, 2002) in cognitive theories, some findings consisted of evidence of dependable gender differed in persuasibility, with women being more easily persuaded than men (Becker, 1986; Eagly & Carli, 1981).

In terms of income differences in leisure and tourists behaviors towards perceptions of online information, previous research findings revealed that those in the low income group (less than RM2500) tend to spend less time than other income groups in accessing online information. Noor Ismawati and Ainin (2005) stated that this probably due to the financial constraints that limit their access to external entertainment. The middle income groups often use online information to access latest information in order to get themselves updated and engaged to the informative society. The high income groups on the other hand, often use online information to meet with the job requirement as being professionals. They tend to access information regardless location and time.

2.1 Research hypotheses

By investigating perceptions of online information in leisure and among tourists behavior with regards to gender and income differences, the hypotheses generated are as followed:

(i) There are differences in perceptions of online information in terms of gender.
(ii) There are differences of intention to use online information in terms of gender.
(iii) There are differences in perceptions of online information in terms of income.
(iv) There are differences of intention to use online information in terms of income.

3 METHODOLOGY

A simple random sampling method was utilized to select 386 local and international tourists attending the Standard Chartered Kuala Lumpur Marathon 2013. Out of 386 respondents, 228 (59.1%) were male whereby 158 (40.9%) were female.

In terms of income, it was reported that 95 (24.6%) respondents were earning RM2000 and lesser, 52 (13.5%) respondents earned RM2001-RM2500, 53 (13.7%) respondents earned RM2501-RM3000, 59 (15.3%) respondents earned between RM3001-RM3500 and 127 (32.9%) respondents earned RM3501 and more. The respondents' ages ranged from 20 to 50 years old and the mean age was 25years 1 month.

This study developed a self-administered questionnaire as a research instrument. Items in the questionnaire were adopted from the measurement scale of Sport Website Acceptance Model (SWAM) developed by Hur, Ko, and Claussen (2011).

4 DATA ANALYSIS

The statistical significance of both hypotheses was tested using t-test to explain differences among gender whereby MANOVA was applied to seek differences among income groups. A significance level $p \leq .05$ was adopted to decide the significance level of each research hypothesis.

5 RESULTS

5.1 Factor analysis and reliability of measurement scale

Results from factor analysis revealed that there were four components of perceptions of online information which include accessibility, flexibility, interactivity, and reliability. The value of item loading for the measurement scale was greater than 0.40, with eigenvalues greater-than-one for the four subscales while the value of item-total correlation for each subscale was more than 0.45. The Coefficient Alpha for the scale of each component was 0.872, 0.844, 0.739 and 0.750 respectively.

5.2 Gender differences towards perceptions of online information

Generally, there are four types of perceptions of online information perceived by sports tourists. Results obtained however indicated that only accessibility component recorded significant differences in perceptions of online information in terms of gender. As referred to Table 1, the particular item was identified as "Access organized collection", $t(384) = 2.635$, $p = .009$, (male: 5.73, female: 5.41). Obviously, the findings showed that male sports tourists perceived higher perceptions on online information in related to accessibility compared to the female sports tourists.

5.3 Gender differences towards intention to use the websites

Result displayed in Table 2 shows that there was no significant difference of intention to use the websites among sports tourists in terms of gender. Thus, it can be concluded that sports tourists perceived the same intentions to use the websites regardless of gender.

Table 1. Mean, standard deviation and significant value of accessibility component on perceptions of online information *(N = 386)*.

Accessibility	Gender	N	Mean	SD	Sig.
Interact with	M	228	5.54	1.28	.201
sport media	F	158	5.37	1.24	
Reduce daily	M	228	5.59	1.16	.304
tasks	F	158	5.46	1.20	
Access organized	M	228	5.73	1.08	.009
collection	F	158	5.41	1.25	
Generate	M	228	5.63	1.14	.730
awareness	F	158	5.67	1.04	
Establish	M	228	5.66	1.02	.757
interactive channel	F	158	5.70	1.10	
Access previous	M	228	5.57	1.08	.990
information	F	158	5.58	1.16	

Table 2. Mean, standard deviation and significant value of intention of online information *(N = 386)*.

Intention	Gender	N	Mean	SD	Sig.
Major source of	M	228	5.68	1.20	.095
information	F	158	6.29	5.29	
Spend more	M	228	5.56	1.04	.335
time	F	158	5.46	1.09	
Continue seeking	M	228	5.64	1.12	.783
information	F	158	5.68	1.17	
Continue purchase	M	228	5.78	4.11	.645
in future	F	158	5.63	1.06	
Recommend to	M	228	5.67	.93	.270
others in future	F	158	5.76	.99	

5.4 Income groups differences toward perceptions of online information

The MANOVA's score for accessibility component on perceptions of online information in leisure and tourist behavior on the five income groups was significant, Wilk's $\wedge$ = 0.900, $F(24, 1313)$ = 1.669, p = .023.

Follow-up ANOVA tests as shown in Table 4 extracted significant results for three items of accessibility reading "Interact with sports media", $F(4, 385)$ = 2.636, p = .034, η^2 = .02, "Generate awareness on particular sporting events and organization", $F(4,385)$ = 2.797, p = .026, η^2=.02 and "Access to previously inaccessible information", $F(4,385)$ = 5.001, p = .001, η^2 = .04.

Findings revealed that sports tourists from the income group of RM3501 and more experienced high perceptions on online information in related to interaction to sport media, generating awareness on particular sports as well as its organization and gain access to previously inaccessible information.

Table 3. Multivariate tests on perceptions of online information in term of income groups.

Multivariate Tests[c]

Effect		Value	F	Hypothesis df	Error df	Sig.
Intercept	Pillai's Trace	.973	2246.092[a]	6.000	376.000	.000
	Wilks' Lambda	.027	2246.092[a]	6.000	376.000	.000
	Hotelling's Trace	35.842	2246.092[a]	6.000	376.000	.000
	Roy's Largest Root	35.842	2246.092[a]	6.000	376.000	.000
income	Pillai's Trace	.102	1.653	24.000	1516.000	.025
	Wilks' Lambda	.900	1.669	24.000	1312.917	.023
	Hotelling's Trace	.108	1.682	24.000	1498.000	.021
	Roy's Largest Root	.074	4.649[b]	6.000	379.000	.000

a. Exact statistic

b. The statistic is an upper bound on F that yields a lower bound on the significance level.

c. Design: Intercept+income

5.5 Income groups differences toward intention to use the websites

Result indicated no significant difference on intention to use the websites in terms of income groups, Wilk's $\wedge$ = 0.925, $F (20, 1251)$ = 1.489, p = .076. Regardless of income, both male and female sports tourists perceived the same intention to use the websites in related to the reasons of "Major source of information", "Spend more time", "Continue seeking information in future", "Continue purchase in future", and "Recommend to others".

6 DISCUSSION

Despite several arguments in leisure and tourism studies that gender differences are not significant (O'Keefe, 2002) in cognitive theories, some findings consisted of evidence of dependable gender differenced in persuasibility, with women being more easily persuaded than men (Becker, 1986; Eagly & Carli, 1981).

The results of this study, however, were contradicted with previous research specifically on the accessibility component of online information. Instead, results of dependable gender differed in accessibility of online information, with men access more online information than women.

Significant differences in perceptions of online information were found among sports tourists in terms of income groups. The middle and high income (RM2500 and more) tourists reported that they perceived high perceptions towards online information particularly on accessibility component than the lower income group did. This finding related to previous study outcome conducted by Noor Ismawati and Ainin (2005) on computer usage and activities. They stated that, these groups of tourists are better able to afford the electricity and Internet connection bills compared to those in lower income groups. Furthermore, middle and high income groups tend to rely on online information to meet with their job requirements.

Table 4. Tests of between-subjects effects on perceptions of online information in related to accessibility in term of income groups.

Source	Dependent variable	Type III sum of squares	df	Mean square	F	Sig.
income	Interact with sports media	16.646	4	4.162	2.636	.034
	Reduce daily tasks	6.243	4	1.561	1.123	.345
	Access organized collections	2.364	4	.591	.435	.783
	Generate awareness	13.240	4	3.310	2.797	.026
	Establish interactive channel	6.234	4	1.558	1.406	.231
	Access to previous information	23.762	4	5.940	5.001	.001
Error	Interact with sports media	601.416	381	1.579		
	Reduce daily tasks	529.749	381	1.390		
	Access organized collections	517.990	381	1.360		
	Generate awareness	450.843	381	1.183		
	Establish interactive channel	422.287	381	1.108		
	Access to previous information	452.559	381	1.188		
	Generate awareness	464.083	385			
	Establish interactive channel	428.521	385			
	Access to previous information	476.321	385			

a R Squared = .027 (Adjusted R Squared = 017)
b R Squared = .012 (Adjusted R Squared = .001)
c R Squared = .005 (Adjusted R Squared = −.006)
d R Squared = .029 (Adjusted R Squared = .018)
e R Squared = .015 (Adjusted R Squared = .004)
f R Squared = .050 (Adjusted R Squared = .040)

Table 5. Multivariate tests on intentions to use the websites in term of income groups.

Multivariate Tests[c]

Effect		Value	F	Hypothesis df	Error df	Sig.
Intercept	Pillai's Trace	.979	3487.541[a]	5.000	377.000	.000
	Wilks' Lambda	.021	3487.541[a]	5.000	377.000	.000
	Hotelling's Trace	46.254	3487.541[a]	5.000	377.000	.000
	Roy's Largest Root	46.254	3487.541[a]	5.000	377.000	.000
income	Pillai's Trace	.076	1.475	20.000	1520.000	.080
	Wilks' Lambda	.925	1.489	20.000	1251.317	.076
	Hotelling's Trace	.080	1.500	20.000	1502.000	.072
	Roy's Largest Root	.061	4.661[b]	5.000	380.000	.000

a. Exact statistic

b. The statistic is an upper bound on F that yields a lower bound on the significance level.

c. Design: Intercept+income

Contributing to the existing literature, this study provides some gender-based insights as well as income group differences for tourism marketers in developing specific strategies for attracting sports tourists, designing respective sports tourism products and thus enhancing the sustainability of sports-based destinations through online information and websites development. We recommend that future studies to further examine the other attributes of sports tourists in relations to online information. To better serve the fast and ever increasing sports tourism market, researchers and industry practitioners should gradually provide concise and precise information on the Internet as it allows sports tourists in decision-making.

7 CONCLUSIONS

The technological revolution through existence of online information and the Internet has changed dramatically the market conditions for sport tourism organizations. Information technology (IT) evolves rapidly providing new tools for sport tourism marketing and management. The development of new and more powerful online information applications empowers both suppliers and consumers to enhance their efficiency and re-engineer their marketing and communication strategies. Thus, future study could focus on other attributes of sports tourism in significant to online information in order to cater different needs of different consumers. Obviously, sports tourism consumers are differ from other products due to its nature which tourism products cannot be tested and controlled in advance. A comprehensive study in sports tourism could offer numerous benefits to its stakeholders.

REFERENCES

Becker, B.J. (1986). Influence again: An examination of reviews and studies of gender differences in social influence. In J.S. Hyde. & M.C. Linn (Eds.). *The Psychology of gender: Advances through meta-analysis* (pp. 178–209). Baltimore, MD: John Hopkins University Press.

Berthon, P., Pitt, L. & Watson, R. (1996). The World Wide Web as an advertising medium: Toward and understanding of conversion efficiency. *Journal of Advertising Research, 36*(1), 432–445.

Bettman, J.R. (1979). Information processing theory on consumer choice. Boston, MA: Addison-Wesley.

Bloch, P.H., Sherrell, D.L. & Ridgway, N.M. (1986). Consumer search: An extended framework. *Journal of Consumer Research, 13*, 119–126.

Cai, L.A., Feng, R. & Breiter, D. (2003). Tourist purchase decision involvement and information preferences. *Journal of Vacation Marketing, 10*(2), 138–148.

Chang, P. & Melbourne, A. (2008). *Drivers and moderators of consumer behavior in the multiple use of mobile phone*. Paper presented at 19th Australasian Conference on Information Systems, 3–5 December.

Collins, D. & Tisdell, C. (2002). Gender and differences in travel life cycles. *Journal of Travel Research, 41*, 133–143.

Eagly, A.H. & Carli, L. (1981). Sex of researchers and sex-typed communications as determinants of sex differences on influenceability: A meta-analysis of social influence studies. *Psychological Bulletin, 90*(1), 1–20.

Fodness, D. & Murray, B. (1998). A typology of tourist information search strategies. *Journal of Travel Research, 37*, 108–119.

Gursoy, D. & Chen, J. (2000). Competitive analysis of cross cultural information search behavior. *Tourism Management, 21*, 583–590.

Heung, V.C.S. (2003). Barriers to implementing E-commerce in the travel industry: A practical perspective. *International Journal of Hospitality Management, 22*(1), 111–118.

Hur, Y., Ko, Y.J. & Claussen, C.L. (2011). Acceptance of sports websites: a conceptual model. *International Journal of Sports Marketing & Sponsorships*, 209–224.

Jang, S.C. (2004). The past, present and future research of online information search. *Journal of Travel & Tourism Marketing, 17*(2/3), 41–47.

Jere, M.G. & Davis, S.V. (2011). An application of uses and gratifications theory to compare consumer motivations for magazine and Internet usage among South African women's magazine readers. *South Africa Business Review, 15*(1), 1–27.

Kleijnen, M.H.P., de Ruyter, K. & Wetzels, M.G.M. (2007). An assessment of value creation in mobile service delivery and the moderating role of time consciousness. *Journal of Retailing, 83*(1), 33–46.

Krugman, H.E. (1966). The measurement of advertising involvement. *Public Opinion Quarterly, 30*, 583–596.

Kurtzman, J. & Zauhar, J. (2003). A wave in time—The sports tourism phenomena. *Journal of Sport & Tourism, 8*, 35–47.

Kwon, H. & Chon, B. (2009). Social influences on terrestrial and satellite mobile TV adoption in Korea: Affiliation, positive self-image and perceived popularity. *International Journal on Media Management, 11*(2), 49–60.

Lake, D. (2001). *American go online for travel information*. Retrieved from http://www.cnn.com/2001/TECH/internet/06/14/travelers.use.net.idg/index.htm

Mircheska, I. & Hristovska, M. (2010). *The Internet marketing: A challenge for fast tourism development*. Paper presented at International Association of Scientific Expert in Tourism.

Moorman, C., Diehl, K., Brinberg, D. & Kiewell, B. (2004). Subjective knowledge, search locations and consumer choice. *Journal of Consumer Research, 31*(3), 673–380.

Moorthy, S., Ratchford, B.T. & Talukdar, D. (1997). Consumer information search revisited: Theory and empirical analysis. *Journal of Consumer Research, 23*, 263–277.

Noor Ismawati, J. & Ainin, S. (2005). Domestic computer usage and activities in West Coast Malaysia: Age and income differences. *Information Development, 21*, 128–136.

O'Keefe, D.J. (2002). *Persuasion: Theory and research* (2nd ed.), Thousand Oaks, CA: Sage.

Peterson, R.A. & Merino, M.C. (2003). Consumer information search behavior and the Internet. *Psychology & Marketing, 20*(2), 99–121.

Radzliyana, R., Khor, P.H. & Lim, K.C. (2012). *Acceptance of online sports marketing among FSR student in UiTM Perlis*. Paper presented at National Postgraduate Colloquium 2012 (NAPAC 2012), December, 2012.

Rafaeli, S. & Sudweeks, F. (1997). Networked interactivity. *Journal of Computer Mediated Communication. 2*(4).

Rosenthal, R. & DePaulo, B.M. (1979). Sex differences in accommodation in nonverbal communication. In R. Rosenthal (Ed.), *Skill in nonverbal communication* (pp. 68–103). Cambridge, MA: Oelgeschlager, Gunn and Hain.

Ritchie, B.W. & Adair, D. (2004). Sport tourism: An introduction and overview. In B.W. Ritchie & D. Adair (Eds.), *Sport tourism: Interrelationships, impacts and issues* (pp. 1–29). Tonawanda, NY: Channel View Publications.

Rogers, E. (2003). *Diffusion of innovations* (7th ed.), New York, NY: The Free Press.

Schertler, W., Schmid, B., Tjoa, A.M. & Werthner, H. (1995). *Information and communication technologies in tourism*. New York, NY: Springer, Wien.

Shilbury, D., Quick, S. & Westerbeek, H. (2003). *Strategic sport marketing* (2nd ed.), Crow's Nest, NSW: Allen & Unwin.

Snepenger, D., Meged, Snelling, M. & Worrall, K. (1990). Information search strategies by destination naïve tourist. *Journal of Travel Research, 29*, 13–16.

Werthner, H. & Klein, S. (1999). Information technology and tourism: A challenging relationship. New York, NY: Springer, Wien.

Zainudeen A., Iqbal, T. & Samarajiva, R. (2010). Who's got the phone? The gendered use of telephones at the bottom of the pyramid. *New Media & Society, 12*, 549–566.

Zwass, V. (1998). Structure and macro-level impacts of electronic commerce: From technological infrastructure to electronic marketplaces. In K.E. Kendalls (Ed.), *Emerging information technologies*. Thousand Oaks. CA: Sage Publications.

Theory and Practice in Hospitality and Tourism Research – Radzi et al. (Eds)
© *2015 Taylor & Francis Group, London, ISBN 978-1-138-02706-0*

A review of family influences on travel decision making

F.N. Osman
School of Graduate Studies, Universiti Putra Malaysia, Malaysia

H. Hashim, H. Nezakati, S.R. Hussin & Y.A. Aziz
Faculty of Economics and Management, Universiti Putra Malaysia, Malaysia

R.N. Raja-Yusof
Putra Business School, Universiti Putra Malaysia, Malaysia

ABSTRACT: Worldwide, there are an increasing number of tourists travelling abroad together as a family, with family members ranging from 3 to 7 persons, and 3 to 50 years of age. These groups of family tourists have illustrated some unique needs which entail further investigation if industry players were to meet and satisfy their needs more effectively. This paper reviews the existing literature on family decision making, paying attention to specific requirements of family needs while travelling. Specific needs of this type of tourists include children-friendly menus, kids clubs, family travel packages, family-friendly environment, and wellness facilities for family members of different age groups etc. The paper reviews existing tourism typologies and concluded that there is a need to develop a new typology for family tourists. This paper proposes three new dimensions for tourist typologies, i.e. cultural-influence (religion), customer characteristic (stage in family life cycle) and destination services (family-friendly facilities).

Keywords: Family decision making, family life cycles, tourist typology, travel decision

1 INTRODUCTION

Tourism is a world-wide socio-economic phenomenon and evolved from developing sectors of world's economy in the past period to emerging sectors in the 21st century. The growth of tourism is an outcome of many factors such as changing of lifestyles, gaining new interests, the growth of income and leisure time, improvement of accessibility and accommodation, and new offer in tourism market (Coccossis & Contantoglou, 2006). According to World Tourism Barometer, international tourist arrivals grew by 5 percent in 2013 and reaching a record of 1,087 million arrivals. In 2014, a 4 percent to 4.5 percent growth was forecasted above the long term projections while demand for international tourism for destinations in Asia and the Pacific are strongest in regional prospect by 5 percent to 6 percent growth where the number of international tourists grew by 14 million to reach 248 million (UNTWO, 2014). Meanwhile, tourism industry in Malaysia is expected to contribute 12.5 percent to the GDP in 2014, a 12 percent increase from 2011 (Malaysia Travel News, 2014). An interesting development in the tourism industry is the growth of tourists travel with their family worldwide. To complement the increase in tourists'

arrival, Malaysia is now adding more destination attraction and facilities that would cater to the need of various groups of tourists all over Malaysia. Apparent example can be found in Iskandar Malaysia, where the expected arrival of family tourist is encouraging to places such as Johor Premium Outlet, LEGOLAND Theme Park, Puteri Harbour Theme Park, Sanrio Hello Kitty Town and Lat's Place (NSTP, 2014).

2 LITERATURE

2.1 *Early studies in family decision making*

Past research examined family role in decision making process. These are summarized as below:

- Jenkins (1978) investigated the vacation decision process for families where it was found that the husband and wife are dominant in influencing vacations decision, while children did not have considerable influences in deciding upon the kinds of vacation activities for the family.
- Cosenza & Davis (1981) focused on dyadic family (husband and wife) as the primary unit of analysis and measured dominance as characteristics of interaction between married partners.

They studied both the influence of family members and the decision structures in travel across stages in the Family Life Cycles.

- Nichols & Snepenger (1988) examined family decision making, tourism behavior, and attitude of tourists specifically on three decisions making mode, namely: husband-dominant, wife-dominant and joint decision making. The finding showed that there were differences between these three groups where majority of the families employed a joint decision making mode compared to the other two modes.
- Fodness (1992) explored the impact of family life cycles on family travel decision making. The result shows that there is a joint decision making process but the wife appears to be most likely to make a decision in families with children while husband only makes a decision in the early stages where there is no presence of kids.

Even though there exists past studies of family decision making over FLC, but the results raise a number of plausible explanations that can be subjected to further empirical research. For example, the numbers of children's participation in the decision making process is still below average. Moreover, it can be said that there has been a revolution on the tourism industry in conformity with the modernization era. We foresee that there would be changes in terms of the influence of different family members at each FLC stages for families today as compared to the decisions back in the 80's and 90's. To name some of decision aspects that may differ because of the time include: the nature of demand for travel among families, product and services offered by the industry players, government policies governing tourism industry and destination development which all opens up a new phase in the tourism industry. In addition, the impacts of family in making decision while travelling will affect in the ways in which it shapes individual host and guest experiences.

2.2 *Family influence in tourist decision making*

Meanwhile, family could be a growing market segment which requires further understanding of the needs and preferences of the family tourists. Factors such as different demographics and behavioural characteristics of different family members at different stage may have different effect of the travel decision making and travelling outcomes. Possible changes of life event and circumstances that happen on individuals and families at different stage of the FLC such as marriage, childbirth, death of spouse, retirement etc. may affect family behaviour and travel motives based on their altered family structure. With such a vast market to research, tourism practitioners must understand and develop

appropriate typology to measure, assess and motivate families to travel (Hong, Fan & Palmer, 2005). According to the study of family decision making in Korea done by Kim, Choi, Agrusa, Wang & Kim (2010), the decision making process of a family is highly influenced by husband and wife, whereas children follow their parent's decisions. This may be true due to the Korean culture itself where children are educated to obey their parents', unlike in the west where children are free to give their opinion. The fact that children are not given voice in decision making process is problematic since they are part of target customers in tourism market (Blichfeldt, Pedersen, Johansen & Hansen, 2011). As assert by Illum (2013), the decision making process in family institution may be influenced by other family members' behaviour and the rise of children involvement also plays an important role in influencing adult decisions. These have been neglected by most of researcher in their studies of family tourist. The specific needs of tourists who travel with families are frequently unmet because of this major gap. In an effort to reflect more complex realities of individual and family needs, this paper proposed family-friendly services for industry business's needs, which emphasize of marketing and product development relative to FLC stages. Family-friendly services play an important role in making family travel more fun as well as influence their decisions of travel arrangement. Family-friendly services could consists of kids clubs, crèches (day care centre, babysitting services), travelling nannies (swimming instructor, sport coaches), family travel packages (bundle or combined family services), all inclusive hotels/resorts (packages of food, dining, accommodation and activities), and five or more family hotels/resorts (flexibility of adjoining rooms) (Bohrer, 2011).

2.3 *Family life cycles*

Lawson (1991) and Hong et al., (2005) posits that family tourists decision making and behaviour can be broken down into cycles known as the Family Life Cycles (FLC). Hong et al., (2005) assert that as individual preferences and the demands placed on resources affected through the progress of FLC, the presence or absence of a spouse, children and their age. Although other variables can be proxies than the FLC, integration of the FLC as one of component in the model will captures some of the relations of family influence in travel decision making. FLC contains eight family cycles which are bachelor, young couples, full nest I, full nest II, full nest III, empty nest, elderly and solitary survivors. These are then further categorizes into; 1) Bachelor: young adults, young singles. 2) Young couples: newly married, no children. 3) Full nest I: young couples with children. 4) Full nest II: growing families with dependent children. 5) Full nest III: older married couple with dependent children. 6) Empty nest: older married couple with no children at home. 7) Elderly: senior citizen, company of other older people. 8) Solitary survivors: widows, widowers. The choice of trip activities for each group may vary because of the characteristics of family itself. Bachelor and young couples group may choose to join adventurous activities as they are young and eager to discover something new as well as they has less commitment compared to family with children. Growing family such as full nest II may be choose leisure activities compared to others as they are travel with their children. Their choice of activities may be based on their children's decisions and to suit their preferences, parents may consider visiting destinations such as theme park and wildlife zoo. Empty nest and elderly group may choose to travel for heritage and nature tourism rather than leisure tourism compared to full nest. The study by Lawson (1991) and (Hong et al., 2005) has shed light into better understanding of family tourists behaviour. Meanwhile, Jang and Wu (2006) studied behaviours of senior tourists. According to them, among travel motives of senior travellers was to seek for knowledge and intellectual enrichment by visiting natural and historical sight. Whereas, travel expenditure for families with children mostly fall in leisure activities (Hong et al., 2005).

3 TOURISTS TYPOLOGY

Typology plays an important role in defining and developing tourism industry as it divides data of one or more attributes into different types by putting data into groups according to how they are similar. On the other hand, tourism also can be a benchmark or a vehicle whereby host countries can locate and present themselves as they would like to be seen by others (Henderson, 2010). In order to successfully manage tourism sector, it is fundamental to understand tourist type's thoroughly. Past researchers grouped tourists by identifying their common characteristics in responding to a certain situation to gain useful knowledge to assure the objectives and goals while trying to satisfy tourists' needs (Jafari 1989, Wickens, 2002; Marwijk & Taczanowska, 2006). Instead of treating tourist as a homogenous group, typology also reflects the multiplicity of tourist experiences, motivations and value which given some space for the family to be the subject of typologies as past research tend to develop typologies based on individualistic figure only with no emphasis of families characteristics.

The development of tourist's demand and supply typologies provides useful information to address the issues. Demand typology is focused on destinations chosen, travel characteristics, travel motivations, and tourist's behaviour while supply typology focus on characteristics of the destination area in terms of the form and extent of supply needs tourism development in the area. Conceptually, Cohen (1972, 1979) was the first researcher who subdivides tourists into a group of different categories. As assert by Cohen conceptual descriptions, tourists can be divided based on their degree of institutionalization and how to differentiate between the drifter, the explorer, the individual mass tourist and the organized mass tourist whereas how tourist react during their travels is therefore determined by the orientation of spiritual or cultural attributes which can be located within or outside their normal region (Arnegger, Woltering & Job, 2010).

One of the current tourist's typology that exists today is the one used for medical tourism. Medical tourists are unlike the other type of tourist since they are seeking activities according their interests rather than choosing destinations first. Instead, they find a suitable destination after deciding on what type of medical services they want (Bookman & Bookman, 2007). Wongkit and McKercher (2013) identified medical tourist typology by combining the trip purpose and decision horizon dimensions. This typology gives rise to four types of tourists: dedicated medical tourist, hesitant medical tourist, holiday medical tourist and opportunistic medical tourist. The dedicated medical tourist makes decision to seek treatment as the main reason and make it decision before departure as important for holiday for pleasure. The hesitant medical tourists are those who have intention to seek treatment but did not make final decision until after arrival. The holiday medical tourist is a tourist who also has intention to seek treatment but identified holiday as main purpose of vacation. Lastly, the opportunistic medical tourist is that person whose main purpose is for vacation and only makes a decision to undergo treatment once in the destination.

Tourist typology can also be expected to adopt supply-sides approach in terms of market segmentation (Arnegger et al., 2010). For example, Pearce (2008) developed a generalized model of tourism distribution emphasized on tourists need in term of time, place, form and possession utilities, timing and location services required by distinguished tourists segments as independent, customized and package. Independent tourists are those who favour choice, flexibility and spontaneity. They are independently arranged of all or the majority of their own travel often not directly from the providers at the destination and market based intermediaries. While customized tourists are those who demand particular features that are not normally included in standard packages and the same time rely on bundled services purchased before departure. Package tourists are those who purchase of all inclusive tours of a single product that sold in a single transaction in the market, which all of the components have been bundled together. Basically, tourists demand three major categories of services like transport, accommodation, attractions and activities while make their travel arrangements before vacationing.

The existing typologies have not considered the decisions or behaviour of children and family travelling as tourists. This is not reflecting the current scenario on worldwide tourism industry where the consumer base of many tourists resort and attractions are formed by families (Obrador, 2012).

4 CONCLUSION

This paper has highlighted some interesting aspect of family tourists as a growing segment in the tourism industry. However there is a need to pay special attention to the needs of this group of customers as initial observations of their travel behaviour indicates that there are several specific needs unmet by current offers in the tourism market.

When tourists travel with their family, travelling decisions may be influenced by various family members. This might lead to specific travelling or purchase behaviour that differs from individual tourists. For example, the head of the family may make decisions for the whole group, thus increasing the quantity of products purchased, customizing types of product to suit different needs of different group members, especially the ones regarding toddlers, teenagers and older people. The paper used the Family Life Cycle Stage to predict tourist behaviour as being influenced by the special needs of tourists at different stages. The paper categorized the specific needs of family into the offers provided by the chosen destination into specific types of product, types of facilities, services and service quality as a starting point in understanding the family tourists. Hence, the paper concluded by proposing a new family tourist typology and recommending that future research consider the needs of family tourist and taking into account the family life cycles characteristics and decision horizon in the new typology.

REFERENCES

Arnegger, J., Woltering, M. & Job, H. (2010). Toward a product-based typology for nature-based tourism: a conceptual framework. *Journal of Sustainable Tourism, 18*(7), 915–928.

Blichfeldt, B.S., Pedersen, B.M., Johansen, A. & Hansen, L. (2011). Tweens on Holidays. In-Situ Decision-making from Children's Perspective. *Scandinavian Journal of Hospitality and Tourism, 11*(2), 135–149.

Bohrer, I. (2011). 7 Family-Friendly Services That Make Family Travel Easier (and More Fun!). BootsnAll. Retrieved from http://www.bootsnall.com/

Bookman, M.Z. & Bookman, K.R. (2007). *Medical Tourism in Developing Countries.*

Coccossis & Contantoglou. (2006). The Use of Typologies In Tourism Planning, (September).

Cosenza, R.M. & Davis, D.L. (1981). Family Vacation Decision Making Over The Family Life Cycle: A Decision And Influence Structure Analysis. *Journal of Travel Research, 20*(2), 17–23.

Fodness, D. (1992). The Impact of Family Life Cycle on the Vacation Decision-making Process. *Journal of Travel Research, 31*(2), 8–13.

Henderson, J.C. (2010). Islam and Tourism. *Bridging Tourism Theory and Practice.* Elsevier.

Hong, G., Fan, J.X. & Palmer, L. (2005). Leisure Travel Expenditure Patterns by Family Life Cycle Stages. *Journal of Travel & Tourism Marketing,* (February 2014), 37–41.

Illum, S.F. (2013). Family tourism: multidisciplinary perspectives. *Journal of Tourism and Cultural Change, 11*(1–2), 126–127.

el motivation and the influential factors: An examination of Taiwanese seniors. *Tourism Management, 27*(2), 306–316.

Jenkins, R.L. (1978). Family Vacation Decision-Making. *Journal of Travel Research, 16*(4), 2–7.

Kim, S.S., Choi, S., Agrusa, J., Wang, K.-C. & Kim, Y. (2010). The role of family decision makers in festival tourism. *International Journal of Hospitality Management, 29*(2), 308–318.

Lawson, R. (1991). Patterns Of Tourist Expenditure And Types Of Vacation Across The Family Life Cycle. *Journal of Travel Research, 29*(4), 12–18.

Malaysia Travel News. (2014). Malaysia has identified four mega-trends to attract tourists. Bernama. Retrieved from http://blis.bernama.com/

Marwijk, R. Van & Taczanowska, K. (2006). Types of Typologies—From Recreationists & Tourists to Artificial Agents, (September), 499–501.

Nichols, C.M. & Snepenger, D.J. (1988). Family Decision Making And Tourism Behavior And Attitudes. *Journal of Travel Research, 26*(4), 2–6.

NSTP. (2014). Malaysia can achieve 26.8 million tourist arrival target. New Straits Times. Retrieved from http://www.nst.com.my/

Obrador, P. (2012). The place of the family in tourism research: Domesticity and thick sociality by the pool. *Annals of Tourism Research, 39*, 401–420.

Pearce, D.G. (2008). A needs-functions model of tourism distribution. *Annals of Tourism Research, 35*(1), 148–168.

UNTWO. (2014). International tourism exceeds expectations with arrivals up by 52 million in 2013 World Tourism Organization UNWTO. Retrieved from http://www2.unwto.org/

Wickens, E. (2002). The Sacred and The Profane: A Tourist Typology. *Annals of Tourism Research, 29*(3), 834–851.

Wongkit, M. & McKercher, B. (2013). Toward a typology of medical tourists: A case study of Thailand. *Tourism Management, 38*, 4–12.

Theory and Practice in Hospitality and Tourism Research – Radzi et al. (Eds)
© *2015 Taylor & Francis Group, London, ISBN 978-1-138-02706-0*

Job performance and employee satisfaction at International Youth Centre, Kuala Lumpur

N.M.A. Ghani & N.S.N. Muhamad-Yunus
Universiti Teknologi MARA, Puncak Alam, Malaysia

ABSTRACT: Surviving and competing successfully in today's unstable economic environment force the organizations to have employees who are proactive and committed while engaging with their job in performing higher standard of job performance. This is parallel with the aims to seek in delivering a customer satisfaction thus establish a good relationship with customers. The nature of customers and employee's interaction constitute the heart of customers' evaluation of the service experience and regarding to that, the employee's role in shaping of customer's satisfaction cannot be overlooked. When the employees have understood their roles and responsibilities, they can successfully perform their job as well as enhance their satisfaction. Therefore, the purpose of this study was to identify the relationship between determinants of job performance; rewards and motivation, empowerment and work environment towards employee satisfaction at International Youth Centre (IYC) in Kuala Lumpur. A total of 60 questionnaires were distributed to the respondents who work at IYC. The degree of the relationship between determinants of job performance and employees satisfaction was investigated through questionnaires distributed. Findings of this study showed that there are positive and significant relationships between the independent variables towards employee satisfaction.

Keywords: Job performance, rewards and motivation, empowerment, work environment, employee satisfaction

1 INTRODUCTION

Employee's job performance is the ability to perform effectively in a job required and understands a complete and up-to-date job description for position; understands the job performance requirements and standards that are expected to meet. According to Bakker and Leiter (2010) in order to survive and compete successfully in today's unstable economic environment, organizations need employees who are proactive, show commitment while engaging with their job and responsibilities and remain committed to performing at higher standard. Organizational agility requires employees who promote high energy and self-confidence and demonstrate strong enthusiasm and passion to their work (Bakker & Schaufeli, 2008). Several factors of job performance will give a big influence to employee's satisfaction. The factors such as rewards and motivation, work environment and empowerment are among the factors that will enhance the employee's satisfaction.

In most organizations, employee satisfaction is important in order to ensure employees are always being loyal with the organization and able to treat external customers with the best services.

Therefore, it is crucial for the organization to have a look at the overall organization environment, whether it is conducive and comfortable enough for employees in performing their works. According to Milliman, Ferguson, Czaplewski and Join (2008), employee satisfaction is how employees feel about the work they are doing and the job performance from that work directly give an impact towards organization's performance and ultimately its stability. Thus, if the employees are highly motivated towards their job, they will perform well and they will be motivated to achieve the goals of the organization. According to Homburg and Stock (2005), the relationship between job satisfaction and job performance is very strong that they affect one another including the relationship with customers. Unfortunately, nowadays some organizations are only concerned about the profits that they get and passionately think about to change towards customer-focused management without focusing on internal customers first.

The factors of job performance in the organizations today were not considered by some of organizations because they were just focus more on customers by ignoring the employee's satisfaction. The employees did not get the desired rewards that worth with the tasks performed and the organizations did not know how to give motivations to the employees directly when they achieve success. Parts from that, most of researches conducted are focused solely on the influence of rewards and motivation thus ignored the importance of measuring other factor such as empowerment and the organization itself. In view of the above discussion, this study appears with the main objective in examining the relationship of each determinant factors of job performance towards employee satisfaction at IYC.

2 LITERATURE REVIEW

2.1 *Job performance*

Job performance is known as the individual efficiency in both quantitative and qualitative aspects of the job expected of an employee and how well those activities were executed and whether the employees are meeting the job duties and the policies and standards of their organization while doing their job. Job performance depends upon the environment of office, work settings and the social relations (Coetzer & Rothmann, 2006). Job performance is an important factor for individual work effectiveness evaluation. Thus, an organization's success or failure depends much on employee's job performance because employees are the heart of the organization.

Munchinsky (2003) stated that job performance is the set of worker's behaviors that can be monitored, measured, and assessed accomplishment in individual level. Thus, job performance is really important not only for the organization's future but also for the employee's future itself. This is because employees with higher job performance have the opportunity to get promoted in view of they have all the competencies required.

2.1.1 *Rewards and motivation*

Rewards are objects or events that make people come back for more. Rewards are crucial for survival and used for behavioral choices that maximize them (Schultz, 2007, 16 April). According to Mahaney and Lederer (2006), organization offers intrinsic and extrinsic rewards to the employees for the purpose of improving employees' performance. An intrinsic reward is an intangible award of recognition or a sense of achievement motivation, and when someone is satisfied with what they had achieved. An extrinsic reward is an award that is tangible or physically given to a person for accomplishing something as recognition of one's efforts. Organizations are responsible in ensuring the rewards are aligned and consistent with organization's strategy in order to motivate

employees in their performance (Allen & Kilmann, 2001). When the employees are satisfied with their performance personally, they will become motivated because it was based on their efforts.

Motivation is defined as the process that initiates, guides and maintains goal-oriented behaviors. Motivation is what causes us to act, whether it is getting a glass of water to reduce thirst or reading a book to gain knowledge (Cherry, 2013, April 16). Therefore, motivation is one of the job performance factors that can lead to employee satisfaction in the long term because motivation comes from the employee's soul and intention to perform the tasks.

2.2 *Empowerment*

The concept of employee's participation has been a focus for research and practiced for many years. It has taken many different forms, which get the employees' involvement and participation in decision-making concepts. Empowerment involves the employees being provided with a greater degree of flexibility and more freedom to make decisions relating to work. Employee empowerment is thought to enhance job satisfaction among the employees. Empowerment can create a positive effect on perceived service quality and job satisfaction (He, Murrmann & Perdue, 2010).

Empowerment is also based on the idea that giving employees skills, all the conveniences that facilitate the employees, the power, chance and motivation as well as holding them responsible and accountable for outcomes of their actions that will contribute to their competencies and satisfaction. On the other hand, empowerment is the combination of the psychological state of a subordinate, which is influenced by the empowering behaviors of supervisors (Lee & Koh, 2001).

2.3 *Work environment*

Working environment can be defined as the surrounding of an employee in a certain work area and can be divided into two categories which are physical and non-physical. Elements of the physical condition include equipment, setting, cubes, tables and other tangible equipment, while the non-physical includes privacy, noise and conversations. When the employees are comfortable in their working place, they will increase their level of improvement in order to treat customers and meet customer's requirement.

Employee's performance in perform their job to meet customers satisfaction have the aspect of innovative and creativity in the working environment. Politis (2005), stated that the employee performs the better work based on their creativity and used all the skill and knowledge that they are represents in making the satisfaction to their customers. Based on the self-management, leadership was relative into the performance on the working environment that leads the creativity and innovation to make the customers satisfied.

2.4 Employee satisfaction

Successful organizations can maintain their benefit only by satisfying the employees and motivate them towards continuous improvement. It is important to satisfy the internal customer if external customers want to be retained and satisfied (Xu & Geodegebuure, 2005). Many studies related demographic variables to job satisfaction has been done and, Lam, Baum & Pine (2001) indicated that job satisfaction was also influenced by the individual's subjective norms. Lam et al. (2001), for instance, found that age, educational level, length of employment, and marital status are related to job satisfaction in a study on Hong Kong hotel employees. Work role and family role variables also have been found to be predictors of job satisfaction in a study in the Turkish Hotel industry (Karatepe & Sokmen, 2006).

Employee's job satisfaction and performance can be more or less related (Judge, Thoresen, Bono & Patton, 2001). Practices which boost employee satisfaction be likely to increase the quality of employees and the level of their performance (Gerhart & Rynes, 2004). According to Wan (2007), satisfaction is a significant goal for organizations to achieve, as it has been publicized that effectiveness, efficiency, employee custody and customer satisfaction are related to employee's happiness. In addition, motivated employees will make higher customer satisfaction and in turn positively influence organization's performance.

2.5 Proposed theoretical framework

In order to conduct this study, rewards and motivation, empowerment and work environment served as three independent variables as the presumed caused and antecedent towards dependent variable. In the other hand, employee satisfaction labeled as dependent variable that presumed effect and consequent. Therefore this study proposes the following framework which illustrated in Figure 1.

Based on the framework above, several hypotheses are tested as follow:

H1: There is a relationship between rewards and motivation and employee satisfaction at IYC.

H2: There is a relationship between work environment and employee satisfaction at IYC.

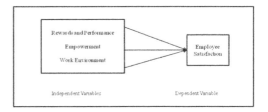

Figure 1. Proposed theoretical framework.

H3: There is a relationship between empowerment and employee satisfaction at IYC.

3 METHODOLOGY

Research design used for this study was correlation in view of it appropriateness to tests statistical relationship between variables. This study has been conducted to determine factors that might influence job performance and employee satisfaction. The data has also been analyzed by using the descriptive analysis.

Respondents involved in this research were from lower level employees in IYC with the total population of 70 employees of lower level management. The researcher choose to involve lower level employees as the respondents in view of these lower level employees were dealing with daily customers service transaction. Simple random sampling was used in this study since each member of population had an equal and independent chance of being selected to be part of the sample. A total of 60 employees participated in this study. According to Krejcie and Morgan (1970), if the total population was 70, the sample size was 59.

In accomplishing this study, a fully structured questionnaire was used as an instrument with the purpose of gathering inclusive information from the respondents. The questionnaire consists of Likert-scale questions which facilitate respondents to express their level of agreement by using five points of response which is the most appropriate when conducting a survey. This set of questionnaire consists of three major sections including Section A which represents respondent's demographic information.

Package for Social Science (SPSS) version 20.0 was used to analyze the collection of data for this study. Descriptive analysis was used to indicate frequency analysis of respondents' demographic profile. Meanwhile, correlational analysis also was used by the researcher to analyze the relationship between the three main independent variables, rewards and motivation; work environment; empowerment and employee satisfaction.

Table 1. Reliability analysis.

Variable	No. of Item	Alpha
Rewards and Motivation	5	.818
Empowerment	5	.663
Work Environment	5	.793

Table 2. Profile of respondents.

Characteristic	%	Characteristic	%
Gender		*Level of Education*	
Male	37	SPM	48
Female	63	STPM	12
Age		Diploma	22
20–30	47	Degree	18
31–40	48	*Employment Status*	
41–50	5	Part-time	20
Marital Status		Full-time	80
Single	35		
Married	65		

4 FINDINGS AND DISCUSSION

4.1 *Reliability analysis*

In this study, Alpha coefficients were calculated in purpose to test three main independent variables which were rewards and motivation, empowerment and work environment. Table 1 below represented the Cronbach's Alpha rating.

Referring to the above table, it indicates the result of the reliability test for variables in purpose to determine the reliability of the data. According to Salkind (2006), a value of .00 or less indicates non reliability. Sekaran (2003) stated when the results show that the number is closer to 1.00, the higher the internal consistent reliability of the variable. The result shows that the highest Alpha coefficient score is rewards and motivation with 0.818 which indirectly represents the most reliable part of the questionnaire.

4.2 *Profile of respondents*

The demographic background for the total number of 60 respondents is presented in Table 2, while the results for the determinant factors that influence employee satisfaction are represented in Tables 3 to 5.

4.3 *Correlation analysis*

Referring to Table 6, it indicates that there are moderate and significant relationship between

Table 3. Rewards and motivation.

Statement	Mean	Std. Deviation
Rewards are distributed rightfully among the employees	3.17	.886
Satisfied with the quality/ quantity of the rewards received	3.05	.852
People acknowledge it when I do my work well	3.38	.825
Rewards have a positive effect on the work atmosphere	3.68	.813
Rewards motivate me to perform well in my job	3.85	.860

Table 4. Empowerment.

Statement	Mean	Std. Deviation
I am involved in making decisions that affect my work	3.48	.873
I am given the opportunity to suggest improvements	3.42	1.078
participate in setting the goals and objectives for my job	3.12	.976
My superior values my suggestions and requests	3.47	.700
I know exactly what I have to do in my work	4.10	.817

Table 5. Work environment.

Statement	Mean	Std. Deviation
I am given an opportunity to learn different tasks	3.78	.922
IYC provide an educative and training environment	3.72	.783
The superiors are willing to share all relevant information with subordinates	3.18	1.017
IYC is always moving forward toward improved ways	3.85	.732
The opportunity for personal growth and development exists in this organization	3.58	.829

Table 6. Correlation coefficient.

	r	P
Rewards and motivation	.488	.000
Empowerment	.644	.000
Work environment	.463	.000

rewards and motivation and employee satisfaction (r=0.488). Thus, *H1* was accepted. Strong and significant relationship is noticed between empowerment and employee satisfaction (r=0.644). Hence, *H2* was accepted. Work environment and employee satisfaction also found has moderate and significant relationship (r=0.463) and these results lead to *H3* was accepted.

5 CONCLUSION AND RECOMMENDATIONS

Based on the researchers' findings, it is found that the hypotheses testing between determinants of job performance; rewards and motivation, empowerment and work environment towards employee satisfaction shows a positive relationship and correlation of these three relationships was also significant. The findings therefore show a support for a positive relationship between *H1, H2* and *H3*. The positive relationship between rewards and motivation and employee satisfaction is consistent with the study conducted by Schultz (2007). In view of positive relationship between empowerment and employee satisfaction also found significant and consistent with He et al. (2010). Finally, positive relationship between work environment and employee satisfaction also consistent with the study conducted by Clark (2009).

Therefore, this study provides good determinant factors for the industry to concern about rewards and motivation, empowerment and work environment in the workplace in order to increase level of employee job performance that leads to employee satisfaction in future.

REFERENCES

Allen, R.S. & Kilmann, R.H. (2001). The role of the reward system for a total quality management based strategy. *Journal of Organizational Change Management, 14*(2), 110–131.

Bakker, A.B. & Leiter, M.P. (2010). *Work engagement: A handbook of essential theory and research.* New York: Psychology Press.

Bakker, A.B. & Schaufeli, W.B. (2008). Positive organizational behavior: Engaged employees in flourishing organizations. *Journal of Organizational Behavior, 29*(2), 147–154.

Cherry, K. (2013, April 16). What is motivation? from http://psychology.about.com/od/mindex/g/motivation-definition.htm

Clark, R.M. (2009). Are we having fun yet? Creating a motivating work environment. *Industrial and Commercial Training, 41*(1), 43–46.

Coetzer, W. & Rothmann, S. (2006). Occupational stress of employees in an insurance company. *S. Afr. J. Bus. Manage, 37*(3), 29–39.

Gerhart, B. & Rynes, S.L. (2004). Compensation: Theory, Evidence, and Strategic Implications. *ILRReview, 57*(3), 81.

He, P., Murrmann, S.K. & Perdue, R.R. (2010). An investigation of the relationships among employee empowerment, employee perceived service quality, and employee job satisfaction in a US hospitality organization. *Journal of Foodservice Business Research, 13*(1), 36–50.

Homburg, C. & Stock, R.M. (2005). Exploring the conditions under which salesperson work satisfaction can lead to customer satisfaction. *Psychology & Marketing, 22*(5), 393–420.

Judge, T.A., Thoresen, C.J., Bono, J.E. & Patton, G.K. (2001). The job satisfaction-job performance relationship: A qualitative and quantitative review. *Psychological bulletin, 127*(3), 376.

Karatepe, O.M. & Sokmen, A. (2006). The effects of work role and family role variables on psychological and behavioral outcomes of frontline employees. *Tourism Management, 27*(2), 255–268.

Krejcie, R.V. & Morgan, D.W. (1970). Determining sample size for research activities. *Educational and psychological measurement, 30*(3), 607–610.

Lam, T., Baum, T. & Pine, R. (2001). Study of managerial job satisfaction in Hong Kong's Chinese restaurants. *International Journal of Contemporary Hospitality Management, 13*(1), 35–42.

Lee, M. & Koh, J. (2001). Is empowerment really a new concept? *International Journal of Human Resource Management, 12*(4), 684–695.

Mahaney, R.C. & Lederer, A.L. (2006). The effect of intrinsic and extrinsic rewards for developers on information systems project success. *Project Management Journal, 37*(4), 42–54.

Milliman, J.F., Ferguson, J.M., Czaplewski, A.J. & Join, A. (2008). Breaking the cycle. *Marketing Management, 17*(2), 14.

Politis, J.D. (2005). Dispersed leadership predictor of the work environment for creativity and productivity. *European Journal of Innovation Management, 8*(2), 182–204.

Salkind, N. (2006). Exploring research (6th ed.). New Jersey: Pearson International Edition Inc.

Schultz, W. (2007, 16 April). from http://www.scholarpedia.org/article/Reward

Sekaran, U. (2003). *Research methods for business: A skill building approach (4th ed.).* New York: John Wiley and Sons Inc.

Wan, H.L. (2007). Human capital development policies: enhancing employees' satisfaction. *Journal of European Industrial Training, 31*(4), 297–322.

Xu, Y. & Geodegebuure, R. (2005). Employee satisfaction and customer satisfaction: Testing the service-profit chain in a Chinese securities firm. *Innovative Marketing Journal, 2.*

Theory and Practice in Hospitality and Tourism Research – Radzi et al. (Eds)
© 2015 Taylor & Francis Group, London, ISBN 978-1-138-02706-0

A conceptual review of homestay community resilience factors

A.A. Ghapar
Ministry of Tourism Malaysia, Putrajaya, Malaysia

N. Othman, S.A. Jamal & A.F. Amir
Faculty of Hotel and Tourism Management, Universiti Teknologi MARA, Shah Alam, Malaysia

ABSTRACT: This conceptual article reviews the literature on resilience at the community level, with the identification of key factors often found in thriving communities. The review concludes with community participation implications and suggestions on how to obtain positive community outcomes. Individual communities, regardless of ethnic status, differ highly in how they experience community strengths, setbacks, and the perceived severity of risk. There is no collective list of key, effective protective and recovery factors, but a review of recent research in the tourism and enterprise survival literature recognizes regular and prominent attributes among resilient, dynamic communities. These factors include: (i) lifestyle values of homestay operators and family members that motivate their participation in the tourism industry; (ii) business age and size of homestay villages; (iii); level of social and human capital; (iv) flexibility of homestay community; (v) relationship with authorities; and (vi) ecological condition.

Keywords: Homestay, community-based tourism, community resilience, social capital, human capital

1 INTRODUCTION

One of the enduring mysteries of community dynamics is why some communities continually thrive and respond optimistically to challenges, whereas others in similar circumstances do not cope well. The successful coping of communities during economic recession, disturbances, various change-related shocks or adversity has been described as community resilience (Mancini & Bowen, 2009; Norris, Steven, Pfefferbaum, Wyche & Pfefferbaum, 2008). The concept of resilience was originally developed by studying the positive adaptation of children under adverse circumstances (Rutter, 1987). Recently, its application has been extended to the study of personal health resilience (Jackson, Firtko & Edenborough, 2007; Worthington & Scherer, 2004), family resilience (Black & Lobo, 2008; Skovholt, Grier & Hanson, 2001; McCubbin, Thompson & McCubbin, 1996; Walsh, 2003), enterprise resilience survival (Bosma, Van Praag, Thurik & De Wit, 2004; Dunne & Hughes, 1994) and social ecological resilience (Berkes & Jolly, 2002; Walker, Holling, Carpenter & Kinzig, 2004; Adger et al., 2005).

In the current phenomenon of globalization, the understanding of the resilience in the vulnerable sectors of social-ecological systems is extremely important (Kaplan, 2002). The community-based homestay tourism sector is highly vulnerable not only to internal challenges such as passive community and leadership problems, but also to external challenges such as economic recession, rivalry from other tourism products and more (Ashikin, Kayat & Mohamed, 2010). However, in the face of large disturbances, the aspects that deliberate resilience to homestay tourism sector have not been studied to date. Indeed, the insufficient studies on resilience in tourism systems furnish conceptual perspectives on the worth of the resilience concept to understanding tourism (Farrell & Twining-Ward, 2004) and qualitative applications of the concept to protected spans and community-based tourism (Ruiz-Ballesteros, 2011; Strickland-Munro). The contribution of this paper is to conceptually discover the components of community resilience of homestay tourism industry in Malaysia to any form of change-related shocks and disturbances.

This conceptual article reviews the literature on resilience at the community level, with the identification of key factors often found in thriving communities. The review concludes with community participation implications and suggestions how to obtain positive community outcomes. Inherent in the community resilience model (Chang & Chamberlin, 2003) is a dual focus of building protective and recovery factors, in addition to reducing ecological risks that threaten family functioning. Skovholt, Grier and Hanson (2001) have argued that an increasingly important realm of community-based homestay tourism business is to identify, enhance, and promote community resiliency. This article describes what is known, identifies conceptual gaps, and proposes how homestay community resilience can be used for family interventions.

2 LITERATURE REVIEW

2.1 Defining community-based homestay tourism and homestay community

Community-based homestay tourism is regarded as a potentially good product in promoting the country as well as getting the community involved in the travel industry (Ibrahim & Rashid, 2010). In Malaysia, apart from being seen as a way to aid in generating income for the community, homestay tourism contributes in achieving the Government's agenda to build job opportunities and eliminate poverty of the communities involved (Pusiran & Xiao, 2013). The concept of community-based tourism is dependent largely on the participation of the local community. Community-based tourism appeared as a potential solution to the mass tourism's negative effects in developing countries. It is also one of the strategies for the community to achieve better living conditions (Russell, 2000).

The main idea is for the community to create a project that presents a sustainable development and promote the relationship between local community and visitors (Ibrahim & Rashid, 2010). To develop a tourism product as such, the core characteristic is to incorporate hotel management, food and beverages, complementary services and tourism management. Not to forget other sub-systems such as infrastructure, health, education and environment (Cioce, Bona & Ribeiro, 2007). It should be noted that community-based tourism is protected and supported by various international organizations such as the World Tourism Organization (2002) and some of the aim is towards achieving a high quality visitor experience, conservation of natural and cultural resources, development of social and economic and community's empowerment and ownership.

Furthermore, the focal benefits of community tourism are the economic impact on communities, improvements of socioeconomic and a more sustainable lifestyle diversification (Manyara & Jones, 2007; Rastegar, 2010). According to the swap of knowledge, analysis and ability among members of the community, it is definitely a useful method of executing policy coordination, attaining synergies, and avoiding disagreement between different actors in tourism (Kibicho, 2008). According to Briedenhann and Wickens in 2004, how the whole community partakes in the development of an area

is important to widen a tourist destination. This is due to the fact that tourism planning will affect the whole community, for example, the awareness of tourism is based on the local community's attitude and their evaluation on the environment, infrastructure and events, and the degree of involvement by the local community exerts a strong influence on the tourist's experience.

According to Ministry of Tourism and Culture (MOTAC, 2013), the homestay program is one of the most successful programs to improve rural-based tourism. A significant increase of 38.4% occupancy rate had been recorded compared to 24.9% in 2011. From January to May 2012, the number of domestic tourists was 110,322 and 23,367 foreigners which make a total of 133,689 tourists who visited homestays in Malaysia. As compared to the previous year in the same period, a 70.7% increase can be seen where it was only 78,333 tourists. For the first five months of 2012, tourist receipts from homestay program boomed to RM 7,376,446.50 from RM 4,817,158.30 in January to May 2011. The improved spending by the tourists helped the rural economy to some extent and this has shown a positive impact to encourage local communities to get involved with rural-based tourism while at the same time maintaining their traditions and identities.

2.2 History of the concept of resilience

The concept of resiliency may be applied to numerous entities, define at different level of aggregation. The term "resilience" will undoubtedly take on different and expanded meaning as it is used by practitioners and other disciplines. For example in the context of organization resiliency, Caralli, Stevens, Willke & Wilson (2004) define operational resiliency as the organization's ability to adapt and to manage risks that originate from day to day operation. Organizations that have resilient operation are able to systematically and transparently cope with disruptive event so that the overall ability of the whole organization to meets its mission is not affected.

In personal health, for example, resiliency is the ability to spring back from and successfully adapt to adversity. An increasing body of research from the field of psychology, psychiatry, and sociology reported that most people including young people who can promptly bounce back from risk, stress, crises and trauma would certainly experience life success (Manyena, 2006).

In physics and engineering, resilience is defined as the capacity of a material to absorb energy when it is deformed elastically and upon unloading to have this energy recovered (Holling, 1996). In other words, it is the maximum energy per volume that can be stored. In the context of system ecology, one of the most familiar definitions is provided by the Resilience Alliance Research Consortium, which defines resilience as the capacity of an ecosystem to tolerate disturbance without collapsing into a qualitatively different state that is controlled by a different set of process (Norris, Steven, Pfefferbaum, Wyche & Pfefferbaum, 2008).

2.3 *Homestay community resilience*

Community resilience has been associated with disaster in many studies done earlier (Mancini & Bowen, 2009; Norris, Steven, Pfefferbaum, Wyche & Pfefferbaum, 2008). The concepts of community resilience refer to the capacity of individual or community to cope with stress, overcome adversity or adapt positively to change (Mancini & Bowen, 2009). The ability to bounce back from negative experiences may reflect the innate qualities of individuals or be the result of learning and experience. Regardless of the origin of resilience, there is evidence to suggest that it can be developed and enhanced to promote greater wellbeing. Resilience is not regarded as a quality that is either present or absent in a person or a group but rather a process may vary across circumstances or time (Luthar, 2003).

In Malaysia, homestay community resilience may not be associated with disaster as severe natural disaster does not happen in the country. In the context of this study, community resilience may be referred as the ability of the homestay community to enhance and sustain in the business. It is timely and vital for the homestay communities to begin exploring for ways to protect and disseminate what is valuable and essential for their survival. At the scale of a village in a homestay tourism system, society could shift from an unstable village economy to a stable economy established generally on homestay tourism or business tourism.

Without spending much on altering the current infrastructure, the communities have been able to develop and maneuver the homestay businesses by seizing supremacy of the existing natural resources, cultural and heritage assets within the community, yet the threats to the ecosystems are quite serious and can put enormous pressure on an area. Such pressures lead to impacts such as soil erosion, increased pollution, natural habitat loss, increased pressure on endangered species and heightened vulnerability to forest fires. The worries facing homestay tourism are not just from the threats facing ecosystems but it is also affected by sociopolitical and economic disturbances such as health concerns, economic recessions, as well as national and local level issues including the regulatory environment (Baker & Coulter, 2007; Gössling & Hall, 2006).

Hence, in the midst of rising global change and related shocks and disturbances, there is a demand to comprehend the factors that allow homestay tourism industry to cope alongside, and make affirmative adaptations. In the face of disturbances and frequently ongoing unpredictable change, a resilience-based approach to understanding and managing homestay tourism is useful as it considers the skill of an arrangement to uphold its useful characteristics and individuality in a coupled social-ecological system (Gunderson, 2001).

2.4 *Prominent homestay community resilience factors*

Individual communities, regardless of ethnic status, differ highly in how they experience community strengths, setbacks, and the perceived severity of risk (Berkes & Jolly, 2002). There is no collective list of key, effective protective and recovery factors, but a review of recent research in the tourism and enterprise survival literature recognizes regular and prominent attributes among resilient, dynamic communities. These factors include: (i) lifestyle values of homestay operators and family members that motivate their participation in the tourism industry; (ii) business age and size of homestay villages; (iii); level of social and human capital; (iv) flexibility of homestay community; (v) relationship with authorities; and (vi) ecological condition.

2.4.1 *Lifestyle values of homestay operators and family members*

It was highlighted that small tourism business owners and operators that are predominantly in non-urban areas are usually driven strongly by non-economic factors such as lifestyle (Ateljevic & Doorne, 2000; Shaw, Williams & Thomas, 2004). Beyond pure economics, lifestyle considerations by small tourism entrepreneurs modify the entry and exit characteristics of enterprises in rural tourism sectors (Ateljevic & Doorne, 2000). Determinants of community resilience in economic sectors such as community-based homestay tourism are interrelated with considerations of lifestyle, sense of place, identity, and cultural values (Sharpley, 2002). Following this, various elements can build up the resilience of homestay tourism in the country and their capability to innovate, reorganize, and adapt.

2.4.2 *Business age and size of homestay villages*

In the early years, businesses are more prone to shut down because a business's age is usually positively related to its future survival (Bosma, Van Praag, Thurik & De Wit, 2004; Dunne & Hughes, 1994). Despite access to funding and healthy revenue and profit levels are essential to business survival and success, the new and young enterprises have less ability in coordinating and roles and developing trust and loyalty among employees as well as in obtaining financial capital and resources (Dunne & Hughes, 1994). However, Fritsch, Brixy, and Falck (2006) proved that the age and success of a business does not necessarily show a direct relationship. Older businesses may display strictness and difficulty to adapt. Ateljevic and Doorne (2000) and Hall, Rusher & Thomas (2004) reported that in general, the ability of business react to crises and change is absolutely important. A lack of experience in dealing with crises has found to be a causal factor in most business failures (Kalleberg & Leicht, 1991).

2.4.3 *The level of social and human capital*
Next, it is fundamental to note the importance of human and social capital for homestay business survival and success. Human capital denotes the expertise and human capacity of an enterprise (Bosma et al., 2004). To provide a valuable services and an unforgettable experience which is adapted and altered according to tourists needs, high level of human capital are necessary. To make sure that the community based tourism truly benefits everyone, it is vital for the community capacity building to be addressed by the stakeholder (Rashid, Hadi Mustafa, Hamzah & Khalifah, 2010). Involvement from the community is one of the important things to look at in building community capacity. It is one of the tools to develop the community by empowering people to get involved.

According to Murphy (1988), getting public participation in tourism planning and management is important. This is because resistance and conflict increase the business cost or abolish the industries prospective when development and planning do not go in line with local objectives and capacities. In enhancing productivity and aiding economic growth and development, social capital is crucial (Uphoff & Wijayaratna, 2000). According to Pretty (2003), social capital refers to the social bonds and customs of mutuality and trust that allows groups and community to function. The ability of individuals and groups to deal with uncertainty and shocks are supported through the bonds and provision enabled by higher levels of social capital (Folke, Hahn, Olsson & Norberg, 2005).

2.4.5 *The flexibility of homestay community*
Flexibility refers to a community's ability to rebound and reorganize in the event of challenges while maintaining a sense of continuity (Mimms, 2010). Mutuality is best achieved when community members possess a clear sense of themselves, both within and outside the context of community. In other words, members identify a connection with the community as well as differentiation from the community. According to MacNeice et. al (2000),

community members tend to prefer stable and orderly patterns, yet function best when a balance is achieved between moderate amounts of structure and flexibility. A moderately structured relationship has a democratic, egalitarian leadership, with negotiations from all members.

It is common for most community members to resist change or loss, yet resilient communities do not view the change with helplessness; rather, roles are reorganized and changes are viewed optimistically as venues to build a new equilibrium (Upham, 2006). Rigid community, on the other hand, tend to operate at the extremes of being either overly inflexible or chaotically unstructured. Members of disengaged community tend to drift off on their own and are unable to find mutual support from the community associate.

2.4.6 *The relationship with authorities*
The relationship between communities and local, state, and national government bodies are included in the extensive outset of social capital. In response to pressures or change, these relationships are an important element of business ability to be adapted and innovated (Hall & Allan, 2008; Hjalager, 2010). Although government bodies can nurture innovation and upkeep ongoing adaptation by business owners, it can also suppress efforts by owners to adapt and change due to the challenges to good governance that include power struggles, corruption, overregulation and institutional inertia (Hall & Allan, 2008).

In Malaysia, through the Ministry of Tourism and Culture (MOTAC), the Government supports the development and growth of homestay programs by granting a specific fund (MOTAC, 2012). The Malaysian Government's increased focus on the development of homestay is significant because it is regarded as a potentially good product in promoting the country as well as getting the community involved in the tourism industry through rural tourism. By taking advantage of the existing natural resources, cultural and heritage assets within the community, communities have been able to develop the homestay product without spending so much on changing the existing infrastructure. Not only is homestay seen as a way to help generate income for the community, it also assists in fulfilling the Government's agenda to eradicate poverty and create job opportunities for the communities involved (Ibrahim & Rashid, 2010).

Much popularity and high reputation had been gained by the homestay program since 1995 and it is considered as a great tool to promote rural tourism. Through the homestay program, the MOTAC is serious in enhancing the economic welfares and incomes to assist the rural communities. Consequently, under the Ninth Malaysia Plan (2006–2010) government had allocated RM40 million and another RM10 million under the Second Stimulus package to upgrade the facilities and infrastructure of all the villages participating in the homestay program (Pusiran & Xiao, 2013). Apart from that, in 2008 the Ministry of Rural and Regional Development had spent another RM6.7 million to improve the infrastructure of rural communities.

2.4.7 *The ecological conditions*

Previous studies have found that tourists' experiences to nature-based attractions such as homestay are influenced by the perceived quality of the primary interest at a destination's features (Deng, King & Bauer, 2002). For example like reef tourism, tourists are less likely to return to a destination after reef degradation or coral bleaching (Kragt, Roebeling & Ruijs, 2009; Uyarra et al., 2005). Similarly, tourists would most likely not return to the same homestay destination if the environment is affected by soil erosion, increased pollution, natural habitat loss, increased pressure on endangered species and such. Therefore, the whole environment and surroundings of the homestay is potentially essential in the community's aptitude to be resilient to shocks. Notably, although visitors to homestays are able to detect some levels of natural degradation at the destination, there is variation in the precision of their perceptions.

3 CONCLUSION

It can be concluded that the community-based homestay tourism sector is highly vulnerable. The way the whole community plays their part in the growth of an area is vital to develop a tourist destination. The ability to bounce back from negative experiences may reflect the essential qualities of individuals or the result of learning and experience. Regardless of the origin of resilience, there is evidence to suggest that it can be developed and enhanced to promote greater wellbeing. To make sure that the community based tourism truly benefits everyone, it is crucial for the stakeholder to address the community capacity building. Areas that could benefit from this research includes professional in the fields of tourism development, tourism planning, community planning, community development, infrastructure planning, economic development, public policy, social welfare policy, and emergency services planning.

ACKNOWLEDGEMENT

This research is funded by Universiti Teknologi MARA through Research Acculturation Grant 600-RMI/RAGS 5/3 (128/2012).

REFERENCES

Adger, W.N., Hughes, T.P., Folke, C., Carpenter, S.R. & Rockström, J. (2005). Social-ecological resilience to coastal disasters. *Science, 309*(5737), 1036–1039.

Ashikin, N., Kayat, K. & Mohamed, B. (2010). The challenges of community-based homestay programme in Malaysia. In *Proceedings of Regional Conference on Tourism Research, Universiti Sains Malaysia, Penang, Malaysia, 13–14 December 2010. The state of the art and its sustainability.* (pp. 66–73). Social Transformation Platform.

Ateljevic, I. & Doorne, S. (2000). 'Staying Within the Fence': Lifestyle Entrepreneurship in Tourism. *Journal of Sustainable Tourism, 8*(5), 378–392.

Baker, K. & Coulter, A. (2007). Terrorism and tourism: the vulnerability of beach vendors' livelihoods in Bali. *Journal of Sustainable Tourism, 15*(3), 249–266.

Berkes, F. & Jolly, D. (2002). Adapting to climate change: social-ecological resilience in a Canadian western Arctic community. *Conservation ecology, 5*(2), 18.

Black, K. & Lobo, M. (2008). A conceptual review of family resilience factors.*Journal of Family Nursing, 14*(1), 33–55.

Bosma, N., Van Praag, M., Thurik, R. & De Wit, G. (2004). The value of human and social capital investments for the business performance of startups. *Small Business Economics, 23*(3), 227–236.

Briedenhann, J. & Wickens, E. (2004). Tourism routes as a tool for the economic development of rural areas-vibrant hope or impossible dream? *Tourism Management,* Vol. 25, No.1, pp.71–79.

Caralli, R.A., Stevens, J.F., Willke, B.J. & Wilson, W.R. (2004). *The critical success factor method: establishing a foundation for enterprise security management* (No. CMU/SEI-2004-TR-010). CARNEGIE-MELLON UNIV PITTSBURGH PA SOFTWARE ENGINEERING INST.

Chang, S.E. & Chamberlin, C. (2003). *Assessing the role of lifeline systems in community disaster resilience.* Multidisciplinary Center for Earthquake Engineering Research.

Cioce, C.A., Bona, M. & Ribeiro, F. (2007). Community tourism: montanha beija-flor dourado pilot project (microbasin of the sagrado river, Morretes, Paraná). *Turismo-Visao e Açao,* Vol. 9, No.2, pp.249–266.

Deng, J., King, B. & Bauer, T. (2002). Evaluating natural attractions for tourism. *Annals of Tourism Research, 29*(2), 422–438.

Dunne, P. & Hughes, A. (1994). Age, size, growth and survival: UK companies in the 1980s. *The Journal of Industrial Economics,* 115–140.

Farrell, B.H. & Twining-Ward, L. (2004). Reconceptualizing tourism. *Annals of Tourism Research, 31*(2), 274–295.

Folke, C., Hahn, T., Olsson, P. & Norberg, J. (2005). Adaptive governance of social-ecological systems. *Annu. Rev. Environ. Resour., 30*, 441–473.

Fritsch, M., Brixy, U. & Falck, O. (2006). The effect of industry, region, and time on new business survival–A multi-dimensional analysis. *Review of Industrial Organization, 28*(3), 285–306.

Gössling, S. & Hall, C.M. (2006). Uncertainties in predicting tourist flows under scenarios of climate change. *Climatic Change, 79*(3–4), 163–173.

Gunderson, L.H. (2001). *Panarchy: understanding transformations in human and natural systems*: Island Press.

Hall, C.M., Rusher, K. & Thomas, R. (2004). Risky
lifestyles? Entrepreneurial characteristics of the New
Zealand bed and breakfast sector. *Small firms in tourism: international perspectives*, 83–97.

Hall, M.C. & Allan, W. (2008). *Tourism and innovation*:
Psychology Press.

Hjalager, A.-M. (2010). A review of innovation research
in tourism. *Tourism Management, 31*(1), 1–12.

Holling, C.S. (1996). Engineering resilience versus ecological resilience.*Foundations of ecological resilience*, 51–66.

Ibrahim, Y. & Rashid, A. (2010). Homestay program and
rural community development in Malaysia. *Journal of
Ritsumeikan Social Sciences and Humanities, 1*(2), 7–24.

Jackson, D., Firtko, A. & Edenborough, M. (2007). Personal resilience as a strategy for surviving and thriving in the face of workplace adversity: a literature
review. *Journal of advanced nursing*, 60(1), 1–9.

Kalleberg, A.L. & Leicht, K.T. (1991). Gender and
organizational performance: Determinants of small
business survival and success. *Academy of management journal, 34*(1), 136–161.

Kaplan, H.B. (2002). Toward an understanding of resilience.
In *Resilience and development* (pp. 17–83). Springer US.

Kibicho, W. (2008). Community-based tourism: A factor-
cluster segmentation approach. *Journal of Sustainable
Tourism*, Vol. 16, No.2, pp.211–231.

Kragt, M.E., Roebeling, P.C. & Ruijs, A. (2009). Effects
of Great Barrier Reef degradation on recreational
reef-trip demand: a contingent behaviour approach*.
*Australian Journal of Agricultural and Resource Eco-
nomics, 53*(2), 213–229.

Luthar, S.S. (Ed.). (2003). *Resilience and vulnerability:
Adaptation in the context of childhood adversities*.
Cambridge University Press.

MacNeice, P., Olson, K.M., Mobarry, C., de Fainchtein,
R. & Packer, C. (2000). PARAMESH: A parallel
adaptive mesh refinement community toolkit.*Computer physics communications, 126*(3), 330–354.

Mancini, J.A. & Bowen, G.L. (2009). Community resilience: A social organization theory of action and
change. *Pathways of human development: Explorations
of change*, 245–265.

Manyara, G. & Jones, E. (2007). Community-based
tourism enterprises development in Kenya: An exploration of their potential as avenues of poverty reduction. *Journal of Sustainable Tourism*, Vol. 15, No.6,
pp.628–644.

Manyena, S.B. (2006). The concept of resilience revisited. *Disasters, 30*(4), 434–450.

McCubbin, H.I., Thompson, A.I. & McCubbin, M.A.
(1996). *Family assessment: Resiliency, coping and
adaptation: Inventories for research and practice*. University of Wisconsin-Madison, Center for Excellence
in Family Studies.

Mimms, M. (2010). *Storied stables: Social engagements
for social sustainability*. (Order No. 1476736, Corc-
oran College of Art + Design). *ProQuest Dissertations
and Theses*, 92. Retrieved from http://search.proquest.
com.ezaccess.library.uitm.edu.my/docview/503429624
?accountid=42518. (503429624).

Ministry of Tourism and Culture (MOTAC). (2012).
Homestay Statistic.

Ministry of Tourism and Culture (MOTAC). (2013).
Homestay Statistic.

Mascarenhas, O.A., Kesavan, R. & Bernacchi, M.(2006).
Lasting customer loyalty: a total customer experience approach. *Journal of Consumer Marketing* 23(7),
387–405.

McNeill, W.H. (2009). *The rise of the West: A history of
the human community*. University of Chicago Press.

Murphy, P.E. (1988). Community driven tourism planning. *Tourism management, 9*(2), 96–104.

Norris, F.H., Stevens, S.P., Pfefferbaum, B., Wyche, K.F.
& Pfefferbaum, R.L. (2008). Community resilience
as a metaphor, theory, set of capacities, and strategy
for disaster readiness. *American journal of community
psychology, 41*(1–2), 127–150.

Pretty, J. (2003). Social capital and the collective management of resources. *Science, 302*(5652), 1912–1914.

Pusiran, A.K. & Xiao, H. (2013). Challenges and Community Development: A Case Study of Homestay in
Malaysia. *Asian Social Science, 9*(5), p1.

Rashid, R., Hadi, M.Y., Mustafa, M.Z., Hamzah, A. &
Khalifah, Z. (2010). Community capacity building
and sustainable

community based tourism in Malaysia. Paper presented
in The National Psychology Seminar. UMS, Sabah,
Malaysia.

Rastegar, H. (2010). Tourism development and residents'
attitude: A case study of Yazd, Iran. *Tourismos*, Vol.
5, No.2, pp. 203–211.

Ruiz-Ballesteros, E. (2011). Social-ecological resilience and
community-based tourism: an approach from Agua
Blanca, Ecuador. *Tourism Management, 32*(3), 655–666.

Russell, P. (2000). Community-based tourism. *Travel &
Tourism Analyst*, (5), 89–116.

Rutter, M. (1987). Psychosocial resilience and protective mechanisms.*American journal of orthopsychia-
try, 57*(3), 316.

Sharpley, R. & Telfer, D.J. (Eds.). (2002). *Tourism and
development: Concepts and issues* (Vol. 5). Channel
View Publications.

Shaw, G., Williams, A.M. & Thomas, R. (2004). From
lifestyle consumption to lifestyle production: changing patterns of tourism entrepreneurship. *Small firms
in tourism: international perspectives*, 99–113.

Skovholt, T.M., Grier, T.L. & Hanson, M.R. (2001).
Career counseling for longevity: Self-care and burnout prevention strategies for counselor resilience.*Journal of Career Development, 27*(3), 167–176.

Strickland-Munro, J.K., Allison, H.E. & Moore, S.A.
(2010). Using resilience concepts to investigate the
impacts of protected area tourism on communities.
Annals of Tourism Research, 37(2), 499–519.

Upham, S.P. (2006). *Communities of innovation: Three
essays on new knowledge development*.(Order No.
3225559, University of Pennsylvania). *ProQuest Dissertations and Theses*, 216–216 p. Retrieved from http://
search.proquest.com.ezaccess.library.uitm.edu.my/do
cview/305278580?accountid=42518. (305278580)

Uphoff, N. & Wijayaratna, C. (2000). Demonstrated benefits from social capital: the productivity of farmer
organizations in Gal Oya, Sri Lanka. *World Development, 28*(11), 1875–1890.

Uyarra, M.C., Cote, I.M., Gill, J.A., Tinch, R.R., Viner,
D. & Watkinson, A.R. (2005). Island-specific prefer-

ences of tourists for environmental features: implications of climate change for tourism-dependent states. *Environmental conservation*, *32*(01), 11–19.

Walker, B., Holling, C.S., Carpenter, S.R. & Kinzig, A. (2004). Resilience, adaptability and transformability in social--ecological systems. *Ecology and society*, *9*(2), 5.

Walsh, F. (2003). Family resilience: A framework for clinical practice. *Family process*, *42*(1), 1–18.

Worthington, E.L. & Scherer, M. (2004). Forgiveness is an emotion-focused coping strategy that can reduce health risks and promote health resilience: Theory, review, and hypotheses. *Psychology & Health*, *19*(3), 385–405.

Theory and Practice in Hospitality and Tourism Research – Radzi et al. (Eds)
© 2015 Taylor & Francis Group, London, ISBN 978-1-138-02706-0

Stock market reaction to major world sporting events

F.F. Cheng & B.T.H. Hwa
University Putra Malaysia, Serdang, Malaysia

ABSTRACT: This paper examined the impact of the international sporting events on hosting country's stock market. A total of 30 international sporting events were used in this study. The events were comprised of five major sporting events, namely Summer Olympics Games, Winter Olympics Games, FIFA World Cup, European Football Championships and Commonwealth Games. This study utilized event study methodology to analyses and investigate the stock's market reaction towards event announcement date. The empirical result indicated that there was short term positive effect on announcement date across all stock markets of event hosting countries except for the FIFA world cup, which may be due to information leakage.

Keywords: Mega sport event, event study, abnormal returns, stock market reaction

1 INTRODUCTION

The international sporting events are usually perceived to create positive effects on the host countries' economy. Thus, in recent decades, there has been an increase in competition among countries to host international sporting events. It is believed that by successfully organizing major international sporting events can bring tremendous tangible and intangible benefits and values to the hosting country. Hence, many developing countries such as South Korea, China, Mexico and South Africa are entering the bidding competition for major international sporting events with the intention to capitalize them as an opportunity to improve city, state, as well as national economic.

The country hosting international sporting events involve substantial investment into various sectors of the economy such as construction, hotels, telecommunications, hospitality, food and tourism. This investment will boost up the economy of the hosting country. Therefore, the economic impact of the international sporting events usually receives broad publicity. However, a full evaluation of this economic impact is rather difficult, since such an international sporting event has short term and long term effects, as well as direct and indirect effects, which complicates the estimation of its costs and benefits.

There is no one simple, direct measure of the impact of hosting international sporting events to a national economy. However, there are many indirect measures where the impact may be gauged from hosting an international sporting event. Among these impact measures or indicators, the most popular one is by observing or studying of the impact through analysis of the movement of hosting countries' stock market movement. The stock market index is a barometer of a country's economy and is commonly believed to reflect the expectations of the economic outlook. Thus, by observing the trends or movement of the stock market within a specified timeframe of the event, it provides some indications of the impact of hosting the international events.

It is often claimed by economist that hosting international sporting events can generate economic benefits to the hosting country. The empirical evidence is quite mixed. Although many countries competed to host international sports events such as the Olympic Games and the World Cup, the benefits associated with the successful bided country hosting the event are uncertain. This study will examine the impact of an international sporting event announcement to the behavior or movement of a hosting country's stock market. Therefore the research questions are: Does the event announcement date have a statistically significant influence on their country's stock market? The objectives of this study are to examine short term impacts of bid winning country's stock market toward the event announcement date by the organizer.

2 LITERATURE REVIEW

Numerous works of academic literature emphasized how international sporting events have had significant impacts on the host nation's economic development. As early as 1995, Humphreys and Plummer (1992) predicted the short-term and long-term effects on Atlanta's economy when hosting the 1996 Summer Olympics Games. The short-term economic impacts were composed of direct, indirect, induced and total impact. The direct impacts involved the expenditure on industries such as the expenditure on broadcasting equipment for international broadcasters. They concluded that there was about US$1.2 billion direct spending during the year of 1991 to 1997. The indirect economic impacts were associated with visitor spending. There was about US$823 million indirect spending by visitors in the 18 days prior to the 1996 Summer Olympic commenced. An earlier study by Kim Rhee, Yu, Koo, and Hong (1989) summarized that there was about 2,382 billion Korean won (US$2.33 billion) invested into Olympic related projects. Moreover, there were about 336,000 jobs opportunities created during year 1982 to 1988. The most impacts of investments were in the infrastructure industry with a 38.8 percent increase followed by the manufacturing industry with a 35 percent increase and then the construction industry with a 32.4 percent increase.

In a recent study, Tziralis, G., Tolis, A., Tatsiopoulos, I., and Aravossis (2006) stated that the Olympics impacted Attica's economy by increasing the labor force with extra employment opportunity and an increased expansion of road networks and improved of existing roads. Moreover, the Games also resulted in the growth of the constructions sectors, hotels and restaurants in a scale larger than the overall growth of the national economy.

Martins and Serra (2007) studied the market reaction to the announcement of the selected country hosting the Summer Olympic Games and Winter Olympic Games, the World Football Cup, the European Football Cup and World specialized exhibition. They concluded no evidence supporting that industries were more likely to extract direct benefit from the events, and there are insignificant cumulative abnormal returns for losing bidders.

Another related study by Willner (1988) examined the impact on the macroeconomic variable Gross Domestic Product (GDP) on the 1998 Seoul Summer Olympics Games by using the ordinary least squares (OLS) model. He found that there was a positive and significant relationship between Olympics and economic growth and investment growth with slightly higher tourist and investment activities in the Olympics year.

Dick and Wang (2008) examined the stock market reactions to the announcement of the Olympics Games host cities during the last three decades. They found a significant and positive announcement effect of hosting the Summer Olympics Games which was reflected in the returns (additional two percent accumulated over the following days). They did not find any significant results for the Winter Olympics Games.

Mirman and Sharma (2010) investigated the stock market impact for the year 1996 to 2010 Olympic Games by comparing the stock market reaction of winners and losers around the announcement date. They revealed the Winter Games announcement, stock markets in winning countries performs significantly worse than in losing countries, while there are insignificant results for the Summer Games.

In the latest article "Commonwealth Games and the Economy" published by Institute International Trade, India, the research analyst Shradha (2010) pointed out that the Commonwealth Games 2010 in India expected to result in an overall economic impact of USD 4.9 billion on India's GDP during a period of four years starting from 2008 and ending 2012, and created employment opportunities for approximately 2.5 million people.

3 DATA AND METHODOLOGY

3.1 Data collection

The data used for this study were categorized into two sets. The first set of data was the daily closing prices for S&P 500 (Standard & Poor 500) and those bid winning countries' index of an event, such as FTSE100 (London), KOPSI 100 Index (Korea), NYSE (New York, USA), TOPIX (Japan), S&P/TSE (Canada), Madrid SE (Spain), S&P/ASX200 (Australia), RTS (Russia), Milan COMIT (Italy), SBF 250 (France), ATHEX Composite Index (Greece), SSE Composite (Shanghai, China), IBOVESPA (Brazil), OSLO (Norway), DAX 30 (Germany), FTSE/JSE (South Africa), NZX 50 Index (New Zealand), KLCI (Malaysia), S&P BSE SENSEX (India), OMXS30 (Sweden), BEL20 (Belgium), PSI20 (Portugal), ATX (Austria), SMI (Switzerland) and WIG (Poland). These set of data was obtained from Thomson Data Stream.

The second set of data consisted of international sporting events announcement date and event starting date from the event organizers such as summer and winter Olympics International Committee, FIFA World Cup, Commonwealth Games Federation and European Football Association. The actual events and the hosting countries are listed in Appendix A. These are international sporting events held globally, and only 30 international sporting events from the five major games were selected for this study after the screening process.

3.2 Event study approach

This study utilized the event study approach to identify the impacts of the international sporting events on hosting country stock market's movement or behavior. This approach is the most common, simple method, widely accepted and frequently used by researchers to analyze the relationship between the effect of a specific event and the market return of a firm or stock market (Brown & Warner, 1985). In general, this event study approach allows the investigation of the stock prices behavior or stock index adjacent to the events and it uses the changes in stock prices or stock index to estimate the effect of an event.

Further, the Market Efficiency Hypothesis (EMH; Fama, Fisher, Jensen & Roll, 1969) test was explored. The hypothesis states that the stock prices on financial market are rapidly and fully reflective to all new available information. Hence, investor cannot make excess return based on the available information they have obtained because when the market is efficient, the stocks are traded at fair and reasonable prices. When the financial market is under equilibrium, investor cannot outperform the market.

The market efficiency test was used to assess the stock market reaction to the relevant information such as events announcement date and events starting month related to the international sporting events. The following are the systematic structure for this event study.

3.3 Event date

This study focused on how the event announcement date and event starting date of an international event affected the host country's stock market. In the events, the announcement date or starting date falls under non-trading day such as weekend or public holiday, where the market is closed, the next available trading date was defined as event date.

3.4 Estimation window and event window

Prior to obtaining data of an events hosting country stock index, it is crucial to identify the estimation period and the event window to be used for this analysis. The estimation window is a time period prior to the event for obtaining α and β in the Market Model (Sharpe, 1982) whereas the event window is the time period around the event of interest which is used to investigate the abnormal returns. Figure 1 shows the event study time line.

In this time line, $t = 0$ represents the event date, $t = -48_{days}$ to $t = +30_{days}$ represents the event windows and $t = -36_{months}$ to $t = -12_{months}$ represents the estimation window. The negative sign means prior to the event date whereas positive sign after the event date.

4 RESULTS AND DISCUSSION

The following results present the cross sectional analysis for the all the stock market's reaction of 30 international sporting events hosting countries and are measured by events hosting country's stock indices to the related information such as events announcement date and events starting month.

4.1 Cumulative average abnormal return on events announcement date

Figure 2 shows the Cumulative Average Abnormal Return (CAAR) across all stock markets of Summer Olympics Games hosting countries. The CAAR trend shows positive return within range of +.23 percent to +2.05 percent prior to the event announcement date. The lowest CAAR is +2.05 percent which is 5 days prior to the event announcement date. However, the CAAR trend shows sharp increase of 1.31 percent from 1.37 percent at event date to 2.68 percent in the following day and continues to increase up to maximum return of 2.97 percent from 1.37 percent at event date to 4.34 percent at day 26 and increase of 2.34 percent from event date till day 30.

Figure 3 shows the CAAR across all stock markets of Winters Olympics Games hosting countries. The CAAR trend shows negative return within the range of −.69 percent to −2.86 percent from day 28 till the event announcement date. However, the CAAR trend shows increasing trend following the event announcement date during which the return increased 2.69 percent from −2.29 percent at event date up to +.40 percent at day 12. From the event

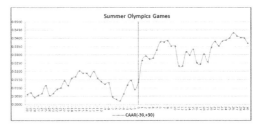

Figure 2. CAAR (−30,+30) across all stock market of Summer Olympics Games hosting countries adjacent to the events announcement date on each of the days in the 30 days symmetric window. *Note*: The actual data can be provided upon request.

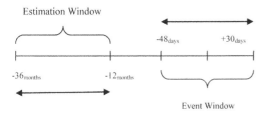

Figure 1. Event study time line.

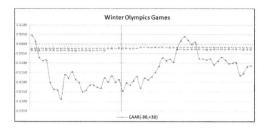

Figure 3. CAAR (−30,+30) across whole stock market of Winter Olympics Games hosting countries adjacent to the events announcement date on each of the days in the 30 days symmetric window. *Note*: The actual data can be provided upon request.

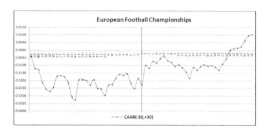

Figure 5. CAAR (−30,+30) across whole stock market of European Football Championships hosting countries adjacent to the events announcement date on each of the days in the 30 days symmetric window. *Note*: The actual data can be provided upon request.

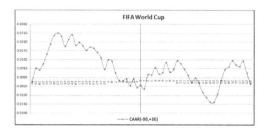

Figure 4. CAAR (−30,+30) across whole stock market of FIFA World Cup hosting countries adjacent to the events announcement date on each of the days in the 30 days symmetric window.
Note: The actual data can be provided upon request.

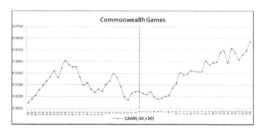

Figure 6. CAAR (−30,+30) across whole stock market of Commonwealth Games adjacent to the events announcement date on each of the days in the 30 days symmetric window. *Note*: The actual data can be provided upon request.

date till day 30, it shows an increase of 1.16 percent in negative zone.

Figure 4 shows the CAAR across whole market of FIFA Would Cup hosting countries. Prior to the event announcement date, the CAAR trend shows maximum return of 2.51 percent at day 23 and is decreasing till negative zone from day 6 till event announcement date. However, the CAAR trend shows increasing trend from negative zone to positive zone following the event announcement date with increase of 1.37 percent from −.42 percent at event date to .95 percent at day 6 but it decreases to −.97 percent from −.42 percent at event date to −1.39 percent at day 20. From the date of announcement till day 30, it shows a small increase at about .06 percent.

Figure 5 shows the CAAR across whole market of European Football Championships hosting countries. The CAAR trend shows negative return of 1.86 percent from −0.40 percent at day 30 to −2.26 percent at event announcement date. The CAAR trend also indicates an increasing trend following the event announcement date which the return increased 1.89 percent from −2.26 percent at event date up to −0.37 percent at day 6 and follows

by decrease to 0.42 percent at day 13. From the day of announcement till day 30, it shows positive return of 3.29 percent.

Figure 6 shows the CAAR across whole market of Commonwealth Games hosting countries. The CAAR trend shows minimum return of 0.85 percent at day 30 and reaches maximum return of 4.06 percent at day 20 but decrease 2.29 percent from 3.74 percent at day 19 till 1.18 percent at event announcement date. However, the CAAR trend shows insignificant and negative increase of 0.32 percent following the event announcement date till day 8. The CAAR trend starts to increase on day 9 onward with maximum return of 4.28 percent from 1.18 percent at the event date to 5.73 percent at day 30.

In sum, the cross sectional CAAR (−30, +30) trend analysis for individual game such as Summer Olympic Games, European Football Championships and Commonwealth Games with positive return in positive zone at 2.34 percent, 3.19 percent and 5.73 percent respectively. However, the CAAR (−30, +30) trend analysis of Winter Olympic Games and FIFA World Cup shows insignificant positive return in negative zone at .06 percent and 1.16 percent after the event announcement date.

Table 1. 30 Events and their hosting countries for five types of international sporting events.

No	Event name	Year hosted	Winning country
1	Summer Olympics	1996	U.S.A.
2	Summer Olympics	2000	Australia
3	Summer Olympics	2004	Greece
4	Summer Olympics	2008	China
5	Summer Olympics	2012	UK
6	Summer Olympics	2016	Brazil
7	Winter Olympics	1992	France
8	Winter Olympics	1994	Norway
9	Winter Olympics	2002	U.S.A.
10	Winter Olympics	2006	Italy
11	Winter Olympics	2010	Canada
12	Winter Olympics	2014	Russia
13	FIFA World Cup	1990	Italy
14	FIFA World Cup	1994	U.S.A.
15	FIFA World Cup	2002	S. Korea
16	FIFA World Cup	2002	Japan
17	FIFA World Cup	2006	Germany
18	FIFA World Cup	2010	S. Africa
19	FIFA World Cup	2014	Brazil
20	Commonwealth Games	1994	Vitoria
21	Commonwealth Games	2002	Manchester
22	Commonwealth Games	2010	Delhi
23	Commonwealth Games	2014	Glasgow
24	Commonwealth Games	2018	Gold Coast
25	European Football	1984	France
26	European Football	1992	Sweden
27	European Football	1996	UK
28	European Football	2000	Belgium
29	European Football	2000	Netherlands
30	European Football	2004	Portugal

5 CONCLUSION

This study investigated and analyzed the impact of international sporting events on events hosting country stock market. A total of 30 international sporting events were utilized in this study where it comprised of five major sporting games namely Summer Olympics Games, Winter Olympics Games, FIFA World Cup, European Football Championships and Commonwealth Games. The data used in this study consisted of daily stock returns of events hosting countries. The market model and t-Statistic test were used to analyses the effect of the events and to test the significant of effect toward the stock market return.

The results of the analysis on the announcement date impact of 30 international sporting events hosting countries' stock market showed evidence that the stock market reacted according to the type of events on event announcement dates, the stock returns during Summer Olympics Games and European Football Championships announcements show positive effect. Whereas, the Winter Olympic Games and Commonwealth Games show weak responses by their hosting countries' stock markets. Lastly, the FIFA World Cup had no effect on the stock market during announcement. This was possible due to leakage of news before the FIFA World Cup announcements.

REFERENCES

Brown, S.J. & Warner, J.B. (1985). Using daily stock returns: the case of event studies. *Journal of Financial Economics, 14 (1)*, 3–31.

Dick, C.D. & Wang, Q. (2008). The economic impact of Olympic Games: evidence from stock markets: ZEW Discussion Papers.

Fama, E.F., Fisher, L., Jensen, M.C. & Roll, R. (1969). The adjustment of stock prices to new information. *International Economic Review, 10*(1), 1–21.

Humphreys, J.M. & Plummer, M.K. (1992). *The economic impact on the state of Georgia of hosting the*

1996 summer Olympic Games: Atlanta Committee for the Olympic Games, Incorporated.

Kim, C.-g., Rhee, S.-w., Yu, J.-C., Koo, K. & Hong, J.C. (1989). Impact of the Seoul Olympic Games on national development: Korea Development Institute Seoul.

Martins, A.M. & Serra, A.P. (2007). *Market impact of international sporting and cultural events*. Journal of Economics and Finance, 35 (4), 382–416.

Mirman, M. & Sharma, R. (2010). Stock market reaction to Olympic Games announcement 1. *Applied Economics Letters, 17 (5), 436–466.*

Sharpe, G.W. (1982). An overview of interpretation. New York, NY: John Wiley & Sons.

Shradha D. (2010). *Commonwealth Games and the Economy', Institute International Trade, India*. Retrieved from http://www.iitrade.ac.in/km/ibank/Commonwealth%20Games%20and%20the%20Economy.pdf

Tziralis, G., Tolis, A., Tatsiopoulos, I. & Aravossis, K. (2006). Economic aspects and sustainability impact of the Athens 2004 Olympic Games. Environmental economics and investment assessment, 21–33.

Willner, J. (1988). Korean Olympics and macro effects: what's there. Economic Growth of Open Economics F, 43.

Theory and Practice in Hospitality and Tourism Research – Radzi et al. (Eds)
© *2015 Taylor & Francis Group, London, ISBN 978-1-138-02706-0*

Performance analysis of hospitality REIT: A case study of YTL hospitality REIT

N.N. Chuweni & S.N.M. Ali
Faculty of Architecture, Planning and Surveying, Universiti Teknologi MARA, Perak, Malaysia

M.N.I. Ismail
Faculty of Hotel and Tourism Management, Universiti Teknologi MARA, Shah Alam, Malaysia

S.H. Ahmad
Faculty of Administrative Science and Policy Studies, Universiti Teknologi MARA, Shah Alam, Malaysia

ABSTRACT: Malaysia currently has 16 Real Estate Investment Trust (M-REITs), of which one is Hospitality REIT. The main focus of this research is ascertain the performance of the Malaysian Hospitality REITs namely YTL Hospitality REITs or formerly known as Starhill REITs. Ratio analysis of financial statement will be used in the research to ascertain the profit margin since 2010 to 2013. The outcome of the research shall be beneficial to the related institutional and potential investors who may consider REITs as another viable alternative investment available in the market. This paper presents the introductory part of the research outlining the performance of YTL Hospitality REITs.

Keywords: Hospitality REITs, performance analysis, financial ratios

1 INTRODUCTION

1.1 *Real Estate Investment Trust*

Real Estate Investment Trust (REITs) or formerly known as Property Trust Fund means unit trust scheme that invests or proposes to invest primarily in income-generating real estate (Securities Commission, 2005). REITs can be divided into three types which are Equity REITs, Mortgage REITs and Hybrid REITs. Before examining the attributes of the performance of REITs, it is perhaps useful to highlight the concept of REITs. The Securities Commission of Malaysia defines REITs as collective investment vehicles (typically in the form of trust funds) which pool money from investors and used this pooled capital to buy, manage and sell real estate. The fund manager will ensure any investment decision made on its property portfolio achieves it optimum return before being distributed to the unit holders. REITs require a trustee to supervise and monitor all activities of the fund management to ensure that it complies with all requirements by Securities Commission.

2 LITERATURE REVIEW

2.1 *Performance of Malaysian Real Estate Investment Trusts*

There are many previous studies conducted on the performance analysis on Malaysian REITs. For instance, previous research indicates that REITs outperform the stock and property market of Malaysia during 2008 Global Financial crisis (GFC) and post GFC period by using the Jensen measurement index (Ong, 2011). The result supports the findings of Yusof and Mohd Nawawi (2012) by using Sharpe and Treynor Ratios, Jensen's Capital Asset Pricing Model (CAPM), Fama-French 3-Factor CAPM as well as Carhart 4-Factor CAPM. The result indicates Malaysian REITs outperform the market index of Kuala Lumpur Composite Index (KLCI). Hence, by looking at these study, Malaysian REIT perform well during and post GFC as compared to other market investment indicating a viable option of diversification in the investment portfolio especially during the economic downturn.

Apart from that, the performance of Malaysian REIT can be measured in the form risk and volatility as compared to the other market alternative investment option. Chai, Choong, Koh and Tham (2011) concludes that Malaysian REITs displayed better performance as compared to other market during the financial crisis period and possess lower degree of overall risk or volatility. Furthermore, the study conducted by Tan (2009) shows the systematic risk of M-REIT is lower than the market portfolio which indicate that M-REITs are less volatile as compared to the stock market. Besides that, the performance of Malaysian REIT can also be seen through correlation with other REIT available in the market especially the Asian REIT. According to Nawawi, Husin, Hadi and Yahya (2010), Malaysian REIT lagged for up two months behind Asian REIT indicating a positive correlation between M-REITs and Asian REIT. However, there is no correlation between Malaysian REIT market and the Asian REITs market for a long period.

Apart from that, in order to understand how well the M-REIT company performs, an analysis of the M-REITs financial and management strength need to be conducted. According to Dynaqueast, the financial strength is one of the most critical measures of the worth of an investment besides stability and growth.

3 RESEARCH METHODOLOGY

Content Analysis is being adopted in the research. The measurement performance of asset performance usually be seen in the form of financial ratios, occupany level, rental rate and growth. However, for the purpose of this study, the performance of the Hospitality REIT will be deployed in term of Ratio Analysis which includes the Profitability Ratios, Operating Ratio and Leverage Ratios. Ratio analysis of financial statement from Annual Report will be used in the research to ascertain the profit margin since 2010 to 2013.

4 A CASE STUDY OF YTL HOSPITALITY REITS

4.1 *Introduction*

YTL Hospitality REIT was listed on 16 December 2005 on the Main Market of Bursa Malaysia Securities Berhad under the name Starhill Real Estate Investment Trust (YTL Hospitality REIT, 2014). YTL Hospitality REIT has a market capitalization of approximately RM1,212 million (as at 30 April 2014) which major investment portfolio focusing on prime hotel and hospitality-related properties.

Table 1. Property portfolio.

Malaysia
JW Marriott Hotel Kuala Lumpur
The Ritz-Carlton Kuala Lumpur
The Pangkor Laut, Tanjong Jara
Cameron Highland resorts
The Vistana chain of hotels in Kuala Lumpur, Penang and Kuantan

International portfolio
Hilton Niseko, Japan
The Sydney Harbour, Australia
Melbourne Marriot hotels, Australia

Figure 1. YTL hospitality REIT's share price.

Table 1 shows the diversified property portfolio owned by the company in optimizing the return of the investors.

4.2 *Performance analysis*

Performance generated from Data Stream is depicted in Figures 1 to 3. It can be seen that market confidence was stable and even increased throughout 2010 and 2013. Share price reached its peak between 2012 and 2013. In the middle 2012, the highest peak of share price was mainly due to two possible reasons: firstly, stable fixed lease rentals arising from its existing property portfolio and, secondly, variable income from the three Marriott hotels located in prime tourist destinations in Australia's major cities will, increasing the potential for distribution per unit growth.

In tandem with the growing performance for the year 2012, the company has also declared a very handsome dividend to its shareholders at the closing year of 2013 with dividend yield of 8.2. Earnings per share between 2012 and 2013 were considerably high at 0.50 and above.

Figure 2. Dividend yield.

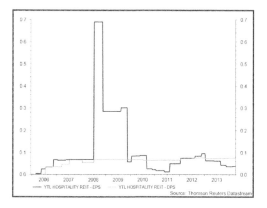

Figure 3. Earnings per share and divided per share.

4.3 Ratio analysis

Table 2 shows a substantial return by the company to its capital employed in 2010 at 46 percent mainly due to the high dependency on equity instead of borrowing. This is reflected in a low gearing ratio of 0.15. The less dependency increased the ability of the company to cover the existing interest payment with the highest times interest earned of 10.34 (see Table 4).

According to Warren Buffet, ROE is one of the important metric in selecting a company. Buffet focuses on ROE rather than on earnings per share (Harper, 2011). Over the year 2011 to 2013, the times interest earned and return on capital employed begun to decrease due to the vigorous project development which caused more capital requirement through borrowing injected into the company. However the ability of the company to repay its interest are still outstanding as reflected in the times interest earned figure which is more than 1 for the year 2011, 2012 and 2013 respec-

Table 2. Profitability ratios.

Ratio	Year			
	2010	2011	2012	2013
Return on assets (%)	1.92	3.67	6.00	1.86
Return on capital employed(%)	46	6.5	4.45	6.23

Table 3. Operating ratios.

Ratio	Year			
	2010	2011	2012	2013
Average collection Period (days)	22.21	17.88	3.84	10.33
Current ratio	4.61x	22.07x	2.32x	1.81x

Table 4. Leverage ratios.

Ratio	Year			
	2010	2011	2012	2013
Debt ratio	0.15	0.14	0.14	0.56
Times interest earned	10.34	8.38	2.01	3.84

tively. This indicates that the company is far from leverage risk.

In terms of efficiency of operation, it is evidenced that the company may be at risk of liquidity due to the declined current ratio from 2010 to 2013. Current ratio of less than 1 is a red flag. The plausible reason behind this figure is the slow debt collection activities which reached 10.33 days in 2013.

5 CONCLUSION

Several sources of individual property investment risk at a portfolio level, it is clear that many are in the nature of the unsystematic risks that can be diversified away by balanced portfolio construction. In conclusion, property as investment is prone to both unsystematic and systematic risks. While former may be diversified away, latter cannot. Although the company is a fixed asset-oriented, and the well-managed of noncurrent asset and liabilities make it far from leverage risk, but liquidity risk may also be observed from time to time. The effect of balanced portfolio construction is therefore to reduce but not abolish the property investment risk. This shows the importance of the studies on the performance analysis on Malaysian REITs especially the Hospitality REIT as another viable investment option for the investors especially in term of diversification of mixed asset allocation portfolio in the long run.

REFERENCES

Chai, M.Y., Choong, Y.J., Koh, C.W. & Tham, W.J. (2011). *Malaysian Real Estate Investment Trusts (M-REITS): A performance and comparative analysis* Final Year Project, UTAR.

Harper, D. (2011). What is Warren Buffet's Investing Style? Retrieved March 30, 2014, from http://www.investopedia.com/articles/05/012705.asp

Nawawi, A.H., Husin, A., Hadi, A.R.A. & Yahya, M.H. (2010). *Relationship and lead-lag effect between Asian Real Estate Investment Trusts (REITs) performance and Malaysian REIT market: Cointegration model-ling.* Paper presented at the International Conference on Science and Social Research (CSSR), 2010, Kuala Lumpur.

Ong, T.S. (2011). A study on the performance of malaysian real estate investment trust from 2005–2010 by using the Net Asset Value Approach. *International Journal of Economics and Research, 2*(1), 1–15.

Securities Commission. (2005). Guidelines on Real Estate In-vestment Trust (REITs).

Tan, S.H. (2009). *Performance of Malaysian REIT stock relative to Bursa Malaysia Stock Index.* (Unpublished Master Thesis), Multimedia University, Malaysia.

YTL Hospitality REIT. (2014). About YTL Hospitality REIT Retrieved May 5, 2014, from http://www.ytl-hospitalityreit.com/overview.html

Yusof, A.Y. & bin Mohd Nawawi, A.H. (2012). *Does Malaysian REITs outperform the equity market?* Paper presented at the International Conference on Statistics in Science, Business, and Engineering (ICSSBE), 2012.

Author index

Printed and bound by CPI Group (UK) Ltd, Croydon, CR0 4YY

18/10/2024

01776219-0014